AF412457

Molecular Imaging with Reporter Genes

Reporter genes have been used for several decades to study intracellular molecular events, such as regulation of gene expression *in vitro*. However, it was little more than a decade ago that a new class of reporter genes was developed for imaging intracellular and intercellular molecular events within living subjects. Studying molecular biology and medicine in the intact and natural environment within living research subjects will enable better translation of knowledge for use in humans. In fact, reporter gene imaging is now possible in humans. This book focuses on this group of imaging reporter genes, starting with detailed descriptions of all reporter genes from different imaging modalities, including optical, MRI, and radionuclide-based imaging. This is followed by a chapter describing all instrumentation used in molecular imaging. Key scientists in the field then explain various applications of the reporter gene imaging technologies.

This is the first comprehensive book on all aspects of reporter gene imaging, detailing what is known in the field and future goals for research. Reporter gene imaging plays a role in basic biomedical science research, in the development of novel therapeutic agents and will soon be an important part of clinical molecular imaging. Therefore, this book will benefit biomedical scientists of all fields in both academia and pharmaceutical/biotechnology industries, assist regulators in forming better policies for assessing safety and efficacy of novel therapeutic agents, and inform physicians of a new molecular imaging tool that will in the near future have clinical applications.

SANJIV SAM GAMBHIR, MD, PhD, is director of the molecular imaging program, division chief of nuclear medicine, and professor in the departments of radiology and bioengineering at Stanford University, Stanford, California.

SHAHRIAR S. YAGHOUBI, PhD, is senior research scientist in the department of radiology's molecular imaging program at Stanford University, Stanford, California.

Cambridge Molecular Imaging Series

Molecular Imaging with Reporter Genes

Edited by

Sanjiv Sam Gambhir

Stanford University, Stanford, California

Shahriar S. Yaghoubi

Stanford University, Stanford, California

CAMBRIDGE UNIVERSITY PRESS
Cambridge, New York, Melbourne, Madrid, Cape Town, Singapore,
São Paulo, Delhi, Dubai, Tokyo

Cambridge University Press
32 Avenue of the Americas, New York, NY 10013-2473, USA

www.cambridge.org
Information on this title: www.cambridge.org/9780521882330

First published 2010

Printed in China by Everbest

A catalog record for this publication is available from the British Library.

Library of Congress Cataloging in Publication data
Molecular imaging with reporter genes / edited by Sanjiv S. Gambhir,
Shahriar S. Yaghoubi.
    p.   cm. – (Cambridge molecular imaging series)
Includes index.
ISBN 978-0-521-88233-0 (hardback)
1. Reporter genes.   2. Molecular probes.   3. Diagnostic imaging.
I. Gambhir, Sanjiv Sam.   II. Yaghoubi, Shahriar S. (Shahriar Shah), 1975–
QH447.8.R47M65   2010
572′.36 – dc22   2009032727

ISBN   978-0-521-88233-0 Hardback

Contents

Contributors

James P. Basilion
Departments of Radiology and Biomedical
Engineering
Case Western Reserve University
Cleveland, Ohio

Dean O. Campbell
Department of Molecular and Medical
Pharmacology
Crump Institute for Molecular Imaging
David Geffen School of Medicine
University of California, Los Angeles
Los Angeles, California

Robert E. Campbell
Department of Chemistry
University of Alberta
Edmonton, Alberta, Canada

Michael Carey
Department of Biological Chemistry
David Geffen School of Medicine
University of California, Los Angeles
Los Angeles, California

Carmel Chan
Molecular Imaging Program
Bio-X Program
Stanford University
Stanford, California

June-Key Chung
Department of Nuclear Medicine
Cancer Research Institute
Tumor Immunity Medical Research Center
Seoul National University College
of Medicine
Seoul, South Korea

Christopher H. Contag
Molecular Imaging Program
Bio-X Program
Departments of Pediatrics, Radiology, and
Microbiology & Immunology
Stanford University
Stanford, California

Michael W. Davidson
National High Magnetic Field Laboratory
Department of Biological Science
Florida State University
Tallahassee, Florida

Timothy C. Doyle
Molecular Imaging Program
Bio-X Program
Stanford University
Stanford, California

Hua Fan-Minogue
Molecular Imaging Program
Bio-X Program
Stanford University
Stanford, California

Marxa L. Figueiredo
Department of Comparative Biomedical Sciences
School of Veterinary Medicine
Louisiana State University
Baton Rouge, Louisiana

Sanjiv Sam Gambhir
Molecular Imaging Program
Division of Nuclear Medicine
Departments of Radiology and Bioengineering
Stanford University
Stanford, California

Isabel J. Hildebrandt
Department of Molecular and Medical
Pharmacology
Crump Institute for Molecular Imaging
David Geffen School of Medicine
University of California, Los Angeles
Los Angeles, California

Joo Hyun Kang
Department of Nuclear Medicine
Cancer Research Institute
Tumor Immunity Medical Research Center
Seoul National University College of Medicine
Laboratory of Nuclear Medicine
Korea Institute of Radiological and
Medical Science
Seoul, South Korea

Keon Wook Kang
Department of Nuclear Medicine
Cancer Research Institute
Tumor Immunity Medical Research Center
Seoul National University College of
Medicine
Seoul, South Korea

Craig S. Levin
Professor of Radiology and Electrical
Engineering
Molecular Imaging Program
Division of Nuclear Medicine
Stanford University School of Medicine
Stanford, California

Tarik F. Massoud
Department of Radiology
University of Cambridge School of Clinical
Medicine
Addenbrooke's Hospital
Cambridge, United Kingdom
Molecular Imaging Program
Bio-X Program
Stanford University
Stanford, California

Evan D. Nair-Gill
Department of Molecular and Medical
Pharmacology
David Geffen School of Medicine
University of California, Los Angeles
Los Angeles, California

Mark D. Pagel
Molecular Imaging Program
Stanford University
Stanford, California

Ramasamy Paulmurugan
Molecular Imaging Program
Bio-X Program
Stanford University
Stanford, California

Iván Peñuelas
Department of Nuclear Medicine
Small Animal Imaging Research Unit CIMA-CUN
University of Navarra
Pamplona, Spain

Vladimir Ponomarev
Molecular Imaging Laboratory
Department of Radiology
Memorial Sloan-Kettering Cancer Center
New York, New York

Felipe Prósper
Hematology and Cell Therapy Area
Foundation for Applied Medical Research
Division of Cancer
University of Navarra
Pamplona, Spain

Caius G. Radu
Department of Molecular and Medical
Pharmacology
Crump Institute for Molecular Imaging
David Geffen School of Medicine
University of California, Los Angeles
Los Angeles, California

Pritha Ray
ACTREC, Tata Memorial Centre
Sector 21, Kharghar
Navi Mumbai, Maharastra

Martin G. Rodriguez-Porcel
Division of Cardiology and Vascular Medicine
Department of Medicine
Mayo Clinic
Rochester, Minnesota

Chengyi J. Shu
Department of Immunology
The Netherlands Cancer Institute
Amsterdam, The Netherlands

Qian Wang
Molecular Imaging Program
Bio-X Program
Departments of Pediatrics, Radiology, and
Microbiology & Immunology
Stanford University
Stanford, California

Owen N. Witte
Department of Microbiology, Immunology, and
Molecular Genetics
David Geffen School of Medicine
Howard Hughes Medical Institute
University of California, Los Angeles
Los Angeles, California

Lily Wu
Departments of Molecular and Medical
Pharmacology, and Urology
David Geffen School of Medicine
University of California, Los Angeles
Los Angeles, California

Shahriar S. Yaghoubi
Molecular Imaging Program
Department of Radiology
Stanford University
Stanford, California

Preface

Multimodality molecular imaging is a combination of imaging strategies that are playing an increasing role in all biological, biomedical, and clinical fields. Molecular imaging can be used to study a whole variety of molecular events in cells, tissues, organs, and the whole body of living organisms. This includes detecting and measuring the levels of mRNA, proteins, enzymes, and protein-protein interactions. Additionally, molecular imaging can be used to detect intracellular metabolic events, the presence and quantity of specific cells within tissues, and changes in cell characteristics through time. Adding to the power of molecular imaging is the fact that many of these techniques can be applied non-invasively in living subjects, allowing repetitive interrogation of molecular events within intact systems.

Reporter genes are among the most powerful tools in molecular imaging. They were originally introduced several decades ago for studying biochemical events *in vitro* including cell/tissue lysates. Later, their use advanced to optical imaging of molecular events within intact cultured cells using microscopes. It was in the early 1990s that imaging reporter genes of several types were developed for non-invasive molecular imaging in *living* subjects. Imaging reporter genes are general tools for imaging gene expression, protein function, protein-protein interactions, and a variety of other molecular events, repetitively and usually non-invasively within living organisms, including humans. Besides their applications in biological research, they have many biomedical applications, including disease diagnosis and optimization of therapeutics.

This is the first book dedicated to teaching all aspects of multimodality molecular imaging of reporter genes.

Imaging reporter genes are now available for optical, magnetic resonance, and radionuclide-based imaging systems. We have gathered international experts in the field to detail reporter gene systems of all imaging modalities (Chapters 1–4). In Chapters 5–6, leading scientists describe available techniques for enhancing the power of imaging reporter genes through multimodality imaging and gene expression amplification strategies. There is also a dedicated chapter on instrumentation for reporter gene imaging in animal models and humans (Chapter 7). Chapters 8–12 were written by molecular imaging experts who have worked extensively to develop many of the current applications of imaging reporter genes. Finally, the book concludes with Chapter 13, which discusses the current progress using imaging reporter genes in cell and gene therapy clinical trials.

This book should help investigators in all biomedical fields to learn about imaging techniques based on reporter genes that are essential for the study of molecular events within intact living systems. We also hope that this knowledge will lead to the development of new applications for imaging reporter genes. Graduate and medical students in all biological and medical fields will also benefit from learning about the technologies described in this book. Investigators in biotechnology and pharmaceutical firms will be able to use the technologies described in this book in their search to optimize novel therapeutics. Finally, clinical investigators, especially in regenerative medicine, cell therapy, and gene therapy will gain tremendous benefits from the knowledge of imaging reporter genes and their applications in clinical therapeutics.

Types of Imaging Reporter Genes

Fluorescent Reporter Proteins

Robert E. Campbell and Michael W. Davidson

INTRODUCTION

For more than a decade the growing class of fluorescent proteins (FPs) defined as homologues of *Aequorea victoria* green FP (avGFP), which are capable of forming an intrinsic chromophore, has almost single-handedly launched and fueled a new era in cell biology. These powerful research tools provide investigators with a means of fusing a genetically encoded optical probe to any one of a practically unlimited variety of protein targets to examine living systems using fluorescence microscopy and related methodology (see Figure 1.1; for recent reviews, see references [1–4]). The diverse array of practical applications for FPs ranges from targeted markers for organelles and other subcellular structures, to protein fusions designed to monitor mobility and dynamics, to reporters of transcriptional regulation (Figure 1.2). FPs have also opened the door to creating highly specific biosensors for live-cell imaging of numerous intracellular phenomena, including pH and ion concentration fluctuations, protein kinase activity, apoptosis, voltage, cyclic nucleotide signaling, and tracing neuronal pathways [5–9]. In addition, by applying selected promoters and targeting signals, FP biosensors can be introduced into an intact organism and directed to specific tissues, cell types, and subcellular compartments to enable monitoring a variety of physiological processes using fluorescence resonance energy transfer (FRET) techniques.

If FPs are the "fuel" for the live-cell imaging revolution, the "engines" are the technical advances in widefield fluorescence and confocal microscopes. Some notable advances include low light level digital charge coupled device (CCD) cameras as well as spinning-disk and swept-field instruments. As of today, avGFP and its color-shifted variants, in conjunction with sophisticated imaging equipment, have demonstrated invaluable service in many thousands of live-cell imaging experiments. One of the most important features of FPs is that they are minimally invasive for living cells, especially compared to many traditional synthetic fluorophores (that are often toxic or photoreactive). The relatively low or nonexistent toxicity of FPs (when expressed at low levels relative to endogenous proteins) permits visualization and recording of time-lapse image sequences for extended periods of time [10, 11]. As we will discuss in this chapter, continued advances in FP engineering technology have enabled the fine-tuning of critical fluorescent imaging parameters, including brightness, spectral profiles, photostability, maturation time, and pH insensitivity, to provide a stream of new and advanced probes for optical microscopy. These structural and functional enhancements have stimulated a wide variety of investigations into protein dynamics and function using FP chimeras imaged at low light intensities for many hours to extract valuable biochemical information.

Today we take the exceptional and revolutionary utility of FPs for granted, and it may be hard for some researchers to imagine research without them. It is therefore somewhat surprising that more than 30 years had to pass between the first scientific report of the isolation of avGFP [12] and its first application as a tool for biological imaging [13]. The first report of fluorescence in the bioluminescent hydrozoan jellyfish species *Aequorea victoria* was recorded more than 60 years ago [14] and a protein extract was independently demonstrated by two investigators to be responsible for this "green" fluorescence in the 1960s and 1970s [12, 15]. It took several more decades to identify the responsible protein, clone the gene encoding the protein, and elucidate the primary amino acid structure [16]. In light of the time span between the original discovery and cloning of avGFP, it is rather remarkable that only 2 years later, an image revealing the fluorescent sensory neurons of the nematode highlighted with the same jellyfish protein was featured on the cover of the journal *Science* [13]. This landmark event unambiguously demonstrated the utility of avGFP as a genetic marker in cells evolutionarily far removed from hydrozoans and ushered in a new

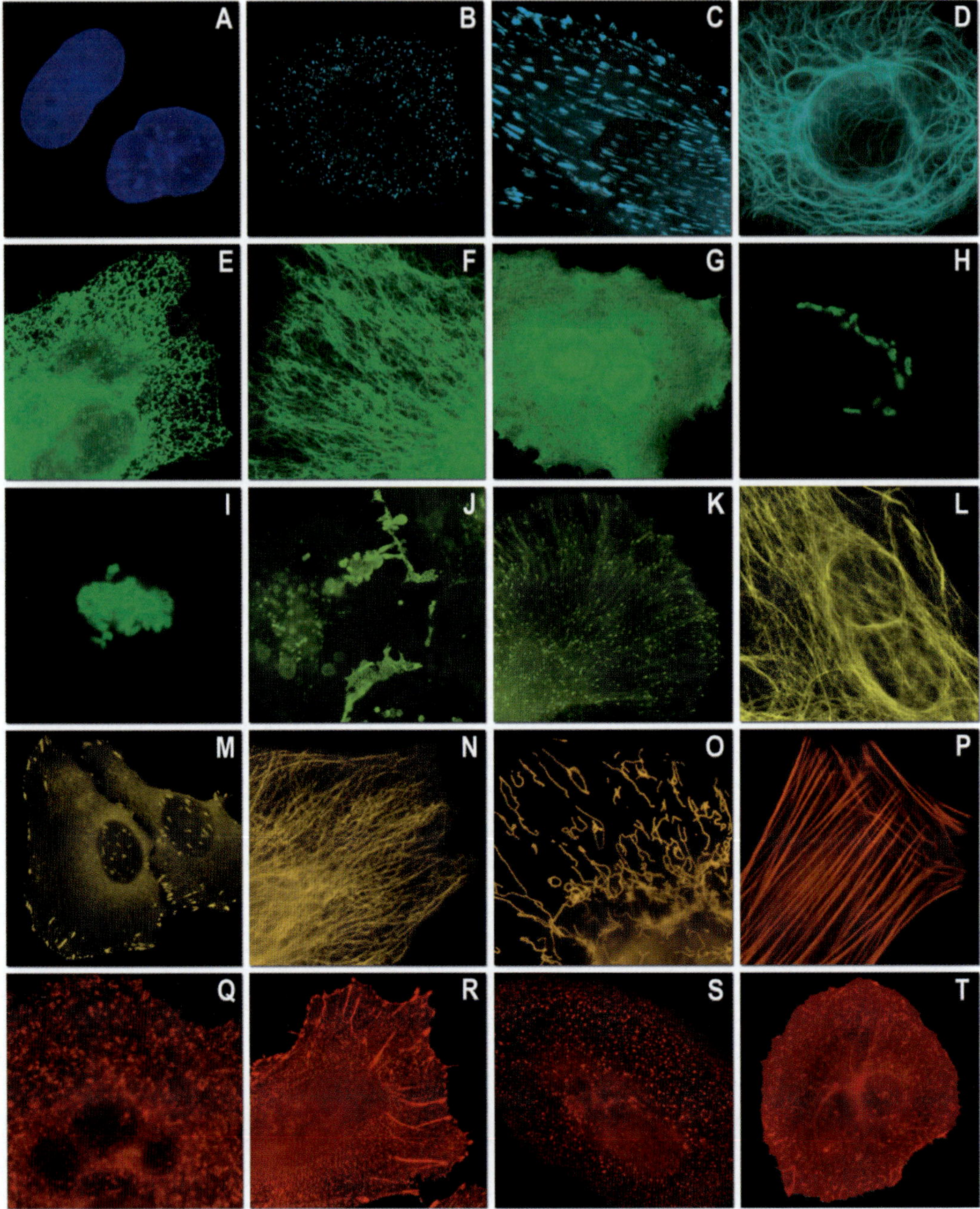

Figure 1.1. Subcellular localization of selected FP fusions (listed in Table 1.1) with targeting proteins imaged in widefield fluorescence. Images are pseudocolored to match the FP emission profile. The FP fusion terminus and number of linker amino acids is indicated after the name of the targeted organelle or fusion protein. The fusion protein and host cell line is given in parentheses (**A**) EBFP2-lamin-B1-N-10 (human lamin B1; nuclear envelope; HeLa); (**B**) ECFP-peroxisomes-C-2 (peroximal targeting signal 1; PTS1; HeLa); (**C**) mCerulean-vinculin-C-23 (human; focal adhesions; Fox Lung); (**D**) mTFP1-keratin-N-17 (human cytokeratin 18; intermediate filaments; HeLa); (**E**) EGFP-endoplasmic reticulum-N-3 (calreticulin signal sequence and KDEL retention sequence; HeLa); (**F**) mEmerald-vimentin-N-7 (human vimentin; intermediate filaments; HeLa); (**G**) mAzami Green-N1 (cloning vector; whole cell fluorescence; HeLa); (**H**) Superfolder avGFP-Golgi-N-7 (N-terminal 81 amino acids of human β-1,4-glactosyltransferase; Golgi complex; HeLa); (**I**) mT-Sapphire-H2B-N-6 (human histone H2B; metaphase; HeLa); (**J**) mVenus-Cx43-N-7 (rat α-1 connexin-43; gap junctions; HeLa); (**K**) YPet-EB3-N-7 (human microtubule-associated protein; RP/EB family; Fox Lung); (**L**) mKusabira Orange-vimentin-N-7 (human; intermediate filaments; Opossum Kidney); (**M**) tdTomato-paxillin-N-22 (chicken; focal adhesions; Fox Lung); (**N**) TagRFP-tubulin-C-6 (human α-tubulin; microtubules; HeLa); (**O**) DsRed2-mitochondria-N-7 (human cytochrome C oxidase subunit VIII; mitochondria; HeLa); (**P**) mStrawberry-actin-C-7 (human β-actin; filamentous actin; Fox Lung); (**Q**) mRFP1-lysosomes-C-20 (rat lysosomal membrane glycoprotein 1; HeLa); (**R**) mCherry-α-actinin-N-19 (human nonmuscle; cytoskeleton; HeLa); (**S**) mKate-clathrin light chain-C-15 (human; clathrin vesicles; HeLa); (**T**) mPlum-farnesyl-C-5 (20-amino acid farnesylation signal from c-Ha-Ras; plasma membrane; HeLa).

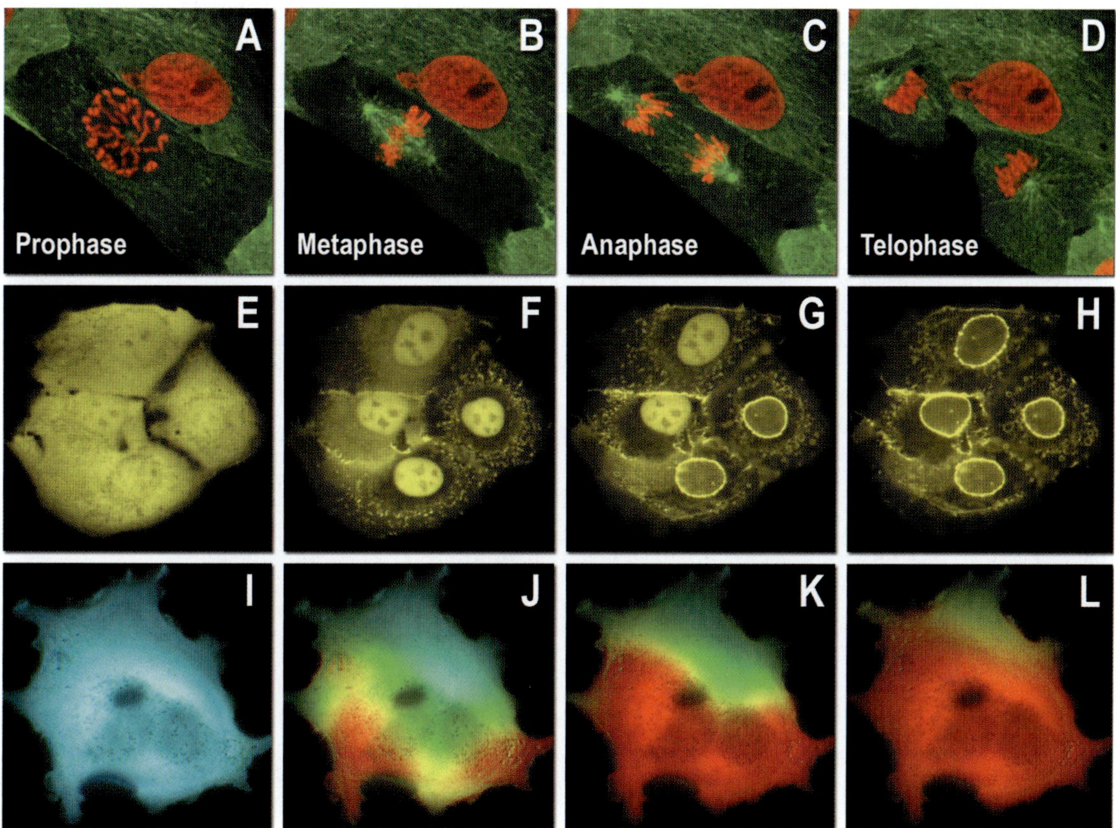

Figure 1.2. Fluorescent protein reporters in action imaged with spinning disk confocal and widefield microscopy. **A–D**: Observing mitosis in dual-labeled normal pig kidney (LLC-PK1 cell line) epithelial cells stably expressing mCherry-H2B-N-6 (histones) and mEmerald-EB3-N-7 (microtubule + end binding protein), **(A)** A cell in prophase (lower) is captured adjacent to a cell in interphase, t = 0; **(B)** The lower cell forms a spindle and enters metaphase. Note the EB3 patterns emanating from the spindle poles and traversing to the plane, t = 20 min; **(C)** During anaphase, the spindle poles translocate to opposite sides of the cell, pulling the condensed chromosomes along, t = 60 min; **(D)** The chromosomes begin to decondense during telophase as the daughter cells recover from cell division (midbody not visible). **E–H**: Spinning disk confocal images selected from a time-lapse series of human cervical adenocarcinoma (HeLa cell line) epithelial cells expressing mKusabira Orange-annexin (A4)-C-12 during ionomycin-induced translocation to the plasma and nuclear membranes, **(E)** A cluster of four cells exhibits expression of the chimera throughout the nucleus and cytoplasm t = 0, ionomycin (10 μM) added; **(F)** Shortly after addition of ionomycin, the annexin chimera begins to translocate to the plasma membrane, clearly revealing the nuclei, t = 3 min; **(G)** The annexin chimera migrates to the membrane in two of the nuclei, time = 5 min; **(H)** The nuclear membranes of all four nuclei display translocated annexin chimera, time = 7 min. **(I–L)** Widefield fluorescence calcium imaging in the cytosol of HeLa cells expressing the circularly permuted cameleon YC3.60; **(I)** Real color image of a single cell, t = 0, histamine (10 μM) added; **(J)** Pseudo-colored ratio image of the HeLa cell as a calcium wave initiates at the two loci on the membrane, t = 10 sec; **(K)** The calcium wave propagates through the cytoplasm, t = 10.5 sec; **(L)** The calcium wave reaches the distant portion of the cell, t = 11.0 sec.

era in biological fluorescence imaging. Through the mid-1990s, a number of genetic variants of the original avGFP nucleotide sequence were developed that featured enhanced green fluorescence (EGFP) [17] and altered fluorescence emission spectral profiles in the blue (BFP) [18, 19], cyan (CFP) [20], and yellow (YFP) [21] regions of the visible spectrum. Perhaps the single most significant advance following the initial cloning and early mutagenesis efforts on wild-type avGFP was the discovery of cyan, green, yellow, orange, and red-fluorescing avGFP homologues in nonbioluminescent reef corals and sea anemones [22]. This discovery not only provided a source of new FPs with new emission colors but also demonstrated that this protein motif can potentially occur in a wide range of classes and species.

FPs have now been discovered in organisms ranging from marine invertebrates to crustaceans and probably exist in many other species [23–26]. In fact, a protein known as nidogen [27], found tucked away in basement membrane of all mammals, has been characterized to have a domain consisting of an 11-stranded β-barrel remarkably similar to the three-dimensional structure of avGFP, despite having only 10% sequence homology.

In nidogen, the amino acid triplet Ile-Gly-Gly (IGG) replaces the chromophore-forming residues Ser-Tyr-Gly (SYG) found in avGFP. In addition, several other residues critical for the generation of a functional chromophore in FPs have been replaced in nidogen by residues that eliminate the possibility of fluorescence. Nevertheless, the β-barrel structure appears to have been evolutionarily conserved for a variety of purposes other than fluorescence, and nature may surprise us again with new sources of chromoproteins and FPs in species previously not considered.

In this chapter, we discuss the basic properties of FPs, including brightness, photostability, color class, oligomerization, folding, and maturation efficiency, and then compare them among themselves and to alternative technologies. In addition, we discuss recent advances in protein engineering strategies as well as improvements to the FP color palette and the development of the current armament of photoactivatable FPs. Finally, we provide suggestions for the best FP choices in single- and multicolor imaging and potential avenues for obtaining the genes encoding these proteins.

COMPARING FLUORESCENT PROTEINS WITH ALTERNATIVE FLUOROPHORES

The single most important advantage of FPs over traditional organic fluorophores and the newer semiconductor quantum dot probes is their widespread compatibility with tissues and intact organisms. In the crowded environment of the cell interior there are thousands of proteins, each with a unique shape, function, and concentration. From the perspective of the cell, expression of the gene encoding an FP (or FP chimera) adds one more relatively benign protein (a perfectly disguised spy!) into this crowded environment. In contrast, a synthetic fluorophore or quantum dot is an unfamiliar and conspicuous entity inside the cell or organism. For example, many synthetic fluorophores are hydrophobic and may bind to exposed hydrophobic patches on other proteins or intercalate into DNA. Furthermore, an FP is created inside the cell from transcription and translation of a gene artificially introduced into the cell's genome. In contrast, synthetic fluorophores and quantum dots are made outside the cell (probably on the lab bench of a chemist) and must breach the cell membrane to reach the cytoplasm, possibly to the detriment of the cell or organism. Other important advantages of FPs include their ability to specifically target fluorescent probes in subcellular compartments and the extremely low or absent levels of phototoxicity. Among the disadvantages of fluorescent proteins are artifacts introduced by delivery of the exogenous nucleic acid, often manifested in high levels of autofluorescence produced by transfection reagents. Overexpression of fluorescent proteins is also a concern

but can be offset by careful selection of clones that stably express the fusion products, at appropriate levels.

The advantages of FPs mentioned previously render them the clear and obvious technology of choice for the study of intracellular protein localization and dynamics in living cells or organisms. Simply put, the fact that FPs are *proteins* and are thus *genetically encoded* is an overwhelming advantage relative to all other fluorescent technologies. However, secondary considerations may or may not impact the choice of technology for certain applications, and these will be addressed in the following paragraphs. A number of reviews comparing synthetic dyes, quantum dots, and FP technology have been published in recent years [28–31]. The following sections, rather than repeating the relative merits of each approach, will focus only on some of the most important issues viewed from an FP-centric perspective.

Brightness

The brightness of a fluorophore is proportional to the product of the fluorescence quantum yield (QY) and the extinction coefficient (EC). The EC (units of $M^{-1}\,cm^{-1}$) describes how effective a molecule is at absorbing light, whereas the QY (a ratio with no units) is the fraction of the absorbed photons subsequently reemitted as fluorescence. By definition, QY values must lie somewhere between 0 (no fluorescence) and 1 (every absorbed photon is emitted as fluorescence). It is not particularly informative to consider either EC or QY in isolation because the actual fluorescent brightness is proportional to the product of these two values. To put some perspective on relative fluorescent brightness, we will arbitrarily pick two fluorophores useful in live- and fixed-cell imaging, one of which is bright and one of which is relatively dim. The bright fluorophore is sulforhodamine 101 (the sulfonyl chloride form of which is known as Texas Red), which has a fluorescent brightness of 125 $mM^{-1}\,cm^{-1}$ (i.e., 139,000 $M^{-1}\,cm^{-1}$ * 0.9) [32]. Note that the units for brightness are arbitrarily provided here as $mM^{-1}\,cm^{-1}$ (as opposed to $M^{-1}\,cm^{-1}$ for EC). The relatively dim fluorophore is the cell tracker dye Lucifer yellow CH, which has a fluorescent brightness of 5 $mM^{-1}\,cm^{-1}$ (24,200 $M^{-1}\,cm^{-1}$ * 0.21) [32]. Nominally, this brightness range of 5 to 125 $mM^{-1}\,cm^{-1}$ is an intuitive and convenient yardstick by which to compare different fluorophores. Due to their high ECs and exceptional QYs [33], quantum dots produce brightness values that typically fall into the range of 100–1000 $mM^{-1}\,cm^{-1}$, depending on excitation wavelength.

In a head-to-head comparison of the brightness of fluorescein and EGFP, two fluorophores with similar excitation and emission wavelength profiles, fluorescein comes out the winner. The brightness of fluorescein (69 $mM^{-1}\,cm^{-1}$) is about double that of EGFP (34 $mM^{-1}\,cm^{-1}$) [3]. This single comparison nicely represents a general trend in comparison of the brightness of FPs and synthetic dyes; FPs are generally dimmer than the highest performance synthetic dyes of similar color. The brightness of FPs spans a broad range with some commercially available proteins, such as mPlum [34] and DsRed-monomer from Clontech (Mountain View, CA), falling near or below the low end of the brightness range provided previously (Table 1.1). At the other extreme, the brightest FPs currently available are YPet at 80 $mM^{-1} cm^{-1}$ [35] and tdTomato at 95 $mM^{-1} cm^{-1}$ [36]. In general, the brightest FPs occur in the green, yellow, and orange color classes, whereas FPs emitting in the blue, cyan, and red spectral regions are generally dimmer. Based on the fact that a number of FPs have ECs approaching 100,000 $M^{-1} cm^{-1}$ and the best have QYs approaching 0.8 [3], it is not unreasonable to expect that it should eventually be possible to engineer an FP color palette where each protein has a brightness of at least 80 $mM^{-1} cm^{-1}$!

Quantitative assessment of EC and QY for an FP is relatively tedious and requires a highly purified and correctly folded protein with, ideally, greater than 95% of the molecules having an active fluorescent chromophore [37]. In addition, for EC determination the total protein concentration must be accurately determined and the measurements of absorption and fluorescence emission performed in reliable, calibrated instrumentation. QY assessment requires the comparison of emission spectra between the FP and an appropriate reference standard having a similar wavelength profile. Investigators should be highly skeptical of purely qualitative FP brightness evaluations (often made by commercial distributors) that lack quantitative information pertaining to the extinction coefficient and quantum yield. It is difficult, if not impossible, to accurately perform brightness comparisons between FPs without knowledge of these critical parameters. Further complicating matters is the fact that even if EC and QY are highly favorable, experimental brightness observed for the FP gene expressed in living cell is intrinsically dependent on the folding and maturation efficiency of the FP (discussed in the following) [37, 38].

Independent of considerations of the intrinsic brightness displayed by a particular FP, the configuration of the imaging equipment is equally and critically important to achieve high signal strength in an imaging experiment. The laser system or arc-discharge lamp coupled to fluorescence filters used to excite the chromophore should strongly overlap the chromophore absorption profile, and the emission filters must have the widest possible bandpass region coinciding with the emission spectrum. In addition, the camera system must be capable of recording images with high quantum efficiency in the fluorescence emission region of interest [39], and the optical system of the microscope should have high throughput in the wavelength regions necessary

for producing excitation and gathering emission. Even with research-level instrumentation, it is often difficult to achieve the maximum potential FP brightness levels in each spectral class unless the fluorescence filter sets are optimized for imaging the proteins. Many multiuser core imaging facilities have limited inventories of filter sets typically designed for traditional synthetic fluorophores rather than FPs. For example, the standard DAPI (4', 6-diamidino-2-phenylindole; ultraviolet excitation), FITC (fluorescein isothiocyanate; cyan–blue excitation), TRITC (tetramethylrhodamine isothiocyanate; green excitation), and Texas Red (yellow excitation) fluorescence filter combinations, often marketed by default with widefield arc-discharge microscopes, are not suitable for many FPs and are less than optimal for others.

Photostability

A commonly cited limitation of FPs relative to other fluorophore technologies is their propensity to photobleach during observation. In other words, illumination of an FP causes it to self-destruct through a series of poorly understood and likely complex mechanisms. Two probable mechanisms for photobleaching of FPs are reaction with a reactive oxygen species (ROS; i.e., singlet oxygen generated by the FP chromophore itself) and photoinduced isomerization [40]. Synthetic dyes are, of course, also susceptible to photobleaching by related mechanisms. One might expect that due to the protective protein shell that holds the FP chromophore rigid and planar and protects it from the bulk environment [41, 42], FPs should be significantly more photostable than a fluorescent dye. Generally speaking, this is not true, and for the FPs considered "best in class," the average photostability is on par with that of the widely used synthetic dye, fluorescein [3]. The most photostable of all currently available monomeric FPs is mEGFP, which is ~33-fold more photostable than fluorescein. The fluorescence of fluorescent nanoparticles (or quantum dots) does not rely on the conjugated systems of double bonds that are the "Achilles heel" of FPs and synthetic dyes with respect to photobleaching. For this reason, nanoparticles have greatly improved photostability over even the best FPs and synthetic dyes [30].

Although there is a high degree of uncorrelated variability between FPs in terms of photostability, most variants listed in Table 1.1 are useful for short-term imaging (from 1 to 25 captures), while several of the more photostable proteins can be employed in time-lapse sequences that span periods of 24 h or longer (in which hundreds to thousands of images are gathered). The long-term stability of any particular protein, however, must be investigated for every illumination scenario (widefield, confocal, multiphoton, swept-field, etc.) because nonlinear differences in photostability are often observed with the

Table 1.1. A compilation of properties of the most useful FP variants. Along with the common name and/or acronym for each FP, the peak excitation (Ex) and emission (Em) wavelengths, molar extinction coefficient (EC), quantum yield (QY), relative brightness, and physiologically relevant quaternary structure are listed (*signifies a weak dimer). The computed brightness values were derived from the product of the molar extinction coefficient and quantum yield, divided by the value for EGFP. This listing was created from scientific and commercial literature resources and is not intended to be comprehensive, but instead represents FP derivatives that have received considerable attention in the literature and may prove valuable in research efforts. The excitation and emission peak values listed may vary in published reports due to the broad spectral profiles. In actual fluorescence microscopy investigations, the experimental brightness of a particular FP may differ (in relative terms) from the brightness provided in this table. Among the many potential reasons for these differences are wavelength-dependent differences in the transmission or reflectance of microscope optics and the efficiency of the camera. Furthermore, the extent of FP folding and maturation will depend on both the particular variant being used as well as the particular characteristics and localization of the fusion partner

Protein (Acronym)	Ex (nm)	Em (nm)	$EC \times 10^{-3}$ M^{-1} cm^{-1}	QY	Quaternary Structure	Relative Brightness (% of EGFP)	Reference
Blue Fluorescent Proteins							
Azurite	384	450	26.2	0.55	Monomer*	43	[91]
EBFP2	383	448	32.0	0.56	Monomer*	53	[57]
mTagBFP	399	456	52.0	0.63	Monomer	98	[100]
Cyan Fluorescent Proteins							
ECFP	439	476	32.5	0.40	Monomer*	39	[185]
TagCFP	458	480	37.0	0.57	Monomer	63	Evrogen
mCerulean	433	475	43.0	0.62	Monomer*	79	[88]
CyPet	435	477	35.0	0.51	Monomer*	53	[35]
AmCyan	458	489	44.0	0.24	Tetramer	31	[22]
Midoriishi Cyan	472	495	27.3	0.90	Dimer	73	[73]
mTFP1	462	492	64	0.85	Monomer	162	[77]
Green Fluorescent Proteins							
EGFP	488	507	56.0	0.60	Monomer*	100	[17]
Emerald	487	509	57.5	0.68	Monomer*	116	[97]
Azami Green	492	505	55.0	0.74	Monomer	121	[72]
mWasabi	493	509	70.0	0.80	Monomer	167	[107]
ZsGreen	493	505	43.0	0.91	Tetramer	117	[22]
TagGFP	482	505	58.2	0.59	Monomer	102	Evrogen
Superfolder avGFP	485	510	83.3	0.65	Monomer*	160	[55]
T-Sapphire	399	511	44.0	0.60	Monomer*	79	[44]
Yellow Fluorescent Proteins							
EYFP	514	527	83.4	0.61	Monomer*	151	[186]
Topaz	514	527	94.5	0.60	Monomer*	169	[60]
Venus	515	528	92.2	0.57	Monomer*	156	[56]
Citrine	516	529	77.0	0.76	Monomer	174	[92]
YPet	517	530	104	0.77	Monomer*	238	[35]
ZsYellow	529	539	20.2	0.42	Tetramer	25	[22]
TagYFP	508	524	64.0	0.60	Monomer	118	Evrogen
mAmetrine	406	526	45.0	0.58	Monomer	78	[187]
Orange Fluorescent Proteins							
Kusabira Orange	548	559	51.6	0.60	Monomer	92	[73]
Kusabira Orange2	551	565	63.8	0.62	Monomer	118	[114]
mOrange	548	562	71.0	0.69	Monomer	146	[36]
mOrange2	549	565	58.0	0.60	Monomer	104	[115]
dTomato	554	581	69.0	0.69	Dimer	142	[36]
dTomato-Tandem	554	581	138	0.69	Pseudo Monomer	283	[36]
DsRed	558	583	75.0	0.79	Tetramer	176	[22]
DsRed-Express (T1)	555	584	38.0	0.51	Tetramer	58	[188]
DsRed-Monomer	556	586	35.0	0.10	Monomer	10	Clontech
TagRFP	555	584	100.0	0.48	Monomer	142	[118]
TagRFP-T	555	584	81.0	0.41	Monomer	99	[115]
Red Fluorescent Proteins							
mRuby	558	605	112.0	0.35	Monomer	117	[126]
mApple	568	592	75.0	0.49	Monomer	109	[115]
mStrawberry	574	596	90.0	0.29	Monomer	78	[36]
AsRed2	576	592	56.2	0.05	Tetramer	8	[22]
mRFP1	584	607	50.0	0.25	Monomer	37	[64]
JRed	584	610	44.0	0.20	Dimer	26	[93]
mCherry	587	610	72.0	0.22	Monomer	47	[36]
HcRed1	588	618	20.0	0.015	Dimer	1	[123]
mRaspberry	598	625	86.0	0.15	Monomer	38	[34]
mKate	588	635	45.0	0.33	Monomer	44	[128]
HcRed-Tandem	590	637	160	0.04	Pseudo Monomer	19	[78]
mPlum	590	649	41.0	0.10	Monomer	12	[34]

same protein when illumination is produced by an arc-discharge lamp versus a laser system. The molecular basis of nonlinear differences in photobleaching of FPs versus light intensity and wavelength is largely an open question that we hope will be addressed in the future. In terms of photostability, the selection of a suitable FP is dictated by numerous parameters, including the illumination conditions, the expression system, and the effectiveness of the imaging setup.

Color Class

What does "color" mean in the context of fluorescence? It could, reasonably, refer to the perceived color of a solution of the fluorophore when viewed in white light. The term could also, reasonably, refer to the perceived color of the solution when illuminated with monochromatic light of a wavelength that corresponds to the absorbance maxima. In practice, any attempt to define fluorescence color by virtue of how it is perceived by eye leads to complications. A more rigorous and practical approach to defining fluorescence color is to say that two fluorophores have different color if their excitation and/or emission maxima and/or peak shapes are significantly different. Defining "significantly different" is troublesome as it depends on the instrumentation available for measuring the shape and maxima of the emission and excitation peaks. For example, spectral imaging can be used to differentiate two colors that could not be differentiated through the use of bandpass filters.

Regardless of whether one considers synthetic fluorophores, FPs, or quantum dots and assuming all other considerations are the same, how will the researcher choose which color to use? An important consideration with respect to color selection is the greater desirability of red-shifted fluorophores [43]. It is generally accepted that excitation with violet or blue light is associated with greater cellular phototoxicity than excitation with green, yellow, or longer wavelength light extending through the near infrared (up to ~1000 nm) but not into the true infrared (where heating due to absorption by water would be problematic for cell viability). Fluorescence excitation and emission hues of FPs are confined to a relatively narrow region of the electromagnetic spectrum (essentially the visible wavelengths) due to protein-imposed restrictions on the possible manipulations of the chromophore structure and environment. In contrast, synthetic dyes and nanoparticles with fluorescence emission tuned to wavelengths that cover the visible and near-infrared regions of the spectrum are available. This spectral limitation of FPs is exacerbated by their relatively broad excitation and emission peaks (ranging up to 100 nm) that further restrict the number of colors that can be distinguished with bandpass filters on a widefield microscope. Practically speaking, the bandwidth of the absorption and emission peaks is

an important consideration in defining the number of colors that are "spectrally distinct." Roughly speaking, there are currently about ten different emission colors of FPs with short Stoke shifts (defined as the distance in nanometers between the absorption and emission peak wavelengths of a fluorophore) and emission maxima spaced every 20 nm between 450 and 650 nm (Table 1.1). These colors include: blue (~450 nm), cyan (~470 nm), teal (~490 nm), green (~510 nm), yellow (~530 nm), yellow–orange (~550 nm), orange (~570 nm), orange–red (~590 nm), red (~610 nm), and far-red (>630 nm). There are a few additional long Stoke shift FPs such as Sapphire [44] and mKeima [45], which, given the definition of fluorescence color provided previously, should be considered additional color classes. However, due to the relatively broad excitation and emission peaks shared by all FPs, it is only really practical to simultaneously image three (Figure 1.3 [46]) or four distinct colors (such as cyan, yellow, and red or blue, green, orange, and far-red) using a bandpass filter-based microscopy system [3]. However, this tenet does not always hold true as the imaging of six distinct colors (CFP, cyan; mMiCy, teal; EGFP, green; YFP, yellow; dKeima570, orange; and mKeima, red) has been achieved using a single laser line for excitation and spectral unmixing of the emission [45].

Hybrid Approaches

This discussion has established that, relative to synthetic dyes and quantum dots, the physical properties of FPs are less than ideal yet more than adequate. Investigators that simply require a fluorophore with high fluorescent brightness, good photostability, and broad color selection would do better with synthetic dyes or quantum dots. However, as mentioned earlier, such superficial comparisons are a disservice to FPs because the fact that these probes are proteins, and are therefore genetically encodable, is their overwhelming advantage for many biological applications. In recent years there has been significant progress in developing "hybrid" technologies for the protein-specific labeling of recombinant proteins in live cells [47–49]. These approaches typically exploit modified dyes (or quantum dots [50]) for noncovalent binding or covalent attachment to a genetically encoded sequence that can be appended to a recombinant protein of interest. Notable examples of such methods include biarsenical xanthene dye-based labeling of tetracysteine motifs [51] and benzylguanine-dye conjugate-based labeling of O6-alkyguanine-DNA alkyltransferase fusion proteins [52], though a number of additional new systems have been reported [47–49]. Although these techniques hold great promise, none of them has yet achieved the versatility and widespread acceptance of FP-based labeling. A major limitation shared by all hybrid methodologies is the nonspecific labeling of intracellular structures with the exogenously applied dye [53]. In

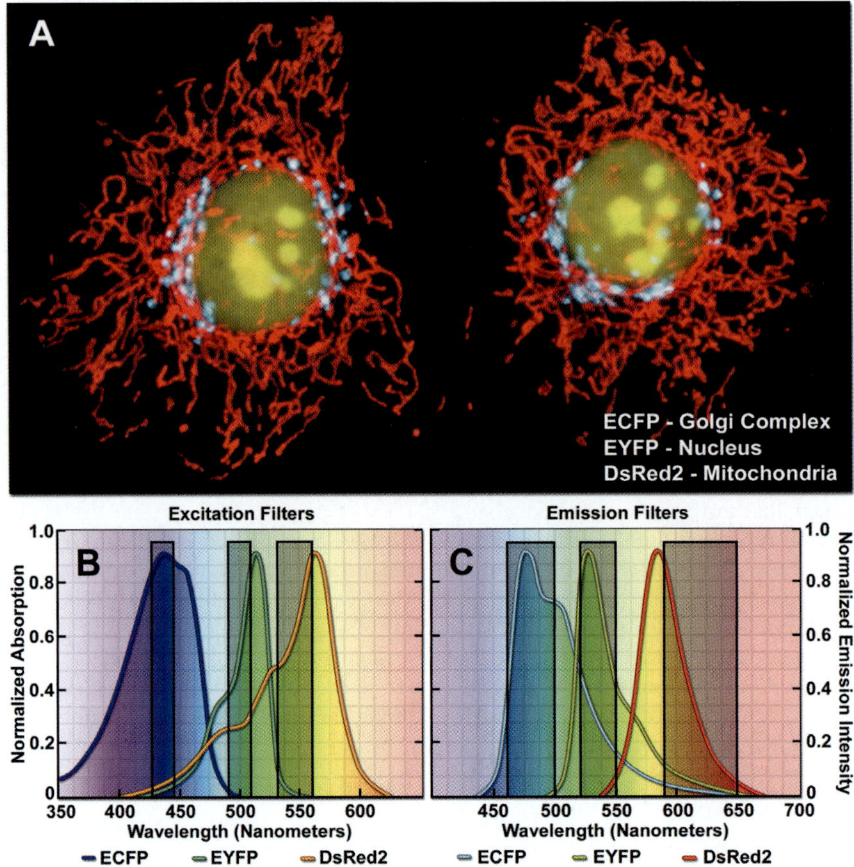

Figure 1.3. Optimized fluorescence filter combinations for multicolor imaging of three FPs spanning the cyan to orange-red wavelength regions; (**A**) Widefield fluorescence image of HeLa cells labeled with ECFP (Golgi complex targeting signal), EYFP (nuclear targeting signal), and DsRed2 (mitochondrial targeting signal); (**B**) Excitation filters optimized for ECFP, EYFP, and DsRed2 FPs having center wavelengths of 436, 500, and 545 nm, respectively. The bandwidth of the ECFP and EYFP excitation filters is 20 nm whereas the bandwidth of the DsRed2 filter is 30 nm; (**C**) Emission filters optimized for the same probes having center wavelengths of 480, 535, and 620 nm with bandwidths of 40, 30, and 60 nm, respectively.

many cases, high levels of nonspecific background staining hampers observation of the targeted structures, and several of the synthetic dyes are sequestered in the mitochondria, lysosomes, and other organelles.

DIRECT COMPARISONS OF FLUORESCENT PROTEINS TO EACH OTHER

For direct comparison of one FP to another, the properties of brightness, photostability, and color remain the three most important criteria. However, there are additional concerns that are direct consequences of the unique experimental designs made possible with FPs. For example, because these probes are proteins, they must undergo efficient transcription, translation, and folding to be functional. Once correctly folded, they then undergo autocatalytic posttranslational chromophore formation, a process informally referred to as "maturation" or "ripening." If the efficiency of any of these steps is compromised, the experimentally observed fluorescence will be diminished or even abolished. Such concerns are not relevant to alternative technologies such as synthetic dyes and quantum dots applied directly to the cells or tissue. In the following sections we list several of the most important criteria that can be used to directly compare FPs and discuss efforts to engineer new variants that are superior by these criteria.

Folding and Maturation Efficiency

Aequorea jellyfish inhabit the cool ocean waters off the coast of Washington and British Columbia. Accordingly, the natural environment of the avGFP protein is one where the temperature hovers around 4–5° C. In contrast, in the unnatural environment (from the FP's perspective) of a transfected cell culture or the cells of a transgenic organism, the avGFP protein will most often experience much higher temperatures than those in which it was evolved to fold, mature, and function. FPs derived from reef corals and sea anemones generally express well at 37° C without genetic selection, presumably because the native species from which the proteins are obtained have evolved in somewhat warmer habitats [54]. The original transposition of wild-type avGFP from jellyfish to cells grown at 37° C substantially decreased the efficiency with which the protein could fold into its proper three-dimensional (tertiary) structure. Clearly this problem needed to be addressed through protein engineering. Indeed, among the first and most substantial improvements to the avGFP protein were realized by selection of variants with more efficient folding at 37° C. Years of progress in this regard have most recently led to a so-called superfolder avGFP with improved folding kinetics, tolerance to circular permutations, high performance in fusions to poorly folding polypeptides, and resistance to denaturation [55]. One of the more interesting and useful aspects of mutations that improve folding efficiency is that they are often translated to different FP colors where they seem to provide similar improvements. The translation of so-called folding mutations to hue-shifted variants has contributed to the excellent folding properties or high brightness of the Venus YFP variant [56], EBFP2 [57], and the series of "super" cyan and yellow FPs [58, 59], among others [55].

The presence of molecular oxygen is also a critical factor in FP chromophore development during the maturation process. During the formation of chromophores in *Aequorea* protein variants, at least one oxygen molecule is required for an oxidation reaction [60, 61], whereas reef coral proteins that emit in the orange–red spectral regions usually require two molecules [62, 63]. In mammalian cell cultures, FP maturation is rarely hampered by a lack of oxygen, but anoxia could become a limiting factor in other systems.

Oligomerization

All of the FPs discovered to date display at least a limited degree of quaternary structure (self-association of individual protein units), exemplified by the weak tendency of native avGFP and its derivatives to dimerize when immobilized at high concentrations [64, 65], as well as the obligate tetrameric structure characteristic of FPs from reef coral and anemones [66, 67]. Oligomerization can be a significant problem for many applications in cell biology, particularly in cases where the FP is fused to a host protein targeted at a specific subcellular location. Once expressed, the formation of dimers and higher-order oligomers induced by the FP portion of the chimera can produce atypical localization, disrupt normal function, interfere with signaling cascades, or restrict the fusion product to aggregation within a specific organelle or the cytoplasm. This effect is particularly

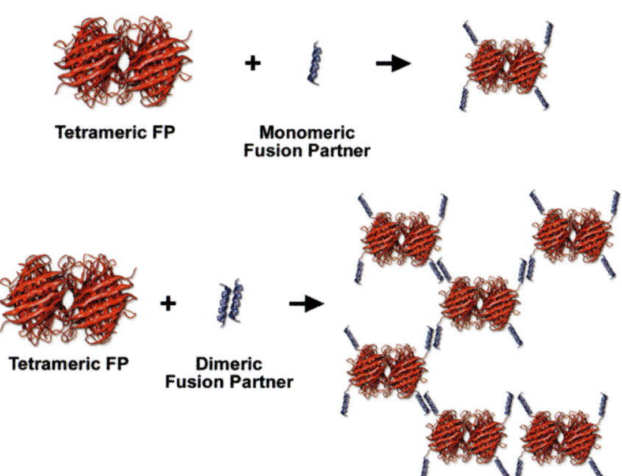

Figure 1.4. Any protein fused to a tetrameric FP will become tetrameric itself.

marked when the FP is fused to partners that participate in natural oligomer formation themselves (see Figure 1.4). Fusion products with proteins that form only weak dimers (i.e., most *Aequorea* variants) may not exhibit aggregation or improper targeting, provided the localized concentration remains low. However, when FPs are targeted to specific cellular compartments, such as the plasma membrane, the localized protein concentration can, in some circumstances, become high enough to permit dimerization.

The basic strategy for overcoming oligomerization artifacts is to modify the FP amino acid sequence to include residues that disrupt intermolecular interactions, a procedure that varies in complexity depending upon the nature and origin of the protein. For many avGFP variants, dimerization can be either significantly reduced or eliminated by replacing the hydrophobic amino acid side chains in the dimer interface with positively charged residues at several key sequence positions [65]. The three most successful mutations, in decreasing order of effectiveness, are A206K, L221K, and F223R, where the nonpolar amino acids alanine, leucine, and phenylalanine are replaced by one of the positively charged hydrophilic amino acids lysine or arginine. In cases where close molecular associations are suspected involving a fusion protein and where quantitative FRET interactions are investigated, it is highly recommended that avGFP variants (i.e., CFP and YFP) be converted into monomers using the A206K point mutation [3, 68].

Creating FP monomers from the tetrameric reef coral and sea anemone proteins is usually far more difficult. Even at exceedingly low concentrations, the original DsRed FP is an obligate tetramer [66] that cannot be dissociated without irreversible denaturation of the polypeptides. In the tetrameric unit, each DsRed protomer interacts with two adjacent neighbors, one through a hydrophobic interface and the other through a hydrophilic interface resulting in a complex assembly

[69, 70]. Other Anthozoa proteins, such as the *Zoanthus* variants and eqFP611, apparently have weaker interactions between the units of the tetramer and may be easier to disrupt into monomers [71]. The most successful approaches [64, 68] have involved the use of site-directed mutagenesis to disrupt the tetrameric interfaces, usually by substitution of hydrophilic or charged amino acids for hydrophobic and neutral moieties. Following the precedent of the process used to break DsRed into a monomeric variant [64], a growing number of coral-derived FPs have now been monomerized. Some notable examples include a green FP from *Galaxeidae* [72]; an orange FP from the *Fungia concinna* [73]; photoconvertible FPs from both *Lobophyllia hemprichii* [74] and *Dendronephthya* [75]; a photoactivatable GFP from *Pectiniidae* [76]; a chromoprotein from *Montipora* [45]; and a cyan FP from *Clavularia* [77].

Another useful technique for preventing oligomerization artifacts for dimeric fluorescent proteins involves creating vectors containing two sequential coding regions separated by a short unstructured linker. Upon expression, the fused FPs, known as "tandem dimers," preferentially bind to each other to form an intramolecular dimeric unit that performs essentially as a monomer although at twice the molecular weight (and size). Tandem dimer constructs have been developed with DsRed derivatives [36, 64], HcRed [78], and a photoconvertible FP known as EosFP [79]. A somewhat different strategy for reducing or eliminating the artifacts associated with FP oligomerization is to simultaneously coexpress FP-tagged proteins with an excess of a nonfluorescent mutant of the same FP [80, 81]. Related to the problem of FP oligomerization is the problem of FP aggregation. Although oligomerization tends to refer to the ability of some FPs to form well-defined quaternary structures, aggregation refers to the tendency of some FPs to act somewhat sticky toward themselves and to form poorly defined complexes of indeterminate stoichiometry. The problem of aggregation seems to be confined to coral-derived FPs and is not generally recognized as a problem with *Aequorea*-derived FPs. An effective strategy for minimizing FP aggregation is the removal of several basic residues that seem to be primarily responsible for the tendency to aggregate from the N-terminus of the FP [82]. Regardless of the specific mechanism employed to overcome FP oligomerization and/or aggregation, the most important point is that experimental results are not compromised by artifacts induced by the existence of quaternary structures.

Although it is generally agreed that a monomeric FP is more desirable than an oligomeric FP, there is at least one case where an oligomeric structure can be advantageous. Dimerization of a FRET pair containing FPs can result in particularly efficient FRET if the chromophores are appropriately oriented. High levels of FRET are desirable in the case of FRET-based sensors of protease activity (e.g., caspase sensors) because cleavage of the substrate (the inter-FP linker) causes dissociation of the FPs and loss of FRET. The higher the FRET level in the initial construct, the greater the change in signal once the linker has been digested. For example, evolutionary optimization of a caspase-3 sensor based on CFP and YFP produced a new FRET pair, known as CyPet and YPet, with greatly improved FRET efficiency [35]. The improved FRET efficiency in the intact state necessarily leads to a substantially improved ratiometric change upon linker cleavage. However, subsequent work has revealed that CyPet and YPet actually have an increased tendency to dimerize in an as yet undetermined orientation compatible with high FRET efficiency [83, 84].

Fusion Tolerance

The most common application for FPs is in the creation of a functional chimera (or fusion) with a second, targeting protein that is of particular interest to a researcher. In the ideal situation, the appended FP would have no effect on the normal folding, localization, biological function, and molecular interactions of the fusion partner. Similarly, the host would ideally have no adverse effect on the folding and maturation of the FP. This requirement poses an important question: In what percentage of fusion proteins are both the FP and the host protein well behaved? Unfortunately, there is no definitive answer. Although the literature is loaded with examples of successful fusions, it is likely that many unsuccessful fusions are never published. In addition, it is possible that in many cases the success or failure of a particular chimera depends on specific details, such as whether it was an N- or C-terminal fusion, the linker length, and the particular identity of the FP. For lack of a better mechanism for addressing this question, we can turn to a study in which avGFP fusions were created with every open reading frame (ORF) of the budding yeast, *Saccharomyces cerevisiae*. Of the 6234 ORFs investigated, 4154 (or 67%) resulted in expression of green fluorescence [85]. Although it is not clear exactly what fraction of these fusions retained full functionality of the host protein, it is also unknown how many of the 33% additional ORFs may have yielded to an alternate fusion topology, linker, or FP variant. With these caveats in mind, two-thirds is probably a reasonably conservative approximation of the fraction of FP fusions that will be well behaved. Multimodality fusion reporter genes are further discussed in Chapter 5.

Fluorescence Lifetime Properties

Although the issues described previously are pertinent to virtually all researchers who employ FPs, the homogeneity of the lifetime decay is of importance to only the growing subset of investigators who employ FPs for fluorescence lifetime imaging microscopy (FLIM) [86].

Whereas standard fluorescence microscopy detects the spatially and wavelength-resolved fluorescence intensity of FPs in cells, FLIM detects the spatially and wavelength-resolved decay of the nanosecond-scale excited state for fluorophores. Thus, in FLIM experiments the nanosecond decay kinetics of the FP chromophore electronic excited state (referred to as the fluorescence lifetime, τ) is determined in spatial coordinates using a specialized microscope detector capable of high-frequency modulation or fast gating.

This approach offers certain advantages over intensity-based methods for cellular imaging, such as being independent of fluorophore concentration and having an exquisite sensitivity toward the chromophore environment. FLIM is particularly useful in combination with FRET for difficult measurements that are not conducive to acceptor photobleaching or sensitized emission methods. However, the very sensitivity that makes FLIM a useful technique also presents additional complexities and technical challenges for data acquisition. The most relevant issue is that many FPs display complex multiple-lifetime decays. Interpretation of FLIM data is therefore nontrivial due to the fact that these decay modes are made even more complex by the heterogeneity of the intracellular environment and the presence of FRET acceptors [87]. This issue could be addressed by development of new FP FRET pairs in which the donor has a homogeneous lifetime decay, and such a goal is widely recognized in FP engineering [77, 88]. Unfortunately, fluorescence lifetime is a good example of a property that is difficult to select in the type of screens typically used in directed evolution of new FPs.

Does Newer Always Mean Better?

With the ever-growing number of new FP variants and the ongoing reengineering of various generations of individual FPs, picking the "best" FP for a particular application is becoming increasingly more confusing. As a general rule of thumb we recommend sticking with those FP variants that are "tried-and-true" rather than simply choosing the most recently published variant. Numerous factors have the potential to negatively impact the performance of an FP in a particular experiment, and new FP variants rarely see testing against all these factors prior to publication. For example, although a new FP variant may behave well in the few standard test fusions attempted by the developers, it might not be practical for all conceivable protein fusions. Even weak residual dimerization for a protein engineered to be monomeric could perturb the localization of some fusions (but not others). Additional subtle factors include the fact that certain FPs are "stickier" (more prone to aggregation) than others and may, under certain conditions, mislocalize due to nonspecific interactions between charged or hydrophobic patches on the surface of the protein. The ultimate validation for an FP is, of course, widespread critical evaluation followed by eventual acceptance by the research community. Using this strictest of criteria, the one FP that stands above the crowd is the avGFP-derived EGFP variant, which exhibits good to excellent performance by all criteria listed previously. Another particularly well-validated and robust FP is the DsRed-derived mCherry, which is an excellent choice for a second color to pair with EGFP.

ENGINEERING IMPROVED FLUORESCENT PROTEINS

As a class, FPs have been subjected to more extensive protein engineering and artificial directed evolution than almost any other class of protein. Why have FPs received so much attention in this regard? The simple answer is that FPs are extremely popular tools in the biological sciences and improved variants can provide huge benefits to researchers. However, this is not the entire story, and there are two additional factors at play. The first factor is that often the very same researchers who employ FPs in their research are the ones who recognize the deficiencies and have the skills to address them. Specifically, researchers who employ FPs are typically experts in molecular biology and fluorescence spectroscopy and microscopy; exactly the tools necessary to undertake engineering of FPs for improved properties. Contrast this to the example of a cell biologist who is dissatisfied with the pharmacological specificity of a particular kinase inhibitor. The cell biologist will almost certainly lack the skills and resources necessary to undertake the synthesis of a potentially more specific inhibitor. The second factor is that, by the very nature of the property that makes them useful tools (i.e., their intrinsic visible fluorescence), identification of FP variants with, for example, improved brightness or a substantial color shift is relatively straightforward.

Fluorescent Protein Engineering and Directed Evolution: General Principles

The defining feature of avGFP is its remarkable ability to autonomously generate a fluorophore within the confines of its distinctive β-barrel structure (see Figure 1.5) [41, 42, 89]. The steric, electrostatic, and hydrogen-bonding environment imposed upon the chromophore by the surrounding residues strongly influences the fluorescence properties. Remarkably and perhaps fortunately, the avGFP chromophore was found to be highly amenable to genetic modification of both its covalent structure and localized environment, and this tolerance has been exploited for the creation of wavelength-shifted variants [60] spanning an 80-nm range. In addition, the avGFP protein is also very tolerant of modifications of residues with side chains that are

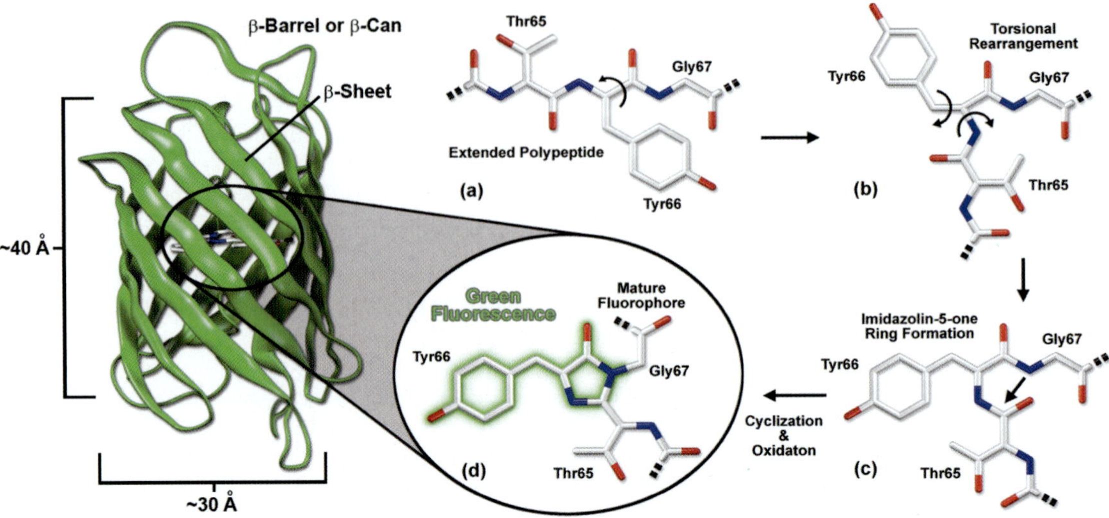

Figure 1.5. Schematic diagram of the β-barrel structure and chromophore formation in maturing enhanced green fluorescent protein (EGFP). (**A**) The prematuration EGFP fluorophore tripeptide amino acid sequence (Thr65-Tyr66-Gly67) is stretched into a linear configuration so that the threonine residue is positioned in the upper left-hand corner of the diagram. The first step in maturation is a series of torsional adjustments (**B**) and (**C**) that relocate the carboxyl carbon of Thr65 so that it is in close proximity to the amino nitrogen of Gly67. The nucleophilic attack of the amide nitrogen of Gly67 on the carboxyl group of Thr65 (**C**), followed by dehydration, results in formation of an imidazolin-5-one heterocyclic ring system. (**D**) Fluorescence occurs when oxidation of the tyrosine α-β carbon bond by molecular oxygen extends electron conjugation of the imidazoline ring system to include the tyrosine residue.

external to the β-barrel, including those on the surface of the barrel and additions (along with limited truncation) at the N- and C-termini of the protein. Efforts to genetically modify the structure of FPs can be broadly classified into two categories: rational modifications and irrational modifications. The former category would contain all FP fusion proteins that are obviously rationally designed and constructed to address a specific biological question. However, when it comes to efforts to modify the FP itself for the purposes of engineering new colors or otherwise improved variants, irrational approaches tend to be more effective than rational approaches.

Generally speaking, our ability to make rational modification of any protein (FPs included) to generate variants with new properties is sadly limited. Accordingly, the number of successful avGFP modifications that probably seemed rational *a priori*, and were ultimately experimentally validated, are relatively few in number. Some important examples of rational mutations of avGFP include Tyr66Trp to create CFP and its descendents [18]; Tyr66His to create BFP and its descendents [18]; introduction of a stacking residue at residue Thr203Tyr to create YFP and its progeny [42]; and the monomerizing mutation Ala206Lys [65] (Figure 1.6). Each of these

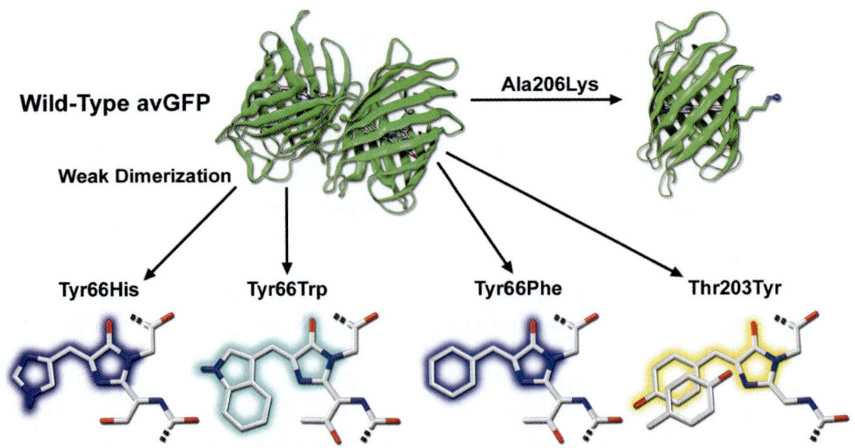

Figure 1.6. Rational modifications of avGFP.

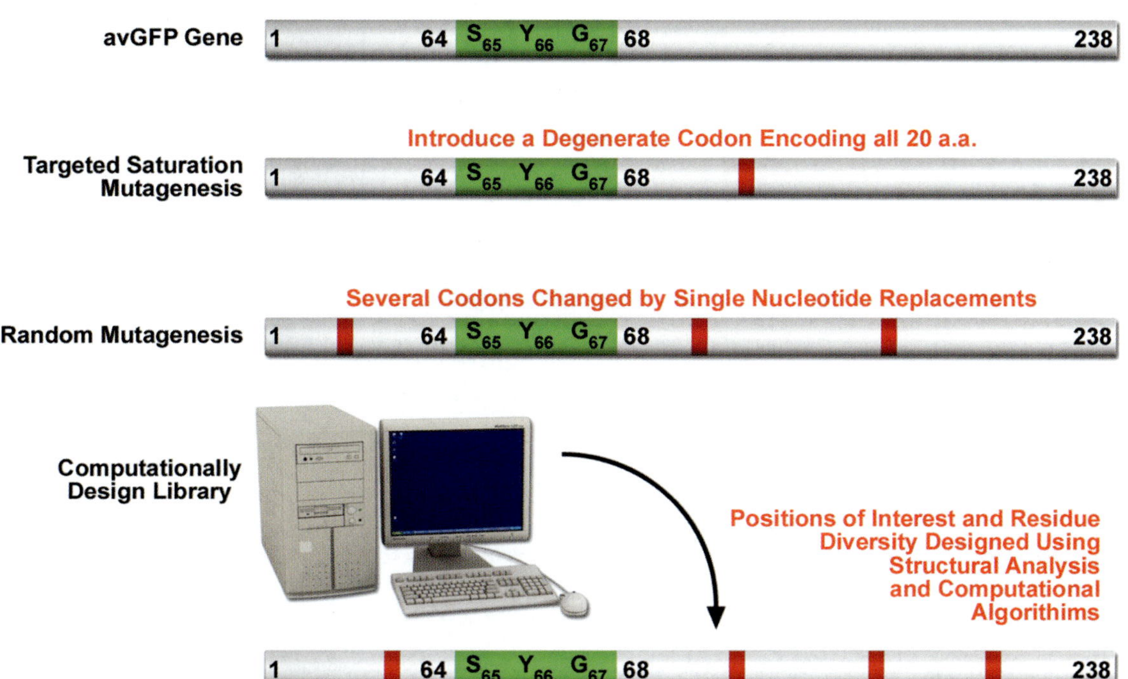

Figure 1.7. Strategies for creation of libraries of FP variants.

modifications involves dramatic changes in the chromophore structure, the chromophore environment, or the nature of oligomeric interactions. Rational modifications of proteins are almost never subtle! However, these dramatic changes are often the critical first steps that ultimately lead to the development of important variants with new colors or properties. From the perspective of protein evolution, rational modifications tend to represent artificially induced evolutionary leaps that stand in marked contrast to the baby-steps that would tend to characterize protein evolution in the wild. Unfortunately, these rational changes are almost always accompanied by substantially decreased fluorescent brightness. To "rescue" the fluorescent brightness of a new variant requires the introduction of compensating mutations compatible with the original rational modification. This is where irrational approaches are most effective. For example, during the breakup of tetrameric DsRed into a functional monomer, the initial mutagenesis efforts designed to disrupt oligomerization resulted in a dramatic reduction in red fluorescence. Subsequent rounds of targeted and random mutagenesis successfully directed the evolution of DsRed into a bright monomer in eight generations.

As they are defined here, irrational approaches attempt to mimic the process of natural protein evolution. That is, the gradual accumulation of mutations that each confer a small, but additive, benefit. These modifications tend to be subtle "tweaks" of the protein structure (far too subtle to be rationally designed), often

involving slight shifts in the packing of hydrophobic cores or in the placement of buried hydrogen bond donors and acceptors or electrostatic charges. Our understanding of protein structure and function is inadequate for rational prediction of which particular mutations might be beneficial. Fortunately, the subtlety and effectiveness of natural protein evolution can be effectively mimicked and even greatly accelerated in the research laboratory. In laboratory-based directed protein evolution, genetic diversity is created through the use of molecular biology (Figure 1.7), and then the resulting library of protein variants is screened to identify variants with improved properties. The primary advantage of this approach is that it does not require a complete understanding of the protein structure and function to be successful. Indeed, it is often difficult to rationalize beneficial mutations identified through this approach.

The key to success for directed evolution of proteins is having an effective high-throughput screen to identify mutant proteins with favorable properties. It should be noted that only those properties being screened for will be caught in the assay, whereas those not being monitored, either beneficial or deleterious, will pass through unnoticed. With respect to directed evolution of FPs, the easiest screen to perform is based on the brightness of the FP when expressed in bacteria. Accordingly, this approach can produce FPs that are brightly fluorescent, but in some cases only when expressed in bacteria. The same variant may or may not be as brightly fluorescent when expressed in other cell types or when expressed as

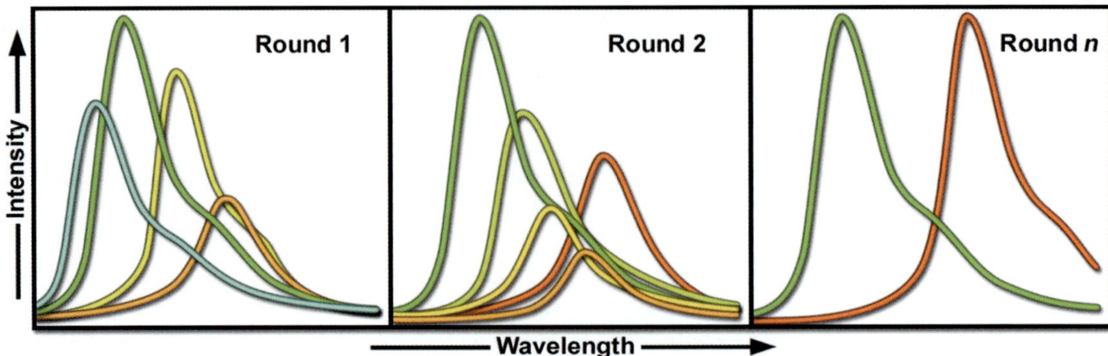

Figure 1.8. Hypothetical representation of the directed evolution of a red-shifted FP.

a fusion to another protein. Consistent with this corollary, FPs evolved only for brightness have not been subjected to selective pressure for photostability, color, or the homogeneity of the lifetime decay, and thus will most likely not be optimal with respect to these important properties. A major challenge for the future is designing library screens that will simultaneously select for variants improved by all relevant criteria. As yet, there has been scant progress in this area beyond the simultaneous screening for brightness and photostability [4].

Figure 1.8 shows a hypothetical series of emission profiles for FPs identified during the process of directed evolution for a new property; in this case a red-shifted emission peak. The intention of this figure is to illustrate general principles of how laboratory evolution of an FP tends to proceed. The green line is the fluorescent emission profile of the progenitor FP. The gene encoding this progenitor FP has undergone diversification by any of a number of strategies (Figure 1.7). In the first round of library screening, a variety of variants with shifted emission maxima, but dimmer fluorescence, were identified. In general, more dramatic color changes (spectral shifts of fluorescence emission) produce more substantial adverse effects on fluorescent brightness. The most red-shifted variant in Figure 1.8 (orange line) was chosen as the template for library creation in the second round. Library screening resulted in the identification of variants with incrementally improved brightness and minor shifts in emission maxima. To be successful, the researcher must choose the variant that is the best compromise of brightness and red shift to carry into the next round of evolution. Fluorescence-activated cell sorting (FACS) is often a method of choice for screening of FP libraries because it enables the rapid screening of millions of variants and with rapid isolation of only the brightest cells [34–36, 90, 91]. Repeating this process through many rounds will often (but not always) result in a new FP color with brightness that equals or exceeds the original protein. For example, Citrine [92] and Venus [56] are red-shifted variants derived from EGFP yet exceed their parent in brightness. It has been demonstrated on

numerous occasions that extensive directed evolution can produce new colors of FP with exceptional brightness. Recent examples include the development of the brightest cyan (or teal) [77] and blue [57] FPs currently available. On the other hand, extensive selection for red-shifted variants was ultimately successful in producing mPlum, which is red-shifted in its emission by 42 nm but only a third as bright as its progenitor, mRFP1 [34].

The Future of Fluorescent Protein Engineering

Through a combination of rational and random mutagenesis, the properties of the most useful FP variants have continued to improve incrementally. With such widespread adoption, one might think of the FPs as mature, optimized tools. However, this is not the case, and the biochemical and photophysical properties of FPs still limit their utility in many applications. It is interesting to note that most FP development to date has focused on just two parent genes: that of the avGFP and the *Discosoma* RFP. The known FPs obtained from phylum Cnidaria are now more than one hundred, and this quantity promises to keep growing [24]. FPs have recently been cloned from organisms of the phylum Arthropoda [25, 93], so protein engineers may soon have access to a new selection of templates from which to evolve exciting new variants.

An obvious direction for future work in developing improved FPs is the engineering of new variants with improved photostability. However, it may not be immediately obvious how one might go about screening FP libraries to identify variants with improved photostability. This is certainly a much more difficult property to screen for than simple fluorescent brightness. One approach proven to be effective in identifying variants with improved photostability is bleaching of FP libraries with an array of intense light emitting diodes (LEDs) [77]. Repeated rounds of selection for variants that bleached the slowest when exposed to intense blue light from an LED array eventually resulted in the identification of mTFP1 (teal FP), which was more than 100-fold more photostable than its progenitor, mTFP0.7. It is possible that applying similar approaches to other FPs could provide similar improvements in photostability. Another approach proven successful is the use of FACS [91], which successfully enriched a highly photostable variant of EBFP from a large library of variants, simply by selecting for the brightest variants. In this case, bleaching during the brief passage of single cells through the intense excitation laser beam may have been significant enough to decrease the overall intensity for the least photostable members of the library. Alternatively, ROS generated during extensive photobleaching may have been toxic to those cells harboring the least photostable members of the library. Despite these few promising examples, it is clear that if dramatic improvements in photostability are to be realized for the most photostable of the FPs (i.e., beyond their already good photostability), much longer exposures and/or much more intense light sources will be required for the FP library bleaching screens.

Other important goals for protein engineers include the development of brighter and more red-shifted RFP variants; ideally with emission extending into the so-called near-infrared window where tissue absorbance is at minimum [43]. Yet another objective will be the development of FPs with homogeneous fluorescence lifetimes. To address these goals, we expect that the design and screening of FP libraries will become more sophisticated as computational methods and multiparameter high-throughput screens become the norm. As more diverse FP sequences are deposited in the nucleotide databases, it is likely that researchers will embrace the creation of fully synthetic FP libraries guided by sequence alignments and consensus FP sequences [94]. Eventually we can expect *de novo* computationally designed libraries [95] in which the excited state dynamics and electronic structure of the chromophore are taken into consideration.

A recurring theme in FP engineering is that there is no one candidate best suited for all applications. Most likely this trend will hold true for the future and when (or if) we eventually do manage to develop FPs that are super-bright, super-photostable, and super-red shifted by today's standards, they will all be separate proteins of diverse origin. That is, no one protein will ever combine the best of all properties, and choosing the right FP for the experiment at hand will necessarily require a compromise with respect to some other property. We are optimistic that these compromises will be fairly minor relative to the benefits. Although an impressive degree of progress in FP development has been made to date, the temptation to say that the current FP palette is "good enough" should be actively resisted. In most cases, current FP variants are good enough to meet the demands of many current applications; in all probability these proteins will not perform as needed in future applications.

THE FLUORESCENT PROTEIN COLOR PALETTE

Over the past decade, a wide variety of new FP variants have been developed featuring fluorescence emission profiles spanning a 200-nm region of the visible light spectrum (~450 nm to ~650 nm), providing useful genetically encoded fluorophores in essentially every color class (Table 1.1) [1, 3, 67]. The fundamental origins of FP emission color have been established and are generally governed by the physical extent of π-orbital conjugation contained within the chromophore [96]. This factor largely determines the general spectral class (i.e., blue, cyan, green, yellow, or red), which can involve a change in the absorption and emission maxima by hundreds of nanometers. Smaller variations in the absorption and emission maxima (ranging from 20 to 40 nm) can be attributed, as discussed previously, to local environmental variables that include the position of charged amino acid residues, hydrogen bonding networks, and hydrophobic interactions within the local chromophore environment. Continued investigations into the photophysics of the FP chromophore will no doubt yield further clues concerning the structure–function relationship with the polypeptide backbone and amino acid side chains, thus rendering the task of engineering more finely tuned color variants and broadening the spectral range much easier.

Blue and Cyan Fluorescent Proteins

Recent advances in developing new FPs in the blue and cyan spectral regions have strengthened the potential for multicolor imaging using proteins that emit in shorter wavelengths. FPs emitting in the blue region (ranging from ~440 nm to 470 nm) were first obtained from site-directed mutagenesis efforts targeted at the tyrosine amino acid residue at position 66 in the avGFP chromophore. Conversion of this residue to histidine (Y66H) produces a blue FP (BFP) that exhibits a broad absorption band in the ultraviolet centered close to 380 nm and an emission maximum at 448 nm [18, 97]. The original protein exhibited only about 15% to 20% of the parent avGFP brightness value due to a low quantum yield and required additional secondary mutations to increase its folding efficiency and expression levels. Subsequent investigations and several additional mutations led to an enhanced BFP version (EBFP) that was still only 25% as bright as EGFP [19] and displayed poor photostability compared to many other FPs.

Caution should be exercised when imaging live cells expressing any of the blue FPs. Aside from limited brightness levels and rapid photobleaching (compared to other FPs), blue FPs also suffer from the fact that they must be excited with ultraviolet light, which is highly phototoxic to mammalian cells, even in limited doses [98,

99]. Furthermore, inherent cellular autofluorescence and high absorption levels by cells and tissues, as well as light-scattering artifacts, often hamper imaging with excitation light in this spectral region. Microscopes operating in the ultraviolet also require specialized light sources, optics, and filter combinations that further complicate imaging. For all of these reasons, the quest for more efficient blue FPs has only recently been renewed.

Using a combination of structurally targeted libraries coupled to random and site-directed mutagenesis, three protein engineering groups have recently reported improved blue *Aequorea* FP variants that feature significantly higher brightness and photostability compared to EBFP [57, 59, 91]. Named Azurite, SBFP2 (strongly enhanced blue FP), and EBFP2, these proteins offer the first real hope for successful long-term imaging of live cells in the blue spectral region (see Table 1.1). The brightest and most photostable of the new blue *Aequorea* FPs, EBFP2 (Figure 1.1a), exhibits typical avGFP-like behavior in fusions and is an excellent FRET donor for proteins in the green spectral class [4]. Recently, an orange-emitting FP derived from coral termed TagRFP (discussed in the following) was subjected to a combination of site-directed and random mutagenesis to produce a blue variant named mTagBFP [100], which exhibits greater brightness and photostability than any previously reported blue FP. The utility of mTagBFP in fusions is similar to other coral-derived FPs (M. W. Davidson, unpublished), and this variant may well emerge as one of the most useful probes in this class. All of the blue FPs can be readily imaged in a fluorescence microscope using standard DAPI filter sets or proprietary BFP sets available from aftermarket optical filter manufacturers.

FPs in the cyan spectral region ($\sim$470 nm to 500 nm) have been widely applied as FRET donors when paired with yellow-emitting FPs [101]. This spectral class was dominated by variants of the original *Aequorea* ECFP until the introduction of a monomeric teal-colored FP, known as mTFP1 [77, 102]. Teal FP exhibits higher brightness and acid stability compared to *Aequorea* CFPs and is far more photostable. Derived from a synthetic gene library built around a *Clavularia* soft coral tetrameric protein, mTFP1 (Figure 1.1d) displays slightly red-shifted spectral characteristics compared to most cyan proteins. In general, members of the cyan FP class contain the amino acid tryptophan at position 66 in the chromophore, but mTFP1 contains the classical tyrosine residue at this location. This amino acid substitution reduces the broad fluorescence emission spectral bandwidth from approximately 60 nm to 30 nm, which is useful in reducing bleed-through in multicolor experiments. The high-emission quantum yield (see Table 1.1) of mTFP1 provides an excellent alternative to the cyan derivatives, such as ECFP [20] and Cerulean (Figure 1.1c; [88, 103]), as a FRET donor when combined with either yellow or orange FPs. For optimal imaging, mTFP1

requires a specialized filter set, but this fluorophore can still produce suitable signal levels with a standard ECFP set. However, mTFP1 is not useful for dual imaging with EGFP due to excessive bleed-through of the teal protein into the green emission channel.

Continued investigation has produced additional useful FPs in the cyan spectral class. Among the improved cyan FPs recently introduced, CyPet [35] and the enhanced cyan variant termed Cerulean [88] show the most promise for use as fusion tags, donors in FRET biosensors, and multicolor imaging. The Cerulean fluorescent probe (named for the sky-blue color) was engineered by site-directed mutagenesis of ECFP (Figure 1.1b) to yield a higher extinction coefficient, improved quantum yield, and a fluorescence lifetime decay having a single exponential component. Cerulean is at least two-fold brighter than ECFP and has been demonstrated to significantly increase contrast as well as the signal-to-noise ratio when coupled with yellow-emitting FPs, such as Venus (see the following), in FRET investigations. The abundance of advantageous features afforded by Cerulean render this protein the most useful all-purpose cyan derivative.

The CFP variant named CyPet (cyan FP for energy transfer) was derived through a unique strategy utilizing FACS to optimize the cyan and yellow pairing for FRET [35]. Libraries were screened for FRET efficiency and the best clones were subjected to several evolutionary cycles consisting of random mutagenesis and synthetic DNA shuffling. A total of seven mutations were accumulated during the directed evolution of the CyPet protein, which features absorption and emission maxima positioned at 435 nm and 477 nm, respectively. CyPet is about half as bright as EGFP and two-thirds as bright as Cerulean, but expresses relatively poorly at 37° C [3]. However, CyPet has a more blue-shifted and narrower fluorescence emission peak than CFP, which greatly increases its potential usefulness for multicolor imaging applications.

The introduction of beneficial "folding" mutations into monomeric variants of ECFP has resulted in the production of new variants featuring enhanced brightness, folding efficiency, solubility, and FRET performance [104]. Termed "super" CFPs (SCFPs), the engineered variants are significantly brighter than the parent protein when expressed in bacteria and almost twofold brighter in mammalian cells. The authors speculate that these high-performance FPs should be useful for fusion tags and in creating new CFP-YFP FRET biosensors exhibiting high dynamic range, and this may well prove true. Another new monomeric cyan FP, TagCFP, was derived from an avGFP-like protein from the jellyfish *Aequorea macrodactyla*. Specific details about the protein are unavailable in the literature, but it is commercially available as mammalian cloning vectors and fusions from Evrogen. The company literature reports TagCFP to be brighter than ECFP and Cerulean, as well as similarly insensitive to physiologically relevant changes in pH.

Several additional potentially useful cyan FPs have been isolated from Anthozoan species. Derived from the reef coral *Anemonia majano*, the AmCyan1 FP [22], which is now commercially available (Clontech), has been optimized with human codons for enhanced expression in mammalian cell systems [105]. Originally named amFP486 (*Anemonia majano* FP with 486 emission maximum) in accordance with a nomenclature scheme [22] devised to simplify the discussion of myriad Anthozoan proteins, this variant exhibits a similar brightness level but a significantly better resistance to photobleaching than CFP. The absorption maximum of AmCyan1 occurs at 458 nm, whereas the fluorescence emission peak resides at 489 nm. Unfortunately, similar to most of the other reef coral proteins, AmCyan1 forms stable tetramers, which will significantly complicate attempts to employ this protein as a fusion tag or FRET biosensor.

First isolated by Miyawaki and associates from an *Acropara* stony coral species, the cyan-emitting Midoriishi-Cyan FP (abbreviated MiCy) [73] was originally designed as the donor in a new FRET combination with the monomeric Kusabira Orange FP (mKO) to generate a biosensor with high spectral overlap (Förster distance of 5.3 nm; mKO is discussed in the section on orange FPs). This protein features the longest absorption and emission wavelength profiles (472 nm and 495 nm, respectively) reported for any probe in the cyan spectral region (see Table 1.1), and similar to mTFP1, could be relegated into a new "teal" class. The high molar extinction coefficient and quantum yield exhibited by MiCy render the protein of equal brightness to Cerulean, although the fluorescence is far more sensitive to pH. Also similar to Cerulean, MiCy features a single exponential lifetime decay component with a time constant of 3.4 ns. An unusual feature of MiCy is that it forms an obligatory dimeric complex rather than the tetrameric variety observed in most coral reef species. A monomeric version of MiCy, known as mMiCy, has been mentioned in the literature [45], but details of its engineering and properties have not been reported.

Green Fluorescent Proteins

The original (wild-type) avGFP isolated from *Aequorea victoria* has been the principal subject of numerous investigations [60] but is not useful in a majority of the practical applications involving FPs due to the bimodal absorption band (395 nm and 475 nm peaks), which is hampered by relatively low extinction coefficients and an absorption maximum in the ultraviolet part of the spectrum. A point mutation replacing the serine residue at position 65 with threonine (S65T) produced a new version of the protein having a well-defined absorption profile with a single peak at 484 nm [17]. This mutation is featured in the most popular variant of avGFP, termed enhanced avGFP (EGFP; Figure 1.1e), which can be imaged using commonly available filter sets designed for fluorescein (FITC) and is among the brightest and most photostable of any FP [3]. These features have rendered EGFP one of the most popular probes and the best choice for most single-label FP experiments.

A large number of proteins emitting in the green (∼500 nm to 525 nm) spectral region have been discovered from a wide range of sources, including different *Aequorea* species [106], copepods [25], amphioxus [26], and coral reefs [22]. However, most of these FPs are naturally oligomeric, and none offers a clear advantage over EGFP. Perhaps the best current choice for live-cell imaging is the avGFP derivative Emerald (available from Invitrogen), which has properties similar to its EGFP parent [97]. Emerald (Figure 1.1f) contains the F64L and S65T mutations featured in EGFP but also has four additional point mutations that improve folding, expression at 37° C, and brightness. Although Emerald is somewhat more efficient than EGFP with respect to maturation and is slightly brighter, it has a fast photobleaching component that might affect quantitative imaging under certain experimental conditions.

The most significant addition to the green spectral region in the past several years has been coined "superfolder" avGFP [55], which is brighter and less sensitive to physiological pH changes than either EGFP or Emerald while retaining similar photostability. Therefore, the superfolder avGFP (Figure 1.1h) should be an excellent candidate for fusions with mammalian proteins, especially those that demonstrate folding problems with standard avGFP derivatives. However, because superfolder avGFP is capable of folding even when fused to insoluble proteins, the potential exists for higher background noise levels when imaging fusions in which a significant portion of the proteins fail to target correctly yet still produce bright fluorescence. In addition to the green superfolder variant, Pédelacq and co-workers also generated blue, cyan, and yellow versions by introducing the appropriate mutations at the chromophore precursor amino acid positions (for the blue and cyan variants) or position 203 (for the yellow variant). Similar to the avGFP derivative, the other superfolder colors also substantially improved fluorescence when fused to poorly folding partners in bacteria. Studies of the superfolder FPs in mammalian cells, either as a reporter for localization or gene expression, have yet to be reported.

Several of the green FP variants mined from reef corals are now commercially available (see Tables 1.1 and 1.3). A brightly fluorescent reporter termed Azami Green (Figure 1.1g; [72]), bearing only a surprisingly scant (less than 6%) sequence homology to EGFP, was isolated from the stony coral *Galaxeidae* and has been demonstrated to mature rapidly during expression in mammalian cell lines. Likewise, one of the original Anthozoa coral reef proteins from *Zoanthus* reported by Matz and co-workers [22] has also been transformed into a commercial product (Clontech) under the name ZsGreen. The probes have absorption maxima at 492 nm and 496 nm and emission peaks at 505 nm and 506 nm, respectively, readily allowing visualization and imaging with standard lasers and filter combinations in confocal and widefield microscopy. However, similar to most of the other proteins isolated in corals, Azami Green and ZsGreen both exist as tetramers in the natural state, which significantly interferes with their use as fusion partners and as a FRET donor or acceptor in biosensors. To overcome the oligomerization problem, site-directed and random mutagenesis efforts were successful in creating a monomeric version of Azami Green (available from MBL International), but this type of effort has not been reported for ZsGreen, although the protein has been reengineered with human codons to optimize expression (resulting in a variant termed ZsGreen1). Because reliable photostability data are lacking, it is unclear whether either of these proteins will outperform EGFP in long-term imaging experiments.

Recently, two bright, monomeric GFPs derived through site-directed and random mutagenesis in combination with library screening in cyan proteins (i.e., mTFP1 and TagCFP) have been reported. Derived from *Clavularia*, mWasabi is a potential alternative green-emitting FRET partner for blue FPs [107] due to negligible absorbance at wavelengths of 400 nm and lower where blue FPs are typically excited. The new green FP is commercially available (Allele Biotechnology) and should be particularly useful in two-color imaging in conjunction with long Stokes shift proteins (such as T-Sapphire) [44] and as a localization tag in fusions with targeting proteins. A derivative of TagCFP, named TagGFP, is a bright and monomeric green variant having an absorption maximum at 482 nm and emission at 505 nm. TagGFP, which is only slightly brighter than EGFP, is available as cloning vectors and fusion tags from Evrogen but has not been thoroughly characterized in literature reports.

The sea pansy, an Anthozoa soft coral, is the source of several green FPs that have been characterized in detail [15, 108, 109]. A protein isolated from *Renilla reniformis* that exhibits properties similar to EGFP is the best characterized of the probes in this class. Having absorption and emission maxima at 485 nm and 508 nm, respectively, in addition to a similar sensitivity to pH, the *Renilla* protein would be an excellent substitute for EGFP were it not for the fact that it is an obligate dimer [37]. Aside from the oligomerization problem, *Renilla* GFPs may be useful in many applications and have been expressed in a wide variety of organisms, including bacteria, fungi, and mammalian cells. Versions with human codon sequences are available from LUX Biotechnology, as are derivatives optimized for expression in other species. There is a general lack of reliable data concerning extinction coefficients, quantum yields, and photostability for the commercial *Renilla* proteins, so valid comparisons to EGFP in terms of brightness and photobleaching are not possible.

Yellow Fluorescent Proteins

Yellow FPs, as a spectral class, are among the most versatile genetically encoded probes yet developed. Ranging in emission wavelength maxima from approximately 525 nm to 555 nm, those proteins residing in the shorter wavelength region actually appear green, rather than yellow, when viewed in a widefield fluorescence microscope. The first member in what has become a rather large family of probes was rationally engineered after the high-resolution crystal structure of avGFP revealed that threonine residue 203 (Thr203) was positioned near the chromophore and potentially able to alter the spectral characteristics upon substitution [42]. Mutations of this aliphatic amino acid to several aromatic moieties were introduced to induce π-orbital stacking and attempt stabilization of the excited state dipole moment of the chromophore. The most successful mutant proved to be tyrosine (T203Y, the original YFP), which resulted in almost a 20-nm shift to longer wavelengths for both the excitation and emission spectra [21, 42, 60].

Several YFP variants were initially constructed to maximize brightness as well as to increase the speed of maturation and optimize expression at 37° C [60, 97]. The variants known as Citrine [92] and Venus (Figure 1.1j; [56]) are currently the most useful proteins in this spectral class (see Table 1.1), but neither is commercially available. Another variant, named after the birthstone topaz, is available from Invitrogen and has been of service in fusion tag localization, intracellular signaling, and FRET investigations [110–112]. A new member of the Evrogen "Tag" commercial series of localization reporter proteins, TagYFP, is a jellyfish-derived (*A. macrodactyla*) monomeric FP that is slightly less bright than EYFP, but an order of magnitude more photostable. Similar to the other members in the "Tag" series, TagYFP (emission peak at 524 nm) has not been characterized in the literature, but can be purchased as mammalian cloning vectors or fusion tags.

During the same FACS-based investigation that led to the generation of CyPet (discussed previously), the evolutionary optimized complementary FRET acceptor, termed YPet (Figure 1.1k), was also obtained [35]. Named after its proficiency in FRET (yellow FP for energy transfer), YPet is the brightest yellow FP variant yet developed and demonstrates very good photostability. The resistance to acidic environments afforded by YPet is superior to Venus and other YFP derivatives, which will enhance the utility of this probe in biosensor combinations targeted at acidic organelles. However, although the optimized CyPet–YPet combination should be the preferred starting point in the development of new FRET biosensors, there remains a serious doubt as to the origin of YPet's increased performance, which is likely due simply to enhanced dimerization with its co-evolved partner, CyPet [83, 84]. Likewise, the suitability of CyPet and YPet in fusion tags for localization experiments, bimolecular complementation analysis, and other routine FP assays has yet to be established. Both proteins exist in solution as relatively weak dimers but presumably could be converted to true monomers using the A206K mutation that has worked so well with other *Aequorea* variants.

Although the potential for new discoveries of yellow and green FPs in Hydrozoan species other than *Aequorea victoria* is significant, only a few viable candidates have surfaced so far. Isolated from the *Phialidium* jellyfish, a protein termed phiYFP [93] is reported to demonstrate very bright yellow fluorescence (absorption and emission at 525 nm and 537 nm, respectively) and to be useful for N-terminal fusion tags. An extraordinary feature of phiYFP is that the naturally occurring protein contains the same mutation at position 64 (leucine) introduced by Venus to increase the folding efficiency [56]. The probe also naturally contains tyrosine at position 203 [42], another site-directed modification of the native avGFP that resulted in yellow fluorescence. This remarkable discovery of a natural similarity between the structure of phiYFP and genetically modified *Aequorea* proteins is a testament to the efficacy of protein engineering efforts directed at avGFP to adjust the spectral properties.

Two monomeric coral reef derivatives with spectral properties falling in the range of *Aequorea* yellow FPs have been created [36]. Named after similarly colored fruits, mHoneydew and mBanana both emit fluorescence in the yellow spectral region. However, a low extinction coefficient and quantum yield render mHoneydew the dimmest member of the monomeric yellow FP cadre, and mBanana is only twice as bright as mHoneydew but features much narrower excitation and emission spectra. Because both proteins exhibit relatively poor photostability, and mBanana is highly pH-sensitive, they probably would not find great utility in imaging experiments. Perhaps the most promising aspect of these probes is that the mere existence of mHoneydew (a Y67W mutant analogous to CFP) demonstrates that the tryptophan-based chromophore of CFP can undergo a further maturation into a longer wavelength-emitting species [36].

ZsYellow (originally referred to as zFP538) is a yellow FP discovered in the Anthozoan button polyp *Zoanthus* during a search in reef corals for naturally occurring avGFP homologs emitting fluorescence in longer-wavelength regions [22, 82, 113]. One of the most unique features of the ZsYellow fluorescence emission spectrum is that the peak (538 nm) occurs almost midway between those of EGFP (508 nm) and DsRed (583 nm), presenting an opportunity to investigate proteins emitting fluorescence in the truly yellow portion of the visible light spectrum. Unfortunately ZsYellow exhibits a marked tendency to form tetramers when expressed *in vivo*, hampering the use of this protein as a fusion partner for localization investigations. Furthermore, the reduced brightness level of ZsYellow when compared to EGFP (~25% of EGFP) also limits the utility of this FP in fluorescence microscopy. The unique emission spectral profile of ZsYellow, however, should encourage the search for genetic modifications that alleviate the tendency to form tetramers while simultaneously increasing the quantum yield and extinction coefficient, an effort that could ultimately yield a high-performance reporter. A human codon-optimized version is commercially available from Clontech as ZsYellow1.

Orange Fluorescent Proteins

In contrast to the relatively large number of FPs engineered in the cyan, green, and yellow spectral classes, only a few promising probes have been developed in the orange portion of the spectrum (ranging from ~555 nm to 590 nm). Even so, all existing orange FPs, which were isolated from coral reef species, have the potential to be useful in a variety of imaging scenarios. Perhaps the most versatile of these is monomeric Kusabira Orange (mKO) [73], a protein originally derived as a tetramer from the mushroom coral *Fungia concinna* (known in Japanese as Kusabira-Ishi). mKO (Figure 1.1) was engineered by site-specific mutagenesis from a cDNA clone of the coral by adding ten amino acids to the N-terminus. The resulting protein has an absorption maximum at 548 nm (ideal for excitation with a 543-nm laser) and emits bright orange fluorescence at 561 nm (Table 1.1). The strategy used to "monomerize" the tetrameric protein was similar to that employed for DsRed to create mRFP1 (discussed in the following) by introducing more than twenty mutations through site-directed and random mutagenesis. The monomeric mKO (commercially available from MBL International) exhibits similar spectral properties to the tetramer and has a brightness value similar to EGFP but is slightly more sensitive than the tetramer to acidic environments. The photostability of this FP, however, is among the best of any FP in all spectral classes, making mKO an excellent choice for long-term imaging experiments. Furthermore, the emission spectral profile is sufficiently well separated from cyan FPs to increase the FRET efficiency in biosensors incorporating mKO, and the probe is useful in multicolor investigations with a combination of cyan, green, yellow, and red FPs. Additional mutagenesis experiments on mKO have yielded a faster folding derivative, termed mKO2 [114], which is slightly brighter (see Table 1.1). Among its obvious applications, mKO2 may also be useful as a FRET partner with rapidly maturing avGFP derivatives in chimeric biosensors.

The mRFP1 derivative, mOrange [36], was derived after four rounds of directed evolution to yield a probe absorbing at 548 nm and emitting orange fluorescence at 562 nm. The mOrange variant is slightly brighter than mKusabira Orange, but has less than 10% the photostability, thus severely compromising its application for experiments requiring repeated imaging. However, mOrange remains one of the brightest proteins in the orange spectral class and is still an excellent choice where intensity is more critical than long-term photostability. In addition, combined with the green-emitting T-Sapphire, mOrange is a suitable alternative to CFP–YFP proteins as a FRET pair to generate longer wavelength biosensors, and can be coupled with FPs in other spectral regions for multicolor investigations. The photostability of mOrange was recently dramatically improved with the introduction of a new strategy to utilize selective pressure for photostability in the directed evolution of FPs [115]. The resulting variant, termed mOrange2, is slightly less bright than mOrange (Table 1.1) but is approximately twenty-five times more photostable.

A novel orange FP isolated from the *Cerianthus* tube anemone [116] is commercially available (cOFP; Stratagene) and has spectral properties similar to mOrange and mKusabira Orange, but like the other anemone proteins isolated to date, exists in solution as a tetramer. The brightness and photostability of cOFP have not been reported so this protein cannot be directly compared to other orange FPs, and its utility will be further limited until it can be converted into a monomer.

The first Anthozoa-derived FP to be extensively investigated was derived from the sea anemone *Discosoma striata* and originally referred to as drFP583, but is now commonly known as DsRed (Figure 1.1o; [22]), although the fluorescence emission is clearly more orange in color than red. Once the protein has fully matured, the emission spectrum of DsRed features a peak at 583 nm, whereas the excitation spectrum has a major peak at 558 nm and a minor peak around 500 nm. Several problems are associated with DsRed in practice. Maturation of DsRed fluorescence occurs slowly and proceeds through an intermediate chromophore stage where a majority of the fluorescence emission is seen in the green region [66]. Termed the "green state," this artifact has proven challenging for multiple labeling experiments in combination with green FPs because of the spectral overlap. In addition, DsRed is an obligate tetramer, an undesirable characteristic that interferes in fusion protein constructs, often leading to poor localization, and increases the tendency to form large protein aggregates in living cells. Although these side effects are not important when the probe is used simply as a reporter for gene expression,

the utility of DsRed as an epitope tag is severely compromised [117]. In contrast to the large *Aequorea* family of proteins employed to successfully tag hundreds of fusion proteins, DsRed fusion proteins have proven far less successful and often exhibit toxic effects.

A bright new monomeric orange protein, named TagRFP (Figure 1.1n) and part of the Evrogen Tag series of FPs, has recently been introduced as a candidate for localization and FRET studies [118]. Derived from the dimer TurboRFP (from the sea anemone *Entacmaea quadricolor*), TagRFP was generated as a result of site-directed mutagenesis to replace several key amino acid residues involved in dimerization while simultaneously performing random mutagenesis to rescue folding properties. In total, seven rounds of semirandom mutagenesis followed by an additional round of random mutagenesis resulted in the final variant, which features excellent photophysical properties and expresses well in a wide variety of fusion tags in mammalian cells. The authors speculate that mTagRFP will be an excellent FRET acceptor when fused to green and yellow donor FPs, but that remains to be demonstrated. During the same investigation that uncovered a highly photostable variant of mOrange [115], similar mutagenesis of TagRFP yielded a single mutation (S158T) that increases the photostability almost tenfold. The resulting FP, named TagRFP-T, is perhaps the most photostable variant yet discovered.

The brightest FP in any spectral class is the tandem version of dimeric Tomato (dTomato), an orange derivative that was one of the original "Fruit" proteins (discussed in more detail in the following) [36]. This FP was derived from an intermediate termed "dimer2" obtained during the directed evolution of mRFP1, which was the first monomeric red FP reported [64], from tetrameric DsRed. The dimeric dTomato protein contains the first and last seven amino acids from avGFP on the N- and C-termini in an effort to increase the tolerance to fusion proteins and reduce potential artifacts in localization. A tandem-dimer version (effectively a "monomer") was created by fusing two copies, head-to-tail, of dTomato with a 23-amino acid linker. Due to the presence of twin chromophores, the resulting tandem dTomato (tdTomato; Figure 1.1m) is extremely bright and has exceptional photostability. A drawback in the use of this protein is the larger size (twice that of a monomeric FP), which may interfere with fusion protein packing in some biopolymers.

Red Fluorescent Proteins

The search for an ideal red-emitting FP has long been the goal for live-cell and whole-animal imaging, primarily due to the requirement for probes in this spectral region in multicolor imaging experiments as well as the fact that longer excitation wavelengths generate less phototoxicity and can probe deeper into biological tissues. As an added convenience, most of the proteins in this wavelength range can be imaged with the common TRITC and Texas Red fluorescence filter sets, as well as common lasers emitting spectral lines at 543 nm, 561 nm, and 594 nm in confocal microscopy. After 5 years of unsuccessful mutagenesis efforts in the avGFP-derived proteins [60], the first real breakthrough occurred with the discovery of potentially fluorescent chromoproteins in nonbioluminescent Anthozoa coral species [22]. To date, a wide spectrum of potentially useful red FPs has been reported (spanning the emission wavelength range of 590–650 nm), many of which still suffer from some degree of the obligatory quaternary structure bestowed by their species of origin (Table 1.1) [1, 3, 67]. Unlike the jellyfish proteins, most of the native and genetically engineered variants of coral reef proteins mature efficiently at 37° C, presumably due to differing water temperatures of their respective host's habitats [1].

Several major problems with DsRed FP have been overcome through site-directed and random mutagenesis efforts, but the construction of truly monomeric variants, as well as monomers from the proteins in other Anthozoa species, has proven to be a difficult task [64]. A total of thirty-three amino acid alterations to the DsRed sequence were required for the creation of the first-generation monomeric red FP (termed mRFP1) [64]. However, this derivative exhibits significantly reduced fluorescence emission compared to the native protein and photobleaches quickly, rendering it much less useful than analogous green and yellow FPs. Extensive mutagenesis research efforts [36], including newly introduced methodology, have successfully been applied in the search for yellow, orange, red, and far-red FP variants that further reduce the tendency of these potentially efficacious biological probes to self-associate while simultaneously pushing emission maxima toward longer wavelengths. The result has been improved monomeric FPs that feature increased extinction coefficients, quantum yields, and photostability, although no single variant has yet been optimized by all criteria. In addition, expression problems with obligate tetrameric red FPs are being overcome by the efforts to generate monomeric variants that are more compatible with biological function.

Perhaps the most substantial developments on this front have been the introduction of a new harvest of FPs derived from monomeric red FP (mRFP1; Figure 1.1q) through directed mutagenesis [36, 119]. The resulting cadre of monomeric FPs exhibit maxima at wavelengths ranging from 560 to 610 nm and have been named in honor of common fruits that bear colors similar to their respective fluorescence emission spectral profiles. Among the potentially efficacious members in the "fruit" series are mStrawberry (Figure 1.1p), mCherry (Figure 1.1r), and tdTomato (discussed previously), all of which have fluorescence emission profiles in the orange and red regions of the spectrum (Table 1.1).

The red proteins, mCherry and mStrawberry (emission peaks at 610 nm and 596 nm, respectively), have brightness levels of approximately 50% and 75% of EGFP, but mCherry is far more photostable than mStrawberry and is the best probe choice to replace mRFP1 for long-term imaging experiments. A variant of mOrange, termed mApple [115], exhibits good photostability and is approximately twice as bright as mCherry (see Table 1.1), making this derivative an exceptional candidate for multicolor imaging. These new proteins essentially fill the gap between the most red-shifted jellyfish FPs (such as YPet) and the multitude of oligomeric coral reef red FPs that have been reported and are commercially available. Although several of these new fluorescent monomeric proteins lack the brightness and photostability necessary for many imaging experiments [3, 120], their existence is encouraging as it suggests the eventuality of bright, stable, monomeric FPs across the entire visible spectrum.

Further extension of the fruit protein spectral class through iterative somatic hypermutation [121] has yielded two new FPs with emission wavelength maxima of 625 nm and 649 nm, representing the first true far-red genetically engineered probes. The most potentially useful probe in this pair was named mPlum (Figure 1.1t), which has a rather limited brightness value (10% of EGFP) but excellent photostability. This monomeric probe should be useful in combination with FPs emitting in the cyan, green, yellow, and orange regions for multicolor imaging experiments and as a biosensor FRET partner with green and yellow proteins, such as Emerald and Citrine. Another far-red FP, termed AQ143, has been derived from mutagenesis efforts on a chromoprotein isolated from the anemone *Actinia equine* [122]. The excitation and emission maxima of AQ143 are 595 nm and 655 nm, respectively, and the brightness is comparable to mPlum. On the downside, the photostability of this protein has not been reported and it forms an obligate tetramer.

Several additional red FPs showing varying degrees of promise have been isolated from the reef coral organisms. One of the first to be adapted for mammalian cell applications is HcRed1 [123], which was isolated from the anemone *Heteractis crispa* and is now commercially available (Clontech). HcRed1 was originally derived from a nonfluorescent chromoprotein through site-directed and random mutagenesis to create a tetrameric red fluorescent species that matures rapidly and efficiently at 37° C (absorption and emission at 588 nm and 618 nm, respectively). Additional mutagenesis efforts resulted in a brighter dimeric variant, but a monomeric version of the protein has not yet been discovered. To generate a variant of the protein that is useful in creating fusion products for localization studies, a tandem dimer expression vector of HcRed similar to tdTomato has been constructed [78].

A red FP, termed eqFP611, was isolated from the sea anemone *Entacmaea quadricolor* and displays one of the largest Stokes shifts and red-shifted fluorescence emission wavelength profiles (excitation and emission maxima at 559 nm and 611 nm, respectively) of any naturally occurring Anthozoan FP [124]. The quantum yield and extinction coefficient of eqFP611 combine to yield a probe approximately as bright as EGFP. In contrast to other Anthozoan FPs, eqFP611 has a reduced tendency to form oligomers at lower concentrations as evidenced through electrophoresis and single-molecule experiments [125], although at high concentrations the protein forms tetramers. Site-directed mutagenesis efforts have yielded functional dimeric variants of eqFP611 [71], and continued efforts have led to a monomeric far-red FP from this species [126].

Two additional reef coral red-emitting FPs, AsRed2 and JRed, are commercially available (Clontech and Evrogen), but these probes form tetrameric and dimeric complexes, respectively, and are less useful than the monomeric proteins described previously. AsRed2 was originally isolated as a chromoprotein from *Anemonia sulcata* [22] and modified through mutagenesis to yield a protein having an absorption maximum at 576 nm and an emission peak at 595 nm [127] with a very modest quantum yield (0.05). Although the protein has been optimized with human codons for expression in mammalian cell lines, it exhibits only about 10% the brightness level of EGFP and the photostability has not been reported. The dimeric protein, JRed, was derived through extensive mutagenesis of a jellyfish chromoprotein [93] to produce a novel red fluorescent marker with peak absorption and emission wavelengths of 584 nm and 610 nm, respectively. JRed is about 25% as bright as EGFP and exhibits limited photostability when illuminated in the 560–580 nm region, but can be successfully employed for long-term imaging experiments when excited with a 543-nm laser.

The application of site-specific and random mutagenesis to TurboRFP variants [128], followed by screening for mutations exhibiting far-red fluorescence, resulted in a dimeric protein named Katushka (emission maxima of 635 nm). Although only two-thirds as bright as EGFP, Katushka exhibits the highest brightness levels of any FP in spectral window encompassing 650–800 nm, a region important for deep tissue imaging. Introduction of the four principal Katushka mutations into TagRFP generated a monomeric, far-red protein named mKate (Figure 1.1s) that has similar spectral characteristics (Table 1.1). The photostability of mKate is reported to be exceptional and the protein displays brightness similar to that of mCherry, which makes it an excellent candidate for localization experiments in the far-red portion of the spectrum.

Large Stokes Shift Fluorescent Proteins

Mutagenesis efforts with FPs have also targeted the separation distance between absorption and emission maxima (termed the Stokes shift) to generate better probes for FRET, fluorescence cross-correlation spectroscopy (FCCS), and multicolor imaging. Substitution of isoleucine for tyrosine at position 203 (T203I) in wtGFP produces a variant, named Sapphire, that eliminates the minor excitation peak at 475 nm [60]. Sapphire exhibits an exceedingly large Stokes shift of 112 nm, with excitation and emission maxima at 399 nm and 511 nm, respectively. An optimized derivative with improved folding and brighter fluorescence, known as T-Sapphire (T for Turbo; Figure 1.1i), was constructed by introducing four additional mutations [44]. These variants should be excellent donors in FRET combinations with orange and red proteins due to their ability to be excited in the ultraviolet region.

Extending the Sapphire strategy to red FPs [45], researchers used a far more rigorous approach to construct the longest Stokes shift FP variant yet developed (180 nm) using a nonfluorescent chromoprotein derived from the *Montipora* stony coral. Mutagenesis of five residues surrounding the chromophore led to a red FP having a bimodal excitation spectrum (peaks at 452 nm and 580 nm) with emission at 606 nm. An additional four mutations substantially reduced the 580 nm peak and blue-shifted the other absorption peak to 440 nm. The resulting derivative, named Keima (after the Japanese chess piece), exhibits an emission maximum at 616 nm but is hampered in most experiments due to obligatory tetramer formation. Additional mutagenesis produced a dimer (dKeima) having similar spectral properties, and a monomer (mKeima; emission = 620 nm) was obtained after continued efforts. mKeima exhibits limited brightness (similar to the value for mPlum) and requires a specialized filter combination for imaging, but it has been demonstrated to be useful in FCCS and multicolor imaging experiments [45].

Although useful fluorophores are now available in every FP spectral class, in most cases there remains no EGFP equivalent in terms of photostability and other critical areas of performance. New additions to the blue and cyan region feature substantially improved brightness and photostability, and any of the orange FPs are excellent choices for long-term multicolor imaging. Although brighter than EGFP, photostability is still suboptimal for most of the yellow FPs, whereas the red FPs are among the dimmest in all spectral classes. Even so, many of the FPs listed in Table 1.1 can be combined for dual- and triple-color imaging to yield excellent results. Given that most of these proteins have only been introduced in the past several years, it is highly likely that in the future, bright and photostable additions will become available for all spectral classes.

Table 1.2. A compilation of properties of the most useful optical highlighter FP reporters. Along with the common name and/or acronym for each highlighter, the peak excitation (Ex) and emission (Em) wavelengths, molar extinction coefficient (EC), quantum yield (QY), relative brightness, and physiologically relevant quaternary structure are listed for both the activated and nonactivated species. The computed brightness values were derived from the product of the molar extinction coefficient and quantum yield, divided by the value for EGFP. Photostability is not listed because little information is available on highlighter FPs. Also listed are references to the original literature sources. [a]Nonactivated species

Protein (Acronym)	Ex (nm)	Em (nm)	$EC \times 10^{-3}$ $M^{-1} cm^{-1}$	QY	Quaternary Structure	Relative Brightness (% of EGFP)	Reference
Photoactivatable FPs							
PA-GFP (NA)[a]	400	515	20.7	0.13	Monomer	8	[131]
PA-GFP (G)	504	517	17.4	0.79	Monomer	41	[131]
PS-CFP2 (C)[a]	400	468	43.0	0.20	Monomer	26	[136]
PS-CFP2 (G)	490	511	47.0	0.23	Monomer	32	[136]
PA-mRFP1 (R)	578	605	10.0	0.08	Monomer	3	[135]
Photoconvertible FPs							
Kaede (G)[a]	508	518	98.8	0.88	Tetramer	259	[130]
Kaede (R)	572	580	60.4	0.33	Tetramer	59	[130]
wtKikGR (G)[a]	507	517	53.7	0.70	Tetramer	112	[137]
wtKikGR (R)	583	593	35.1	0.65	Tetramer	68	[137]
mKikGR (G)[a]	505	515	49.0	0.69	Monomer	101	[139]
mKikGR (R)	580	591	28.0	0.63	Monomer	53	[139]
mEosFP (G)[a]	505	516	67.2	0.64	Monomer	128	[74]
mEosFP (R)	569	581	37.0	0.62	Monomer	68	[74]
mEos2FP (G)[a]	506	519	56.0	0.84	Monomer	140	[140]
mEos2FP (R)	573	584	46.0	0.66	Monomer	90	[140]
tdEosFP (G)[a]	506	516	84.0	0.66	Monomer	165	[79]
tdEosFP (R)	569	581	33.0	0.60	Monomer	59	[79]
Dendra2 (G)[a]	490	507	45.0	0.50	Monomer	67	[75]
Dendra2 (R)	553	573	35.0	0.55	Monomer	57	[75]
Photoswitchable FPs							
Kindling – KFP1 (R)	580	600	59.0	0.07	Tetramer	12	[132]
Dronpa (G)	503	518	95.0	0.85	Monomer	240	[76]

OPTICAL HIGHLIGHTERS: TOOLS FOR THE STUDY OF PROTEIN DYNAMICS

A special class of FPs known as "optical highlighters" includes a subset of the color palette with photophysical characteristics that enable the activation or conversion of fluorescent properties by controlled illumination [11, 129]. Table 1.2 presents a compilation of physical data for the current palette of optical highlighter FPs that display significant potential in applications as *in vivo* probes targeting cellular structure and function. Optical highlighters can be further divided into three classes based on whether they are photoactivated, photoconverted, or photoswitched. FPs that can be activated to initiate fluorescence emission from a dark or quiescent state are termed "photoactivatable," whereas those that can

be optically converted from one fluorescence emission bandwidth to another are "photoconvertible." A third class, FPs that can be reversibly toggled between dark and light states, is referred to as "photoswitchable." Synthetic fluorophores with optical highlighter-like properties have been known for decades, but these attributes have only recently been discovered in genetically encoded FPs [130–132]. Photoactivated FPs generally exhibit little or no initial fluorescence under excitation at the imaging wavelength, but dramatically increase their fluorescence intensity after activation by irradiation at a different (usually lower) wavelength (Figure 1.9). In contrast, photoconversion involves a change in the fluorescence emission bandwidth profile (generally from shorter to longer wavelengths) upon optically induced changes to the FP chromophore. Photoswitchable FPs can be turned

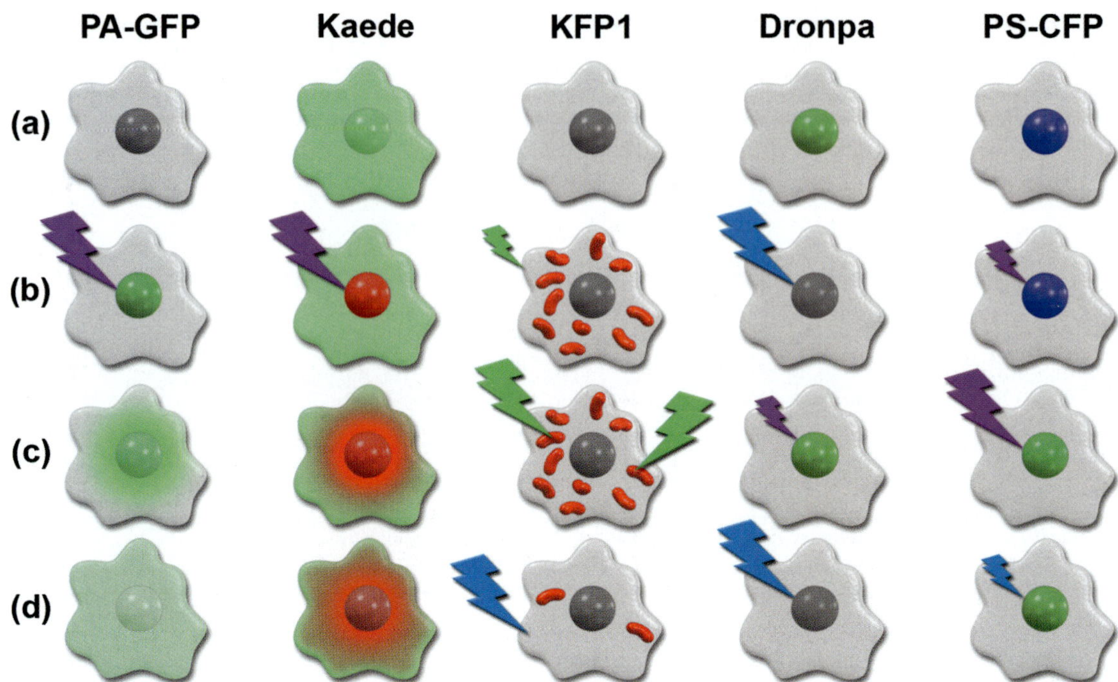

Figure 1.9. Cartoons illustrating the photoconversion mechanisms for the most useful optical highlighters developed to date. The photoactivation, photoconversion, or photoswitching sequence for each highlighter is outlined in a single column, with a repetitive cartoon drawing representing an individual cell containing a central nucleus being employed for each step in the sequence. Illumination at high intensities (photoconversion light levels) is indicated by large lightning bolts, while low-intensity illumination (imaging light levels) is represented by smaller bolts. The fluorescence excitation wavelengths are similar to the colors of the lightning bolts, whereas the emission colors appear in the nucleus and/or cytoplasm of the cell cartoons. To examine a sequence, start with the uppermost cell drawing and proceed down the column. For example, a single cell containing photoactivatable green fluorescent protein (PA-GFP) before conversion is illustrated at the top of the first column and appears nonfluorescent (grayscale). After illumination of the nucleus with 405-nm light (purple lightning bolt), the PA-GFP protein exhibits green fluorescence throughout the nucleus, which slowly diffuses into the cytoplasm, as illustrated by the lower two cell cartoons ((**C**) and (**D**)) in the first column. The other columns contain sequences that symbolize the photoconversion mechanisms for selected optical highlighters. A cell expressing green fluorescent native Kaede (second column) is photoconverted to red in the nuclear region, which slowly diffuses into the cytoplasm. Highlighting mitochondria with KFP1 (third column) enables these organelles to be visualized transiently or permanently, whereas Dronpa (fourth column) can be turned on and off with alternating 488- and 405-nm lasers. PS-CFP (fifth column) can be visualized with weak 405-nm illumination or photoconverted from cyan to green with intense 405-nm illumination.

"on" or "off," in some cases with timescales in the millisecond range, by simply changing the illumination wavelength. Optical highlighters represent perhaps the most promising approach to the *in vivo* investigation of protein dynamics [11] and have recently become useful for high-resolution microscopy techniques that break the classical diffraction barrier [133, 134].

The ideal optical highlighter FP should be readily photoconvertible or photoactivatable (through the process of fluorescence activation and/or emission wavelength shifts) to produce a high level of contrast. It should also be monomeric for optimum expression in the target system. These probes will be especially useful in experiments paralleling results obtained with photobleaching techniques, such as recovery (FRAP) and loss (FLIP) of photobleaching, because they have the advantage that measurements are not influenced by newly synthesized or nonconverted proteins, which either remain invisible or continue to emit the original wavelengths [11]. Also, by repeated excitation in the region of interest, optical highlighters can be continuously photoconverted at a specific intracellular location. This technique is more efficient than FLIP because the translocation of activated proteins can be directly imaged. In addition, time required for photoactivation (a few seconds) is often much less than the time required to completely photobleach a similar region. Investigations involving extremely rapid cellular processes will clearly benefit from such improvements in temporal resolution.

Photoactivatable Fluorescent Proteins

The first useful optical highlighter designed specifically for photoactivation studies is a variant of avGFP, termed PA-GFP (Figure 1.10a–c). This photoactivatable version of avGFP was developed by improving on the

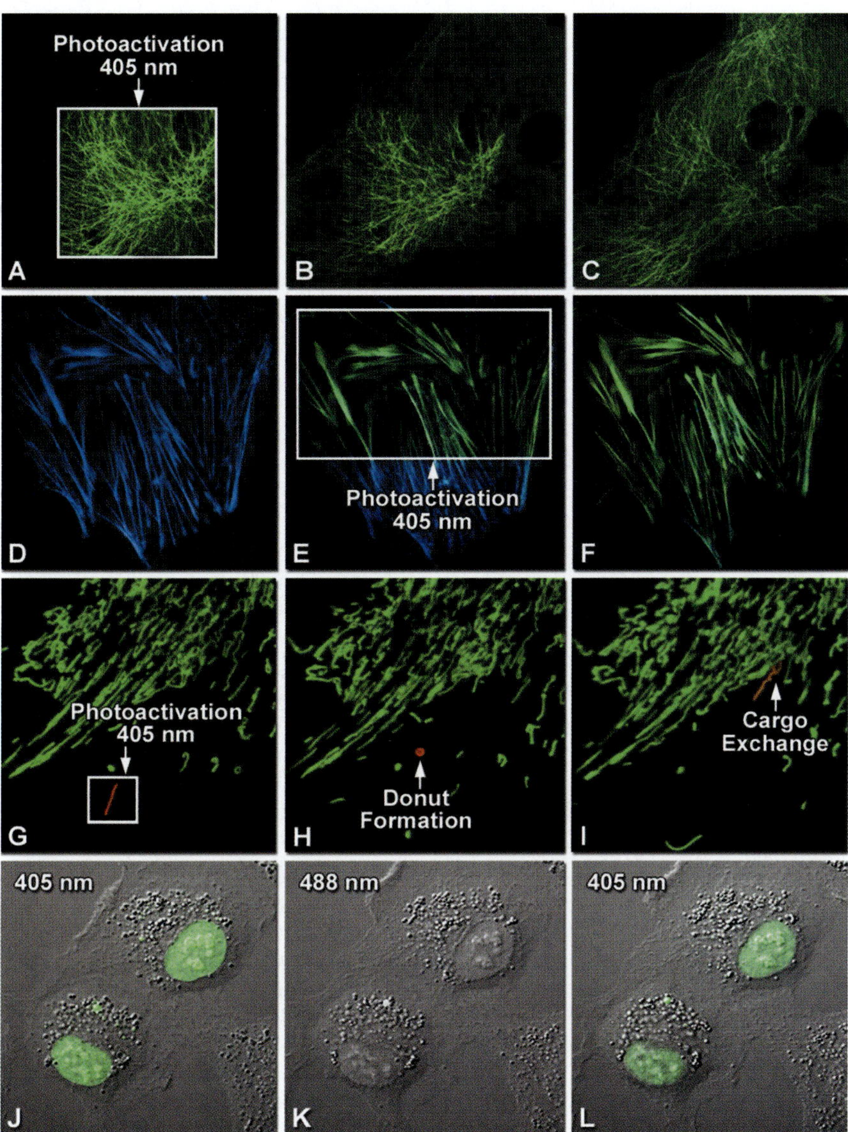

Figure 1.10. Optical highlighter FP reporters in action imaged with laser scanning confocal microscopy. **A–C:** Photoactivation of mPA-GFP-tubulin-C-6 in normal African green monkey kidney (CV-1 cell line) fibroblast cells; (**A**) Rectangular region of interest (white box) selected with Olympus FV1000 SIM scanner is illuminated at 405 nm for 5 sec, t = 0; (**B**) The photoactivated tubulin chimera initially translocates to microtubules throughout the cytoplasm as fluorescence intensity decreases in the activated region, t = 15 min; (**C**) The microtubule network gains more intensity at t = 60 min. **D–F:** Photoconversion of PS-CFP2-actin-C-7 in gray fox lung (FoLu cell line) fibroblast cells; (**D**) Single FoLu cell expressing the actin chimera imaged with a 405-nm laser; (**E**) Photoconversion of labeled actin filaments in the lower portion of the viewfield in a selected region of interest (white box) with 405 nm illumination at t = 0; (**F**) Photoconverted PS-CFP2 in the actin chimera has translocated into most of the filaments at t = 30 min. **G–I**: Tracking of mitochondria labeled with tdEos-mito-N-7 in rabbit kidney (RK-13 cell line) epithelial cells; (**G**) Photoconversion of a single mitochondrion (red) in a selected region (white box) with 405 nm illumination at t = 0; (**H**) Formation of a donut by the photoconverted mitochondrion at t = 2 min; (**I**) Cargo exchange between mitochondria (arrow) at t = 3 min. **J–L:** Photoswitching of the histone H2B with Dronpa-H2B-N-6 in opossum kidney (OK cell line) epithelial cells; (**J**) Labeled nuclei in adjacent cells imaged with 488 nm laser with differential interference contrast, t = 0; (**K**) After completely photoswitching the labeled histones "off" at 488 nm, the nuclei now appear devoid of fluorescence, t = 3 min; (**L**) Dronpa label in nucleus, reactivated with illumination at 405 nm, appears dimmer due to photobleaching after forty rounds of photoswitching.

photoconversion efficiency of the natural wild-type protein chromophore from a predominately neutral form to a species that is anionic in character [131]. Substitution of histidine for threonine at position 203 (T203H) produced a variant with negligible absorbance in the region between 450 nm and 550 nm, thus enhancing the contrast between the nonactivated and activated species. After photoactivation with violet light (~405 nm), the absorption maximum at 504 nm in PA-GFP increases approximately 100-fold. This event evokes high contrast differences between the converted and unconverted pools of PA-GFP and is useful for tracking the dynamics of molecular subpopulations within a cell (Figure 1.9, Column 1). On the downside, intracellular targets expressing PA-GFP are not easily distinguishable prior to being photoactivated, thus making the definition of regions for observation difficult.

Several new photoactivatable proteins have been produced using site-directed mutagenesis of a monomeric red-shifted reef coral FP. The monomeric derivative of DsRed FP, mRFP1, has been converted to a probe photoactivated by either green or violet irradiation [135]. This FP, termed PA-mRFP1, exhibits a 70-fold increase of fluorescence intensity upon activation by wavelengths between 380–400 nm. Unfortunately, the relatively low level of fluorescence intensity of PA-mRFP1 in the photoactivated form (3 percent of EGFP) renders it significantly less useful than PA-GFP for live-cell investigations. Clearly, this class of optical highlighters would benefit from efforts to engineer the mFruit proteins (such as mCherry and mPlum) into photoactivatable probes.

A novel photoconvertible optical highlighter, termed photoswitchable cyan FP (PS-CFP2; Figure 1.10d–f), derived from the *Aequorea coerulescens* green FP variant, aceGFP, has been observed to transition from cyan-to-green fluorescence upon illumination at 405 nm (Figure 1.9, Column 5). The PS-CFP2 highlighter was generated by site-directed mutagenesis of aceGFP [136] and is expressed as a monomer *in vivo*. Among the advantages of PS-CFP2 is the significant level of cyan fluorescence present before photoconversion, a factor that allows investigators to track and selectively illuminate specific intracellular regions or entire cells for study. However, the dynamic range of PS-CFP2 is significantly lower than that of PA-GFP, and the probe is inferior to highlighters in the green-to-red spectral class in terms of photoconversion efficiency.

Photoconvertible Fluorescent Proteins

Several potentially useful green-to-red photoconvertible optical highlighters have been developed in FPs cloned from reef coral and sea anemone species. One of the first and most important examples, a tetrameric FP isolated from the stony Open Brain coral, *Trachyphyllia geoffroyi*, has been found to photoconvert from green to red fluorescence emission in ultraviolet light (Figure 1.9, Column 2; [130]). The unusual color transition prompted investigators to name the protein Kaede after the leaves of the Japanese maple tree, which turn from green to red in the fall months. Illumination of Kaede with ultraviolet or violet light results in a spectral shift of the native (green) species from 508 nm (absorption) and 518 nm (emission) to longer wavelength peaks at 572 nm and 582 nm, respectively. Upon photoconversion, Kaede exhibits a dramatic increase in the red-to-green fluorescence ratio (approximately 2000-fold, considering both the decrease in green and the increase in red emission). The photoconversion is stable and irreversible under aerobic conditions, and the red fluorescent state of the Kaede chromophore is comparable to the green in terms of brightness and photostability.

The stony coral *Favia favus* has yielded a promising tetrameric derivative that exhibits efficient photoconversion from green to red fluorescence emission wavelengths (similar to Kaede) upon irradiation with near ultraviolet or violet light [137]. Engineering efforts based on structural analysis of this protein produced a variant, termed KikGR, which is several-fold brighter than Kaede in both the green and red states and features a wider separation of green and red emission maxima than Kaede (75 nm vs. 54 nm). Another tetrameric stony coral FP, EosFP, emits bright green fluorescence at 516 nm that shifts to orange–red (581 nm) when illuminated with light at wavelengths in the near-ultraviolet region of 390–405 nm [74]. Two single point mutations have been employed to split the wild-type tetramer into dimeric subunits and a combination of both single point mutations yields a true monomeric protein (mEosFP). The monomer can be incorporated into functional biological chimeras to serve as a marker in live-cell imaging [138], although the monomer fusions are only efficiently expressed at temperatures below 30° C (limiting their applications in mammalian systems). An EosFP tandem dimer (Figure 1.10g–i; [74, 79]) exhibits far better maturation at 37° C and is much brighter, making this variant one of the best choices in its spectral class. Recently, monomeric versions of KikGR [139] and Eos (mEos2; [140]) have been reported. Initial studies using both of these derivatives (Davidson, unpublished) indicate that they have significant potential in a wide variety of fusions for investigations of live-cell dynamics and superresolution microscopy.

The first useful monomeric green-to-red photoconvertible FP was derived from a soft coral and originally named Dendra [75] but was shortly followed by an improved commercial version (Evrogen) known as Dendra2. Capable of being photoconverted by ultraviolet, violet, and blue wavelengths (in decreasing order of effectiveness), Dendra2 exhibits a high dynamic range (up to 4000) and matures rapidly at 37° C. The monomeric nature of Dendra2 makes this FP ideal for sensitive fusions and FRET investigations, and the ability to photoconvert with a common argon-ion laser at 488 nm reduces the phototoxic effects induced by shorter wavelengths. On the downside, the green species of Dendra2 is less than half as bright as the tetrameric analogs or tandem dimer EosFP, and the photoconverted species may not be as photostable as originally reported.

Other proteins similar to Kaede, KikGR, Dendra2, and EosFP that are capable of being photoconverted by violet and ultraviolet illumination have been discovered in the Great Star coral (mcavRFP) and the mushroom coral (rfloRFP) [23]. All of these highlighters contain a common chromophore derived from the tripeptide His-Tyr-Gly (HYG) that initially emits green fluorescence until driven into a red state by strong ultraviolet illumination. Irradiation induces cleavage between the amide nitrogen and *alpha* carbon atom in the histidine residue with subsequent formation of a highly conjugated dual imidazole ring system, a process requiring catalysis by the intact protein and resulting in the dramatic shift of fluorescence emission to red wavelengths [141, 142]. The unconventional chemistry involved in this chromophore transition should give engineers an excellent foundation upon which to develop more advanced highlighters.

Photoswitchable Fluorescent Proteins

Although the phenomenon of photochromism (the ability to switch fluorescence on and off) has been observed in the wild-type and several yellow FP derivatives of avGFP at the single molecule level [143, 144], none has demonstrated this phenomenon when measured in bulk. In these studies, during illumination at 488 nm, the molecules exhibited fluorescence for several seconds, followed by an equally short time interval without emission, followed later by resumption of emission. Termed blinking behavior [145], this on-and-off switching sequence can be repeated a number of times before each avGFP molecule ultimately photobleaches. Unfortunately, photoswitching in most of the FPs described previously cannot be done successfully in quantitative experiments.

A new generation of specialized optical highlighters with reversible on–off switching capabilities was created by the introduction of Dronpa (Figure 1.10j–l), a monomeric FP derived from the *Pectiniidae* coral [76]. Named after a fusion of the ninja term for vanishing (dron) and photoactivation (pa), Dronpa exhibits unusual behavior due to its ability to toggle fluorescence on and off by illumination with two different wavelengths (Figure 1.9, Column 4). Dronpa was engineered using both directed and random mutagenesis to yield a monomeric version of the wild-type oligomeric FP having a major absorption maximum at 503 nm and a minor peak at 390 nm. The absorption peak at 503 nm is due to the deprotonated species of the protein, whereas the smaller peak at 390 nm arises from the protonated form.

When irradiated at 488 nm, the fluorescence emission of the deprotonated species has a maximum at 518 nm with a relatively high quantum yield of 0.85 (Table 1.2). In contrast, the protonated form of the protein is almost nonfluorescent. Photoswitching of Dronpa occurs by interconversion between the deprotonated and protonated forms [146]. Upon irradiation at 488 nm, Dronpa is driven to the protonated species with a commitment decrease in fluorescence to produce a dim (off) state in which the 390-nm absorption peak predominates. The dim state is readily converted to the original fluorescent (on) deprotonated state with minimal illumination at 405 nm. Similar behavior has been reported for a teal FP, termed mTFP0.7 [40], in which the dark and fluorescent states have been characterized by crystallography.

Another potentially useful photoswitchable optical highlighter, the Kindling FP (KFP1), has been developed from a nonfluorescent chromoprotein isolated in *Anemonia sulcata* [23, 132, 147] and is now commercially available (Evrogen). Kindling FP does not exhibit emission until illuminated with green or yellow light in the region between 525 nm and 580 nm. Low-intensity light results in transient red fluorescence (kindling) with excitation and emission maxima at 580 nm and 600 nm, respectively, which slowly decays in the dark as the protein relaxes to its initial nonfluorescent state. Irradiation with intense blue light quenches the kindled fluorescence immediately and completely, allowing tight control over fluorescent labeling (Figure 1.9, Column 3). Note that both kindling with low-intensity green light and quenching of fluorescence by blue light are reversible processes for the wild-type protein. In contrast, high-intensity illumination or continued irradiation at moderate levels results in irreversible kindling with a fluorescence intensity approximately 30-fold greater than that of the nonactivated protein. Irreversibly kindled molecules do not lose their fluorescence and are not quenched by illumination with blue light. This feature allows for stable long-term highlighting of cells, tissues, and organelles similar to PA-GFP and other highlighter proteins. The major drawback of kindling protein is its tendency to aggregate into tetramers, which seriously affects the potential for use as a protein fusion tag without some degree of disturbance to normal biological processes. However, the kindling protein is an excellent candidate for bulk photolabeling and tracking of individual organelles and cells within a large population.

Investigations into the underlying mechanism of FP photoswitching [40, 148–151] suggest that *cis–trans* isomerization of the tyrosine chromophore is a key event in the process. The *cis* conformation represents the bright fluorescent state, whereas the *trans* isomer is adopted by the chromophore in the nonfluorescent, or dark, state. The conformational changes are thought to be accompanied by varied chromophore protonation states that also contribute to the determination of the fluorescent properties. Furthermore, photoswitching is probably a manifestation of chromophore planarity and structural rearrangements of internal amino acid side chains within the chromophore cavity. These collective features may constitute a fundamental mechanism common to all photoactivatable and reversibly photoswitchable FP derivatives.

The potential for optical highlighters remains far greater than the current realization of useful derivatives in this category. Among the photoactivatable probes, PA-GFP is still the best choice and is far superior in terms of dynamic range to the only red variant yet reported, PA-mRFP1 [135]. The only choice for cyan-to-green photoconversion, PS-CFP2 [136], exhibits monomeric character but is compromised by low brightness levels and the artifact of continued photoconversion during imaging. In the green-to-red class, the best performers in terms of brightness and conversion efficiency are tetramers. Thus, they are not useful for a majority of experiments. The monomeric variant, Dendra2 [75], is probably the best choice for fusions and FRET studies but suffers from rapid photobleaching of the red species and is less than half as bright as the tetramers. A tandem dimer of the green-to-red highlighter named EosFP [74, 79] behaves better than Dendra2 in terms of brightness and photostability, but is twice as large. Clearly, there is a significant need for better performers in all of the optical highlighter categories.

ADDITIONAL PRACTICAL CONSIDERATIONS

Recent advances in FP technology have turned many biologists into experts in live-cell imaging and microscopy. Investigators who want to track their favorite protein simply fuse the gene to the cDNA for an FP of choice and transfer the resulting recombinant vector into a host cell or whole organism. In the best cases, the new chimera enables the host protein to be normally involved in its routine cellular duties while the piggyback FP contributes a fluorescent molecular beacon to report on the position of the conjoined pair. FRET biosensors are slightly more complicated, but the general aspects are the same. Unfortunately, the best case scenario is not universally observed in practice, leaving the investigator puzzled over whether a different FP would produce the optimum result.

Assuming the host cells are initially in log phase and healthy, and provided the transfection protocol does not produce excessive levels of trauma, the most common problems encountered in working with FPs are aggregation, incorrect localization, nonfunctional fusions, and suppression of the expected fluorescence intensity. Once these problems are corrected, the investigator must choose between examining the cells in transient FP transfections, where expression levels are highly varied, or in taking more time to select stably expressing cell populations that often produce superior results. It is often tempting (and much quicker) to transiently express the FP fusion and search for cells exhibiting low levels of fluorescence intensity that may correspond to expression levels minute enough to not interfere with normal function (see Figure 1.11). However, producing stable cell lines presents an opportunity for a quantitative comparison of fusion expression to that of the endogenous protein and is a much safer bet. Alternatively, placing the fusion construct into a vector having an inducible promoter enables control in modulating the level of expression.

In some cases, the FP fusion exhibits unexpectedly low levels of fluorescence when expressed. This problem can be due to a number of factors, including the localized environment (primarily pH) of the target organelle, improper folding of the chimera, lack of expression, or a highly unstable fusion complex. Many organelles have internal pH ranges that differ widely from that of the cytoplasm, and this can interfere with FP detection in some fusions. Successfully targeting FPs to the Golgi, endosomes, lysosomes, secretory granules, and other acidic organelles requires the use of FPs with relatively low pKa values. For example, EGFP and Venus are more likely to experience a loss of fluorescence in the lumen of the Golgi complex than is TagRFP, which has a much lower pKa. In general, fine-tuning the target environment with suitable FP parameters will ensure that intensity problems must be assigned to other variables.

Improper folding of either the FP, the host fusion protein, or the entire chimeric complex is another source of poor fluorescence intensity. Provided each of these entities folds properly when expressed separately, the first candidate for examination is the linker between the FP and its host, which should be optimized for every application. If the linker is not sufficiently long and flexible enough to separate the two protein domains, steric hindrance can lead to folding interference in one or both of the proteins. The first choice for linker components is the amino acid glycine, which has the smallest side chain and bestows the greatest degree of flexibility to any peptide. Interspersing several glycine residues with serine improves solubility and should be considered for longer linkers. Additionally, the use of many other amino acids in FP fusion linkers has been reported, indicating that there is a wide tolerance to residue choice. Usually, a linker length between two and ten amino acids is sufficient, but the optimum size also depends upon whether the fusion occurs at the amino (N) or carboxy (C) terminus of the FP. EGFP and other avGFP variants have a flexible C-terminal stretch of approximately ten amino acids, which considerably shortens the required linker length. In contrast, the N-terminal region of these probes is much less tolerant to fusions and requires a longer linker. The opposite is true for several coral FPs, but the

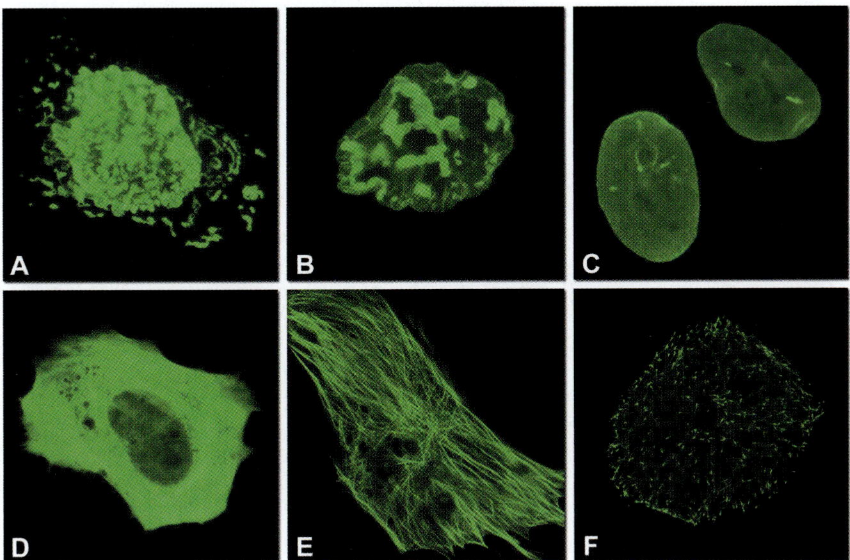

Figure 1.11. Varied FP expression levels in transient transfections. **A–C:** mEmerald-lamin B1-C-10 expressed in HeLa cells; **(A)** Extreme overexpression is manifested by FP background fluorescence in the Golgi complex, aggregates in the nucleus, and distortion of nuclear envelope structures; **(B)** Moderate overexpression distorts the nucleus and produces aggregates in the nucleoplasm and nuclear envelope; **(C)** Low expression levels exhibit evenly distributed lamin fusion protein and distinct foci that inhabit the nucleoplasm; **D–F:** mEGFP-EB3-N-7 expressed in Gray fox lung (FoLu cell line) fibroblast cells; **(D)** Extreme overexpression pervades the cytoplasm with little definition of microtubules, but the fusion protein is apparently too large to enter the nucleus; **(E)** Moderate overexpression resembles outlines in the microtubule network, but individual (+) ends are obscured; **(F)** Low expression levels clearly reveal defined EB3-labeled microtubule (+) termini, which can be followed as they migrate through the cytoplasm with time-lapse imaging. Stable transfectants should be selected from cells expressing FP fusions at levels similar to those represented in **(C)** and **(F)**. Images were pseudocolored to correspond to the respective FP emission wavelength maximum.

crystal structure for many proteins remains undetermined, so the investigator should err on the side of caution and choose longer linkers in the absence of structural data.

In cases where the FP fusion is not expressed or is highly unstable, the focus of troubleshooting efforts should be on the position of the fusion or the molecular integrity of the construct. If the protein is expressed but unstable, the chimera will often perform better if the FP is placed in a different position, such as the N-terminus when the C-terminal fusion behaves poorly. Although rare, some host proteins will not tolerate a FP fusion to either end, leaving only the choice of inserting the probe somewhere in the middle of the host sequence. The most tolerant regions are highly flexible loops, but other sequences where the amino acids are relatively disordered may work as well. Another possibility is to use circularly permuted FP derivatives [152, 153] where the original amino and carboxy termini have been linked with a short spacer and new terminal ends established within the β-barrel. A total lack of expression usually indicates a problem with the construct design (such as the FP being inserted out of frame). After ensuring that the nucleotide sequence is correct and contains a Kozak initiation site, check to determine if the flanking sequences of the fusion gene may be interfering with transcription or translation. The remedy may be to excise the entire fusion gene and insert it into a multiple cloning site of a standardized expression vector.

Poor localization, which can occur for a variety of reasons, is perhaps the single largest source of problems when examining FP fusion chimeras. In many cases, the primary cause is interference with the host protein's normal biological function, but aggregation and oligomerization artifacts from the FP itself, as well as excessively high expression levels, can also produce the same result (see Figures 1.11 and 1.12). The same approach of checking fusion termini and modulating linker lengths discussed previously is the first place to start in troubleshooting localization problems, provided the FP is believed to be monomeric and does not aggregate when expressed alone. In situations where the fusion host protein itself forms a biopolymer or intermolecular complex (such as actin, tubulin, and the histones), any degree of oligomerization can be disruptive to proper localization and should be suspected. In general, many newly developed monomeric orange and red proteins from coral species should be suspected of contributing artifacts if aggregation or improper localization is observed. The best approach when using these FPs is to compare the distribution of new fusion chimeras with the distribution of the native protein fused to a well-studied EGFP

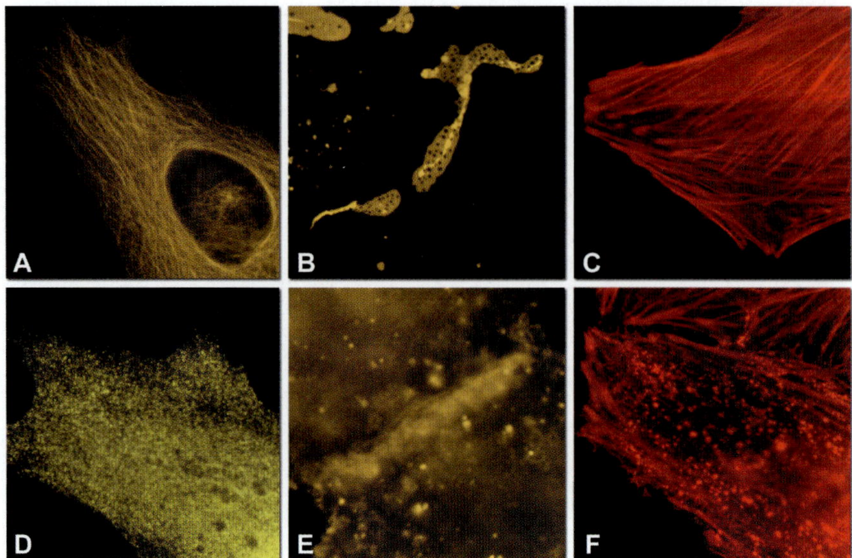

Figure 1.12. Examples of good and poor localization and cytoplasmic aggregation artifacts. (**A**) mTagRFP-tubulin-C-6 in HeLa cells demonstrates excellent morphology in defining the microtubule network; (**D**) mKO-tubulin-C-6 in HeLa cells displays much poorer localization than the identical mTagRFP construct; (**B**) mTagRFP-Cx43-N-7 forms well-defined gap junctions in HeLa cells; (**F**) tdTomato-Cx43-N-7 forms inferior gap junctions and is difficult to distinguish from overlapping cell membranes; (**C**) mPlum-actin-C-7 localizes very nicely to the filamentous network in many cases, but can also form aggregates throughout the cytoplasm (**G**), often centered around the nucleus in the area of the Golgi complex. Images were pseudocolored to correspond to the respective FP emission wavelength maximum.

variant or to subsequently verify proper localization using immunofluorescence.

A final note on applying new FP variants in fusion constructs serves to underscore the fact that expression levels can often be maximized by redesigning the nucleic acid sequence to coincide with codon preferences of the host organism [68]. Neither the native jellyfish nor coral reef native codon usage is optimal for mammalian cells, and translation of any protein in eukaryotes can be further assisted by including the proper initiation sequence [154]. Installing a new codon beginning with a G immediately after the start codon (Met, ATG) is sufficient to produce the Kozak site but introduces an extra amino acid (preferably Val or Ala) into the sequence. In avGFP variants the N-terminal region is tolerant to such additions, but the same may not hold true for other FPs.

Commercial Availability

Obtaining new FPs is often a major impediment to investigators who want to begin work in this area or transition into multicolor imaging experiments. Fortunately, many variants listed in Tables 1.1 and 1.2 are now available through commercial distributors, and new FPs are continuously being added to their existing inventories. Table 1.3 lists the current (spring 2009) choices for commercial sources of FPs. These probes are usually available as plasmid "cloning" vectors that have been optimized with silent base-pair changes to generate codons ideal

for mammalian or bacterial expression and often contain the Kozak sequence. Cloning vectors contain a specialized region of 80–100 nucleotides, termed a "multiple cloning" site (MCS) positioned either N- or C-terminal to the FP, which houses a host of popular restriction endonuclease sites for convenient insertion of fusion host proteins. Genes inserted into the MCS will be expressed as fusions provided they are in the same reading frame as the FP and there are no intervening stop codons. Other features common to many commercial cloning vectors include a polyadenylation signal downstream of the fusion site to aid in processing of the 3′ terminus of the fusion mRNA and origins of replication for both bacterial and mammalian hosts. Antibiotic resistance cassettes are usually included to allow for selection in bacteria with kanamycin or ampicillin and in mammalian cells with G-418 or another antibiotic.

FLUORESCENT PROTEIN APPLICATIONS

Initially, FPs were used almost exclusively for targeting fusions to specific subcellular locations using proteins and signal peptides in mammalian cell cultures, a task for which they are well suited. However, as FP technology has matured, these ubiquitous fluorescent probes are finding uses in an increasing number of applications as diverse as superresolution microscopy [133, 155, 156], single-molecule imaging [157, 158], neurobiology

Table 1.3. Commercial Sources of FP and Optical Highlighter Vectors

Company Name	Web Site Address	Postal Address	Contact Information	Fluorescent Protein Products
Addgene	http://www.addgene.org	Addgene Inc. 1 Kendall Square Cambridge, MA 02139 USA	Tel: (617) 225–9000 Fax: (888) 734–0533 E-mail: info@addgene.org	Nonprofit plasmid archive for research scientists. Distributes numerous fluorescent protein vectors
Allele Biotechnology	http://www.allelebiotech.com	Allele Biotech 9924 Mesa Rim Road San Diego, CA 92121 USA	Tel: (858) 587–6645 Toll Free: 800 991-7624 Fax: (858) 587–6692 E-mail: oligo@allelebiotech.com	Alleleustrous Line: mTFP1, mWasabi
Amaxa Biosystems	http://www.amaxa.com	Amaxa Inc. 205 Perry Parkway, Suite 7 Gaithersburg, MD 20877 USA	Tel: (888) 632–9110	Amaxa pmaxFP Line:pmaxFP-Green, pmaxFP-Yellow, pmaxFP-Yellow-m, pmaxFP-Red.
BD Biosciences	http://www.bdbiosciences.com	BD Biosciences 2350 Qume Drive San Jose, CA 95131 USA	Tel: (877) 232–8995 Fax: (800) 325–9637 E-mail: facservice@bd.com	Baculovirus Transfer Vectors with BFP and YFP variants
Clontech Laboratories	http://www.clontech.com	Clontech Laboratories, Inc. 1290 Terra Bella Ave. Mountain View, CA 94043 USA	Tel: (800) 662–2566 Fax: (800) 424–1350 E-mail: orders@clontech.com	Living Colors Line: AcGFP1, AmCyan1, AsRed2, DsRed2, DsRed-Express, DsRed-Monomer, HcRed1, ZsGreen1, ZsYellow1, mFruits
Evrogen	http://www.evrogen.com	Evrogen Joint Stock Company Miklukho-Maklaya str, 16/10, 117997, Moscow, Russia	Tel: +7(495) 429–8020 Fax: +7(495) 429–8520 E-mail: evrogen@evrogen.com	Turbo and Tag Line: TurboGFP, YFP, RFP, FP602TagCFP, GFP, YFP, RFP; PhiYFP, JRed, PS-CFP2, Dendra2, KFP-Red, HyPer, KillerRed
Invitrogen	http://www.invitrogen.com	Invitrogen Corporation 1600 Faraday Avenue Carlsbad, CA 92008 USA	Tel: (760) 603–7200 Fax: (760) 602–6500 E-mail: catalog@invitrogen.com	Vivd Colors Line: Emerald, Topaz, CFP, BFP Cycle 3 GFP
LUX Biotechnology	http://luxbiotech.com	LUX Biotechnology Ltd BioSpace,King's Buildings EdinburghUnited Kingdom EH9 3JF	Tel: +44(0)131–662–3350 Fax: +44(0)131-662-3396	NanoLight Line (UK): *Renilla Mullerei* GFP, *Ptilosarcus* GFP, *Renilla Reniformis* GFP
MBL International Corporation	http://www.mblintl.com	MBL International 15 B Constitution Way Woburn, MA 01801 USA	Tel: 800 200–5459 Fax: (781) 939–6963 E-mail: info@mblintl.com	CoralHue Line: mAzami Green, mKusabira Orange, Dronpa, Kaede, Kikume Green-Red, Keima Red, Midoriishi-Cyan,
NanoLight Technology	http://www.nanolight.com	Bruce Bryan, MD 163 W. White Mountain Blvd. Pinetop, AZ 85935 USA	Tel: (928) 367–1200 Fax: (928) 367–1205 Lab Tel: (928) 333–2001 E-mail: bruce@prolume.com	NanoLight Line (USA): *Renilla Mullerei* GFP, *Ptilosarcus* GFP, *Renilla Reniformis* GFP
PerkinElmer	http://las.perkinelmer.com	PerkinElmer Life and Analytical Sciences 940 Winter StreetWaltham, MA 02451 USA	Tel: (781) 663–6900 E-mail: customer-careus@perkinelmer.com	BRET^2 Assay Vectors Line: GFP^2 Humanized Codon Cloning Vectors
Promega	http://www.promega.com	Promega Corporation 2800 Woods Hollow Road Madison, WI 53711 USA	Tel: 608-274-4330 Fax: 608-277-2516 E-mail: custserv@promega.com	Monster Green Line: phMGFP
Stratagene	http://www.stratagene.com	Stratagene11011 N. Torrey Pines Road La Jolla, CA 92037 USA	Tel: (858) 373–6300	Vitality Vectors: hrGFP and hrGFPII Nuc, Mito, Golgi, Peroxy

[9, 159], visualization of mRNA localization and trafficking [160, 161], FRET biosensors [162–164], bioluminescence resonance energy transfer or BRET [165–167], plant cell biology [168–170], bimolecular fluorescence complementation [171, 172], fluorescence correlation spectroscopy [173], *in vivo* imaging [174–176], drug discovery [177, 178], and in correlative morphological investigations of optical fluorescence techniques in combination with electron microscopy [29, 179].

The combined methodologies being developed with the aid of FPs are impressive and will no doubt continue to expand in the coming years, spanning the range from high-resolution single-molecule experiments *in vitro* to the behavior of entire cell populations in living animals and even beyond. Furthermore, the introduction of new and smaller genetically encoded fluorescent reporters, such as the light-, oxygen-, or voltage-sensing domain of the plant blue light receptor phototropin (iLOV; [180]), should enable even more advanced fusions that potentially exhibit a wide spectrum of new properties and are not hampered by larger steric constraints of traditional FPs. Additionally, directed molecular engineering of endogenous proteins to modify or create unusual fluorescent properties holds promise as an emerging technique [181]. Underpinning all current and future achievements using FPs are the rapid advances in imaging technologies that emerge from the commercial sector [182–184]. Advanced new FP-related fluorophores coupled with highly sensitive imaging systems will afford new opportunities to examine age-old problems in biology and medicine.

CONCLUSIONS

The current thrust of FP development strategies centers on fine-tuning the current palette of blue-to-yellow FPs derived from the *Aequorea victoria* jellyfish while simultaneously developing monomeric FPs emitting in the orange to far-red regions of the visible light spectrum. Progress toward these goals has been substantial, and it is not inconceivable that near-infrared emitting FPs loom on the horizon. The latest efforts in jellyfish variants have resulted in new and improved monomeric probes for the blue, cyan, green, and yellow regions, whereas the search for a bright monomeric and fast-maturing red FP has yielded a host of excellent candidates spanning longer wavelengths. Continuing efforts in protein engineering of existing FPs, coupled with advanced new technologies, should further expand the color palette and ultimately provide proteins in every spectral class that mature rapidly and are bright and photostable. As the development of optical highlighters continues, FPs useful for optical marking should evolve toward brighter monomeric derivatives with high contrast that can be easily photoconverted and display a wide spectrum of emission colors. For example, proteins capable of reversible photoactivation, red-to-green photoconversion, improved expression at elevated temperatures, and derivatives emitting in the far-red or near-infrared regions of the spectrum would be useful.

REFERENCES

1 Chudakov, D. M., Lukyanov, S., Lukyanov, K. A. (2005). Fluorescent proteins as a toolkit for in vivo imaging. *Trends Biotechnol* 23: 605–613.

2 Miyawaki, A., Nagai, T., Mizuno, H. (2005). Engineering fluorescent proteins. *Adv Biochem Eng Biotechnol* 95: 1–15.

3 Shaner, N. C., Steinbach, P. A., Tsien, R. Y. (2005). A guide to choosing fluorescent proteins. *Nat Methods* 2: 905–909.

4 Shaner, N. C., Patterson, G. H., Davidson, M. W. (2007). Advances in fluorescent protein technology. *J Cell Sci* 120: 4247–4260.

5 Zhang, J., Campbell, R. E., Ting, A. Y., Tsien, R. Y. (2002). Creating new fluorescent probes for cell biology. *Nat Rev Mol Cell Biol* 3: 906–918.

6 Meyer, T., Teruel, M. N. (2003). Fluorescence imaging of signaling networks. *Trends Cell Biol* 13: 101–106.

7 Miyawaki, A., Mizuno, H., Nagai, T., Sawano, A. (2003). Development of genetically encoded fluorescent indicators for calcium. *Methods Enzymo* 360: 202–225.

8 Zaccolo, M., Cesetti, T., Di Benedetto, G., Mongillo, M., Lissandron, V., Terrin, A., Zamparo, I. (2005). Imaging the cAMP-dependent signal transduction pathway. *Biochem Soc Trans* 33: 1323–1326.

9 Livet, J., Weissman, T. A., Kang, H. N., Draft, R. W., Lu, J., Bennis, R. A., Sanes, J. R., Lichtman, J. W. (2007). Transgenic strategies for combinatorial expression of fluorescent proteins in the nervous system. *Nature* 450: 56–62.

10 Day, R. N., Schaufele, F. (2005). Imaging molecular interactions in living cells, *Mol Endocrinol* 19: 1675–1686.

11 Lippincott-Schwartz, J., Patterson, G. H. (2003). Development and use of fluorescent protein markers in living cells. *Science* 300: 87–91.

12 Shimomura, O., Johnson, F. H., Saiga, Y. (1962). Extraction, purification and properties of Aequorin, a bioluminescent protein from luminous Hydromedusan, Aequorea. *J Cell Comp Physiol* 59 (1962): 223–239.

13 Chalfie, M., Tu, Y., Euskirchen, G., Ward, W. W., Prasher, D. C. (1994). Green fluorescent protein as a marker for gene expression. *Science* 263: 802–805.

14 Davenport, D., Nicol, J. A. C. (1955). Luminescence in Hydromedusae. *Proc R Soc Lond, B, Biol Sci* 144: 399–411.

15 Morin, J. G., Hastings, J. W. (1971). Energy transfer in a bioluminescent system. *J Cell Physiol* 77: 313–318.

16 Prasher, D. C., Eckenrode, V. K., Ward, W. W., Prendergast, F. G., Cormier, M. J. (1992). Primary structure of the Aequorea victoria green-fluorescent protein. *Gene* 111: 229–233.

17 Heim, R., Cubitt, A. B., Tsien, R. Y. (1995). Improved green fluorescence. *Nature* 373: 663–664.

18 Heim, R., Prasher, D. C., Tsien, R. Y. (1994). Wavelength mutations and posttranslational autoxidation of green fluorescent protein. *Proc Natl Acad Sci U S A* 91: 12501–12504.

19 Yang, T. T., Sinai, P., Green, G., Kitts, P. A., Chen, Y. T., Lybarger, L., Chervenak, R., Patterson, G. H., Piston, D. W., Kain, S. R. (1998). Improved fluorescence and dual color detection with enhanced blue and green variants of the green fluorescent protein. *J Biol Chem* **273**: 8212–8216.

20 Heim, R., Tsien, R. Y. (1996). Engineering green fluorescent protein for improved brightness, longer wavelengths and fluorescence resonance energy transfer. *Curr Biol* **6**: 178–182.

21 Wachter, R. M., Elsliger, M. A., Kallio, K., Hanson, G. T., Remington, S. J. (1998). Structural basis of spectral shifts in the yellow-emission variants of green fluorescent protein. *Structure* **6**: 1267–1277.

22 Matz, M. V., Fradkov, A. F., Labas, Y. A., Savitsky, A. P., Zaraisky, A. G., Markelov, M. L., Lukyanov, S. A. (1999). Fluorescent proteins from nonbioluminescent Anthozoa species, *Nat Biotechnol* **17**: 969–973.

23 Labas, Y. A., Gurskaya, N. G., Yanushevich, Y. G., Fradkov, A. F., Lukyanov, K. A., Lukyanov, S. A., Matz, M. V. (2002). Diversity and evolution of the green fluorescent protein family. *Proc Natl Acad Sci U S A* **99**: 4256–4261.

24 Matz, M. V., Lukyanov, K. A., Lukyanov, S. A. (2002). Family of the green fluorescent protein: journey to the end of the rainbow. *Bioessays* **24**: 953–959.

25 Masuda, H., Takenaka, Y., Yamaguchi, A., Nishikawa, S., Mizuno, H. (2006). A novel yellowish-green fluorescent protein from the marine copepod, Chiridius poppei, and its use as a reporter protein in HeLa cells. *Gene* **372**: 18–25.

26 Deheyn, D. D., Kubokawa, K., McCarthy, J. K., Murakami, A., Porrachia, M., Rouse, G. W., Holland, N. D. (2007). Endogenous green fluorescent protein (GFP) in amphioxus. *Biol Bull* **213**: 95–100.

27 Hopf, M., Gohring, W., Ries, A., Timpl, R., Hohenester, E. (2001). Crystal structure and mutational analysis of a perlecan-binding fragment of nidogen-1. *Nat Struct Biol* **8**: 634–640.

28 Jaiswal, J. K., Simon, S. M. (2004). Potentials and pitfalls of fluorescent quantum dots for biological imaging. *Trends Cell Biol* **14**: 497–504.

29 Giepmans, B. N. G., Adams, S. R., Ellisman, M. H., Tsien, R. Y. (2006). Review – the fluorescent toolbox for assessing protein location and function. *Science* **312**: 217–224.

30 Sapsford, K. E., Pons, T., Medintz, I. L., Mattoussi, H. (2006). Biosensing with luminescent semiconductor quantum dots. *Sensors* **6**: 925–953.

31 Michalet, X., Pinaud, F. F., Bentolila, L. A., Tsay, J. M., Doose, S., Li, J. J., Sundaresan, G., Wu, A. M., Gambhir, S. S., Weiss, S. (2005). Quantum dots for live cells, in vivo imaging, and diagnostics. *Science* **307**: 538–544.

32 McNamara, G., Boswell, C. (2006). PubSpectra; http://home.earthlink.net/~pubspectra/.

33 Dabbousi, B. O., Rodriguez-Viejo, J., Mikulec, F. V., Heine, J. R., Mattoussi, H., Ober, R., Jensen, K. F., Bawendi, M. G. (1997). (CdSe)ZnS core-shell quantum dots: synthesis and characterization of a size series of highly luminescent nanocrystallites. *J Phys Chem B* **101**: 9463–9475.

34 Wang, L., Jackson, W. C., Steinbach, P. A., Tsien, R. Y. (2004). Evolution of new nonantibody proteins via iterative somatic hypermutation. *Proc Natl Acad Sci U S A* **101**: 16745–16749.

35 Nguyen, A. W., Daugherty, P. S. (2005). Evolutionary optimization of fluorescent proteins for intracellular FRET. *Nat Biotechnol* **23**: 355–360.

36 Shaner, N. C., Campbell, R. E., Steinbach, P. A., Giepmans, B. N., Palmer, A. E., Tsien, R. Y. (2004). Improved monomeric red, orange and yellow fluorescent proteins derived from Discosoma sp. red fluorescent protein. *Nat Biotechnol* **22**: 1567–1572.

37 Ward, W. W. (2006). Biochemical and physical properties of green fluorescent protein. In: Chalfie, M., & Kain S. R. (Eds.), *Green Fluorescent Protein: Properties, Applications, and Protocols*, 2nd ed., Wiley-Interscience: New York, pp. 39–65.

38 Lakowicz, J. R. (2006). *Principles of Fluorescence Spectroscopy*, 3rd ed., Springer: New York.

39 Hazelwood, K. L., Olenych, S. G., Griffin, J. D., Cathcart, J. A., Davidson, M. W. (2007). Entering the portal: Understanding the digital image recorded through a microscope. In: Shorte, S. L., & Frischknecht, F. (Eds.), *Imaging Cellular and Molecular Biological Functions*. Springer: Berlin, pp. 3–43.

40 Henderson, J. N., Ai, H. W., Campbell, R. E., Remington, S. J. (2007). Structural basis for reversible photobleaching of a green fluorescent protein homologue. *Proc Natl Acad Sci U S A* **104**: 6672–6677.

41 Yang, F., Moss, L. G., Phillips, G. N., Jr. (1996). The molecular structure of green fluorescent protein. *Nat Biotechnol* **14**: 1246–1251.

42 Ormo, M., Cubitt, A. B., Kallio, K., Gross, L. A., Tsien, R. Y., Remington, S. J. (1996). Crystal structure of the Aequorea victoria green fluorescent protein. *Science* **273**: 1392–1395.

43 Weissleder, R. (2001). A clearer vision for in vivo imaging. *Nat Biotechnol* **19**: 316–317.

44 Zapata-Hommer, O., Griesbeck, O. (2003). Efficiently folding and circularly permuted variants of the Sapphire mutant of GFP. *BMC Biotechnol* **3**: 5.

45 Kogure, T., Karasawa, S., Araki, T., Saito, K., Kinjo, M., Miyawaki, A. (2006). A fluorescent variant of a protein from the stony coral Montipora facilitates dual-color single-laser fluorescence cross-correlation spectroscopy. *Nat Biotechnol* **24**: 577–581.

46 Xia, J., Kim, S. H., Macmillan, S., Truant, R. (2006). Practical three color live cell imaging by widefield microscopy. *Biol Proced Online* **8**: 63–68.

47 Chen, I., Ting, A. Y. (2005). Site-specific labeling of proteins with small molecules in live cells. *Curr Opin Biotechnol* **16**: 35–40.

48 Miller, L. W., Cornish, V. W. (2005). Selective chemical labeling of proteins in living cells. *Curr Opin Chem Biol* **9**: 56–61.

49 Gronemeyer, T., Godin, G., Johnsson, K. (2005). Adding value to fusion proteins through covalent labeling. *Curr Opin Biotechnol* **16**: 453–458.

50 Howarth, M., Takao, K., Hayashi, Y., Ting, A. Y. (2005). Targeting quantum dots to surface proteins in living cells with biotin ligase. *Proc Natl Acad Sci U S A* **102**: 7583–7588.

51 Griffin, B. A., Adams, S. R. Tsien, R. Y. (1998). Specific covalent labeling of recombinant protein molecules inside live cells. *Science* **281**: 269–272.

52 Keppler, A., Gendreizig, S., Gronemeyer, T., Pick, H., Vogel, H., Johnsson, K. (2003). A general method for the covalent

labeling of fusion proteins with small molecules in vivo. *Nat Biotechnol* 21: 86–89.

53 Martin, B. R., Giepmans, B. N., Adams, S. R., Tsien, R. Y. (2005). Mammalian cell-based optimization of the biarsenical-binding tetracysteine motif for improved fluorescence and affinity. *Nat Biotechnol* 23: 1308–1314.

54 Salih, A., Larkum, A., Cox, G., Kuhl, M., Hoegh-Guldberg, O. (2000). Fluorescent pigments in corals are photoprotective. *Nature* 408: 850–853.

55 Pedelacq, J. D., Cabantous, S., Tran, T., Terwilliger, T. C., Waldo, G. S. (2006). Engineering and characterization of a superfolder green fluorescent protein. *Nat Biotechnol* 24: 79–88.

56 Nagai, T., Ibata, K., Park, E. S., Kubota, M., Mikoshiba, K., Miyawaki, A. (2002). A variant of yellow fluorescent protein with fast and efficient maturation for cell-biological applications, *Nat Biotechnol* 20: 87–90.

57 Ai, H.-w., Shaner, N. C., Cheng, Z., Tsien, R. Y., Campbell, R. E. (2007). Exploration of new chromophore structures leads to the identification of improved blue fluorescent proteins. *Biochemistry* 46: 5904–5910.

58 Kremers, G. J., Goedhart, J., vanMunster, E. B., Gadella, T. W. J. (2006). Cyan and yellow super fluorescent proteins with improved brightness, protein folding, and FRET Forster radius. *Biochemistry* 45: 6570–6580.

59 Kremers, G. J., Goedhart, J., vandenHeuvel, D. J., Gerritsen, H. C., Gadella, T. W. J. (2007). Improved green and blue fluorescent proteins for expression in bacteria and mammalian cells. *Biochemistry* 46: 3775–3783.

60 Tsien, R. Y. (1998). The green fluorescent protein. *Annu Rev Biochem* 67: 509–544.

61 Wachter, R. M. (2007). Chromogenic cross-link formation in green fluorescent protein. *Acc Chem Res* 40: 120–127.

62 Gross, L. A., Baird, G. S., Hoffman, R. C., Baldridge, K. K., Tsien, R. Y. (2000). The structure of the chromophore within DsRed, a red fluorescent protein from coral. *Proc Natl Acad Sci U S A* 97: 11990–11995.

63 Zimmer, M. (2002). Green fluorescent protein (GFP): applications, structure, and related photophysical behavior. *Chem Rev* 102: 759–781.

64 Campbell, R. E., Tour, O., Palmer, A. E., Steinbach, P. A., Baird, G. S., Zacharias, D. A., Tsien, R. Y. (2002). A monomeric red fluorescent protein. *Proc Natl Acad Sci U S A* 99: 7877–7882.

65 Zacharias, D. A., Violin, J. D., Newton, A. C., Tsien, R. Y. (2002). Partitioning of lipid-modified monomeric GFPs into membrane microdomains of live cells. *Science* 296: 913–916.

66 Baird, G. S., Zacharias, D. A., Tsien, R. Y. (2000). Biochemistry, mutagenesis, and oligomerization of DsRed, a red fluorescent protein from coral. *Proc Natl Acad Sci U S A* 97: 11984–11989.

67 Verkhusha, V. V., Lukyanov, K. A. (2004). The molecular properties and applications of Anthozoa fluorescent proteins and chromoproteins. *Nat Biotechnol* 22: 289–296.

68 Zacharias, D. A., Tsien, R. Y. (2006). Molecular biology and mutation of green fluorescent protein. *Methods Biochem Anal* 47: 83–120.

69 Yarbrough, D., Wachter, R. M., Kallio, K., Matz, M. V., Remington, S. J. (2001). Refined crystal structure of DsRed,

a red fluorescent protein from coral, at 2.0-A resolution. *Proc Natl Acad Sci U S A* 98: 462–467.

70 Wall, M. A., Socolich, M., Ranganathan, R. (2000). The structural basis for red fluorescence in the tetrameric GFP homolog DsRed. *Nat Struct Biol* 7: 1133–1138.

71 Wiedenmann, J., Vallone, B., Renzi, F., Nienhaus, K., Ivanchenko, S., Rocker, C., Nienhaus, G. U. (2005). Red fluorescent protein eqFP611 and its genetically engineered dimeric variants. *J Biomed Opt* 10: 14003.

72 Karasawa, S., Araki, T., Yamamoto-Hino, M., Miyawaki, A. (2003). A green-emitting fluorescent protein from Galaxeidae coral and its monomeric version for use in fluorescent labeling. *J Biol Chem* 278: 34167–34171.

73 Karasawa, S., Araki, T., Nagai, T., Mizuno, H., Miyawaki, A. (2004). Cyan-emitting and orange-emitting fluorescent proteins as a donor/acceptor pair for fluorescence resonance energy transfer. *Biochem J* 381: 307–312.

74 Wiedenmann, J., Ivanchenko, S., Oswald, F., Schmitt, F., Rocker, C., Salih, A., Spindler, K. D., Nienhaus, G. U. (2004). EosFP, a fluorescent marker protein with UV-inducible green-to-red fluorescence conversion. *Proc Natl Acad Sci U S A* 101: 15905–15910.

75 Gurskaya, N. G., Verkhusha, V. V., Shcheglov, A. S., Staroverov, D. B., Chepurnykh, T. V., Fradkov, A. F., Lukyanov, S., Lukyanov, K. A. (2006). Engineering of a monomeric green-to-red photoactivatable fluorescent protein induced by blue light. *Nat Biotechnol* 24: 461–465.

76 Ando, R., Mizuno, H., Miyawaki, A. (2004). Regulated fast nucleocytoplasmic shuttling observed by reversible protein highlighting. *Science* 306: 1370–1373.

77 Ai, H.-w., Henderson, J. N., Remington, S. J., Campbell, R. E. (2006). Directed evolution of a monomeric, bright and photostable version of Clavularia cyan fluorescent protein: structural characterization and applications in fluorescence imaging. *Biochem J* 400: 531–540.

78 Fradkov, A. F., Verkhusha, V. V., Staroverov, D. B., Bulina, M. E., Yanushevich, Y. G., Martynov, V. I., Lukyanov, S., Lukyanov, K. A. (2002). Far-red fluorescent tag for protein labelling. *Biochem J* 368: 17–21.

79 Nienhaus, G. U., Nienhaus, K., Holzle, A., Ivanchenko, S., Renzi, F., Oswald, F., Wolff, M., Schmitt, F., Rocker, C., Vallone, B., Weidemann, W., Heilker, R., Nar, H., Wiedenmann, J. (2006). Photoconvertible fluorescent protein EosFP: biophysical properties and cell biology applications. *Photochem Photobiol* 82: 351–358.

80 Gavin, P., Devenish, R. J., Prescott, M. (2002). An approach for reducing unwanted oligomerisation of DsRed fusion proteins. *Biochem Biophys Res Commun* 298: 707–713.

81 Bulina, M. E., Verkhusha, V. V., Staroverov, D. B., Chudakov, D. M., Lukyanov, K. A. (2003). Hetero-oligomeric tagging diminishes non-specific aggregation of target proteins fused with anthozoa fluorescent proteins, *Biochem J* 371: 109–114.

82 Yanushevich, Y. G., Staroverov, D. B., Savitsky, A. P., Fradkov, A. F., Gurskaya, N. G., Bulina, M. E., Lukyanov, K. A., Lukyanov, S. A. (2002). A strategy for the generation of non-aggregating mutants of Anthozoa fluorescent proteins. *FEBS Lett* 511: 11–14.

83 Vinkenborg, J. L., Evers, T. H., Reulen, S. W., Meijer, E. W., Merkx, M. (2007). Enhanced sensitivity of FRET-based

protease sensors by redesign of the GFP dimerization interface. *Chembiochem* **8**: 1119–1121.

84 Ohashi, T., Galiacy, S. D., Briscoe, G., Erickson, H. P. (2007). An experimental study of GFP-based FRET, with application to intrinsically unstructured proteins. *Protein Sci* **16**: 1429–1438.

85 Ghaemmaghami, S., Huh, W. K., Bower, K., Howson, R. W., Belle, A., Dephoure, N., O'Shea, E. K., Weissman, J. S. (2003). Global analysis of protein expression in yeast. *Nature* **425**: 737–741.

86 Wallrabe, H., Periasamy, A. (2005). Imaging protein molecules using FRET and FLIM microscopy. *Curr Opin Biotechnol* **16**: 19–27.

87 Millington, M., Grindlay, G. J., Altenbach, K., Neely, R. K., Kolch, W., Bencina, M., Read, N. D., Jones, A. C., Dryden, D. T., Magennis, S. W. (2007). High-precision FLIM-FRET in fixed and living cells reveals heterogeneity in a simple CFP-YFP fusion protein. *Biophys Chem* **127**: 155–164.

88 Rizzo, M. A., Springer, G. H., Granada, B., Piston, D. W. (2004). An improved cyan fluorescent protein variant useful for FRET. *Nat. Biotechnol* **22**: 445–449.

89 Shimomura, O. (1979). Structure of the chromophore of Aequorea green fluorescent protein. *FEBS Lett* **104**: 220–222.

90 Cormack, B. P., Valdivia, R. H., Falkow, S. (1996). FACS-optimized mutants of the green fluorescent protein (GFP). *Gene* **173**: 33–38.

91 Mena, M. A., Treynor, T. P., Mayo, S. L., Daugherty, P. S. (2006). Blue fluorescent proteins with enhanced brightness and photostability from a structurally targeted library. *Nat Biotechnol* **24**: 1569–1571.

92 Griesbeck, O., Baird, G. S., Campbell, R. E., Zacharias, D. A., Tsien, R. Y. (2001). Reducing the environmental sensitivity of yellow fluorescent protein. Mechanism and applications. *J Biol Chem* **276**: 29188–29194.

93 Shagin, D. A., Barsova, E. V., Yanushevich, Y. G., Fradkov, A. F., Lukyanov, K. A., Labas, Y. A., Semenova, T. N., Ugalde, J. A., Meyers, A., Nunez, J. M., Widder, E. A., Lukyanov, S. A., Matz, M. V. (2004). GFP-like proteins as ubiquitous metazoan superfamily: evolution of functional features and structural complexity. *Mol Biol Evol* **21**: 841–850.

94 Dai, M., Fisher, H. E., Temirov, J., Kiss, C., Phipps, M. E., Pavlik, P., Werner, J. H., Bradbury, A. R. (2007). The creation of a novel fluorescent protein by guided consensus engineering. *Protein Eng Des Sel* **20**: 69–79.

95 Treynor, T. P., Vizcarra, C. L., Nedelcu, D., Mayo, S. L. (2007). Computationally designed libraries of fluorescent proteins evaluated by preservation and diversity of function. *Proc Natl Acad Sci U S A* **104**: 48–53.

96 Remington, S. J. (2006). Fluorescent proteins: maturation, photochemistry and photophysics. *Curr Opin Struct Biol* **16**: 714–721.

97 Cubitt, A. B., Woollenweber, L. A., Heim, R. (1999). Understanding structure-function relationships in the Aequorea victoria green fluorescent protein. *Methods Cell Biol* **58**: 19–30.

98 Stephens, D. J., Allan, V. J. (2003). Light microscopy techniques for live cell imaging. *Science* **300**: 82–86.

99 Khodjakov, A., Rieder, C. L. (2006). Imaging the division process in living tissue culture cells. *Methods* **38**: 2–16.

100 Subach, O. M., Gundorov, I. S., Yoshimura, M., Subach, F. V., Zhang, J. H., Gruenwald, D., Souslova, E. A., Chudakov, D. M., Verkhusha, V. V. (2008). Conversion of red fluorescent protein into a bright blue probe. *Chem Biol* **59**: 1116–1124.

101 Miyawaki, A., Llopis, J., Heim, R., McCaffery, J. M., Adams, J. A., Ikura, M., Tsien, R. Y. (1997). Fluorescent indicators for Ca2+ based on green fluorescent proteins and calmodulin. *Nature* **388**: 882–887.

102 Day, R. N., Booker, C. F., Periasamy, A. (2008). The characterization of an improved donor fluorescent protein for Förster resonance energy transfer microscopy. *J Biomed Optics.*

103 Rizzo, M. A., Piston, D. W. (2005). High-contrast imaging of fluorescent protein FRET by fluorescence polarization microscopy. *Biophys J* **88**: L14–L16.

104 Kremers, G. J., Goedhart, J., van Munster, E. B., Gadella, T. W. Jr. (2006). Cyan and yellow super fluorescent proteins with improved brightness, protein folding, and FRET Forster radius. *Biochemistry* **45**: 6570–6580.

105 Richards, B., Zharkikh, L., Hsu, F., Dunn, C., Kamb, A., Teng, D. H. (2002). Stable expression of Anthozoa fluorescent proteins in mammalian cells. *Cytometry* **48**: 106–112.

106 Luo, W. X., Cheng, T., Guan, B. Q., Li, S. W., Miao, J., Zhang, J., Xia, N. S. (2006). Variants of green fluorescent protein GFPxm. *Mar Biotechnol* **8**: 560–566.

107 Ai, H.-w., Olenych, S. G., Wong, P., Davidson, M. W., Campbell, R. E. (2008). Hue-shifted monomeric variants of Clavularia cyan fluorescent protein: identification of the molecular determinants of color and applications in fluorescence imaging. *BMC Biol* **6**: 13.

108 Cormier, M. J., Eckroade, C. B. (1962). Studies on the bioluminescence of Renilla reniformis. III. Some biochemical comparisons of the system to other Renilla species and determinations of the spectral energy distributions. *Biochim Biophys Acta* **64**: 340–344.

109 Miyawaki, A. (2002). Green fluorescent protein-like proteins in reef Anthozoa animals. *Cell Struct Funct* **27**: 343–347.

110 Cornea, A., Conn, P. M. (2002). Measurement of changes in fluorescence resonance energy transfer between gonadotropin-releasing hormone receptors in response to agonists. *Methods* **27**: 333–339.

111 Sturman, D. A., Shakiryanova, D., Hewes, R. S., Deitcher, D. L., Levitan, E. S. (2006). Nearly neutral secretory vesicles in Drosophila nerve terminals. *Biophys J* **90**: L45–47.

112 Tavare, J. M., Fletcher, L. M., Welsh, G. I. (2001). Using green fluorescent protein to study intracellular signalling. *J Endocrinol* **170**: 297–306.

113 Gurskaya, N. G., Savitsky, A. P., Yanushevich, Y. G., Lukyanov, S. A., Lukyanov, K. A. (2001). Color transitions in coral's fluorescent proteins by site-directed mutagenesis. *BMC Biochem* **2**: 6.

114 Sakaue-Sawano, A., Kurokawa, H., Morimura, T., Hanyu, A., Hama, H., Osawa, H., Kashiwagi, S., Fukami, K., Miyata, T., Miyoshi, H., Imamura, T., Ogawa, M., Masai, H., Miyawaki, A. (2008). Visualizing spatiotemporal dynamics of multicellular cell-cycle progression. *Cell* **132**: 487–498.

115 Shaner, N. C., Lin, M. Z., McKeown, M. R., Steinbach, P. A., Hazelwood, K. L., Davidson, M. W., Tsien, R. Y.

(2008). Improving the photostability of bright monomeric orange and red fluorescent proteins. *Nat Methods* **5**: 545–551.

116 Ip, D. T., Chan, S. H., Allen, M. D., Bycroft, M., Wan, D. C., Wong, K. B. (2004). Crystallization and preliminary crystallographic analysis of a novel orange fluorescent protein from the Cnidaria tube anemone Cerianthus sp. *Acta Crystallogr D Biol Crystallogr* **60**: 340–341.

117 Rizzo, M. A., Piston, D. W. (2005). Fluorescent protein tracking and detection. In: Goldman, R. D. & Spector D. L. (Eds.), *Live Cell Imaging: A Laboratory Manual.* Cold Spring Harbor Laboratory Press: Cold Spring Harbor: 3–23.

118 Merzlyak, E. M., Goedhart, J., Shcherbo, D., Bulina, M. E., Shcheglov, A. S., Fradkov, A. F., Gaintzeva, A., Lukyanov, K. A., Lukyanov, S., Gadella, T. W., Chudakov, D. M. (2007). Bright monomeric red fluorescent protein with an extended fluorescence lifetime. *Nat Methods* **4**: 555–557.

119 Patterson, G. H. (2004). A new harvest of fluorescent proteins. *Nat Biotechnol* **22**: 1524–1525.

120 Tsien, R. Y. (2005). Building and breeding molecules to spy on cells and tumors. *FEBS Lett* **579**: 927–932.

121 Wang, C. L., Yang, D. C., Wabl, M. (2004). Directed molecular evolution by somatic hypermutation. *Protein Eng Des Sel* **17**: 659–664.

122 Shkrob, M. A., Yanushevich, Y. G., Chudakov, D. M., Gurskaya, N. G., Labas, Y. A., Poponov, S. Y., Mudrik, N. N., Lukyanov, S., Lukyanov, K. A. (2005). Far-red fluorescent proteins evolved from a blue chromoprotein from Actinia equina. *Biochem J* **392**: 649–654.

123 Gurskaya, N. G., Fradkov, A. F., Terskikh, A., Matz, M. V., Labas, Y. A., Martynov, V. I., Yanushevich, Y. G., Lukyanov, K. A., Lukyanov, S. A. (2001). GFP-like chromoproteins as a source of far-red fluorescent proteins. *FEBS Lett* **507**: 16–20.

124 Wiedenmann, J., Schenk, A., Rocker, C., Girod, A., Spindler, K. D., Nienhaus, G. U. (2002). A far-red fluorescent protein with fast maturation and reduced oligomerization tendency from Entacmaea quadricolor (Anthozoa, Actinaria). *Proc Natl Acad Sci U S A* **99**: 11646–11651.

125 Schenk, A., Ivanchenko, S., Rocker, C., Wiedenmann, J., Nienhaus, G. U. (2004). Photodynamics of red fluorescent proteins studied by fluorescence correlation spectroscopy. *Biophys J* **86**: 384–394.

126 Kredel, S., Oswald, F., Nienhaus, K., Deuschle, K., Rocker, C., Wolff, M., Heilker, R., Nienhaus, G. U., Wiedenmann, J. (2009). A monomeric eqFP611 variant for labeling of subcellular structures. *PLoS ONE* **3**.

127 Lukyanov, K. A., Fradkov, A. F., Gurskaya, N. G., Matz, M. V., Labas, Y. A., Savitsky, A. P., Markelov, M. L., Zaraisky, A. G., Zhao, X., Fang, Y., Tan, W., Lukyanov, S. A. (2000). Natural animal coloration can be determined by a nonfluorescent green fluorescent protein homolog. *J Biol Chem* **275**: 25879–25882.

128 Shcherbo, D., Merzlyak, E. M., Chepurnykh, T. V., Fradkov, A. F., Ermakova, G. V., Solovieva, E. A., Lukyanov, K. A., Bogdanova, E. A., Zaraisky, A. G., Lukyanov, S., Chudakov, D. M. (2007). Bright far-red fluorescent protein for whole-body imaging. *Nat Methods* **4**: 741–746.

129 Lukyanov, K. A., Chudakov, D. M., Lukyanov, S., Verkhusha, V. V. (2005). Innovation: Photoactivatable fluorescent proteins. *Nat Rev Mol Cell Biol* **6**: 885–891.

130 Ando, R., Hama, H., Yamamoto-Hino, M., Mizuno, H., Miyawaki, A. (2002). An optical marker based on the UV-induced green-to-red photoconversion of a fluorescent protein. *Proc Natl Acad Sci U S A* **99**: 12651–12656.

131 Patterson, G. H., Lippincott-Schwartz, J. (2002). A photoactivatable GFP for selective photolabeling of proteins and cells. *Science* **297**: 1873–1877.

132 Chudakov, D. M., Belousov, V. V., Zaraisky, A. G., Novoselov, V. V., Staroverov, D. B., Zorov, D. B., Lukyanov, S., Lukyanov, K. A. (2003). Kindling fluorescent proteins for precise in vivo photolabeling. *Nat Biotechnol* **21**: 191–194.

133 Betzig, E., Patterson, G. H., Sougrat, R., Lindwasser, O. W., Olenych, S., Bonifacino, J. S., Davidson, M. W., Lippincott-Schwartz, J., Hess, H. F. (2006). Imaging intracellular fluorescent proteins at nanometer resolution. *Science* **313**: 1642–1645.

134 Schwentker, M. A., Bock, H., Hofmann, M., Jakobs, S., Bewersdorf, J., Eggeling, C., Hell, S. W. (2007). Widefield subdiffraction RESOLFT microscopy using fluorescent protein photoswitching. *Microsc Res Tech* **70**: 269–280.

135 Verkhusha, V. V., Sorkin, A. (2005). Conversion of the monomeric red fluorescent protein into a photoactivatable probe. *Chem Biol* **12**: 279–285.

136 Chudakov, D. M., Verkhusha, V. V., Staroverov, D. B., Souslova, E. A., Lukyanov, S., Lukyanov, K. A. (2004). Photoswitchable cyan fluorescent protein for protein tracking. *Nat Biotechnol* **22**: 1435–1439.

137 Tsutsui, H., Karasawa, S., Shimizu, H., Nukina, N., Miyawaki, A. (2005). Semi-rational engineering of a coral fluorescent protein into an efficient highlighter. *EMBO Rep* **6**: 233–238.

138 Ivanchenko, S., Rocker, C., Oswald, F., Wiedenmann, J., Nienhaus, G. U. (2005). Targeted green-red photoconversion of EosFP, a fluorescent marker protein. *J Biol Phys* **31**: 249–259.

139 Habuchi, S., Tsutsui, H., Kochaniak, A. B., Miyawaki, A., van Oijen, A. M. (2008). mKikGR, a monomeric photoswitchable fluorescent protein. *PLoS ONE* **3**: e3944.

140 McKinney, S. A., Murphy, C. S., Hazelwood, K. L., Davidson, M. W., Looger, L. L. (2008). A bright and photostable photoconvertible fluorescent protein for fusion tags. *Nature Meth* **6**.

141 Mizuno, H., Mal, T. K., Tong, K. I., Ando, R., Furuta, T., Ikura, M., Miyawaki, A. Photo-induced peptide cleavage in the green-to-red conversion of a fluorescent protein. *Mol Cell* **12**: 1051–1058.

142 Nienhaus, K., Nienhaus, G. U., Wiedenmann, J., Nar, H. (2005). Structural basis for photo-induced protein cleavage and green-to-red conversion of fluorescent protein EosFP. *Proc Natl Acad Sci U S A* **102**: 9156–9159.

143 Cinelli, R. A. G., Pellegrini, V., Ferrari, A., Faraci, P., Nifosi, R., Tyagi, M., Giacca, M., Beltram, F. (2001). Green fluorescent proteins as optically controllable elements in bioelectronics. *Appl Phys Lett* **79**: 3353–3355.

144 McAnaney, T. B., Zeng, W., Doe, C. F. E., Bhanji, N., Wakelin, S., Pearson, D. S., Abbyad, P., Shi, X. H., Boxer, S. G., Bagshaw, C. R. (2005). Protonation, photobleaching, and photoactivation of yellow fluorescent protein (YFP 10C): A unifying mechanism. *Biochemistry* **44**: 5510–5524.

145 Dickson, R. M., Cubitt, A. B., Tsien, R. Y., Moerner, W. E. (1997). On/off blinking and switching behaviour of single molecules of green fluorescent protein. *Nature* **388**: 355–358.

146 Habuchi, S., Cotlet, M., Gensch, T., Bednarz, T., Haber-Pohlmeier, S., Rozenski, J., Dirix, G., Michiels, J., Vanderleyden, J., Heberle, J., De Schryver, F. C., Hofkens, J. (2005). Evidence for the isomerization and decarboxylation in the photoconversion of the red fluorescent protein DsRed. *J Am Chem Soc* **127**: 8977–8984.

147 Chudakov, D. M., Feofanov, A. V., Mudrik, N. N., Lukyanov, S., Lukyanov, K. A. (2003). Chromophore environment provides clue to "kindling fluorescent protein" riddle. *J Biol Chem* **278**: 7215–7219.

148 Andresen, M., Wahl, M. C., Stiel, A. C., Grater, F., Schafer, L. V., Trowitzsch, S., Weber, G., Eggeling, C., Grubmuller, H., Hell, S. W., Jakobs, S. (2005). Structure and mechanism of the reversible photoswitch of a fluorescent protein. *Proc Natl Acad Sci U S A* **102**: 13070–13074.

149 Andresen, M., Stiel, A. C., Trowitzsch, S., Weber, G., Eggeling, C., Wahl, M. C., Hell, S. W., Jakobs, S. (2007). Structural basis for reversible photoswitching in Dronpa. *Proc Natl Acad Sci U S A* **104**: 13005–13009.

150 Henderson, J. N., Remington, S. J. (2006). The kindling fluorescent protein: a transient photoswitchable marker. *Physiology (Bethesda)* **21**: 162–170.

151 Stiel, A. C., Trowitzsch, S., Weber, G., Andresen, M., Eggeling, C., Hell, S. W., Jakobs, S., Wahl, M. C. (2007). 1.8 A bright-state structure of the reversibly switchable fluorescent protein Dronpa guides the generation of fast switching variants. *Biochem J* **402**: 35–42.

152 Baird, G. S., Zacharias, D. A., Tsien, R. Y. (1999). Circular permutation and receptor insertion within green fluorescent proteins. *Proc Natl Acad Sci U S A* **96**: 11241–11246.

153 Topell, S., Hennecke, J., Glockshuber, R. (1999). Circularly permuted variants of the green fluorescent protein. *FEBS Lett* **457**: 283–289.

154 Kozak, M. (1987). An analysis of 5'-noncoding sequences from 699 vertebrate messenger RNAs. *Nucleic Acids Res* **15**: 8125–8148.

155 Fernandez-Suarez, M., Ting, A. Y. (2008). Fluorescent probes for super-resolution imaging in living cells. *Nat Rev Mol Cell Biol* **9**: 929–943.

156 Lippincott-Schwartz, J., Manley, S. (2009). Putting super-resolution fluorescence microscopy to work. *Nat Methods* **6**: 21–23.

157 Seefeldt, B., Kasper, R., Seidel, T., Tinnefeld, P., Dietz, K. J., Heilemann, M., Sauer, M. (2008). Fluorescent proteins for single-molecule fluorescence applications. *J Biophoton* **1**: 74–82.

158 Douglass, A. D., Vale, R. D. (2008). Single-molecule imaging of fluorescent proteins. *Methods Cell Biol* **85**: 113–125.

159 Miyawaki, A. (2005). Innovations in the imaging of brain functions using fluorescent proteins. *Neuron* **48**: 189–199.

160 Querido, E., Chartrand, P. (2008). Using fluorescent proteins to study mRNA trafficking in living cells. *Methods Cell Biol* **85**: 273–292.

161 Dahm, R., Zeitelhofer, M., Gotze, B., Kiebler, M. A., Macchi, P. (2008). Visualizing mRNA localization and local protein translation in neurons. *Methods Cell Biol* **85**: 293–327.

162 Piston, D. W., Kremers, G. J. (2007). Fluorescent protein FRET: the good, the bad and the ugly. *Trends Biochem Sci* **32**: 407–414.

163 Miyawaki, A. (2003). Visualization of the spatial and temporal dynamics of intracellular signaling. *Dev Cell* **4**: 295–305.

164 Lalonde, S., Ehrhardt, D. W., Frommer, W. B. (2005). Shining light on signaling and metabolic networks by genetically encoded biosensors. *Curr Opin Plant Biol* **8**: 574–581.

165 Ciruela, F. (2008). Fluorescence-based methods in the study of protein-protein interactions in living cells. *Curr Opin Biotechnol* **19**: 338–343.

166 Pfleger, K. D. G., Eidne, K. A. (2006). Illuminating insights into protein-protein interactions using bioluminescence resonance energy transfer (BRET). *Nat Methods* **3**: 165–174.

167 Xu, X. D., Soutto, M., Xie, Q., Servick, S., Subramanian, C., von Arnim, A. G., Johnson, C. H. (2007). Imaging protein interactions with bioluminescence resonance energy transfer (BRET) in plant and mammalian cells and tissues. *Proc Natl Acad Sci U S A* **104**: 10264–10269.

168 Dixit, R., Cyr, R., Gilroy, S. (2006). Using intrinsically fluorescent proteins for plant cell imaging. *Plant J* **45**: 599–615.

169 Mathur, J. (2007). The illuminated plant cell. *Trends Plant Sci* **12**: 506–513.

170 Haseloff, J., Siemering, K. R. (2006). The uses of green fluorescent protein in plants. *Methods Biochem Anal* **47**: 259–284.

171 Kerppola, T. K. (2006). Visualization of molecular interactions by fluorescence complementation. *Nat Rev Mol Cell Biol* **7**: 449–456.

172 Kerppola, T. K. (2008). Biomolecular fluorescence complementation (BiFC) analysis as a probe of protein interactions in living cells. *Annu Rev Biophys* **37**: 465–487.

173 Langowski, J. (2008). Protein-protein interactions determined by fluorescence correlation spectroscopy. *Methods Cell Biol* **85**: 471–484.

174 Hoffman, R. M., Yang, M. (2006). Whole-body imaging with fluorescent proteins. *Nat Protoc* **1**: 1429–1438.

175 Barth, A. L. (2007). Visualizing circuits and systems using transgenic reporters of neural activity. *Curr Opin Neurobiol* **17**: 567–571.

176 Hadjantonakis, A. K., Dickinson, M. E., Fraser, S. E., Papaioannou, V. E. (2003). Technicolour transgenics: Imaging tools for functional genomics in the mouse. *Nat Rev Genet* **4**: 613–625.

177 Michnick, S. W., Ear, P. H., Manderson, E. N., Remy, I., Stefan, E. (2007). Universal strategies in research and drug discovery based on protein-fragment complementation assays. *Nat Rev Drug Discov* **6**: 569–582.

178 Wolff, M., Wiedenmann, J., Nienhaus, G. U., Valler, M., Heilker, R. (2006). Novel fluorescent proteins for high-content screening. *Drug Discov Today* **11**: 1054–1060.

179 Gaietta, G. M., Giepmans, B. N. G., Deerinck, T. J., Smith, W. B., Ngan, L., Llopis, J., Adams, S. R., Tsien, R. Y., Ellisman, M. H. (2006). Golgi twins in late mitosis revealed by genetically encoded tags for live cell imaging and correlated electron microscopy. *Proc Natl Acad Sci U S A* **103**: 17777–17782.

180 Chapman, S., Faulkner, C., Kaiserli, E., Garcia-Mata, C., Savenkov, E. I., Roberts, A. G., Oparka, K. J., Christie, J. M. (2008). The photoreversible fluorescent protein iLOV outperforms GFP as a reporter of plant virus infection. *Proc Natl Acad Sci U S A* **105**: 20038–20043.

181 Berndt, A., Yizhar, O., Gunaydin, L. A., Hegemann, P., Deisseroth, K. (2008). Bi-stable neural state switches. *Nat Neurosci* **12**(2) 229–234.

182 Sluder, G., Wolf D. W. (Eds.) (2007). *Digital Microscopy*, 3rd ed., Elsevier: New York.

183 Shorte, S. L., Frischknecht, F. (Eds.) (2007). *Imaging Cellular and Molecular Biological Functions.* Springer-Verlag: Berlin.

184 Tsien, R. Y. (2003). Imagining imaging's future. *Nat Rev Mol Cell Biol*: Ss16–Ss21.

185 Cubitt, A. B., Heim, R., Adams, S. R., Boyd, A. E., Gross, L. A., Tsien, R. Y. (1995). Understanding, improving and using green fluorescent proteins. *Trends Biochem Sci* **20**: 448–455.

186 Miyawaki, A., Griesbeck, O., Heim, R., Tsien, R. Y. (1999). Dynamic and quantitative Ca2+ measurements using improved cameleons. *Proc Natl Acad Sci U S A* **96**: 2135–2140.

187 Ai, H.-w., Hazelwood, K. L., Davidson, M. W., Campbell, R. E. (2008). Fluorescent protein FRET pairs for ratiometric imaging of dual biosensors. *Nat Methods* **5**: 401–403.

188 Bevis, B. J., Glick, B. S. (2002). Rapidly maturing variants of the Discosoma red fluorescent protein (DsRed). *Nat Biotechnol* **20**: 83–87.

Revealing Biomolecular Mechanisms Through *In Vivo* Bioluminescence Imaging

2

Timothy C. Doyle, Qian Wang, and Christopher H. Contag

Since the first publication describing *in vivo* bioluminescence imaging (BLI) [1], this molecular imaging strategy has been adapted to investigate a range of biological questions in a variety of fields. This imaging modality has been used to investigate mammalian physiology, disease mechanisms, and response to therapy as well as development of new therapeutic agents. This reporter gene imaging approach was enabled by the development of bioluminescent reporter genes (luciferases) as transcriptional reporters in cultured cells [2–7] and small transparent organisms [8–11]. As such, expression of luciferases has been used to create light-emitting cells, which can be studied in correlative culture assays and then used in animal models where a low intrinsic background signal from the host animal provides significant signal-to-noise ratios. BLI has the advantage of being relatively inexpensive and easy to use, and because it uses relatively nontoxic substrates, it is ideally suited to small animals, such as mice and rats. In addition, BLI avoids hazards of ionizing radiation. Laboratory rodents are small enough to allow light originating from luciferase-expressing cells deep in the body to be transmitted to the body surface where the photons can then be detected by sensitive camera systems based on charge-coupled devices (CCDs, see Figure 2.1).

The response of the cells expressing luciferase, or the expression of luciferase by a promoter of interest, can thus be observed in the complex environment of the living body. As with other molecular imaging modalities outlined in this book, bioluminescence imaging in small animals offers the opportunity to repeatedly and noninvasively image the same animals over extended periods of time, thus decreasing the number of animals required for a study relative to other approaches that use tissues acquired by biopsy or after necropsy. Furthermore, because the animals can be used as their own controls, the quality of data obtained is improved by comparing changes within a subject as well as between subjects. One of the greatest advantages of BLI is low intrinsic background from tissues; this allows detection of small numbers of labeled cells through detection of low levels of luciferase activity. Optical imaging tools in general, including BLI, are relatively inexpensive, rapid, and sensitive. However, as with all reporter gene imaging strategies, BLI is usually limited by the necessity to genetically engineer cells or host animals to express the reporter. Due to the absorbing and scattering properties of mammalian tissues [12], transmission of light in the visible region of the spectrum is relatively poor, and thus optical tools are well suited for use in small animals such as mice and rats, with limited applications in larger animals and humans. Mice and rats are, however, widely used to model biological pathways and diseases of humans, and BLI techniques have been applied broadly to such models, allowing a greater understanding of these processes. BLI has tremendous utility in the study of infection, both monitoring the pathogen and the host response, tumor biology, stem cell biology, tissue transplantation, gene regulation, protein–protein interactions, and drug and gene delivery strategies (reviewed in [13]). This chapter describes BLI from the biochemistry and physics of bioluminescence imaging to the required instrumentation and then highlights selected applications.

FIRST PRINCIPLES

Imaging in the Optical Domain Using Biological Sources of Light: the Basics

Luciferases are a class of enzymes that generate light in the presence of a substrate (known as a luciferin), an energy source (which may also be provided by the luciferin itself), and oxygen [14, 15]. Many types of organisms have evolved enzymes that generate light, and there are a handful of well-characterized proteins and substrates that have been used as biological reporters. Luciferase activity can be detected over many orders of magnitude; the inherently low background enables a broad dynamic

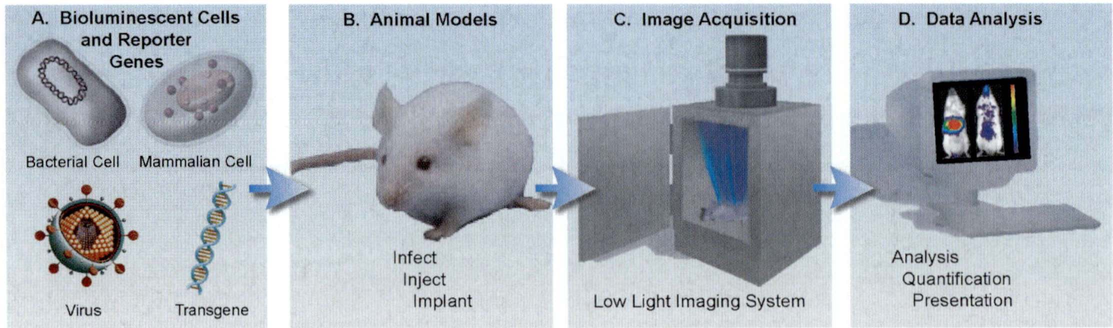

Figure 2.1. Overview of *in vivo* bioluminescence imaging. Cells, viruses, or genes of interest are genetically engineered to express a luciferase gene (**A**), which are then introduced to the animal model (typically mice or rats (**B**)). The animals are then imaged (**C**) with a sensitive CCD camera mounted on a light-tight imaging chamber, and the resulting images (**D**) can then be quantified. Figure adapted from Doyle et al. [47].

range beginning at extremely low levels (in the attamolar range) in biochemical assays. In most cases, the substrates themselves do not produce light in the absence of luciferase activity, which is one reason for the low background signal in biochemical assays, and coupled with a near absence of autoluminescence in mammalian tissues is the reason for extremely low background *in vivo*. This provides a significant advantage over radioactive imaging strategies, such as PET and SPECT, in which the unbound or unincorporated injected radioactive tracer produces a relatively high background in the host animal. However, the photons in the visible region of the spectrum are of relatively low energy compared to those emitted by radionuclides used in PET and SPECT, and are thus subject to significant attenuation by animal tissues, being both scattered and absorbed. A reasonable depth of penetration for imaging with photons in the visible region of the spectrum is roughly 1–2 cm with a maximum of 6–10 cm [12, 16]. For BLI, a maximum tissue depth is on the order of 1–2 cm for reasonable sensitivity, and thus BLI is most often used in small animals such as mice and rats, or in specific applications where the source is close to the surface.

Luciferase Biochemistries

Several different classes of luciferases have been cloned and used as reporter genes. Each class is defined by a different substrate and energy source. Three main classes of luciferases have been used *in vivo*: the aldehyde-utilizing enzyme encoded by the *lux* operon from marine and terrestrial bacteria (e.g., *Vibrio spp.* and *Photorhabdus luminescens*) [17–21], the coelenterazine-utilizing luciferases from marine organisms (e.g., jellyfish and sea pansies) [4, 5], and the benzothiazole-utilizing luciferases from terrestrial beetles (e.g., fireflies, click beetles, and railroad worms) [2, 3, 6, 7, 22]. Many of the genes encoding these enzymes and modified versions of the enzymes are commercially available and have been codon optimized for

specific expression in mammals or Gram+ or Gram− bacteria. Because the host cells do not typically generate the luciferin substrates (the exception being bacteria and the Lux operon), these compounds are also available commercially and have been widely used in animal imaging with no overt toxicities.

The most commonly used luciferase in cell biology and BLI is from the firefly, *Photonis pyloris*, a member of the beetle family that also includes other bioluminescent insects such as railroad worms and click beetles. The luciferases from these organisms (termed Luc for the enzyme, and *luc* for the gene) are a single polypeptide related to the CoA ligase family of proteins [23]. They use a benzothiazole luciferin, adenosine triphosphate (ATP), and oxygen as substrates and emit in the range of 560 to 614 nm light (green to red), depending on the luciferase and physiological conditions [24]. The firefly luciferase (fluc), for example, will emit a yellow–green light with an emission peak at ∼560 nm at room temperature (25° C) but will emit red light (λmax ∼605 nm) at 37° C, whereas commercial variants of the click beetle enzyme will emit either green (λmax ∼540 nm, CBGr luc) or red (λmax ∼615 nm, CBR luc), neither of which is spectrally affected by temperature [24]. Because these enzymes require both ATP and oxygen, bioluminescence is greatest in metabolically active cells in an aerobic environment. Nonetheless, fluc has been used as a hypoxia sensor and has been reported to have activity even in low oxygen environments [25].

Bioluminescence is common in many marine organisms, and genes for luciferases from the sea pansy *Renilla* (RLuc), the jellyfish *Aequorea* (ALuc), and the marine copepod *Gaussia* (GLuc) have all been cloned and expressed in mammalian cells and used in BLI [26–29]. All these enzymes have unique amino acid sequences, but each utilizes the same energy-rich substrate, coelenterazine, and oxygen to generate light. The hydrozoan *Aequorea* enzyme (aequorin) also requires a calcium ion

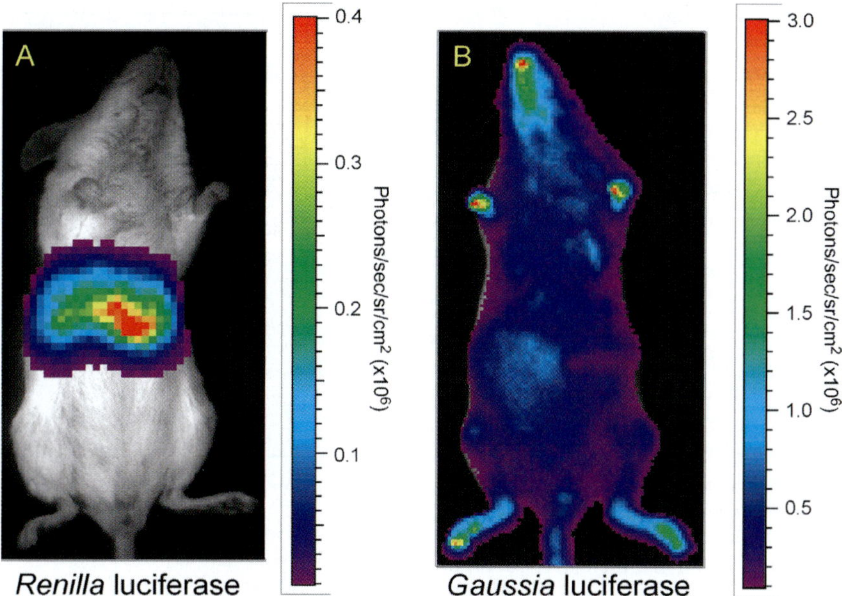

Figure 2.2. Luciferase activity in mice expressing either *Renilla* luciferase (**A**) or *Gaussia* luciferase (**B**) in the liver. Hempatocytes were transfected in living mice using s10μg plasmid DNA with either the *Renilla* or *Gaussia* luciferase genes, and imaged 24 h later following coelenterazine injection (IV). The cytoplasmic RLuc enzyme remains within the liver, whereas the GLuc protein is secreted into the blood pool, causing the entire mouse to bioluminesce. Bioluminescent signal from the *Renilla*-expressing mouse was 0.8×10^7 photons/sec, whereas the *Gaussia*-expressing mouse was 14×10^7 photons/sec (unpublished data).

to generate light and has been used as a measure of calcium ion release in cells and tissues [30, 31]. All enzymes in this class that require coelenterazine naturally generate blue light (λmax ~475 nm) with the native substrate, and none requires a cellular cofactor as an energy source because the coelenterazine substrate itself provides the necessary energy. Mutant forms of the RLuc enzyme and the use of different substrates can yield different wavelengths of emission [32, 33].

The *Gaussia* enzyme is unique in this family in that it is an excreted homopolymer in the *Gaussia* organism and along with its substrate generates a cloud of bioluminescence thought to distract potential predators. This has led to the use of the *Gaussia* enzyme as a secreted reporter in mammalian systems [29] (Figure 2.2). The *Gaussia* luciferase is a small (185 amino acid) monomeric protein, and a mammalian-optimized version of the protein is nontoxic to the host mammalian cell and is naturally secreted. Further, it generates a significantly higher bioluminescent signal in mammalian cells than either the *Renilla* or firefly luciferases. Because the enzyme is secreted, it is also possible to quantify labeled cells in animals by measuring its activity in blood samples [29]. The fact that these marine luciferases function in the absence of cellular cofactors has enabled their use as fusions to other proteins such as extracellular antibodies [28, 34] or attached to fluorescent nanoparticles to create self-illuminating particles [35] that when injected into animals will bioluminesce in the presence of substrate. Such

agents remain extracellular and can thus be targeted to surface antigens on cells or vascular walls and subsequently imaged because no cellular cofactor is required for bioluminescence. The native substrate is, however, relatively unstable in aqueous solutions, and its breakdown is catalyzed by serum proteins that can generate light independent of the luciferase activity [36]. Therefore a higher background can be anticipated from this type of reporter. Because the substrates for the marine and beetle luciferases do not cross-react with each other, it is possible to use both reporters in the same tube, cell culture, or animal, either imaging each enzyme sequentially following administration of each substrate or taking advantage of different spectral emission peaks to differentiate the two enzymes.

Several bacteria use a heterodimeric luciferase (Lux) to generate light using reduced flavine mononucleotide (FMNH$_2$), the long chain fatty aldehyde decanal, and oxygen to produce a blue–green light (λmax ~490 nm). In the bacterial systems characterized to date, the genes encoding the luciferase (*luxA* and *luxB*) as well as three enzymes (*luxC–E*) that synthesize the aldehyde substrate are encoded in a single operon (the Lux operon consisting of the genes *luxCDABE*), which may also contain other genes encoding a fluorescence protein that binds the LuxAB protein or regulator proteins that control expression of the operon [37, 38]. Two systems have been well characterized, from the marine *Vibrio spp.* and the terrestrial *P. luminescens*, and because the latter retains significant activity at 37° C [38, 39], this has been used in BLI studies in mammals. The *luxCDABE* operon has been successfully expressed in many Gram-negative bacteria such as *Escherichia coli* and *Salmonella* spp. for studies in pathogenesis [1] and food safety [40]. The operon has been modified for expression in Gram-positive organisms. The major limitation to expression in Gram-positive bacteria was the more restrictive canonical transcriptional start sites (Shine-Delgano sequences), and these needed to be introduced into the construct to generate a Lux operon that could be functionally expressed in Gram-positive bacteria such as *Stapholo-coccus aureus* [41] and *Listeria monocytogenes* [42]. In these modified operons the gene order was also changed to *luxABCDE*; this was not only to list the genes in alphabetical order but also to arrange the genes encoding the heterodimeric luciferase first in the gene order with the intent of producing more enzyme and generating more light per cell. The bacterial system has also been successfully expressed in yeast [43] and mammalian cells [44] by expressing each component separately and/or as fusions from different promoters. The main advantage of this luciferase system is that no exogenous substrate needs be added to the cells to generate light, although for maximal light production the cells must be in aerobic conditions and be metabolically active to generate light, as both oxygen and $FMNH_2$ must be present. These luciferases may function in anaerobic or microaerophilic environments, but the data to date are anecdotal. Slow-growing bacteria and those that thrive in anaerobic conditions in mammalian infection models are likely to produce significantly less signal, but there may be sufficient signals to detect these bacteria even in oxygen-limited environments.

Light Propagation Through Mammalian Tissues and Use of Optical Reporter Genes

Bioluminescence is easily observed in nature due to either the small size of the organisms expressing them (such as the firefly) or the transparent nature of the host (such as jellyfish or bacteria living in symbiosis with fish), in addition to the optimal conditions within the organism to produce a bright signal and to control timing and signal intensities. Application of these optical reporter genes in biology, especially in imaging in mammals, requires modifications and adaptation for use in vertebrate organisms that are both larger and less transparent. The codon optimization and expression from highly modified vectors has led to extremely bright signals from engineered organisms. For *in vivo* imaging applications, engineered cells that express reporter genes introduce into their hosts and can be localized to any number of anatomic sites. Because mammalian tissues attenuate optical signals through both scattering and absorption of light in the visible region of the spectrum, the signals are surface

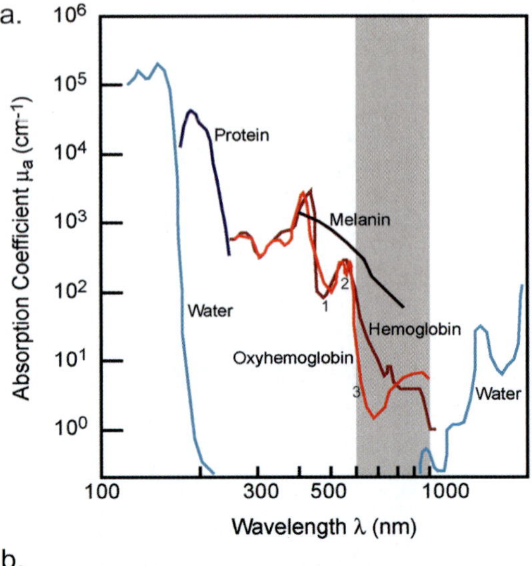

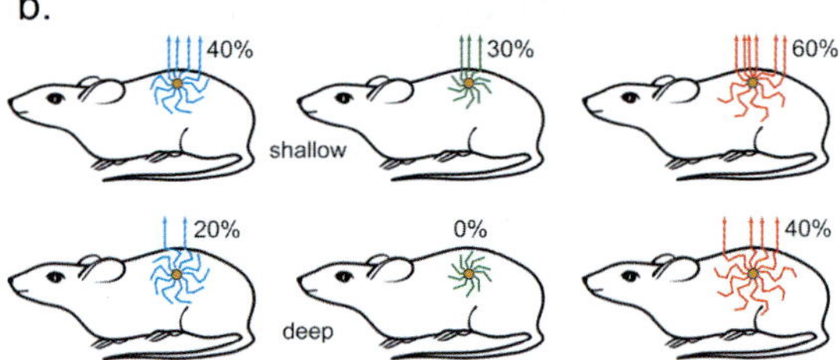

Figure 2.3. (A) Absorption properties of the principal chromophores found in mammalian tissue. An "optical window" between 600–1000 nm (shaded in gray) represents the optimal wavelengths to visualize photon emission from within an animal, where hemoglobin (red), melanin (black), and water (blue) absorptions are smallest. Spectral peaks of *Renilla* (1), click beetle green (2), and firefly luciferases are shown. **(B)** Simplified representation of photon propagation through tissue at different wavelengths and depths. Ten photons with scattered paths are shown for each of blue, green, and red wavelengths, where green photons (middle) are absorbed most strongly (and hence travel the least distance) and red photons (right mice) absorbed the least. As the light source is moved deeper into the animal (lower mice), fewer photons are emitted from the surface to be detected by a CCD camera.

weighted and an understanding of tissue optics is essential for interpreting optical imaging data. Scattering of photons in the visible and near-infrared region of the spectrum is due to changes in refractive index at cell membranes and cell organelles, as well as at tissue boundaries [45, 46]. Photon absorption can vary with tissue type and is largely due to levels of hemoglobin, which absorbs strongly in the blue–green region of the visible spectrum (Figure 2.3a).

The extent of absorption by hemoglobin is significantly lower in the red and near-infrared region of the spectrum (wavelengths longer than 600 nm). Thus, when a white flashlight is held behind your fingers, the blue and green components are strongly absorbed, and the red light component of the white light is apparent [47]. Photon scattering is relatively unaffected by wavelength in the visible range, and above 600 nm (in the red region) scattering becomes a main attenuator of light passing through tissue. The selection of an appropriate optical reporter, luciferase or other reporter, is therefore based on wavelength of emission in addition to other parameters to be discussed in the following, and longer-wavelength emitters generally have greater tissue penetration. Although each luciferase described previously has a different λmax ranging from blue (*Renilla* RLuc) through red (firefly at 37° C and click beetle red Luc; CBR), the spectral emission profiles for each is very broad, and at least some emitted photons are above 600 nm (RLuc ~5% at 37° C, CBR Luc ~60%) [24]. Therefore, a simple *in vitro* comparison of different luciferases in tissue culture dishes may lead to the selection of a bright luciferase *in vitro* that works poorly *in vivo* due to attenuation of the signal. This would be the case in comparing the click beetle green (λmax ~540 nm) and red (λmax ~615 nm) luciferases, where the green-emitting click beetle enzyme appears brighter in the absence of overlying tissue, but this enzyme performs poorly *in vivo* [24].

In animal models where reporter cells are introduced close to the surface of the animal, such as tumor cells injected subcutaneously, the absorption can dramatically reduce signal intensities; nonetheless, emission from many enzymes that emit shorter wavelengths are of sufficient intensity that they can be used (Figure 2.3b). This is true for the RLuc luciferase, which performs well in these models, although its utility at deeper tissue sites is limited by absorption of blue light [24]. Although mouse strains differ little internally with regard to their optical properties, further consideration is needed when designing imaging experiments in the visible regime. Because melanin is another major absorber of visible light, dark fur and skin color can result in signal attenuation. The use of nude (*nu/nu*) and other hairless mice can improve signal collection, and because nude mice are immunodeficient, they are widely used in oncology models that use human tumor cells, xenografts. Albino mice are also widely used in BLI studies; white fur is scattering and will affect optical signals relative to hairless mice. The effect is modest, and if necessary the fur can be removed by shaving or applying depilatory creams. Light scattering at the surface of hairless mice can be reduced by applying a hyperosmotic agent to the skin, and it has been shown that application of 50% glycerol to the skin of mice can reduce the skin scattering coefficient by more than an order of magnitude while having little effect on the absorption coefficient [48].

Patterns of surface radiance will differ between weak surface signals and bright signals deep in the animal due to the optical properties of mammalian tissues. Signals from deeper sources will be attenuated, and radiance patterns at the surface will appear more diffuse and will be projected over a larger area of the animal. The surface-emitted spectral profile will also be significantly altered, with only those photons >600 nm being detectable outside the body (see Figure 2.3b). Although this can make identifying the light source origin within the animal difficult, these differences can be utilized to improve quantification by calculating the source depth and to enable reconstruction of 3D modeling of optical signals including those obtained with bioluminescent reporters [49]. In preclinical imaging the subjects are typically sacrificed during or after the experiment, and therefore tissue analysis after necropsy can verify the origin of the luciferase-expressing cells using biochemical assays. This improves the data sets because many 3D reconstructions are based on assumptions that enable predictions of the origin of the source.

Gene Transfer Tools for BLI

The luciferase protein can be introduced into animals in a number of ways, primarily by direct expression within the host cell [13] or, more recently, by the injection of purified luciferase, usually tagged to a second moiety, into the mouse [28, 34, 35]. Transfer of the reporter gene is typically performed on cells in culture that are then selected and transferred into the animal models. Alternatively, the gene can be integrated into the mouse genome to create transgenic mice that express the luciferase from regulated or constitutive promoters. It is also possible to introduce the luciferase gene directly into naïve animals, using viral vectors, naked DNA, or other nonviral techniques (see below). The development of labeled cell lines in culture has the advantage of being able to selectively enrich for the cells that express the reporter, and these can be used as clones or selected lines that can be pooled to create a population in which the heterogeneity of the parental line is preserved.

Genetic reporters can be engineered with a promoter designed to express luciferase genes constitutively. These gene constructs can be incorporated into cells *ex vivo* and used to image trafficking of these cells in small living animals. Alternatively, a regulatable promoter induced under developmental, physiological, or environmental influences can be used to reveal patterns of gene expression. In most cases, a second selectable marker gene such as those that encode drug resistance proteins or fluorescent proteins, is also introduced along with the luciferase gene to allow rapid selection of transfected/transformed cells to create clonal populations [50]. It is possible, however, to select for transgenic cells by virtue of their bioluminescence [51]. Cell lines can also be engineered with promoterless luciferase genes using a random integration strategy that selects for insertions downstream of actively expressed endogenous proteins [52]. This strategy has been successful in the development of bacterial lines, in which the *lux* operon is linked to a drug resistance gene and integrated into the genome using a transposon strategy. Screening these cells for resistance to the drug allows identification of transposition events in which the luciferase-resistance reporters are expressed from a native bacterial promoter, and these cells are further screened for their bioluminescent properties. Libraries of such promoter-trapped luciferase lines can then be screened in mice to determine expression levels of genes *in vivo*, and the promoters are subsequently identified by molecular techniques.

The development of transgenic reporter mice that incorporate luciferase has also been widely employed, and mice have been engineered to express the luciferase from a number of different native promoters, as well as promoters that can be controlled with exogenous regulators, such as the tet-system [53–58]. The creation of a mouse expressing luciferase from a hybrid actin/beta-globulin promoter has proven useful [59–61], because most cells in this animal strongly express the luciferase, and cells and organs can be transplanted to naïve mice to study cell and organ transplant models [62]. As with any transgenic mouse line, it is important to ensure that the expression profile of the introduced luciferase gene accurately mimics that of the native promoter in the host. Because generation and screening of transgenic animals can be slow and costly, the ability to screen reporter constructs in mice prior to creating the transgenic line can help choose the appropriate constructs [63–65]. To this end, it is possible to introduce purified nucleic acid directly into hepatocytes of living mice using high-pressure injections. In this approach the nucleic acid is injected rapidly into the mouse tail vein in a volume of saline approximately equal in volume to 9% of the animal's body weight [66]. DNA is then taken up by cells in the liver, where expression peaks at 24 h and can be expressed for up to several weeks following the procedure, allowing screening of the functionality of the cassette. Expression of the luciferase is controlled by the promoter sequence and in most cases does not appear to be regulated by hepatic factors, and thus constructs can be selected using this strategy to commit to the generation of transgenic mice [67]. The luciferase gene can also be introduced into mice using viral vectors into which the expression cassette has been introduced [68]. This may allow the study of the viral infection process in the host.

In Vivo Bioluminescence and Fluorescence Imaging: Signal-to-Noise Benefits

Luciferases have been used as reporters for several decades in cells and transparent animals, and more recently fluorescent proteins have provided another powerful reporter strategy with tremendous potential for studying cells in culture. Fluorescent proteins have been expressed in cells to allow simple identification via microscopy [69, 70] and cytometery or fused to native proteins and used to visualize subcellular localization using fluorescent microscopy. The proteins can be visualized by illuminating the cells with a bright light source either spectrally filtered for optimal fluorescent excitation or using a suitable laser, and then they are viewed with another filter that selects for fluorescently emitted photons. Excitation wavelengths for these proteins typically exist in the blue-to-yellow-green range. The transition from fluorescent proteins in cells to their use in animals began with green fluorescent protein (GFP). Relative to *in vivo* bioluminescence, *in vivo* imaging of fluorescent proteins [71–73] required an understanding of two additional factors; the adsorption and scatter of the excitation light source and the intrinsic autofluorescence of tissue, which is significant in animals. These two factors limit the visualization of GFP in mice to a depth of only a few millimeters from the surface unless the overlying tissues are removed [74]. Greater sensitivity and deeper tissue penetration can be obtained using the red-emitting fluorescent proteins [75, 76] and, more recently, a far-red fluorescent protein [77]; this is due to greater penetration of the excitation wavelength and to reduced autofluorescence. *In vivo* fluorescence offers the benefits of microscopic detection and analysis by flow cytometery and microscopy after removal of the tissues and cells. Therefore, combinations of bioluminescent and fluorescent reporters in dual function constructs can provide greater opportunities for analysis [59, 60].

The ability to detect a bioluminescent signal is also affected by the absorption of photons by absorbing particles in the tissue, but there is significantly less background noise in mammals, which allows the detection of very low signals relative to fluorescence. Signals from fluorescent proteins can, in general, be brighter than bioluminescent signals because extremely bright excitation sources can be used. The very low background for bioluminescence imaging usually results in significant signal-to-background-noise ratios. Comparing cells that express both firefly luciferase and the *ds*Red protein implanted subcutaneously on the back of a mouse, a 100-fold brighter signal was observed under fluorescent illumination compared to the bioluminescent signal, but the autofluorescent signal from the animal was 100,000-fold higher than the autobioluminescent signal [78]. The signal-to-background ratio was thus significantly better for the bioluminescent signal (7500) compared to the fluorescent signal (7.8 and 150 with background subtraction), allowing detection of as few as 400 bioluminescent cells in this model compared to 400,000 cells by fluorescence. It should be noted, however, that if the cells of interest are close to the surface of the animal or externalized by surgery, it is possible to detect single fluorescently labeled cells using intravital microscopes, and cells have been visualized moving through surface capillaries in real time [79]. It is improbable that a similar level of sensitivity to detect single bioluminescent cells will be possible, and thus these reporters have different applications in molecular imaging, with bioluminescence excelling at macroscopic imaging and fluorescence at microscopic imaging.

Fluorescence imaging in animal models is not, however, limited to fluorescent proteins. A number of near-infrared fluorescent dyes have been successfully used to label and visualize them *in vivo*. Autofluorescent background levels of mammalian tissues drop significantly as the wavelength increases, and thus use of fluorescent dyes that excite in the 650–800-nm range has proven successful in visualizing deep within animals [80–82]. Again, this allows a synergy of strategies to image both bioluminescent cells and those labeled with near-infrared dyes. One source of fluorescence background is due to chlorophyll in regular mouse chow, which fluoresces with excitation wavelengths up to ~700 nm. Feeding the animals with a suitable chlorophyll-free diet can remove this background signal [78].

LUCIFERASE OPTIMIZATION AND ADVANCES

Each of the luciferases used as reporter genes have been cloned from nonmammalian sources, and as such their native forms may not be optimal for expression in mammalian host cells or may not have optimal properties for *in vivo* bioluminescence imaging. Each luciferase used *in vivo* has had changes made to its genetic sequences to better optimize functionality. The enzymes used in BLI to date are described in the following.

Beetle Luciferases

The firefly luciferase gene was first commercialized with no changes to the DNA sequence and contained peptide-targeting sequences that directed the enzyme to the peroxisome organelle in the cell. A modified version was later released (termed Luc+) in which the peroxisome-targeting sequence (lys-arg-leu-C-term) was changed to the nontargeting sequence ile-ala-val-C-term, as well as having a few nucleotide changes in the coding region that removed frequently used restriction sites and some sequences that may permit transcription factor binding proteins to affect mammalian transcription. This gene allowed significantly improved luciferase expression in mammalian cells and has been used for much of the luciferase data reported to date. Recently, further modifications to the nucleotide sequence were incorporated, and the resulting luciferase gene ("luc2") has changes to many of the codons so that optimal expression of the gene can be obtained in mammalian cells ("codon optimization"), which resulted in a tenfold increase in enzyme

activity compared to the luc+ enzyme, without changing the peptide sequence. These modifications increase the sensitivity of BLI, allowing detection of small numbers of cells and perhaps single-cell detection in some animal models.

Sequence modifications have been made by many groups to better understand the biochemistry of the light-producing reaction, as well as change the properties of the enzymes. Site-directed mutagenesis around the active site of the enzyme has allowed changes in the emission spectrum of the enzyme, resulting in both red and green shifts of the peak [83], as well as affecting the *p*H affect on wavelength (firefly luciferase shows a significant red shift as the *p*H becomes more acidic). Shapiro et al. [84] have reported two mutations that have a λmax ~609 nm at both *p*H 5.5 and 7.0 and that also generate more light above 600 nm than the native enzyme, although neither are as bright over the entire emission spectrum. Nakatsu et al., using the Japanese firefly luciferase, generated a single mutation (S286N) that showed a similar red shifting of the emission spectrum, which they used to help elucidate the mechanism of green versus red light production by the luciferase enzyme [85]. Baggett et al. screened luciferase mutations to identify enzymes with improved thermostability [86], the native enzyme being heat labile and having an *in vitro* half-life of only 3 min at 37° C, and identified several residues that when mutated resulted in up to a 25-fold higher half-life than the native enzyme. Increasing transcription and translation of luciferase, creating a more stable enzyme, and spectrally shifting their emission to generate photons above 600 nm all further increase the potential sensitivity of BLI with the beetle enzymes. It is also important to note that the firefly luciferase is red-shifted as the temperature increases; however, the spectral peaks are usually reported at 25° C. At 37° C, the native enzyme has an emission peak at 605 nm, again favoring its use in BLI approaches.

The railroad worm and click beetle luciferases, although closely related to the firefly enzyme, have *p*H- and temperature-insensitive emission spectra. Further, these insects naturally show significant variation in their color, and indeed one railroad worm (a beetle) expresses both green and red luciferases, with two red spots at its head and green spots along each flank [22]. These railroad worm enzymes were reported to have activity when expressed at 20° C but have negligible activity at 37° C, limiting their use as *in vivo* reporters until a temperature stable variant has been generated. The click beetle enzymes have been subjected to mutagenesis to generate two variants that are commercially available that have spectral peaks separated by over 70 nm, the green variant (CBGr luc) having a peak at 540 nm, and the red enzyme (CBR luc) that emits at 610 nm [7]. The two genetic sequences were then codon-optimized for mammalian expression, and can be co-expressed in cell lines, and the relative signals differentiated *in vitro* by using spectral filters [87, 88]. However, although the green variant is significantly brighter *in vitro*, the red enzyme has better properties for small-animal imaging, since over 60% of its spectrum is emitted above 600 nm [24], the start of the "optical window" for tissue imaging (see Figure 2.3).

Renilla Luciferase Optimization

In a similar fashion to the optimization of the firefly and click beetle luciferase, a humanized *Renilla* luciferase (hRLuc) has been generated, which is commercially available and has been used extensively as a reporter gene both *in vitro* and in murine models. In its natural state, the RLuc enzyme is associated with a green fluorescent protein, and this protein absorbs the blue photons emitted by the luciferase and itself emits a green light (in a process known as bioluminescent resonant energy transfer, BRET). The luciferase has not, therefore, been under evolutionary pressure to shift its spectral properties, and has been subjected to random and targeted mutagenesis to create mutants that have improved imaging properties. The first improvement reported for this enzyme was the generation of a brighter, thermally stable variant (Rluc8), which is eight fold brighter [32]. However, since it still emitted in the blue region of the spectrum, had poor deep-tissue imaging properties, this was subjected to further rounds of random mutation to red shift the emission spectrum from 486 nm to >540 nm, with an increase in the spectrum above 600 nm from 3% to over 20% [33]. Such variants will be useful for allowing imaging of two reporters in the same animal, because the firefly and *Renilla* enzymes use substrates that do not cross-react with each other [26].

Bacterial Luciferases

The bacterial Lux operon has been used to genetically engineer many bacteria to be autonomously bioluminescent, and no optimization of the codon usage has been necessary to express this reporter in Gram-negative bacteria. However, as discussed previously, although the native operon worked well in many species of Gram-negative bacteria, this was not the case for Gram-positive organisms. It was necessary to introduce Gram-positive-derived transcriptional start sites (Shine-Dalgarno sequences) between each open reading frame in the operon for expression in Gram-positive bacteria such as *Staphylococcus* and *Listeria* [41]. In addition, in the modified constructs the gene order was changed from *LuxCDABE* to *LuxABCDE*. This was not to have the genes appear in alphabetical order, but it was done with the intent of optimizing bioluminescence in Gram-positive bacteria. By placing the genes encoding the heterodimeric luciferase first in the operon sequence, the levels of expression and hence bioluminescence may be

greater [41]. Since the temperature optima for many bacterial luciferases are significantly less than that of the mammalian host, the Lux operon of *Photorhabdus luminescens*, from which the expressed luciferase is functional at temperatures greater than 35Å°C, was selected for these genetic modifications.

To functionally express the five bacterial *lux* genes in the yeast *Saccharomyces cerevisiae*, each open reading frame was cloned downstream of separate promoters, although the *luxD* gene encoding the acyl-ACP transferase involved in substrate generation was fused to the yeast internal ribosome entry sequence (IRES) downstream of the *luxC* gene [43]. Bioluminescent signals from yeast were poor until a flavin oxidoreductase gene was co-expressed to supply sufficient concentrations of the $FMNH_2$ cofactor. Expression of the bacterial luciferase in mammalian cells has also been reported [44], although only the *luxA* and *luxB* genes were expressed and the substrate was provided in the media. Codon optimization of the bacterial luciferase genes resulted in a six-fold increase in light production over the native sequence. Expressing the other three genes in mammalian cells would provide a useful additional reporter system for optical imaging of luciferases in mice, as the decanal substrate is likely to be toxic if injected into mice.

Extracellular Luciferases

The coelenterazine-utilizing luciferases do not require cellular cofactors to generate light, and because the energy is provided by coelenterazine, these enzymes can function as extracellular reporters. These enzymes can therefore be exogenously expressed, purified, and injected into animals and still remain active without being internalized into cells. Both the *Renilla* Rluc8 and *Gaussia* enzymes have been expressed as fusions to antibody fragments (diabodies) to the carcinoembryonic antigen (CEA) [28, 34]. When these fusions were injected into mice that had CEA-positive LS174T human colon carcinoma cell xenographs, bioluminescence could be detected from this tumor with a tumor-to-background ratio of 6.0 at 6 h following protein injection, whereas no signal indicative of binding was observed on control xenografts. Colabeling the Rluc8 fusion protein with Iodine-124 allowed microPET imaging of the fusion and confirmed that the bioluminescent signal was due to the antibody-mediated localization of the luciferase to the target tumor. Similarly, the *Gaussia* fusion to the same diabody showed specific tumor targeting of the fusion, and a fifteen amino acid N-terminal truncation of the luciferase protein resulted in a monomeric form of the fusion with increased bioluminescence. Such fusions will allow simple, nonradioactive imaging of tumors without the need to express the protein in the mammalian host, but instead using bacterial expression systems.

This opens up the possibility of using bioluminescent reporters in human subjects.

The *Renilla* Rluc8 luciferase has also been conjugated to quantum dots, which can be injected into mice and imaged following administration of coelenterazine [35]. Quantum dots are small fluorescent particles that efficiently absorb light at blue wavelengths, including the emission from luciferases that emit at the appropriate wavelengths. The luciferase-quantum dot conjugates emit light at wavelengths determined by the size of the particle, with larger particles emitting in the near infrared (800 nm). Because the bioluminescent light is generated adjacent to the dot, resonant energy transfer is high, and different quantum dot sizes can be easily differentiated in a mouse model by using appropriate filters. Using an external light source to excite the dots is therefore unnecessary, which means there is no autofluorescent signal observed from mice with these self-illuminating quantum dots. By further conjugating diabodies to these Rluc-quantum dots, it may be possible to label and differentiate multiple cell-surface markers with low backgrounds.

Substrates for Bioluminescence Reactions

The different substrates for each of the luciferases in optical imaging have already been discussed, and the delivery of each substrate to the mouse and modifications to improve their functionality are extensive. The substrates can be provided by different routes of administration depending on the class of luciferase used and the intended target tissue. The most convenient means of substrate delivery occurs with the bacterial luciferin, decanal, which can be synthesized *in situ* by the bacterial cell, which obviates the need for exogenous addition of substrate. Bacteria expressing the *lux* operon are thus autonomously bioluminescent in the murine host, provided the cells are actively growing and exist in an aerobic, or microaerophilic, environment. Although bacterial luciferase has been expressed in eukaryotic cells, attempts to image these cells in mice have not been reported, and there is no specific data on substrate administration.

The benzothiazole luciferin substrate for the beetle luciferases is a relatively stable, water-soluble compound that following delivery to the animal can pass through most tissue and cellular barriers in the animal, including the blood–brain and placental barriers and will also appear in the milk of lactating females (Contag and Bachmann, unpublished results). Further, this substrate can pass through the skin with a suitable organic carrier [89] and is absorbed through the gut if delivered orally (Contag and Bachmann, unpublished results). The substrate is typically delivered by an interperitoneal injection (at ~150mg/kg, in saline) and distributes throughout the animal within approximately 10 min. The optimal window for imaging is, for most tissues, between 10 and 30 min, at which time the substrate levels are fairly constant. Signals are apparent for several hours, although signal intensity typically falls off after about 40 minutes depending on the site of expression of luciferase. The skin and brain appear to be the tissue sites that receive the least amount of substrate after interperitoneal injections, and imaging at these locations benefits from intravenous injections (see the following). Cohorts of mice can be injected quickly and imaged in groups of up to five to ten, with data acquisition times of several seconds to 5 or 10 min. At these substrate concentrations, no toxicity has been observed, and the substrate and products are rapidly cleared and excreted in the urine. Injection through the tail vein has also been used, which allows a rapid distribution through the animal but is rarely used because this technique is more time-consuming, results in rapid clearance time from most organs, and cannot be repeatedly used on the animals due to damage to the vein. Biodistribution studies with a ^{14}C-radiolabeled luciferin molecule [90] following intravenous injection confirm that luciferin can be detected in all organs as soon as 1-min postinjection, but that different organs show markedly different uptake, with the liver, kidneys, and intestines having a high uptake, whereas organs such as the brain have a lower uptake. The use of an implanted osmotic pump has also been reported [91], and this implantable device delivers a constant infusion of the substrate resulting in reliable repeated imaging over short periods without the need to inject fresh substrate at each time point. Injecting luciferin into a lactating female will result in luciferin delivery to her pups, which can subsequently be imaged, although screening of transgenic pups for luciferase is more typically done by direct injection of the substrate (Bachmann, Ravaridan, Contag, unpublished results).

The luciferin substrate can be modified to allow other types of biochemical reactions to be monitored using bioluminescence as a readout (Figure 2.4). The addition of moieties to a hydroxyl group in the substrate effectively "cages" the luciferin, prevents binding to the luciferase, and must first be cleaved to release the free luciferin, which will react to produce light. Attaching a galactose sugar to the luciferin by a glycosidic link (Figure 2.4a) to generate a luciferin–galactoside conjugate ("lugal") generates such a substrate that will not bind to the luciferase until the sugar is removed [92]. Cleavage of this bond can be achieved by a bacterial beta-galactosidase enzyme (ß-gal, encoded by the *lacZ* gene), an enzyme that mammals do not normally express. This lugal substrate can thus be used as a reporter of *lacZ* expression in transgenic mice in the presence of luciferase-expressing cells. Cells expressing both luciferase and ß-gal will thus bioluminesce in the presence of the lugal substrate, while cells that express either enzyme alone will not generate light. The *lacZ* gene is a common reporter gene widely used in both mammalian cell and transgenic mice

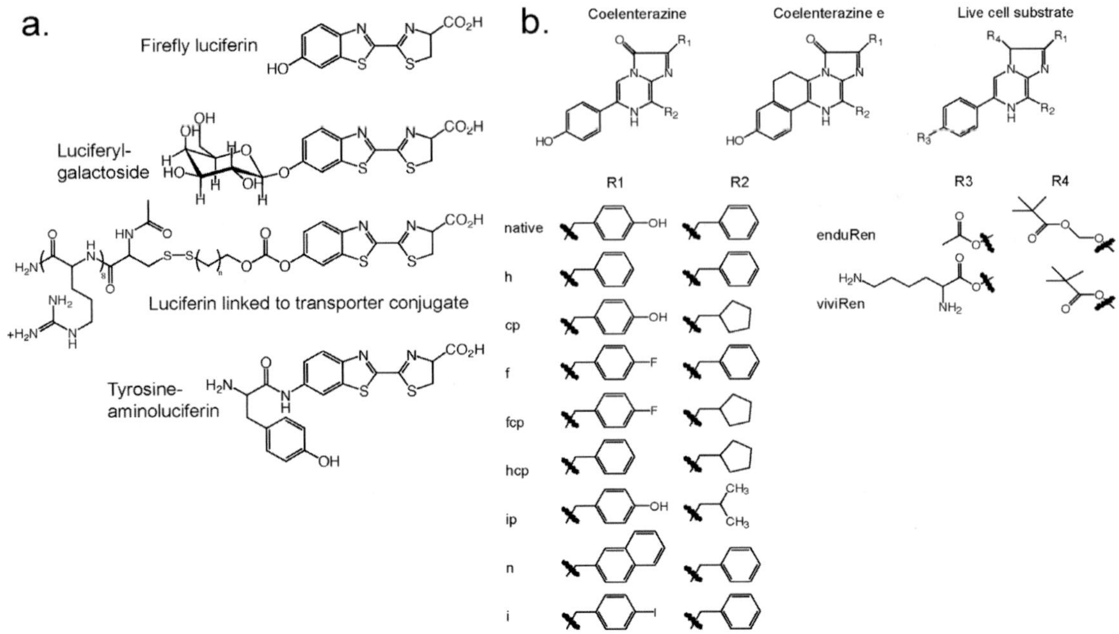

Figure 2.4. Chemical structures of firefly (**A**) and *Renilla* (**B**) substrates and derivatives used in molecular imaging. Firefly luciferin (a, 4,5-dihydro-2-(6-hydroxy-2benzothiazolyl)-4-thiazolecarboxylate) has been modified with a sugar ("lugal"), a polyarginine transporter conjugate, and addition of amino acids to generate substrates that require enzymatic activation before reacting with luciferase. Many derivatives of the *Renilla* luciferin (**B**), coelenterazine, also a substrate for the *Gaussia* enzyme) are commercially available, including the enduRen and viviRen compounds that have been optimized for *in vivo* imaging.

lines, and use of this caged luciferin allows imaging of a large number of established models. An ester linkage at this hydroxyl group has also been employed to introduce a disulphide-linked cargo to the luciferin molecule (Figure 2.4a).

An octa-arginine moiety has also been added to luciferin using a disulphide ester linkage [89, 93, 94], and this polyarginine peptide allows transport of the luciferin complex through tissue barriers and the cell membrane into the cytoplasms, where cytoplasmic glutathione cleaves the disulphide bond releasing the luciferin substrate that is then free to react with luciferase to generate light. Such compounds were used in transgenic luciferase reporter mice to noninvasively determine the biodelivery of luciferin through the skin by the polyarginine transporter and release of the luciferin cargo in real time [89].

The hydroxyl group of luciferin can also be changed to an amino group, and the amino-luciferin compound is still a substrate for the luciferase, albeit with reduced light output. This amino group can be modified with the addition of amino acids or peptides via a peptide bond; the resulting caged amino-luciferin peptide is again no longer an effective substrate for the luciferase reaction, although low levels of activity can be detected. Cleavage of the peptide by proteases may then release the amino-luciferin that is free to react with the luciferase and generate light [95]. A number of caspase-specific peptides have been added to the amino group and provide a

bioluminescent reporter substrate to monitor these enzymes *in vivo* [96–99]. The amino group has also been used as a target for PEGylation [100], which extends the circulatory half-life of the substrate to 3–4 h, compared to the rapid clearance time (40 min) observed with the native substrate. Signal flux was similar to the poorer substrate, amino-luciferin, but is still significantly lower than that observed for luciferin.

The coelenterazine substrate is an energy-rich molecule hydrolyzed in water and is typically stored in ethanol or methanol prior to dilution in saline immediately before injection into the mouse [26]. The substrate is unstable in the host animal, reacting with serum proteins (especially albumins) to generate low levels of bioluminescence and is rapidly broken down and cleared by the animal [36]. The coelenterazine substrate is therefore injected via the tail vein immediately prior to imaging, and individual animals are imaged. The substrate will disseminate quickly through the animal and easily enter cells by the multidrug resistance *MDR1* P-glycoprotein transporter [101]. The bioluminescent signal from the luciferase can typically be measured for several minutes. Recently, the caged coelenterazine compounds enduRen^{TM} and vivoRen^{TM} have been developed commercially that have small groups added to reactive hydroxyl groups that can be cleaved by cellular lipases and esterases (Figure 2.4b). These caged compounds show significantly improved stability in aqueous solutions, and brighter, longer-lasting bioluminescent signals can be obtained following IV injection into mice compared to the native substrates [102]. These substrates may therefore enable effective use and perhaps improved quantification of the *Renilla* luciferase activity *in vivo*, although the cost of the substrates may limit such utility at the present time.

Many variants of the native coelenterazine are also commercially available (Figure 2.4b), some of which show improved light generation *in vivo* [36] and others having altered spectral emissions. One such compound ("Clz400") shows light emission significantly blue shifted to 400 nm, which when used with GFP has permitted sensitive BRET measurements in animals [103, 104].

IMAGING INSTRUMENTATION

The Basics

Part of the power of *in vivo* bioluminescence imaging is the simplicity of the instrumentation used to detect the photons emitted from animals as a result of the luciferase reaction. Essentially the animals are placed in a light-tight imaging chamber with a sensitive CCD camera mounted on one side (typically the top) and a computer controller to acquire the images and allow image analysis. Although many groups have built their own imaging systems, a number of companies sell systems capable of detecting low light levels emitted from the animals, including systems from ART Technologies, Berthold Technologies, BioSpace, Fujifilm, Roper, and Xenogen (now part of CaliperLS) [16]. The authors of this chapter have experience with the Xenogen IVIS Imaging Systems that were designed specifically to image luciferases in mice and are well suited for this purpose.

In vivo bioluminescence imaging is performed by transferring luciferase-labeled cells or using gene transfer tools to label cells in mice and injecting the luciferin substrate into the animals. The animals are then anesthetized and placed into a light-tight imaging chamber equipped with an appropriate ultrasensitive imaging device. Typically, a grayscale reference image or photograph is taken under low light illumination, and then an image of the bioluminescent signal is acquired in complete darkness, which may be acquired over a few seconds or several minutes, depending on the magnitude of the signal. The bioluminescence data is then represented in pseudocolor to indicate the signal intensities, and this image is superimposed over the grayscale reference image using appropriate software for imaging processing. The magnitude of the bioluminescent signal can be appreciated by the pseudocolored pattern or otherwise assessed by drawing regions of interest over user-specified locations. Animals may be imaged from several different views (dorsal, ventral, or lateral) to best detect the origin of the bioluminescent reporter, but in each case the two-dimensional image acquired represents the projected photons from the surface of the animal that will have been scattered as they pass through the tissue.

Selection of the cooled charge-couple device (CCD) camera is critical for *in vivo* imaging of small animals, and intensified and integrating cooled cameras have typically been used. A review by Rice et al. compares these systems [16], in which they present an argument for the use of cooled, back-illuminated, integrating CCD cameras, but recent developments in cooled intensified CCD cameras now offer the option of using these instead, and indeed one company now mounts such a camera and can perform imaging at video rates (33 frames/sec) of bioluminescence in mice. Biological applications of this fast system include calcium sensing through Aequorin luciferase: a Ca^{++} sensitive enzyme. Intensified CCD cameras typically suffer low quantum efficiency, especially at wavelengths above 600 nm. Further, intensified CCD detectors have a limited dynamic range and methods for controlling signal intensity must be applied. Integrating CCD cameras, however, have a higher quantum yield and are sensitive from the blue to near-infrared spectral regions. Cooling of the chip (-90 to $-105°C$) can reduce the background dark current signals, although the read noise on the detector can still present a problem with extremely weak signals, and background measurements must be taken and subtracted from experimental data.

Alternative camera architecture includes the electron-multiplying charge-coupled device (EMCCD) that allows for on-chip multiplication of electron counts. This device has increased sensitivity and detection rates, relative to back-thinned cooled CCDs, enabling video rate data acquisition. The largest current back-illuminated EMCCD camera chip, however, is only 512×512 pixels; cooled CCD detectors for low light imaging are typically 2048×2048 pixels. EMCCD devices tend to be more expensive than cooled CCDs and the signal-to-noise ratios on these detectors are less than that of other detectors. The signal-to-noise ratios can be compensated for with background subtraction, and EMCCDs can be cooled to reduce the noise; however, cooling slows the collection rate. Recently, these fast readout cameras have been used to image bioluminescent signals from mice, and an elegant experiment using a calcium-dependent aequorin luciferase expressing luciferase mouse allows real-time imaging of calcium signaling in free-roaming animals [31]. Following coelenterazine injection, bioluminescent signals could be detected as the muscles contracted during motion, as reported in this study.

Signal intensity can be increased on CCD detectors by increasing the effective pixel size; this is referred to as "binning," in which photons exciting groups of 5×5 pixels, for example, on the CCD chip, that are summed can be collected on one "superpixel," in this case 25-fold larger than the single pixel. Sensitivity is thus greatly increased at the cost of image resolution, but because photons passing through mammalian tissue are significantly scattered and resolution lost, optical resolution on the detector is often not a priority. By adjusting pixel binning and acquisition time, the user can obtain data sets where the signal is within the working range of the camera and above the noise of the system (typically 2–95% of the dynamic range of the camera, corresponding to ~1,000–60,000 counts on a 16-bit camera). Care is also required to ensure that the camera pixels are not "saturated." Saturation occurs when the number of photons, converted to electrons, hitting a given pixel exceeds the pixel, or superpixel, capacity, and is thus not collected. Some commercial systems have warning signals built into the software to prevent the use of saturated images. However, in systems not equipped with this feature, saturation is obvious when signals from what should be an increasing situation plateau at a fixed count. Unfortunately, this plateau is often only apparent after the fact, and the experiment will need to be repeated. If the camera is suitably calibrated, the data can be converted from the arbitrary counts to a scale that accounts for acquisition time, binning, and distance of the animals from the camera (e.g., photons/sec/cm^2/steradian), which will allow direct comparison of data acquired using different parameters and systems. Well-designed systems are often calibrated against standards approved by the National Institutes of Standards and Technology (NIST) to both facilitate calibration of individual systems and ensure that all images obtained on these devices are comparable. In this way images obtained at one institution can be compared to those of a similar animal model obtained at another institution. Well-developed user interfaces are present on many systems such that imaging parameters can be readily adjusted while retaining acceptable ease of use. An early objective in the development of *in vivo* bioluminescence imaging was to make molecular imaging accessible to all biologists [105]; one aspect of this objective was to develop easy-to-use instruments.

Because integration times may last several minutes, images obtained with long data acquisition times are improved by anesthetizing the animals before imaging and keeping them anesthetized during the image acquisition process. Although this is not always necessary, if data acquisition time is fast, immobilizing the subject is usually desirable. Anesthetics are known to influence bioluminescence from bacterial [106] and firefly luciferase [107–109] and will therefore affect signal intensity in BLI. A similar inhibition of the coelenterazine-utilizing luciferase by anesthetics is probable, although this has not yet been reported. It has also been observed that low-level chemiluminescent signals can be observed in the nasal area and lungs of mice anesthetized with isoflurane in oxygen (unpublished observation), which may affect the ability to observe low-level bioluminescent signals from these tissues. Because anesthetics differ in their effects on biology, understanding the effects of a given anesthetic on the process and the reporter is important for all imaging studies where the subject is anesthetized. Most often gaseous anesthesia such as isoflurane is used because it is generally well tolerated by the animals and can be used to obtain repeated measures at relatively fine temporal resolution. When gas anesthesia is used, the animals are maintained under anesthesia during imaging by placing their noses in a manifold on the instrument stage that can accommodate several nose cones fed with fresh anesthetic. It is also important to keep the animals warm during imaging, as mice rapidly lose body heat when anesthetized, which not only can harm the animal but also affect the efficiency of the bioluminescent reaction. Many instruments designed for small-animal imaging have heated stages to keep the animals warm, and use of stages heated to 37° C will maintain core body temperature, which will keep the animals from going into shock and stabilize their biology, leading to more consistent results. If the stage of the instrument can be moved up and down relative to the camera, different fields of view may be obtained, allowing sequential high-throughput imaging of multiple animals or close-up, focused high-resolution imaging of a small region of a selected animal. These features, built into many of the imaging systems, are essential for optimal image data acquisition, reproducible results, and safe animal handling.

Immediately before imaging, mice are injected, if necessary, with luciferin and then anesthetized. They are then placed on the stage of the imaging chamber under continued anesthesia, and the door is closed. Injectable anesthetics can be used for imaging, and users should consult their veterinary services or appropriate Web sites, for optimal dosages and dosing schedules. After acquiring the reference image, bioluminescence imaging begins, ranging from fractions of a second to 1 sec and up to 5 min, and if necessary, sequential images can be acquired to follow changes in the signal due to physiological or kinetic changes (such as the movement of bacteria in the gut or uptake of chemical substrates into various tissues of the animal). Animals can be repeatedly imaged using this strategy over many days or weeks, depending on the experimental model and expected temporal changes, and therefore each animal can serve as its own control, and the data are strengthened by the power of repeated measures. Care must also be taken to ensure suitable control groups are included within a study. Animals whose tissues do not express any luciferases, but are injected luciferase substrates can be used as negative controls and to determine signal background. This is more relevant to *in vivo* fluorescence imaging where the background levels can be high. These control groups provide a reference as to the lower limit of detection for a given experiment. Because autoluminescent and background fluorescent signals from mice can be significant, especially if the unstable coelenterazine substrate is used, these controls can provide important insights.

Spectral Imaging and 3D Imaging

Due to low levels of light typically emitted during imaging, often all photons emitted from a single view (dorsal, ventral, or lateral) detected by the CCD camera are counted, and a simple two-dimensional image is acquired. However, this does not use any spectral information contained in the images. The spectral data can provide information about the depth within the animal of the bioluminescent source and can be used to distinguish two or more spectrally distinct reporters [110]. Because red photons are less attenuated by hemoglobin than green or blue photons, by comparing the ratios of these wavelengths it is possible to determine tissue depth of the source. Several groups have been using this strategy to provide three-dimensional reconstructions of bioluminescent sources in the mouse, either acquiring data with spectral band-pass filters or using beam splitters to acquire images in the green and red spectral regions [111, 112]. Kuo et al. have generated a simple 3D surface map of a mouse by using structured light analysis, and then acquired multiple spectrally filtered images (up to six 20 nm wide bandpass filters every 20 nm from 560 to 660 nm) and applied forward Green's function kernels to map photon propagation incorporating absorption and scattering properties of tissue at each wavelength [49]. They then generated 3D reconstructions of luminescent sources from images acquired from one view, with the spectral data contributing the third dimension. By placing the animal on a transparent gantry that can be rotated around a mirror to allow similar imaging of the mouse from multiple views, they can generate a more accurate 3D reconstruction, and systems are available commercially that can perform such single view (IVIS 200) or multiview (IVIS 3D) analysis (see Figure 2.5). These images have also been coregistered with MRI data as a source of anatomical reference image [113]. An alternative approach to obtain multiple views in an IVIS 200 imaging system without moving the specimen mouse was reported by Chaudhari et al., where they placed a number of mirrors around the mouse, which allows simultaneous imaging from top, both sides, and bottom [114]. Accurate localization of the bioluminescent source was obtained in these studies. Strategies have also been proposed to generate 3D tomographic data from monochromatic measurements [115, 116], and models of imaging with an array of six photodiode detectors arranged around a specimen capable of detecting both bioluminescent and positron photons have been proposed [117, 118] and have been refined to include mouse anatomical information to account discrepancies arising from different optical properties of different tissues.

APPLICATIONS

Bioluminescence imaging has been applied to many different animal models of human diseases and development. These have been extensively reviewed [13, 119], and a few illuminating examples are presented here to indicate the broad applicability of *in vivo* bioluminescence imaging in rodent models.

Infectious Disease

Animal models of viral [120], bacterial [42], fungal [121], and parasitic [122] infections have all been adapted by engineering pathogens to be bioluminescent (reviewed in [47, 123, 124] and have provided a powerful refinement to animal models of these diseases that permit pathogen load and localization to be determined noninvasively. This not only significantly reduces the number of animals required for such a study because many infection models require the sacrifice and necropsy to determine pathogen load and location, but has also allowed a more thorough analysis of the course of infection, as individual animals are tracked and can be compared over time, rather than the snapshot of infection within animals that necropsy provides. Further, a better understanding of an infection can be determined, which was elegantly demonstrated by Hardy et al., who infected mice with bioluminescent *Listeria monocytogenes* [42]. Using sublethal doses, they observed that some mice showed a persistent bioluminescent signal from the abdomen but no physiological symptoms. The signal was maintained over many days but varied in intensity, and by careful *ex vivo* imaging of the abdominal organs it was determined that the bacteria were residing in the lumen of the gall bladder, a site of infection that had not been previously suspected. These studies also revealed the source of variability by studying the bacterial load relative to time of eating and contraction of the gall bladder. These investigations showed that fasted animals had a larger signal from this organ, and that feeding the animals or injecting cholecystokinin resulted in a contraction of the gall bladder, which the team could observe by ultrasound imaging, accompanied by a decrease in the bioluminescent signal from the gall bladder and the appearance of bioluminescent signals in the intestines (see Figure 2.6). The signal in the intestine was observed by sequential imaging to move through the abdominal cavity as the bolus of bacteria from the gall bladder transited the gut [125]. The studies showed that the bacteria reside extracellularly in the lumen of the gall bladder, and that during feeding as the bile is released into the intestine; the bacteria are similarly released and subsequently egested with fecal matter. This may be a significant source of infection for other animals and may have relevance to the human population.

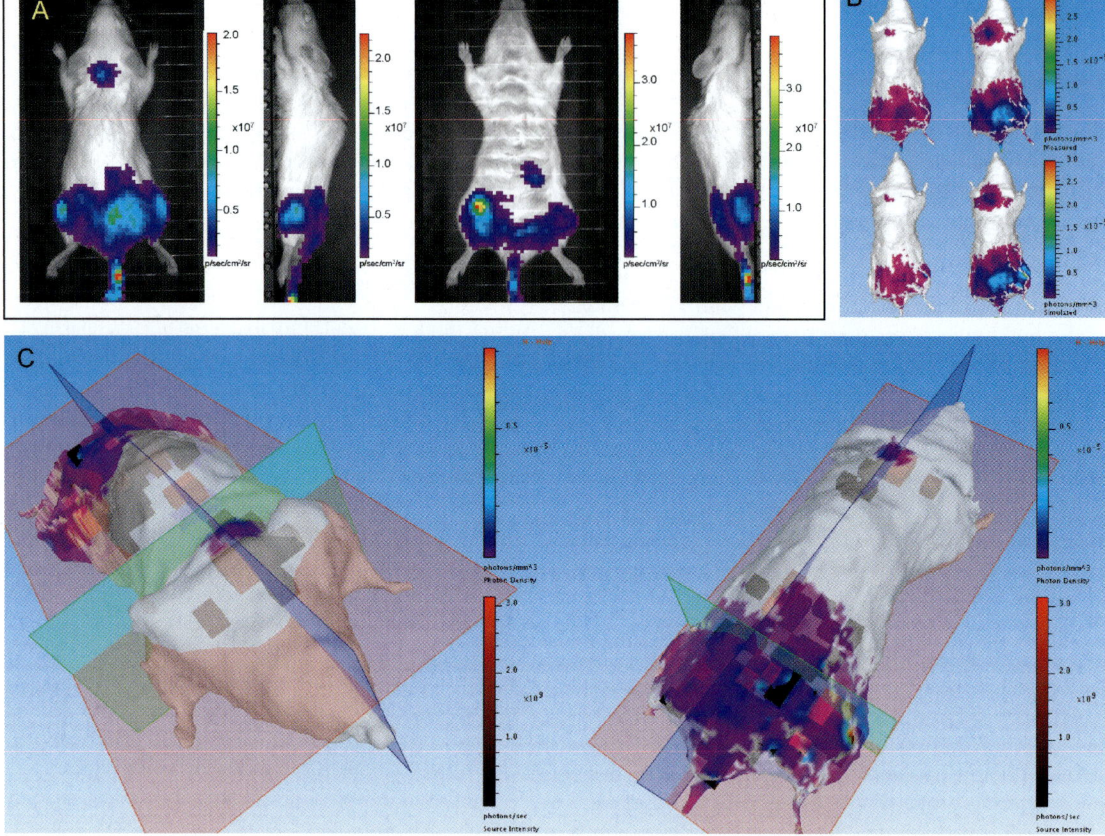

Figure 2.5. Multiview bioluminescence imaging of a mouse with metastatic breast cancer (4T1-luciferase cells) growths. The animal was imaged at 90° increments at two wavelengths (600 ± 10 nm shown (**A**). A surface map of the mouse is generated from multiple structured light and photographic images (not shown) onto which derived surface bioluminescence projections can be compared against measured signals (**B**). A three-dimensional representation of the animal with bioluminescent sources and intensities can be shown (**C**), in this case showing growths in the hind limbs and neck of the mouse (unpublished data).

The response of animals to infection has also been studied using transgenic animals expressing luciferase under the control of specific promoters from regulated genes such as serum amyloid A [126] and NF-κB [127]. In these studies an infection simulation was performed using a lipopolysaccharide (LPS) that led to an increase in luciferase expression. Treatments of the animals could then be performed to alter this induced response to help determine the mechanisms of gene regulation.

It is possible to sequentially image both bacterial infections using Lux bioluminescence and the host animal response using the firefly luciferase driven by a host response gene. In one example, a pneumococcal meningitis infection with *Streptococcus pneumoniae* expressing the *lux* operon was established in a transgenic mouse containing an inducible firefly luciferase gene under the transcriptional control of the mouse glial fibrillary acid protein (GFAP) promoter [110]. Luciferases could be differentiated in the animal by spectral imaging and also by substrate specificity of the two enzymes; the bacterial

luciferase was imaged with a 500 ± 10 nm bandpass filter without injection of substrate (the bacterial operon produces a spectral emission peaking at 490 nm and generates its own substrate). Subsequent injection of the firefly substrate and imaging through a 620 ± 10 nm filter permitted the imaging of the host response by the GFAP expression. Antibiotic treatment of infected animals quickly killed the bacteria with resultant loss of the short wavelength signal, but GFAP response could be measured for several days after the infection had cleared. From *in vivo* images, a significantly brighter and more diffuse signal from the head (brain) and back (spinal column) could be measured from the firefly reporter, either reflecting the ability to more easily detect the longer wavelength emitted by this enzyme or an infection localized to discrete patches eliciting a response from the whole organ. *Ex vivo* imaging of the infected brain revealed that the latter was the case, with a strong bacterial infection where it was injected in the brain but firefly luciferase expression over the entire organ. The

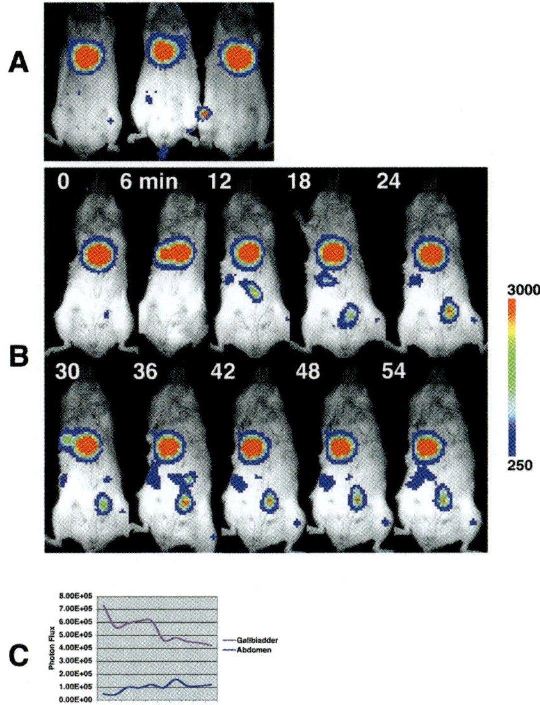

Figure 2.6. Bioluminescent *Listeria monocytogenes* bacteria infection model in mice, showing signals from the gall bladder in three representative animals (**A**) following intravenous infection. Subsequent feeding of the first mouse showed the bacteria leaving the gall bladder and entering the intestine (**B**), which can be quantified (**C**). Figure taken from Hardy et al. [125].

significance of a broad organ-level response to a focal infection that lasts beyond the time of infection also is relevant to human disease and postinfection sequelae.

Studies of infection demonstrate many of the strengths of *in vivo* bioluminescent imaging and its broad applicability in the study of the etiologic agent and the host response. Studying disease models where subtlety and nuance can be assessed noninvasively improves these models and reveals new features of the disease previously either not apparent or too difficult to assess in the body such that their study was prevented. Thus, we are learning more about disease states, chronic infections, host response, and therapeutic intervention than was possible through nonimaging approaches to the study of disease.

Cancer

Traditional oncology models were quickly adapted to using bioluminescent tumor cell lines to observe the progression of disease in mouse models. Other noninvasive methods of studying tumor burden over time have relied on measurement of tumor size by calipers, assessment of tumor volume by MRI and CT, or radionuclide uptake via microPET or microSPECT. These tools are useful for following the tumor progression *in vivo*, but each of these strategies relies on relatively large tumor size (at least 1 mm) before they can be observed, at which point the tumor is mature [128]. The ability to detect tumors at their earliest stages when tumor burden is low, or to visualize the appearance of metastases away from the primary tumor, are crucial to improve treatment approaches. Neither strategy can noninvasively detect these events, at least until the new growth is well established in the animal. The tremendous sensitivity of bioluminescence imaging has allowed visualization of tumors that are much smaller, and tumors or metastases as small as a few hundred cells can now be detected, even deep within animals, where calipers cannot be used or radionuclide approaches may be affected by nonspecific uptake of adjacent organs [129–132]. Animal models can thus be adapted to the development of earlier drug intervention, and the ability to monitor tumor regression permits investigation of minimal residual disease states. This is an important unmet clinical need in oncology, and now tumor models that enable the study of small numbers of tumor cells will lead to therapies that remove residual disease and prevent relapse.

Because the entire disease course can be followed with BLI in oncology models, we have the opportunity to evaluate early events in the onset of cancers, mechanisms of disease progression and metastasis, mediators of the immune response, kinetics of the therapeutic response, the nature of minimal residual disease, and rates of relapse. Thus, studies of animal models should not be limited to large subcutaneous tumors initiated with cell lines that represent end-stage disease. The field, in general, should move to evaluating aspects of cancer not previously accessible and developing intervention strategies directed at key steps over the full range of disease states. With the emerging paradigm of the cancer stem cell being substantiated in many different cancers, imaging will play a key role revealing the nature of these cells and their behavior *in vivo*. Sensitive *in vivo* assays will be essential for these studies, and BLI is ideally suited for this purpose. Use of BLI as a guide for the selection of times and tissues to evaluate refines our studies of cancer, and when imaging is coupled to advanced *in vitro* and *ex vivo* assays the animal models are more informative and cancer studies are more insightful. The use of reporter genes offers these links between cell culture assays, and *in vivo* measures, and investigators should be taking advantage of such aspects of BLI and other modalities as we increase the sophistication of our cancer studies. Imaging requires fewer animals without compromising the statistical validity of a given study while providing more information.

Implantation of tumor cells in mice allows modeling of tumor growth and studying the effect of drugs on this growth, but it does not allow investigation of

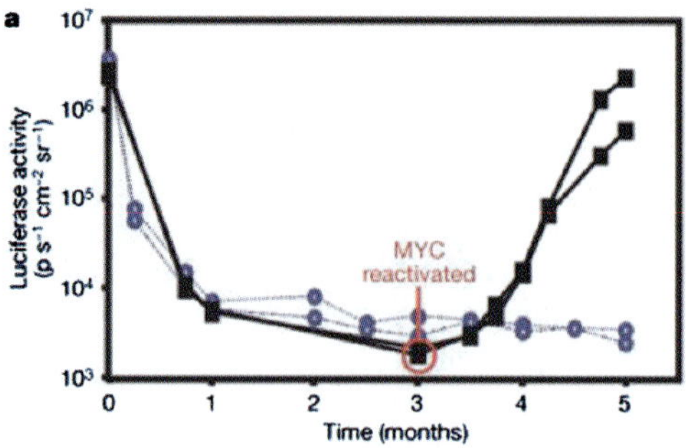

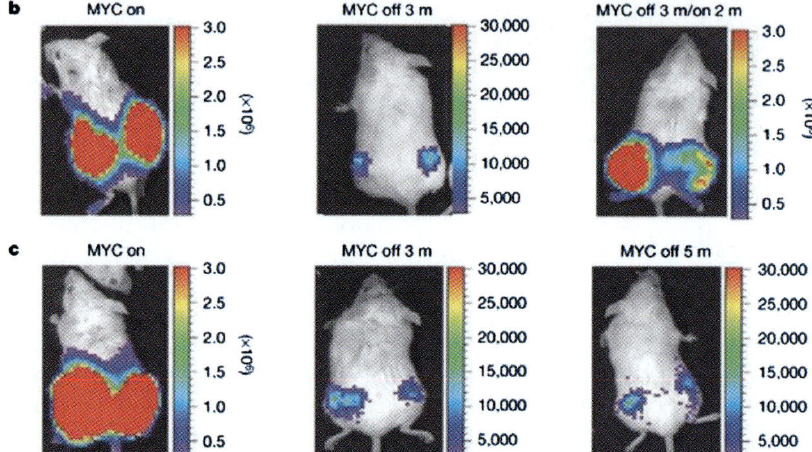

Figure 2.7. Tumor dormancy observed in liver tumors labeled with luciferase. The MYC oncogene is expressed under the control of the tet-response element and can be turned off by feeding the mice doxycycline. (**A**) Kinetics of tumor regression and regrowth monitored by bioluminescent imaging. Mice with tumors (MYC expressed) are treated with doxycycline, and the tumors recede to a minimal residual level of bioluminescent signal, and after 3 months tumors will grow upon removal of doxycycline (squares in **A**, and representative mouse in **B**), whereas mice where the MYC expression is continually repressed for 5 months (circles in **A**, representative mouse in **C**) show no recurrence. Figure taken from Shachaf et al. [134].

the earliest events in tumor biology as the normal cell changes physiology to become a tumor cell. Models of such early events, and the ability to study them, are essential to advance cancer therapies. An elegant transgenic animal line was developed in which a gene known to cause such a switch, the *Kras2^{v12}* gene, was engineered to be expressed only after a Cre recombinase-directed chromosomal rearrangement had been induced [133]. By placing the luciferase gene under the same genetic switch in this mouse, chromosomal rearrangement by the recombinase would result in both tumorigenic and bioluminescent cells. Nonsmall cell lung carcinomas-like lung adenocarcinomas were then induced in the mice following lung intubation of adenoviral particles expressing the Cre recombinase, and the development of tumor in the lungs monitored by BLI. Of course, the recombination event is a one-way switch, and the oncogene cannot be turned off in this model once induced, but control of the activation of malignancy in this model is nonetheless important.

Another example of *in vivo* modeling and imaging of cancer was developed by placing the Myc proto-oncogene under the control of the tetracycline-inducible Tet-promoter in a transgenic mouse line [134]. Two mouse lines were initially developed, one in which the MYC gene is driven by the Tet-response element (TRE-MYC), and a second in which the liver activator protein (LAP) promoter drives the expression of the tetracycline transactivating protein (tTA) in liver cells. Progeny of breeding between these lines that possess both transgenes express MYC, but expression could be stopped by oral treatment of mice with doxycycline that prevents tTA protein binding to the Tet-response element. Mice continuously treated with doxycycline show no tumor onset, but removal of the drug resulted in tumor development with a mean latency of 12 weeks. Reintroduction of doxycycline leads to tumor remission, with tumor cells differentiating into hepatocytes and biliary cells (see Figure 2.7). By breeding these mice with CMV-GFP-Luc mice, tumor cells could be transplanted

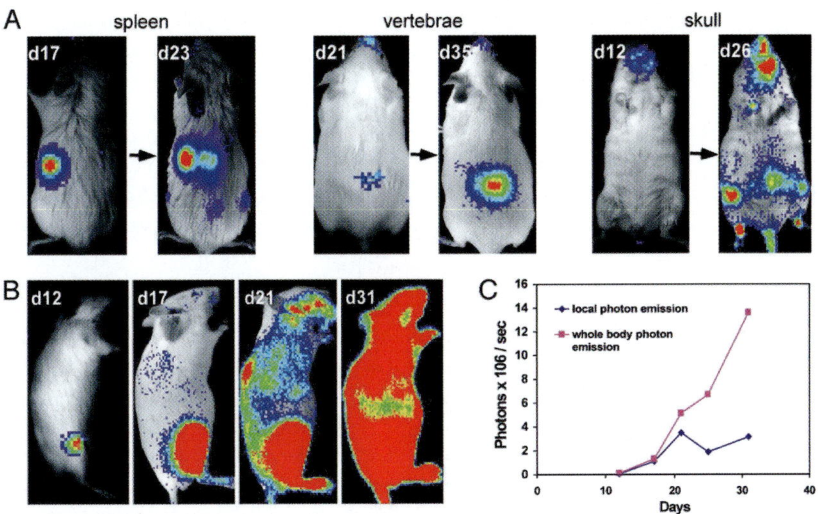

Figure 2.8. Engraftment and expansion of single luciferase-labeled hematopoietic stem cell in mice monitored by bioluminescent imaging. **(A)** Representative images of engraftment of single cells in the spleen, vertebrae, or skull of different mice. Foci expanded over time, and in some cases new detectable sites became apparent over time. **(B)** Expansion of foci from right femur of one animal that resulted in a significant degree of hematopoietic reconstitution from single cell. **(C)** Kinetics of reconstitution observed in mouse from B over the entire body, compared to the constant signal from the site of the initial focus. Figure taken from Cao et al. [60].

to naïve mice, and the size of the tumor correlated with the light emitted by luciferase activity, with as few as 1000 cells being detectable. These cells remained detectable in mice in the presence of doxycycline for many months, but after removal of the repressor, cells immediately regained their capacity for proliferation, which was detectable by increased luciferase activity and eventual appearance of palpable tumors. Reintroduction of the doxycycline resulted in tumor regression to a minimal luciferase-expressing level, and these cells returned to a more normal phenotype with restoration of liver-specific markers and loss of tumor antigens. Moreover, these cells began to differentiate back into hepatocytes and to form liver-like tissue architecture. This demonstrated how the oncogene inactivation reverses tumorigenesis in clinically aggressive cancer models and shows the pluripotent capacity of tumor cells to differentiate into normal cells while still maintaining their latent potential to become cancerous. This model may therefore allow a greater understanding of a minimal residual disease and the stem cell nature of cancer cells that may persist in patients following molecularly targeted therapy.

Imaging has provided us with tremendous opportunities to study cancers at many different stages of disease and to probe the molecular regulators of these stages. Building animal models representative of human disease and building into these models reporter genes that provide a handle with which to study them will lead to greater insight and improved therapies. Visible animal models that are relevant and predictive of human disease may be the greatest contribution to the control of cancer that has arisen from the advances in molecular imaging.

Regenerative Medicine and Organ and Stem Cell Transplantation

The creation of transgenic mice in which the transgene is the luciferase gene expressed from a strong constitutive promoter (e.g., the hybrid chicken-actin/beta-globulin promoter) has provided a powerful tool to probe cell and organ transplantation biology [60]. Transplanting organs or purified cell lines from such a donor animal into a naïve unlabeled mouse allows the fate of the transplanted cells and tissues to be assessed over the lifetime of the host animal or until the transplanted tissue is rejected. The use of BLI to monitor such cell transfers was first reported in a study to investigate the early events and dynamics of hematopoietic reconstitution in living mice [60]. Hematopoietic stem cells (HSC) are well-characterized multipotent cells that when introduced into a lethally irradiated host can stably reconstitute the hematopoietic system for the lifetime of the host, but determining the success of these transfers had only been possible weeks or months following transplantation and required tissue removal. Thus the early events of cell homing to spleen or bone and subsequent expansion was inaccessible in most studies. BLI provides an opportunity to monitor these events (see Figure 2.8). Bone marrow cells were isolated from a luciferase-expressing donor mouse line, and c-kit$^+$Thy1.1^{lo}Lin$^-$Sca-1$^+$ (KTLS) HSC were isolated by double fluorescence-activated cell sorting, and 10, 50 or 250 KTLS HSC were injected into lethally irradiated, nontransgenic syngenic animals, along with unlabeled whole bone marrow cells to allow animals to recover.

HSC engraftment to the bone marrow or spleens of these animals could be detected within 6 d of transfer as discrete foci, and expansion of these cells, as they reconstituted the hematopoietic system, could be monitored over time as the bioluminescent signal expanded from the initial foci in the bone marrow and spleen to the whole animal as circulating blood cells. Indeed, a single labeled KTLS HSC could be injected with 3×10^5 radio-protective unlabeled bone marrow cells and result in the appearance of a bioluminescent focus that could be detected within 12 d, although the fate of these foci could vary between animals, with some showing expansion to the whole animal, while others display a bioluminescent signal that lessened over time and remained within the original focal location.

This transgenic mouse line was further used as a source of cells to monitor bone marrow cell homing and engraftment in ischemic myocardium, a possible treatment for ischemic heart disease [61]. Animal and clinical trials have suggested that bone marrow cells may improve restoration of cardiac function after infarcts, leading to speculation that these stem cells may release angiogenic factors, protect cardiomyocytes from apoptosis, and recruit and induce proliferation of cardiomyocytes and cardiac stem cells. Naïve mice were subjected to sham or ischemic cardiac injury and subsequently received 5 million luciferase-labeled HSC and allowed to recover. BLI showed preferential homing of the labeled cells to injured heart compared to those that received the sham surgery, and further, echocardiography demonstrated improved cardiac function in animals that received cells compared to control animals with similar injury, suggesting that the systemically delivered HSC home to, and are retained by, the injured myocardium, where they may help recovery.

Embryonic stem (ES) cell therapy has also been monitored in rodent models by luciferase tagging the cells and introducing them into mice or rats [59, 135]. In these studies, human or mouse embryonic stem cells were transduced with luciferase by a lentiviral-mediated gene delivery as part of a triple reporter construct (with a fluorescent protein and thymidine kinase reporter for microPET studies). Delivery of 10 million undifferentiated murine ES cells to rat myocardium allowed detection of cells by both BLI and microPET studies, and subsequent increase in signals over the following weeks was due to formation of intra- and extracardiac teratomas, which indicates the limitations of using undifferentiated ES cells for therapy [59]. Treatment of animals with ganciclovir, a suicide substrate for the thymidine kinase gene, prevented the development of teratomas, with a loss of both bioluminescent and PET signals. Using neural precursors to differentiate luciferase-labeled human ES cells, and delivering 50,000–175,000 cells to the brain of immunocompromised animals, bioluminescence imaging could be used to follow the cell fate for over 2 months

postgraft, without the appearance of teratomas [135]. Similar grafts in immunocompetent mice resulted in a loss of signal due to cell rejection within 2 weeks. The ability to monitor ES cells and the consequences of transplantation into animals will hopefully allow a better understanding of the risks and great potential that these cells offer to disease therapies.

Luciferase-labeled organs can also be transplanted to host animals and the survival or rejection monitored noninvasively. Two groups have used pancreatic islet transplantation as a model to detect transplanted tissue and monitor potential graft rejection [62, 136–138]. Islets isolated either from transgenic mice expressing the luciferase from the rat insulin promoter (RIP; [136, 137] or transduced with an adenovirus that encoded both luciferase and the green fluorescent protein [138] were implanted in the kidney capsule or liver. Bioluminescent signal detected from these animals correlated with the number of islets transplanted, with as few as ten being detectable. Stable long-term expression (up to 18 months) could be followed in isografts, but rejection in allografts occurred within several days, as determined by both loss of bioluminescent signal and *ex vivo* histological investigation. One group went on to show that orthotopic implantation of the islets was not necessary for long-term engraftment and demonstrated that streptozotocin-induced hyperglycemic mice engrafted islets in the epidermal fat pad with similar efficiency as those with islets engrafted in the liver [136]. Transplanted islets could be identified in the fat pad or liver by both BLI and histological analysis for over 5 months, and animals had a similar tolerance to glucose, confirming correct insulin secretion. BLI can thus provide an important tool to understand transplantation models, which may be used in clinical studies.

Immune Cell Trafficking

A key aspect of immunity is immune cell migration, and the ability to monitor cell trafficking patterns *in vivo* is necessary to understand immune function, as well as cell–cell and cell–tissue interactions. A number of studies have used BLI to demonstrate the tempo of immune cell migration using various cell types in disease models to understand the immune response or to determine optimal means of using immune cells to deliver therapeutic genes and proteins to target sites in the body. These studies have been performed in animal models of immune responses to exogenous insult or in mouse models of human autoimmune disease.

Rheumatoid arthritis (RA) is an autoimmune disease characterized by inflammatory synovitis and subsequent destruction of the articular tissue, and CD4+ T-cell-medicated autoimmune responses are thought to play a critical role in the pathogenesis of RA. By selecting a type II collagen-specific CD4+ T-cell hybridoma or primary mouse CD4+ T cells and transfecting them with a luciferase–GFP retrovirus, cells could be injected into a collagen-induced arthritis mouse model of RA, and the cell trafficking monitored over time [139]. Immediately following intravenous injection, cells could be visualized in the lungs of mice, with subsequent localization (within 24 h) to the lymph nodes, and 2–4 d later they appeared in the sites of infection, specifically joints inflamed with the collagen, where they remained for the following week. Using T cells derived from mice transgenic for a T-cell receptor for myelin basic protein (MBP), cells also transiently traffic to inflammation sites, indicating nonspecific migration to such sites, but they were not retained there and were cleared from the animal within a week. Injection of luc-labeled MBP-specific T cells to mice previously immunized with MBP showed that bioluminescent signals could be measured in the central nervous system. The objective in these studies of autoimmune disease was to understand cell homing to the sites of tissue destruction with the intent of using the immune cells that would otherwise cause disease to transport immune modulators to control disease progression.

The application of bone marrow-derived dendritic cells transduced with a lentiviral vector expressing the cytokine IL4 to treat nonobese diabetic (NOD) mice has been shown to delay or prevent the onset of diabetes in these animals [140, 141]. By tagging these same T cells with a GFP–firefly luciferase fusion, the fate of the injected cells could be monitored over time, and were found to appear initially in the lungs and spleen of the mice following intravenous injection, and 3 d postinjection the cells had cleared the lungs and were visualized in the pancreatic lymph nodes and spleen (see Figure 2.9). Subsequent *ex vivo* analysis confirmed that these were the only fate for these cells – no bioluminescence was detected in the mesenteric, inguinal, or lumbar lymph nodes, or the thymus. Further optimization of the luciferase gene sequence [142] to remove cryptic splice sites and underrepresented codons has resulted in a greater than 100-fold enhancement of luciferase expression in T cells, allowing detection of as few as three cells implanted subcutaneously, or the tracking of fewer than 30,000 injected T cells into a mice with pre-established tumors. Application of such modified luciferases will greatly improve the sensitivity of such studies, allowing use of fewer injected labeled T cells.

Cytokine-induced killer (CIK) cells are a population of lymphocytes that can be enriched for therapeutic purposes, and these cells bear phenotypic markers of natural killer (NK) and T cells (CD3+) and express the natural killer group 2D (NKG2D) receptor. These cells mediate killing of tumor cells through the recognition of a class of stress-associated ligands expressed on the surface of the tumor cell called NKG2D ligands. CIK cells can be induced from splenocytes by growth under

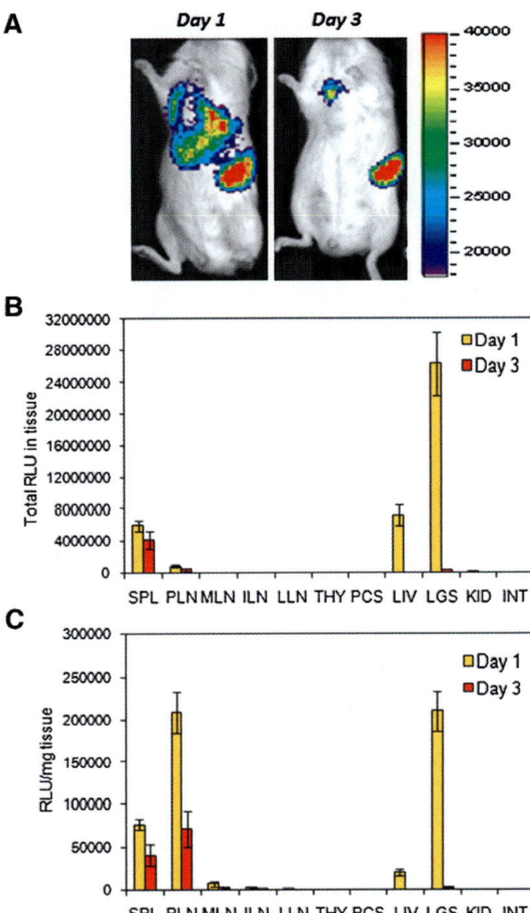

Figure 2.9. Biodistribution of luciferase-labeled bone marrow-derived dendritic cells in nonobese diabetic mice observed by bioluminescence. (**A**) Bioluminescence is observed in the lungs and spleen of living mice 1 d postintravenous injections, with subsequent trafficking to the pancreatic lymph node by day 3, with similar signal visible from the spleen. (**B**) and (**C**). *Ex vivo* analysis of luciferase activity in different tissues in representative animals either 1 or 3 d postimplantation. Early trafficking to lungs (LGS) and liver (LIV) is followed by targeting of the spleen (SPL) and pancreatic lymph nodes (PLN). Little trafficking is observed to the mesenteric lymph nodes (MLN), inguinal LN (ILN), thymus (THY), pancreas (PCS), kidneys (KID), or ileum (INT). Figure taken from Creusot et al. [140].

appropriate anti-CD3 antibody stimulation and cytokine (IL2) induction, and when tagged with the luciferase gene by retroviral transduction, could be tracked by BLI to an implanted tumor in mice with CIK cells arriving at the tumor site within hours; by 3 d the tumor was the only tissue site where signals were detected. The signal from the CIK cells remained detectable at the tumor site for least 12 d after transfer [143, 144]. Although the CIK cells are tumoricidal and are retained within the tumor for 14–21 d, they are capable of eliminating many cancer cells, but not all malignant cells are killed after a single

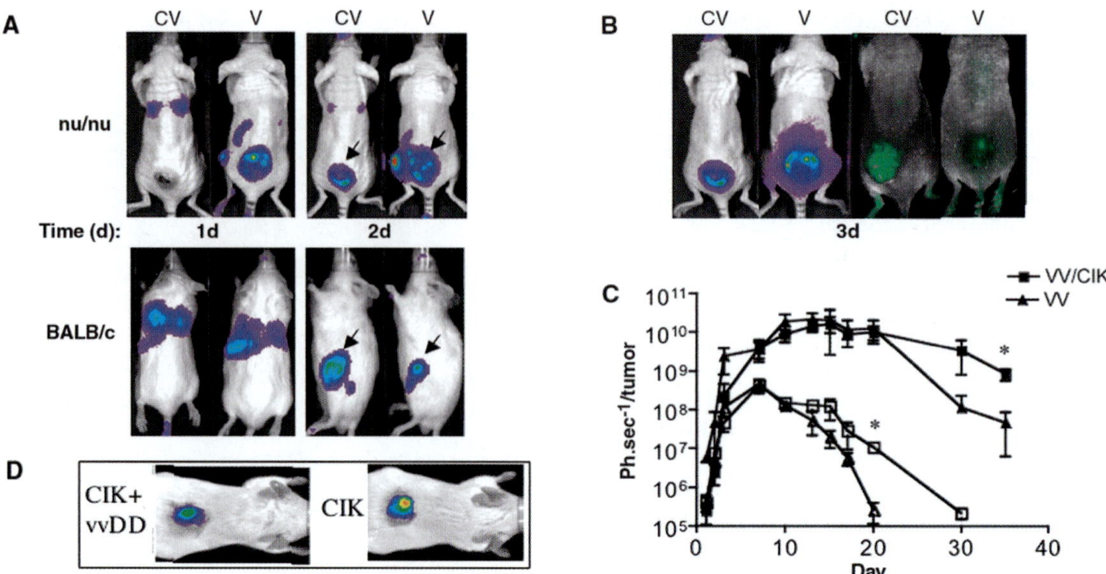

Figure 2.10. Trafficking of luciferase-labeled vaccinia virus (V) or cytokine-induced killer (CIK) cells preinfected with the same virus (CV) following intravenous injection to UCI-101 human ovarian tumors in nu/nu mice (**A**, top) or mouse 4T1 breast cancer tumors in BALB/c mice (**A**, bottom). Three days postinfection (**B**) bioluminescence is confined to the tumors in mice treated with the combination CV therapy compared to the virus alone. Fluorescent signals were also measured from vaccinia virus encoding the green fluorescent protein showing a uniform distribution of expression in the tumor after viral delivery in CIK cells. (**C**) Quantification of bioluminescent signal from tumors in nu/nu (closed symbols) or BALB/c mice (open symbols) following treatment with virus alone (triangles) or CIK cells preinfected with the virus (squares). (**D**) Labeling of CIK cells with Cy5.5 dye confirms trafficking of CIK cells is not affected by infection with the vaccinia virus. Figure taken from Thorne et al. [145].

administration of CIK cells. Attempts have been made to increase the potency of CIK cells with the objective of using a patient's own CIK cells as anticancer therapy. One strategy is to use the CIK cells as a "Trojan horse" to traffic a cargo of oncolytic virus to the tumor, where the virus can be released by the CIK cells within the tumor to specifically target the cancer cells and evade the host immune response [145]. In developing virotherapies the effects of the immune system on the engineered virus and the off-target effects are of prime concern. An elegant strategy to follow the efficacy of such a virus is to tag it with a firefly luciferase gene, such that expression and bioluminescence can only occur in infected cells, because the virus itself cannot by itself express the reporter gene or provide the necessary ATP and cofactors. By infecting unlabeled CIK cells with such a reporter-tagged oncolytic virus, the CIK cells will be bioluminescent when the viral genes are expressed and can be tracked in the animal as infected cells traffic to the tumor (see Figure 2.10). Following viral release from the carrier CIK cell and infection of the tumor, these tumor cells then bioluminesce, and their infection and subsequent destruction can be monitored.

The nature of oncolytic viruses is that they are engineered to retain lytic activity in malignant cells but not to replicate in normal cells. Thorne et al. utilized an oncolytic vaccinia virus (the vaccine strain used as the small pox vaccine) in which two crucial genes necessary for viral replication were deleted [146]. This double deleted virus had the thymidine kinase (TK) [147] and viral growth factor (VGF) [148] genes removed. The viral TK knockout is complemented only in tumor cells where cellular TK activity is elevated to provide a pool of thymidine nucleides for accelerated cell division. These cells also complement the viral VGF defect with mutations in the Ras/mitogen-activated protein kinase/extracellular signal-regulated kinase pathway. Oncolytic viruses are essentially nontargeted particles that circulate after injection with some virus ending up in the tumor [149]. Injection of the double deleted vaccinia virus alone into tumor-bearing mice resulted in some (30%) increased survival of mice with xenograft tumors, and bioluminescence signals were detected from the tumor. In contrast to the inefficient delivery of virions to tumor targets, use of CIK cells to deliver the virus to the tumor target dramatically increased survival of tumor-bearing mice. BLI enabled a directed study of the infection of CIK cells with oncolytic vaccinia virus and revealed synergistic effects with the combined viro- and cellular therapy. A thorough study of the vaccinia replication in CIK cells revealed that following infection there was a prolonged eclipse period of 48–72 h, compared to 2–8 h in cell lines. During the eclipse period there is no infectious virus and the infection is hidden from the host immune system. The tempo of CIK migration to tumor targets fortuitously matched the time of the eclipse period and virus was released from the CIK cells after a period of time when a majority of CIK localized to the tumor. Synchronous release of the virus in the tumor target with roughly a 1000-fold amplification and subsequent amplification of the virus in the first round of infection of tumor cells led to a rapid destruction of the tumor.

Any combination therapy requires optimization of delivery, and this study of dual biotherapy demonstrates that imaging can provide the insight necessary to optimize the combinations and timing that can inform the next study and improve the design of clinical studies. Thus, in many cases it is the information gained through imaging that translates to the clinic and not the imaging modality itself; this comprises one of the most significant contributions of imaging to medicine. That is, refinement and acceleration of preclinical studies, and this carries over into many areas of involved therapeutic intervention strategies.

Bone marrow transplantation, or hematopoietic stem cells, has proven to be an effective therapy for a variety of life-threatening malignancies in which the donor cells being given to the patient both reconstitute the host bone marrow and in a graft versus tumor effect destroy residual tumor cells that may have survived radiation or chemotherapy. However, in such transplants in humans there are minor immunological mismatches and the allogenic, or mismatched, immune cells can recognize the host cells as foreign and destroy the cells and tissues of the bone marrow recipient. This disease termed graft-versus-host disease (GvHD) can limit the application of this therapy, and careful immunological matching of donor and host is always performed prior to transplantation.

Despite transplantation of the best-matched bone marrow in the clinical setting GvHD remains a significant health risk. To better understand this disease, animal models were developed in which bioluminescent bone marrow and splenocytes were purified and transplanted into allogenic or syngenic irradiated recipient mice, and the trafficking patterns of these cells were monitored [150]. By mixing cell populations from the donor mice and assessing disease outcome in the recipients the cells responsible for GvHD could be identified and studied. T-cell-depleted bone marrow cells allow long-term survival of either syngeneic or allogeneic recipients, but the addition of splenocytes resulted in host death due to GvHD in the allogeneic setting. Use of bioluminescent splenocytes revealed that these cells were detected initially in the secondary lymphoid tissue (spleen, Peyer patches, and mesenteric lymph nodes, confirmed by *ex vivo* imaging of excised tissues) before infiltrating the intestine at day 4, and the liver and skin by day 6. Subsequent analysis of the animals with obvious T-cell proliferation indicative of GvHD revealed changing patterns of T-cell

subpopulations in the affected tissues. There was initial proliferation of the CD4+ T cells followed by CD8+ cells in the secondary lymphoid tissues, with subsequent trafficking to the other organs. When purified naïve T cells were used in the allogeneic transplantation model, a similar outcome was observed, but in contrast transplanted CD4+ effector memory T cells did not proliferate in the secondary lymphoid tissue and did not cause GvHD.

BLI served as an indicator of disease status and provided a useful guide for times and tissue samples in the *ex vivo* studies that revealed the specific cellular involvement and effector populations. From these studies it can be concluded that a cell transplant consisting of well-defined cell populations could provide the needed function and reduce the risk of GvHD. Transplants consisting of bone marrow-derived stem cells, memory T cells, and enriched CIK cells would lead to reconstitution of the hematopoietic system (by stem cells), temporary immunity to common antigens (by memory T cells), and an effective graft versus tumor effect (by the CIK cells).

Gene Discovery

Animal models provide the ability to understand disease progression in relation to systemic changes in gene expression where tissue barriers exist and organ systems are intact. This is in contrast to the isolated and unnatural environments that cell cultures provide. The ability to monitor disease progression in an animal and then assess gene expression profiles at specific anatomic sites within an animal will allow us to better understand the disease and potentially target drugs to proteins expressed under specific circumstances; for example, the study of the genes that are increased, or decreased, in expression when a primary tumor metastasizes to other locations in the host. By further interrogating the cells that show different physiological responses, clusters of co-regulated genes may be identified that are associated with, or lead to, a specific outcome. Such a screen was applied using BLI of breast cancer cells in a mouse metastasis model [151, 152]. To demonstrate the *in vivo* gene screen, breast cancer cells obtained from a human patient were labeled, individual cells from this population were transduced with a luciferase reporter and transferred to mice. The single-cell clones injected into mice displayed markedly different abilities to metastasize to the lung, bone, or adrenal medulla.

Transcriptional profiling by microarray expression analysis of human carcinomas has identified gene expression profiles that when present in the bulk of the tumor cells predict a poor prognosis for patients (5-year metastasis-free survival), and the presence of such signatures predicts metastasis of the primary tumor. Transcriptional profiling of cell lines that showed a preference to metastasize to the bones showed a similar gene expression profile to the parental line, but those that showed a preference to colonize the lungs showed a different profile. When a cell line metastasized to the lung (as observed by BLI) was recovered from the mouse, reintroduced to another animal, and the process repeated to a third round of animal imaging and recovery, the isolated cell lines from these animals were analyzed by transcriptional profiling and showed a marked difference from the parental and bone-specific cell lines. Ninety-five genes were identified that showed a threefold difference in expression level between these and the parental cells, of which forty-eight were over-expressed and forty-seven were under-expressed, and this gene set was largely distinct from the bone metastasis cell lines.

To identify which genes were directly linked to lung metastasis, each was over-expressed by retroviral infection in the parental cell line, either individually, in groups of three, or groups of six. Only one gene alone showed a small increase in the lung BLI signal, whereas combinations of three genes led to aggressive lung but not bone metastasis. Similarly, inhibition of these genes using small inhibitory RNAs (siRNA) in cell lines obtained from lung "enrichment" decreased their lung metastatic activity. Using a combined BLI imaging and gene profiling strategy, a group of genes that mediates breast cancer metastasis to the lung was identified, which when compared to clinical samples showed a correlation, and many of the genes identified had not been previously linked to metastasis. By performing such screens, it may be possible to better profile the genes involved in different metastatic fates of cancers, which in turn will lead to a more personalized treatment regime for a specific patient.

Transgenic Luciferase Reporter Mice

The generation of transgenic mice with the luciferase gene under the control of specific gene promoter sequences has been widely used with many different mouse models, some of which have been discussed previously. The generation of such animals allows the visualization of specific gene expression profiles in the context of the whole animal by simply observing the light emitted from the mouse. As with the generation of any transgenic animal, care must be taken in the selection of the founder breeding animals that the expression of the luciferase gene closely mimics the endogenous protein, and reporter expression profiles can be affected both by the extent of the flanking regions used to make the reporter construct and the chromosomal insertion site of the expression cassette. Typically, luciferase expression profiles are compared with *ex vivo* analysis of endogenous protein expression or activity of the gene of interest. It should also be noted that such reporter animals reflect the expression of the gene of interest, not necessarily the presence of the gene product of interest, as the stability of the luciferase enzyme or any other reporter protein

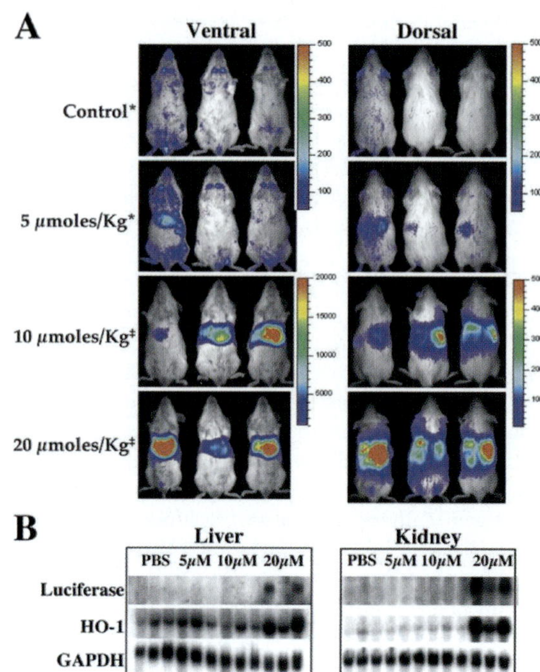

Figure 2.11. Expression of luciferase under the control of the heme oxygenase 1 (HO1) promoter in a transgenic mouse. Dose-dependent increases in luciferase activity are apparent in the liver and kidneys 9 h after different doses of cadmium chloride. (**A**) Ventral and dorsal views of three female transgenic mice with control (phosphate-buffered saline), 5, 10, or 20 μmol/kg doses of CdCl₂. (**B**) Northern blot analysis of liver and kidney total RNA with probes for luciferase, HO1, or control (GAPDH) confirm increased luciferase expression is observed with increased HO-1 expression following sacrifice of animals in (**A**). Figure taken from Zhang et al. [65].

used in these approaches may be different than that of the targeted protein.

Transgenic luciferase reporter mice provide a powerful tool to model and understand the expression patterns of genes that are either developmentally regulated or altered in response to challenges to the animal. Developmentally expressed genes, such as the bone morphogenic protein 4 (BMP4) gene, are expressed strongly during embryogenesis and fetus development, but after birth the expression levels drop until only expression is seen in the paws at adulthood [64]. A heme-oxygenase 1 (HO1) promoter-luciferase mouse model has also been used to monitor HO1 promoter activity during fetal and postnatal development [63–65], as well as being used to monitor chemical toxicity of compounds such as cadmium chloride (see Figure 2.11), doxorubicin, and thioacetamide [153]. Expression of luciferase in these HO1-luc mice was observed primarily in the liver, but other organs showed elevated expression levels in response to chemical insults. Although luciferase expression levels did not reflect the extent of tissue damage induced by the toxin, this reporter mouse may prove to be a useful model to screen animals for chemical toxicity.

The expression of an endogenous gene targeted for study may not be high enough to allow bioluminescence imaging of a luciferase gene directly linked to the promoter of interest, and amplification strategies have been used that permit the low-level expression from the gene of interest to be amplified by an intermediate protein to result in high cellular expression of the luciferase gene (see Chapter 6). By expressing chimeric transcriptional activator proteins composed of a yeast GAL4 DNA-binding domain and viral VP16 activation domains from the weakly induced gene of interest, the few molecules of this protein can bind to yeast *GAL4* DNA-binding sequences engineered upstream of a luciferase gene, which will then permit higher expression of the luciferase gene [154, 155]. Using this two-step genetic amplification strategy, expression of the luciferase protein was over 800-fold greater than if the luciferase had been linked directly to the weaker promoter (in this case the prostate-specific antigen), and indeed the induced expression level was more than twenty times greater than the constitutive cytomegalovirus promoter. Such strategies, when used in conjunction with destabilizing factors may provide a sensitive *in vivo* assay of gene expression in animals.

Nucleic Acid-Based Therapies

The field of gene therapy to treat human genetic disorders holds great potential for treating genetic disorders resulting from expression of nonfunctional proteins or reducing levels of adherently expressed genes. A variety of strategies have been adopted to deliver the genetic material to the host, and it is important to establish that the DNA is being delivered to the correct tissue, as well as confirming that the delivered gene is correctly expressed. Animal models are key to developing such delivery systems, and the luciferase gene provides a sensitive reporter to confirm that the genetic material is delivered to the correct organ and is correctly expressed at this site. Modified viruses offer an attractive mechanism to deliver genes to a host, and by engineering a luciferase expression cassette into the viral genome it is possible to model the efficiency of payload delivery and long-term stability of the therapy in mice. Delivery of the luciferase gene in adeno-associated virus (AAV) serotypes 2 and 5 *in utero* confirmed that both could sustain luciferase expression for more than 15 months and demonstrated that the AAV5 serotype resulted in greater expression [156]. Lentiviral vector-mediated gene therapy studies have also been successfully tested in mouse and rat models using luciferase as a reporter gene. Stereotactic injection of lentiviral vectors containing a luciferase reporter into the brain of rodents resulted in long-term expression from the brain, with an initial high expression falling 80% within the first

month, but stably expressed thereafter for more than a year [68].

In addition to providing a strategy to monitor gene therapy, it may also be possible to cointroduce neurodegenerative genes to the animal to model such diseases noninvasively. Direct injection of plasmid DNA has also been proposed as a method to deliver genes to tissue, but there is limited long-term stability using this approach as little of the transferred DNA integrates into the host tissue. Phage integrases have been shown to catalyze the integration of plasmids into mammalian genomic DNA if the plasmids contain specific sequences. The ϕC31 integrase can catalyze integration of plasmids containing an attB recognition sequence into "pseudo-attP" sites of mammalian sequences, and when plasmids containing this attB recognition sequences are coinjected with plasmids expressing the ϕC31 integrase, the resulting expression of the plasmid cargo was both stable and tenfold higher than in the absence of the integrase. Using luciferase as the cargo expression system, plasmid DNA was injected into the muscle of mice with or without the integrase plasmid, followed by electroporation of the tissue [157]. Mice were injected in both thighs with the luciferase plasmid, with only one thigh injected with the integrase plasmid, the other thigh getting a control construct. Luciferase expression could be detected as soon as 4 d postinjection in both thighs of the mice, and was detectable only in the thigh getting the integrase after more than a year. Genetic analysis confirmed that presence of the integrase plasmid resulted in stable incorporation in the chromosomal DNA, whereas extrachromosomal copies observed in the absence of the integrase were undetectable after 90 d. Further experiments using plasmids expressing the dystrophin gene injected with the luciferase gene confirmed that a therapeutic construct would be similarly expressed in muscle tissue that also received the integrase gene.

The use of siRNA to inhibit the expression of genes has allowed an understanding of the functions of genes in culture by selectively silencing them without the need to generate genetic knockouts. The clinical applications of such a strategy to turn off undesired genes, either from adherent overexpression observed with oncogenes in cancers or targeting essential genes of pathogens, holds the promise of exciting new therapies. The delivery of the small RNA species to target tissues remains the most significant impediment to effective therapies based on RNAi. Animal models where the efficacy of a therapy can be noninvasively assessed provide powerful tools for the development of novel delivery strategies. One of the earliest proof-of-principle studies that RNA inhibition might work in an animal system was demonstrated using luciferase as the target gene of the RNA molecules, and mice were injected with luciferase expression plasmids by a hydrodynamic transfection protocol that resulted in high liver expression in mice that could be detected by

BLI [158, 159]. A similar injection of a small inhibitory RNA (siRNA) or a plasmid that expressed short hairpin RNA (shRNA) molecules that inhibited luciferase expression resulted in a significant decrease in the detectable light from the liver, whereas scrambled siRNA or shRNA sequences had no effect on the BLI signal. RNA interference has been proposed as a therapeutic approach to treat hepatitis C virus (HCV) infections, targeting the internal ribosome entry site (IRES) sequence within the HCV genome that would control viral protein expression in the host cells. By engineering the luciferase gene downstream of the virus, bioluminescence can be a sensitive assay of IRES function. In a similar hydrodynamic transfection study as shown previously, it was shown that direct injection of shRNA sequences that target the IRES sequence decrease liver luciferase expression within 24 h and maintain this silencing for at least 5 d [160]. Imaging approaches have the potential to revolutionize drug discovery and drug development, and BLI is a cornerstone technology in the molecular imaging toolbox that has and will refine and accelerate the development of novel therapies (see Chapter 9 for a more detailed discussion of reporter gene imaging or therapeutic transgenes).

CONTRIBUTIONS OF *IN VIVO* BIOLUMINESCENCE TO HUMAN HEALTH

Unlike other molecular imaging modalities such as PET and SPECT, it is unlikely that bioluminescence imaging will find many opportunities for translation to the clinic due to the significant attenuation of visible light by mammalian tissues and other requirements of this modality. However, discoveries made at the preclinical level using BLI will provide insight that can be applied to the clinical arena. Translation from preclinical studies could be comprised of new technologies, as with PET, SPECT, and MRI, or new information. BLI will continue to generate new information that will be translated to clinical studies and refine and improve these studies and lead to better trials and approaches. Animal models can be rapidly screened with BLI without the need for large cohorts to generate statistically relevant data, as each animal in a study can act as its own control as the experiment proceeds. Thus, it should be possible to allow more drug candidates to pass along the development pipeline to preclinical studies, which may permit identification of candidates that show less than promising results at an *in vitro* level but may fare better in the physiologically more relevant animal model. Similarly, promising drugs can be more extensively investigated using altered treatment regimes that might refine and improve established therapies. The high cost of conducting these studies without molecular imaging strategies may preclude such studies, but BLI and other such strategies open these avenues to researchers.

BLI also offers improved sensitivity to animal models unmatched by other modalities, and this may help better understand diseases where small residual disease may be otherwise undetectable. The example of minimal residual disease in cancer treatment is an ideal example of this, and the ability to test drugs in such animal models may help identify new treatments. Similarly, bioluminescence imaging in gene screens may help identify new therapeutic targets and create new opportunities for therapy. Likewise, as stem cell biology translates to the clinic, the insights gained through visible animal models provided by BLI will significantly impact how we use these cells in the clinic.

THE FUTURE OF *IN VIVO* BIOLUMINESCENCE IMAGING

In vivo bioluminescence imaging is broadly utilized in a variety of fields in biomedical research. New applications of BLI are ever increasing as new variants and modifications of luciferase reporters are described and new models are developed. BLI relies on luciferases that emit in the visible region of the spectrum and are therefore influenced by absorption and scattering. This limitation may be addressed by the development of luciferase mutants that emit light at longer wavelengths and perhaps in the near infrared. This would improve detection and sensitivity in rodents and may allow imaging in larger animals. Significant wavelength shifts have already been achieved with both *Renilla* and beetle luciferases, and based on the incredible shifts obtained in fluorescent proteins, it may be possible to generate luciferases that emit light at longer wavelengths.

Of course, translation of BLI to larger animals would result in the need for more substrate administration so the identification of genes involved in the production of the luciferin substrate would greatly improve the utility of BLI, especially if eukaryotic cells could be made to autobioluminesce in a similar fashion to bacterial cells. To these ends, the luciferin-regenerating enzyme has been cloned [161, 162], an enzyme that mature fireflies use to recycle the luciferin substrate, but to date, the enzymes from the insect that actually synthesize the substrate have not been reported. Other luciferases may also be identified with their substrates that will allow monitoring of many events in animals – fungi, earthworms, and insects have been studied with different luciferase/luciferin chemistries that have yet to be characterized.

Further improvements to the instrumentation might be expected, especially with respect to 3D tomographic capabilities in BLI. Current instrumentation provides intriguing insight of what might be possible – the modeling of photon flux through animals that accounts for tissue properties may be in the near future, and multimodality instruments that provide anatomical, bioluminescent, and/or radionuclide imaging capabilities will greatly increase the information content possible from bioluminescence imaging. If all of these developments can be achieved, it may be possible to translate BLI to the large animal and possibly the clinical arena. But regardless of the extent of these advances we will continue to see improvements in BLI of mice and rats, and even without the spectacular advances intimated previously, BLI will continue to inform us of mammalian biology and guide our animal studies.

REFERENCES

1. Contag, C.H., et al. (1995). Photonic detection of bacterial pathogens in living hosts. *Mol Microbiol.* **18**(4): 593–603.

2. de Wet, J.R., et al. (1987). Firefly luciferase gene: structure and expression in mammalian cells. *Mol Cell Biol.* **7**(2): 725–37.

3. de Wet, J.R., et al. (1985). Cloning of firefly luciferase cDNA and the expression of active luciferase in Escherichia coli. *Proc Natl Acad Sci U S A.* **82**(23): 7870–3.

4. Liu, J. and A. Escher. (1999). Improved assay sensitivity of an engineered secreted Renilla luciferase. *Gene.* **237**(1): 153–9.

5. Liu, J., D.J. O'Kane, and A. Escher. (1997). Secretion of functional Renilla reniformis luciferase by mammalian cells. *Gene.* **203**(2): 141–8.

6. Wood, K.V., et al. (1989). Bioluminescent click beetles revisited. *J Biolumin Chemilumin.* **4**(1): 31–9.

7. Wood, K.V., et al. (1989). Complementary DNA coding click beetle luciferases can elicit bioluminescence of different colors. *Science.* **244**(4905): 700–2.

8. Brandes, C., et al. (1996). Novel features of drosophila period Transcription revealed by real-time luciferase reporting. *Neuron.* **16**(4): 687–92.

9. Plautz, J.D., et al. (1997). Quantitative analysis of Drosophila period gene transcription in living animals. *J Biol Rhythms.* **12**(3): 204–17.

10. White, M.R., et al. (1995). Real-time analysis of the transcriptional regulation of HIV and hCMV promoters in single mammalian cells. *J Cell Sci.* **108**(Pt 2): 441–55.

11. White, M.R., C.D. Wood, and A.J. Millar. (1996). Real-time imaging of transcription in living cells and tissues. *Biochem Soc Trans.* **24**(3): 411S.

12. Mobley, J. and T. Vo-Dinh. (2003). Optical Properties of Tissue. in *Biomedical Photonics Handbook*, T. Vo-Dinh, Editor. CRC Press LLC: Boca Raton. p. 2-1-2-75.

13. Contag, C.H. and M.H. Bachmann. (2002). Advances in vivo bioluminescence imaging of gene expression. *Annual Review of Biomedical Engineering.* **4**: 235–260.

14. Hastings, J.W. (1996). Chemistries and colors of bioluminescent reactions: a review. *Gene.* **173**(1 Spec No): 5–11.

15. Wilson, T. and J.W. Hastings. (1998). Bioluminescence. *Annu Rev Cell Dev Biol.* **14**: 197–230.

16. Rice, B.W., M.D. Cable, and M.B. Nelson. (2001). In vivo imaging of light-emitting probes. *J Biomed Opt.* **6**(4): 432–40.

17. Baldwin, T.O., et al. (1984). Cloning of the luciferase structural genes from Vibrio harveyi and expression of bioluminescence in Escherichia coli. *Biochemistry.* **23**(16): 3663–7.

18. Cohn, D.H., et al. (1983). Cloning of the Vibrio harveyi luciferase genes: use of a synthetic oligonucleotide probe. *Proc Natl Acad Sci U S A.* **80**(1): 120–3.

19. Craney, A., et al. (2007). A synthetic luxCDABE gene cluster optimized for expression in high-GC bacteria. *Nucleic Acids Res.* **35**(6): e46.

20. Manukhov, I.V., et al. (2000). [Cloning and expression of the lux-operon of Photorhabdus luminescens, strain Zm1: nucleotide sequence of luxAB genes and basic properties of luciferase]. *Genetika.* **36**(3): 322–30.

21. Miyamoto, C.M., et al. (1985). Polycistronic mRNAs code for polypeptides of the Vibrio harveyi luminescence system. *J Bacteriol.* **161**(3): 995–1001.

22. Viviani, V.R., E.J. Bechara, and Y. Ohmiya. (1999). Cloning, sequence analysis, and expression of active Phrixothrix railroad-worms luciferases: relationship between bioluminescence spectra and primary structures. *Biochemistry.* **38**(26): 8271–9.

23. Schroder, J. (1989). Protein sequence homology between plant 4-coumarate:CoA ligase and firefly luciferase. *Nucleic Acids Res.* **17**(1): 460.

24. Zhao, H., et al. (2005). Emission spectra of bioluminescent reporters and interaction with mammalian tissue determine the sensitivity of detection in vivo. *J Biomed Opt.* **10**(4): 41210.

25. Safran, M., et al. (2006). Mouse model for noninvasive imaging of HIF prolyl hydroxylase activity: assessment of an oral agent that stimulates erythropoietin production. *Proc Natl Acad Sci U S A.* **103**(1): 105–10.

26. Bhaumik, S. and S.S. Gambhir. (2002). Optical imaging of Renilla luciferase reporter gene expression in living mice. *Proc Natl Acad Sci U S A.* **99**(1): 377–82.

27. Tannous, B.A., et al. (2005). Codon-optimized Gaussia luciferase cDNA for mammalian gene expression in culture and in vivo. *Mol Ther.* **11**(3): 435–43.

28. Venisnik, K.M., et al. (2007). Fusion of Gaussia luciferase to an engineered anti-carcinoembryonic antigen (CEA) antibody for in vivo optical imaging. *Mol Imaging Biol.* **9**(5): 267–77.

29. Wurdinger, T., et al. (2008). A secreted luciferase for ex vivo monitoring of in vivo processes. *Nat Methods.* **5**(2): 171–3.

30. Curie, T., et al. (2007). Red-shifted aequorin-based bioluminescent reporters for in vivo imaging of Ca2 signaling. *Mol Imaging.* **6**(1): 30–42.

31. Rogers, K.L., et al. (2007). Non-invasive in vivo imaging of calcium signaling in mice. *PLoS ONE.* **2**(10): e974.

32. Loening, A.M., et al. (2006). Consensus guided mutagenesis of Renilla luciferase yields enhanced stability and light output. *Protein Eng Des Sel.* **19**(9): 391–400.

33. Loening, A.M., A.M. Wu, and S.S. Gambhir. (2007). Red-shifted Renilla reniformis luciferase variants for imaging in living subjects. *Nat Methods.* **4**(8): 641–3.

34. Venisnik, K.M., et al. (2006). Bifunctional antibody-Renilla luciferase fusion protein for in vivo optical detection of tumors. *Protein Eng Des Sel.* **19**(10): 453–60.

35. So, M.K., et al. (2006). Self-illuminating quantum dot conjugates for in vivo imaging. *Nat Biotechnol.* **24**(3): 339–43.

36 Zhao, H., et al. (2004). Characterization of coelenterazine analogs for measurements of Renilla luciferase activity in live cells and living animals. *Mol Imaging.* **3**(1): 43–54.

37 Meighen, E.A. (1991). Molecular biology of bacterial bioluminescence. *Microbiol Rev.* **55**(1): 123–42.

38 Meighen, E.A. (1993). Bacterial bioluminescence: organization, regulation, and application of the lux genes. *Faseb J.* **7**(11): 1016–22.

39 Szittner, R. and E. Meighen. (1990). Nucleotide sequence, expression, and properties of luciferase coded by lux genes from a terrestrial bacterium. *J Biol Chem.* **265**(27): 16581–7.

40 Karsi, A., et al. (2008). Development of bioluminescent Salmonella strains for use in food safety. *BMC Microbiol.* **8**: 10.

41 Francis, K.P., et al. (2000). Monitoring bioluminescent Staphylococcus aureus infections in living mice using a novel luxABCDE construct. *Infect Immun.* **68**(6): 3594–600.

42 Hardy, J., et al. (2004). Extracellular replication of Listeria monocytogenes in the murine gall bladder. *Science.* **303**(5659): 851–3.

43 Gupta, R.K., et al. (2003). Expression of the Photorhabdus luminescens lux genes (luxA, B, C, D, and E) in Saccharomyces cerevisiae. *FEMS Yeast Res.* **4**(3): 305–13.

44 Patterson, S.S., et al. (2005). Codon optimization of bacterial luciferase (lux) for expression in mammalian cells. *J Ind Microbiol Biotechnol.* **32**(3): 115–23.

45 Chance, B. (1991). Optical method. *Annu Rev Biophys Biophys Chem.* **20**: 1–28.

46 Chance, B., et al. (1995). Effects of solutes on optical properties of biological materials: models, cells, and tissues. *Anal Biochem.* **227**(2): 351–62.

47 Doyle, T.C., S.M. Burns, and C.H. Contag. (2004). In vivo bioluminescence imaging for integrated studies of infection. *Cell Microbiol.* **6**(4): 303–17.

48 Jansen, E.D., et al. (2006). Effect of optical tissue clearing on spatial resolution and sensitivity of bioluminescence imaging. *J Biomed Opt.* **11**(4): 041119.

49 Kuo, C., et al. (2007). Three-dimensional reconstruction of in vivo bioluminescent sources based on multispectral imaging. *J Biomed Opt.* **12**(2): 024007.

50 Duda, J., et al. (2007). Methods for imaging cell fates in hematopoiesis. *Methods Mol Med.* **134**: 17–34.

51 Doyle, T.C., et al. (2006). Expression of firefly luciferase in Candida albicans and its use in the selection of stable transformants. *Microb Pathog.* **40**(2): 69–81.

52 Francis, K.P., et al. (2001). Visualizing pneumococcal infections in the lungs of live mice using bioluminescent Streptococcus pneumoniae transformed with a novel gram-positive lux transposon. *Infect Immun.* **69**(5): 3350–8.

53 Dhawan, J., et al. (1995). Tetracycline-regulated gene expression following direct gene transfer into mouse skeletal muscle. *Somat Cell Mol Genet.* **21**(4): 233–40.

54 Furth, P.A., et al. (1994). Temporal control of gene expression in transgenic mice by a tetracycline-responsive promoter. *Proc Natl Acad Sci U S A.* **91**(20): 9302–6.

55 Gossen, M. and H. Bujard. (1992). Tight control of gene expression in mammalian cells by tetracycline-responsive promoters. *Proc Natl Acad Sci U S A.* **89**(12): 5547–51.

56 Gossen, M., et al. (1995). Transcriptional activation by tetracyclines in mammalian cells. *Science.* **268**(5218): 1766–9.

57 Kistner, A., et al. (1996). Doxycycline-mediated quantitative and tissue-specific control of gene expression in transgenic mice. *Proc Natl Acad Sci U S A.* **93**(20): 10933–8.

58 Weinmann, P., et al. (1994). A chimeric transactivator allows tetracycline-responsive gene expression in whole plants. *Plant J.* **5**(4): 559–69.

59 Cao, F., et al. (2006). In vivo visualization of embryonic stem cell survival, proliferation, and migration after cardiac delivery. *Circulation.* **113**(7): 1005–14.

60 Cao, Y.A., et al. (2004). Shifting foci of hematopoiesis during reconstitution from single stem cells. *Proc Natl Acad Sci U S A.* **101**(1): 221–6.

61 Sheikh, A.Y., et al. (2007). Molecular imaging of bone marrow mononuclear cell homing and engraftment in ischemic myocardium. *Stem Cells.* **25**(10): 2677–84.

62 Cao, Y.A., et al. (2005). Molecular imaging using labeled donor tissues reveals patterns of engraftment, rejection, and survival in transplantation. *Transplantation.* **80**(1): 134–9.

63 Contag, C.H. and D.K. Stevenson. (2001). In vivo patterns of heme oxygenase-1 transcription. *J Perinatol.* **21** Suppl 1: S119–24; discussion S125–7.

64 Zhang, W., et al. (2001). Rapid in vivo functional analysis of transgenes in mice using whole body imaging of luciferase expression. *Transgenic Res.* **10**(5): 423–34.

65 Zhang, W., et al. (2002). Selection of potential therapeutics based on in vivo spatiotemporal transcription patterns of heme oxygenase-1. *J Mol Med.* **80**(10): 655–64.

66 McCaffrey, A.P., et al. (2002). Determinants of hepatitis C translational initiation in vitro, in cultured cells and mice. *Mol Ther.* **5**(6): 676–84.

67 Tolar, J., et al. (2005). Real-time in vivo imaging of stem cells following transgenesis by transposition. *Mol Ther.* **12**(1): 42–8.

68 Deroose, C.M., et al. (2006). Noninvasive monitoring of long-term lentiviral vector-mediated gene expression in rodent brain with bioluminescence imaging. *Mol Ther.* **14**(3): 423–31.

69 Chalfie, M. (1995). Green fluorescent protein. *Photochem Photobiol.* **62**(4): 651–6.

70 Chalfie, M., et al. (1994). Green fluorescent protein as a marker for gene expression. *Science.* **263**(5148): 802–5.

71 Hoffman, R. (2002). Green fluorescent protein imaging of tumour growth, metastasis, and angiogenesis in mouse models. *Lancet Oncol.* **3**(9): 546–56.

72 Hoffman, R.M. (2002). In vivo imaging of metastatic cancer with fluorescent proteins. *Cell Death Differ.* **9**(8): 786–9.

73 Hoffman, R.M. (2002). Green fluorescent protein imaging of tumor cells in mice. *Lab Anim (NY).* **31**(4): 34–41.

74 Li, C.Y., et al. (2000). Initial stages of tumor cell-induced angiogenesis: evaluation via skin window chambers in rodent models. *J Natl Cancer Inst.* **92**(2): 143–7.

75 Shaner, N.C., et al. (2004). Improved monomeric red, orange and yellow fluorescent proteins derived from Discosoma sp. red fluorescent protein. *Nat Biotechnol.* **22**(12): 1567–72.

76 Wang, L., et al. (2004). Evolution of new nonantibody proteins via iterative somatic hypermutation. *Proc Natl Acad Sci U S A.* 101(48): 16745–9.

77 Shcherbo, D., et al. (2007). Bright far-red fluorescent protein for whole-body imaging. *Nat Methods.* 4(9): 741–6.

78 Troy, T., et al. (2004). Quantitative comparison of the sensitivity of detection of fluorescent and bioluminescent reporters in animal models. *Mol Imaging.* 3(1): 9–23.

79 Hoffman, R.M. (2008). In vivo real-time imaging of nuclear-cytoplasmic dynamics of dormancy, proliferation and death of cancer cells. *APMIS.* 116(7–8): 716–29.

80 Mahmood, U., et al. (1999). Near-infrared optical imaging of protease activity for tumor detection. *Radiology.* 213(3): 866–70.

81 Ntziachristos, V., et al. (2002). In vivo tomographic imaging of near-infrared fluorescent probes. *Mol Imaging.* 1(2): 82–8.

82 Weissleder, R., et al. (1999). In vivo imaging of tumors with protease-activated near-infrared fluorescent probes. *Nat Biotechnol.* 17(4): 375–8.

83 Branchini, B.R., et al. (2007). Thermostable red and green light-producing firefly luciferase mutants for bioluminescent reporter applications. *Anal Biochem.* 361(2): 253–62.

84 Shapiro, E., C. Lu, and F. Baneyx. (2005). A set of multicolored Photinus pyralis luciferase mutants for in vivo bioluminescence applications. *Protein Eng Des Sel.* 18(12): 581–7.

85 Nakatsu, T., et al. (2006). Structural basis for the spectral difference in luciferase bioluminescence. *Nature.* 440(7082): 372–6.

86 Baggett, B., et al. (2004). Thermostability of firefly luciferases affects efficiency of detection by in vivo bioluminescence. *Mol Imaging.* 3(4): 324–32.

87 Gammon, S.T., et al. (2006). Spectral unmixing of multicolored bioluminescence emitted from heterogeneous biological sources. *Anal Chem.* 78(5): 1520–7.

88 Michelini, E., et al. (2008). Spectral-resolved gene technology for multiplexed bioluminescence and high-content screening. *Anal Chem.* 80(1): 260–7.

89 Wender, P.A., et al. (2007). Real-time analysis of uptake and bioactivatable cleavage of luciferin-transporter conjugates in transgenic reporter mice. *Proc Natl Acad Sci U S A.* 104(25): 10340–5.

90 Berger, F., et al. (2008). Uptake kinetics and biodistribution of (14)C-D: -luciferin-a radiolabeled substrate for the firefly luciferase catalyzed bioluminescence reaction: impact on bioluminescence based reporter gene imaging. *Eur J Nucl Med Mol Imaging.* 35(12): 2275–2285.

91 Gross, S., et al. (2007). Continuous delivery of D-luciferin by implanted micro-osmotic pumps enables true real-time bioluminescence imaging of luciferase activity in vivo. *Mol Imaging.* 6(2): 121–30.

92 Wehrman, T.S., et al. (2006). Luminescent imaging of beta-galactosidase activity in living subjects using sequential reporter-enzyme luminescence. *Nat Methods.* 3(4): 295–301.

93 Goun, E.A., et al. (2006). Intracellular cargo delivery by an octaarginine transporter adapted to target prostate cancer cells through cell surface protease activation. *Bioconjug Chem.* 17(3): 787–96.

94 Jones, L.R., et al. (2006). Releasable luciferin-transporter conjugates: tools for the real-time analysis of cellular uptake and release. *J Am Chem Soc.* 128(20): 6526–7.

95 Shinde, R., J. Perkins, and C.H. Contag. (2006). Luciferin derivatives for enhanced in vitro and in vivo bioluminescence assays. *Biochemistry.* 45(37): 11103–12.

96 Liu, J.J., et al. (2005). Bioluminescent imaging of TRAIL-induced apoptosis through detection of caspase activation following cleavage of DEVD-aminoluciferin. *Cancer Biol Ther.* 4(8): 885–92.

97 O'Brien, M.A., et al. (2005). Homogeneous, bioluminescent protease assays: caspase-3 as a model. *J Biomol Screen.* 10(2): 137–48.

98 O'Brien, M.A., et al. (2008). Homogeneous, bioluminescent proteasome assays. *Methods Mol Biol.* 414: 163–81.

99 Shah, K., et al. (2005). In vivo imaging of S-TRAIL-mediated tumor regression and apoptosis. *Mol Ther.* 11(6): 926–31.

100 Chandran, S.S., S.A. Williams, and S.R. Denmeade. (2009). Extended-release PEG-luciferin allows for long-term imaging of firefly luciferase activity in vivo. *Luminescence.* 24(1): 35–38.

101 Pichler, A., J.L. Prior, and D. Piwnica-Worms. (2004). Imaging reversal of multidrug resistance in living mice with bioluminescence: MDR1 P-glycoprotein transports coelenterazine. *Proc Natl Acad Sci U S A.* 101(6): 1702–7.

102 Otto-Duessel, M., et al. (2006). In vivo testing of Renilla luciferase substrate analogs in an orthotopic murine model of human glioblastoma. *Mol Imaging.* 5(2): 57–64.

103 De, A. and S.S. Gambhir. (2005). Noninvasive imaging of protein-protein interactions from live cells and living subjects using bioluminescence resonance energy transfer. *Faseb J.* 19(14): 2017–9.

104 De, A., A.M. Loening, and S.S. Gambhir. (2007). An improved bioluminescence resonance energy transfer strategy for imaging intracellular events in single cells and living subjects. *Cancer Res.* 67(15): 7175–83.

105 Contag, P.R., et al. (1998). Bioluminescent indicators in living mammals. *Nat Med.* 4(2): 245–7.

106 Curry, S., W.R. Lieb, and N.P. Franks. (1990). Effects of general anesthetics on the bacterial luciferase enzyme from Vibrio harveyi: an anesthetic target site with differential sensitivity. *Biochemistry.* 29(19): 4641–52.

107 Franks, N.P., et al. (1998). Structural basis for the inhibition of firefly luciferase by a general anesthetic. *Biophys J.* 75(5): 2205–11.

108 Moss, G.W., W.R. Lieb, and N.P. Franks. (1991). Anesthetic inhibition of firefly luciferase, a protein model for general anesthesia, does not exhibit pressure reversal. *Biophys J.* 60(6): 1309–14.

109 Moss, G.W., N.P. Franks, and W.R. Lieb. (1991). Modulation of the general anesthetic sensitivity of a protein: a transition between two forms of firefly luciferase. *Proc Natl Acad Sci U S A.* 88(1): 134–8.

110 Kadurugamuwa, J.L., et al. (2005). Reduction of astrogliosis by early treatment of pneumococcal meningitis measured by simultaneous imaging, in vivo, of the pathogen and host response. *Infect Immun.* 73(12): 7836–43.

111 Dehghani, H., et al. (2006). Spectrally resolved bioluminescence optical tomography. *Opt Lett.* 31(3): 365–7.

112 Wang, G., et al. (2008). Overview of bioluminescence tomography–a new molecular imaging modality. *Front Biosci.* **13**: 1281–93.

113 Allard, M., et al. (2007). Combined magnetic resonance and bioluminescence imaging of live mice. *J Biomed Opt.* **12**(3): 034018.

114 Chaudhari, A.J., et al. (2005). Hyperspectral and multispectral bioluminescence optical tomography for small animal imaging. *Phys Med Biol.* **50**(23): 5421–41.

115 Slavine, N.V., et al. (2006). Iterative reconstruction method for light emitting sources based on the diffusion equation. *Med Phys.* **33**(1): 61–8.

116 Soloviev, V.Y. (2007). Tomographic bioluminescence imaging with varying boundary conditions. *Appl Opt.* **46**(14): 2778–84.

117 Alexandrakis, G., F.R. Rannou, and A.F. Chatziioannou. (2006). Effect of optical property estimation accuracy on tomographic bioluminescence imaging: simulation of a combined optical-PET (OPET) system. *Phys Med Biol.* **51**(8): 2045–53.

118 Rannou, F.R., et al. (2004). Investigation of OPET Performance Using GATE, a Geant4-Based Simulation Software. *IEEE Trans Nucl Sci.* **51**(5): 2713–2717.

119 Contag, C.H. (2006). Molecular imaging using visible light to reveal biological changes in the brain. *Neuroimaging Clin N Am.* **16**(4): 633–54, ix.

120 Cook, S.H. and D.E. Griffin. (2003). Luciferase imaging of a neurotropic viral infection in intact animals. *J Virol.* **77**(9): 5333–8.

121 Doyle, T.C., et al. (2006). Visualizing fungal infections in living mice using bioluminescent pathogenic Candida albicans strains transformed with the firefly luciferase gene. *Microb Pathog.* **40**(2): 82–90.

122 Hitziger, N., et al. (2005). Dissemination of Toxoplasma gondii to immunoprivileged organs and role of Toll/interleukin-1 receptor signalling for host resistance assessed by in vivo bioluminescence imaging. *Cell Microbiol.* **7**(6): 837–48.

123 Hutchens, M. and G.D. Luker (2007). Applications of bioluminescence imaging to the study of infectious diseases. *Cell Microbiol.* **9**(10): 2315–22.

124 Piwnica-Worms, D., D.P. Schuster, and J.R. Garbow. (2004). Molecular imaging of host–pathogen interactions in intact small animals. *Cell Microbiol.* **6**(4): 319–31.

125 Hardy, J., J.J. Margolis, and C.H. Contag. (2006). Induced biliary excretion of Listeria monocytogenes. *Infect Immun.* **74**(3): 1819–27.

126 Zhang, N., et al. (2005). Serum amyloid A-luciferase transgenic mice: response to sepsis, acute arthritis, and contact hypersensitivity and the effects of proteasome inhibition. *J Immunol.* **174**(12): 8125–34.

127 Zhang, N., et al. (2005). NF-kappaB and not the MAPK signaling pathway regulates GADD45beta expression during acute inflammation. *J Biol Chem.* **280**(22): 21400–8.

128 Lyons, S.K. (2005). Advances in imaging mouse tumour models in vivo. *J Pathol.* **205**(2): 194–205.

129 Jenkins, D.E., et al. (2005). Bioluminescent human breast cancer cell lines that permit rapid and sensitive in vivo detection of mammary tumors and multiple metastases in immune deficient mice. *Breast Cancer Res.* **7**(4): R444–54.

130 Jenkins, D.E., et al. (2003). Bioluminescent imaging (BLI) to improve and refine traditional murine models of tumor growth and metastasis. *Clin Exp Metastasis.* **20**(8): 733–44.

131 Jenkins, D.E., et al. (2003). In vivo monitoring of tumor relapse and metastasis using bioluminescent PC-3M-luc-C6 cells in murine models of human prostate cancer. *Clin Exp Metastasis.* **20**(8): 745–56.

132 Scatena, C.D., et al. (2004). Imaging of bioluminescent LNCaP-luc-M6 tumors: a new animal model for the study of metastatic human prostate cancer. *Prostate.* **59**(3): 292–303.

133 Lyons, S.K., et al. (2003). The generation of a conditional reporter that enables bioluminescence imaging of Cre/loxP-dependent tumorigenesis in mice. *Cancer Res.* **63**(21): 7042–6.

134 Shachaf, C.M., et al. (2004). MYC inactivation uncovers pluripotent differentiation and tumour dormancy in hepatocellular cancer. *Nature.* **431**(7012): 1112–7.

135 Bradbury, M.S., et al. (2007). Optical bioluminescence imaging of human ES cell progeny in the rodent CNS. *J Neurochem.* **102**(6): 2029–39.

136 Chen, X., et al. (2007). The epididymal fat pad as a transplant site for minimal islet mass. *Transplantation.* **84**(1): 122–5.

137 Chen, X., et al. (2006). In vivo bioluminescence imaging of transplanted islets and early detection of graft rejection. *Transplantation.* **81**(10): 1421–7.

138 Fowler, M., et al. (2005). Assessment of pancreatic islet mass after islet transplantation using in vivo bioluminescence imaging. *Transplantation.* **79**(7): 768–76.

139 Nakajima, A., et al. (2001). Antigen-specific T cell-mediated gene therapy in collagen-induced arthritis. *J Clin Invest.* **107**(10): 1293–301.

140 Creusot, R.J., et al. (2008). Tissue-targeted therapy of autoimmune diabetes using dendritic cells transduced to express IL-4 in NOD mice. *Clin Immunol.* **127**(2): 176–87.

141 Creusot, R.J., et al. (2009). Lymphoid tissue-specific homing of bone marrow-derived dendritic cells. *Blood.* **113**(26): 6638–47.

142 Rabinovich, B.A., et al. (2008). Visualizing fewer than 10 mouse T cells with an enhanced firefly luciferase in immunocompetent mouse models of cancer. *Proc Natl Acad Sci U S A.* **105**(38): 14342–6.

143 Edinger, M., et al. (2003). Revealing lymphoma growth and the efficacy of immune cell therapies using in vivo bioluminescence imaging. *Blood.* **101**(2): 640–8.

144 Hardy, J., et al. (2001). Bioluminescence imaging of lymphocyte trafficking in vivo. *Exp Hematol.* **29**(12): 1353–60.

145 Thorne, S.H., R.S. Negrin, and C.H. Contag. (2006). Synergistic antitumor effects of immune cell-viral biotherapy. *Science.* **311**(5768): 1780–4.

146 McCart, J.A., et al. (2001). Systemic cancer therapy with a tumor-selective vaccinia virus mutant lacking thymidine kinase and vaccinia growth factor genes. *Cancer Res.* **61**(24): 8751–7.

147 Hengstschlager, M., et al. (1994). Different regulation of thymidine kinase during the cell cycle of normal versus DNA tumor virus-transformed cells. *J Biol Chem.* **269**(19): 13836–42.

148 Andrade, A.A., et al. (2004). The vaccinia virus-stimulated mitogen-activated protein kinase (MAPK) pathway is

required for virus multiplication. *Biochem J.* **381**(Pt 2): 437–46.

149 Yu, Y.A., et al. (2004). Visualization of tumors and metastases in live animals with bacteria and vaccinia virus encoding light-emitting proteins. *Nat Biotechnol.* **22**(3): 313–20.

150 Beilhack, A., et al. (2005). In vivo analyses of early events in acute graft-versus-host disease reveal sequential infiltration of T-cell subsets. *Blood.* **106**(3): 1113–22.

151 Minn, A.J., et al. (2005). Genes that mediate breast cancer metastasis to lung. *Nature.* **436**(7050): 518–24.

152 Minn, A.J., et al. (2005). Distinct organ-specific metastatic potential of individual breast cancer cells and primary tumors. *J Clin Invest.* **115**(1): 44–55.

153 Malstrom, S.E., et al. (2004). In vivo bioluminescent monitoring of chemical toxicity using heme oxygenase-luciferase transgenic mice. *Toxicol Appl Pharmacol.* **200**(3): 219–28.

154 Iyer, M., et al. (2005). Non-invasive imaging of a transgenic mouse model using a prostate-specific two-step transcriptional amplification strategy. *Transgenic Res.* **14**(1): 47–55.

155 Zhang, L., et al. (2002). Molecular engineering of a two-step transcription amplification (TSTA) system for transgene delivery in prostate cancer. *Mol Ther.* **5**(3): 223–32.

156 Lipshutz, G.S., et al. (2003). Comparison of gene expression after intraperitoneal delivery of AAV2 or AAV5 in utero. *Mol Ther.* **8**(1): 90–8.

157 Bertoni, C., et al. (2006). Enhancement of plasmid-mediated gene therapy for muscular dystrophy by directed plasmid integration. *Proc Natl Acad Sci U S A.* **103**(2): 419–24.

158 McCaffrey, A., M.A. Kay, and C.H. Contag. (2003). Advancing molecular therapies through in vivo bioluminescent imaging. *Mol Imaging.* **2**(2): 75–86.

159 McCaffrey, A.P., et al. (2002). RNA interference in adult mice. *Nature.* **418**(6893): 38–9.

160 Wang, Q., et al. (2005). Small hairpin RNAs efficiently inhibit hepatitis C IRES-mediated gene expression in human tissue culture cells and a mouse model. *Mol Ther.* **12**(3): 562–8.

161 Gomi, K., K. Hirokawa, and N. Kajiyama. (2002). Molecular cloning and expression of the cDNAs encoding luciferin-regenerating enzyme from Luciola cruciata and Luciola lateralis. *Gene.* **294**(1–2): 157–66.

162 Gomi, K. and N. Kajiyama. (2001). Oxyluciferin, a luminescence product of firefly luciferase, is enzymatically regenerated into luciferin. *J Biol Chem.* **276**(39): 36508–13.

3. Reporter Gene Imaging with PET/SPECT

June-Key Chung, Joo Hyun Kang, and Keon Wook Kang

Molecular imaging (MI) allows *in vivo* visualization of normal and abnormal cellular processes at the molecular and genomic levels, rather than at the anatomical level [1]. MI is a relatively new biomedical discipline that enables cellular and subcellular biologic processes within living subjects to be visualized, characterized, and quantified. MI combines molecular biology and medical imaging and is increasingly attracting research attention in the molecular cell biology, chemistry, genetics, biomedical physics, engineering, and medical fields. It can be used to study genomics, proteomics, metabolomics, various intracellular processes, and cell–cell interactions. A major focus of MI is genetic imaging, that is, "molecular–genetic imaging," and imaging reporter genes are set to play a leading role in molecular–genetic imaging.

Conventionally, gene expression levels can be determined by assaying reporter gene expression. To achieve this, a recombinant plasmid is constructed that expresses simultaneously a gene of interest and a reporter gene in a correlated manner, then it is transfected into target cells. When transcription and translation of the gene of interest and reporter gene occur simultaneously, by assaying reporter protein activity, gene expression can be indirectly evaluated in transfected cells. Conventional reporter genes include β-galactosidase, alkaline phosphatase, luciferases, and green fluorescent protein, but the conventional techniques of assaying them often required tissue sampling. More recent techniques with imaging reporter genes allow noninvasive and repetitive determination of transgene expression studies in living animals. If the gene of interest is a therapeutic or suicide gene, "imaging reporter gene"-transfected cells can emit specific imaging signals that enable visualizing the locations, durations, and magnitudes of transgene expressions and render gene therapy monitoring *in vivo* [2, 3].

REPORTER GENE IMAGING WITH RADIONUCLIDE-BASED REPORTER GENES

Radionuclides that emit positrons and gamma rays are used routinely in nuclear medicine. Positron emission tomography (PET) scanners are used to visualize positron emitters like [^{18}F], [^{11}C], and [^{124}I], and conventional gamma cameras and single photon emission computed tomography (SPECT) instruments can produce planar and tomographic images of gamma emitters like [^{131}I], [^{123}I], [^{111}In], and [^{99m}Tc].

Several small-animal PET cameras (i.e., microPET cameras) have been developed for basic research to meet the need for higher spatial resolution in small-animal studies, for example, a recently commercialized microPET scanner has a spatial resolution of 1 mm [5]. In addition, small high-resolution animal SPECT systems with pinhole collimators have also been developed for imaging gamma emitters with a spatial resolution of 0.35 mm [6]. Furthermore, advances in imaging technology have now resulted in the development of fused imaging modalities such as PET/CT, SPECT/CT, and PET/MRI [7, 8]. These fusion modalities allow us to obtain biological information in combination with precise anatomical localization in a single imaging session [9].

In parallel with these advances in instrumentation, various imaging reporter genes and positron and gamma emitting tracers (probes) have been developed (Figure 3.1). However, because they have unique advantages and disadvantages, given situations require that appropriate choices be made. Table 3.1 lists some common reporter gene/probes systems.

The Herpes Simplex Virus-1 Thymidine Kinase (HSV1-tk) System

Imaging reporter genes have been investigated in combination with high-resolution PET [2, 3, 4, 10].

Table 3.1. Radionuclide-Based Reporter Gene/Probe Systems

Reporter Gene	Radionuclide Probes	References
Wild-type or mutant herpes simplex virus type 1 thymidine kinase (HSV1-tk,HSV1-sr39tk or HSV1-A167Y-sr39tk)	[18F]ganciclovir, [18F]penciclovir, [18F]FHBG, [124I]FIAU, [18F]FEAU	2, 3, 4, 11, 12, 18, 21, 22, 24, 26
Human mitochondrial thymidine kinase (△hTK2)	[18F]FHBG, [124I]FIAU, [18F]FEAU	31, 32
Dopamine 2 receptor (D2R)	[18F]fluoroethylspiperone	10, 32, 33
Sodium/iodide symporter (NIS)	Radioiodines, 99mTc, 188Re	35, 41, 42, 78, 79
Somatostatin receptor (SSTr-2)	[123I]Tyr3-octereotide 111In-DTPA-D-Phe-octereotide 94mTc-Tyr3-octereotide	45, 46, 47, 49
Norepinephrine transporter (NET)	[123I], or [124I]MIBG [11C]ephedrine	52, 54, 56
Estrogen receptor (ER)	[18F]estradiol	57
Varicella–Zoster Virus thymidine kinase	[123I], [18F], [11C] bicyclic fluoropyrimidine deoxynucleoside analogus (BCNA)	61–63

The HSV1-tk gene has been used most commonly, and in transfected cells it is transcribed to HSV1-tk mRNA, which is then translated to HSV1-TK protein enzyme. The expressed TK then phosphorylates its substrate, for example, fluorine-18 labeled 9-(4-fluoro-3-hydroxymethylbutyl)-guanine ([18F]FHBG), which cannot traverse the cell membrane, whereas the unphosphorylated probe can, thereby phosphorylated [18F]FHBG gets trapped. [18F]FHBG emits positrons that can be visualized by PET. Thus, PET can localize

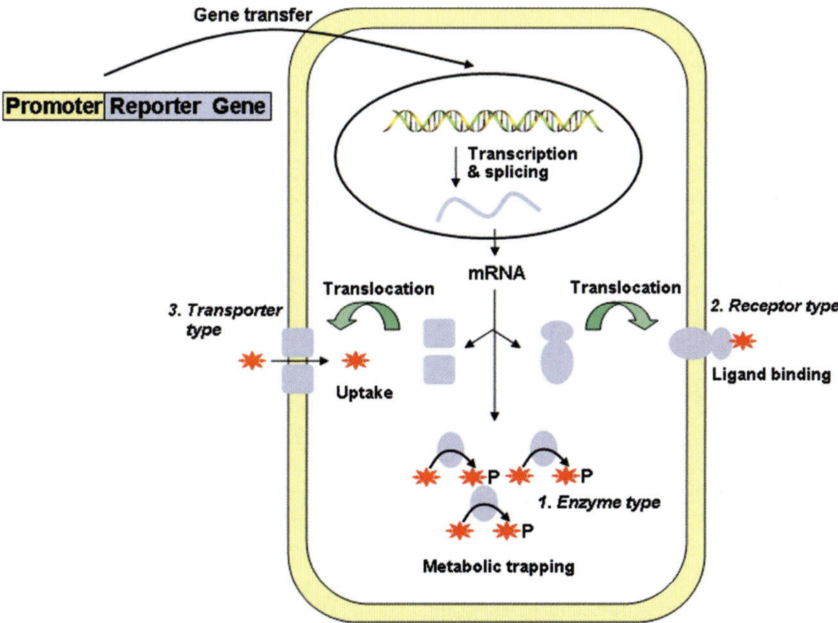

Figure 3.1. Schematic illustration of the three types of reporter gene expression imaging mechanisms used in nuclear medicine. HSV1-tk (oval shape) and NIS (squares) are representative of the enzyme and transporter-based mechanisms, respectively. The receptors used are D2R, NET, and SSTr-2. Stars represent radioisotope-labeled substrates (enzyme), ligands (receptor), or radionuclides (transporter type).

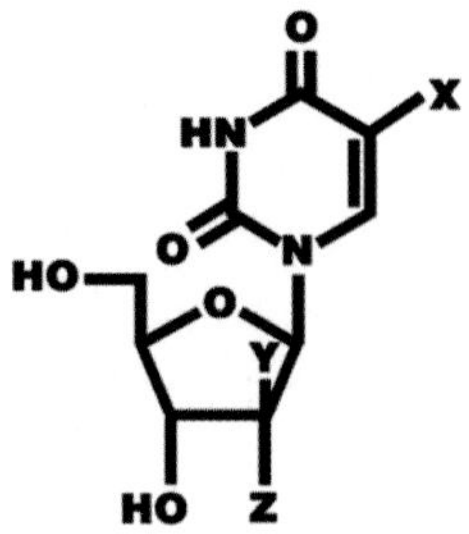

X	Y	Z	
CH₃	H	H	<TdR>
I	F	H	<FIAU>
CH₃	F	H	<FMAU>
CH₃CH₂	F	H	<FEAU>

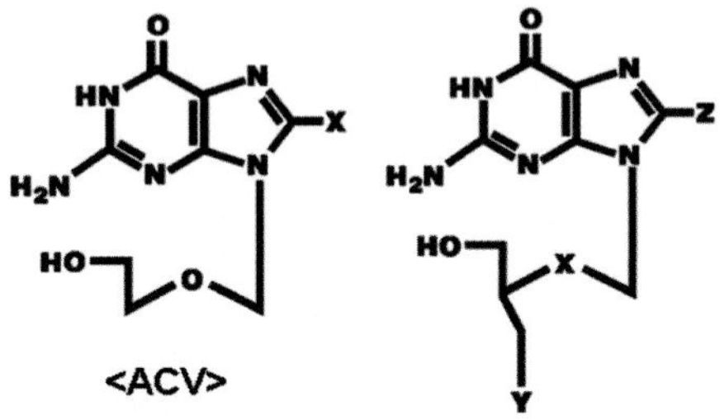

X	Y	Z	
O	OH	H	<GCV>
O	OH	F	<FGCV>
CH₂	OH	F	<FPCV>
CH₂	F	H	<FHBG>

<ACV>

Figure 3.2. Structures of thymidine and HSV1-TK reporter gene substrates. TdR (thymidine), 5-methyluracil-2′-deoxyribose; FIAU, 2′-fluoro-2′-deoxy-1-β-D-arabinofuranosyl-5-iodo-uracil; FMAU, 2′-fluoro-2′-deoxy-5-methyl-1-β-D-arabinofuranosyl-uracil; FEAU, 2′-fluoro-2′-deoxyarabinofuranosyl-5-ethyluracil; ACV, 9-[(2-hydroxy-1-ethoxy)methyl]guanine (acyclovir); GCV, 9-[(2-hydroxy-1-(hydroxymethyl)ethoxy)methyl]guanine (ganciclovir); FGCV, 8-fluoro-9-[(2-hydroxy-1-(hydroxymethyl)ethoxy)methyl]guanine (fluoroganciclovir); FPCV, 8-fluoro-9-[4-hydroxy-3-(hydroxymethyl)butyl]guanine (fluoropenciclovir); FHBG, 9-(4-fluoro-3-hydroxymethylbutyl)guanine.

HSV1-tk reporter gene expression, and the magnitude of reporter probe accumulation, in this case radioactivity, reflects HSV1-tk enzyme activity and HSV1-tk gene expression.

HSV1-tk enzyme has two types of substrates, pyrimidine nucleoside derivatives (e.g., 2′-fluoro-2′-deoxy-1-β-D-arabinofuranosyl-5-iodouracil (FIAU), 2′-fluoro-2′-deoxy-5-methyl-1-β-D-arabinofuranosyluracil (FMAU), and 2′-fluoro-2′-deoxy-5-ethyl-1-β-D-arabinofuranosyluracil (FEAU)), and acycloguanosine derivatives (acyclovir, ganciclovir, penciclovir, and FHBG) [2, 3] (Figure 3.2). Unlike mammalian thymidine kinase, HSV1-tk is less substrate specific, and phosphorylates these derivatives as well as thymidine, which is why derivatives such as Penciclovir (PCV) can be used for antiherpetic therapy without harming normal organs in humans. ¹⁸F-labeled acyclovir, ganciclovir, penciclovir, and FHBG have been tested *in vitro* and *in vivo* for reporter gene imaging sequentially, and [¹⁸F]FHBG has been found to accumulate better in cells expressing HSV1-tk relative to other acycloguanosine derivatives [11]. [¹⁸F]FHBG is a side chain fluorine-18 labeled analog of the antiherpetic drug penciclovir. In a reporter probe comparison study Min et al. [12] found that accumulation of [¹⁸F]FHBG in rat Morris hepatoma nude mouse tumor xenografts stably expressing HSV1-tk (MH3924A-stb-tk+) is 0.7% injected dose/gram tissue (%ID/g). The level of [¹⁸F]FHBG accumulation in the livers of mice injected with

adenoviruses carrying HSV1-tk was about 5.2%ID/g. As we will discuss in the following, sensitivity of [¹⁸F]FHBG imaging has now been improved by using mutant HSV1-tk reporter genes.

To improve substrate uptake, several different reporter probes (substrates) have been tested. Since pyrimidine nucleoside derivatives are structurally similar to natural thymidine, they have been suggested to be more sensitive probes than acycloguanosine derivatives for HSV1-tk reporter gene imaging. Pyrimidine nucleoside derivatives are readily phosphorylated by mammalian cytosolic and mitochondrial thymidine kinase and by HSV1-tk. However, because of their mitochondrial toxicities and lethal effects on neurons and liver, as determined by clinical studies, FMAU and FIAU are unsuitable antiviral therapeutics [13–15]. However, by taking advantage of the high sensitivities of PET and SPECT, it is possible to administer trace doses of versions of these drugs labeled with positron- or gamma-emitting radioisotopes as probes for HSV1-tk expression [16]. FIAU accumulates well in cells expressing HSV1-tk (% ID/g tissue was 1.1 in rat glioma cells transduced with the retrovirus cells and 1.6 in MH3924A-stb-tk+ cells in xenografted mice) [12, 17]. Moreover, the net accumulation of FMAU in HSV1-tk transfected cells (TK+) was higher than those of FEAU and FHBG, but its TK+/TK- uptake ratio was lower [18, 19]. On the other hand, FEAU had high TK+ uptake and a high TK+/TK- uptake ratio, whereas [¹²⁴I]FIAU suffered deiodination and label loss *in vivo.* High affinity between mammalian thymidine kinase and FMAU makes it more appropriate as a cell proliferation marker, that is, like [^{18}F]fluoro-L-thymidine (FLT) rather than a reporter gene probe [18]. Because of its high sensitivity and selectivity FEAU has been suggested to be a promising reporter probe for HSV1-tk [18, 19].

Another strategy of improving substrate accumulation is using more efficient enzymes for acycloguanosine derivatives. Mutant herpes simplex virus type 1 thymidine kinase (HSV1-sr39tk), a mutant enzyme with six amino acid substitutions [20], was originally developed to increase the effect of gene therapy. Of 426 mutants examined HSV1-sr39tk produced the most stable mammalian cell transfectants, which were 43-fold more sensitive to ganciclovir and 20-fold more sensitive to acyclovir. Tumor uptake of [^{18}F]penciclovir was 5.3% ID/g for C6 rat glioma cells stably transfected with mutant HSV1-sr39tk (C6-stb-sr39tk1+), which was 3.7 times higher than that of C6 cells transfected with wild-type HSV1-tk (C6-stb-tk1+) [20]. FHBG accumulation was 7.6% ID/g for C6-stb-sr39tk1+ and only 2.8% ID/g for FIAU [12]. This mutant HSV1-TK enzyme more effectively uses fluorinated acycloguanosines as substrates. MicroPET imaging of mice carrying tumor C6-stb-tk+ and C6-stb-sr39tk+ xenografts showed that C6-stb-tk+ tumors accumulated more [^{18}F]FEAU than [^{18}F]FHBG (0.48 vs 0.15% ID/g), whereas C6-stb-sr39tk+ tumors accumulated 75% more [^{18}F]FHBG than [^{18}F]FEAU (8.8 vs 2.8% ID/g) [22]. Consequently, HSV1-sr39tk/[^{18}F]FHBG offers the more effective PET reporter gene/reporter probe combination.

Recently, Degrève et al. [23] showed that the Ala-to-Tyr mutation at position 167 of HSV1-tk (HSV1-A167Y-tk) increases the relative affinity of HSV1-tk for purine over pyrimidine nucleosides. They developed HSV1-A167Y-tk to reduce endogenous competition with natural substrates dThd and favor ganciclovir phosphorylation. Likar et al. [24] introduced the Ala-to-Tyr mutation of a mutant HSV1-sr39tk and tested a new HSV1-tk supermutant, HSV1-A167Ysr39tk. It lacks the ability to phosphorylate pyrimidine-based nucleoside derivatives but exhibits high phosphorylation activity with acycloguanosine analogs. Small-animal PET imaging of transduced U87 xenografts showed no [^{18}F]FEAU accumulation in HSV1-A167Ytk–expressing and HSV1-A167Ysr39tk–expressing tumors. [^{18}F]FHBG accumulation in HSV1-A167Ysr39tk–expressing tumors was 2.5-fold higher than those in U87/wild-type HSV1-tk and U87/HSV1-A167Ytk tumors. But the highest level of accumulation of [^{18}F]FHBG was still observed in U87/HSV1-sr39tk tumors (twofold higher than in U87/HSV1-A167Ysr39tk). The supermutant HSV1-A167Y-sr39tk can be used for PET with [^{18}F]FHBG and suicidal gene therapy protocols with ganciclovir in patients receiving pyrimidine-based antiviral treatment.

A good reporter probe for *in vivo* imaging should confer high sensitivity, be highly selective for HSV1-tk/sr39tk, and have favorable *in vivo* pharmacokinetic profiles as well. [^{18}F]FHBG is excreted through the hepatobiliary and renal systems, which makes it hard to interpret reporter gene expression in the abdominal and pelvic areas in mice. However, [^{18}F]FEAU is predominantly cleared via the renal system and produces very low gastrointestinal signals.

Substrate metabolites, if present in sufficiently high concentrations, will kill cells, and HSV1-tk has been used for suicide gene therapy. Yaghoubi et al. [25] performed preclinical safety studies of FHBG in rats and rabbits and found that acute treatment with FHBG at one hundred times the conceived human dose did not harm organ functions or tissues. This group also performed human pharmacokinetic and dosimetry studies of [^{18}F]FHBG using PET in ten healthy volunteers, and the pharmacologic dose of [^{18}F]FHBG was estimated to be 160,000-fold less than that of a single dose of ganciclovir (5 mg/kg) [26]. [^{18}F]FHBG is rapidly cleared from blood through the renal and hepatobiliary systems. Initial levels of [^{18}F]FHBG in liver and kidneys are high, but it is also rapidly cleared from these organs. The maximum injected dose has been estimated at 530 MBq (14 mCi), as the radiation exposure of urinary bladder should be kept less than the limit of 0.05 Gy. At this injected level, doses to testes and ovaries were well below the 0.03 Gy limit issued by the Food and Drug Administration (FDA). [^{18}F]FHBG does not cross the blood–brain barrier, and [^{18}F]FHBG has been approved by the FDA as an investigational drug (IND #61,880).

In fact, the metabolites generated by an enzyme reaction may not accurately represent HSV1-tk expression. Green et al. [27] showed that tracer kinetic modeling of PET imaging estimates the rate of phosphorylation by thymidine kinase and that this is strongly correlated with actual TK activity. The cellular uptakes of thymidine derivatives and acycloguanosines are mediated by nucleoside transporters, equilibrative nucleoside transporters (ENTs), and concentrative nucleoside transporters (CNTs) [28]. Therefore, the cellular retention of these tracers is dependent on transporter levels on target cells, the ability of a tracer to utilize these transporters, and on HSV1-tk expression. The tracer retention is also affected by blood circulation of the body and blood flow in the tissues during *in vivo* imaging.

To optimize reporter gene systems, several factors should be considered. The pharmacokinetics of nucleosides and their analogs may differ. For example, Choi et al. [29] demonstrated that only the D-isomer of FIAU is useful for imaging HSV1-tk. Changes in regional blood flow to target tissues and the permeability of cell membranes to nucleosides also cause uptake differences. Of course, transport mechanisms, CNT, and ENT influence tracer accumulation, and after phosphorylation, the metabolic fates of nucleosides may alter tracer retention and uptake. Carrier-level and endogenous ligand competition (i.e., thymidine) are also important [29].

The immunogenicity of nonhuman-derived reporter protein is a major limitation of reporter gene translation in clinical practice. This is especially important when repetitive reporter gene administration or long-term monitoring of transgene expression is required. Riddell et al. [30] described an immune response specific for HSV1-tk protein in five of six patients who received anti-HIV cytotoxic HSV1-tk transduced T lymphocytes, and to overcome this limitation, Ponomarev et al. [31] developed a different human-derived gene encoding a nonimmunogenic endogenous enzyme, human mitochondrial thymidine kinase type 2 (hTK2). hTK2 phosphorylates deoxythymidine, deoxycytidine, and deoxyuridine, as well as several antiviral nucleoside analogs [32, 33]. ΔhTK2 was produced by truncating the N-terminal of human mitochondrial thymidine kinase to prevent its nuclear localization, which causes it to be expressed throughout the cytoplasm. In addition, these researchers found that [^{18}F]FEAU is a more specific substrate than [^{18}F] or [^{124}I]FIAU, whereas [^{18}F]FHBG is not phosphorylated by ΔhTK2 [32].

The transduced/wild-type uptake ratio in ΔhTK2-transduced cells (seventy-five for FEAU and ten for FIAU) is less than that in HSV1-tk-transduced cells (540 for FEAU and 30 for FIAU) or HSV1-sr39tk-transduced cells (460 for FEAU and 430 for PCV) [34]. For the uptake (% dose/ml) *in vivo*, high values were obtained with the HSV1-sr39tk reporter (7.6 for FHBG and 2.8 for FIAU), and lower values were obtained with the ΔhTK2 reporter (0.32 for FEAU and 0.25 for FIAU) [34]. An explanation is that thymidine competes with the PET imaging probe (e.g., FEAU) in case of ΔhTK2 and that the mutant HSV1-sr39tk has the advantage that it does not have as high an affinity for thymidine as the mammalian thymidine kinase. Looking at the aspect of reporter gene/probe combination, the viral HSV1-tk/[^{18}F]FEAU and the HSV1-sr39tk/[^{18}F]FHBG or HSV1-sr39tk/[^{18}F]FEAU systems are still more likely to play a role in future clinical studies because they show imaging characteristics better than those of ΔhTK2/[^{18}F]FEAU.

Genetic or functional cellular alterations after reporter gene transfection are also possible problems and are likely to become major issues when imaging reporter gene technology is applied to stem cell therapy. Wu et al. [35] examined the effects of a triple fusion gene, consisting of firefly luciferase, red fluorescent protein, and HSV1-sr39tk, on embryonic stem cell transcriptional profiles. Microarray studies showed the downregulation of genes related to the cell cycle, cell death, and protein and nucleic acid metabolism, and the upregulation of genes related to homeostasis and antiapoptosis. However, despite these changes, the expressions of triple reporter genes were found to have no significant effects on stem cell viability, proliferation, and differentiation capability.

The Dopamine 2 Receptor (D$_2$R) Gene System

D$_2$R is also used as a reporter imaging gene because of the availability of the well-established radiolabeled probe, [^{18}F]fluoroethyl spiperone (FESP). PET signals of [^{18}F]FESP in D$_2$R-expressing adenovirus-infected nude mice are well correlated with tritium-labeled spiperone binding and D$_2$R gene expression indicated by mRNA level [10]. Quantitative *in vivo* assays of FESP accumulation and *in vitro* assays of hepatic D$_2$R levels have demonstrated that noninvasive microPET analysis of the accumulated radioactivity in target tissue accurately reflects gene expression levels [10].

However, this technique has potential problems, such as the occupancy of ectopic D$_2$R by endogenous natural ligands. Moreover, high levels of reporter receptor expression cannot be achieved on cell membranes because of competition by other receptors for cell membrane sites. In addition, when a ligand activates D$_2$R, cellular levels of cyclic adenosine mono-phosphate (cAMP) may be affected, which could have physiologic consequences. For this reason, mutant strains of D$_2$R have been developed that do not activate the signaling pathway [36]. The binding of [^{3}H]spiperone by lysate of cells transduced with D$_2$R or mutant D$_2$R has been reported to be at the 500–7000 pmol/mg of protein level [37].

The Sodium/Iodide Symporter (NIS) Gene System

Conventional PET imaging reporter genes often require the synthesis of expensive probes in cyclotrons located close by and the use of expensive PET equipment. This has encouraged a number of investigators to examine the use of gamma emitters for reporter gene imaging [38]. Of these, the simplest and most applicable is the NIS gene system, and others and our group have shown that NIS gene can be used as an imaging reporter gene [39, 42, 46]. Iodine enters thyroid cells with sodium through a specific transporter – sodium/iodide symporter (NIS). This iodine uptake is driven by the sodium ion transmembrane concentration gradient, which is generated and maintained by the sodium–potassium pump (Na$^+$–K$^+$ ATPase).

The NIS gene was identified in 1996 by Carrasco in rats [40], and its human equivalent (hNIS) was isolated and cloned using the complementary DNA sequence of rat NIS [41]. The hNIS gene contains fifteen exons interrupted by fourteen introns and codes for a 3.9-kb mRNA transcript. Moreover, NIS protein undergoes post-translational modifications, which include glycosylation and phosphorylation. NIS is an intrinsic membrane protein with thirteen putative transmembrane domains, an extracellular amino terminal, and an intracellular carboxy terminal domain [43]. Using site-directed mutagenesis, three putative glycosylation sites in asparagine residues were identified. Furthermore, N-linked glycosylation was found not to be essential for NIS activity or stability [43]. Five phosphorylation sites in NIS were identified by mass spectrometry [44], phosphorylation of NIS modulates its protein stability and iodine transport rates. NIS cotransports two sodium ions and one iodide ion. In addition to iodide, several other anions are transported by NIS, that is, in order of transport rates, $ClO_4^- > ReO_4^- > I^- \geq SCN^- > ClO_3^- > NO_3^-$ [45]. As might be expected, the finding that Tc-99m pertechnetate and Re-188 perrhenate are also transported by NIS is important in terms of nuclear medicine imaging and radionuclide therapy.

The expression of functional NIS protein in target cells enables cells to concentrate radioiodine from plasma. Although the iodine taken up by NIS-expressing cell is not organified, more rapid uptake than efflux results in higher iodine concentration than those of extracellular space or plasma [39, 42, 46]. NIS has many advantages as an imaging reporter gene because of the wide availability of its substrates, that is, radioiodines and [^{99m}Tc], and the well-understood metabolism and clearance of its substrates from the body. NIS imaging does not require close proximity of radiochemistry laboratory facilities when radioiodine or [^{99m}Tc] are used. In addition, NIS is unlikely to perturb the underlying cell biochemistry. Iodine is not metabolized in most tissues, and although sodium influx may be a concern, no adverse effects have been observed to date. Lee et al. [46] showed that adenovirus-mediated NIS gene delivery to the myocardium does not cause significant injury or adversely affect cardiac function, which suggests that this system is safe in living subjects. Importantly, reporter gene imaging may be easier with NIS because all nuclear medicine departments have access to a gamma camera, SPECT, radioiodines, and [^{99m}Tc].

Another important merit of NIS is that NIS of human origin is available, which cannot provoke an immune response. Another general advantage of this transporter system is that the genes are expressed on cell surfaces, and therefore can be reached easily by reporter probes. In addition, radioiodines and [^{99m}Tc] pertechnetate are highly selective for NIS-expressing cells and are rapidly cleared from background tissues, which reduces interference due to background radioactivity [39, 46]. Moreover, the continuous active transport of radiotracers is possible in NIS-based systems, which amplifies the signals produced.

Nevertheless, NIS also has its limitations. First, NIS occurs naturally at high concentrations in the thyroid and stomach, and each substrate of NIS is excreted through the urinary tract, and this can sometimes cause image interpretation difficulties. Second, the NIS system is hampered by the rapid efflux of radionuclides from cells. In normal thyroid cells, iodine is organified by peroxidase, which traps intracellular iodine, and thus cotransfection with the thyroid peroxidase gene may improve radioiodine retention in target cells [47].

Vadysirisack et al. [48] demonstrated that NIS cell-surface levels are proportional to total NIS protein levels in several types of cultured cells, and that radioiodide uptake is also proportional to total NIS protein level. However, radioiodide uptake could not be increased by increasing NIS cell-surface levels beyond a certain limit, which indicated that the presence of factors, such as the availability of substrates iodide or sodium ions, limit NIS-mediated radioiodide uptake. The quantitative analysis of NIS expression is complicated because NIS protein must be properly localized on the cellular membrane before it can uptake radiolabeled substrates. The localization of NIS reporter protein in plasma membrane requires posttranslational modifications, such as its glycosylation and dimer formation, and proper membrane trafficking.

The Somatostatin Receptor (SSTr) Gene System

The somatostatin receptors are G-protein-linked seven membrane pass receptors. The expression of one of the six SSTr genes, SSTr2, is basically restricted to the pituitary, which can reduce background activity [49]. Although other tissues express lower quantities of this protein. Octreotide is a somatostatin analog peptide and is bound with high affinity by SSTr2 and SSTr5. Radiolabeled octreotides have been developed, that is, [^{123}I] labeled Tyr3-octreotide [50] and [^{111}In] DTPA-D-Phe-octreotide [51], and have been approved for clinical use in Europe and the United States [52]. [^{94m}Tc] Tyr3-octreotate has also been developed for PET imaging [53]. Human ovarian tumor xenografts expressing human SSTr2 with adenoviral vector have been monitored by [^{111}In] DTPA-D-Phe-octreotide-binding *in vivo* and *in vitro* [49]. Zinn et al. [54] showed that adenovirus-mediated expression of SSTr2 in subcutaneous and intraperitoneal tumors is detectable by gamma camera imaging.

However, hSSTr2 is naturally expressed in a number of normal tissues, and ligand binding to hSSTr2 reporter gene potentially could perturb normal physiology. To solve undesirable signaling effects of hSSTr2-ligand binding, generation of mutant hSSTr2 in which ligand binding is uncoupled from signal transduction may be required as in the case of mutant D_2R [36]. An alternative attempt was constructing a model epitope-tagged receptor in an adenoviral vector expressing hemagglutinin (HA) sequence of the influenza virus fused to the extracellular N-terminus of hSSTr2 gene [55]. The HA epitope is novel and not found in normal tissues, and therefore should be a specific marker for determining gene expression of HA-fused hSSTr2 using [^{99m}Tc] anti-HA antibody *in vitro* and *in vivo* tumor xenograft model.

The Human Norepinephrine Transporter (hNET) Gene System

hNET is a transmembrane protein involved in the transport of norepinephrine analogs into cells at adrenergic nerve terminals. It has been used clinically for imaging myocardial sympathetic innervation and neural crest tumors using [^{123}I], [^{131}I] MIBG, and [^{11}C] ephedrine [56–59]. hNET has several advantages as a reporter gene. First, it is a human protein that should eliminate immunogenicity concerns. Second, its radiolabeled probes are currently used clinically, and thus the safety of the probes has been proven [56, 60]. Moreover, the small size (<2 kb) of the hNET gene allows it to be easily incorporated in the expression cassette of the delivery vehicle [60]. *In vivo* experiments performed with nude mice bearing both hNET-expressing and wild-type tumors showed a tenfold higher accumulation of [^{131}I] MIBG in transfected tumors [56].

Buursma et al. [58] evaluated the feasibility of hNET as a reporter gene using [^{11}C]ephedrine, and managed to visualize one of three hNET-transduced tumors. However, Doubrovin et al. [59] reported that as few as 10^4 transduced T cells injected into tumors could be visualized by SPECT or PET after injecting [^{123}I]MIBG or [^{124}I]MIBG. Moroz et al. [60] suggested that a "late imaging" paradigm be used at 1, 2, and 3 d after injecting [^{124}I]MIBG, which takes advantage of the slow clearance of tracer from gene-transfected tumors.

The Estrogen Receptor Ligand (ERL) Gene System

Furukawa et al. [61] designed a new reporter gene imaging system that uses [^{18}F]-labeled estradiol (FES) and a human estrogen receptor ligand (hERL) binding domain. This system takes advantage of the fact that FES is used in human studies and that it has access to a wide range of tissues due to its lipophilic nature, including the brain. hERL was chosen as the reporter because it is a human protein with very low endogenous expression in tissues other than the uterus, ovaries, and mammary glands. hERL lacking N-terminal activation and DNA-binding domains loses its ability to bind target DNA and can no longer work as a transcription factor, which is an important reason for using a receptor as an imaging reporter [61]. They carried out basic studies to evaluate its potential for gene therapy monitoring with human thymidylate phosphorylase as therapeutic gene.

The radioligand uptake in hERL-transfected cells was one hundred times higher than that of control cells, and this result indicates that hERL-expressing cells efficiently uptake and accumulate radioligand and hERL-FES can work well as a PET reporter system to detect gene expression. However, estrogen concentration in plasma must be considered in female subjects, and mutated hERL with altered ligand specificity is required as mentioned by the authors.

Mintun et al. [62] reported that estrogen receptor concentrations and FES tumor uptakes are well correlated in thirteen breast cancer patients, and that cells expressing estrogen receptors at a level of 3 fmol/mg protein may be visualized *in vivo*.

The Varicella–Zoster Virus Thymidine Kinase (VZV-tk) Gene System

VZV-tk system has been applied to cancer gene therapy for treatment of hepatocellular carcinoma by conversion of nontoxic prodrug 6-methoxypurine arabinonucleoside to cytotoxic adenine arabinonucleoside triphosphate [63]. The cytotoxic effects of VZV-tk-transfected human osteosarcoma with pyrimidine nucleoside analogs is also higher than that of nontransfected tumor cells [64]. Recently, a new class of antiviral compounds, the bicyclic fluoropyrimidine deoxynucleoside analogs (BCNAs), has been developed [65]. These compounds are highly potent and selective inhibitors of VZV replication. Like HSV1-tk, VZV thymidine kinase efficiently phosphorylates BCNAs to the corresponding monophosphates and then further converts them to diphosphates by thymidylate kinase. Phosphorylated BCNAs are negative charged and remain trapped in cells. [^{123}I]-, [^{18}F]-, and [^{11}C]-labeled BCNAs have been developed and advocated as a new reporter gene system for use with SPECT and PET [66, 67]. Radioisotope-labeled BCNA showing high lipophilic values can cross the blood–brain barrier (BBB), and VZV-tk/radiolabeled BCNA system could evaluate the gene expression in brain and cell trafficking to brain region [67].

COMPARISONS OF RADIONUCLIDE IMAGING WITH OTHER MODALITIES

Radionuclide imaging strategies have several advantages (Table 3.2); they are highly sensitive and can detect at the 10^{-10}–10^{-12} mol/L radiotracer level, and they are highly quantitative, which means that dynamic studies and kinetic modeling can be easily performed. In general, radionuclide imaging modalities are well suited for the quantitative assessments of radiotracer/drug pharmacokinetics and metabolism. Moreover, nuclear medicine imaging is free of the attenuation problem that limits

Table 3.2. Comparison of Radionuclide and Optical Imaging Modalities

	Optical Imaging	Radionuclide Imaging
Spatial resolution	3–5 mm	1–2 mm
Temporal resolution	seconds to minutes	10 sec to min
Tissue penetration	1 cm (variable)	No limit
Sensitivity	10^{-15} to 10^{-17} mol/L (bioluminescence) or 10^{-9} to 10^{-12} mol/L (fluorescence)	10^{-10} to 10^{-12} mol/L
Radiation hazard	No	Yes
Costs	Low	High
Reporter gene transfer	Yes	Yes
Probes	Yes (bioluminescence) or No (fluorescence)	Radiolabeled tracers
Background	Low (bioluminescence) or moderate (fluorescence)	High
Chemical synthesis or labeling labors	No	Yes
Potential to clinical uses	Very low	High

optical imaging. A wide range of molecular imaging agents already have been approved for clinical use, and many novel agents are currently undergoing preclinical testing.

The ability to translate data from small-animal SPECT and PET imaging studies to patients is of great benefit. Moreover, rapid advances in imaging technology have resulted in the development of fused imaging modalities, such as PET/CT and SPECT/CT. Fusion modalities allow us to understand biological information in the context of precise anatomical localization in single imaging sessions [7].

Optical imaging modalities, such as fluorescence and bioluminescence imaging, are simpler, cheaper, more convenient, and more user friendly than other imaging modalities. Another advantage, especially for bioluminescence imaging, is that they are highly sensitive and allow the detection of low levels of gene expression [68]. Moreover, biological hypotheses can be tested rapidly in living experimental models, as demonstrated by their wide use in many *in vitro* reporter gene assays. However, the attenuation of light photons is a basic problem. About 90% of bioluminescence signal flux is lost per centimeter of tissue, and thus photon intensities detected by CCD cameras may not proportionately or sufficiently reflect endogenous reporter gene expression in the inner organs of even small animals.

MRI has two obvious advantages: high spatial resolution (micrometers) and the ability to extract physiologic and anatomic conformation simultaneously. However, MRI has poor sensitivity in terms of imaging enhancement agents and molecular reactions, requires highly trained personnel, and has high associated capital costs [69]. Moreover, despite continued efforts for more than a decade to develop suitable MR reporter genes, the field is still in its early stage [70].

MULTIMODALITY IMAGING

Because all the various imaging technologies have unique advantages and disadvantages, researchers have developed gene transfer vehicles, probes, and detectors that are compatible with several imaging modalities. The development of multimodality noninvasive imaging reporter genes will allow us to choose imaging technologies that best suit specific biological problems. Clinical applications of noninvasive reporter gene technologies should also be facilitated by multimodality reporter genes.

Several strategies are being used to link the expressions of multiple genes (Figure 3.3). The bicistronic approach to gene linkage involves the incorporation of an internal ribosomal entry site (IRES) sequence between two genes [37, 71]. Both genes are then transcribed into a single mRNA and translated into two different proteins. To enhance the expression of downstream gene of IRES, Wang et al. [72] developed a super IRES expression vector using ten copies of nine-nucleotide Gtx sequence that resulted in improved imaging signal for downstream gene expression. Another strategy involves the use of a fusion gene vector, whereby two genes are connected in such a way that their coding sequences are in the same reading frame to generate a single protein [73–75]. A third strategy makes use of bidirectional

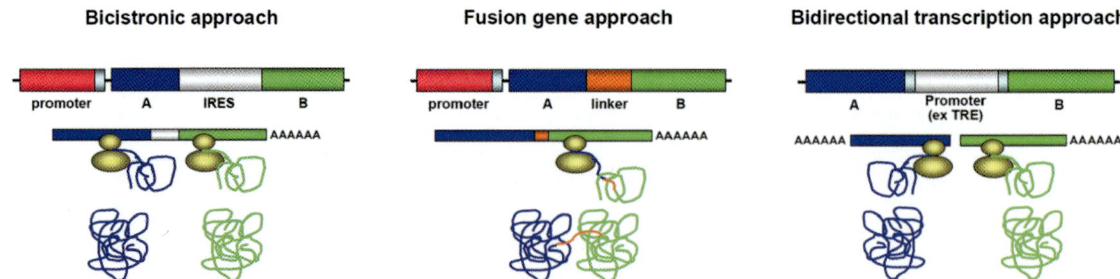

Figure 3.3. Several strategies to link the expressions of multiple genes in a single DNA construct. The bicistronic approach to gene linkage involves the incorporation of an internal ribosomal entry site (IRES) sequence between two genes. Both genes are then transcribed into a single mRNA and translated into two different proteins. A fusion gene approach, whereby two genes are connected in such a way that their coding sequences are in the same reading frame to generate a single protein. The bidirectional transcription approach uses a centrally located promoter element such as tetracycline-responsive element in which two genes are coexpressed in a bidirectional manner. AAAAAA: polyA tail in mature RNA, dumbell: ribosome, thread bundles: protein

transcription. For example, HSV1-sr39tk and D_2R were coexpressed in a bidirectional manner with centrally located tetracycline-responsive promoter elements [76]. As a result of these developments, chimeric fusion genes or bicistronic vectors that can be monitored either by bioluminescence and fluorescence [75], microPET and fluorescence [74], microPET and bioluminescence [77], and by microPET, fluorescence, and bioluminescence [78] have been used for the noninvasive imaging of reporter gene expression. PET imaging coupled with optical imaging appears to be the most amenable technology. In fact, PET can provide 3D images and allows quantitative analyses of reporter expression. On the other hand, optical bioluminescence imaging can easily and rapidly produce bidimensional images with high sensitivity [68].

Similarly, multimodality microPET/microCT and microSPECT/microCT instruments have been developed [79], and instruments that permit concurrent, coregistered optical imaging and radionuclide imaging, using common detector systems, are under development. These instruments should provide convenient and sensitive means of bioluminescent noninvasive reporter gene imaging with the resolution, quantifying ability, and three-dimensional properties of microPET imaging.

APPLICATION OF REPORTER GENE IMAGING

Reporter genes are genetic markers that encode easily detectable proteins. Once located downstream of a specific promoter, these markers become extraordinary tools for determining the activities of specific promoters and of factors that regulate their activities. Although the originally devised reporter genes have been widely utilized to study *in vitro* cell biology, recent technical developments allow the direct visualization and analysis of gene expression and regulation in living subjects [80].

Most genomic DNA are involved in the regulation of gene expression, which can be exercised at the transcriptional or posttranslational level. Many genes contain their own promoters, which control the expression of downstream genes, and the activity of a particular promoter may be specific to a disease process. These types of promoters are of particular interest, first because they are part of the molecular signature of the pathological process concerned, and second because they are potentially useful as specific promoters for use in gene therapy. By placing an imaging reporter gene under the control of such a promoter, dynamic visualization of promoter activity can be achieved [73].

Gene Therapy and Targeted Gene Expression

Gene-based therapy is a promising and flexible therapeutic approach to the management of various diseases. The lack of any convincing therapeutic success of current gene therapy protocols can be attributed in part to an inability to monitor gene expression at targeted sites in living subjects. Combining molecular imaging with gene therapy will enable real-time assessment of therapies and modification of treatment protocols [9]. Thus an imaging reporter gene delivered and expressed in conjunction with a therapeutic gene could become a general approach for monitoring of therapeutic transgene expression in living subjects.

Expression of a therapeutic gene can be visualized by gene imaging. HSV1-thymidine kinase converts acycloguanosine prodrugs, such as acyclovir, ganciclovir, and penciclovir, into their therapeutically active compounds. Thus, HSV1-TK was designed to kill target cells in the presence of pharmacologic concentrations of these prodrugs (suicide gene therapy). The location and magnitude of HSV1-tk gene expression can be monitored repeatedly by PET, using either [^{18}F]FHBG [81] or [^{124}I]FIAU. Moreover, radiolabeled somatostatin analogs, such as Y-90 octreotide, have been used as therapeutics to treat neuroendocrine tumors that over-express the SSTr2 gene [82], which can be monitored using [^{111}In]octreotide. Furthermore, NIS expression has been used to both image thyroid cancer and to concentrate [^{131}I] in cancer cells. NIS has also been investigated as an exogenous therapeutic gene that kills target cells in the presence of pharmacologic concentrations of [^{131}I] or [^{188}Re] (radionuclide gene therapy) [83, 84]. Exogenous NIS gene expression can be determined with radioiodines or [^{99m}Tc] pertechnetate.

In terms of the use of imaging reporter genes in clinical practice, Jacobs et al. [85] reported the first human PET image of HSV1-tk expression in glioma patients. They performed HSV1-tk suicide gene therapy and undertook to determine whether the gene was functionally active in human glioma tissues. Accordingly, they performed reporter gene imaging using [^{124}I]FIAU before and after HSV1-tk gene transfer and confirmed the successful transfection of the HSV1-tk gene into a human tumor *in vivo*. More recently, [^{18}F]FHBG PET imaging visualized HSV1-TK activity 2 d after intratumoral injection of HSV1-tk-containing adenovirus in patients with hepatocellular carcinoma at the injection site [86]. One week after treating patients with the prodrug ganciclovir (day 9) specific [^{18}F]FHBG accumulation was not observed within the tumor nodule region, which may indicate suicide killing of HSV1-tk-expressing cells. Furthermore, reinjection of adenovirus did not result in increased activity of [^{18}F]FHBG at the injected tumor site, which may indicate immune system rejection of the adenovirus. These clinical studies demonstrate the application of whole-body therapeutic transgene imaging using reporter probes in planning appropriate gene therapy protocols. Chapter 13 provides a more detailed discussion of clinical reporter gene imaging.

Transcription is initiated by the activation of a gene promoter, a specific DNA sequence that can be targeted by both transcription factors and RNA polymerase. Promoters specifically activated in certain tissues or cells (tissue/cell-specific promoters) are often used for gene therapy [73]. Adams et al. [87] used the tissue-restricted expression of prostate-specific antigen (PSA) in prostate cancer cells, and we also developed hepatoma-specific NIS gene expression using an alpha fetoprotein promoter [88].

Endogenous Gene Expression

Several investigators have designed specific reporter gene constructs under the control of upstream promoter/enhancer elements possessing binding sites for specific transcription factors. These promoters/enhancers can be activated by specific endogenous transcription factors and subsequently associate with specific endogenous genes. This strategy is referred to as "the cis-promoter/enhancer reporter gene system." Once a promoter/enhancer element has been activated due to the expression or activation of an endogenous gene product, imaging reporter gene expression occurs, thus enabling visualization.

One example is provided by the visualization of endogenous p53, an important antioncogenic gene. Doubrovin et al. [89] used the HSV1-tk gene, whereas Kim et al. [90] used the NIS gene for reporter gene imaging. In Doubrovin's study, a method for monitoring endogenous p53 activation was developed with retroviral vector system-expressing dual-reporter gene of HSV1-TK and GFP fusion controlled by artificial cis-acting p53 enhancer (p53RE). They have used N, N′-bis(2-chloroethyl)-N-nitrosourea for p53 activation and exhibited clear PET images with [^{124}I] FIAU by p53 activation *in vivo*. The p53RE-hNIS reporter system, in which the hNIS reporter gene is expressed under the control of a p53RE, was transfected into a human hepatoma cell line. Adriamycin was used to induce the enhanced expression of endogenous p53, and adriamycin-treated cells were found to accumulate more [^{125}I] than nontreated cells. Moreover, intracellular [^{125}I] uptake increased as the adriamycin dose was increased, and this was found to correlate significantly with p53 levels as determined by Western blotting with p53-specific antibody (Figure 3.4). Xenografted tumors of these cells also showed increased radionuclide accumulation after adriamycin treatment.

Intracellular Biologic Phenomenon

Using cis-enhancer-imaging reporter genes, some intracellular biological events, such as the activation of specific signal transduction pathways and nuclear receptors, can be visualized. In particular, the intracellular signaling pathway of TGF-beta receptor has been imaged [91]. After TGF-beta binds to its receptor, a specific intracellular signal transduction pathway is activated, which results in the production of several Smad proteins. In this study, HSV1-tk/GFP fusion retroviral reporter vector controlled by a promoter with a Smad binding site was constructed. This DNA construct was then transfected into target cancer cells, and *in vivo* images were acquired using a mouse xenograft model. [^{124}I] FIAU images visualized target tumors after injecting TGF-beta, indicating the presence of Smads and the successful signal transduction of TGF-beta and its receptor in tumors.

We also imaged the activities of estrogen and retinoic acid nuclear receptors, using a cis-enhancer reporter imaging system [92, 93]. NIS and luciferase genes were linked with IRES to express simultaneously two reporter genes and placed under the control of a cis-acting retinoic acid responsive element (RARE). In human hepatoma cells expressing this DNA construct, [^{125}I] uptake and bioluminescent intensity increased after

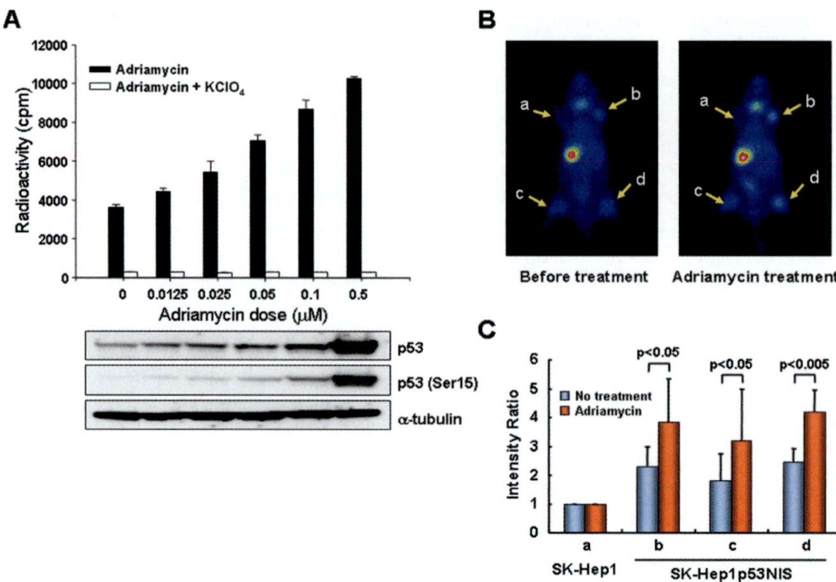

Figure 3.4. Relation between p53 gene expression and the accumulation of radioiodines in SK-Hep1 cells expressing p53RE-hNIS (SK-Hep1p53NIS) after adriamycin treatment. (**A**) Western blot analysis and radioiodine uptake. After treating SK-Hep1p53NIS cells with adriamycin, ^{125}I uptake was found to increase with adriamycin dose. Expression levels of total p53 and activated p53 (p53 Ser15) proteins increased dose-dependently in SK-Hep1p53NIS cells, and this increase was completely inhibited by KClO₄. (**B**) Scintigraphic image of endogenous p53 activation. Tumors were xenografted into nude mice at four different sites: 1×10^7 of SK-Hep1 (a, negative control); SK-Hep1p53NIS, (b, 5×10^6; c, 1×10^7; d, 2×10^7). Scintigraphic images of ^{99m}Tc were obtained before and after adriamycin treatment in same mice. Scintigraphy showed increased radioactivity in test tumors (SK-Hep1p53NIS) after adriamycin treatment versus untreated control tumors (SK-Hep1). (**C**) Radioactivity ratio of SK-Hep1p53NIS to SK-Hep1 in mice before and after adriamycin treatment. After adriamycin treatment, significantly higher uptake was observed in SK-Hep1p53NIS tumors than in SK-Hep1 tumors or in nontreated SK-Hep1p53NIS tumors.

retinoic acid treatment. The enhanced expression of NIS and luciferase genes induced by retinoic acid treatment was also demonstrated in an animal tumor model by using scintigraphic and optical bioluminescent images (Figure 3.5).

To acquire the image of a reporter gene controlled by a weak promoter or enhancer, it is sometimes necessary to augment its transcriptional activity, and an elaborate two-step transcriptional amplification (TSTA) system has been designed for this purpose. In this system, a weak promoter or enhancer controls the expression of a potent transcriptional activator like GAL4-VP16. Accordingly, an expressed activator binds to the binding sites of GAL4, which then drives the expressions of a reporter or therapeutic gene. Iyer et al. [94] demonstrated 50- and 12-fold enhancements of firefly luciferase and HSV1-tk, respectively, by prostate-specific antigen (PSA) promoter using the TSTA approach. They also developed transgenic mouse model to investigate tissue specificity and regulation of luciferase gene expression by bioluminescence imaging with TSTA system for PSA promoter [95]. To image the neuronal differentiation process, weak neuron-specific enolase promoter activity could be overcome by TSTA strategies. Hwang et al. [96] prepared a DNA construct expressing the NIS-Luc gene controlled by neuron-specific enolase (NSE) promoter,

which is specific to neuronal cells, and transfected this into neural stem cells. After these cells had been treated *in vitro* with db-cAMP, which induces neural differentiation, radioiodine was observed to accumulate and optical signals from cells increased.

The molecular imaging of interacting protein partners is an important application of reporter gene imaging and could provide a means of performing functional proteomic studies in living animals and provide a tool for the whole-body evaluations of new pharmaceuticals targeted to modulate protein–protein interactions. To monitor protein–protein interaction, the elaborate method was developed [97]. Luciferase gene expression was controlled by GAL4-VP16 transcriptional activation in which GAL4 and VP16 is expressed as fusion protein forms of MyoD and ID, respectively. In this system, luciferase expression is induced by GAL4-VP16-mediated transcription in GAL4-binding site through MyoD and ID interaction.

Luker et al. [98] confirmed the feasibility of imaging protein–protein interactions. The interaction of the tumor suppressor p53-GAL4 fusion protein with large T-antigen-VP16 hybrid protein was employed to activate the cellular expression of a mutant HSV1-tk reporter gene under control of a promoter containing a GAL4-binding site. Tumors modified genomically to express

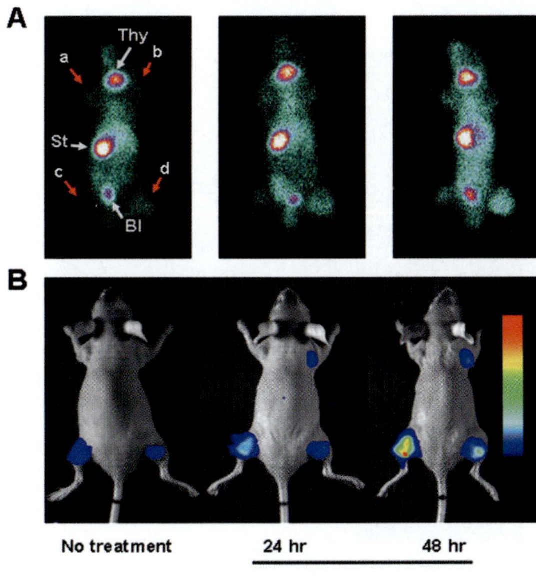

Figure 3.5. ^{99m}Tc scintigraphy (**A**) and bioluminescent (**B**) images of nude mice bearing SK-HEP1 and SK-HEP1/RARE/NIS-Luc tumors before and after all transretinoic acid (ATRA) treatment. Xenograft tumors derived from SK-HEP1 (a, 1×10^7 wild-type) and SK-HEP1/RARE/NIS-Luc cells (b, 1×10^6; c, 1×10^7; d, 1×10^8) were grown in nude mice. (**A**) Serial planar gamma camera images of the same mouse showing increased radioactivity after ATRA treatment in an SK-HEP1/RARE/NIS-Luc tumor, but not in an SK-HEP1 tumor (Thy = thyroid; St = stomach; Bl = bladder). (**B**) Serial optical images of the same mouse showing an increased optical signal after ATRA treatment in SK-HEP1/RARE/NIS-Luc tumors, but not in SK-HEP1 tumors.

two fusion proteins in the presence of doxycycline were grown subcutaneously in mice, and after treatment with doxycycline, PET imaging using F-18 FHBG detected and allowed the activation of reporter gene to be measured. Paulmurugan et al. [99] produced the splitted firefly luciferase as N-terminal and C-terminal halves that are fused into ID and MyoD, respectively (Nfluc-ID and MyoD-Cfluc). When Nfluc-ID and MyoD-Cfluc are simultaneously expressed, full luciferase activity could be recovered through MyoD and ID interaction. Therefore protein–protein interaction can be monitored using split reporter *in vivo* as well as *in vitro*.

Monitoring Tumor Mass

Noninvasive reporter gene imaging offers excellent opportunities to understand cancer progression, metastasis, and therapy in whole animals. Individual animals can be visually monitored for tumor burden at primary sites, differences in tumor progression rates can be distinguished, the possibility of metastases can be investigated, individual responses to alternative therapies can be repeatedly monitored, and therapies can be altered and the consequences of these alterations observed [100].

A large number of models of engrafted tumors have been marked with optical or nuclear medicine reporter genes and then used to monitor tumor burden and response to therapy. For example, we transfected hNIS and Fluc (firefly luciferase) genes under constitutive cytomegalovirus (CMV) promoter into human hepatoma cells. Using animal models, the effects of anticancer therapeutic regimens was easily monitored using a gamma camera system and radioiodine or Tc-99m. We found an excellent correlation ($R^2 = 0.99$) between accumulated radioiodine activity in cells and number of viable cancer cells [71]. In this study, reporter imaging using the NIS gene reflected viable cancer cell numbers and allowed changes in cell numbers to be detected after anticancer treatment. Furthermore, *in vivo* scintigraphic imaging using [^{99m}Tc] showed a moderate correlation ($R^2 = 0.87$) between tumor weight and radioactivity (Figure 3.6). However, it is likely that radioactivity reflects viable cancer cell numbers more accurately than tumor weight because tumor tissues contain immune cells, necrotic, and fibrous tissues as well as cancer cells. In addition, our group developed a lentiviral vector system carrying sodium iodide symporter (hNIS) gene under UbC promoter to establish stable and long-term gene expression *in vitro* and *in vivo* and subsequently demonstrated a marked therapeutic effect by radioiodine gene therapy using [^{131}I] in a tumor xenograft model [101].

Cell Therapy

Molecular imaging using imaging reporter genes can also be applied to the monitoring of *in vivo* distributions of target cells or therapy cells, such as immune and stem cells. The imaging of targeted T-cell trafficking using optical luciferase bioluminescence imaging has been demonstrated in several models of autoimmunity, including collagen-induced arthritis [102] and experimental autoimmune encephalomyelitis [103].

It would be of great value if the antigen-dependent activation states of T cells could be monitored noninvasively clinically and in animal models. Ponomarev et al. [104] transfected HSV1-tk/GFP fusion protein under the control of an NFAT-responsive promoter into a human T-cell line. In this study, cells were selected by flow cytometry for the NFAT induction of GFP expression in response to T-cell activation. Expression from NFAT promoter was measured by PET using I-124 FIAU as a noninvasive measure of T-cell receptor-dependent gene expression.

The transplantation of cells, such as stem cells or progenitor cells, into damaged tissues has tremendous potential for the treatment of a number of disorders. After stem cells have been administered systemically or locally, they may be able to migrate and repopulate pathologic sites. Reporter gene imaging techniques have

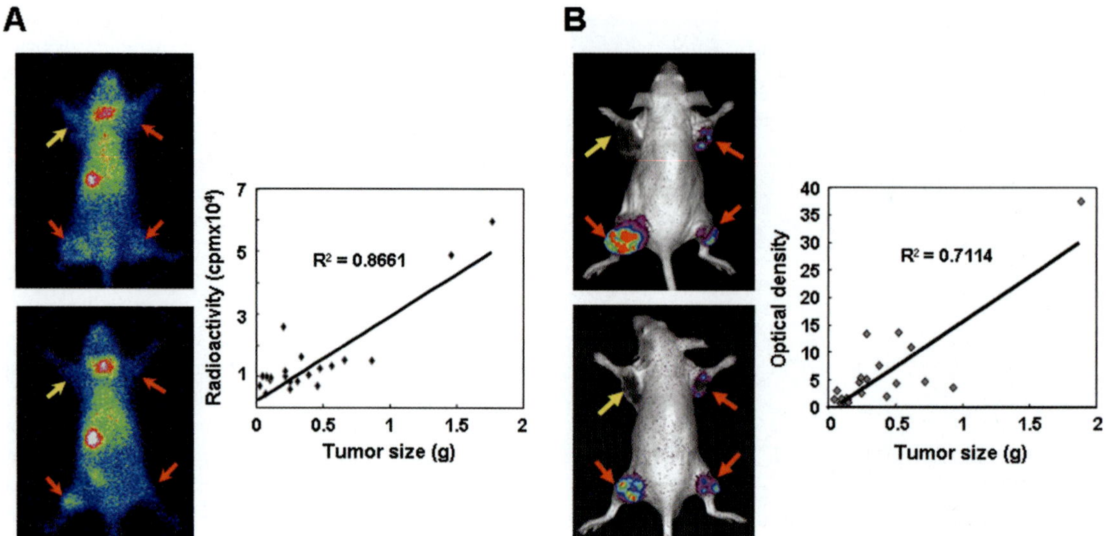

Figure 3.6. Correlation between imaging signals from tumors and tumor weight. SK-HEP1 and SK-HEP1/NIS-Luc cells simultaneously expressing the NIS and luciferase genes were xenografted into left shoulders (SK-HEP1; 5×10^6 cells), right shoulders (SK-HEP1/NIS-Luc; 5×10^6 cells), left thighs (SK-HEP1/NIS-Luc; 1×10^7 cells), and right thighs (SK-HEP1/NIS-Luc; 5×10^6 cells). (**A**) ^{99m}Tc scintigram showing increased radioactivity in SK-HEP1/NIS-Luc tumors, but not in SK-HEP1 tumors. After acquiring scintigraphic images, mice were sacrificed and radioactivity biodistributions were examined. The correlation found between tumor weight and radioactivity is shown ($r^2 = 0.8661$). (**B**) Bioluminescent image showing the viable tumor burden of SK-HEP1/NIS-Luc tumors, but not that of SK-HEP1 tumors. A moderate correlation was found between bioluminescence signals and tumor weights ($r^2 = 0.7114$).

been used to monitor cells and can provide information on three important aspects of cellular implants, that is, cell tracking, cell viability, and cell numbers [16]. After the stable transfection of cells containing an imaging reporter gene, a suitable imaging probe can visualize the distribution of stem cells and permit the longitudinal monitoring of their survival and distribution.

The locations, numbers, and survival duration of embryonic cardiomyoblasts have been monitored non-invasively using HSV1-tk and luciferase gene as imaging reporters [105]. We transfected the NIS and luciferase genes into F-3 human neural stem cells and increased imaging signals from cells, which suggests the possibility of visualizing stem cell migration [106]. Another important issue in stem cell therapy concerns the successful differentiation of stem cells and mature functional cells, which can also be visualized using reporter genes [96].

Transgenic Animals

Transgenic mice expressing a specific reporter gene can be produced by injecting a gene fragment of interest into fertilized eggs and selecting positive founders by southern blotting or genomic polymerase chain reaction (PCR) to ensure that the injected gene is incorporated into the genome.

The major advantage of reporter animals is that they provide measurable endpoints for evaluating drug efficacy in all tissues, which would radically change preclinical experimentation strategies, provide us an opportunity to better understand physiological and pathological mechanisms, and allow the effects of therapies in healthy animals or models of specific diseases to be evaluated. In the future, efforts will be aimed at devising animal testing modalities that allow pharmacodynamic, pharmacokinetic, and possibly toxicologic aspects to be investigated simultaneously, which would lead to more efficient and rational candidate drug screening protocols and considerable cost and time reductions [107].

The ERE-Luc transgenic mouse, which ubiquitously expresses the luciferase reporter gene driven by an estrogen-responsive element, is one example of a reporter animal that can be used for drug development [108]. This mouse model was specifically developed to study ligands that activate via nuclear estrogen receptor binding, and can report on the generalized state of estrogen receptor activity and be used in studies on the dynamics of estrogen receptor activity during physiological fluctuations of estrogen. It has been demonstrated that this model is sensitive enough to detect responses to low and physiological concentrations of various ligands [109].

To obtain repetitively PET images of endogenous albumin gene expression, transgenic mice in which HSV1-tk reporter gene expression is regulated by albumin promoter were generated [110]. Transgenic mice showed a relatively high $[^{18}F]$FHBG accumulation in liver region, and this result supports that the endogenous gene expression of albumin can be investigated by PET imaging with [18F]FHBG.

Transgenic mice can also be used for stem cell therapy studies. Cao et al. [111] produced transgenic mice expressing firefly luciferase controlled by β-actin promoter and monitored engrafted hematopoietic stem cells in irradiated recipient mice with real-time bioluminescence imaging.

Furthermore, the differentiation of stem cells to mature functional cells in target tissues is critical for successful therapy. We developed a transgenic mouse model to image cardiomyocyte differentiation using alpha-myosin heavy chain (MHC) promoter as a differentiation marker and NIS as a reporter gene [112]. Alpha-MHC promoter is a differentiated cardiomyocyte-specific regulator of gene expression. Nuclear medicine images using [131I] and [124I] in transgenic mice showed rapid and intense radioiodine uptake only in myocardium, indicating functional activation of alpha-MHC (Figure 3.7). This transgenic mouse model should be useful for myocardial stem cell differentiation studies. By injecting bone marrow-derived stem cells obtained from these transgenic animals into other animals with myocardial infarct, radioiodines or [99mTc] scintigraphy and PET could be used to confirm the differentiation of stem cells into mature cardiomyocytes.

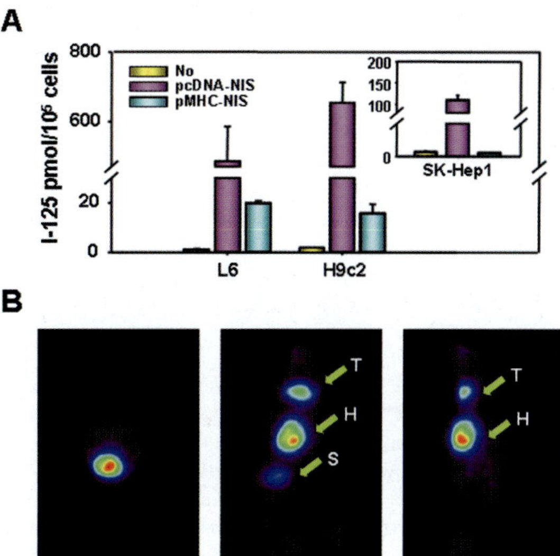

Figure 3.7. Development of NIS-transgenic mice for cardiomyocyte-specific reporter gene expression. Upper panel: To confirm the cell-type specific expression of pMHC-NIS vector, pMHC-NIS was transfected into two rat myoblast cell lines, L6 and H9c2, and into a human hepatoma cell line, SK-Hep1. Transfected L6 and H9c2 cells showed 20- and 10-fold higher 125I uptakes than their parental cells without change in hepatoma cells (Inset), thus confirming that MHC promoter was active only in cardiomyocytes. Lower panel: MicroPET image of a transgenic mouse using I-124 (left; Transverse, center; Coronal, right; Sagittal section). Transgenic mice showed higher 124I uptake in heart myocardium (H) than in the thyroid (T) or stomach (S).

CONCLUSION

Molecular imaging embraces proteomic, metabolic, cellular biologic processes, and genetic imaging. Imaging reporter genes are set to play a leading role in molecular–genetic imaging, and of the various imaging modalities, nuclear medicine-based methods have the advantage of high sensitivity and of allowing data obtained from small-animal imaging studies to be applied to patients. Several imaging reporter genes and radioactive probes have been developed, and successful transitions from bench to bedside have already occurred.

By accurately characterizing tumor properties or biological processes, molecular–genomic imaging using imaging reporter genes may play a unique role in patient management, that is, in diagnosis, staging, determination of therapeutic targets, monitoring therapy, and evaluation of prognoses. In addition, they will undoubtedly increase our knowledge of the critical biological pathways involved in disease progression. In addition, molecular imaging has become an indispensable tool in basic research [9].

The remarkable efforts currently focused on molecular imaging technologies amply demonstrate its potential importance and range of applications. Continued improvements in instrumentation, the identification of novel targets and genes, and the availability of improved imaging probes indicate that molecular imaging is likely to play an increasingly important role in clinic as well as basic research. Furthermore, dual- or triple-reporter constructs for optical and nuclear medicine imaging provide opportunities for multimodality imaging (discussed in Chapter 5) and should ease the transition between laboratory studies and clinical application. In particular, the clinical applications of reporter gene imaging will expand as new reporter systems, both transgenes and probes – become "humanized" and more widely available. In parallel with these developments, we should strive to develop more efficient, safer vector systems to deliver and target transgenes to selected organs and tissues.

The collaborative efforts of physicists, biologists, molecular biologists, biochemists, chemists, bioengineers, computer scientists, and clinicians have led to substantial recent advances in instrumentation and detection devices, the generation of disease-specific reporter animal models, and the developments of target-specific probes and genetically encoded reporters. Moreover, molecular imaging is expected to play an increasingly more important role in the personalized health care and therapy of tomorrow. To achieve this vision, the multidisciplinary team approach becomes increasingly important, and effective communication and cooperation within such teams are essential requirements of success [80].

REFERENCES

1 Massoud, T. F., Gambhir, S. S. (2007). Integrating non-invasive molecular imaging into molecular medicine: an evolving paradigm. *Trends Mol Med* 13: 183–191.

2 Tjuvajev, J. G., Stockhammer, G., Desai, R. et al. (1995). Imaging the expression of transfected genes in vivo. *Cancer Res* 55: 6123–6132.

3 Gambhir, S. S., Barrio, J. R., Wu, L. et al. (1998). Imaging of adenoviral-directed herpes simplex virus type 1 thymidine kinase reporter gene expression in mice with radiolabeled ganciclovir. *J Nucl Med* 39: 2003–2011.

4 Gambhir, S. S., Barrio, J. R., Phelps, M. E. et al. (1999). Imaging adenoviral-directed reporter gene expression in living animals with positron emission tomography. *Proc Natl Acad Sci USA* 96: 2333–2338.

5 Kim, J. S., Lee, J. S., Im, K. C. et al. (2007). Performance measurement of the microPET Focus 120. *J Nucl Med* 48: 1527–1535.

6 Beekman, F., van der Have, F. (2007). The pinhole: gateway to ultra-highresolution three-dimensional radionuclide imaging. *Eur J Nucl Med Mol Imaging* 34: 151–161.

7 Townsend, D. W., Carney, J. P., Yap, J. T., Hall, N. C. (2004). PET/CT today and tomorrow. *J Nucl Med* 45(Suppl 1): 4S–14S.

8 Pichler, B. J., Judenhofer, M. S., Catana, C. et al. (2006). Performance test of an LSO-APD detector in a 7-T MRI scanner for simultaneous PET/MRI. *J Nucl Med* 47: 639–647.

9 Iyer, M., Sato, M., Johnson, M. et al. (2005). Application of molecular imaging in cancer therapy. *Curr Gene Ther* 5: 607–618.

10 MacLaren, D. C., Gambhir, S. S., Satyamurthy, N. et al. (1999). Repetitive, noninvasive imaging of the dopamine D2 receptor as a reporter gene in living animals. *Gene Ther* 5: 785–791.

11 Gambhir, S. S., Herschman, H. R., Cherry, S. R. et al. (2000). Imaging transgene expression with radionuclide imaging technologies. *Neoplasia* 2: 118–138.

12 Min, J. J., Iyer, M., Gambhir, S. S. (2003). Comparison of F-18 FHBG and C-14 FIAU for imaging of HSV1-tkk reporter gene expression: adenoviral infection vs stable transfection. *Eur J Nucl Med Mol Imaging* 30: 1547–1560.

13 Abbruzzese, J. L., Schmidt, S., Raber, M. N. et al. (1989). Phase I trial of 1-(2'-deoxy-2'-fluoro-1-beta-D-arabinofuranosyl)-5-methyluracil (FMAU) terminated by severe neurologic toxicity. *Invest New Drugs* 7: 195–201.

14 Cui, L., Yoon, S., Schinazi, R. F., Sommadossi, J. P. (1995). Cellular and molecular events leading to mitochondrial toxicity of 1-(2-deoxy-2-fluoro-1-beta-D-arabinofuranosyl)-5-iodouracil in human liver cells. *J Clin Invest* 95: 555–563.

15 Wang, J., Eriksson, S. (1996). Phosphorylation of the anti-hepatitis B nucleoside analog 1-(2'-deoxy-2'-fluoro-1-beta-D-arabinofuranosyl)-5-iodouracil (FIAU) by human cytosolic and mitochondrial thymidine kinase and implications for cytotoxicity. *Antimicrob Agents Chemother* 40: 1555–1557.

16 Acton, P. D., Zhou, R. (2005). Imaging reporter genes for cell tracking with PET and SPECT. *Q J Nucl Mol Imaging* 49: 349–360.

17 Tjuvajev, J. G., Doubrovin, M., Akhurst, T. et al. (2002). Comparison of radiolabeled nucleoside probes (FIAU, FHBG, and FHPG) for PET imaging of HSV1-tk gene expression. *J Nucl Med* 43: 1072–1083.

18 Kang, K. W., Min, J. J., Chen, X., Gambhir, S. S. (2005). Comparison of [14C]FMAU, [3H]FEAU, [14C]FIAU, and [3H]PCV for monitoring reporter gene expression of wild type and mutant herpes simplex virus type 1 thymidine kinase in cell culture. *Mol Imaging Biol* 7: 296–303.

19 Buursma, A. R., Rutgers, V., Hospers, G. A., Mulder, N. H., Vaalburg, W., de Vries, E. F. (2006). 18F-FEAU as a radiotracer for herpes simplex virus thymidine kinase gene expression: in-vitro comparison with other PET tracers. *Nucl Med Commun* 27: 25–30.

20 Black, M. E., Newcomb, T. G., Wilson, H. M. P., Lobe, L. A. (1996). Creation of drug-specific herpes simplex virus type 1 thymidine kinase mutant for gene therapy. *Proc Nat Acad Sci USA* 93: 3525–3529.

21 Gambhir, S. S., Bauer, E., Black, M. E. et al. (2000). A mutant herpes simplex virus type 1 thymidine kinase reporter gene shows improved sensitivity for imaging reporter gene expression with positron emission tomography. *Proc Natl Acad Sci USA* 97: 2785–2790.

22 Chin, F. T., Namavari, M., Levi, J., Subbarayan, M., Ray, P., Chen, X., Gambhir, S. S. (2008). Semiautomated radiosynthesis and biological evaluation of [18F]FEAU: a novel PET imaging agent for HSV1-tk/sr39tk reporter gene expression. *Mol Imaging Biol* 10: 82–91.

23 Degrève, B., Esnouf, R., De Clercq, E., Balzarini, J. (2000). Selective abolishment of pyrimidine nucleoside kinase activity of herpes simplex virus type 1 thymidine kinase by mutation of alanine-167 to tyrosine. *Mol Pharmacol* 58: 1326–1332.

24 Likar, Y., Dobrenkov, K., Olszewska, M. et al. (2008). A new acycloguanosine-specific supermutant of herpes simplex virus type 1 thymidine kinase suitable for PET imaging and suicide gene therapy for potential use in patients treated with pyrimidine-based cytotoxic drugs. *J Nucl Med* 49: 713–720.

25 Yaghoubi, S., Couto, M. A., Chen, C. C. et al. (2006). Preclinical safety evaluation of F-18 FHBG: a PET reporter probe for imaging herpes simplex virus type 1 thymidine kinase (HSV1-tk) or mutant HSV1-sr39tk's expression. *J Nucl Med* 47: 706–715.

26 Yaghoubi, S., Barrio, J. R., Dahlbom, M. et al. (2001). Human pharmacokinetic and dosimetry studies of [(18)F]FHBG: a reporter probe for imaging herpes simplex virus type-1 thymidine kinase reporter gene expression. *J Nucl Med* 42: 1225–1234.

27 Green, L. A., Nguyen, K., Berenji, B. et al. (2004). A tracer kinetic model for F-18 FHBG for quantitating herpes simplex virus type 1 thymidine kinase reporter gene expression in living animals using PET. *J Nucl Med* 45: 1560–1570.

28 Cabrita, M. A., Baldwin, S. A., Young, J. D., Cass, C. E. (2002). Molecular biology and regulation of nucleoside and

nucleobase transporter proteins in eukaryotes and prokaryotes. *Biochem Cell Biol* **80**: 623–638.

29 Choi, S. R., Zhuang, Z. P., Chacko, A. M. et al. (2006). SPECT imaging of herpes simplex virus type 1 thymidine kinase gene expression by I-123 FIAU. *Acad Radiol* **12**: 798–805.

30 Riddell, S., Elliott, M., Lewinshon, D. et al. (1996). T-cell mediated rejection of gene-modified HIV-specific cytotoxic T lymphocytes in HIV-infected patients. *Nat Med* **2**: 216–223.

31 Ponomarev, V., Doubrovin, M., Shavrin, A. et al. (2007). A human-derived reporter gene for noninvasive imaging in humans: Mitochondrial thymidine kinase type 2. *J Nucl Med* **48**: 819–826.

32 Wang, J., Eriksson, S. (1996). Phosphorylation of the anti-hepatitis B nucleoside analog 1-(2′-deoxy-2′-fluoro-1-beta-D-arabinofuranosyl)-5-iodouracil (FIAU) by human cytosolic and mitochondrial thymidine kinase and implications of cytotoxicity. *Antimicrob Agents Chemother* **40**: 1555–1557.

33 Piskur, J., Sandrini, M. P., Knecht, W., Munch-Petersen, B. (2004). Animal deoxyribonucleoside kinases: 'forward' and 'retrograde' evolution of their substrate specificity. *FEBS Lett* **560**: 3–6.

34 Serganova, I., Ponomarev, V., Blasberg, R. (2007). Human reporter genes: potential use in clinical studies. *Nucl Med Biol* **34**: 791–807.

35 Wu J. C., Spin, J. M., Cao, F. et al. (2006). Transcriptional profiling of reporter genes used for molecular imaging of embryonic stem cell transplantation. *Physiol Genomics* **25**: 29–38.

36 Liang, Q., Satyamurthy, N., Barrio, J. R. et al. (2001). Noninvasive, quantitative imaging in living animals of a mutant dopamine D2 receptor reporter gene in which ligand binding is uncoupled from signal transduction. *Gene Ther* **8**: 1490–1498.

37 Liang, Q., Gotts, J., Satyamurthy, N. et al. (2002). Noninvasive, repetitive, quantitative measurement of gene expression from a bicistronic message by positron emission tomography, following gene transfer with adenovirus. *Mol Ther* **6**: 73–82.

38 Rogers, B. E., Zinn, K. R., Buchsbaum, D. J. (2000). Gene transfer strategies for improving radiolabeled peptide imaging and therapy. *Q J Nucl Med* **44**: 208–223.

39 Chung, J. K. (2002). Sodium/iodide symporter; its role in nuclear medicine. *J Nucl Med* **43**: 1188–1200.

40 Dai, G., Levy, O., Carrasco, N. (1996). Cloning and characterization of the thyroid iodide transporter. *Nature* **379**: 458–460.

41 Smanik, P. A., Lui, Q., Furminger, T. L. et al. (1996). Cloning of the human sodium Iodide symporter. *Biochem Biophys Res Commun* **226**: 339–345.

42 Shin, J. H., Chung, J-K., Kang, J. H. et al. Feasibility of sodium/iodide symporter (NIS) gene as a new imaging reporter gene: comparison with HSV1-tk. *Eur J Nucl Med Mol Imaging* **31**: 425–432.

43 Levy, O., De la Vieja, A., Ginter, C. S., Riedel, C., Dai, G., Carrasco, N. (1998). N-linked glycosylation of the thyroid Na+/I- symporter (NIS). Implications for its secondary structure model. *J Biol Chem* **273**: 22657–22663.

44 Vadysirisack, D. D., Chen, E. S., Zhang, Z., Tsai, M. D., Chang, G. D., Jhiang, S. M. (2007). Identification of in vivo phosphorylation sites and their functional significance in the sodium iodide symporter. *J Biol Chem* **282**: 36820–36828.

45 van Sande, J., Massart, C., Beauwens, R. et al. (2003). Anion selectivity by the sodium iodide symporter. *Endocrinology* **144**: 247–252.

46 Lee, K. H., Bae, J. S., Lee, S. C. et al. (2006). Evidence that myocardial Na/I symporter gene imaging does not perturb cardiac function. *J Nucl Med* **47**: 1851–1857.

47 Huang, M., Batra, R. K., Kogai, T. et al. (2001). Ectopic expression of the thyroidperoxidase gene augments radioiodide uptake and retention mediated by the sodium iodide symporter in non-small cell lung cancer. *Cancer Gene Ther* **8**: 612–618.

48 Vadysirisack, D. D., Shen, D. H., Jhiang, S. M. (2006). Correlation of Na+/I− symporter expression and activity: implications of Na+/I− symporter as an imaging reporter gene. *J Nucl Med* **47**: 182–190.

49 Rogers, B. E., McLean, S. F., Kirkman, R. L. et al. (1999). In vivo localization of [111In]-DTPA-D-Phe1-octreotide to human ovarian tumor xenografts induced to express the somatostatin receptor subtype 2 using an adenoviral vector. *Clin Cancer Res* **5**: 383–393.

50 Krenning, E. P., Bakker, W. H., Breeman, W. A. et al. (1989). Localisation of endocrine-related tumours with radioiodinated analogue of somatostatin. *Lancet* **863**: 242–244.

51 Bakker, W. H., Albert, R., Bruns, C. et al. (1991). [111In-DTPA-D-Phe1]-octreotide, a potential radiopharmaceutical for imaging of somatostatin receptor-positive tumors: synthesis, radiolabeling and in vitro validation. *Life Sci* **49**: 1583–1591.

52 Van Den Bossche, B., Van de Wiele, C. (2004). Receptor imaging in oncology by means of nuclear medicine: current status. *J Clin Oncol* **22**: 3593–3607.

53 Rogers, B. E., Parry, J. J., Andrews, R., Cordopitis, P., Nock, B. A., Maina, T. (2005). MicroPET imaging of gene transfer with a somatostatin receptor-based reporter gene and Tc-94m Demotate 1. *J Nucl Med* **46**: 1889–1897.

54 Zinn, K. R., Chaudhuri, T. R. (2002). The type 2 human somatostatin receptor as a platform for reporter gene imaging. *Eur J Nucl Med Mol Imaging* **29**: 288–399.

55 Rogers, B. E., Chaudhuri, T. R., Reynolds, P. N., Della Manna, D., Zinn, K. R. (2003). Non-invasive gamma camera imaging of gene transfer using an adenoviral vector encoding an epitope-tagged receptor as a reporter. *Gene Ther* **10**: 105–114.

56 Altmann, A., Kissel, M., Zitzmann, S. et al. (2003). Increased MIBG uptake after transfer of the human norepinephrine transporter gene in rat hepatoma. *J Nucl Med* **44**: 973–980.

57 Anton, M., Wagner, B., Haubner, R. et al. (2004). Use of the norepinephrine transporter as a reporter gene for noninvasive imaging of genetically modified cells. *J Gene Med* **6**: 119–126.

58 Buursma, A. R., Beerens, A. M. J., de Vries, E. F. J. et al. (2005). The human norepinephrine transporter in combination with C-11-m-hydroxyephedrine as a reporter gene/probe for PET of gene therapy. *J Nucl Med* **46**: 2068–2075.

59 Doubrovin, M. M., Doubrovin, E. S., Zanzonico, P., Sadelain, M., Larson, S. M., O'Reilly, R. J. (2007). In vivo imaging and quantitation of adoptively transferred human antigen-specific T cells transduced to express a human norepinephrine transporter gene. *Cancer Res* 67: 11959–11969.

60 Moroz, M. A., Serganova, I., Zanzonico, P. et al. (2007). Imaging hNET reporter gene expression with I-124 MIBG. *J Nucl Med* 48: 827–836.

61 Furukawa, T., Lohith, T. G., Takamatsu, S., Mori, T., Tanaka, T., Fujibayashi, Y. (2006). Potential of the FES-hERL PET reporter gene system – basic evaluation for gene therapy monitoring. *Nucl Med Biol* 33: 145–151.

62 Mintun, M. A., Welch, M. J., Siegel, B. A. et al. (1988). Breast cancer: PET imaging of estrogen receptors. *Radiology* 169: 45–48.

63 Huber, B. E., Richards, C. A., Krenitsky, T. A. (1991). Retroviral-mediated gene therapy for the treatment of hepatocellular carcinoma: an innovative approach for cancer therapy. *Proc Natl Acad Sci U S A* 88: 8039–8043.

64 Degreve, B., Andrei, G., Izquierdo, M. et al. (1997). Varicella-zoster virus thymidine kinase gene and anti-herpetic pyrimidine nucleoside analogues in a combined gene/chemotherapy treatment for cancer. *Gene Ther* 4:1107–1114.

65 McGuigan, C., Barucki, H., Carangio, A. et al. (2000). Highly potent and selective inhibition of varicella-zoster virus by bicyclic furopyrimidine nucleosides bearing an aryl side chain. J Med Chem 43: 4993–4997.

66 Chitneni, S. K., Deroose, C. M., Fonge, H. et al. (2007). Synthesis and biological evaluation of an I-123 labeled bicyclic nucleoside analogue (BCNA) as potential SPECT tracer for VZV-tk reporter gene imaging. *Bioorg Med Chem Lett* 17: 3458–3462.

67 Chitneni, S. K., Deroose, C. M., Balzarini, D. J. et al. (2007). Synthesis and preliminary evaluation of F-18 or C-11 labeled bicyclic nucleoside analogue (BCNA) for imaging Varicella-Zoster virus thymidine kinase gene expression using positron emission tomography. *J Med Chem* 50: 1041–1049.

68 Kang, J. H., Chung, J. K. (2008). Molecular-genetic imaging based on reporter gene expression. *J Nucl Med* 49: 164s–179s.

69 Gross, S., Piwnica-Worms, D. (2006). Molecular imaging strategies for drug discovery and development. *Curr Opin Chem Biol* 10: 334–342.

70 Gilad, A. A., Winnard, P. T. Jr., Van Zijl, P. C., Bulte, J. W. (2007). Developing MR reporter genes: promises and pitfalls. *NMR Biomed* 20: 275–290.

71 Shin, J. H., Chung, J. K., Kang, J. H. et al. (2004). Noninvasive imaging for monitoring of viable cancer cells using dual-imaging reporter gene. *J Nucl Med* 45: 2109–2115.

72 Wang, Y., Iyer, M., Annala, A. J. et al. (2005). Noninvasive monitoring of target gene expression by imaging reporter gene expression in living animals using improved bicistronic vectors. *J Nucl Med* 46: 667–674.

73 Lily, W., Mai, J., Makoto, S. (2003). Transcriptionally targeted gene therapy to detect and treat cancer. *Trends Mol Med* 9: 421–429.

74 Ponomarev, V., Doubrovin, M., Serganova, I. et al. (2003). Cytoplasmically retargeted HSV1-tk/GFP reporter gene mutants for optimization of noninvasive molecular-genetic imaging. *Neoplasia* 5: 245–254.

75 Wang, U., Yu, Y. A., Shabahang, S., Wang, G., Szalay, A. A. (2002). Renilla luciferase-Aequorea GFP (Ruc-GFP) fusion protein, a novel dual reporter for real-time imaging of gene expression in cell cultures and in live animals. *Mol Genet Genomics* 268: 160–168.

76 Sun, X., Annala, A. J., Yaghoubi, S. S. et al. (2001). Quantitative imaging of gene induction in living animals. *Gene Ther* 8: 1572–1579.

77 De, A., Lewis, X. A., Gambhir, S. S. (2003). Noninvasive imaging of lentiviral-mediated reporter gene expression in living mice. *Mol Ther* 7: 681–691.

78 Ray, P., Tsien, R., Gambhir, S. S. (2007). Construction and validation of improved triple fusion reporter gene vectors for molecular imaging of living subjects *Cancer Res* 67: 3085–3093.

79 Goertzen, A. L., Meadors, A. K., Silverman, R. W., Cherry, S. R. (2002). Simultaneous molecular and anatomical imaging of the mouse in vivo. *Phys Med Biol* 47: 4315–4328.

80 Ottobrini, L., Ciana, P., Biserni, A. et al. (2006). Molecular imaging: a new way to study cellular processes in vivo. *Mol Cell Endo* 246: 69–75.

81 Yaghoubi, S. S., Barrio, J. R., Namavari, M., Satyamurthy, N., Phelps, M. E., Herschman, H. R., Gambhir, S. S. Imaging progress of herpes simplex virus type 1 thymidine kinase suicide gene therapy in living subjects with positron emission tomography. *Cancer Gene Ther* 12: 329–339.

82 Bushnell, D., Menda, Y., O'Dorisio, T. et al. (2004). Effects of intravenous amino acid administration with Y-90 DOTA-Phe1-Tyr3-Octreotide (SMT487[OctreoTher) treatment. *Cancer Biother Radiopharm* 19: 35–41.

83 Dadachova, E., Carrasco, N. (2004). The Na/I symporter (NIS): imaging and therapeutic applications. *Semin Nucl Med* 34(1): 23–31.

84 Kang, J. H., Chung, J. K., Lee, Y. J. et al. (2004). Establishment of a human hepatocellular carcinoma cell line highly expressing sodium iodide symporter for radionuclide gene therapy. *J Nucl Med* 45: 1571–1576.

85 Jacobs, A., Voges, J., Reszka, R. et al. (2001). Positron-emission tomography of vector-mediated gene expression in gene therapy for gliomas. *Lancet* 9283: 727–729.

86 Penuelas, I., Mazzolini, G., Boan, J. F. et al. (2005). Positron emission tomography imaging of adenoviral-mediated transgene expression in liver cancer patients. *Gastroenterology* 128: 1787–1795.

87 Adams, J. Y., Johnson, M., Sata, M. et al. (2002). Visualization of advanced human prostate cancer lesions in living mice by a targeted gene transfer vector and optical imaging. *Nat Med* 8: 891–897.

88 Jin, Y. N., Chung, H. K., Kang, J. H. et al. (2007). Hepatoma-targeted human sodium iodide symporter gene expression using alpha fetoprotein enhancer/promoter [abstract]. *J Nucl Med* 48(Suppl): 328.

89 Doubrovin, M., Ponomarev, V., Beresten, T. et al. (2001). Imaging transcriptional regulation of p53-dependent genes with positron emission tomography in vivo. *Proc Natl Acad Sci USA* 98: 9300–9305.

90 Kim, K. I., Chung, J. K., Kang, J. H. et al. (2005). Visualization of endogenous p53-mediated transcription in vivo

using sodium iodide symporter. *Clin Cancer Res* 11: 123–128.

91 Kang, Y., He, W., Tulley, S. et al. (2005). Breast cancer bone metastasis mediated by the Smad tumor suppressor pathway. *Proc Natl Acad Sci USA* 102: 13909–13914.

92 So, M. K., Kang, J. H., Chung, J. K. et al. (2004). In vivo imaging of retinoic acid receptor activity using a sodium/iodide symporter and luciferase dual imaging reporter gene. *Mol Imaging* 3: 163–171.

93 Kang, J. H., Chung, J. K., Lee, Y. J. et al. (2006). Evaluation of transcriptional activity of the oestrogen receptor with sodium iodide symporter as an imaging reporter gene. *Nucl Med Commun* 27: 773–777.

94 Iyer, M., Wu, L., Carey, M., Wang, Y., Smallwood, A., Gambhir, S. S. (2001). Two-step transcriptional amplification as a method for imaging reporter gene expression using weak promoters. *Proc Natl Acad Sci USA* 98: 14595–14600.

95 Iyer, M., Salazar, F. B., Lewis, X. et al. (2005). Non-invasive imaging of a transgenic mouse model using a prostate-specific two-step transcriptional amplification strategy. *Transgenic Res* 14: 47–55.

96 Hwang, do W., Kang, J. H., Jeong, J. M. et al. (2008). Noninvasive in vivo monitoring of neuronal differentiation using reporter driven by a neuronal promoter. *Eur J Nucl Med Mol Imaging* 35: 135–145.

97 Ray, P., Pimenta, H., Paulmurugan, G. et al. (2002). Noninvasive quantitative imaging of protein-protein interactions in living subjects. *Proc Natl Acad Sci U S A* 99: 3105–3110.

98 Luker, G. D., Sharma, V., Pica, C. M. et al. (2002). Non-invasive imaging of protein-protein interactions in living animals. *Proc Natl Acad Sci USA* 99: 6961–6966.

99 Paulmurugan, R., Umezawa, Y., Gambhir, S. S. (2002). Non-invasive imaging of protein-protein interactions in living subjects by using reporter protein complementation and reconstitution strategies. *Proc Natl Acad Sci USA* 99: 15608–15613.

100 Herschman, H. R. (2004). Noninvasive imaging of reporter gene expression in living subjects. *Adv Cancer Res* 92: 29–80.

101 Kim, H. J., Jeon, Y. H., Kang, J. H. et al. (2007). In vivo long-term imaging and radioiodine therapy by sodium iodide symporter gene expression using a lentiviral system containing ubiquitin C promoter. *Cancer Biol Ther* 6: 1130–1135.

102 Nakajima, A., Seroogy, C. M., Sandora, M. R. et al. (2001). Antigen-specific T cell-mediated gene therapy in collagen-induced arthritis. *J Clin Invest* 107: 1293–1301.

103 Costa, G. L., Sandora, M. R., Nakajima, A. et al. (2001). Adoptive immunotherapy of experimental autoimmune encephalomyelitis via T cell delivery of the IL-12 p40 sub-unit. *J Immunol* 167: 2379–2387.

104 Ponomarev, R., Doubrovin, M., Lyddane, C., Beresten, T., Balatoni, J. G. (2001). Imaging TCR-dependent NFAT-mediated T-cell activation with positron emission tomography in vivo. *Neoplasia* 3: 480–488.

105 Wu, J. C., Chen, I. Y., Sundaresan, G. et al. (2003). Molecular imaging of cardiac cell transplantation in living animals using optical bioluminescence and positron emission tomography. *Circulation* 108: 1302–1305.

106 Shin, J. H., Chung, J. K., Roh, J. K. et al. (2002). Monitoring of neural stem cell using sodium/iodide symporter gene [abstract]. *J Nucl Med* 43: 238.

107 Adriana, M., Luisa, O., Andrea, B. et al. (2005). Techniques: reporter mice-a new way to look at drug action. *Trends Pharm Sciences* 25: 337–342.

108 Ciana, P., Luccio, G., Belcredito, S. et al. (2001). Engineering of a mouse for the in vivo profiling of estrogen receptor activity. *Mol Endocrinol* 15: 1104–1113.

109 Ciana, P., Raviscioni, M., Mussi, P. et al. (2003). In vivo imaging of transcriptionally active estrogen receptors. *Nat Med* 9: 82–86.

110 Green, L. A., Yap, C. S., Nguyen, K. et al. (2002). Indirect monitoring of endogenous gene expression by positron emission tomography (PET) imaging of reporter gene expression in transgenic mice. *Mol Imaging Biol* 4: 71–81.

111 Cao, Y. A., Wagers, A. J., Beilhack, A. et al. (2004). Shifting foci of hematopoiesis during reconstitution from single stem cells. *Proc Natl Acad Sci USA* 101: 221–226.

112 Kang, J. H., Lee, D. S., Paeng, J. C. et al. (2005). Development of a sodium/iodide symporter (NIS)-transgenic mouse for imaging of cardiomyocyte-specific reporter gene expression. *J Nucl Med* 46: 479–483.

Noninvasive Imaging of Gene Expression with Magnetic Resonance Imaging and Magnetic Resonance Spectroscopy

Mark D. Pagel and James P. Basilion

INTRODUCTION

Magnetic resonance imaging (MRI) has developed from an intriguing research project initially conceived in 1973 to an essential diagnostic method in the armamentarium of clinical radiologists [1]. An estimated 26.6 million MRI procedures were performed in 2006 in the United States that generated approximately $20 billion in service revenue [2]. The demand for clinical MRI diagnoses is expected to increase by 30% by 2020. This projected growth is due in part to the rising prevalence of age-related pathologies of soft tissues that can be conveniently monitored with MRI, such as the anatomy of pathologies in the cardiopulmonary system (e.g., regions of myocardial infarcts), neurological system (e.g., regions of cerebral infarcts, morphological changes during multiple sclerosis), and musculoskeletal system (e.g., tears in ligaments, tendons, and cartilage). MRI offers advantages relative to optical imaging methods limited to making diagnoses only near tissue surfaces, and relative to PET, SPECT, CT, and X-ray imaging methods that use potentially harmful ionizing radiation. Unlike these other imaging modalities, MRI also provides excellent spatial resolution at or smaller than 1 mm^3 for clinical diagnostics and approaching 0.1 mm^3 for small-animal research studies. MRI can also assess physiological function, such as the function of the cardiopulmonary system (e.g., MR angiography of vasculature), neurological system (e.g., fMRI of brain activity), renal system (e.g., perfusion imaging of kidney function), musculoskeletal system (e.g., MR elastography of connective tissues), and cancer lesions (e.g., dynamic contrast enhancement MRI of angiogenic tumors). MRI methods that diagnose and monitor anatomy and function can be translated from biochemistry studies, *in vitro* cell culture studies, and small-animal *in vivo* studies to clinical studies, which all greatly facilitate the development of MRI for human health diagnoses.

Although MRI has many advantages for assessing the anatomy and function of soft tissues, it suffers from poor detection sensitivity relative to most other imaging modalities. Clinical MRI is primarily limited to detecting the ^{1}H atoms in water and fat molecules, which are the most abundant molecules in most tissues. Water and fat molecules are typically present at concentrations greater than 100 mM, which is much greater than the 10–10,000 μM detection threshold of MRI and MRS/I [3, 4]. Yet some pathological (e.g., diseased) tissues and nonpathological tissues may possess similar compositions of water or fat, which may lead to misdiagnoses based on MR images. The poor specificity of MRI for distinguishing pathological and nonpathological tissues is being addressed by detecting the influence of other molecules on water and subsequent changes to water MRI signals. These new MRI methods attempt to indirectly detect molecular compositions that serve as early biomarkers of pathologies and therefore comprise a major role in the paradigm of molecular imaging [5, 6].

Magnetic resonance spectroscopy (MRS) has also evolved from an analytical chemistry tool initially demonstrated in 1946 to an established method for clinical radiology [7, 8]. MRS has many of the same advantages as MRI and has the additional advantages of directly detecting types of molecules other than water and fat and detecting atomic nuclei such as ^{31}P, ^{13}C, or ^{19}F. These advantages have been exploited to detect and monitor metabolism in many tissues such as muscle, brain, cartilage, and cancer tumors. The MRS capability for direct detection of molecules places it directly within the realm of molecular imaging, which may provide excellent specificity for diagnosing pathological tissues at an early stage. Unfortunately, the detection sensitivity of MRS is even worse than MRI. To compensate for this weak sensitivity, the MR signals of a relatively large number of molecules must be detected. In practice, the average MR signal from the entire sample is used for

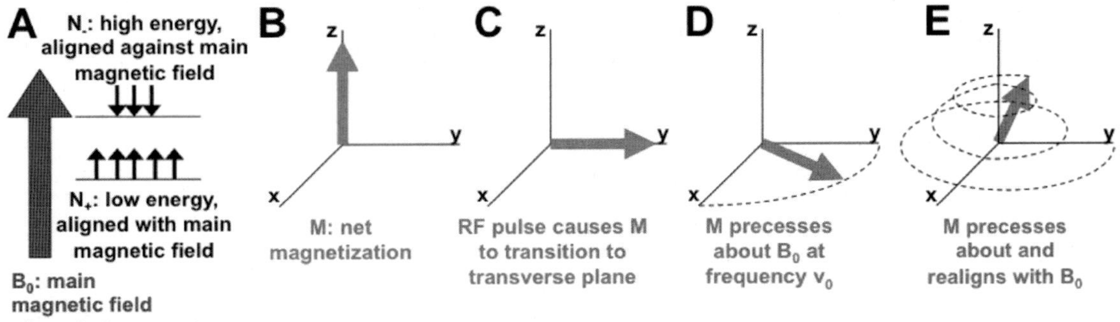

Figure 4.1. Evolution of sample magnetization in an MR magnetic field. (A) A strong magnetic field, B_0, causes the magnetization of individual atoms to align with B_0 in a low-energy state N_+, or to align against B_0 in a high-energy state, N_-. **(B)** The summation of the magnetizations of individual atoms produces a net magnetization M aligned with B_0 at equilibrium. **(C)** A radio frequency (RF) pulse of energy perturbs the equilibrium and causes M to transition partway from the low-energy state to the high-energy state, and can "tip" to align in the transverse plane. **(D)** When the RF pulse is terminated, M precesses about B_0, rotating in the transverse plane. **(E)** As the system relaxes to equilibrium, M continues to precess about B_0 as it realigns with B_0, tracing a helical path.

MRS to detect MR signals from all molecules in the sample. As a compromise between detection sensitivity and spatial resolution, MR spectroscopic imaging (MRSI) is accomplished with a coarse spatial resolution of several mm^3 that contains enough molecules for sufficient detection of MR signals. This poor sensitivity and coarse spatial resolution has limited the utility of MR spectroscopy or spectroscopic imaging (MRS/I) to clinical applications that lack any other adequate analysis technique. Yet MRS/I continues to be vigorously explored as a preclinical and clinical research method, which may expand the utility of this method for more clinical diagnoses.

The poor sensitivity of MRI and MRS/I is a major impediment to the noninvasive imaging of genes and gene expression. Only one or two copies of a gene or specific DNA sequence is typically present in each cell, and expression of a gene usually results in less than 100–10,000 copies of a specific RNA sequence and 10^2–10^8 copies of a specific protein within the cell or on the cell surface. Depending on cellular size and density, nucleic acids are typically present at concentrations of 1–1000 pM, and intracellular or cell-surface proteins are typically present at concentrations of 1 to 1000 nM. These concentrations are at least two orders of magnitude lower than the 10–10,000 μM detection threshold of MRI and MRS/I [3, 4]. Therefore, signal amplification strategies are critical for detecting genes and gene expression with MRI and MRS/I.

An interdisciplinary approach is needed to generate these signal amplification strategies. In particular, expertise in radiology and medical physics, chemistry and biochemistry, and cell biology and physiology must be combined to image genes and gene expression with MRI and MRS/I. This chapter is arranged to address each of these disciplines. Section 2 reviews the concepts of MRI and MRS to provide a perspective of radiology and medical physics. Section 3 describes the physiochemistry of exogenous contrast agents. Section 4 presents an overview of the interactions between exogenous contrast agents and endogenous tissues and biochemicals. Sections 5 and 6 present a comprehensive survey of MRI/S applications for imaging genes and gene expression, which demonstrate how the material in Sections 2 through 4 can be combined for these applications. Future directions are highlighted in Section 7.

PHYSICS OF MAGNETIC RESONANCE IMAGING AND SPECTROSCOPY

Concepts of MR Spectroscopy

All atomic nuclei consist of a combination of protons and neutrons. Nuclei with an odd number of protons, such as ^1H, ^{19}F, ^{13}C, and ^{31}P, have an uneven spatial distribution of protons within the nucleus, which creates a net ionic charge distribution. In addition, each nucleus has a quantum mechanical property of nuclear spin. As with all spinning charged particles, these charged atomic nuclei act as a spinning magnet [9, 10]. When placed in a magnetic field of strength and direction B_0, the nuclear spinning magnet will align with B_0 in a low-energy state or will oppose B_0 in a high-energy state. The summation of spins in the sample yields a net macroscopic magnetization vector, M, aligned with B_0 (Figures 4.1a and 4.1b).

M can transition from the low-energy state to a high-energy state by absorbing a photon with the same spinning frequency. In practice, a radio frequency (RF) photon is used to produce an oscillating magnetic field in a plane perpendicular to B_0. This oscillating frequency, v_0, depends on the gyromagnetic ratio, γ, and defines the

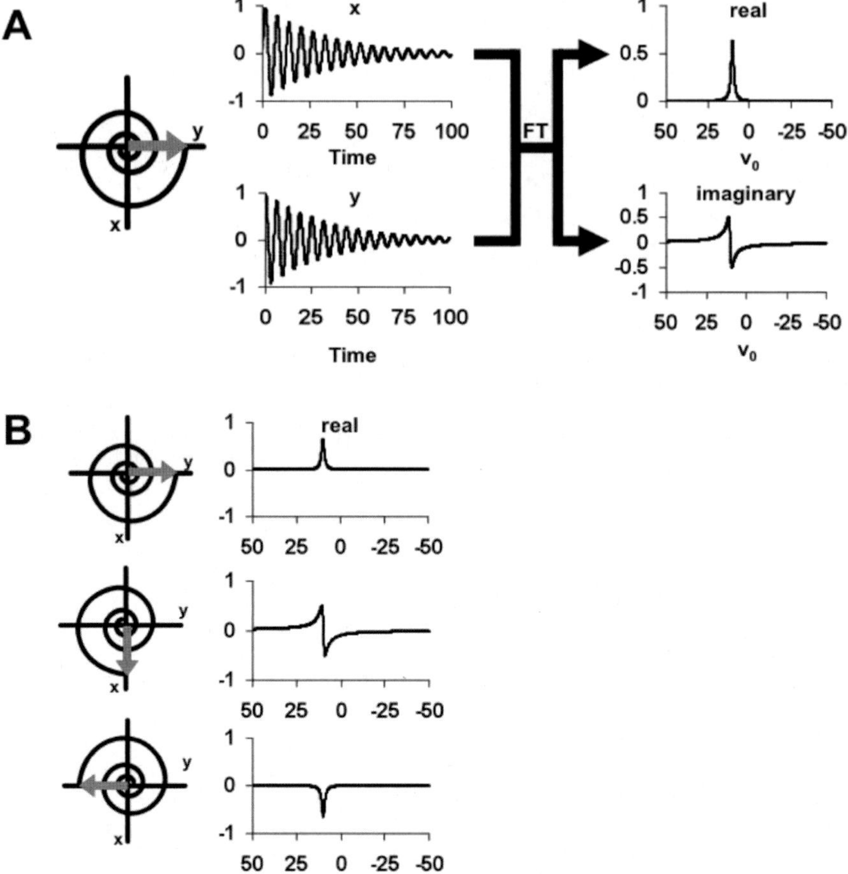

Figure 4.2. Fourier transformation. (A) After being excited from equilibrium, the net magnetization of the sample (M, gray arrow) traces a spiral path in the transverse (xy) plane (Figure 4.1e). The magnetization along the x-axis and y-axis is detected over time to form a pair of time-domain signals. This signal pair can be Fourier transformed to produce two frequency-domain spectra, which more easily shows the number of signals, frequencies of each signal, and uncertainties of these frequency measurements (width of each spectral peak). **(B)** M can be rapidly oriented in specific directions in the transverse plane (prior to tracing a spiral path). This initial orientation defines the phase of the signal, which determines the relative phase of the peak in the frequency spectrum. Only the real spectrum is shown for each example. By convention, MR frequencies from samples that are faster than the MR frequency of the spectrometer (known as the Larmor frequency) are considered to be positive in sign and are plotted on the left side of the MR spectrum.

difference in energy between the low-energy and high-energy states (ΔE) (Equation 4.1).

$$v_0 = \gamma B_0 \tag{4.1a}$$

$$^1H\gamma = 42.58 \text{ MHz/T}$$
$$^{13}C\gamma = 10.71 \text{ MHz/T}$$
$$^{19}F\gamma = 40.08 \text{ MHz/T}$$
$$^{31}P\gamma = 17.25 \text{ MHz/T}$$

$$\Delta E = h v_0, \text{ where } h \text{ is Planck's constant} \tag{4.1b}$$

An interesting quantum mechanical phenomenon is created when a brief RF pulse causes M to transition partway from the low-energy state to the high-energy state, to lie in the transverse plane perpendicular to B_0

(Figure 4.1c). This transverse M will precess around B_0 at the Larmor frequency (Figure 4.1d), which generates a radio frequency photon that can induce an electrical current in a coil of wire placed perpendicular to the axis of rotation. Without further RF stimulation, this transverse magnetization will eventually decay as M returns to equilibrium and realigns with B_0 (as described in section 4.3a; Figure 4.1e). The detected signal is termed a Free Induction Decay (FID) as a reminder that the signal is "free" to precess, causes an "induction" of electromagnetic frequency in the detector coil, and will "decay" from the transverse plane.

The FID is detected along two orthogonal axes to create a pair of oscillating sinusoidal signals that evolve over time, as usually measured in units of seconds (Figure 4.2). Because the information within the FID is difficult to visualize in this format, a Fourier Transform is performed on the pair of sinusoids to convert the signals from the time domain to an MR spectrum in the frequency domain in units of sec^{-1} or Hertz. The resulting frequency spectrum clearly shows peaks that represent the characteristics of signals that comprise the FID, including the number of signals, the relative frequency difference between signals in Hertz (Hz), and the uncertainty in measuring the exact frequency of each signal (width of the peak).

Another interesting quantum mechanical phenomenon is created by the electrons surrounding each atomic nucleus. Electrons are also spinning charged particles and therefore also act as a spinning magnet. The magnetism created by the electrons causes the nucleus to be shielded from B_0, with greater electron density causing greater magnetic shielding. The effective magnetic field experienced by each nucleus, B_{eff}, is a combination of the main magnetic field, B_0, and the magnetic shielding. The density of electrons about each nucleus is dependent on chemical bonds and chemical electrostatic interactions between atoms so that each type of atomic nucleus within a molecule possesses a different magnetic shielding from the surrounding electrons. Therefore, according to Equation 4.1, each type of nucleus in a molecule will possess a unique MR frequency v. Because v scales with B_0, and MR research investigators use a variety of magnets with different B_0 magnetic field strengths, MR frequencies for each type of nucleus in a molecule are usually normalized to B_0 and reported in units of "parts per million" or ppm (Equation 4.2). This relative MR frequency of each type of nucleus is termed the "chemical shift" as a reminder that these unique frequency shifts arise from chemical properties.

$$parts\ per\ million = (v - v_0)/v_0 \qquad (4.2a)$$

$$= (B_{eff} - B_0)/B_0 \qquad (4.2b)$$

In practice, the unique chemical bonds and electrostatic interactions within each molecule create a unique "fingerprint" of MR frequencies for each molecule. These MR frequency "fingerprints" can be used to identify types of molecules within a complex mixture, such as metabolites within biological tissues (Figure 4.3; see also Figure 4.17).

MR Sensitivity and Hyperpolarization

The population ratio of the number of atomic nuclei in the low-energy state (N_+) and high-energy state (N_-) is determined by the Boltzman distribution (Equation 4.3). The difference in energy between these two states, ΔE, is relatively small, and therefore MRI has the advantage of being a relatively safe and nondestructive imaging method. Yet ΔE is small relative to body temperature, T,

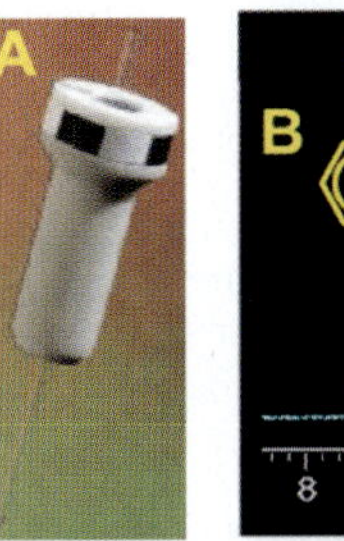
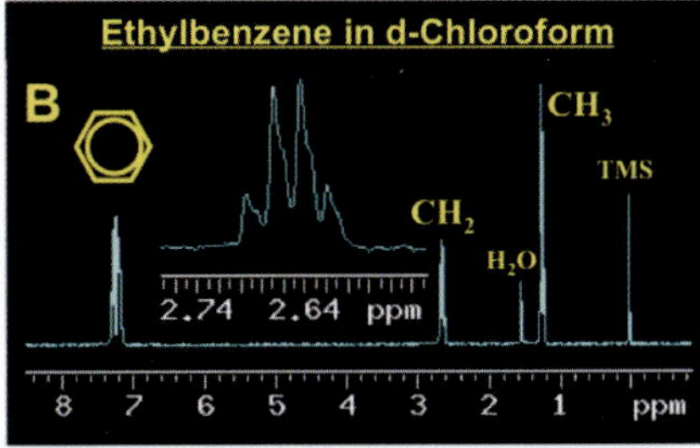

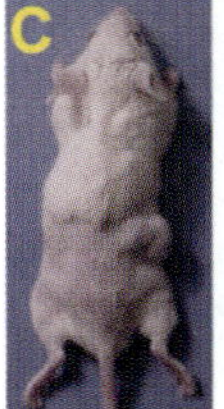
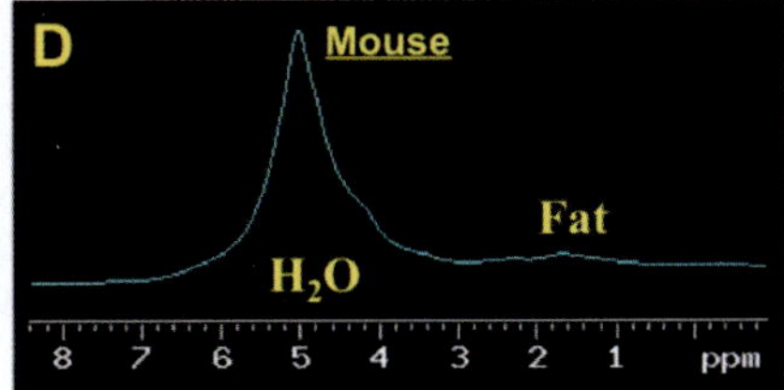

Figure 4.3. Examples of ^{1}H MR spectra. (A) Solutions of small molecules can be packaged in test tubes and a sample tube holder to create samples with excellent homogeneity. **(B)** This homogeneity leads to MR spectra with good detection sensitivity and excellent spectral resolution. **(C)** For comparison, a mouse has relatively poor homogeneity throughout the body. **(D)** This poor homogeneity leads to poor detection sensitivity and poor resolution. In practice, only the MR chemical shift frequencies of water and fat can be detected and resolved from each other in the MR spectrum of the mouse.

so that the difference in population between N_+ and N_- that creates the net magnetization M_0 is relatively small. Therefore MRI has the disadvantage of being a relatively insensitive imaging method (Table 4.1).

$$M_0 = N_+ - N_-$$

$$\frac{N_-^o}{N_+^o} = \exp\left(-\frac{\Delta E}{kT}\right) \qquad (4.3)$$

N_-^o: the thermal equilibrium populations of protons in the m = $-\frac{1}{2}$ states
N_+^o: the thermal equilibrium populations of protons in the m = $+\frac{1}{2}$ states
ΔE: the energy difference between the two states
k: the Boltzmann constant.
T: the absolute temperature in Kelvin of the surrounding lattice

Among the stable atomic nuclei, ^{1}H has the best MR sensitivity due to its high gyromagnetic ratio (Equation 4.1). Fortunately, ^{1}H is also particularly abundant in biological soft tissues with high water content, so that MRS of water in biological tissues is feasible (Figure 4.3; Table 4.1). Aliphatic components of molecules in cell membranes and adipose tissues are also abundant, so that MRS of "fat" in biological tissues is also feasible. Unfortunately, other molecules such as metabolites, proteins, and nucleic acids are much less abundant in biological tissues, so that detecting weak ^{1}H MR signals

Table 4.1. Relative sensitivity of net magnetization

Nucleus	ΔE^a	Net Spin[b] $(N_+ - N_-)/(N_+ + N_-)$	Isotope Abundance	Relative Sensitivity Net Spin* Isotope Abund.	Biological Abundance
^{1}H	1.66×10^{-24} J	0.019%	99.99%	0.019%	63.0%
^{13}C	0.42×10^{-24} J	0.0049%	1.11%	0.000054%	9.4%
^{19}F	1.56×10^{-24} J	0.018%	100.00%	0.018%	~0.0%
^{31}P	0.68×10^{-24} J	0.0079%	100.00%	0.0079%	0.24%

[a] Determined using Equation 4.1 at 9.40T B_0 magnetic field strength.
[b] Determined using Equation 4.3 at 37.0° C.

of these other molecules is arduous relative to the overwhelming ^{1}H signals from water and fat. For comparison, molecules with naturally abundant ^{31}P atoms, or exogenous molecules with ^{19}F atoms or isotopically enriched ^{13}C atoms, can be detected in biological tissues with MRS because water and fat do not contain these atoms at high concentrations. The greater number of electrons surrounding these atoms (relative to ^{1}H) causes a greater range of magnetic shielding, which expands the range of MR frequencies from these atoms and greatly facilitates their identification through their MR frequency "fingerprints." However, the lower gyromagnetic ratios of these atoms relative to ^{1}H generates less net MR signal (Equations 4.1 and 4.3), so that only the molecules most abundant with ^{31}P, ^{19}F, or ^{13}C can be detected with MRS (Table 4.1).

To address this deficiency, atomic nuclei can be hyperpolarized to create a tremendously large MR signal [11, 12]. Hyperpolarization exploits the relatively high gyromagnetic ratio of electrons, γ_e, which results in a very large difference in low-energy and high-energy magnetic states for the electron, ΔE_e. According to Equation 4.3, this large energy difference causes a higher population in the low-energy state, N_{e+}, relative to the high-energy state, N_{e-}. To prepare the electrons in a state that can transfer their hyperpolarization to atomic nuclei, the atoms are exposed to a discharge of RF and a heated alkali metal such as rubidium or other materials with an unpaired electron. The transfer of the hyperpolarization is then most often accomplished via optical laser pumping for ^{3}H and ^{129}Xe, and ^{13}C can be hyperpolarized with radio frequency pumping or microwave pumping to trigger the polarization transfer [13–14]. As much as 1 liter of ^{3}He or ^{129}Xe gas can reach 20–60% polarization levels, and as much as 20–40% of a solid containing ^{13}C can be hyperpolarized.

Care must be taken to avoid loss of hyperpolarized magnetization during each process, as lost polarization cannot be recovered without repeating the process. In particular, hyperpolarized ^{3}He and ^{129}Xe gases undergo T1 relaxation when gas molecules collide with container walls or impurities, but hyperpolarization can last for more than 10 h with proper engineering controls [15].

The greater asymmetry of electrons around a carbon nucleus causes even faster T1 relaxation, which typically limits the T1 relaxation time to 4–40 s [16]. This limited lifetime of the ^{13}C magnetization requires rapid MRI acquisition methods that avoid further destruction of the hyperpolarization. Although this temporal limitation is a major hurdle, the potential 10^5 improvement in ^{13}C MRI sensitivity drives interest in further refining this technique for biomedical applications, including applications in imaging genes and gene expression.

Research studies of hyperpolarized ^{3}H and ^{129}Xe gases have primarily focused on evaluations of lung function [17–19]. Similar research studies of hyperpolarized ^{13}C have focused on evaluations of vascular perfusion to study coronary disease [20–22]. The detection of the metabolic products that contain hyperpolarized ^{13}C has been used to assess metabolism in the heart [23, 24] and cancerous tumors [25–27]. Overall, these evaluations of metabolism or pulmonary and vascular function cannot be easily linked to specific genes and gene expression, so the use of hyperpolarized gases and fleeting hyperpolarized metabolites appears to have limited utility for directly detecting genes. More recently, hyperpolarized nuclei have also been used to detect enzymes and cell receptors [28–30].

These approaches have more promise for reporter gene imaging, as discussed in Sections 6a and 6c.

Concepts of MR Imaging

MRS analyses assume that chemical shift frequency measurements are due only to chemical effects and do not arise from differences in B_0 throughout the sample. Indeed, great efforts are taken to generate a homogeneous magnetic field throughout the sample volume. For example, superconducting magnets detached from fluctuating power supplies are often used for MRS, despite their costs ranging from $100,000 to $1 million. Magnets are often placed on antivibration platforms that further stabilize the magnetic field, and laboratories with MR magnets are often shielded from external electromagnetic fields with a metal cage of copper mesh or iron within the walls. Prior to each study, the magnetic field is electronically

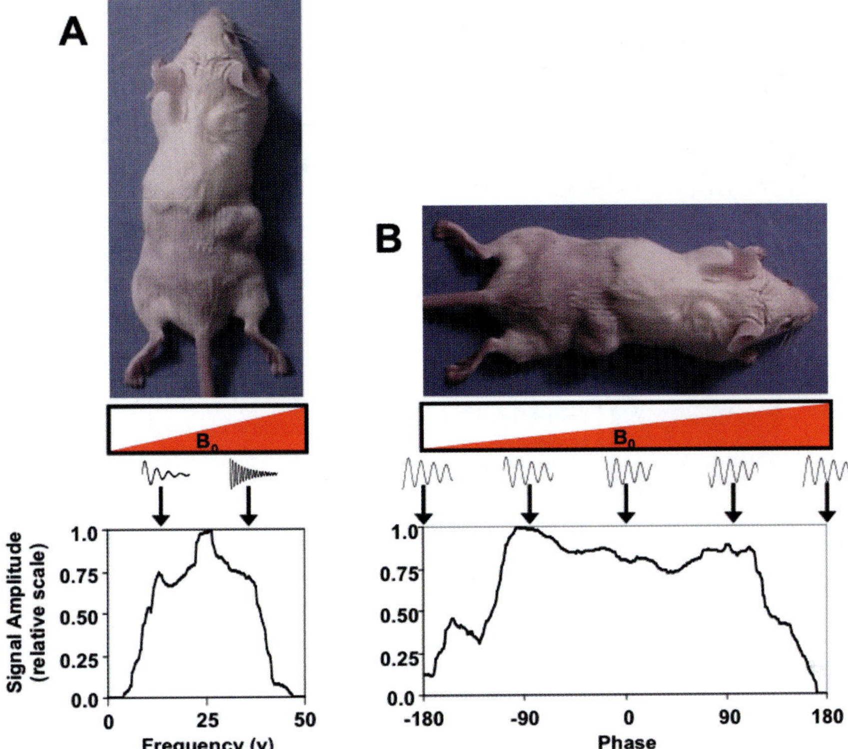

Figure 4.4. Effect of B_0 gradients on MR signals. (A) A B_0 gradient applied during the acquisition of MR frequencies causes magnetization at different locations across the sample to possess different frequencies. The signal amplitudes with each frequency can be used to construct a profile of the sample along the spatial dimension of the gradient. **(B)** A brief B_0 gradient applied before the acquisition of MR frequencies causes magnetization at different locations across the sample to possess different phases. The signal amplitudes with each frequency can be used to construct a profile of the sample along the spatial dimension of the gradient.

"shimmed" to remove small fluctuations caused by noncylindrical chemical samples or patients within the magnet.

As with many inspirational paradigm shifts in science and engineering, the woeful disadvantage of an inhomogeneous B_0 for MRS has been exploited as a tremendous advantage to create the concept of MRI [1, 31]. A magnetic field can be generated that increases in strength across the distance of the sample so that each spatial increment of the sample experiences a different value of B_0. This in turn generates MR signals with different frequencies from each increment of the sample (Equation 4.1). If B_0 is applied as a linear gradient across the sample, the Fourier transform of the MR signals yields the projection of signal amplitudes of the sample along the direction of the B_0 gradient (Figure 4.4a).

Oscillating sinusoidal MR signals also possess the property of a phase in addition to a frequency (Figure 4.2b). To measure a spatial dimension by using phase information, a linear B_0 gradient may be applied for a brief period along the dimension, which causes each increment along the sample to have MR signals with different frequencies (Figure 4.4b). This causes M_0 at each increment of the sample to briefly rotate to a unique position in the transverse plane. When the linear B_0 gradient is turned off, the phase of each oscillating sinusoidal MR signal can be used to identify the position of the signal along the spatial dimension. In practice, measuring the phases of oscillating sinusoidal MR signals can be relatively inaccurate, so that this experiment is repeated many times with B_0 gradients applied for incrementally longer periods that cause incrementally greater amounts of phase changes to the sinusoidal signals. Alternatively, the B_0 gradient can be applied with incrementally stronger powers, which has the same effect of creating incrementally greater amounts of phase changes for each sinusoidal signal. The rate of change in phase of each sinusoidal signal can then be used to accurately measure the initial phase of the signal, which can then be used to determine the projection of signal amplitudes along the direction of the B_0 gradient.

These two types of measurements can be combined to obtain a 2D MR image (Figure 4.5). First, a brief linear B_0 gradient is applied along the "phase encode" dimension to generate signals with different phases. Next, a linear B_0 gradient is applied along the "frequency encode" dimension to generate signals with different frequencies. This process is repeated many times to generate a series

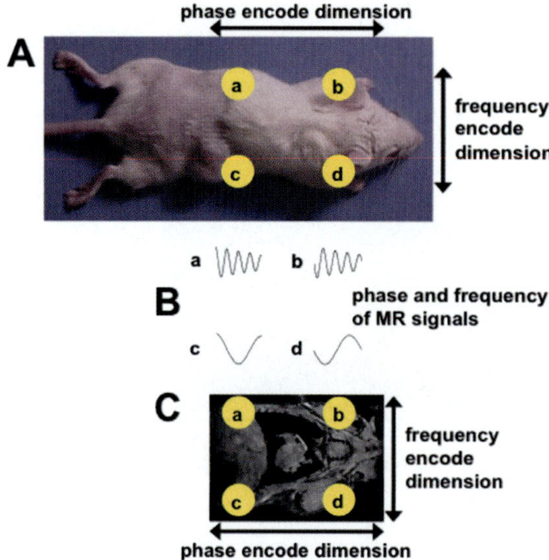

Figure 4.5. Phase- and frequency-encoding. (A) A brief linear B_0 gradient is applied along the "phase encode" dimension, followed by a linear B_0 gradient applied along the "frequency encode" dimension. **(B)** MR signals with unique phases and frequencies corresponding to the spatial location of the atoms that generate each MR signal. **(C)** A Fourier transformation of signals along each dimension produces a 2D image, where the amplitude at each spatial position within the image is a reflection of the signal amplitude with the specific frequency and phase at each image position. The resulting MR image shows a 1-mm-thick image slice through the center of the mouse's body, which contains the trachea, heart, lungs, and liver.

of signals with incrementally greater phase changes. A Fourier transformation of signals along each dimension produces a 2D image, where the "brightness" or amplitude at each position within the image is a reflection of the sinusoidal signal amplitude with the specific frequency and phase at each image position.

Phase encoding can be applied along two dimensions to create a 3D image. Similar to the process of acquiring a 2D image, the first dimension is phase-encoded, then the second dimension is also phase-encoded; finally, the third dimension is frequency-encoded. This process is repeated many times to generate a 2D matrix of signals with incrementally greater phase changes along both phase-encoded dimensions. A Fourier transformation of signals along each dimension produces a 3D image. MRSI is a special example of 3D MRI, in which the third dimension is *not* frequency-encoded with a linear B_0 gradient, and instead the frequencies are recorded in a homogeneous B_0 field so that the frequencies reflect the chemical shifts of the molecules within the sample. A Fourier transformation of signals along each dimension produces a 2D spatial image, with a 1D chemical shift spectrum at each image position. In general, 3D MRI requires long acquisition times to repeat the phase encodings along two dimensions or suffers from relatively coarse resolution along one dimension to reduce the total

acquisition time. A proper balance between spatial resolution, spectral resolution, and acquisition times greatly depends on the goals of each diagnostic application.

MAGNETIC RESONANCE CONTRAST MECHANISMS AND CONTRAST AGENTS

Although MRI is primarily limited to the detection of the atomic nuclei of ^{1}H in water (and fat), the surrounding molecular environment can affect the physicochemical properties of water molecules, which leads to changes in the MR properties of the water. The MRI signals from water can be detected using a variety of methods that exploit these MR properties to generate images with different bright/dark contrast. Exogenous contrast agents can be introduced that alter the physicochemical properties of water, which can be used to indirectly detect the contrast agents via changes in water MRI signals. A brief review of these physicochemical properties is required to relate bright/dark contrast in MR images to the molecular environment and molecular contrast agents.

T1 Contrast

T1 relaxation, also known as longitudinal or spin-lattice relaxation, represents the "relaxation" or realignment of the component of M_0 in the transverse plane with B_0 in an equilibrium state. Due to the small ΔE between MRI energy states, this relaxation is not spontaneous, and instead M_0 must be stimulated to release energy to the "lattice" of surrounding material through appropriate quantum-mechanical mechanisms. In the case of *in vivo* ^{1}H MRI, the predominant mechanisms are dipole–dipole interactions between a hydrogen nucleus and another nucleus with a magnetic dipole rotationally tumbling at the same ν (Equation 4.4), and scalar interactions between a hydrogen nucleus in contact with an unpaired electron. Therefore, the T1 relaxation time of water is a consequence of the rotational tumbling time of a nearby magnetic dipole, the accessibility of water to the dipole, or the electron spin state of the dipole (Figure 4.6).

$$\frac{1}{T1_{DD}} = \gamma^4 r^6 \left(\frac{2\tau_c}{1 + \omega^2 \tau_c^2} \right)$$

$$\times \left(\frac{\tau_c}{1 + \omega^2 \tau_c^2} + \frac{4\tau_c}{1 + 4\omega^2 \tau_c^2} \right) \tag{4.4}$$

γ = gyromagnetic ratio
r = distance between nuclei
ω = resonance frequency
τ_c = correlation time
$\quad = \dfrac{4\pi \eta a^3}{3kT}$
η = viscosity of surrounding medium
a = average molecular radius
k = Boltzmann constant
T = temperature

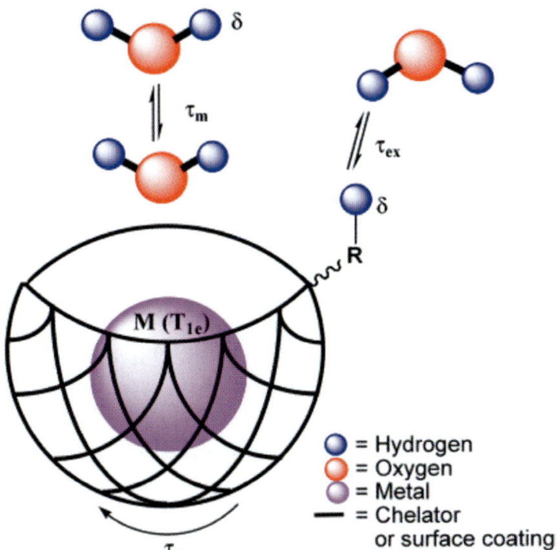

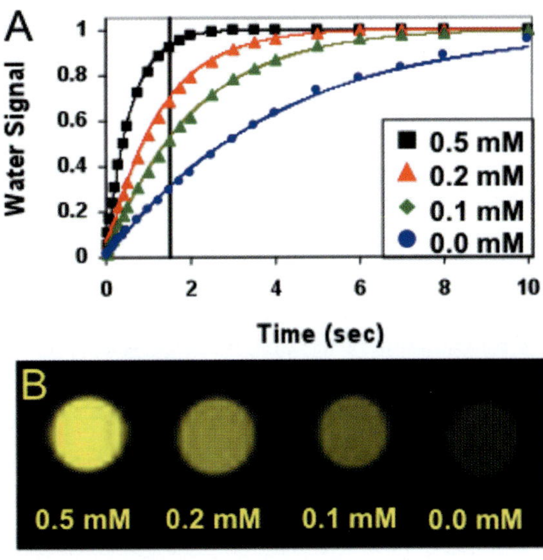

Figure 4.6. Schematic of responsive mechanisms for MR contrast agents. The metal (M) is shown as partially encapsulated in a basket-like chelator or coating to represent the chemical moieties of most MRI contrast agents, such as the clinically approved Gd-DTPA chelate and dextran-coated iron oxide. However, some contrast agents don't have a chelate or coating, such as the manganese ion. The water accessibility (τ_m), rotational tumbling time (τ_r), electron spin state (T_{1e}), chemical exchange rate (τ_{ex}), and MR frequency (δ) are shown, but RF inhomogeneity is not shown in this scheme.

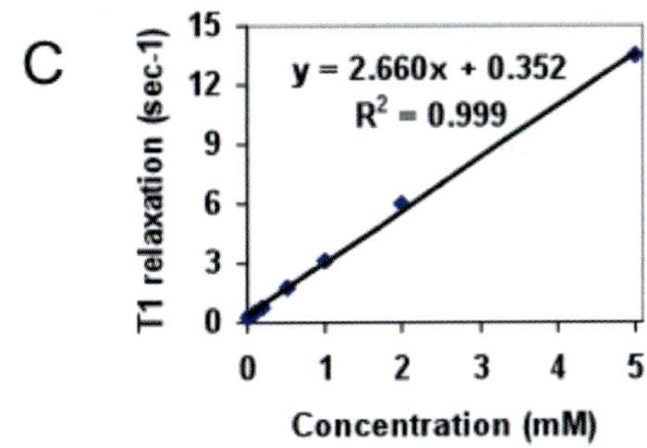

Figure 4.7. Effects of T1 relaxation. Solutions of GdDTPA were prepared at different concentrations and analyzed using a 9.4T (400 MHz) MR scanner at 37° C. (**A**) Magnetization recovers at different rates according to T1 relaxation times of different concentrations of Gd-DTPA (Eq. 4.5). (**B**) An MR image shows the effects of T1 relaxation on image contrast. The vertical line at 1.5 sec in graph A corresponds to the timing used to acquire the image. (**C**) The T1 relaxivity, or efficacy of T1 relaxation per unit of concentration, was determined to be 2.66 mM^{-1} sec^{-1} for GdDTPA at 9.4T and 37° C. The T1 relaxivity can be used to correlate the T1 relaxation effects of the agent with the concentration of the agent.

Exogenous T1 relaxation contrast agents typically contain gadolinium, which has a large spin-7/2 magnetic dipole moment and seven unpaired electrons, and therefore has excellent characteristics to decrease the T1 relaxation time of water molecules near the gadolinium metal ion (Figure 4.7). Because gadolinium is toxic, the agent consists of a polyamine with carboxylate ligands that tightly binds to the gadolinium but that still allows water molecules to directly access the gadolinium ion. The T1 relaxation can be further improved by covalently conjugating or noncovalently encapsulating the gadolinium chelate to a nanosized carrier, such as a linear polymer, dendrimer, micelle, liposome, antibody, or protein, which reduces the rotational tumbling time of the water–Gd–chelate complex. Other metals such as manganese can also act as T1 relaxation agents.

The T1 relaxation time constant indicates the speed at which M_0 realigns with B_0 (Equation 4.5). The amplitudes of the water MR signals that are detected are dependent on the ratio of the relaxation time period used in the MRI experiment, TR, and the T1 relaxation time of the sample (including the endogenous T1 relaxation time of the sample and contributions to further relaxation from T1 relaxation contrast agents). Therefore, by using an appropriate TR time in an MRI acquisition scheme, the signal amplitudes of the MR image can reflect differences in T1 relaxation times for different parts of the sample (Figure 4.7). If MR signals are acquired just one time, then signals with long T1 relaxation times should have the greatest amplitude. However, MR signals detected during subsequent acquisitions can only detect magnetization that has reached equilibrium and realigned with the main B_0 magnetic field at the end of the TR time period. Therefore, signals with short T1 relaxation times and fast realignment with B_0 will have the greatest amplitude. This counterintuitive result is evident in Equation 4.5, which indicates that large TR/T1 ratios produce the greatest relative signal amplitudes (M) from T1-weighted MRI experiments.

These T1-weighted MR images are particularly useful for identifying biological tissues with different viscosities, which cause the endogenous water molecules in each tissue to have different rotational tumbling times, such as the differences between normal brain tissues versus necrotic brain tissues. T1-weighted MRI images can also distinguish biological tissues that contain different magnetic dipoles, such as tissues that contain different salt contents, for example, normal kidneys versus impaired kidneys. The T1 relaxation rate (1/T1) caused by a contrast agent is directly proportional to the concentration of the contrast agent, which can be exploited to measure the concentration of the agent in the sample.

$$M = \text{Mo} \sin\theta \frac{1 - e^{-TR/T_1}}{1 - \cos\theta\, e^{-TR/T_1}} \qquad (4.5a)$$

where TR is the relaxation time period
to allow M to realign with B_0.
With a 90-degree excitation pulse ($\theta = 90°$),

$$M = \text{Mo}\left(1 - \exp\left(-\frac{TR}{T_1}\right)\right) \qquad (4.5b)$$

T2 Contrast

The T2 relaxation time, also known as time for transverse or spin–spin relaxation, represents the loss of excited magnetization within the detection plane that is transverse to the main static field of the MRI magnet. This loss of magnetization from the transverse detection plane occurs as excited magnetization returns to equilibrium during the T1 relaxation time, so that the T2 relaxation time can be no slower than the T1 relaxation time. Furthermore, the excited magnetization detected in the transverse plane represents a net sum of small magnetic moments from many nuclei, which must have the same MR frequency to produce a coherent net signal. Perturbations to nuclei that cause changes in MR frequency will cause a loss of coherence from among these small magnetic moments, which contributes to the loss of the net coherent magnetization from the transverse detection plane. These perturbations arise from a "spin–spin" exchange of energy between two dipolar hydrogen nuclei, which become more efficient as the rotational tumbling time of the two nuclei becomes slower (Equation 4.6). These perturbations also arise from spatial inhomogeneities in the main magnetic field that cause differences in MR frequencies from hydrogen nuclei in different tissue locations but are otherwise in homogeneous environments (the inclusion of this type of transverse relaxation is often termed T2* relaxation). Therefore, the T2* relaxation time is dependent on the rotational tumbling time of a magnetic dipole

and the local inhomogeneities in the main B_0 magnetic field (Figure 4.6).

$$\frac{1}{T2_{DD}} = \gamma^4 r^6 \left(\frac{2\tau_c}{1 + \omega^2\tau_c^2}\right)$$
$$\times \left(3\tau_c + \frac{5\tau_c}{1+\omega^2\tau_c^2} + \frac{2\tau_c}{1 + 4\omega^2\tau_c^2}\right) \qquad (4.6)$$

γ = gyromagnetic ratio
r = distance between nuclei
ω = resonance frequency
τ_c = correlation time
$\quad = \dfrac{4\pi\eta a^3}{3kT}$
η = viscosity of surrounding medium
a = average molecular radius
k = Boltzmann constant
T = temperature

Local magnetic field inhomogeneities are usually considered a woeful detriment to MRI. However, this apparent detriment has been exploited as a tremendous advantage to create T2* relaxation contrast agents, such as superparamagnetic iron oxide nanoparticles. These particles cause local magnetic field inhomogeneities that decrease the T2* relaxation time of water molecules that are near the iron oxide nanoparticle (Figure 4.8). Iron oxide is an essential metabolite for biological function, and therefore large amounts of iron oxide are nontoxic. Iron oxide nanoparticles are typically coated with a biomaterial such as dextran to further enhance biocompatibility and provide opportunities for further derivatization. The T2 relaxation is directly related to the size of the nanoparticle, although the ability to deliver higher amounts of smaller nanoparticles to biological tissues can offset the lower T2 relaxations of these smaller agents. Other metals such as dysprosium can also act as T2 relaxation agents.

The amplitudes of the water MR signals detected starting at time TE are dependent on the T2 relaxation time, including the endogenous T2 time and contributions from T2 relaxation contrast agents (Equation 4.7). Therefore, by using an appropriate TE time in an MRI acquisition scheme, the signal amplitudes are strongest for portions of samples with long T2 relaxation times (Figure 4.8). These T2-weighted MR images are particularly useful for identifying biological tissues with different magnetic field inhomogeneities, such as tissues with different amounts of air–tissue interfaces such as inhomogeneous lung tissues versus homogeneous lung tumors. As with T1 agents, the T2 relaxation rate (1/T2) caused by a contrast agent is directly proportional to the concentration of the contrast agent, which can be

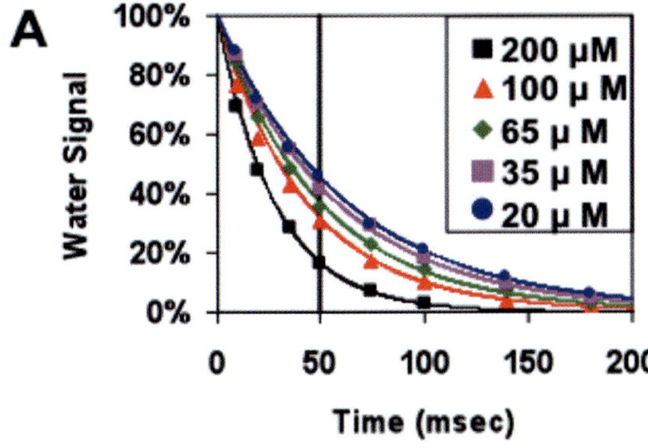

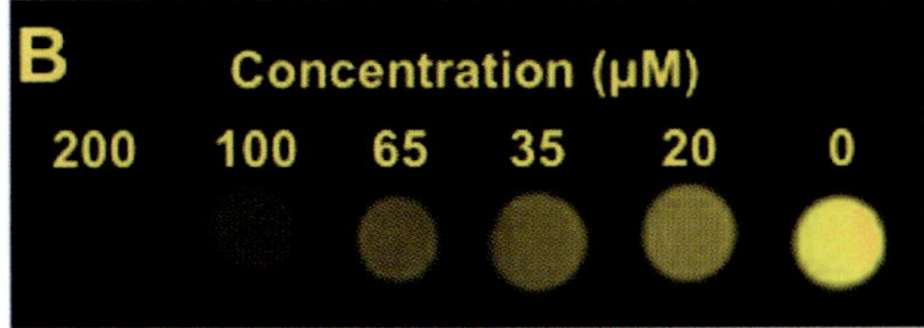

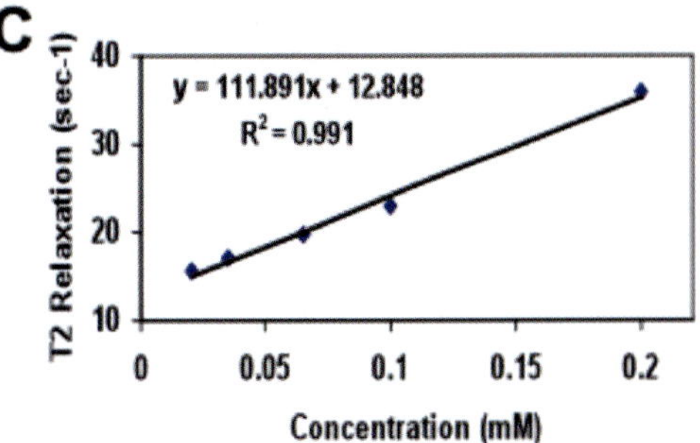

Figure 4.8. Effects of T2 relaxation. Solutions of dextran-coated iron oxide nanoparticles were prepared at different concentrations and analyzed using a 9.4T (400 MHz) MR scanner at 37° C (Eq. 4.7). (**A**) Magnetization decays at different rates according to T2 relaxation times of different concentrations of the iron oxide. (**B**) An MR image shows the effects of T2 relaxation on image contrast. The vertical line at 50 msec in graph A corresponds to the timing used to acquire the image. (**C**) The T2 relaxivity, or efficacy of T2 relaxation per unit of concentration, was determined to be 112 mM^{-1} sec^{-1} for the dextran-coated iron oxide nanoparticles at 9.4T and 37° C. The T2 relaxivity can be used to correlate the T2 relaxation effects of the agent with the concentration of the agent.

exploited to measure the concentration of the agent in the sample.

$$M = M_o \sin\theta \, e^{-TR/T_2} \tag{4.7a}$$

where TR is the relaxation time period to allow M to realign with B_0.

With a 90-degree excitation pulse ($\theta = 90°$),

$$Mxy = M_o \, e^{(-TR/T_2)} \tag{4.7b}$$

CEST Contrast

A fundamentally new type of exogenous MRI contrast agent can be detected through the mechanism of chemical exchange saturation transfer (CEST) [32]. These contrast agents contain hydrogens that undergo chemical exchange with hydrogens of surrounding water molecules. These hydrogens may lie within functional groups in the covalent structure of the contrast agent, such as amide, amine, or hydroxyl groups. Selective saturation can be applied at the specific MR frequency of the exchangeable hydrogens, which reduces the detectable magnetization from these hydrogens. Chemical exchange with water causes a transfer of reduced detectable magnetization to the water signal (Equation 4.8) [33]. Therefore, CEST is dependent on the MR frequency of the hydrogens that exchange with water and the hydrogen exchange rate with water (Figure 4.6).

$$\frac{Ms}{Mo} = \frac{1}{1 + \dfrac{n_{CA}[CA]T_{1sat}}{n_{H_2O}[H_2O]\tau_M}} \tag{4.8}$$

A CEST image can be created by acquiring MR images with and without selective saturation at the MR frequency of the exchangeable hydrogens and subtracting one image from the other. CEST images are useful for identifying biological tissues with a high concentration of exchangeable hydrogens that resonate at one MR frequency, such as liver tissues with high glycogen concentrations [34].

The chemical exchange rate (k_{ex}) must be faster than the T1 relaxation rate (1/T1) so that the reduced magnetization of the water signal may be detected before the magnetization relaxes from the transverse plane ($k_{ex} \gg$ 1/T1). Yet k_{ex} must be slower than the difference in MR frequency between the hydrogen on the contrast agent and the hydrogen on a water molecule ($\Delta\omega$) so that the hydrogens have magnetically distinct frequencies that can allow for selective saturation of the agent's MR frequency to generate a CEST effect ($k_{ex} \ll \Delta\omega$). Although this limited range of k_{ex} was initially a woeful disadvantage in developing new CEST agents, this limitation has also been exploited as a tremendous advantage to create CEST agents that are responsive to their molecular environment. For example, k_{ex} is sensitive to pH so that the extracellular pH can be mapped with CEST agents [35, 36].

An exogenous MRI contrast agent with a paramagnetic europium ion has been developed that showed relatively tight binding to a single water molecule [37]. This tight water binding of this agent, a negative for T1-weighted imaging, has been exploited as a tremendous advantage to create a PARAmagnetic CEST (PARACEST) contrast agent. The lanthanide ion shifts

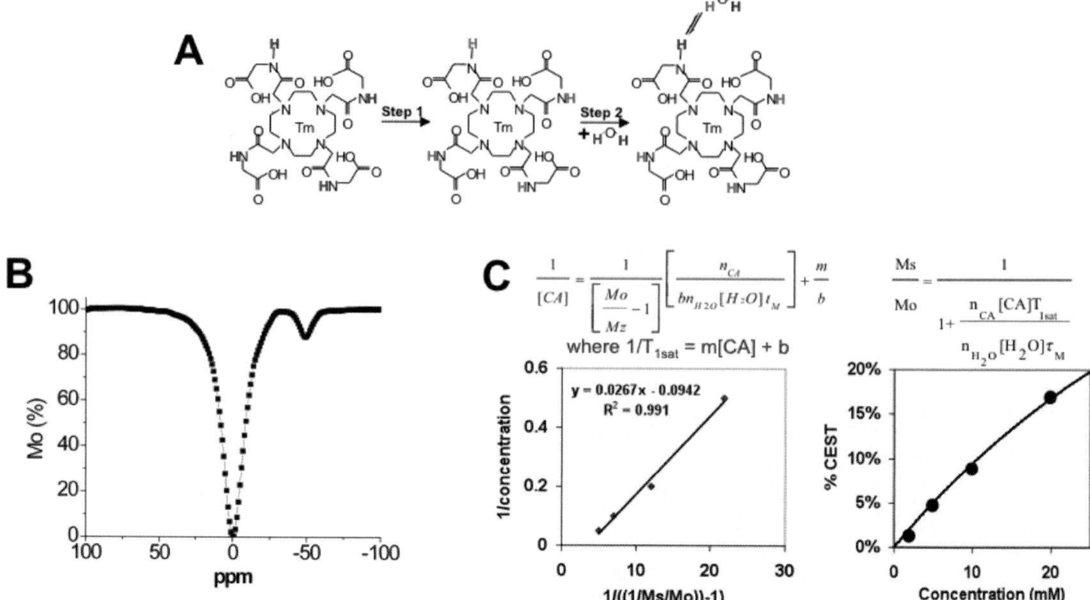

Figure 4.9. An example of a PARAmagnetic chemical exchange saturation transfer (PARACEST) MRI contrast agent. (A) Tm-DOTAMGly has amide hydrogens that have a unique chemical shift (Eq. 4.2) at −51 ppm, due to their proximity to the Tm lanthanide ion. Selective saturation at the −51 ppm chemical shift frequency (Step 1) causes the coherent magnetization of these amide hydrogens to be "spoiled" so that they have no net magnetization (i.e., N_+ and N_- are equal, so that $M_0 = 0$ according to Eq. 4.2). Chemical exchange of the hydrogens with hydrogens on water molecules (Step 2) transfers this loss of net magnetization to the water. (B) The PARACEST effect of 10 mM of Tm-DOTAMGly can be shown in a CEST spectrum, which records the net water magnetization M_0 (y-axis) immediately after selective saturation at each chemical shift frequency (x-axis). Selective saturation at −51 ppm creates the PARACEST effect. Selective saturation at 0 ppm, the chemical shift frequency of water, causes direct saturation of the water magnetization. A Varian Inova 600 MHz NMR spectrometer was used to record this CEST spectrum, with a modified presaturation pulse sequence that included a continuous wave saturation pulse, saturation pulse power of 4.95 μT, saturation delay of 4 sec and in 1-ppm increments from 100 ppm to −100 ppm. (C) The PARACEST effect can be correlated with the concentration of the agent, assuming a two-site exchange model with complete saturation of the agent and no direct saturation of water [16, 38].

the MR frequency of the hydrogens on the bound water, which greatly facilitates selective excitation (Figure 4.9). The greater $\Delta\omega$ between the bound water and bulk water also allows for a faster exchange rate, k_{ex}, which improves detection sensitivity (Equation 4.8) [38]. Additional PARACEST agents have been developed that show a variety of unique MR frequencies from amides, amines, and hydroxyl groups within the covalent structure of the chelate as well as from the tightly bound water. This variety expands the flexibility in designing PARACEST agents, resulting in agents that can detect enzymes [39], metabolites [40–42], metal ions [43], and temperature [44].

PHARMACOKINETICS AND PHARMACODYNAMICS OF EXOGENOUS CONTRAST AGENTS

The relative insensitivity of MRI impairs the direct detection of endogenous molecular compositions within *in vivo* tissues. Except for molecular imaging studies that detect an accumulation of endogenous iron oxide, most MR molecular imaging studies critically depend on exogenous MRI agents that change image contrast in response to these endogenous molecules. The MRI contrast agent must be designed to interact with the intended molecular target without also interacting with unintended targets. The agent must also be delivered to the biological site of the intended molecular target and avoid adverse biological effects before, during, and after reaching the target site. These biological and physiological criteria pose daunting problems for developing new exogenous MRI contrast agents, particularly for imaging genes and gene expression within cells. As shown in the following, some of these woeful problems can also be exploited to provide tremendous advantages for molecular imaging.

Reversible and Irreversible Responsive MRI Contrast Agents

Many exogenous MRI contrast agents are designed to bind to endogenous molecules in a reversible manner that only involves noncovalent interactions, such as peptides that bind to cell receptors or extracellular matrix components [45]. Yet exogenous MRI contrast agents are relatively insensitive, requiring a minimum threshold of 10–10,000 μM for adequate detection during *in vivo* applications [7, 8]. Endogenous metabolites present at concentrations higher than the detection threshold are feasible targets for this approach. Nucleic acids are typically present at concentrations of 1–1000 pM, and intracellular proteins are typically present at concentrations of 1 to 1000 nM, therefore these reversible responsive agents have been difficult to use to target intracellular proteins and nucleic acids [46, 47]. Extracellular proteins can exist at typical concentrations of 0.01–100 μM so that the most abundant proteins may possibly serve as targets for reversibly binding MRI contrast agents [48]. Newer approaches that load MRI contrast agents onto nanoparticles may overcome this sensitivity problem, but at the expense of altered pharmacokinetics and potential toxicities [49, 50]. In addition, the specificity of the noncovalent binding for the desired target molecule relative to other potential targets must be carefully evaluated to confidently interpret the response of the reversible MRI contrast agent. Therefore, reversible responsive MRI contrast agents are problematic for molecular imaging.

Alternatively, responsive MRI contrast agents may be designed to undergo irreversible covalent changes during interactions with endogenous molecules, such as agents that are substrates for enzymes. Irreversible responsive contrast agents can exploit a high catalytic turnover rate from a relatively low concentration of target enzyme to generate a high concentration of responsive agent that is above the MRI detection threshold. The high specificity of the enzyme reaction may also lend confidence when interpreting the response of an irreversible MRI contrast agent. Therefore, irreversible responsive MRI contrast agents have advantages for molecular imaging studies.

Tissue and Intracellular Delivery

To translate changes in image contrast to these changes in T1 relaxation time, T2* relaxation time, or the PARACEST effect, other characteristics that change image contrast must be constant or monitored by other methods. In particular, the concentration of the contrast agent affects image contrast, and *in vivo* pharmacokinetics rarely allow for constant tissue concentrations of contrast agents during the MRI scan session, so that the concentration of the contrast agent must be monitored. This may be accomplished by adding a second, unresponsive contrast agent that has identical pharmacokinetics or that is covalently linked to the responsive contrast agent [51, 52]. This poses a daunting problem for MRI contrast agents that depend on T1 and T2* relaxation times, because the T1 or T2* contrast can only monitor one agent during the MRI scan session, and the two

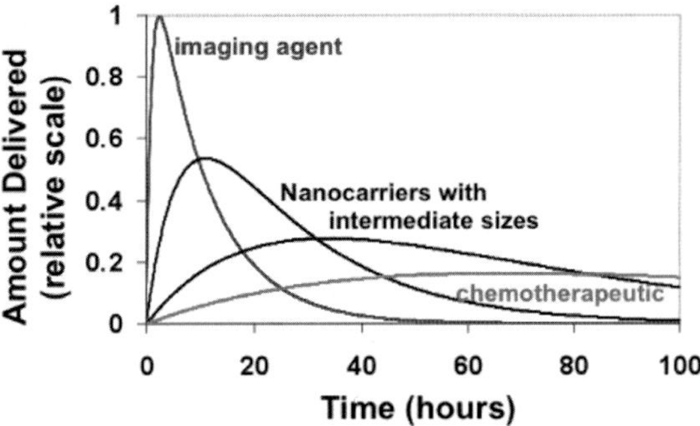

Figure 4.10. Schematic of the tissue deliveries of MRI contrast agents and chemotherapeutics. MRI contrast agents are typically detected within the first hours after the initial administration to the subject, while the effects of a chemotherapeutic may occur many hours or days after initial administration. The insensitivity of MRI requires that high amounts of contrast agents must be delivered to tissues, whereas a much lower amount of chemotherapeutic can be delivered to tissues to cause a therapeutic effect. Nanocarriers that deliver MRI contrast agents or chemotherapeutics must be designed to deliver appropriate amounts at appropriate times for each application.

types of relaxation times are too correlated to distinguish one effect from the other during the same MRI scan session. In practice, the responsive and unresponsive agents are administered in series, and the pharmacokinetics of each agent are assumed to be identical. The addition of an unresponsive contrast agent during the same scan session is feasible with PARACEST MRI, because PARACEST agents can be selectively detected via different saturation frequencies [53–55].

Due to limited sensitivity of MRI, relatively high concentrations of exogenous MRI contrast agents must be delivered to *in vivo* tissues. Methods to accomplish this delivery have followed many of the same paradigms developed to deliver chemotherapeutics to intended pathological tissues. Examples have included the use of nanocarriers that covalently conjugate or noncovalently encapsulate their cargo, such as a linear polymer [56], dendrimer [57], micelle [58], liposome [59], or antibody [60]. However, exogenous MRI contrast agents are typically detected within the first hours after the initial administration to the subject, while the effects of a chemotherapeutic may occur many hours or days after initial administration (Figure 4.10). Also, a high percentage of the injected MRI contrast agent should accumulate in the tissue of interest for adequate detection, while only a small percentage of an injected chemotherapeutic is needed to cause a therapeutic effect in the tissue of interest. These differences in timing and amount to be delivered require smaller nanocarriers for MRI contrast agents that can rapidly permeate the vascular endothelial walls and extravasate into the extracellular tissue space. In general, a nanocarrier that delivers MRI contrast agents to tissues should have a spherical diameter no larger than ~7 nm, which roughly corresponds to a

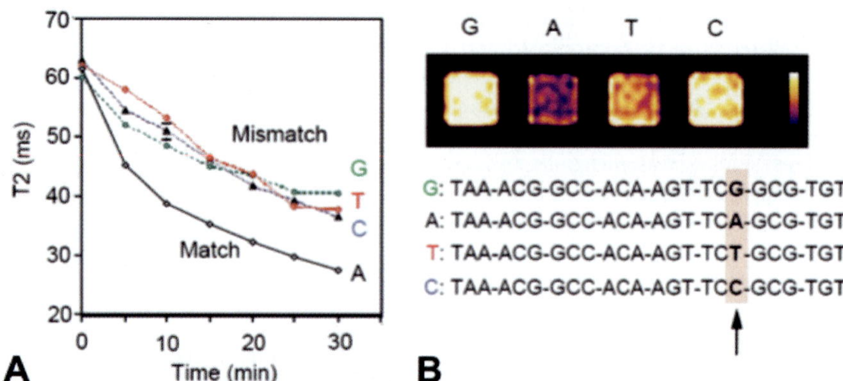

Figure 4.11. An example of a DNA-responsive MRI contrast agent. The contrast agent consisted of a complementary oligonucleotide conjugated to an iron oxide nanoparticle (10 µg Fe/mL). This agent was added to solutions of the target oligonucleotide (containing A) and other solutions of oligonucleotides containing single-nucleotide mismatches G, T, and C (53 fmol) in 25 mM KCl, 50 mM Tris-HCl, pH 7.4. The perfect target sequence is clearly distinguished from single-nucleotide mismatches by benchtop measurements of the temporal changes in T2 relaxation times (**A**) and T2 weighted MRI (**B**) of samples. (Reproduced with permission of Nature Publishing Group from Perez J., et al. (2002 *Nature Biotech* 20, 816–820)).

protein with a molecular weight of ∼70 kD [61]. This is in contrast to nanocarriers that deliver chemotherapeutics, with molecular weights ranging from 40–700 kD that can successfully deliver ∼3% of chemotherapeutics to pathological tissues [62]. These size estimates are approximate and greatly depend on the permeability and surface area of the vascular endothelium, the nonspherical shape of the nanocarrier, and other physicochemical properties of the nanocarrier such as hydrophilicity and electronic charge.

Exogenous MRI contrast agents designed to directly detect genes must also cross the membranes of the cell and cell nucleus. Other MRI contrast agents that detect gene expression may also need to cross cell membranes to interact with their RNA or protein targets. These additional hurdles can be daunting challenges for most MRI contrast agents because these agents are usually charged and hydrophilic and cannot permeate hydrophobic membranes. Nonspecific intracellular and intranuclear delivery has been investigated by coupling a cell-penetrating peptide (CPP) to the MRI contrast agent [63]. Although initial research studies indicated that CPPs can directly deliver cargos to the cytosol and nucleus, growing evidence indicates that CPPs are first endocytosed and processed through the Golgi complex, and only a fraction of CPPs are finally delivered to the cytosol and cell nucleus [64, 65]. CPPs are a rapidly developing "hot topic" for the delivery of imaging agents and therapeutics, and new CPPs with improved delivery to the cytosol and nucleus may be developed in the near future.

Although the cellular and nuclear membranes present hurdles for exogenous MRI contrast agents, this woeful problem can be exploited as a tremendous advantage for delivering agents only to cells that exhibit the desired gene expression. As described in Section 6b, intracellular transporters only expressed by the cells of interest can deliver MRI contrast agents into these cells. The high accumulation of contrast agents in these cells of interest, and/or the removal of contrast agents that have not permeated into these cells, can create strong concentration differences that can generate excellent image contrast.

IMAGING OF GENES WITH MRI TARGETING AGENTS

The strong interactions between complementary nucleic acid sequences can provide high selectivity for detecting a specific nucleic acid target. Because a gene normally consists of a long nucleic acid sequence, multiple short nucleic acid sequences can be designed to complement segments of the long gene. These short complementary sequences can be coupled to iron oxide nanoparticles, which aggregate only when the short sequences bind to the long gene sequence. The aggregation of the iron oxide nanoparticles increases local field inhomogeneities and generates T2* image contrast. The "magnetic relaxation switch" method has been successfully applied to detect the DNA sequence for green fluorescence protein in biochemical solutions within well plates, versus DNA sequences with single mismatches relative to the target DNA sequence (Figure 4.11) [66, 67]. Several types of peptide–nucleic acid sequences conjugated to gadolinium chelates have been used to target overexpressed RNAs within *in vitro* cell cultures [46, 68, 69]. Yet directly detecting a specific nucleic acid sequence within the *in vivo* context is a daunting challenge for MRI, primarily due to the extremely low concentration of specific DNA and RNA sequences present at ∼1 and

10–100 copies within each cell, respectively. The intranuclear and intracellular locations of nucleic acids pose the additional problem of intracellular delivery and trafficking of the nucleic acid-targeting contrast agent. Therefore, the "magnetic relaxation switch" method appears to be limited to biochemical screening studies of turbid media (which are not amenable to evaluation by more efficient and well-established optical imaging methods).

IMAGING GENE EXPRESSION WITH REPORTER GENE SYSTEMS

A more promising approach is the use of a reporter gene imaging strategy to study gene expression during *in vitro* or *in vivo* studies [70, 71]. This strategy requires transgenic manipulation of the cell or animal model to include two genes that are coexpressed. One is the "target gene," the focus of the gene expression study. The other gene represents the "reporter gene" that encodes for a "reporter protein" that can be detected by an MRI contrast agent that responds to the function of the reporter protein. The change in MR image contrast caused by the responsive agent thus reports on the expression of the "target gene." Reporter gene imaging with optical and radionuclide imaging modalities has become tremendously successful (see Chapters 2 and 3), and similar successes with MRI reporter gene strategies are likely to emerge in the near future.

Cell Receptors and Polypeptides

Due to the relative insensitivity of MRI, responsive contrast agents are generally limited to targeting proteins that are present at high concentrations, such as proteins that contribute to connective tissues and proteins that reside in the blood pool at high concentrations. Highly overexpressed cell receptors may also be detected with MRI. For example, antibodies labeled with iron oxide nanoparticles can be used to detect the overexpression of ICAM-1 or the H2K^k antigen, and biotin labeled with iron oxide can be used to detect the overexpression of an avidin-cell receptor fusion protein [72–74]. Other similar examples couple iron oxide to a targeting moiety to create a T2 contrast agent, because these agents can be detected at lower concentrations relative to T1 agents or CEST agents. However, these demonstrations have been limited to *in vitro* cell suspensions, and these preliminary studies indicate that this strategy is not promising for accumulating sufficient concentrations for *in vivo* MRI detection of gene expression.

A similar approach has been used to detect proteins by conjugating a protein-targeting moiety to a nanocage containing hyperpolarized ^{129}Xe [30]. After the ^{129}Xe-labeled agent binds to the target protein, the complex shows a change in MR frequency for the ^{129}Xe nucleus.

Although *in vitro* studies have shown small yet detectable changes in MR frequencies after receptor-binding, these examples have yet to be translated to *in vivo* studies in which small changes in MR frequencies may be more difficult to detect. In addition, the MR frequencies are dependent on other environmental conditions such as pH and temperature, which may further complicate *in vivo* evaluations. Yet the boost in sensitivity offered by hyperpolarization may be able to detect dilute concentrations of cell receptors *in vivo*, so that this approach still warrants investigation.

The direct targeting of intracellular proteins with MRI contrast agents faces the same problems with generating sufficient sensitivity for detection. A clever exception exploits the CEST effect that can be detected from expressed polypeptides that rapidly exchange hydrogens with water (Figure 4.12). For example, the expression of a poly-L-lysine polypeptide that consists of 200 lysine amino acid residues has a unique MR frequency at 3.76 ppm from the water resonance, which can be selectively saturated to generate a CEST effect detected *in vitro* and *in vivo* [75]. The homogeneous composition of the polypeptide is required to generate a single MR frequency. Other homogeneous polypeptides with unique MR frequencies may also be expressed by other cells, which may allow for the simultaneous detection of multiple gene expressions during a single MRI scan session. Care must be taken to ensure that the CEST detection accounts for other effects such as magnetization transfer, T2* relaxation, and magnetic susceptibilities, which can be especially problematic during *in vivo* studies. Furthermore, cationic polymers such as polylysine may alter the gene expression profile of the cell and can even be toxic at sufficiently high concentrations.

Intracellular Delivery and Accumulation of Iron Oxide

The delivery and accumulation of high concentrations of iron oxide nanoparticles can alter T2* MR image contrast. The transferrin receptor has been extensively studied as a mechanism to promote the intracellular delivery of iron oxide (Figure 4.13) [76–79]. This receptor binds two transferrin molecules each containing two paramagnetic iron atoms, internalizes this cargo into cells, and releases the iron within acidic endosomes. Significant levels of the transferrin receptor can be overexpressed on a cell membrane through a reporter gene system so that a high amount of iron oxide can be accumulated within the cells to generate substantial T2* image contrast. However, a relatively high amount of receptor must be present to produce sufficient intracellular delivery for MRI detection. Interestingly, as a result of malignant transformation, imageable levels of the transferrin receptor are expressed (fivefold or greater overexpression) in 40% of breast cancers, making this a potential

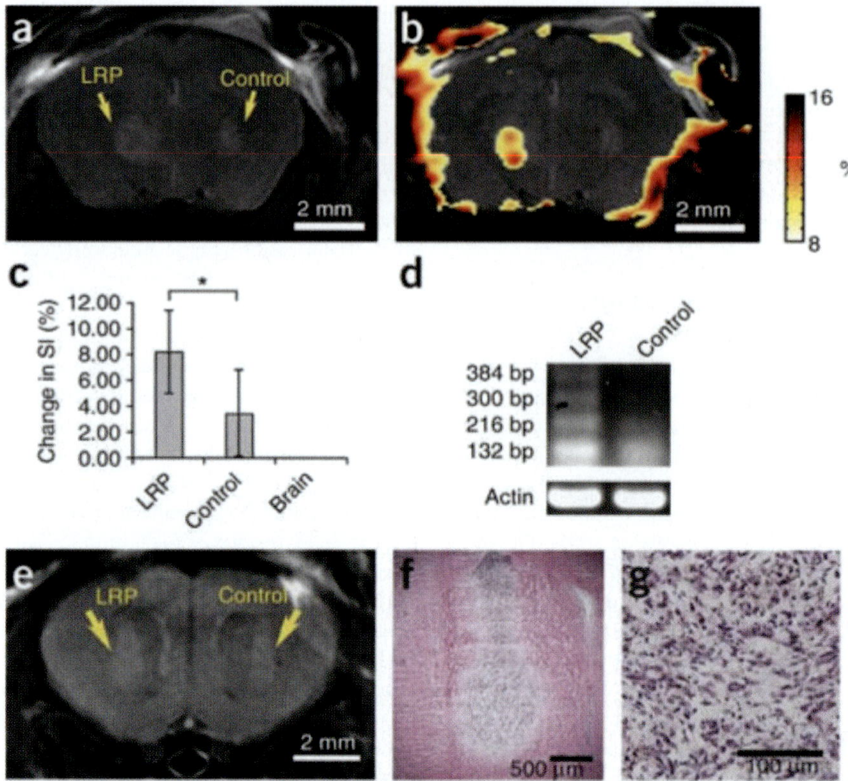

Figure 4.12. CEST imaging of a reporter polypeptide. (A) Anatomical MR image of rat glioma cells transfected with a lysine-rich protein (LRP) or "control" glioma cells without LRP. **(B)** CEST signal intensity–difference map overlaid on the anatomical image distinguishes LRP-expressing and control xenografts; **(C)** signal-intensity differences (mean s.d.; six mice, each containing two xenografts; *, $P = 0.03$, two-tailed, unpaired t-test); to compare different mice, signal-intensity changes were normalized to make signal-intensity change of normal brain equal to zero. The proper adjustment of field homogeneity could only be done inside the brain, leading to some artifacts at brain edges. **(D)** RT-PCR of xenografts from rat brains, showing expression of LRP only in the LRP xenograft. **(E)** Anatomical photo of the same tissue during postmortem analysis. **(F)** Eosin–hematoxylin stain of a frozen section off the tissue corresponding to LRP tumor in E. **(G)** Magnification of F shows a uniform tumor mass. Scale bars: 2 mm for A, B, and E, 500 µm for F and 100 µm for G. (Reproduced with permission of Nature Publishing Group from Gilad et al. (2007 *Nature Biotech* 25, 217–219)).

target for probe development for this cancer [80]. Probe development has progressed to detect threefold to five-fold changes in receptor levels *in vitro*, although this may not always be the case for the tissue of interest [79]. Furthermore, the accumulated iron oxide may reside within the cell after the cell has died so that this reporter gene system cannot report on cell viability. Lastly, unnaturally high concentrations of iron oxide have been shown to alter gene expression patterns, which may complicate the analysis of gene expression.

Another approach receiving extensive study exploits the overexpression of ferritin to store high amounts of iron oxide within cells (Figure 4.14). Ferritin serves as the body's iron depot and creates an antiferromagnetic iron crystal, which acts as a weak T2* contrast agent that can generate sufficient T2* image contrast only when highly overexpressed [81, 82]. Although this approach has been used to detect gene expression *in vitro* [57] and *in vivo* [83–86], the accumulation of iron oxide within

cells suffers from the same problems as listed previously for the transferrin receptor reporter gene system. In addition, the time required for sufficient iron accumulation compromises the temporal response, and the process of loading iron oxide into ferritin causes fluctuations in T2* image contrast. More recently, an analogous approach has been developed that employs MagA, a protein found in the *M. magneticum* magnetotactic bacterium involved in the production of magnetosomes [87]. The *in vitro* and *in vivo* expression of a magA reporter gene leads to natural synthesis of magnetosomes that contain iron oxide crystals, which generate 20–60% more T2* image contrast than ferritin.

A second approach has been developed to accumulate iron oxide within cells. A transgenic tyrosinase reporter gene system can produce melanin, which can sequester intracellular iron and enhance T1 image contrast [88]. However, this reporter gene system also produces highly reactive oxygen species that can exhibit toxic effects,

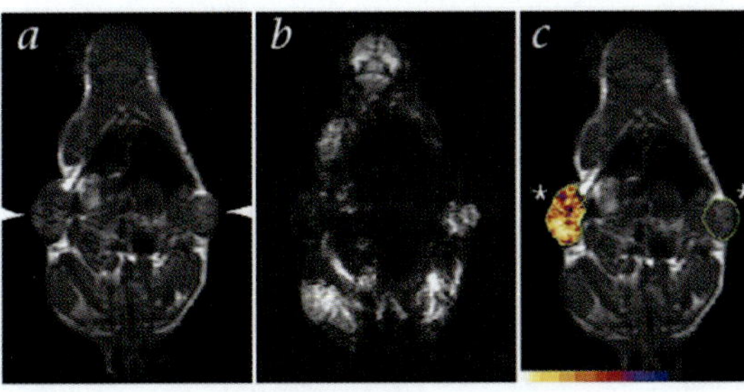

Figure 4.13. Tracking transferrin reporter gene expression. *In vivo* MR imaging of a single mouse with a flank tumor containing an engineered transferrin receptor (ETR+, left arrowheads) and a second flank tumor without the engineered transferrin receptor (ETR-, right arrowheads). (**A**), T1-weighted coronal SE image (imaging time, 3.5 min; voxel resolution, 300 × 300 × 3000 μm). ETR- and ETR+ tumors have similar signal intensities. (**B**), T2-weighted gradient-echo image corresponding to the image in (**A**), showing substantial differences between ETR- and ETR+ tumors (imaging time, 8 min; voxel resolution, 300 × 300 × 3000 m). As expected, ETR-mediated cellular accumulation of the superparamagnetic probe decreases signal intensity. These differences in MR signal intensity were most pronounced using T2- and T2*-weighted imaging pulse sequences, consistent with the increased transverse relaxation rate (1/T2) after cellular internalization. (**C**) Composite image of a T1-weighted spin-echo image obtained for anatomic detail with superimposed 1/T2 changes after Tf-MION administration, as a color map. *, difference in 1/T2 changes between the ETR+ and ETR- tumors. Scale bar (bottom left) represents 10 mm, *n* = 1. (Reproduced with permission of Nature Publishing Group from Weissleder, R., et al. (2000 *Nature Med* 6, 351–354)).

which may eventually limit this application to the study of targeted cancer gene therapy.

Enzyme Activities

Targeting enzymes provides several advantages for the design of responsive MRI contrast agents. First, the high catalytic rate of a relatively low concentration of enzyme can develop a relatively high concentration of altered contrast agent so that MRI sensitivity is less problematic for enzyme detection. Second, the specificity of enzymatic reactions is usually high so that a change in MRI contrast can often be confidently attributed to the specific targeted enzyme. Lastly, enzymatic activity can cause a variety of irreversible responses in a contrast agent, which can be exploited to develop many types of responsive MRI contrast agents. As with all reporter gene systems, the reporter enzyme must have good specificity for the MRI contrast agent and must not exhibit unintended biological changes in the cell or tissue of interest.

Enzymes can cleave a ligand of an MRI contrast agent, which can then change MR image contrast. For example, β-galactosidase can cleave a galactopyranose ligand from a contrast agent, which can increase the accessibility of water for a chelated gadolinium ion and shorten the T1 relaxation time [89, 90], alter the MR frequency of an ^{19}F galactopyranose derivative [91, 92], or alter the T2* relaxation time of an iron-based galactopyranose derivative [93]. The change in T1 or T2* relaxation times

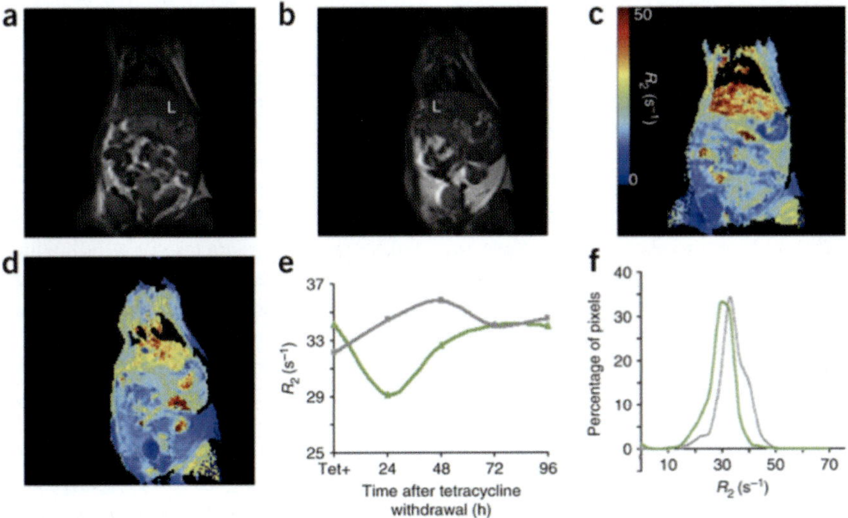

Figure 4.14. Tracking ferritin reporter gene expression. (**A,B**) MRI of sTG tet-hfer mouse (**A**) and a dTG liver-hfer sibling, overexpressing h-ferritin in liver hepatocytes (**B**). L, liver. (**C,D**) Corresponding R2 (1/T2) relaxation maps of sTG tet-hfer (**C**) and dTG liver-hfer (**D**) mice. H-ferritin overexpression resulted in a significant decrease in liver R2. (**E**) Kinetic MRI detection of the induction of h-ferritin expression. The changes in R2 induced by transgene expression were determined by consecutive MRI analysis of mice following tetracycline withdrawal. STG tet-hfer mice (gray; n = 6) showed small variation in R2; in contrast, a significant transient decrease in R2 was detected in dTG liver-hfer mice (green; n = 5; P = 0.035). (**F**) A histogram describing the spatial distribution of R2 in the liver of sTG (gray; n = 4) and dTG (green; n = 5) mice. Overexpression of h-ferritin resulted in reduced R2. (Reproduced with permission of Nature Publishing Group from Cohen B et al. (2007 *Nature Med* 13, 498–503)).

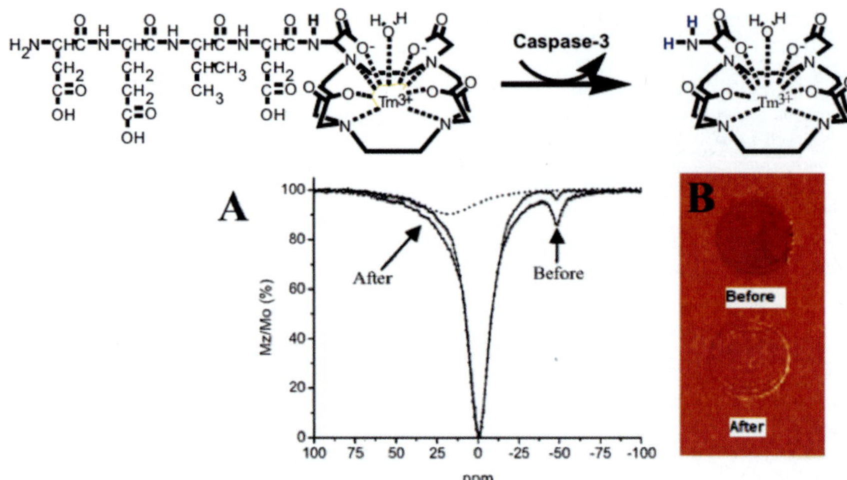

Figure 4.15. Enzyme detection with PARACEST MRI. PARACEST spectra (**A**) and MR parametric map (**B**) of DEVD-(Tm-DOTA) amide before and after addition of caspase-3. PARACEST spectra were acquired using a Varian Inova 600 MHz NMR spectrometer with a concentration of 25 mM at 37° C and pH 7.4, with a continuous-wave saturation pulse applied at 31 μT for 4 s. The deconvoluted PARACEST spectrum of the product after reaction, showing a PARACEST effect at +8 ppm, is also shown. MR images were acquired at 37° C and pH 7.4, with a Bruker Biopsin 9.4 T MR scanner. A MSME T1 method was used with TR/TE = 1623/10.9 ms and a train of Gaussian-shaped saturation pulses applied at 50 μT for 1.106 s, and with saturation offsets at −51 and +51 ppm. The parametric map was obtained by subtracting the MR image with a saturation offset at −51 ppm from the MR image with saturation offset at +51 ppm. This parametric map shows a 14.5% decrease in water MR signal before the enzymatic reaction, and no significant change in water MR signal after reaction. The magnitude of the scale of the original MR images was used as the scale for this parametric map to properly represent the practical difference in MRI contrast obtained with different saturation offsets. (Reproduced with permission of Nature Publishing Group from Yoo, B. and Pagel, M. D. (2006 *J Am Chem Soc* 6, 351–354)).

can be modest, which compromises the interpretation of these results. For comparison, the MR frequency of a PARACEST agent has been shown to be drastically altered after a peptidyl ligand of the agent is cleaved by caspase-3 (Figure 4.15) [39]. Only 3.4 nM of caspase-3 was required to generate sufficient MRI contrast in a practical time frame, and the specificity of caspase-3 for the PARACEST contrast agent substrate was shown to be excellent [55].

The cleavage of a ligand can also alter the chemical shift of the contrast agent, which can be detected using hyperpolarized MR spectroscopy. For example, the chemical shift of hyperpolarized ^{13}C in glutamine undergoes a change as glutaminase converts glutamine to glutamate [28]. The chemical shift of hyperpolarized ^{129}Xe coupled to a peptide ligand undergoes a change as matrix metalloproteinase-7 cleaves the peptide ligand [29]. Further refinement of these methods is required to detect small changes in chemical shifts within *in vivo* tissues and to account for changes in chemical shifts caused by pH and temperature. Furthermore, time required for pharmacokinetic delivery and enzyme catalysis of a sufficient concentration of contrast agent for adequate detection may be longer than the lifetime of the hyperpolarization.

Enzymes can alter the rotational tumbling times of substrates, which can alter T1 or T2* MR image contrast. For example, the rotational tumbling time can be changed after a phosphate monoester ligand of a contrast agent is hydrolyzed to an alcohol by alkaline phosphatase, which can then more easily bind to human serum albumin (HSA) [94]. Similarly, a lysine-containing ligand of a contrast agent can be cleaved by thrombin-activatable fibrinolysis inhibitor, which facilitates interactions between the contrast agent and HSA and causes an increase in rotational tumbling time [95]. The rotational tumbling time of a contrast agent can also be changed by polymerizing monomeric agents. Polymerization of phenolic contrast agents has been exploited to detect several peroxidises [96], including myeloperoxidase and oxidoreductase [97–100], which can cause up to a threefold decrease in T1 relaxation time. Peroxidase-induced polymerization of phenolic contrast agents can also assemble dextran-coated iron oxide nanoparticles to create an even stronger superparamagnetic particle that decreases the T2* relaxation time [101]. The polymerization also helps retain the contrast agent at the target site, either through slower pharmacokinetics of the polymeric form of the agent or by cross-linking the agent to the local extracellular matrix. As another example, a stearic acid ligand of a contrast agent can be cleaved by an esterase, which leads to spontaneous polymerization of the cleaved chelate [102]. Degradation of a polymer can also change the rotational tumbling time of a contrast agent. This mechanism is typically exploited by conjugating contrast agents with a linker cleaved by a specific enzyme. For example, a glucuronide linker is cleaved by

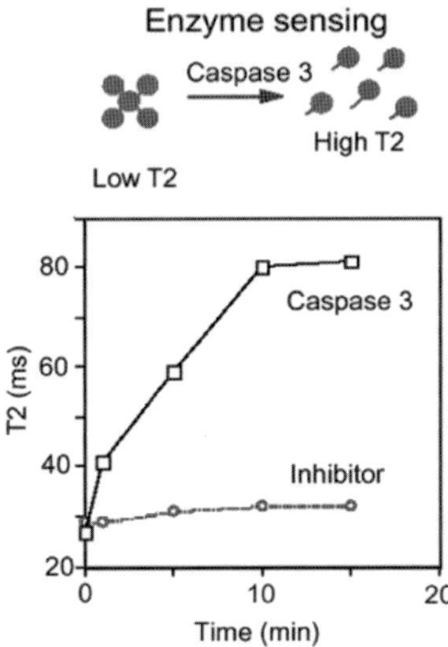

Figure 4.16. Detection enzymatic reactions. Using a DEVD peptide containing a "magnetic resonance switch" nanoassembly, caspase 3 activity can be detected by an increase in T2. This effect is not observed when a specific caspase 3 inhibitor (*N*-acetyl-DEVD-CHO) is added. The nanoassembly was formed by incubating avidin − P1 with biotin − DEVD-SS-P2. This is a generic platform by which any biotinylated peptide or protein cleavage site may be sensed. (Reproduced with permission of Nature Publishing Group from Perez et al. (2002 *Nature Biotech* 20, 816–820)).

glucuronidase [103], and a hyaluronan linker is cleaved by hyaluronidase [104, 105], which releases gadolinium chelate from the polymer. Caspase-3-induced degradation [66] or MMP-2-induced degradation [106] of a short peptide sequence linking multiple avidin-biotin-T2* contrast agents can increase the T2* relaxation time (Figure 4.16). The relatively rapid pharmacokinetics of monomeric contrast agents can accelerate removal of the agent from the site of degradation, which can further enhance the response of the agent to degradative enzyme activity.

Metabolites

A reporter gene may possibly be designed that modulates metabolite concentrations so that the detection of metabolites can be used to detect reporter gene expression. Many metabolites exist at high concentrations within an *in vitro* or *in vivo* environment, which greatly facilitates their detection by responsive MRI contrast agents. The variety of molecular structures provides opportunities to exploit several mechanisms employed by responsive MRI contrast agents. Yet this variety of molecular structures is a challenge when designing responsive MRI contrast agents so that first-generation agents may only weakly interact with the target metabolite. In addition, care must be taken to selectively detect

one of many similar metabolites to ensure that the MRI response can be attributed to the target metabolite.

PARACEST agents are insensitive and require high concentrations of their target, but they can be responsive to weak interactions. Therefore, PARACEST agents are well suited to detect metabolites. The hydrogen exchange rate of a PARACEST contrast agent has been shown to change when the agent noncovalently binds to a sugar, although selectivity for specific sugars is marginal [40, 41]. The MR frequency of another PARACEST contrast agent changes when the agent noncovalently binds to L-lactate [42]. Both the hydrogen exchange rate and the MR frequency of a PARACEST contrast agent change when the agent undergoes an irreversible covalent reaction with nitric oxide [107]. In each case, care must be taken to ensure that interactions between high concentrations of the PARACEST agent and the metabolite do not perturb the biological system. Furthermore, care must be taken to create reporter gene systems that generate sufficiently high concentrations of these metabolites for MRI detection, but not sufficiently high to alter the biology or physiology of the subject.

Fewer mechanisms have been exploited to change T1 or T2* relaxation times in response to metabolites. Water accessibility can be altered when glucose binds to a gadolinium chelate [108]. Degradation of thiol-containing polymeric contrast agents can be caused by dithiothreitol or primary radicals OH˙, H˙, HO$_2^{\cdot}$, H$_3$O$^+$, O$_2^-$, or e$_{aq}^-$ [109]. Further research to improve the strength of interactions between contrast agents and specific metabolites is required to exploit additional responses in T1 or T2* relaxation times.

The high abundance of metabolites can also overcome the insensitivity of MRS techniques. In particular, phosphorylated metabolites generated through reporter gene systems can be detected with ^{31}P MRS. For example, creatine kinase is an endogenous enzyme that can be transgenically overexpressed, resulting in high levels of phosphocreatine that can be detected with ^{31}P MRS (Figure 4.17) [110–114]. A similar strategy has been demonstrated by the transgenic expression of arginine kinase, an enzyme found only in invertebrates that enzymatically generates phosphoarginine [115, 116]. This metabolite product is normally not present in vertebrate cells, so that detection of phosphoarginine by ^{31}P MRS is aided by low endogenous background signal. However, both of these ^{31}P-based reporter gene systems depend on ^{31}P energy metabolism that can vary between tissues and over time, which complicates the interpretation of results [117]. An alternative MRS approach can detect changes in the ^{19}F MR frequency of a fluorinated derivative of cytosine that is converted to a fluorocil derivative by cysteine deaminase [118]. The absence of endogenous fluorine greatly facilitates the detection of these metabolites by ^{19}F MRS. Yet the low spatial resolutions of MRS and MRSI are detrimental to these approaches.

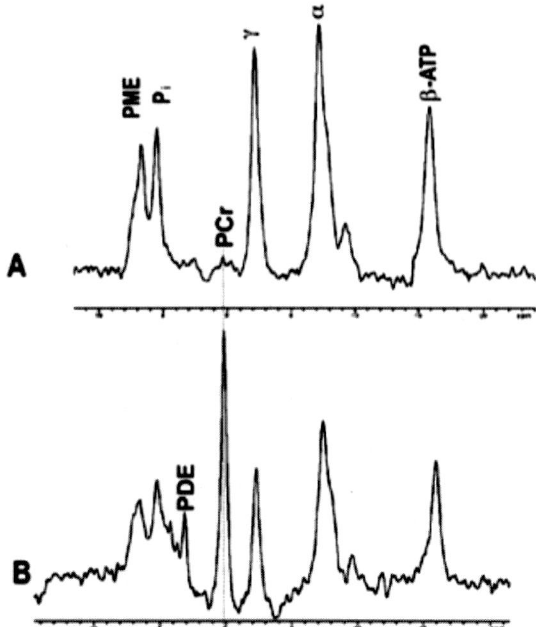

Figure 4.17. ^{31}P MRS to detect a metabolite produced by a reporter protein. A unique phosphocreatine (PCr) resonance is detected in liver tissue that was transduced with an Ad-CMV-hLDLr-IRES-ckb gene (**B**), but is absent in liver transduced with the control virus AdCMV-Empty gene (**A**). The ckb gene produces creatine kinase that generates PCr. Other resonances are phosphomonoesters (PME), inorganic phosphate (Pi), phosphodiesters (PDE), and the phosphate groups of adenosine triphosphate (ATP). (Reproduced with permission of Nature Publishing Group from Auricchio et al. (2001 *Proc Natl Acad Sci USA* 98: 5205–5210)).

STRATEGIES FOR CLINICAL TRANSLATION

Several studies in small animals have demonstrated that *in vitro* loading of stem cells with untargeted iron oxides (by exploiting their physiologic propensity to take up the particles *in vitro*), followed by introducing the loaded cells into an *in vivo* system, is a sensitive way to track the migration of the cells *in vivo* [81, 119–121]. The iron oxides remain cell associated and detectable for at least 4 weeks and are therefore suitable to make many conclusions about the migration of the cells. Others have exploited perfluoropolyether (PFPE) agents to load cells *in vitro* and track cells *in vivo* [122]. However, these studies are not able to address cell viability. The MR imaging of gene expression can directly report on cell viability to fully assess the migration of viable stem cells or other viable cells *in vivo*, which provides additional motivation for clinical translation [81, 120, 121].

It is likely that translation into the clinic will be sequential. For example, the use of untargeted iron oxides as imaging agents has been demonstrated in humans for liver imaging and has been exploited to interrogate lymph nodes for cancer in prostate cancer patients [122]. In each case these applications do not *per se* image specific genes, but rather a multigene process – iron oxide nanoparticle uptake. These successes are, however, encouraging and advance the acceptance of MRI as a molecular imaging tool capable of being used clinically to image gene expression.

There is clearly a significant gap between basic biological research of MRI reporter genes and their translation to the clinic. Sections 5 and 6 list these disadvantages for each research design. Technologies are now available to close that gap. Once these technologies have been reined in for clinical use, then how should one select the target to image? Potential targets may be selected from a variety of molecular markers based on their relevance to various physiologic or pathophysiologic processes. Because pathologies often exhibit modest changes in the expression, production, delivery, or activity of various biomarkers, emphasis should be placed on the ability to quantitatively translate the response in MR image contrast to the concentration of the biomarker. The absolute quantification of MR image contrast is a daunting challenge because many characteristics of the clinical patient, the instrumentation, and the choice of acquisition parameters and image processing methods can alter image contrast. Ratiometric quantifications are more feasible, in which the response caused by a molecular biomarker is the only effect allowed to vary between two MR images or between two or more regions of the same image. PARACEST may facilitate ratiometric quantifications because two or more PARACEST agents can be selectively detected within the same sample volume by saturating their respective CEST MR frequencies.

Target selection may be informed by genomic or proteomic screens that identify markers or groups of markers associated with a particular disease, physiology, or biochemical pathway under investigation in the clinic. Often the information derived from genomic analysis of disease does not identify single markers but rather groups of markers whose variable expression as a cohort is diagnostic. Therefore, in the future, molecular imaging probes and technologies will have to be designed to survey the expression of several targets simultaneously to meaningfully report on the biological state. The selective detection of each PARACEST agent also provides opportunities to detect more than one molecular target during a single MRI scan session, which may lead to the diagnoses of "molecular signatures," or multiple molecular biomarkers of pathological tissues. Additional multistep mechanisms such as posttranslational protein processing, protein–protein interactions, and cell–cell signaling are likely to be employed with future responsive MRI contrast agents for reporter gene imaging. Therefore, interdisciplinary research combining molecular biology with radiology and chemistry will be critical for further developing responsive MRI contrast agents that can provide value in the clinic. Finally, it should be considered that multimodality imaging devices will become standard in clinical care. Anatomical resolution and biochemical data obtained with MRI can be combined with biochemical information obtained by nuclear or optical imaging such that a 3D graphic/biochemical image can represent the disease state.

As we move forward with molecular imaging toward the goal of earlier detection of disease, we should expect that the results will modify if not dramatically alter the face of medicine. Medical practices will have to be rethought, as imaging based on biochemical rather than anatomical differences between tissues enables detection of developing disease before patients become symptomatic. The practice of medicine will potentially change from medical intervention to a clinical pretreatment/ prevention paradigm.

REFERENCES

1 Lauterbur, P. C. (1973). Image formation by induced local interactions: Examples employing nuclear magnetic resonance. *Nature* **242**: 190–191.

2 2006 MRI Market Summary Report. (2007). IMV International: Des Plaines, Ill.

3 Ahrens, E. T., Rothbacher, U., Jacobs, R. E., Fraser, S. E. (1998). A model for MRI contrast enhancement using T1 agents. *Proc Nat Acad Sci USA* **95**(15): 8443–8448.

4 Mills, P. H., Ahrens, E. T. (2007). Theoretical MRI contrast model for exogenous T2 agents. *Magn Res Med* **57**(2): 442–447.

5 Massoud, T. F., Gambhir, S. S. (2003). Molecular imaging in living subjects: seeing fundamental biological processes in a new light. *Genes & Dev* **17**: 545–580.

6 Weissleder, R., Mahmood, U. (2001). Molecular imaging. *Radiology* **219**: 316–333.

7 Bloch, F., Hansen, W. W., Packard, M. (1946). The nuclear induction experiment. *Phys Rev* **70**: 474–485.

8 Purcell, E.M., Torrey, H. C., Pound, R. V. (1946). Resonance absorption by nuclear magnetic moments in a solid. *Phys Rev* **69**: 37–38.

9 Abragam A. (1961). *Principles of Nuclear Magnetism.* Oxford University Press: New York.

10 Schlicter, C. P. (1978). *Principles of Magnetic Resonance.* Springer-Verlag: New York.

11 Conradi, M. S., Saam, B. T., Yablonsky, D. A., Woods, J. C. (2006). Hyperpolarized ^{3}He and perfluorocarbon gas diffusion MRI of lungs. *Prog NMR Spect* **48**: 63–83.

12 Mansson, S., Johansson, E., Magnusson, P., Chai, C.-M., Hansson, G., Petersson, J. S., Stahlberg, F., Golman, K. (2006). ^{13}C imaging – a new diagnostic platform. *Eur Radiol* **16**: 57–67.

13 Wu, X. (2003). Optical pumping and hyperpolarized spin relaxation. Proceedings of SPIE-The International Society for Optical Engineering (Photonics and Imaging in Biology and Medicine) **5254**: 97–107.

14 Golman, K., Olsson, L. E., Axelsson, O., Mansson, S., Karlsson, M., Petersson, J. S. (2003). Molecular imaging using hyperpolarized 13C. *Brit J Radiol* **76**(Spec. *Iss.* 2): S118–S127.

15 Hersman, F. W., Ruset, I. C., Ketel, S., Muradian, I., Covrig Silviu, D., Distelbrink, J., Porter, W., Watt, D., Ketel, J., Brackett, J., Hope, A., Patz, S. (2008). Large production system for hyperpolarized 129Xe for human lung imaging studies. *Acad Radiol* **15**(6): 683–692.

16 Mansson, S., Johansson, E., Magnusson, P., Chai, C.-M., Hansson, G., Petersson, J. S., Stahlberg, F., Golman, K. (2006). 13C imaging – a new diagnostic platform. *European Radiol* **16**(1): 57–67.

17 Dugas, J. P., Garbow, J. R., Kobayashi, D. K., Conradi, M. S. (2004). Hyperpolarized 3He MRI of mouse lung. *Magn Reson Med* **52**(6): 1310–1317.

18 van Beek, E. J. R., Wild, J. M., Kauczor, H.-U., Schreiber, W., Mugler, J. P., de Lange, E. E. (2004). Functional MRI of the lung using hyperpolarized 3-helium gas. *J Magn Reson Imaging* **20**(4): 540–554.

19 Driehuyst, B., Cofer, G. P., Pollaro, J., Mackel, J. B., Hedlund, L. W., Johnson, G. A. (2006). Imaging alveolar-capillary gas transfer using hyperpolarized 129Xe MRI. *Proc Nat Acad Sci USA* **103**(48): 18278–18283.

20 Johansson, E., Mansson, S., Wirestam, R., Svensson, J., Petersson, J. S., Golman, K., Stahlberg, F. (2004). Cerebral perfusion assessment by bolus tracking using hyperpolarized 13C. *Magn Reson Med* **51**(3): 464–472.

21 Olsson, L. E., Chai, C.-M., Axelsson, O., Karlsson, M., Golman, K., Petersson, J. S. (2006). MR coronary angiography in pigs with intraarterial injections of a hyperpolarized 13C substance. *Magn Reson Med* **55**(4): 731–737.

22 Ishii, M., Emami, K., Kadlecek, S., Petersson, J. S., Golman, K., Vahdat, V., Yu. J., Cadman, R. V., MacDuffie-Woodburn, J., Stephen, M., Lipson, D. A., Rizi, R. R. (2007). Hyperpolarized 13C MRI of the pulmonary vasculature and parenchyma. *Magn Reson Med* **57**(3): 459–463.

23 Golman, K., Petersson, J. S., Magnusson, P., Johansson, E., Aakeson, P., Chai, C.-M., Hansson, G., Maansson, S. (2008). Cardiac metabolism measured noninvasively by hyperpolarized 13C MRI. *Magn Reson Med* **59**(5): 1005–1013.

24 Schroeder, M. A., Cochlin, L. E., Heather, L. C., Clarke, K., Radda, G. K., Tyler, D. J. (2008). In vivo assessment of pyruvate dehydrogenase flux in the heart using hyperpolarized carbon-13 magnetic resonance. *Proc Nat Acad Sci USA* **105**(33): 12051–12056.

25 Golman, K., in't Zandt, R., Lerche, M., Pehrson, R., Ardenkjaer-Larsen, J. H. (2006). Metabolic imaging by hyperpolarized 13C magnetic resonance imaging for in vivo tumor diagnosis. *Cancer Res* **66**(22): 10855–10860.

26 Chen, A. P., Albers, M. J., Cunningham, C. H., Kohler, S. J., Yen, Y-F., Hurd, R. E., Tropp, J., Bok, R., Pauly, J. M., Nelson, S. J., Kurhanewicz, J., Vigneron, D. B. (2007). Hyperpolarized C-13 spectroscopic imaging of the TRAMP mouse at 3T-initial experience. *Magn Reson Med* **58**(6): 1099–1106.

27 Day, S. E., Kettunen, M. I., Gallagher, F. A., Hu, D.-E., Lerche, M., Wolber, J., Golman, K., Ardenkjaer-Larsen, J. H., Brindle, K. M. (2007). Detecting tumor response to treatment using hyperpolarized 13C magnetic resonance imaging and spectroscopy. *Nature Med* **13**(11): 1382–1387.

28 Gallagher, F. A., Kettunen, M. I., Day, S. E., Lerche, M., Brindle, K. M. (2008). 13C MR spectroscopy measurements of glutaminase activity in human hepatocellular carcinoma cells using hyperpolarized 13C-labeled glutamine. *Magn Reson Med* **60**(2): 253–257.

29 Wei, Q., Seward, G. K., Hill, P. A., Patton, B., Dimitrov, I. E., Kuzma. N. N., Dmochowski, I. J. (2006). Designing 129Xe NMR biosensors for matrix metalloproteinase detection. *J Am Chem Soc* **128**(40): 13274–13283.

30 Schroder, L., Lowery, T. J., Hilty, C., Wemmer, D. E., Pines, A. (2006). Molecular imaging using a targeted magnetic resonance hyperpolarized biosensor. *Science* **314**: 446–449.

31 Haacke, E. M., Brown, R. W., Thompson, M. R., Venkatesan. R. (1999). *Magnetic Resonance Imaging: Physical Principles and Sequence Design.* John Wiley & Sons: New York.

32 Ward, K. M., Aletras, A. H., Balaban, R. S. (2000). A new class of contrast agents for MRI based on proton chemical exchange dependent saturation transfer (CEST). *J Magn Reson* **143**: 79–87.

33 Woessner, D. E., Zhang, S., Merritt, M. E., Sherry, A. D. (2005). Numerical solutions of the bloch equations provides insights into the optimum design of PARACEST agents for MRI. *Magn Reson Med* **53**: 790–799.

34 van Zijl, P. C. M., Jones, C. K., Ren, J., Malloy, C. R., Sherry, A. D. (2007). MRI detection of glycogen in vivo by using chemical exchange saturation transfer imaging (glycoCEST). *Proc Nat Acad Sci USA* **104**(11): 4359–4364.

35 Ward, K. M., Balaban, R. S. (2000). Determination of pH using water protons and chemical exchange dependent saturation transfer. *Magn Reson Med* **44**: 799–802.

36 van Zijl, P. C. M., Duyn, J. H., Bryant, L. H., Bulte, J. W. M. (2003). The use of starburst dendrimers as pH contrast agents. *Proc Int Soc Magn Reson Med* **9**: 878.

37 Zhang, S., Winter, P., Wu, K., Sherry, A. D. (2001). A novel europium(III)-based MRI contrast agent. *J Am Chem Soc* **123**: 1517–1518.

38 Zhang, S., Merritt, M., Woessner, D. E., Lenkinski, R. E., Sherry, A. D. (2003). PARACEST agents: modulating MRI contrast via water proton exchange. *Acc Chem Res* **36**: 783–790.

39 Yoo, B., Pagel, M. D. (2006). A PARACEST MRI contrast agent to detect enzyme activity. *J Am Chem Soc* **128**: 14302–14303.

40 Zhang, S., Trokowski, R., Sherry, A. D. (2003). A paramagnetic CEST agent for imaging glucose by MRI. *J Am Chem Soc* **125**(50): 15288–15289.

41 Trokowski, R., Zhang, S., Sherry, A. D. (2004). Cyclen-based phenylborate ligands and their Eu^{3+} complexes for sensing glucose by MRI. *Bioconj Chem* **15**: 1431–1440.

42 Aime, S., Castelli, D. D., Fedeli, F., Terreno, E. (2002). A paramagnetic MRI-CEST agent responsive to lactate concentration. *J Am Chem Soc* **124**: 9364–9365.

43 Trokowski, R., Ren, J., Kalman, F. K., Sherry, A. D. (2005). Selective sensing of zinc ions with a PARACEST contrast agent. *Angew Chem Int Ed* **44**: 6920–6923.

44 Zhang, S., Malloy, C. R., Sherry, A. D. (2005). MRI thermometry based on PARACEST agents. *J Am Chem Soc* **127**: 17572–17573.

45 Artemov, D. (2003). Molecular magnetic resonance imaging with targeted contrast agents. *J Cell Biochem* **90**: 518–524.

46 Heckl, S., Piplorn, R., Waldeck, W., Spring, H., Jenne, J., von der Lieth, C-W., Corban-Wilhelm, H., Debus, J., Braun, K. (2003). Intracellular visualization of prostate cancer using magnetic resonance imaging. *Cancer Res* **63**: 4766–4772.

47 Artemov, D., Bhujwalla, Z. M., Bulte, J. W. M. (2004). Magnetic resonance imaging of cell surface receptors using targeted contrast agents. *Curr Pharm Biotech* **5**: 485–494.

48 Flacke, S., Fischer, S., Scott, M. J., Fuhrhop, R. J., Allen, J. S., McLean, M., Winter, P., Sicard, G. A., Gaffney, P. J., Wickline, S. A., Lanza, G. M. (2001). Novel MRI contrast agent for molecular imaging of fibrin. Implications for detecting vulnerable plaques. *Circulation* **104**: 1280–1285.

49 Caruthers, S. D., Winter, P. M., Wickline, S. A., Lanza, G. M. (2006). Targeted magnetic resonance imaging contrast agents. *Methods Molec Med* **124**: 387–400.

50 Torchilin, V. P. (2000). Polymeric contrast agents for medical imaging. *Curr Pharm Biotech* **1**(2): 183–215.

51 Raghunand, N., Howison, C., Sherry, A. D., Zhang, S., Gillies, R. J. (2003). Renal and systemic pH imaging by contrast-enhanced MRI. *Magn Reson Med* **49**: 249–257.

52 Garcia-Martin, M. L., Martinez, G. V., Raghunand, N., Sherry, A. D., Zhang, S., Gillies, R. J. (2006). High resolution pHe imaging of rat glioma using pH-dependent relaxivity. *Magn Reson Med* **55**(2): 309–315.

53 Aime, S., Barge, A., Castelli, D. D., Fedeli, F., Mortillaro, A., Nielsen, F. U., Terreno, E. (2002). Paramagnetic lanthanide(III) complexes as pH-sensitive chemical exchange saturation transfer (CEST) contrast agents for MRI applications. *Magn Reson Med* **47**: 639–648.

54 Terreno, E., Castelli, D. D., Cravotto, G., Milone, L., Aime, S. (2004). Ln(III)-DOTAMGly complexes: a versatile series to assess the determinants of the efficacy of paramagnetic chemical exchange saturation transfer agents for magnetic resonance imaging applications. *Invest Radiol* **39**(4): 235–243.

55 Yoo, B., Raam, M., Rosenblum, R., Pagel, M. D. (2007). Enzyme-responsive PARACEST MRI contrast agents: A new biomedical imaging approach for studies of the proteasome. *Contrast Media Molec Imag* **2**(4): 189–198.

56 Nori, A., Kopecek, J. (2004). Intracellular targeting of polymer-bound drugs for cancer chemotherapy. *Adv Drug Delivery Rev* **57**: 609–636.

57 Lee, C. C., MacKay, J. A., Frechet, J. M. J., Szoka, F. C. (2005). Designing dendrimers for biological applications. *Nature Biotech* **23**(12): 1517–1526.

58 Gaucher, G., Dufresne, M.-H., Sant, V. P., Kang, N., Maysinger, D., Leroux, J.-C. (2005). Block copolymer micelles : preparation, characterization and application in drug delivery. *J Control Rel* **109**(1–3): 169–188.

59 Drummond, D. C., Meyer, O., Hong, K., Kirpotin, D. B., Papahadjopoulos, D. (1999). Optimizing liposomes for delivery of chemotherapeutic agents to solid tumors. *Pharm Rev* **51**(4): 691–743.

60 Coiffier, B. (2004). Chemotherapy combined with monoclonal antibodies in the treatment of patients with diffuse large B-cell lymphoma. In *Progress in Oncology.* Jones and Bartlett Publishers: Sudbury, MA, pp. 220–235.

61 Taylor, J. S., Tofts, P. S., Port, R., Evelhoch, J. L., Knopp, M., Reddick, W. E., Runge, V. M., Mayr, N. (1999). MR imaging of tumor microcirculation: Promise for the new millennium. *J Magn Reson Imag* **10**(6): 903–907.

62 Maeda, H., Wua, J., Sawa, T., Matsumura, Y., Hori, K. (2000). Tumor vascular permeability and the EPR effect in macromolecular therapeutics: a review. *J Control Rel* **65**: 271–284.

63 Allen, M. J., MacRenaris, K. W., Venkatasubramanian, P. N., Meade, T. J. (2004). Cellular delivery of MRI contrast agents. *Chem & Biol* **11**: 301–307.

64 Fischer, R., Kohler, K., Fotin-Mlezek, M., Brock, R. (2004). A stepwise dissection of the intracellular fate of cationic cell-penetrating peptides. *J Biol Chem* **279**(13): 12625–12635.

65 Melikov, K., Chernomordik, L. V. (2005). Arginine-rich cell penetrating peptides: from endosomal uptake to nuclear delivery. *Cell Mol Life Sci* **62**: 2739–2749.

66 Perez, J. M., Josephson, L., O'Loughlin, T., Hogemann, D., Weissleder, R. (2002). Magnetic relaxation switches capable of sensing molecular interactions. *Nature Biotech* **20**: 816–820.

67 Perez, J., O'Loughlin, T., Simeone, F. J., Weissleder, R., Josephson, L. (2002). DNA-based magnetic naoparticle assembly acts as a magnetic relaxation nanoswitch allowing screening of DNA-cleaving agents. *J Am Chem Soc* **124**: 2856–2857.

68 Su, W., Mishra, R., Pfeuffer, J., Wiesmueller, K.-H., Ugurbil, K., Engelmann, J. (2007). Synthesis and cellular uptake of a MR contrast agent coupled to an antisense peptide nucleic acid – cell-penetrating peptide conjugate. *Contrast Media Molec Imag* **2**(1): 42–49.

69 Tian, X., Chakrabarti, A., Amirkhanov, N., Aruva, M. R., Zhang, K., Cardi, C. A., Lai, S., Thakur, M. L. Receptor-mediated internalization of chelator-PNA-peptide hybridization probes for radioimaging or magnetic resonance imaging of oncogene mRNAs in tumours. *Biochem Soc Trans* **35**(1): 72–76.

70 Bremer, C., Weissleder, R. (2001). In vivo imaging of gene expression: MR and optical technologies. *Acad Radiol* **8**: 15–23.

71 Hogemann, D., Basilion, J. P. (2002). "Seeing inside the body:": MR imaging of gene expression. *Eur J Nucl Med* **29**: 400–408.

72 Bulte, J. W. M., Verkuyl, J. M., Herynek, V., Katsanis, E., Brocke, S., Holla, M., Frank, J. A. (1998). Magnetoimmunodetection of (transfected) ICAM-1gene expression. *Proc Int Soc Magn Reson Med* **6**: 307.

73 So, P. W., Hotee, S., Herlihy, A. H., Bell, J. D. (2005). Generic method for imaging transgene expression. *Magn Reson Med* **54**: 218–221.

74 Mantyla, T., Hakumaki, J. M., Huhtala, T., Narvanen, A., Yla-Herttuala, S. (2006). Targeted magnetic resonance imaging of Scavidin-receptor in human umbilical vein endothelial cells in vitro. *Magn Reson Med* **55**: 800–804.

75 Gilad, A. A., McMahon, M. T., Walczak, P., Winnard, P. T., Raman, V., van Laarhoven, H. W. M., Skoglund, C. M., Bulte, J. W. M., van Zijl, P. C. M. (2007). Artificial reporter gene providing MRI contrast based on proton exchange. *Nature Biotech* **25**(2): 217–219.

76 Koretsky, A., Lin, Y.-J., Schorle, H., Jaenisch, R. (1996). Genetic control of MRI contrast by expression of the transferrin receptor. *Proc Int Soc Magn Reson Med* **4**: 69.

77 Ichikawa, T., Hoegemann, D., Saeki, Y., Tyminski, E., Terada, K., Weissleder, R., Chiocca, E. A., Basilion, J. P. (2002). MRI of transgene expression: correlation to therapeutic gene expression. *Neoplasia* **4**(6): 523–530.

78 Hoegemann, D., Josephson, L., Weissleder, R., Basilion, J. P. (2000). Improvement of MRI probes to allow efficient detection of gene expression. *Bioconj Chem* **11**(6): 941–946.

79 Weissleder, R., Moore, A., Mahmood, U., Bhorade, R., Benveniste, H., Chiocca, E. A., Basilion, J. P. (2000). In vivo magnetic resonance imaging of transgene expression. *Nature Med* **6**(3): 351–354.

80 Hogemann-Savellano, D., Bos, E., Blondet, C., Sato, F., Abe, T., Josephson, L., Weissleder, R., Gaudet, J., Sgroi, D., Peters, P. J., Basilion, J. P. (2003). The transferrin receptor: a potential molecular imaging marker for human cancer. *Neoplasia* **5**(6): 495–506.

81 Gilad, A. A., Winnard, P. T., van Zijl, P. C. M., Bulte, J. W. M. (2007). Developing MR reporter genes: promises and pitfalls. *NMR Biomed* **20**: 275–290.

82 Bulte, J. M., Vymazal, J., Brooks, R. A., Pierpaoli, C., Frank, J. A. (1993). Frequency dependence of MR relaxation times. II. Iron oxides. *J Magn Reson Imag* **3**: 641–648.

83 Gottesfeld, Z., Neeman, M. (1996). Ferritin effect on the transverse relaxation of water: NMR microscopy at 9.4 T. *Magn Reson Med* **35**: 514–520.

84 Cohen, B., Dafni, H., Meir, G., Harmelin, A., Neeman, M. (2005). Ferritin as an endogenous MRI reporter for non-invasive imaging of gene expression in C6 glioma tumors. *Neoplasia* **7**: 109–117.

85 Cohen, B., Ziv, K., Plaks, V., Israely, T., Kalchenko, V., Harmelin, A., Benjamin, L. E., Neeman, M. (2007). MRI detection of transcriptional regulation of gene expression in transgenic mice. *Nature Med* **13**(4): 498–503.

86 Genove, G., Demarco, U., Xu, H., Goins, W. F., Ahrens, E. T. (2005). A new transgene reporter for in vivo magnetic resonance imaging. *Nat Med* **11**: 450–454.

87 Zurkiya, O., Chan, W. S., Hu, X. (2008). MagA is sufficient for producing magnetic nanoparticles in mammalian cells, making it an MRI reporter. *Magn Reson Med* **59**: 1225–1231.

88 Weissleder, R., Simonova, M., Bogdanova, A., Bredow, S., Enochs, W. S., Bogdanov, A. MR imaging and scintigraphy of gene expression through melanin induction. *Radiology* **204**: 425–429.

89 Moats, R. A., Fraser, S. E., Meade, T. (1997). A "smart" magnetic resonance imaging agent that reports on specific enzyme activity. *Angew Chem Intl Edn Engl* **36**: 726–728.

90 Louie, A. Y., Huber, M. M., Ahrens, E. T., Rothbacher, U., Moats, R., Jacobs, R. E., Fraser, S. E., Meade, T. J. (2000). In vivo visualization of gene expression using magnetic resonance imaging. *Nat Biotech* **18**: 321–325.

91 Cui, W., Otten, P., Li, Y., Koeneman, K. S., Yu, J., Mason, R. P. (2004). Novel NMR approach to assessing gene transfection: 4-fluoro-2-nitrophenolbeta-d-galactopyranoside as a prototype reporter molecule for β-galactosidase. *Magn Reson Med* **51**: 616–620.

92 Kodibagkar, V. D., Yu, J., Liu, L., Hetherington, H. P., Mason, R. P. (2006). Imaging beta-galactosidase activity using (19)F chemical shift imaging of LacZ gene-reporter molecule 2-fluoro-4-nitrophenolbeta-d-galactopyranoside. *Magn Reson Imaging* **24**: 959–962.

93 Cui, W., Liu, L., Adam, A., Yu, J., Li, X., Mason, R. P. (2005). Detection of beta-galactosidase activity in a human tumor

xenograft by 1H MRI in vivo using S-Gal. *Proc Int Soc Magn Reson Med* 13: 2593.

94 Lauffer, R., McMurry, T. J., Dunham, S. O., Scott, D. M., Parmelee, D. J., Dumas, S. (1997). Bioactivated diagnostic imaging contrast agents. *WIPO Patent Application* 97/36619.

95 Nivorozhkin, A., Kolodziej, A. F., Caraban, P., Greenfield, M. T., Lauffer, R. B., McMurry, T. J. (2001). Enzyme activated Gd^{3+} magnetic resonance imaging contrast agents with a prominent receptor induced magnetization enhancement. *Angew Chem Int Ed* 40: 2903–2906.

96 Bogdanov, A., Matuszewski, L., Bremer, C., Petrovski, A., Weissleder, R. (2002). Oligomerization of paramagnetic substrates result in signal amplification and can be used for MR imaging of molecular targets. *Molec Imag* 1: 16–23.

97 Chen, J., Pham, W., Weissleder, R., Bogdanov, A. Human myeloperoxidase: a potential target for molecular MR imaging in atherosclerosis. *Magn Reson Med* 52: 1021–1028.

98 Chen, J., Querol, M., Bogdanov, A., Weissleder, R. (2006). Imaging myeloperoxidase in mice by using novel amplifiable paramagnetic substrates. *Radiology* 240: 473–481.

99 Querol, M., Chen, J. W., Weissleder, R., Bogdanov, A. (2005). DTPA-bisamide-based MR sensor agents for peroxidase imaging. *Org Lett* 7: 1719–1722.

100 Querol, M., Chen, J. W., Bogdanov, A. (2006). A paramagnetic contrast agent with myeloperoxidase-sensing properties. *Org Biomol Chem* 4: 1887–1895.

101 Perez, J., Simeone, F. J., Tsourkas, A., Josephson, L., Weissleder, R. Peroxidase substrate nanosensors for MR imaging. *Nanolett* 4: 119–122.

102 Aime, S., Cabella, C., Colombatto, S., Crich, S. G., Gianolio, E., Maggioni, F. (2002). Insight into the use of paramagnetic Gd(III) complexes in MR molecular imaging investigations. *J Magn Reson Imag* 16: 394–406.

103 Duimstra, J., Meade, T. J. (2005). *Self-immolative magnetic resonance imaging contrast agents sensitive to beta-glucuronidase.* WIPO Patent Application 05/115105.

104 Shiftan, L., Israely, T., Cohen, M., Frydman, V., Dafni, H., Stern, R., Neeman, M. (2005). Magnetic resonance imaging visualization of hyaluronidase in ovarian carcinoma. *Cancer Res* 65: 10316–10323.

105 Shiftan, L., Neeman, M. (2006). Kinetic analysis of hyaluronidase activity using a bioactive MRI contrast agent. *Contrast Media Molec Imag* 1: 106–112.

106 Zhao, M., Josephson, L., Tang, Y., Weissleder, R. (2003). Magnetic sensors for protease assays. *Angew Chem Int Ed* 43: 1375–1378.

107 Liu, G., Lu, Y., Pagel, M. D. (2007). Design and characterization of new irreversible responsive PARACEST MRI contrast agent that detects nitric oxide. *Magn Reson Med* 58: 1249–1256.

108 Rohovec, J., Maschmeyer, T., Aime, S., Peters, J. A. (2003). The structure of the sugar residue in glycated human serum albumin and its molecular recognition by phenylboronate. *Chem Eur J* 9: 2193–2199.

109 Glogard, C., Stensrud, G., Aime, S. (2003). Novel radical-responsive MRI contrast agent based on paramagnetic liposomes. *Magn Reson Chem* 41(8): 585–588.

110 Koretsky, A. P., Brosnan, M. J., Chen, L. H., Chen, J. D., Van Dyke, T. (1990). NMR detection of creatine kinase expressed in liver of transgenic mice: determination of free ADP levels. *Proc Natl Acad Sci USA* 87: 3112–3116.

111 Auricchio, A., Zhou, R., Wilson, J. M., Glickson, J. D. (2001). In vivo detection of gene expression in liver by 31P nuclear magnetic resonance spectroscopy employing creatine kinase as a marker gene. *Proc Natl Acad Sci USA* 98: 5205–5210.

112 Li, Z., Qiao, H., Lebherz, C., Choi, S. R., Zhou, X., Gao, G., Kung, H. F., Rader, D. J., Wilson, J. M., Glickson, J. D., Zhou, R. (2005). Creatine kinase, a magnetic resonance-detectable marker gene for quantification of liver-directed gene transfer. *Hum Gene Ther* 16: 1429–1438.

113 Askenasy, N., Koretsky, A. P. (2002). Transgenic livers expressing mitochondrial and cytosolic CK: mitochondrial CK modulates free ADP levels. *Am J Physiol Cell Physiol* 282: C338–C346.

114 Walter, G., Barton, E. R., Sweeney, H. L. (2000). Noninvasive measurement of gene expression in skeletal muscle. *Proc Natl Acad Sci USA* 97: 5151–5155.

115 Landis, C. S., Yamanouchi, K., Zhou, H., Mohan, S., Roy-Chowdhury, N., Shafritz, D. A., Koretsky, A., Roy-Chowdhury, J., Hetherington, H. P., Guha, C. Noninvasive evaluation of liver repopulation by transplanted hepatocytes using 31P MRS imaging in mice. *Hepatology* 44(5): 1250–1258.

116 Ki, S., Sugihara, F., Kasahara, K., Tochio, H., Okada-Marubayashi, A., Tomita, S., Morita, M., Ikeguchi, M., Shirakawa, M., Kokubo, T. (2006). A novel magnetic resonance-based method to measure gene expression in living cells. *Nucleic Acids Res* 34: e51.

117 Weiss, R. G., Gerstenblith, G., Bottomley, P. A. (2005). ATP flux through creatine kinase in the normal, stressed, and failing human heart. *Proc Natl Acad Sci USA* 102: 808–813.

118 Stegman, L. D., Rehemtulla, A., Beattie, B., Kievit, E., Lawrence, T. S., Blasberg, R. G., Tjuvajev, J. G., Ross, B. D. (1999). Noninvasive quantitation of cytosine deaminase transgene expression in human tumor xenografts with in vivo magnetic resonance spectroscopy. *Proc Natl Acad Sci USA* 96: 9821–9826.

119 Kraitchman, D. L., Bulte, J. W. (2008). Imaging of stem cells using MRI. *Basic Res Cardiol* 103(2): 105–113.

120 Walczak, P., Kedziorek, D. A., Gilad, A. A., Barnett, B. P., Bulte, J. W. M. (2007). Applicability and limitations of MR tracking of neural stem cells with asymmetric cell division and rapid turnover: the case of the shiverer dysmyelinated mouse brain. *Magn Reson Med* 58: 261–269.

121 Serganova, I., Blasberg, R. (2005). Reporter gene imaging: potential impact on therapy. *Nucl Med Bio* 32: 763–780.

122 Ahrens, E. T., Flores, R., Xu, H., Morel, P. A. (2005). In vivo imaging platform for tracking immunotherapeutic cells. *Nature Biotech* 23: 983–987.

123 Harisinghani, M. G., Barentsz, J., Hahn, P. F., Deserno, W. M., Tabatabaei, S., van de Kaa, C. H., de la Rosette, J., Weissleder, R. (2003). Noninvasive detection of clinically occult lymph-node metastases in prostate cancer. *N Engl J Med* 348(25): 2491–2499.

Enhancing Reporter Gene Imaging Techniques

5

Multimodality Imaging of Reporter Genes

Pritha Ray and Sanjiv Sam Gambhir

INTRODUCTION

Reporter genes (RGs), an integral part of molecular imaging, have become essential tools for studying biology in living subjects noninvasively. Currently, molecular imaging techniques can be broadly classified into five categories based on the spectrum and source of energy used for detection [1]. These are optical imaging (fluorescence and bioluminescence imaging), radionuclide imaging (positron emission tomography (PET) and single photon emission computed tomography (SPECT), X-ray computed tomography imaging (CT), magnetic resonance imaging (MRI), and ultrasound (US) imaging. Excluding CT and US, a variety of reporter genes have been developed for the remaining three categories, which can be used to study specific biological processes (such as promoter activation, transcription, translation, protein–protein interaction) and monitor disease progression and therapy (Figure 5.1). Reporter genes therefore are also categorized into different groups based on their usage for different imaging techniques [1].

REPORTER GENES

Optical Reporter Genes

By definition, an optical reporter protein can emit light in the visible range (300 nm–600 nm) either by interacting with specific substrates (luminescence) or by being excited with light of specific wavelength (fluorescence). The emitted light can then be captured in a sensitive charge coupled device (CCD) camera and presented as an optical signature [2]. Both luminescence and fluorescence reporter genes have advantages and disadvantages that carry over to their *in vivo* imaging instrumentation and their application to noninvasive imaging. The luminescent reporter genes are commonly known as luciferases. Luciferase proteins (translated from luciferase genes) were originally isolated from different beetles, bacteria, and marine organisms. Of the four luciferase reporter genes (Firefly, Renilla, click beetle, and Gaussia), the Firefly and Renilla luciferases are most commonly used for *in vitro* research. Recently, both these genes have been adapted and successfully employed for *in vivo* imaging in smaller animals [2–4]. These genes are still being improved by directed mutagenesis for higher light output, longer wavelength emission (for better tissue penetration), and better stability [5–9]. Due to low photon generation by the bioluminescence reaction, luciferase genes had limited use for imaging molecular events from single live cell; however, invention of highly sophisticated optical cameras is now facilitating the imaging of single live cells by bioluminescence [10, 11]. In contrast, fluorescence reporter genes are already ideal and have been used extensively for decades to image molecular events in single live cell due to the high level of photon output from fluorescence reactions [12]. These fluorescence reporters are primarily isolated from marine organisms and can be classified into two major groups (green fluorescent and red fluorescent proteins) based on their fluorophores. Currently, only a few highly sophisticated optical cameras are available (Xenogen, Maestro, eOPtics, etc.), equipped with specialized filters for whole-body fluorescence imaging. In contrast to bioluminescence imaging, fluorescence imaging always suffers from higher background signal due to the auto-fluorescing nature of the biomolecules [13]. Researchers are actively mutating these existing fluorescent proteins for higher brightness, photostability, monomerization, and longer emission wavelength [14]. These optimized fluorescent reporters would have greater potential to be imaged over tissue autofluorescence and therefore would be more suitable for imaging in living animals. (Please see Chapters 1 and 2 for detailed teachings on the topics of fluorescence and bioluminescence RGs, respectively.)

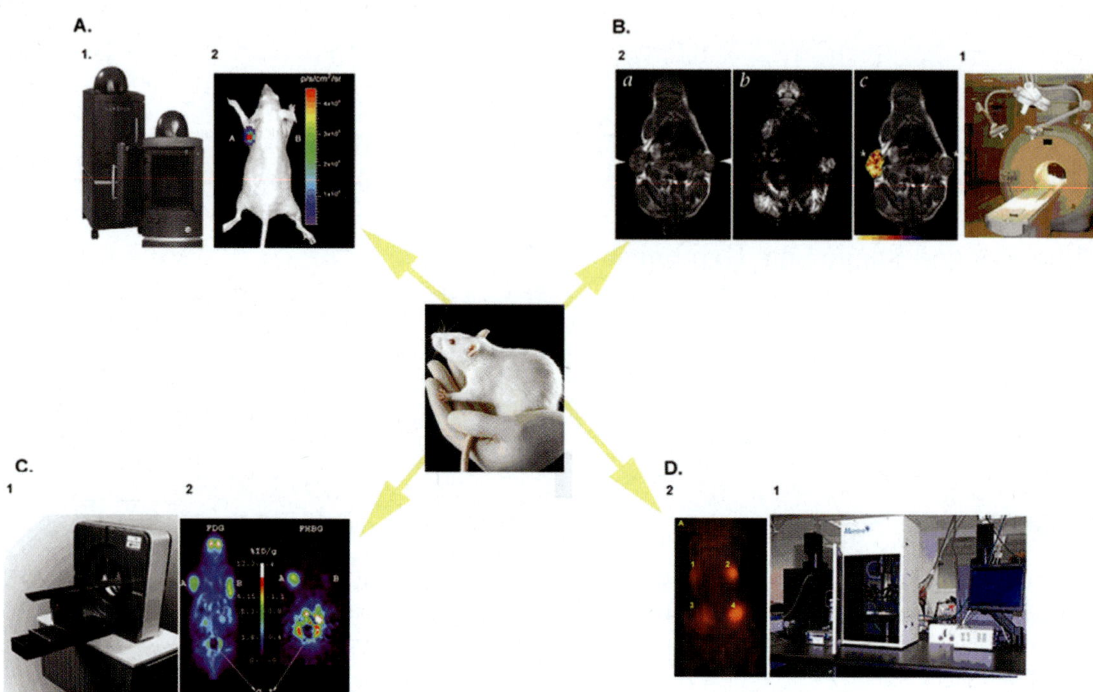

Figure 5.1. Multiple imaging modalities are available for small-animal molecular imaging. Shown are views of typical instruments available and illustrative examples of the variety of images obtained with these modalities. (A) Bioluminescence imaging: (A.1) The Xenogen IVIS-200 cooled CCD camera. **(A.2)** An optical bioluminescence image of a mouse with a subcutaneous xenograft expressing Renilla luciferase in the left shoulder region, after tail-vein injection of the substrate coelenterazine. Images were obtained using a cooled CCD camera. The color image of visible light is superimposed on a photographic image of the mouse with a scale in photons per second per square centimeter per steradian (sr) (reprinted from Ref. [51] with permission). **(B) Magnetic resonance imaging. (B.1)** A microMRI from General Electric. **(B.2)** *In vivo* MR imaging of a single mouse with ETR+ (left arrowheads) and ETR− (right arrowheads) flank tumors. *a*, T1-weighted coronal SE image (imaging time, 3.5 min; voxel resolution, $300 \times 300 \times 3000$ μm). ETR− and ETR+ tumors have similar signal intensities. *b*, T2-weighted gradient-echo image corresponding to the image in *a* showing substantial differences between ETR− and ETR+ tumors (imaging time, 8 min; voxel resolution, $300 \times 300 \times 3000$ μm). As expected, ETR-mediated cellular accumulation of the superparamagnetic probe decreases signal intensity. These differences in MR signal intensity were most pronounced using T2- and T2*-weighted imaging pulse sequences, consistent with the increased transverse relaxation rate (R2) after cellular internalization. *c*, Composite image of a T1-weighted spin-echo image obtained for anatomic detail with superimposed R2 changes after Tf-MION administration, as a color map. *, difference in R2 changes between the ETR+ and ETR− tumors. Scale bar (bottom left) represents 10 mm, n = 1 (reprinted from Ref. [87] with permission). **(C) microPET Imaging: (C.1)** A representative image of a microPET scanner (R4) from Siemens. **(C.2)** MicroPET imaging of a nude mouse carrying tumor stably expressing a fusion ($tk_{20}rl$) protein **(A)** and a control tumor **(B)**. The mouse was imaged by microPET using FHBG (*left panel*) and again using FDG on the following day (*right panel*). The tumor formed by $tk_{20}rl$-expressing cells shows high FHBG accumulation in comparison to the control tumor. The FDG image represents the viability of both $tk_{20}rl$ and control tumor. Nonspecific accumulation of tracer was found in the gastrointestinal tracts (GI), bladder in case of FHBG (attributable to clearance of tracer), and in GI tract and brain in case of FDG (attributable to high metabolic activity) (reprinted from Ref. [51] with permission). **(D) Fluorescence imaging: (D.1)** A representative image of Maestro (CRI) *in vivo* fluorescence imaging camera. **(D.2)** Fluorescence image of a nude mouse implanted with 2 and 4 million 293T cells transiently transfected with *jred* (*left*) and *mrfp1* (*right*) reporters in fusion backbone, respectively (sites *A1* and *A2* have 2 million cells and sites *A3* and *A4* have 4 million cells) using Maestro system. Note that the JRED reporter exhibits lesser fluorescence than mRFP1 reporter in the fusion backbone (reprinted from Ref. [7] with permission).

Radionuclide Reporter Genes

This class of reporter genes encodes for proteins that interact with small molecules labeled with a radioisotope resulting in the accumulation of their probes within or on the surface of cells that express the reporter genes. Depending on the rays emitted by those radioisotopes (either high-energy positron or gamma rays), PET or SPECT scanners are used to detect the location of the reporter probes from any depth within the animal [1]. For example, the ^{99}Tc-labeled probe for somatostatin receptor (SSTr2) is suitable for SPECT imaging, and the ^{18}F-labeled probe for HSV1-thymidine kinase (HSV1-tk) is imageable by a PET scanner [1]. (Please see Chapter 3.)

Magnetic Resonance (MR) Reporter Genes

Compared to radionuclide or optical reporter genes, the development of MR reporter genes is still in its early stage and needs more improvement and validation. Nevertheless, significant progress has been made during the last few years. In general, three classes of reporter genes can generate MR signals or contrast: (1) enzyme-based cleavage of functional groups that block water (proton) exchange or protein binding of MR contrast agents (e.g., β-galactosidase and gadolinium-based substrate (EgadMe)); (2) expression of surface receptors that enable binding of specific MR contrast agents (e.g., transferrin receptors); (3) expression of para or antiferromagnetic (metallo) proteins involved with iron metabolism (e.g., ferritin and tyrosinase). (15) (Please see Chapter 4.)

Comparison of Molecular Imaging Modalities

High sensitivity, high resolution, ability to obtain tomographic images, cost, and ease of use are the five main criteria researchers consider when choosing reporter gene systems to apply in their specific research area. No single existing reporter gene imaging system is superior to the others in all these criteria. Although bioluminescence imaging has excellent sensitivity (possibly $10^{-15}-10^{-17}$ mole/L) and is user friendly, the current imaging instrumentation cannot obtain tomographic images. Furthermore, bioluminescence imaging does not allow volumetric quantification, provides limited depth information, and suffers from low resolution. Currently available fluorescence imaging techniques are not sensitive due to absorption of excitation photons prior to reaching the fluorescent protein, light scattering, attenuation of emission photons by tissues, and high background signal from living mice. However, currently fluorescence imaging is the best modality for imaging molecular events in cultured live cells with high resolution and sensitivity. Both bioluminescence and fluorescence *in vivo* imaging techniques need mass levels of substrates (micrograms to milligrams) to generate sufficient signals. PET or SPECT imaging has lower sensitivity ($10^{-11}-10^{-12}$ mole/L) but higher resolution than bioluminescence imaging. They generate tomographic images and do not suffer from significant tissue attenuation and require trace amounts of probe (picograms to nanograms) to generate sufficient signal over background. MRI is probably the least used imaging technique for small animals but has higher resolution than optical imaging and does not suffer any attenuation in tissues. Low sensitivity of detection ($10^{-3}-10^{-5}$ mole/L), and need for mass doses of contrast agents/substrates (micrograms to milligrams) are some limitations of MR imaging.

MULTIMODALITY MOLECULAR IMAGING

A new imaging approach has emerged in the last decade to overcome the shortcomings of each individual modality either by combining several reporter genes in a single vector or by integrating different modalities in a single instrument. This chapter will focus only on the construction and application of these unique multimodality reporter vectors (Figure 5.2). By definition, a multimodality-imaging vector consists of several reporter genes that can be inserted in a specific cell type or a viral or plasmid vector and then be injected in small animals and monitored by various imaging modalities. The motivation to build these multimodality vectors is to take advantage of the strengths of each imaging modality and to allow the biologist to not worry about a particular modality but instead focus on the biology and obtain comprehensive knowledge. There are four main strategies to build these multimodality imaging vectors.

Multimodality Fusion Reporters

Two or more reporter genes, joined together with short linkers controlled by a single promoter, which gives rise to a single transcript and a single polypeptide. The length of the linker and orientation of each gene are crucial factors to retain the properties of each component protein. Sometimes one or more reporter proteins lose their activity when fused together. However, even if partial functionality can be retained, the fusion reporter vector is the best multimodality vector because decoupling (as seen in the dual promoter-based approach) or differential attenuation of gene expression in different cell types (as observed in the IRES-based approach) is unlikely to occur [16]. Composition and length of the linkers and orientation of the polypeptides are the critical parameters to construct a successful fusion protein. Proteins like green fluorescent protein and its mutants are flexible enough to tolerate direct fusion with many different proteins at either N or C terminal ends without losing their fluorescence activity [17, 18]. However, a large number of proteins are sensitive to fusions and need careful strategies to preserve their function. Several studies on the linker selections suggest that length, flexibility [19–21], and nature of amino acids [22] are all important factors for preserving the functions of protein domains. A short linker (2–5 amino acid long) often creates a fusion protein with reduced or total loss of activity due to physical interference of the protein domains [23]. Conversely, a long linker (40–100 amino acid long) could potentially act as a short polypeptide that could affect the active quaternary folding of the fusion protein and exert undesirable effect on the function of the fusion protein. A linker sequence with high propensity for forming α-helical or β-strand structures might limit the

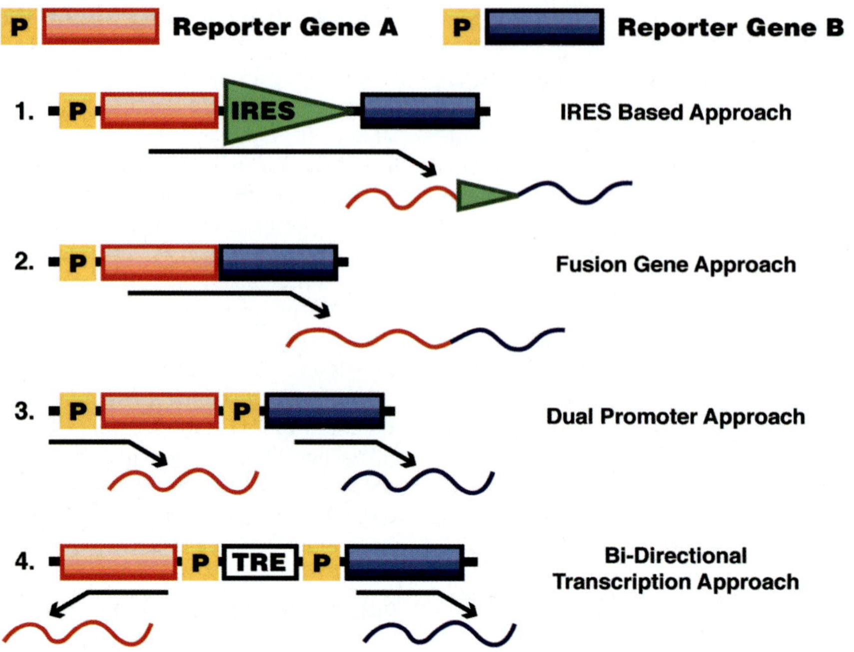

Figure 5.2. Approaches for multimodality reporter gene construct. This schematic diagram demonstrates the four different strategies of building a vector for multimodality imaging. The second reporter gene or a therapeutic gene can be replaced with another reporter gene suitable for a different imaging modality. Currently approaches 1 and 3 are used more frequently than the other two strategies.

flexibility of the fusion protein and consequently affect functional activity [24]. Similarly, presence of highly charged amino acids like arginine, lysine, or proline that potentially forms a kink in the peptide bond, or cysteine that forms a very reactive disulphide bond, could potentially form secondary structures among the protein domains [20]. Care should also be taken to avoid the presence of proteolytic sensitive sites of the common proteases in the linkers. (Unfortunately, there are no universal linkers available to fuse every combination of protein to generate a moderately functional fusion protein.) Another important criterion affecting the functionality of a fusion protein is the orientation of critical (functional) amino acids in the polypeptide. Often amino acids responsible for catalysis, substrate binding, or phosphorylation cluster on either N or C end of the polypeptide. Fusion of another polypeptide on that end can restrict those amino acids to function freely and decrease the catalytic rate or binding of substrate and other properties. Changing the orientation of the polypeptides in these fusion proteins often results in gaining activity [25].

Bi-cistronic or IRES-based Multimodality Reporter

The internal ribosome entry site (IRES) is a small (450–800 bp) DNA sequence found in polioviruses and encephalomyocarditis viruses that was further improved as a new gateway for expressing two DNA sequences from a single cassette [26]. In an IRES-mediated bi-cistronic vector, one single transcript but two different polypeptides are produced. The expression of the downstream gene in an IRES-based vector usually suffers attenuation, and addition of a second or third reporter through an IRES cassette significantly decreases the expression. Moreover, an IRES may have different behavior in different cell lines, and the expression of both genes may not be well correlated [27]. Chappell et al. have recently demonstrated that the 196-nucleotide (nt) 5' UTR of the Gtx homeodomain mRNA contains an IRES and that multiple nonoverlapping fragments of this 5' UTR possess IRES activity [28]. They identified a 9-nt segment in one of these fragments that functions independently as an IRES, and multiple copies of this segment lead to a synergistic increase in IRES activity. Wang et al. (2005) directly compared the efficiency of this 9-nt segment (also termed as super IRES or sIRES) with the EMCV-IRES by small-animal imaging using FL and TK reporters and showed that sIRES can lead to 4–6 times higher expression in small living animals. [29].

Dual Promoter-based Multimodality Reporter

Two different genes expressed from distinct promoters within a single vector (e.g., pCMV-D2R-pCMV/

Ubiquitin-HSV1sr39tk) can potentially be a useful way to couple the expression of two genes. However, presence of two promoters increases the size of the vector, and both promoters often compete for the same transcription factors, resulting in reduction and decoupling of the expression levels of both genes. The expression levels of the genes also uncouple if either of the promoter activity gets altered by accumulation of mutations over time.

Bidirectional Multimodality Reporter

When regulated expression of two different reporter genes is sought, the bidirectional approach with both genes expressed under a common promoter and regulatory sequences (e.g., tetracycline-inducible or repressing sequence) is undertaken. This approach leads to bidirectional transcription of two mRNAs and subsequently two proteins. The gene expression level can be modulated by a chemical regulator (e.g., tetracycline) and can be employed in specific researches [30, 31].

Comparison of Different Coupling Strategies

Each of the strategies described previously has strengths and weaknesses and thus could be utilized in specific research problems. The fusion strategy works well with small, monomeric reporter proteins that have less chance of losing activity after fusing them to other polypeptides. This strategy also provides best-correlated expression of all the component proteins. While building multimodality vectors with large, di, or tetrameric reporter proteins, one must follow the IRES-based strategy to obtain best expression level from all polypeptides. Even though the second reporter protein would suffer attenuation in expression level, this strategy will work better than the other strategies mentioned previously. Dual promoter-based strategy is ideal when one wants to track two reporter genes governed by two different promoters to monitor two different biological events in the same cell or same tissue. Finally, the bidirectional strategy is useful for imaging biological events under inducible condition.

Among these four types of multimodality vectors, the fusion reporter vector can incorporate more than two reporter genes without significant loss of expression. For the other three strategies, expression level gets significantly affected with the third or subsequent reporter coupled. In addition, this strategy usually provides best-correlated expression of all component proteins because a single fusion protein is made from the multimodality vector, whereas in other strategies individual reporter proteins are formed from those multimodality vectors. Consequently, any change at the transcriptional or translation level of these vectors would alter the correlation.

Therefore, to date the fusion gene strategy is the most well-liked method of building multimodality reporters in the biomedical research community (Table 5.1).

APPLICATIONS OF MULTIMODALITY MOLECULAR IMAGING

Gene Therapy

Gene therapy is an alternative to traditional pharmaceutical agents as an intended cure for many diseases. It is also probably the most direct procedure to address hereditary or genetically altered diseases. Gene therapy is still in its infancy and will need much validation before large-scale human application [32–34]. Thus the monitoring of gene therapy with molecular imaging, specifically multimodality molecular imaging, will be important, with reporter genes playing a crucial role in gene therapy research. In 2003, Jacobs et al. constructed a series of vectors using fusion and IRES-based approaches using *HSV1-tk* and enhanced green fluorescence protein (*egfp*) as reporter genes and cytosine deaminase (*cd*) as a therapeutic gene, and validated extensively both in cell culture and in living animals. They used the retroviral vector pHSV-GN (a basic retroviral amplicon backbone) mediated transduction of these multimodality vectors in tumors of mice to assess dose-dependent therapeutic effects of the cytosine deaminase (*cd*) gene by micro PET and fluorescence microscopy imaging. So et al. (2004) developed an hNIS-IRES-fl (NIS-sodium iodide symporter gene) bi-cistronic reporter under an artificial cis-acting retinoic acid responsive element (pRARE/NL) and showed increased bioluminescence imaging and higher Tc-99m uptake in retinoid acid-treated tumors [35]. We and other groups have been applying and validating several reporter gene-based gene therapy approaches. To validate the lentiviral-mediated gene therapy in molecular imaging, De et al. (2003) constructed a lentiviral vector (HSV1-tk-IRES-fl) carrying two reporter genes separated by an IRES and imaged neuroblastoma tumors transduced with the virus with bioluminescence and microPET imaging modalities [36]. Wang et al. (2005) validated a new IRES molecule (ten linked copies of GTx (homeodomain protein) for constructing a multimodality vector (*fl-SIRES-HSV1-tk*) that exhibits higher expression and better correlation between two reporter genes (*fl* and *sr39tk*) [29]. Bioluminescence and microPET imaging of *fl* and HSV1-tk reporter gene expression in nude mice bearing N2a (murine neuroglial) tumor xenografts shows the gene expression mediated by SIRES to be fourfold and eightfold higher, respectively, than EMCV IRES. This improved vector will be suitable for many applications in gene therapy, specifically for monitoring efficacy where one of the genes is replaced with a therapeutic gene for long-term monitoring. In another approach, our lab has actively pursued the

Table 5.1. Multimodality Reporter Vectors

Multimodality Fusion Vector	Study	Imaging Modality	Reference
HSV1-TK-GFP	Cell culture	Fluorescence Microscope	(17)
HSV1-TK-GFP	Cell culture	Fluorescence Microscope	(62)
HSV1-TK-GFP	Cell culture and *in vivo* tumors	Fluorescence Microscope, microPET	(50)
HSV1-TK-GFP	Cell culture	Fluorescence Microscope	(63)
HSV1-TK-GFP	Cell trafficking	Fluorescence Microscope, microPET	(57, 64)
HSV1-TK-GFP and DsRed2-XPRT	Hypoxic tumor xenograft model	Fluorescence Microscope, microPET	(49)
HSV1-GFP-TK	Promoter activation	Fluorescence Microscope, microPET	(65)
HSV1-TK-GFP, humanized mitochondrial TK type2-GFP	Transgene activation in lung, tumor xenograft	Fluorescence Microscope, microPET	(66, 67)
RL-GFP	Promoter activation in cell culture and *in vivo*	Fluorescence Microscope, bioluminescence	(68)
FL-GFP	Cell culture, imaging of embryonic stem cell	Fluorescence microscope, Bioluminescence	(38–41, 69)
FL-TK	Cell culture and tumor xenograft	Bioluminescence	(70)
Sr39TK-RL	Cell culture and tumor xenograft	Bioluminescence microPET	(51)
Triple reporter (tk-egfp-fl)	Cell culture, tumor xenograft, hypoxic hepatocarcinoma model, breast metastasis	Fluorescence Microscope, microPET, bioluminescence, and fluorescence camera	(54, 71–74)
Triple reporter (hrl-mrfp-ttk)	Cell culture and lung metastasis, promoter analysis	Fluorescence Microscope, microPET, bioluminescence, and fluorescence camera	(52, 53, 75)
Triple reporter (fl-mrfp-ttk)	Imaging of embryonic stem cell for cardiac application and skeletal regeneration	Fluorescence Microscope, microPET, bioluminescence, and fluorescence camera	(43, 45, 76)
DsRed-xanthine phosphoribosyltransferase-xanthine	Tumor xenograft	Fluorescence microscope and microPET	(77)
Triple reporter (hrl-mrfp-ttk)	Cell trafficking	Bioluminescence and microPET	(58, 59), (60)
dsRed-xanthine phosphoribosyltransferase-xanthine and HSV1-tk/GFP fusion	Tumor xenograft for HIF-1 promoter activity	Fluorescence microscope and microPET	(49)
Cytosine deaminase and uracil phosphoribosyltransferase fused with mDsRed	Tumor xenograft	Fluorescence microscope, 19(F) MR spectroscopy	(78)

Multimodality IRES-Based Vector	Study	Imaging Modality	Reference
NorEpinephrine transporter (hNET)-IRES-GFP	Tumor xenograft	Fluorescence microscope, gamma camera, SPECT, and microPET	(79)
Firefly luciferase-IRES-GFP	Cardiac cell implantation	Fluorescence microscopy, bioluminescence imaging	(80)
Cytosine deaminase-IRES-tkgfp or tkgfp-IRES-cytosine deaminase	Tumor xenograft	Fluorescence microscopy and microPET	(81)
Sodium iodide symporter (NIS)-IRES-firefly luciferase	Tumor xenograft	Bioluminescence and scintigraphic imaging, gamma camera imaging	(82) (35) (83)
TK-IRES-FL	Tumor xenograft	Bioluminescence and PET imaging	(36)

Multimodality Inducible Bidirectional Vector	Study	Imaging Modality	Reference
EGFP-TRE-Ferritin	Transgenic mice	MRI and fluorescence microscopy	(84, 85)
RFP-TRE-sr39TK	Tumor xenograft	Fluorescence microscopy and microPET	(86)

bidirectional approach for surviving-mediated gene therapy in liver metastasis [37] using an adenoviral construct. Currently, this vector carries a bioluminescence reporter (*fl*) in one direction and survivin in another, and the effect is being monitored by bioluminescence imaging after intraliver injection of the vector in living mice. In the future, a multimodality triple-fusion cassette will replace the *fl* gene, therefore, the therapeutic effect could be monitored with multimodality imaging. Multimodality reporter vectors have also emerged as important tools for monitoring gene therapy/stem cell therapy for cardiac applications. The Wu Laboratory at the Stanford University is extensively exploiting the power of multimodality fusion reporter vectors for monitoring cardiac infarction and the effect of stem cell therapy in the infracted heart of small animals. Many important insights such as spatial and temporal effect of stem cell implantation in heart, cell migration, teratoma-forming capacity of embryonic stem cells, and immunogenic properties of stem cells have emerged from their study [38–44]. Lee et al. (2008) have recently investigated the power of adult stem cells (adipose) as therapeutic reagents for skeletal regeneration [45]. Many other investigators are actively monitoring adoptive therapy with tumor-targeted T lymphocytes, adult and embryonic stem cell therapy, and targeted gene therapy in different cancer models with noninvasive molecular imaging that would not have been possible without these multimodality reporter vectors (see Table 5.1).

Protein–Protein Interaction

Progression of life relies upon interactions of cellular proteins at the right place and at the right moment. Rapid advancements in genomics and proteomics enable us to assemble valuable information on the interactions of cellular proteins involved in different signaling pathways. Imaging interactions of these proteins in living subjects are therefore promising, and currently a few studies are under way using multimodality imaging techniques. While we were developing a novel approach for imaging protein–protein interactions with the two-hybrid system in living animals by bioluminescence imaging using the firefly luciferase reporter [46], Luker et al. (2001) were working on a similar approach with a bifusion (HSV1-TK-GFP) fusion protein using fluorescence microscopy and microPET imaging. They showed that interaction of two proteins (p53 and T antigen of SV40 virus) could be imaged via a two-hybrid system in a living cell by fluorescence microscopy and in living mice by microPET. In the absence of specific interacting partners, no fluorescence or microPET signals were produced [47]. Later they extended this study to a doxycycline titratable system to obtain more quantitative data and analyze the time kinetics of p53 and Tag interaction using microPET imaging and fluorescence microscopy in living mice [48].

Further details of imaging protein–protein interactions may be found in Chapter 8.

Cancer Metastasis Imaging

Multimodality imaging of reporter genes has most extensively been applied in studying the progression of cancer and the effects of drug therapy in small-animal models. Among all the vectors described in Section II, the fusion gene approach is clearly the most preferred method for multimodality imaging of cancer. In the history of development of fusion reporter gene imaging, bifusion vectors were initially investigated for PET-fluorescence, PET-bioluminescence or bioluminescence–fluorescence technologies (see Table 5.1). The bifusion TK-eGFP or eGFP-TK proteins were developed independently by many investigators using different length and composition of linkers (see Table 5.1). Most of them were tested only in cell/tissue culture, but a few of them were applied for imaging in living animals often using microPET and fluorescence microscopy. Researchers at MSKCC developed a HSV1-TK-GFP fusion vector and after initial validation in cell culture applied this fusion reporter to image the temporal and spatial dynamics of hypoxia-inducible factor-1 (HIF-1)-mediated activity in tumors of living mice [49]. Ponomarev et al. at MSKCC (2003) later modified the HSV1-tk part of this fusion by either deleting or mutating the nuclear localization signal or by adding a nuclear export signal (NES) to retarget the fusion gene into the cytoplasm [50]. This retargeted fusion vector had higher access to the substrates of TK and thus was more efficient for *in vivo* imaging.

While the MSKCC group was constructing and validating the PET-fluorescence fusion reporter, our lab started building the first PET-bioluminescence reporter vector for multimodality imaging. Ray et al. (2003) first described a renilla luciferase-HSV1-thymidine kinase fusion protein ($TK_{20}RL$) and demonstrated optical bioluminescence and microPET imaging of tumor xenografts of N2a cells stably expressing $tk_{20}rl$ fusion gene in living mice [51]. Soon thereafter, we reported several triple-fusion reporter vectors combining a fluorescence, bioluminescence, and PET reporter gene [52] (Figure 5.3). The novelty of these triple-fusion vectors is enabling the imaging of gene expression in single cultured living cells using fluorescence microscopy, in living small animals using a cooled CCD camera (bioluminescence and fluorescence), and in animals using a microPET scanner (PET imaging). The ability to noninvasively report molecular information ranging from a single living cell to the multicellular environment of a living animal holds tremendous potential for studying cancer metastasis, efficacy of drug therapy, and many other applications in a preclinical setting. Ray et al. (2004) have followed the progression of metastatic melanoma using

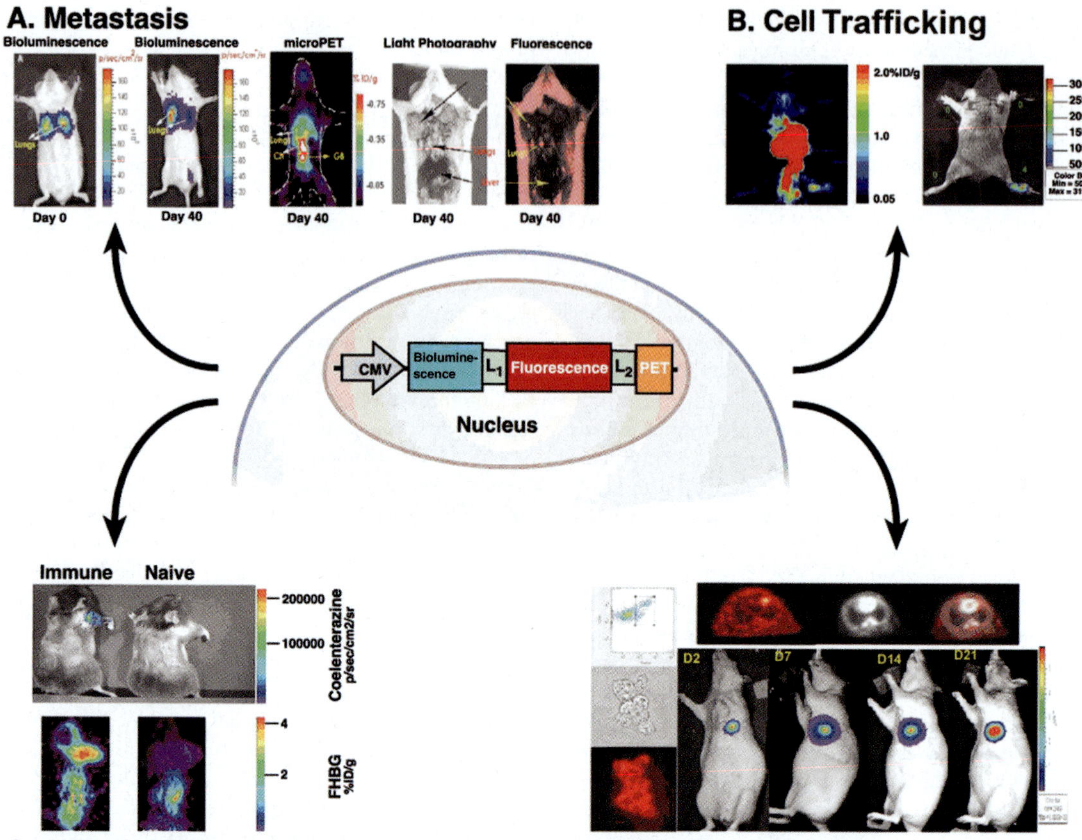

Figure 5.3. **Application of the multimodality triple-fusion reporter vector in biomedical research: Shown here are examples of multimodality imaging of molecular events using fluorescence microscopy, bioluminescence, fluorescence, and microPET modalities**. Shown in the middle is a schematic diagram of a multimodality triple-fusion vector carrying a fluorescence (monomeric red fluorescent protein or enhanced green fluorescent protein), a bioluminescence (firefly luciferase or Renilla luciferase), and a PET (HSV1-sr39 thymidine kinase or HSV1-thymidine kinase) reporter genes connected by two small peptide linkers. Transcription of this fusion vector yields a single mRNA and subsequent translation leads to a single polypeptide capable of retaining partial, if not full, activities of the three proteins fused. **A. Multimodality imaging of metastasis of A375M cells stably expressing the** *hrl-mrfp-ttk* **fusion reporter gene in living mice**: 7×10^5 A375M cells stably expressing the triple fusion were injected via tail vein in an SCID mouse and imaged for bioluminescence signal following tail vein injection of coelenterazine at day 0 and day 40. Prominent bioluminescence signal was found from the region of both the lungs ($1.3–1.5 \times 10^5$ max (p/sec/cm^2/sr)) at day 0 and from left lung (2×10^5 max (p/sec/cm^2/sr)) at day 40. Following a bioluminescence scan, the mouse was imaged in microPET using 18F-FHBG. A strong signal ($\sim$0.78 %ID/g) was present from the chest region (Ch) with lower signal (0.35% ID/g) from the lung region. The stronger PET signal was found to be from a metastatic tumor present deep inside the body as evident from the fluorescence photograph. Note the gallbladder (GB) retains FHBG so background signal from the GB is also seen in the microPET images. Next panel shows the light photograph of the same SCID mouse after sacrifice and organ exposure and rightmost panel demonstrates the whole-body fluorescence imaging of the same SCID mouse showing fluorescing metastatic tumors in lung and chest regions that correspond with the bioluminescence and PET images (reprinted from Ref. (52) with permission). **(B) Multimodality imaging of immune cell trafficking in collagen-induced arthritis mice**. 26×10^6 A2-TFR cells (joint targeting T-cell hybridomas transduced with a lentivirus carrying the hRluc-mRFP-HSV1-tsr39tk triple-fusion gene) were injected into the tail vein of collagen-induced arthritic mice after inflammation had developed in their paws. The mouse was imaged for coelenterazine activity (right panel) and [18F]FHBG distribution (left panel) using the Xenogen CCD camera and the Concorde MicroPET R4, respectively, the day after injection of the cells (reprinted from Ref. [60] with permission). **(C) Multimodality imaging of antitumor lymphocytic infiltrate expressing the** *hrl-egfp-ttk* **fusion reporter by optical bioluminescence and microPET**: Antitumor lymphocytes were transduced with lentivirus containing the triple-fusion reporter gene (*hrl-gfp-ttk*) and adoptively transferred. Day 6 images are shown. For CCD imaging analysis, coelenterazine was injected intravenously and each mouse was immediately imaged for 5 min. [18F]FHBG PET imaging was performed on the same mice on the same day. Only the tumors in the immune mice showed bioluminescence (upper panel) signal and FHBG uptake (lower panel). Tumor in naïve mouse did not show significant bioluminescence or PET signal (reprinted from Ref. [58] with permission). **(D) Multimodality imaging of murine embryonic stem cells (mESCs) in nude rat myocardium**. After lentiviral transduction of mESCs with the *fl-mrfp1-ttk* fusion reporter, about 28% of mESC cells were positive for RFP on FACS, and the cell morphology was unchanged under bright field and fluorescence microscopy (left vertical panel). The upper panel shows the microPET images of transplanted cells scanned with [18F]FHBG (left), [18F]FDG (middle), and fusion image (right) in rat myocardium. Noninvasive bioluminescence imaging of mESCs in rat heart at Day 2, 7, 14, and 21 (from left to right in lower panel) were performed by injecting D-luciferin intraperitoneally (reprinted from Ref. [43] with permission).

all three imaging modalities [52]. Later Deroose et al. (2007) followed the progression of these melanoma cells extensively by bioluminescence, 18F-FHBG-PET, 18F-FDG-PET, and coregistered with anatomical CT images [53]. In their study, the 18F-FHBG PET/CT allowed better detection and localization of lesions present in the thorax and the head that were not seen on CT because of poor contrast resolution and were not seen on 18F-FDG PET because of higher background uptake relative to 18F-FHBG. Subsequently, various investigators have applied these triple-fusion vectors for imaging telomerase promoter expression, immune cell trafficking, cardiac cell therapy, and in many other research arenas (see Table 5.1). The MSKCC group also independently developed a triple-fusion (*HSV1-tk-egfp-fl*) and validated noninvasively by bioluminescence, fluorescence, and gamma camera imaging in a tumor xenograft model of nude mice [54]. To avoid the localization of the fusion protein in nucleus (driven by the nuclear localization signal (NLS) present in the HSV1-tk gene), MSKCC group constructed two HSV1-tk mutants (a deletion mutant and an HSV1-tk with an added nuclear export signal (NES)) and fused with *egfp* and *fl* reporter genes. Both the fusion constructs showed similar levels of activity for *fl*, *egfp*, and HSV1-tk reporter in cell culture and in *in vivo* imaging experiments. To achieve cytoplasmic retargeting, we have generated a deletion mutant of HSV1-tk and fused it with several fluorescent and bioluminescent reporter genes to develop a small library of triple-fusion vectors. This small library offers us a range of combinations of triple-fusion reporter genes suitable for different research applications.

Recently, a multimodality fusion vector (*mrfp1-DEVD-tk-DEVD-fl*) has been developed to image caspase 3 activation in apoptosing cells with fluorescence, bioluminescence, and microPET imaging [55]. The ability to report apoptosis noninvasively from living cells to living animals will be useful in studying many pathological disease models. In this vector three different reporter (fluorescent, bioluminescent, and PET) genes are joined together through a small peptide linker (DEVD) that can be recognized and cleaved by active caspase-3. In the fused form, all three reporter proteins have markedly attenuated activity. However, when activated caspase-3 recognizes and cleaves the fusion protein into its individual components, each reporter protein shows an increase in activity. This unique caspase sensor vector therefore can report the event and levels of caspase-3 activation indirectly by an increase in reporter protein signal. Using melanoma cells stably expressing this sensor and treated with protein kinase C inhibitors, the authors could image the activation of caspase 3 in living cells by fluorescence and bioluminescence imaging and in tumors of living mice by bioluminescence and microPET imaging. This multimodality caspase sensor vector could effectively and noninvasively monitor caspase-3 activation from single live cells to a multicellular tumor environment and thus would be a valuable tool for drug screening in preclinical models and future patient cell-based therapy.

Cell Trafficking and Cell Therapy Imaging

The ability to noninvasively monitor cell trafficking *in vivo* in a longitudinal fashion is a pressing need for emerging cellular therapeutic strategies. Cell trafficking or cell therapy imaging therefore is an emerging area of modern biomedicine especially in the context of stem cell therapy. However, sensitivity is always a challenging issue due to the low numbers of cells that ultimately reach the target tissues (e.g., tumors). A recombinant protein with fluorescent and bioluminescent characteristics might take advantage of the strengths of both reporters. For example, a TK-GFP dual-reporter gene was used to monitor NFAT-mediated transcriptional activation in human Jurkat cells [56]. Later this TK-GFP fusion was also used to monitor the migration of adoptively transferred T-cells in human tumor xenograft bearing nude mice by serial microPET imaging [57]. Recently Kim et al. (2004) showed that migration of 1.5×10^6 T cells stably expressing the *hrl-egfp-ttk* gene could be imaged in the tumors by both microPET and bioluminescence [58]. Later Shu et al. (2005) described a noninvasive and tomographic method to visualize primary antitumor immune response with microPET using bone marrow chimeric mice generated by engraftment of hematopoietic stem and progenitor cells transduced with the *hrl-egfp-ttk* reporter gene [59]. A study by our laboratory showed that dendritic cells carrying a GFP–luciferase fusion indeed traffic to inflamed joints, and at least some of them persist for 14 d by bioluminescent imaging (unpublished data). In another study, Yaghoubi et al. have also demonstrated that T-cell hybridomas expressing *hrl-mrfp-ttk* fusion reporter proteins migrate to inflamed paws of collagen induced arthritic (CIA) mice [60]. Quantified data from optical images demonstrate that the higher the degree of inflammation, the greater the number of T-cell hybridomas trafficking into the paws [60]. However, increasing the number of cells injected intravenously did not result in greater number of cells trafficking to the paws.

Stem cells are primal, undifferentiated cells that have the unique potential to produce any kind of cell in the body. Stem cell therapy is an exciting and challenging area with the potential to revolutionize medicine by repairing the tissue damage in many pathological diseases. Molecular imaging may play an important role in this exciting field, especially to follow the fate of transplanted stem cells labeled with reporter genes, and thus predict the outcome of therapy as demonstrated in a study by Cao et al. (2006) [43]. In an elegant study, they labeled murine embryonic stem cells (mES cells) with a triple-fusion reporter (*fl-mrfp-ttk*), injected them into the myocardium of adult nude rats, and followed the fate of the cells with both bioluminescence and microPET imaging. This study clearly demonstrated the potency of the embryonic stem cells to differentiate into teratomas, a challenging side effect of stem cell therapy. In a later study the same group demonstrated the usefulness of the therapeutic characteristic of a reporter gene (thymidine kinase with ganciclovir treatment) to monitor therapy simultaneously with multimodality imaging [44]. They transduced the same mES cells with two different multimodality fusion reporters (a triple-fusion reporter (TF) as described previously and a bifusion reporter (DF) consisting of firefly luciferase and green fluorescence protein) and implanted them in shoulders of adult female mice. Cell survival was followed noninvasively by both bioluminescence and microPET imaging, and teratomas were found in all mice implanted with mES-TF and mES-DF cells. Administration of ganciclovir (GCV) targeting the HSV-ttk gene resulted in selective ablation of teratomas arising from the ES-TF cells but not ES-DF cells. This study demonstrates an important aspect of cell therapy where imaging property of a reporter can be coupled with its therapeutic effect [44].

FUTURE OF MULTIMODALITY IMAGING

Noninvasive multimodality imaging specifically with fusion reporter vectors is rapidly becoming the method of choice for many applications in biomedical science research. A few significant studies with small-animal imaging have been described in this chapter, and examples of more applications (validated mostly in cell culture) can be found elsewhere. Improvement in the existing vectors by using better (mutated) reporter genes, in combination with better reporter probes and detection systems, will help maximize sensitivity and specificity. Reporter genes/proteins currently in use have probably gone through several rounds of improvements for higher light output, longer wavelength, better substrate utilization, efficient transcription and translation in mammalian cells. However, we still lack the ideal reporter gene for each imaging modality. For example, after several rounds of improvement of the red fluorescent proteins, we still require a fluorescent protein with emission spectra in the near-infrared (NIR-650–900 nm) region to overcome tissue attenuation and absorption issues for *in vivo* imaging. Similarly, HSV1-thymidine kinase gene, the most validated PET reporter gene, has been mutated for selective substrate (either purine, for example, or pyrimidine-based nucleoside analogs) to develop as suicide gene in combination with GCV and as a nuclear imaging reporter gene with appropriate reporter probe. Many of the purine- (gancyclovir developed by Roche and acyclovir developed by GlaxoSmithKline) and pyrimidine (bromovinyldeoxyuridine or

BVdU developed by ZOSTEX)-based drugs are used in clinic for antiviral therapy. Scientists are now actively engaged to develop HSV1-tk mutants that will have least specificity for either purine- or pyrimidine-based analogs while having high-level phosphorylating activity for the other group of analogs [11]. This will be advantageous for patients undergoing suicide gene therapy with mutant HSV1-tk without being affected by drugs applied for antiviral therapy. One such mutant has already been constructed and validated in small-animal model by microPET imaging, and many more are in the pipeline. There is a need to explore new reporter genes with less immunogenicity, new PET reporter gene/probe systems to work behind the blood–brain barrier suitable for brain applications, and finally development of multimodality reporters incorporating all these new reporter genes and their mutated (improved) versions.

Until today multimodality reporter vectors were constructed mostly with combinations of optical (fluorescence and bioluminescence) and optical and PET (fluorescence and PET or bioluminescence or PET or fluorescence, bioluminescence and PET) reporter genes and imaged with fluorescence microscope, optical imaging, and PET imaging. Some other functional imaging modalities (such as MR or photo acoustics) are also taking advantage of reporter genes to image fine molecular changes required for medical applications, including early detection and treatment of disease and basic pharmaceutical development. Therefore it would be desirable to have a multireporter system in which one of the reporter components will be replaced by an MR (e.g., ferritin) or photo acoustic (e.g., β-galactosidase) reporter gene. In this way, the high resolution of MRI could be combined with the sensitivity and straightforward whole-body hot spot interpretation of PET or bioluminescent imaging, as well as histological validation with GFP or RFP. Photo acoustic imaging, a hybrid biomedical imaging modality, is probably one of the most powerful future imaging techniques with strong clinical potential. In photo acoustics imaging, nonionizing laser pulses are delivered to biological tissues, and the signal from a photo acoustic agent/reporter gene is imaged in the form of ultrasonic waves. Photo acoustic imaging of living subjects offers higher spatial resolution and allows deeper tissues to be imaged compared with most optical imaging techniques [61]. A future multimodality reporter vector combining PET and fluorescence imaging with photo acoustic imaging will likely contribute to a much better understanding of complex molecular pathways from single living cell with higher sensitivity and from living subjects with higher sensitivity, specificity in a tomographic and quantitative manner. Multimodality imaging of these multimodality reporter genes will facilitate rapid translation of approaches developed in cells to preclinical models and clinical applications.

ACKNOWLEDGEMENT

We would like to thank Drs. Shahriar Yaghoubi and Abhijit De for their comments on the manuscript. Financial support from Canary Foundation, National Cancer Institute and Doris Duke Foundation are duly acknowledged.

REFERENCES

1 Massoud, T., Gambhir, S. (2003). Molecular imaging in living subjects: seeing fundamental biological processes in a new light. *Genes Dev* 17: 545–580.

2 Contag, C. H., Bachmann, M. H. (2002). Advances in in vivo bioluminescence imaging of gene expression. *Annu Rev Biomed Eng* 4: 235–260.

3 Sato, A., Klaunberg, B., Tolwani, R. (2004). In vivo bioluminescence imaging. *Comp Med* 54: 631–634.

4 Bhaumik, S., Gambhir, S. S. (2002). Optical imaging of Renilla luciferase reporter gene expression in living mice. *Proc Natl Acad Sci U S A* 99: 377–382.

5 Braggett, B., Roy, R., Momen, S., Morgan, S., Tisi, L., Morse, D., Gillies, R. J. (2004). Thermostability of firefly luciferase affects efficiency by in vivo bioluminescence. *Molecular Imaging* 3: 324–332.

6 Branchini, B. R., Ablamsky, D. M., Murtiashaw, M. H., Uzasci, L., Fraga, H., Southworth, T. L. (2007). Thermostable red and green light-producing firefly luciferase mutants for bioluminescent reporter applications. *Anal Biochem* 361: 253–262.

7 Ray, P., Tsien, R., Gambhir, S. S. (2007). Construction and validation of improved triple fusion reporter gene vectors for molecular imaging of living subjects. *Cancer Res* 67: 3085–3093.

8 Loening, A. M., Fenn, T. D., Wu, A. M., Gambhir, S. S. (2006). Consensus guided mutagenesis of Renilla luciferase yields enhanced stability and light output. *Protein Eng Des Sel* 19: 391–400.

9 Loening, A. M., Wu, A. M., Gambhir, S. S. (2007). Red-shifted Renilla reniformis luciferase variants for imaging in living subjects. *Nat Methods* 4: 641–643.

10 Gammon, S. T., Leevy, W. M., Gross, S., Gokel, G. W., Piwnica-Worms, D. (2006). Spectral unmixing of multicolored bioluminescence emitted from heterogeneous biological sources. *Anal Chem* 78: 1520–1527.

11 Bradbury, M. S., Panagiotakos, G., Chan, B. K., Tomishima, M., Zanzonico, P., Vider, J., Ponomarev, V., Studer, L., Tabar, V. (2007). Optical bioluminescence imaging of human ES cell progeny in the rodent CNS. *J Neurochem* 102: 2029–2039.

12 Muller-Taubenberger, A., Anderson, K. I. (2007). Recent advances using green and red fluorescent protein variants. *Appl Microbiol Biotechnol* 77: 1–12.

13 Troy, T., Jekic-McMullen, D., Sambucetti, L., Rice, B. (2004). Quantitative comparison of the sensitivity of detection of fluorescent and bioluminescent reporters in animal models. *Mol Imaging* 3: 9–23.

14 Shanerm, N. C., Steinbach, P. A., Giepmans, B. N., Palmer, A. E., Tsien, R. Y. (2004). Improved monomeric red, orange and yellow fluorescent proteins derived from Discosoma sp. red fluorescent protein. *Nat Biotechnol* 22, 1567–1572.

15 Gilad, A. A., Winnard, P. T., Jr., van Zijl, P. C., Bulte, J. W. (2007). Developing MR reporter genes: promises and pitfalls. *NMR Biomed* 20, 275–290.

16 Ray, P., Bauer, E., Iyer, M., Barrio, J. R., Satyamurthy, N., Phelps, M. E., Herschman, H. R., Gambhir, S. S. (2001). Monitoring gene therapy with reporter gene imaging. *Semin Nucl Med* 31: 312–320.

17 Jacobs, A., Dubrovin, M., Hewett, J., Sena-Esteves, M., Tan, C. W., Slack, M., Sadelain, M., Breakefield, X. O., Tjuvajev, J. G. (1999). Functional coexpression of HSV-1 thymidine kinase and green fluorescent protein: implications for noninvasive imaging of transgene expression. *Neoplasia* 1: 154–161.

18 Hoggarth, J., Jones, A. E., Meredith, D. M. (2004). Functional expression of thymidine kinase in human leukaemic and colorectal cells, delivered as EGFP fusion protein by herpesvirus saimiri-based vector. *Cancer Gene Therapy* 11, 613–624.

19 Crasto, C. J., Feng, J. A. (2000). LINKER: a program to generate linker sequences for fusion proteins. *Protein Engineering* 13, 309–312.

20 Robinson, C. R., Sauer, R. T. (1998). Optimizing the stability of single-chain proteins by linker length and composition mutagenesis. *Proc Nat Acad Sci USA* 95, 5929–5934.

21 Gustavsson, M., Lethao, S. D., Teeri, T. T., Hult, K., Martinelle, M. (2001). Stable linker peptides for a cellulose-binding domain-lipase fusion protein expressed in Pichia pastoris. *Protein Engineering* 14, 711–715.

22 Witte, V., Wolf, H. D. (1996). Clostripain linker deletion variants yield active enzyme in Escherichia coli: A possible function of the linker pepetide as intramolecular inhibitor of Clostripain automaturation. *Curr Microbiol* 33: 281–286.

23 Maeda, Y., Ueda, H., Kazami, J., Kawano, G., Suzuki, E., Nagamune, T. (1997). Engineering of functional chimeric protein G-Vargula luciferase. *Anal Biochem* 249: 147–152.

24 Arai, R., Ueda, H., Kitayama, A., Kamiya, N., T Nagamune. (2001). Design of linkers which effectively separate domains of a bifunctional fusion protein. *Protein Engineering* 14: 529–532.

25 Soling, A., Theiss, C., Jungmichel, S., Rainov, N.G. (2004). A dual function fusion protein of Herpes simplex virus type 1 thymidine kinase and firefly luciferase for noninvasive in vivo imaging of gene therapy in malignant glioma. *Genet Vaccines* 2: 7.

26 Stoneley, M., Willis, A. E. (2004). Cellular internal ribosome entry segments: structures, trans-acting factors and regulation of gene expression. *Oncogene* 23: 3200–3207.

27 Ngoi, S. M., Chien, A. C., Lee, C. G. (2004). Exploiting internal ribosome entry sites in gene therapy vector design. *Current Gene Therapy* 4: 15–31.

28 Chappell, S. A., Edelman, G. M., and Mauro, V. P. (2004). Biochemical and functional analysis of a 9-nt RNA sequence that affects translation efficiency in eukaryotic cells. *Proc Natl Acad Sci U S A* 101: 9590–9594.

29 Wang, Y., Iyer, M., Annala, A. J., Chappell, S., Mauro, V., Gambhir, S. S. (2005). Noninvasive monitoring of target gene expression by imaging reporter gene expression in

living animals using improved bicistronic vectors. *J Nucl Med* **46**: 667–674.

30 Sun, X., Annala, A. J., Yaghoubi, S. S., Barrio, J. R., Nguyen, K. N., Toyokuni, T., Satyamurthy, N., Namavari, M., Phelps, M. E., Herschman, H. R., Gambhir, S. S. (2001). Quantitative imaging of gene induction in living animals. *Gene Ther* **8**: 1572–1579.

31 Gafni, Y., Pelled, G., Zilberman, Y., Turgeman, G., Apparailly, F., Yotvat, H., Galun, E., Gazit, Z., Jorgensen, C., Gazit, D. (2004). Gene therapy platform for bone regeneration using an exogenously regulated, AAV-2-based gene expression system. *Mol Ther* **9**: 587–595.

32 Rueger, M. A., Winkeler, A., Thomas, A. V., Kracht, L. W., Jacobs, A. H. (2008). Molecular imaging-guided gene therapy of gliomas. *Handb Exp Pharmacol* 341–359.

33 Gray, S. J., Samulski, R. J. (2008). Optimizing gene delivery vectors for the treatment of heart disease. *Expert Opin Biol Ther* **8**: 911–922.

34 Briat, A., Vassaux, G. (2006). Preclinical applications of imaging for cancer gene therapy. *Expert Rev Mol Med* **8**: 1–19.

35 So, M. K., Kang, J. H., Chung, J. K., Lee, Y. J., Shin, J. H., Kim, K. I., Jeong, J. M., Lee, D. S., Lee, M. C. (2004). In vivo imaging of retinoic acid receptor activity using a sodium/iodide symporter and luciferase dual imaging reporter gene. *Mol Imaging* **3**: 163–171.

36 De, A., Lewis, X. Z., Gambhir, S. S. (2003). Noninvasive imaging of lentiviral-mediated reporter gene expression in living mice. *Mol Ther* **7**: 681–691.

37 Ray, S., Paulmurugan, R., Patel, M. R., Ahn, B. C., Wu, L., Carey, M., Gambhir, S. S. (2008). Noninvasive imaging of therapeutic gene expression using a bidirectional transcriptional amplification strategy. *Mol Ther* **16**: 1848–1856.

38 van der Bogt, K. E., Sheikh, A. Y., Schrepfer, S., Hoyt, G., Cao, F., Ransohoff, K. J., Swijnenburg, R. J., Pearl, J., Lee, A., Fischbein, M., Contag, C. H., Robbins, R. C., Wu, J. C. (2008). Comparison of different adult stem cell types for treatment of myocardial ischemia. *Circulation* **118**: S121–S129.

39 Swijnenburg, R. J., Schrepfer, S., Cao, F., Pearl, J. I., Xie, X., Connolly, A. J., Robbins, R. C., Wu, J. (2008). In vivo imaging of embryonic stem cells reveals patterns of survival and rejection following transplantation. *Stem Cells and Development* **17**: 1023–1029.

40 Swijnenburg, R. J., Schrepfer, S., Govaert, J. A., Cao, F., Ransohoff, K., Sheikh, A. Y., Haddad, M., Connolly, A. J., Davis, M. M., Robbins, R. C., Wu, J. C. (2008). Immunosuppressive therapy mitigates immunological rejection of human embryonic stem cell xenografts. *Proc Natl Acad Sci U S A* **105**: 12991–12996.

41 Li, Z., Suzuki, Y., Huang, M., Cao, F., Xie, X., Connolly, A. J., Yang, P. C., Wu, J. C. (2008). Comparison of reporter gene and iron particle labeling for tracking fate of human embryonic stem cells and differentiated endothelial cells in living subjects. *Stem Cells* (Dayton, Ohio) **26**: 864–873.

42 Cao, F., Drukker, M., Lin, S., Sheikh, A. Y., Xie, X., Li, Z., Connolly, A. J., Weissman, I. L., Wu, J. C. (2007). Molecular imaging of embryonic stem cell misbehavior and suicide gene ablation. *Cloning and Stem Cells* **9**: 107–117.

43 Cao, F., Lin, S., Xie, X., Ray, P., Patel, M., Zhang, X., Drukker, M., Dylla, S. J., Connolly, A. J., Chen, X., Weissman, I. L., Gambhir, S. S., Wu, J. C. (2006). In vivo visualization of embryonic stem cell survival, proliferation, and migration after cardiac delivery. *Circulation* **113**: 1005–1014.

44 Cao, F., van der Bogt, K. E., Sadrzadeh, A., Xie, X., Sheikh, A. Y., Wang, H., Connolly, A. J., Robbins, R. C., Wu, J. C. (2007). Spatial and temporal kinetics of teratoma formation from murine embryonic stem cell transplantation. *Stem Cells and Development* **16**: 883–891.

45 Lee, S. W., Padmanabhan, P., Ray, P., Gambhir, S. S., Doyle, T., Contag, C., Goodman, S. B., Biswal, S. (2008). Stem cell-mediated accelerated bone healing observed with in vivo molecular and small animal imaging technologies in a model of skeletal injury. *J Orthop Res*.

46 Ray, P., Pimenta, H., Paulmurugan, R., Berger, F., Phelps, M. E., Iyer, M., Gambhir, S. S. (2002). Noninvasive quantitative imaging of protein-protein interactions in living subjects. *Proc Nat Acad Sci USA* **99**: 3105–3110.

47 Luker, G. D., Sharma, V., Pica, C. M., Dahlheimer, J. L., Li, W., Ochesky, J., Ryan, C. E., Piwnica-Worms, H., Piwnica-Worms, D. (2002). Noninvasive imaging of protein-protein interactions in living animals. *Proc Natl Acad Sci U S A* **99**: 6961–6966.

48 Luker, G. D., Sharma, V., Pica, C. M., Prior, J. L., Li, W., Piwnica-Worms, D. (2003). Molecular imaging of protein-protein interactions: controlled expression of p53 and large T-antigen fusion proteins in vivo. *Cancer Res* **63**: 1780–1788.

49 Serganova, I., Doubrovin, M., Vider, J., Ponomarev, V., Soghomonyan, S., Beresten, T., Ageyeva, L., Serganov, A., Cai, S., Balatoni, J., Blasberg, R., Gelovani, J. (2004). Molecular imaging of temporal dynamics and spatial heterogeneity of hypoxia-inducible factor-1 signal transduction activity in tumors in living mice. *Cancer Research* **64**: 6101–6108.

50 Ponomarev, V., Doubrovin, M., Serganova, I., Beresten, T., Vider, J., Shavrin, A., Ageyeva, L., Balatoni, J., Blasberg, R., Tjuvajev, J. G. (2003). Cytoplasmically retargeted HSV1-tk/GFP reporter gene mutants for optimization of noninvasive molecular-genetic imaging. *Neoplasia* **5**: 245–254.

51 Ray, P., Wu, A. M., Gambhir, S. S. (2003). Optical bioluminescence and positron emission tomography imaging of a novel fusion reporter gene in tumor xenografts of living mice. *Cancer Res* **63**: 1160–1165.

52 Ray, P., De, A., Min, J. J., Tsien, R. Y., Gambhir, S. S. (2004). Imaging tri-fusion multimodality reporter gene expression in living subjects. *Cancer Res* **64**: 1323–1330.

53 Deroose, C. M., De, A., Loening, A. M., Chow, P. L., Ray, P., Chatziioannou, A. F., Gambhir, S. S. (2007). Multimodality imaging of tumor xenografts and metastases in mice with combined small-animal PET, small-animal CT, and bioluminescence imaging. *J Nucl Med* **48**: 295–303.

54 Ponomarev, V., and M. Doubrovin, I. S., J. Vider, A. Sharvin, T. Beresten, A. Ivanova, L. Ageyeva, V. Tourkova, J. Balatoni, W. Bornmann, R. Blasberg, J. Gelovani Tjuvajev. (2004). A novel triple-modality reporter gene for whole-body fluorescent, bioluminescent, and nuclear noninvasive imaging. *Eur J Nucl Med Mol Imaging* **31**: 740–751.

55 Ray, P., De, A., Patel, M., Gambhir, S. S. (2008). Monitoring caspase-3 activation with a multimodality imaging sensor in living subjects. *Clin Cancer Res* **14**: 5801–5809.

56 Ponomarev, V., Dubrovin, M., Lyddane, C., Beresten, T., Balatoni, J., Bornman, W., Finn, R., Akhurst, T., Larson,

S., RBlasberg, R., Sadelain, M., Tjuvajev, J.G. (2001). Imaging TCR-dependent NFAT-mediated T-cell activation with positron emission tomography in vivo. *Neoplasia* 3: 480–488.

57 Koehne, G., Doubrovin, M., Doubrovina, E., Zanzonico, P., Gallardo, H. F., Ivanova, A., Balatoni, J., Teruya-Feldstein, J., Heller, G., May, C., Ponomarev, V., Ruan, S., Finn, R., Blasberg, R. G., Bornmann, W., Riviere, I., Sadelain, M., O'Reilly, R. J., Larson, S. M., Tjuvajev, J. G. (2003). Serial in vivo imaging of the targeted migration of human HSV-TK-transduced antigen-specific lymphocytes. *Nat Biotechnol* 21: 405–413.

58 Kim, Y., Dubey, P., Gambhir, S.S., Witte, O.N. (2004). Multimodality imaging of lymphocytic migration using lentiviral-based transduction of a tri-fusion reporter gene. *Molecular Imaging Biology* 6: 331–340.

59 Shu, C. J., Guo, S., Kim, Y. J., Shelly, S. M., Nijagal, A., Ray, P., Gambhir, S. S., Radu, C. G., Witte, O. N. (2005). Visualization of a primary anti-tumor immune response by positron emission tomography. *Proc Natl Acad Sci U S A* 102: 17412–17417.

60 Yaghoubi, S. S., Creusot, R. J., Ray, P., Fathman, C. G., Gambhir, S. S. (2007). Multimodality imaging of T-cell hybridoma trafficking in collagen-induced arthritic mice: image-based estimation of the number of cells accumulating in mouse paws. *J Biomed Opt* 12: 064025.

61 De la Zerda, A., Zavaleta, C., Keren, S., Vaithilingam, S., Bodapati, S., Liu, Z., Levi, J., Smith, B. R., Ma, T. J., Oralkan, O., Cheng, Z., Chen, X., Dai, H., Khuri-Yakub, B. T., Gambhir, S. S. (2008). Carbon nanotubes as photoacoustic molecular imaging agents in living mice. *Nat Nanotechnol* 3: 557–562.

62 Loimas, S., Wahlfors, J., Jänne, J. (1998). Herpes simplex virus thymidine kinase-green fluorescent protein fusion gene: new tool for gene transfer studies and gene therapy. *Biotechniques* 24: 614–618.

63 Uch, R., Gerolami, R., Faivre, J., Hardwigsen, J., Mathieu, S., Mannoni, P., Bagnis, C. (2003). Hepatoma cell-specific ganciclovir-mediated toxicity of a lentivirally transduced HSV-TkEGFP fusion protein gene placed under the control of rat alpha-fetoprotein gene regulatory sequences. *Cancer Gene Therapy* 10: 689–695.

64 Ponomarev, V., Doubrovin, M., Lyddane, C., Beresten, T., Balatoni, J., Bornman, W., Finn, R., Akhurst, T., Larson, S., Blasberg, R., Sadelain, M., Tjuvajev, J. G. (2001). Imaging TCR-dependent NFAT-mediated T-cell activation with positron emission tomography in vivo. *Neoplasia* 3: 480–488.

65 Luker, G. D., Luker, K. E., Sharma, V., Pica, C. M., Dahlheimer, J. L., Ocheskey, J. A., Fahrner, T. J., Milbrandt, J., Piwnica-Worms, D. (2002). In vitro and in vivo characterization of a dual-function green fluorescent protein–HSV1-thymidine kinase reporter gene driven by the human elongation factor 1 alpha promoter. *Mol Imaging* 1: 65–73.

66 Richard, J.C., Factor, P., Welch, L.C., Schuster, D.P. (2003). Imaging the spatial distribution of transgene expression in the lungs with positron emission tomography. *Gene Ther* 10: 2074–2080.

67 Ponomarev, V., Doubrovin, M., Shavrin, A., Serganova, I., Beresten, T., Ageyeva, L., Cai, C., Balatoni, J., Alauddin, M., Gelovani, J. (2007). A human-derived reporter gene for noninvasive imaging in humans: mitochondrial thymidine kinase type 2. *J Nucl Med* 48: 819–826.

68 Yu, Y., Szalay, A. (2002). A Renilla luciferase-Aequorea GFP (ruc-gfp) fusion gene construct permits real-time detection of promoter activation by exogenously administered mifepristone in vivo. *Molecular Genetics Genomics* 268: 160–168.

69 Xie, X., Chan, K. S., Cao, F., Huang, M., Li, Z., Lee, A. C., Weissman, I. L., Wu, J. C. (2008). Imaging of STAT3 signaling pathway during mouse embryonic stem cell differentiation. *Stem Cells and Development* 18: 205–214.

70 Soling, A., Simm, A., Rainov, N. (2002). Intracellular localization of Herpes simplex virus type 1 thymidine kinase fused to different fluorescent proteins depends on choice of fluorescent tag. *FEBS Lett* 527: 153–158.

71 Brader, P., Riedl, C. C., Woo, Y., Ponomarev, V., Zanzonico, P., Wen, B., Cai, S., Hricak, H., Fong, Y., Blasberg, R., Serganova, I. (2007). Imaging of hypoxia-driven gene expression in an orthotopic liver tumor model. *Mol Cancer Ther* 6: 2900–2908.

72 Likar, Y., Dobrenkov, K., Olszewska, M., Vider, E., Shenker, L., Cai, S., Pillarsetty, N., Hricak, H., Ponomarev, V. (2008). A new acycloguanosine-specific supermutant of herpes simplex virus type 1 thymidine kinase suitable for PET imaging and suicide gene therapy for potential use in patients treated with pyrimidine-based cytotoxic drugs. *J Nucl Med* 49: 713–720.

73 Dobrenkov, K., Olszewska, M., Likar, Y., Shenker, L., Gunset, G., Cai, S., Pillarsetty, N., Hricak, H., Sadelain, M., Ponomarev, V. (2008). Monitoring the efficacy of adoptively transferred prostate cancer-targeted human T lymphocytes with PET and bioluminescence imaging. *J Nucl Med* 49: 1162–1170.

74 Minn, A. J., Kang, Y., Serganova, I., Gupta, G. P., Giri, D. D., Doubrovin, M., Ponomarev, V., Gerald, W. L., Blasberg, R., Massague, J. (2005). Distinct organ-specific metastatic potential of individual breast cancer cells and primary tumors. *The Journal of Clinical Investigation* 115: 44–55.

75 Padmanabhan, P., Otero, J., Ray, P., Paulmurugan, R., Hoffman, A. R., Gambhir, S. S., Biswal, S., Ulaner, G. A. (2006). Visualization of telomerase reverse transcriptase (hTERT) promoter activity using a trimodality fusion reporter construct. *Journal of Nuclear Medicine* 47: 270–277.

76 Wu, J. C., Spin, J.M., Cao, F., Lin, S., Xie, X., Gheysens, O., Chen. I. Y., Sheikh, A. Y., Robbins, R. C., Tsalenko, A., Gambhir, S. S., Quertermous, T. (2006). Transcriptional profiling of reporter genes used for molecular imaging of embryonic stem cell transplantation. *Physiological Genomics* 25: 29–38.

77 Doubrovin, M., Ponomarev, V., I. Serganova, I., Soghomonian, S., Myagawa, T., Beresten, T., Ageyeva, L., Sadelain, M., Koutcher, J., Blasberg, R. G., Tjuvajev, J. G. (2003). Development of a new reporter gene system—dsRed/xanthine phosphoribosyltransferase-xanthine for molecular imaging of processes behind the intact blood-brain barrier. *Molecular Imaging* 2: 93–112.

78 Xing, L., Deng, X., Kotedia, K., Ackerstaff, E., Ponomarev, V., Clifton Ling, C., Koutcher, J. A., Li, G. C. (2008). Non-invasive molecular and functional imaging of cytosine deaminase and uracil phosphoribosyltransferase fused with red fluorescence protein. *Acta Oncologica* (Stockholm, Sweden) 47: 1211–1220.

79 Moroz, M. A., Serganova, I., Zanzonico, P., Ageyeva, L., Beresten, T., Dyomina, E., Burnazi, E., Finn, R. D., Doubrovin, M., Blasberg, R. G. (2007). Imaging hNET reporter gene expression with 124I-MIBG. *J Nucl Med* **48**: 827–836.

80 Kutschka, I., Chen, I. Y., Kofidis, T., von Degenfeld, G., Sheikh, A. Y., Hendry, S. L., Hoyt, G., Pearl, J., Blau, H. M., Gambhir, S. S., Robbins, R. C. (2007). In vivo optical bioluminescence imaging of collagen-supported cardiac cell grafts. *J Heart Lung Transplant* **26**: 273–280.

81 Jacobs, A. H., Winkeler, A., Hartung, M., Slack, M., Dittmar, C., Kummer, C., Knoess, C., Galldiks, N., Vollmar, S., Wienhard, K., Heiss, W. D. (2003). Improved herpes simplex virus type 1 amplicon vectors for proportional coexpression of positron emission tomography marker and therapeutic genes. *Hum Gene Ther* **14**: 277–297.

82 Shin, J. H., Chung, J. K., Kang, J. H., Lee, Y. J., Kim, K. I., So, Y., Jeong, J. M., Lee, D. S., Lee, M. C. (2004). Noninvasive imaging for monitoring of viable cancer cells using a dual-imaging reporter gene. *J Nucl Med* **45**: 2109–2115.

83 Che, J., Doubrovin, M., Serganova, I., Ageyeva, L., Zanzonico, P., Blasberg, R. (2005). hNIS-IRES-eGFP dual reporter gene imaging. *Mol Imaging* **4**: 128–136.

84 Cohen, B., Dafni, H., Meir, G., Harmelin, A., Neeman, M. (2005). Ferritin as an endogenous MRI reporter for noninvasive imaging of gene expression in C6 glioma tumors. *Neoplasia* **7**: 109–117.

85 Cohen, B., Ziv, K., Plaks, V., Israely, T., Kalchenko, V., Harmelin, A., Benjamin, L. E., Neeman, M. (2007). MRI detection of transcriptional regulation of gene expression in transgenic mice. *Nat Med* **13**: 498–503.

86 Winkeler, A., Sena-Esteves, M., Paulis, L. E., Li, H., Waerzeggers, Y., Ruckriem, B., Himmelreich, U., Klein, M., Monfared, P., Rueger, M. A., Heneka, M., Vollmar, S., Hoehn, M., Fraefel, C., Graf, R., Wienhard, K., Heiss, W. D., Jacobs, A. H. (2007). Switching on the lights for gene therapy. *PLoS ONE* **2**: e528.

87 Weissleder, R., Moore, A., Mahmood, U., Bhorade, R., Benveniste, H., Chiocca, E. A., Basilion, J. P. (2000). In vivo magnetic resonance imaging of transgene expression. *Nat Med* **6**: 351–355.

Cell-Specific Imaging of Reporter Gene Expression Using a Two-Step Transcriptional Amplification Strategy

6

Marxa L. Figueiredo, Sanjiv Sam Gambhir, Michael Carey, and Lily Wu

INTRODUCTION

The monitoring of reporter gene expression allows measurement of the location(s), magnitude, and time variation of gene transcription in living animals and humans. Several imaging modalities can be employed for repetitive, noninvasive monitoring of reporter gene expression. The most common methods include positron emission tomography (PET), single photon emission computed tomography (SPECT), magnetic resonance imaging (MRI), and optical imaging by bioluminescence (e.g., Firefly luciferase, Fluc, or luc) or fluorescence (e.g., green fluorescent protein, GFP). The strengths of each imaging modality are reviewed in Chapters 1–4. Noninvasive imaging has been applied extensively to monitor gene therapy, to detect cell migration and metastasis, and finally to monitor endogenous gene expression (by the use of transgenic mice expressing a reporter gene).

A common and successful means to target imaging of reporter gene expression to a particular tissue is to employ a transcriptional targeting strategy. Transcriptional targeting refers to the use of a cell-specific regulatory element (promoter or promoter/enhancer) to restrict gene expression to a particular tissue or cell type. A pitfall in using tissue- or tumor-specific promoters (TSPs) is that the relatively weak transcriptional activity of a cellular promoter could in principle greatly limit imaging sensitivity due to low levels of reporter gene expression *in vivo*. This contrasts with the potent but non-tissue–specific viral promoters like the simian virus 40 (SV40) early promoter or the cytomegalovirus (CMV) enhancer/promoter. Transcriptional targeting can be of great benefit to several experimental and preclinical applications, for example, by improving the safety and specificity of cancer gene therapy or by achieving high-level specific gene expression in transgenic animal models. In cancer gene therapy applications, a cytotoxic gene is often the therapeutic payload. The use of a TSP restricts cytotoxic gene expression to the targeted tumor or tissue. Even if inadvertent delivery of the cytotoxic gene to vital organs should occur, the TSP will remain silent in normal, nontargeted tissues. Similar to the reporter gene imaging applications, TSP-driven gene therapy is hampered either by low potency and/or specificity of TSP, resulting in limited therapeutic efficacy. Thus, a prerequisite to achieve more effective cell-specific gene imaging and therapy is to boost the potency of TSP activity, reaching magnitudes comparable to common viral promoter/enhancers while maintaining specificity. Transcriptional targeting strategies also can amplify reporter gene expression from a relatively weak promoter to achieve high-level tissue-specific targeting of gene expression in transgenic animal models. When coupled with noninvasive imaging technology, transcriptional targeting may aid noninvasive examination of gene expression over time during a biological process. For example, the biological processes of oncogenesis, cancer progression, or wound healing can be monitored in a highly sensitive manner. For either gene therapy or transgenic mouse model applications, the augmentation of TSP activity cannot be indiscriminant; it needs to be accomplished in conjunction with maintaining cell-selective capacity.

We have explored several strategies to augment the intrinsic activity of TSP. Some productive approaches will be discussed in this chapter in the context of the prostate-specific antigen (PSA) promoter [1]. Fruitful strategies include creating more potent and compact promoters by eliminating DNA sequences that do not contribute to promoter strength [2] while retaining specificity and combining the strengths of transcriptional regulatory elements from different promoters specific for the same tissue to attain more potent synthetic chimeric promoters [2, 3, 4]. In our experience, however, the most effective method is to use recombinant transcriptional activators in an approach termed the two-step transcriptional amplification, or TSTA.

The TSTA system consists of a potent transcriptional activator driven by a cell-specific promoter, which acts on a second "expression cassette" (made up of one or more genes and the sequences controlling their expression), which encodes the reporter/therapeutic component. This binary approach results in cell-specific amplification of gene expression (Figure 6.1a). The activator in our method is a fusion protein termed GAL4-VP16, comprising the DNA-binding domain from the yeast transcription activator GAL4 and the transactivation domain from the herpes simplex virus 1 (HSV-1) activator viral protein 16 (VP16) [5]. GAL4-VP16 binds specifically to 17-base pair GAL4 DNA-binding sites and recruits the transcription machinery via the acidic VP16 transactivation domain. Although the initial feasibility of this TSTA approach for prostate cancer therapy was demonstrated with a 5.3-kb intact PSA regulatory element driving a firefly luciferase (*fl* or *luc*) reporter gene and the toxic polyglutamine gene [6], the subsequent improvements in the TSTA methodology have expanded its utility to a wide range of applications, including imaging of exogenous reporter/therapeutic genes (viral or nonviral vectors) for therapeutic applications (gene therapy), *in vivo* imaging of endogenously expressed reporter genes (transgenic mice), or amplification of signal utilizing different tissue-specific promoters. In this chapter we will discuss the TSTA and the refinement schemes that have emerged over the past few years. These schemes have shown considerable promise for highly regulated, tissue-specific imaging to visualize gene expression and functionality of signaling pathways important for cancer progression *in vivo*.

STRATEGIES TO OPTIMIZE THE TWO-STEP TRANSCRIPTIONAL AMPLIFICATION (TSTA) EXPRESSION SYSTEM

Given the bipartite nature of transcriptionally amplified gene regulation systems, there are several variables that can be modulated to achieve an optimal TSTA system. These variables include: (1) the potency of the tissue- or cell-specific promoter driving expression of an effector (e.g., GAL4-VP16); (2) the number of GAL4 DNA-binding sites proximal to the *fl* or other reporter genes, which determines how many effector proteins can bind to DNA and therefore affects the potency of activation; (3) the potency of the GAL4-VP16 derivative; and (4) the configuration of the effector and reporter genes on the same plasmid or viral vector (Figure 6.1b).

PSA Promoter Optimization

Many prostate-specific gene regulatory regions are well characterized, and several of them have been assessed in preclinical and clinical therapeutic studies (reviewed in [7]). The most extensively studied prostate-specific promoter to date is derived from the prostate-specific antigen (PSA or hK3) gene, which encodes a serine protease [8]. Because PSA is expressed in normal prostate, as well as at all stages of prostate carcinogenesis including the metastatic stage, it remains an excellent candidate for directing therapeutic gene expression specifically toward prostate tissue. Extensive studies have been performed to boost the inherently weak but specific activity of the native PSA promoter and enhancer by developing chimeric PSA promoters (Figure 6.1c) [2, 9]. Enhancers are nucleotide sequences located as many as several thousand base pairs away in either direction from the target gene, which enhances transcription of the target gene. Although a 6-kb enhancer/promoter region upstream of the PSA gene transcriptional start site directed prostate-specific expression in transgenic mice [10], the transcriptional elements required to achieve prostate-restricted expression are confined to smaller regions. Briefly, the PSA regulatory regions have been defined and contain a proximal promoter (−541 to +12) comprising a TATA box (a consensus sequence found in the promoter region of most genes transcribed by eukaryotic RNA polymerase II) and two functionally important binding sites for the transcription factor androgen receptor (AR; ARE I and II) at its enhancer region. The PSA promoter exhibits a degree of tissue-specific expression *in vitro*, although it is insufficient to direct prostate-specific expression *in vivo* [11, 12]. A 440-bp core segment of the PSA enhancer is responsible for the majority of the enhancer activity and plays a major role in androgen responsiveness due to the presence of high affinity ARE (AREIII), to which AR binds when complexed to androgen [9]. The enhancer core also contains sites for other transcription factors and is active in both androgen-dependent and -independent prostate cancer cells [13]. Multiple dimers of AR bind cooperatively to a cluster of AREs encompassing AREIII in the core enhancer and synergistically activate transcription in the presence of dihydrotestosterone (DHT) in normal cells and androgen-dependent prostate cancer.

Relevant to the improvement of PSA-based promoter sequences for targeted gene expression is that in many recurrent cancers, the AR gene is either amplified and/or overexpressed but remaining able to function under low DHT levels or by adrenal androgen activation [14]. Furthermore, AR has been shown to retain functionality in androgen-dependent (AD) and androgen-independent (AI) prostate cancers *in vitro* and *in vivo* [15, 16]. Therefore, the initial strategy for augmenting the specificity and activity of the PSA promoter, by creating chimeric PSA-based promoters, exploited the synergistic nature of the AR action. The starting construct was the PSE, bearing the 2.4-kb enhancer and 541-bp promoter fragments from the PSA gene [17]. In the first promoter optimization approach, we fused an artificial regulatory element (ARE4) comprising four tandem copies of a high-affinity class I ARE (derived from 15-bp AREI site

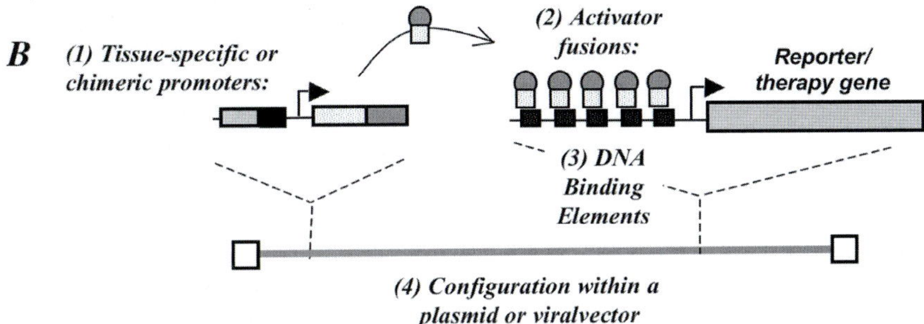

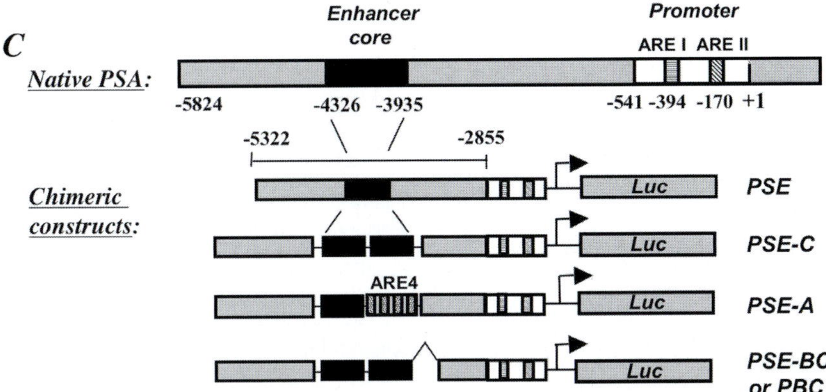

Figure 6.1. Evolution of the chimeric prostate-specific PBC promoter for use in the two-step amplification (TSTA) system for gene therapy and imaging applications. (A) *The two-step transactivation (TSTA) system.* In the first step, the artificial GAL4-VP16 activator is expressed in a tissue-specific manner by virtue of regulation by a tissue- or cancer-specific promoter (PSE-BC). In the second step, the GAL4-VP16 protein binds to five GAL4 DNA-binding sites upstream of a minimal promoter and activates expression of a reporter or therapeutic gene. **(B)** *The modular nature of two-step gene regulation systems allows for a high level of flexibility in vector design and construction.* The objective is to maximize specificity and delivery efficiency for *in vivo* applications. Native or chimeric tissue- or cancer-specific promoters can be introduced with the *activator* component (with optimal number of activation domains) and the *target/effector* cassette (containing optimal number of DNA-binding elements and a reporter/therapeutic gene) within the same vector for maximum delivery efficiency. **(C)** *Promoter optimization can enhance the activity of a tissue-specific promoter for optimal expression of reporter genes.* To exploit the synergistic nature of AR action, chimeric enhancers were generated harboring either four copies of the proximal ARE I element, (2) duplication of enhancer core, or (3) removal of intervening sequences (−3744 to −2855) between the enhancer and promoter. The most efficacious chimeric promoter was PSE-BC (duplication of core), which contained the PSA enhancer (−5322 to −2855) fused to the proximal promoter (−541 to +12). The PSE-BC exhibited 18.9-fold higher activity than the baseline construct PSE. Both PSE and PSE-BC were found to be highly androgen-inducible and tissue-specific.

of PSA promoter) to the native PSA enhancer and promoter construct, creating the chimeric PSE-A promoter [3] (Figure 6.1c). Other strategies to increase gene expression from chimeric PSA promoters included duplication of the core enhancer and removal of intervening sequences (-3744 to -2855) to position the enhancer elements closer to the proximal promoter to achieve more potent activation of gene expression [3]. These alternate chimeric constructs improved activity (>tenfold) versus the PSE yet retained androgen inducibility and tissue specificity [3]. Indeed, the use of streamlined promoter/enhancers smaller than the native versions has the advantage of allowing the constructs to fit into gene therapy vectors with stringent size limitations.

The construct with the highest specific activity *in vitro* was the PSE-BC (PBC) configuration, which deleted an 890-bp intervening sequence between the enhancer and promoter (-3743 to -2855) and contained the PSA enhancer with a duplicated 390-bp core region (Figure 6.1c). These modifications augmented androgen-responsive expression 20-fold in cell culture [3]. Importantly, the augmented activity of the PBC construct was validated further in cell culture using an adenoviral vector driving FL expression (AdPBC) in LNCaP cells [18]. The AdPBC was further tested *in vivo* for specificity of reporter gene expression and responsiveness to androgen in xenograft tumor models. The AdPBC vector targeted efficient reporter gene expression restricted to both AD and AI LAPC4 prostate tumors in live mice, whereas there was no reporter gene expression in the liver, indicating tissue specificity of the vector (Figure 6.2a, upper panels). This example demonstrates the successful use of a noninvasive imaging modality in therapeutic and diagnostic strategies for prostate cancer. Interestingly, the AdPBC vector also specifically detected distal lung and spinal metastatic lesions in xenograft models using the cooled charge coupled device (CCD) optical imaging system over a 3-week period following adenovirus systemic injection (Figure 6.2a, lower panels) [18]. LAPC-4 prostate tumors were shown to progress from AD to AI while still retaining AR and PSA expression (Figure 6.2b), despite the absence of testicular androgen (upon castration), providing initial crucial evidence that the PBC regulatory element is androgen-regulated *in vivo*. Transcriptional activity of AdPBC was tenfold higher in AI than in AD tumors (Figure 6.2b), supporting the possibility that a prostate-targeted vector could be developed to treat patients with advanced prostate cancer. One drawback of this vector, however, was that overall activity of the AdPBC was only 5% of AdCMV, the benchmark for CCD imaging studies, making it difficult to dynamically monitor the androgen response. This study was the basis for further development of an Ad-TSTA-*fl*, which could achieve higher tissue-specific transcriptional activity while retaining specificity and proper androgen regulation [15].

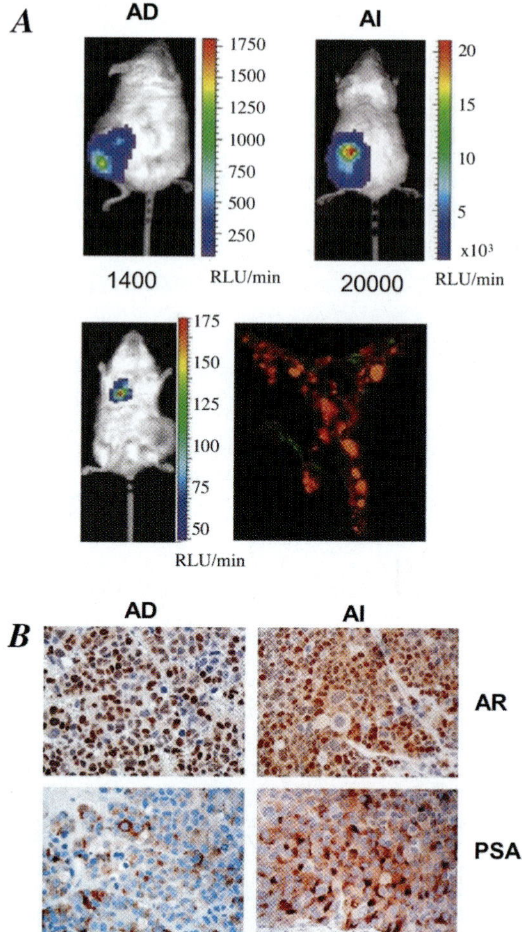

Figure 6.2. Chimeric promoter activity *in vivo* using noninvasive bioluminescence imaging. (A) *Detection of PSE-BC tissue-specific activity in vivo.* The PSE-BC promoter was utilized to examine androgen-regulated gene expression in an adenoviral vector configuration (AdPBC). The upper panel shows CCD luciferase imaging of LAPC-4 AD (left) or AI (right) tumor-bearing mice 11 days following intratumoral injection, and the AdPBC vector-targeted expression of a luciferase reporter (*fl*) specifically to the tumors. Interestingly, the AdPBC vector could also detect prostate cancer metastasis to lung at 12 d post systemic injection (tail vein) of 3.6×10^7 infectious units in an LAPC-4 AD tumor-bearing mouse. The lower panel shows micrometastases in a lung section using anticytokeratin (red/orange) and confocal microscopy. Vessels were visualized by lectin (green), 400x. **(B)** *Prostate tumors retain AR and PSA expression.* Endogenous AR (top) and PSA (bottom) expression (brown color) in AD (left) and AI (right) LAPC-4 tumors. PSA expression in the AI tumor appears to be elevated compared with the AD LAPC-4 tumor. AR expression appears less well localized to the nucleus in the AI tumor.

Optimization of Effector Constructs

The modular nature of gene expression (activator and target/effector components) renders TSTA systems highly flexible, and many combinations can be systematically manipulated to achieve maximal activity and specificity. Importantly, the chimeric promoters with the highest specificity and activity should be chosen prior to transitioning a system into TSTA for *in vivo* gene delivery. The relative efficacy of PSE versus the PBC promoter to drive expression of the activator construct (GAL4-VP16) was examined. This was described in the previous section. Next, the effect of optimizing the effector component was examined by either varying the number of GAL4 binding sites or VP16 activation domains. Lastly, we examined the effect of combining the optimal activator and effector components within one vector (Figure 6.3).

Titration of effector components was first achieved by varying the number of activation domains on the GAL4-derived activator fusion (GAL4-VP16) and GAL4-binding sites on the reporter plasmid. The reporter templates were constructed to contain either one, two, five, or nine copies of the 17-bp GAL4 binding sites positioned 23-bp upstream of a minimal promoter containing the adenovirus E4 gene TATA box and driving *fl* reporter gene expression (G1-, G2-, G5- or G9-*fl*) (Figure 6.2a). The PSE or PBC chimeric promoters were used to express recombinant GAL4-VP16 variants to generate a series of effector plasmids that displayed a gradient of activities. First, the number of VP16 transactivation domains and their effect on transcriptional strength was evaluated in LNCaP cells. The PSE was used to express a GAL4-VP16 fusion consisting of the 147-amino acid GAL4 DNA-binding domain (DBD) plus one copy of the 42-amino acid VP16 activation subdomain (aa 413–454), PSE-VP1. PBC was used to examine expression of fusion proteins containing the GAL4 DBD plus one, two, or four copies of the VP16 subdomain (VP1, VP2, or VP4) (Figure 6.3a). The vector with highest fold synergy and absolute levels of activation contained two VP16 domain copies (PBC-VP2). By using a fixed G5-*fl* reporter plasmid, the PBC-VP2 construct exhibited threefold greater activity than with one VP16 subdomain. Interestingly, four VP16 domains had a modest reduction in expression, possibly due to a phenomenon termed "transcriptional squelching," whereby overexpression of a transcription factor suppresses transcription possibly by sequestration of a limiting cofactor. Controls included parental plasmids containing the SV40 enhancer and either GAL4-VP1, -VP2, or -VP4, and a construct with *fl* driven directly by the CMV enhancer (CMV-*fl*) (Figure 6.3b, left panel). In another study, the specificity and activity of PSE-VP1 was confirmed using two different reporter genes, *fl* and *sr39tk*, an HSV1-tk mutant with enhanced imaging sensitivity. Using G5-*fl* and G5-*sr39tk* cassettes, ~50-fold (*fl*) and ~12-fold (*sr39tk*) gene expression enhancement was observed with the PSE-VP1 two-step approach in prostate cancer cells [19].

Second, varying the number of GAL4 binding sites contributed to the titratability of the system. The activity with PSE-VP1 increased ~eightfold when using two GAL4 binding sites (G2-*fl*), but the best activity was obtained with G5-*fl*, with an additional 60-fold increase in activity. The activity appeared to saturate at five GAL4 sites because a construct bearing nine sites (G9-*fl*) exhibited the same activity as G5-*fl*. Therefore, transcription activation increased synergistically with the increase in the number of activator binding sites. The optimal combination for the activator construct was the PBC promoter-bearing GAL4 fused to two copies of VP16 (PBC-VP2) and for the target construct, five GAL4 DNA-binding sites upstream of *fl* (G5-*fl*). These constructs then were combined in the TSTA system (Figure 6.3a). Finally, the TSTA constructs were evaluated in side-by-side comparison with a SV40-driven *fl* construct by cotransfection assays into androgen-responsive prostate cancer lines and control lines to evaluate cell specificity (Figure 6.3b, right panel). These optimized components (PBC-VP2 and G5-*fl*) were evaluated either in cotransfection (two construct) or single transfection (cloned into a single plasmid; Figure 6.3c). The combination of PBC-VP2 and G5-FL-augmented activity and androgen inducibility compared to PBC driving *fl* directly, indicating that the TSTA system can successfully amplify *fl* expression by 250-fold while retaining high specificity in a transfection scheme. Further, the TSTA system retained strong androgen responsiveness. Overall, the best construct leading to the generation of the TSTA prostate-specific system was the PBC-VP2 activator and G5-*fl* effector combination, with activated expression levels exceeding those of both CMV promoter and SV40 promoter with G5-*fl* (Figure 6.3b). An 800-fold augmentation from the weakest one-step to the strongest two-step system was observed, and the PBC-VP2/G5-*fl* construct combination was selected (Figure 6.3c) to further evaluate tissue and cell specificity of the TSTA system *in vitro* and *in vivo* (Figure 6.3d).

Despite the promise of the system and its applications, some limitations can be observed, including high background gene expression in some cases. This two-tiered amplification methodology now has been extensively tested and validated in many configurations to address this limitation. In two studies performed using adenoviruses to deliver TSTA to prostate cancer cells, the positioning of the two-step expression elements dramatically affected the magnitude of background gene expression, and these studies will be discussed in the following section. These adenoviral vectors contain all TSTA components in a contiguous DNA insert and display activity levels higher than the CMV promoter in prostate tumors while succeeding in reducing background activity in prostate and nonprostatic cell types [6, 9, 16, 19–23].

Configurations of the TSTA System

Overall, the TSTA system (PBC-VP2 and G5-*fl* combination) performed well to achieve high activity, inducibility,

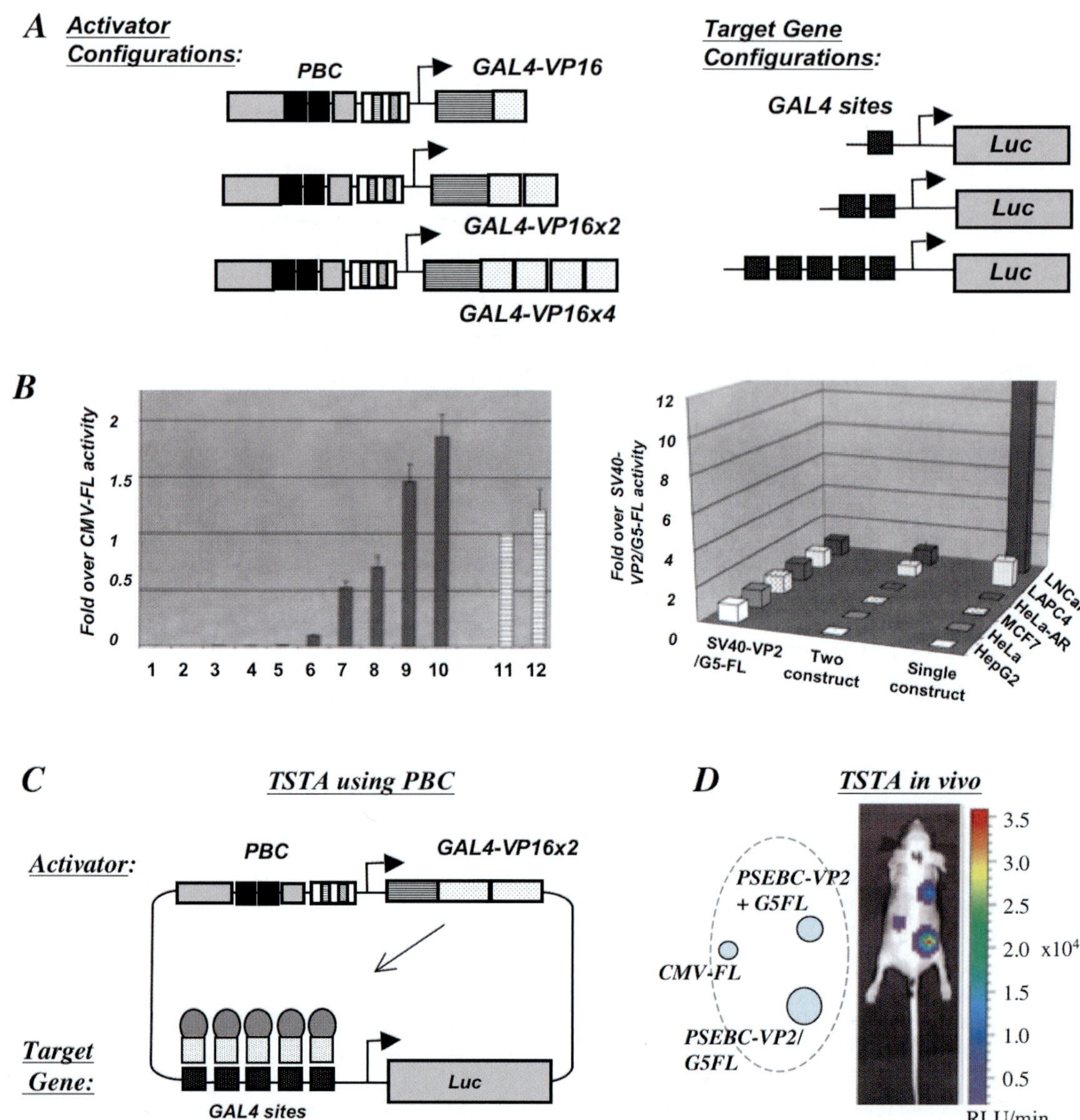

Figure 6.3. Flexibility of modular gene expression amplification systems. (A) *Optimizing tissue-specific reporter gene expression by titration of transcriptional amplification components.* Several systematic modifications of the number of *activator* (GAL4-VP16) or *target/effector* (GAL4 binding sites) were examined for their ability to enhance tissue-specific reporter gene expression. From this analysis, the TSTA system was derived to comprise the activator construct (PBC GAL4 plus two VP16 domains), plus a target construct (5 GAL4 DNA-binding sites upstream of the firefly luciferase (*fl*) reporter). **(B)** *The spectrum of fl expression activity generated by TSTA as compared with CMV- or SV40-driven constructs.* The left panel compares cell-specific expression when TSTA components are present within a single or two constructs. Shown here are six human cell lines transfected with either SV40-VP2 + G5-*fl*, PBC-VP2 + G5-*fl* control constructs or a single PSE-BC-VP2/G5-*fl* construct. In the right panel, measurements are normalized to CMV-*fl* activity in the presence of R1881, a synthetic androgen. CMV-*fl* was assigned a value of 1. The samples are ordered by their activities: (1) PSE-*fl*; (2) PSE-VP1/G1-*fl*; (3) PSE-BC-VP4/G1-*fl*; (4) PSE-BC-*fl*; (5) PSE-VP1/G2-*fl*; (6) PSE-BC-VP4/G2-*fl*; (7) PSE-VP1/G5-*fl*; (8) PSE-BC-VP1/G5-*fl*; (9) PSE-BC-VP4/G5-*fl*; (10) PSE-BC-VP2/G5-*fl*; (11) CMV-*fl*; and (12) SV40-VP4/G5-*fl*. **(C)** *The TSTA system using a prostate cancer-specific promoter (PSE-BC).* PSEBC was used to activate expression of firefly luciferase (*fl* or *luc*) reporter gene in this TSTA system application. **(D)** *Efficiency of a one-construct TSTA vector.* The one-construct TSTA system was more efficient than CMV or two-construct TSTA plasmids at inducing specific and strong gene expression of reporter gene *in vivo* following transfection. Bioluminescent color images of imaging in athymic nude mice superimposed on the gray-scale mouse photographs. The color scale is in units of RLU/min. A map representing the dorsal surface of the mice is on the left; the circles denote the relative position of the three injection spots, with the transfected plasmids labeled over each circle. A description of the group is shown on top of each panel and the acquisition time of the CCD camera for each image generated is in parentheses.

and specificity in plasmid transfection experiments. To examine the gene transfer capability of the TSTA system in animal models, we incorporated the initial nonviral plasmid constructs into more efficient adenoviral vectors. In the context of the larger and more complex adenoviral genome, the binary TSTA system could be inserted in different configurations to influence transcriptional regulation. Hence, we have investigated the activation, androgen-responsiveness, and tissue specificity of different TSTA viral constructs. In the original Ad configuration, both the activator and reporter components were inserted into a single vector (AdTSTA) in a head-to-head configuration in the deleted E1 (ΔE1) region of the adenovirus (single system), whereas in another they were placed at the ΔE1 in two separate viruses (separate system, designated AdPBC-VP2 and AdG5-*fl*) (Figure 6.4a) [21]. The capability of single-system AdTSTA to direct prostate-specific expression was evaluated in comparison to the separate system by infection of a variety of cell lines, including androgen-responsive AR-positive and AR-negative cells. The *fl* activity in AR-negative prostate cancer cell lines was negligible and was not induced by androgen at a multiplicity of infection or number of infectious viral particles of one per cell infected (m.o.i. = 1). At a higher m.o.i. of ten, however, the cell specificity became lower, even in the absence of androgen, an effect hypothesized to be due to increased basal or background transcriptional activity. Interestingly, the single-system AdTSTA consistently expressed an elevated level of GAL4-VP2 activator as detected by Western blot. This observation suggested that because this particular vector configuration positioned the activator and reporter genes in close proximity in the adenoviral genome in a head-to-head manner, the multiple GAL4-VP2 activators could also stimulate transcription in the direction of the PBC promoter. Therefore, a feed-forward loop was hypothesized to be involved in the single-system TSTA activation mechanism in an enhancer-like manner. When the activator and reporter components were placed in two separate Ads, androgen induction was more robust than for the single AdTSTA ($\sim$six- to ninefold higher in LNCaP cells). Therefore, each TSTA configuration presented particular advantages, with higher androgen induction and higher specificity with the separate system but higher activity with the single system. For these reasons, additional configurations of the TSTA system were generated to achieve improved functional separation of the activator and reporter component while retaining in a single-vector system, which is highly advantageous for vector delivery *in vivo*.

One important consideration of prostate-targeted adenoviral expression systems is to overcome constitutive expression of the reporter or therapeutic gene in the liver because adenoviral vectors naturally display strong hepatic tropism following leakage into systemic circulation [18]. This constitutive promoter expression of AdCMV

in liver may lead to inadvertent hepatotoxicity [22]. *In vivo* activity of the single AdTSTA exceeded that of a CMV promoter-driven vector following intraprostatic vector delivery of 10^7 pfu (Figure 6.4b) while maintaining tissue specificity. Systemic administration of AdTSTA resulted in undetectable expression in the liver and low-level signals in the lung. In contrast, AdCMV was highly expressed in liver (Figure 6.4c). The weak signal in the lung after systemic administration of AdTSTA could be attributed to the presence of epithelial-specific transcription factors in lung that are competent to activate the PSA promoter and the preferred distribution of the adenovirus to lung via the intravenous injection route [23]. The absence of prostatic signal when AdTSTA is delivered systemically may be due to liver sequestration of the Ad and the relatively low dose of 10^7 pfu. When a higher dose Ad was administered, a prostatic signal could be detected [18, 21].

Our recent findings suggested that the orientation and placement of the two components of TSTA in the adenoviral genome can impact on the amount of transcriptional activation [21, 25]. Several new configurations of the two-step TSTA system in the same viral vector are illustrated in Figure 6.5. We showed that separating the two components by significant genomic distances reduces background activity (Figure 6.5a) [25], likely because it interrupts the feed-forward loop described for the parental head-to-head configured AdTSTA (Figure 6.4a). The advantage of bidirectional (Figure 6.5b) and multiple-gene (Figure 6.5c) configurations is that simultaneous expression of several genes can be regulated by the same stringent tissue-specific promoter [26]. These improved constructs can be applied to direct more effective multigene combined therapy or coupled imaging and therapeutic strategies in a single vector.

A recent development of the TSTA system included a bidirectional configuration to enhance expression of therapeutic gene TNF-a related apoptosis inducing ligand (TRAIL) and the reporter gene *fl* from a survivin promoter [27]. An Ad carrying the bidirectional cassette under the control of eight GAL4 DNA-binding sites (AdpSurv-TRAIL-G8-*fl*) displayed tenfold higher expression of both therapeutic and reporter genes in cells and animals. The TSTA Ad maintained promoter specificity in nontarget tissues following systemic administration. This configuration illustrates the potential of the TSTA system to achieve high levels of gene expression from a weak promoter while preserving specificity and the ability to noninvasively image expression of therapeutic/reporter gene cassettes.

APPLICATIONS OF TSTA SYSTEM

Highly sensitive noninvasive molecular imaging technology using reporter genes has been used to evaluate

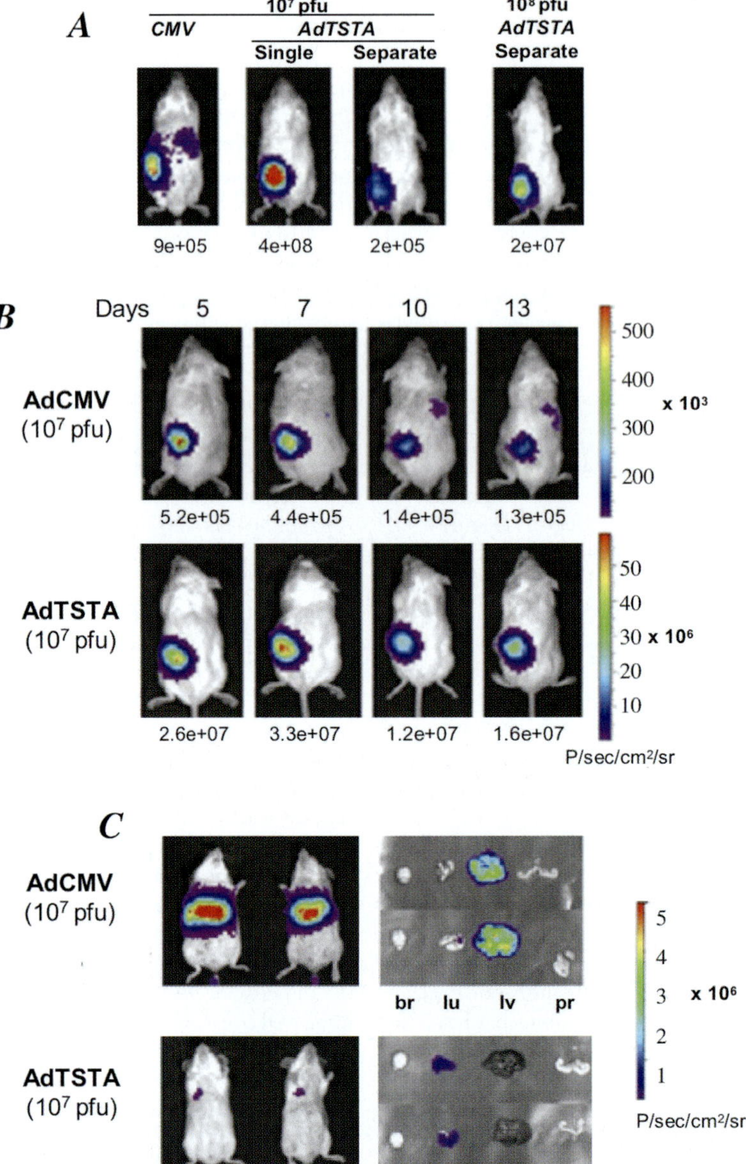

Figure 6.4. The TSTA system achieves highly specific gene expression levels. The engineering of higher efficiency vectors for TSTA delivery utilized the adenovirus vector genome as a prototype. (**A**) *An improved configuration of the TSTA system enhances vector delivery efficiency and gene expression in vivo*. Typically, the TSTA system consists of two separate Adenoviral vectors (Ad) to deliver the activator and target gene components *in vitro* and *in vivo*. An improved vector configuration was constructed, where both activator and reporter are inserted into the E1 region of the same Ad in a head-to-head orientation. This new configuration optimized gene delivery using the Ad vector and improved *fl* reporter gene expression levels *in vivo*. Optical signals after injections of the respective Ads in LAPC-4 tumors. 10^7 or 10^8 pfu of Ads were administered. The injection of separate Ads denotes the coadministration of both AdBC-VP2 and AdG5-fl at the specified dosage. CCD images of representative animals analyzed at 4 d postinjection were shown. Kinetics of *fl* expression in LAPC-9 tumors. 10^7 pfu of AdCMV or AdTSTA were injected intratumorally. Optical signals were monitored on the specified days after viral injection. The number below each image represents the maximal signal over the tumor. The graph on the right represents the averaged serum PSA level measured in the animals at the specified days postviral injection. (**B**) *Comparing specificity and activity of TSTA with CMV promoter in adenoviral vectors*. The single-system TSTA (AdTSTA) harbors activator and reporter inserted into the E1 region of the same adenovirus in a head-to-head configuration, while in the separate system (AdPBC-VP2 and AdG5-fl), activator and reporter components were incorporated into the E1 region of two separate adenoviruses. *In vivo fl* expression mediated by TSTA or CMV Ads driving luciferase expression in LAPC-4 xenografts 4 d postintratumoral injections of the respective Ads. The AdTSTA (AdTSTA) displayed higher sensitivity than the CMV-driven vector. Numbers below the images are the maximal activities in the region of interest as photons (p) acquired per second per square centimeter per steradian (sr). (**C**) *The TSTA system achieves highly specific gene expression compared to CMV promoter in vivo*. Optical signals monitored by CCD bioluminescence imaging *in vivo* and *ex vivo* on the specified days after 10^7 pfu (plaque-forming units = pfu or infectious units) of Ad (Days 3, 6, and 22). Robust liver signals were noted in the AdCMV-fl (AdCMV)-injected animals starting at Day 3 and increasing from that point onward. The AdTSTA-injected animals remained transcriptionally silent until Day 22, when a weak signal was noted in the lung. *Ex vivo* imaging (excised organs) revealed that the liver is the predominant site of expression in AdCMV-injected animals. Low level of expression in the lung was observed in the AdTSTA-injected animals (br, brain; lu, lung; lv, liver; and pr, prostate).

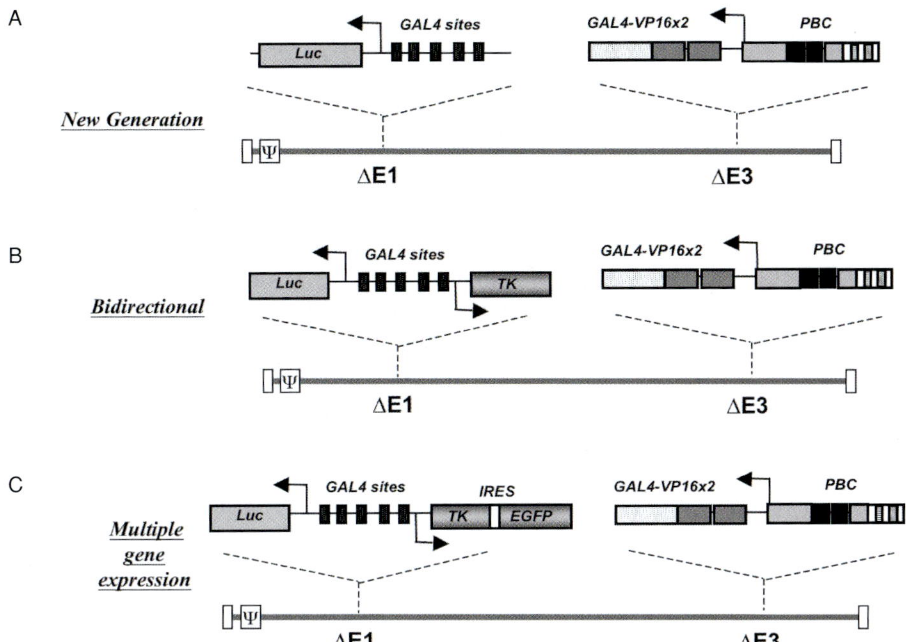

Figure 6.5. Current and future TSTA configurations for therapeutic applications. The engineering of higher efficiency and safer vectors for TSTA delivery and expression has promoted the design of alternative configurations for potential *in vivo* applications. **(A)** *New-generation AdTSTA:* the components are separated by ~30 kb of adenoviral genome, with the activator inserted in the E3 region and the reporter inserted into the E1 region. In the bidirectional variation AdTSTA-TK: the activator is inserted in the E3 region, and the reporter/therapeutic gene is inserted in a head-to-head orientation in the E1 region to achieve simultaneous expression of two genes. **(B)** *Bidirectional configuration:* the components are placed in the new-generation configuration, with two target genes placed in the deleted E1 region in a bidirectional manner flanking G5 sites. **(C)** *Multiple gene expression AdTSTA-TK-ires-EGFP configuration:* the activator is inserted in the E3 region, and a combination of genes can be expressed in a head-to-head fashion. The number of genes expressed can be increased by virtue of internal ribosome entry site (IRES) sequences placed in between them. ψ denotes the packaging signal of adenovirus and open rectangles at both termini denote inverted terminal repeats of the viral genome.

current TSTA applications and has allowed us to image tumors and metastases to monitor the therapeutic efficacy of drugs, cell trafficking, gene delivery, and expression, as well as the study of transgenic models. Following optimization of the components of the TSTA system in transfection conditions, the system was adapted to incorporate more efficient gene delivery vectors for gene therapy applications or long-term gene expression *in vivo*. The design of improved configurations of the TSTA system within the Ad vector improved the delivery efficiency and activity of gene expression even further. The functionality of AR was confirmed in all TSTA studies in AD and AI prostate tumor models.

To assess long-term expression of TSTA-driven reporter gene *in vivo* and to further test specificity, two models were developed and examined over time for continued expression of a reporter gene. First, a transgenic mouse (Tg-TSTA-*fl*) model revealed that the TSTA system was well tolerated and that the system remains functional and androgen-responsive *in vivo* for extended periods of time. This model can be useful to help detect and track malignancy progression if used in the context of a cross into a second, oncogenic transgenic model. In

a second model, we developed a mouse model where a lentiviral TSTA system was injected intraprostatically. Although the most commonly used gene therapy viral vector has been the adenovirus in the clinic, lentiviral vectors are emerging as an attractive choice for transducing and maintaining stable, long-term expression of transgenes in target tissues. Therefore, extending the applicability of the TSTA system *in vivo* has included Lv-TSTA-*fl* development and successful testing in a systemic approach.

Finally, the TSTA system was adapted to direct gene expression for imaging and gene therapy applications. Of notice is the application to Ad-TSTA-*tk*, which has enabled us to examine transduction efficiency and therapy efficacy following prodrug administration *in vivo* using microPET/CT imaging. The approaches outlined previously will be discussed in more detail in the following sections.

Tissue-Specific Imaging of Reporter Genes

The success of gene therapy approaches lies in the development of tools for efficient gene transfer, prolonged expression, and gene expression monitoring. To achieve more stable tumor-specific targeting, the TSTA system was adapted to a lentivirus-based vector to noninvasively monitor the prostate-specific expression of *fl* reporter gene in living mice [28]. The utility of Lv-TSTA-*fl* (Lv-TSTA) was first demonstrated in cell culture, in which LNCaP cells showed a six- to eightfold induction of FL activity in the presence of androgen, with minimal levels of FL activity in nonprostate cell lines. To demonstrate the potency of the TSTA system to amplify reporter gene expression in an Lv context, *fl* expression was compared between LvTSTA and Lv-PBC-*fl* (LvPBC) vectors in LAPC9 tumors (Figure 6.6a). The bioluminescence signals observed with LvTSTA were 100-fold greater. Mice displayed the *fl* expression signals primarily in the tumor whereas other organs showed basal levels of gene expression. The androgen dependency was confirmed by visualizing decreasing FL signal following androgen withdrawal by surgical castration. Injection of LvTSTA into the prostate of nude mice revealed efficient and long-term *fl* gene expression for up to 3 months (Figure 6.6b). These studies demonstrated that the LvTSTA could also confer high levels of tissue-specific gene expression from a weak promoter while preserving cell-type specificity and the ability to noninvasively image the sustained, long-term expression of reporter genes in living animals. In another study, systemic administration of LvTSTA resulted in specific targeting of prostate LAPC9 tumors *in vivo* in severe combined immunodeficient (SCID) mice [29]. There were significant levels of transduction at the tumor site when compared with other organs, and gene expression was sustained for up to 3 weeks, although low levels of transduction were also observed in the spleen and liver. Although not as robust a delivery vector as the Ad, the results supported the use of TSTA-based lentiviral vectors for prostate tumor targeting after systemic delivery.

Further applications of the TSTA system have included adaptations to examine endogenous expression within transgenic animal models. Noninvasive imaging of bioluminescence reporter genes is being extensively used to study the onset and progression of disease in small-animal models. Two TSTA transgenic models have been generated carrying either the PBC or vascular endothelial growth factor (VEGF) tissue-specific promoters and have provided valuable insight into the regulation of promoter activities *in vivo*. The prostate-specific TSTA mouse was engineered in a single cassette to express the *fl* reporter gene. The transgenic mice show prostate-specific expression as early as 3 weeks of age (Figure 6.6c), while female animals did not display expression of the *fl* gene. The bioluminescence signal in the prostate is significantly higher than in other organs. We also demonstrated that blocking androgen availability could downregulate *fl* expression in the prostate. The transgenic mice displayed normal physical characteristics and

developmental behavior, indicating that the high level of GAL4-VP2-driven expression was well tolerated. These findings suggested that the GAL4-VP16 transactivator can be used to amplify reporter gene expression from a relatively weak promoter in a transgenic mouse model. Persistent prostate-selective expression was observed and no discernible toxic effect was observed in the prostate secondary to the expression of the potent GAL4-VP16 transactivator. Future studies combining an oncogenic model and the prostate-targeted imaging model described in this study would significantly aid in investigating oncogenesis and the progression of prostate cancer. Also, the strategies developed could be adapted for transgenic research in general by allowing for higher amplification of tissue-specific transgene expression.

Another interesting application of the TSTA system has been the generation of a transgenic mouse model to monitor endogenous VEGF expression during wound healing as a model for monitoring VEGF levels in the early angiogenic process. Transgenic mice carried the VEGF promoter (pVEGF) driving the TSTA system [30], and pVEGF-TSTA-*fl* (VEGF-TSTA) induced *fl* expression in the wound lesions by Days 4 and 5 and peaking on Days 15–22 postwounding (Figure 6.6d). The bioluminescence signals showed correlation with the endogenous VEGF protein levels in the wound tissue. In a mammary tumor model, *fl* expression was detected at Day 3, peaked at Day 17, and declined thereafter. These results supported the notion of noninvasive bioluminescence imaging for longitudinally monitoring a gene induction in response to biological processes such as a wound insult and tumor progression *in vivo*. This study has important implications for several applications in which it would be critical to noninvasively examine gene expression *in vivo* over time during a biological process or development.

Imaging Signaling Pathway Function using TSTA *in vivo*

AR Function in Prostate Cancer Progression and Drug Response

One of the original goals of the TSTA system had been to gain the ability to monitor the status of AR-mediated transcription over time during the AD to AI transition within a tumor by utilizing noninvasive *in vivo* imaging. An important application of imaging is to determine whether a drug is reaching its intended target within a living subject and to determine whether the efficacy of target inhibition or activation correlates with the therapeutic effect. A second application of imaging is to identify periods when a drug is displaying maximal efficacy to perform invasive analyses to understand the mechanism of inhibition. These approaches will be discussed in detail in the present section.

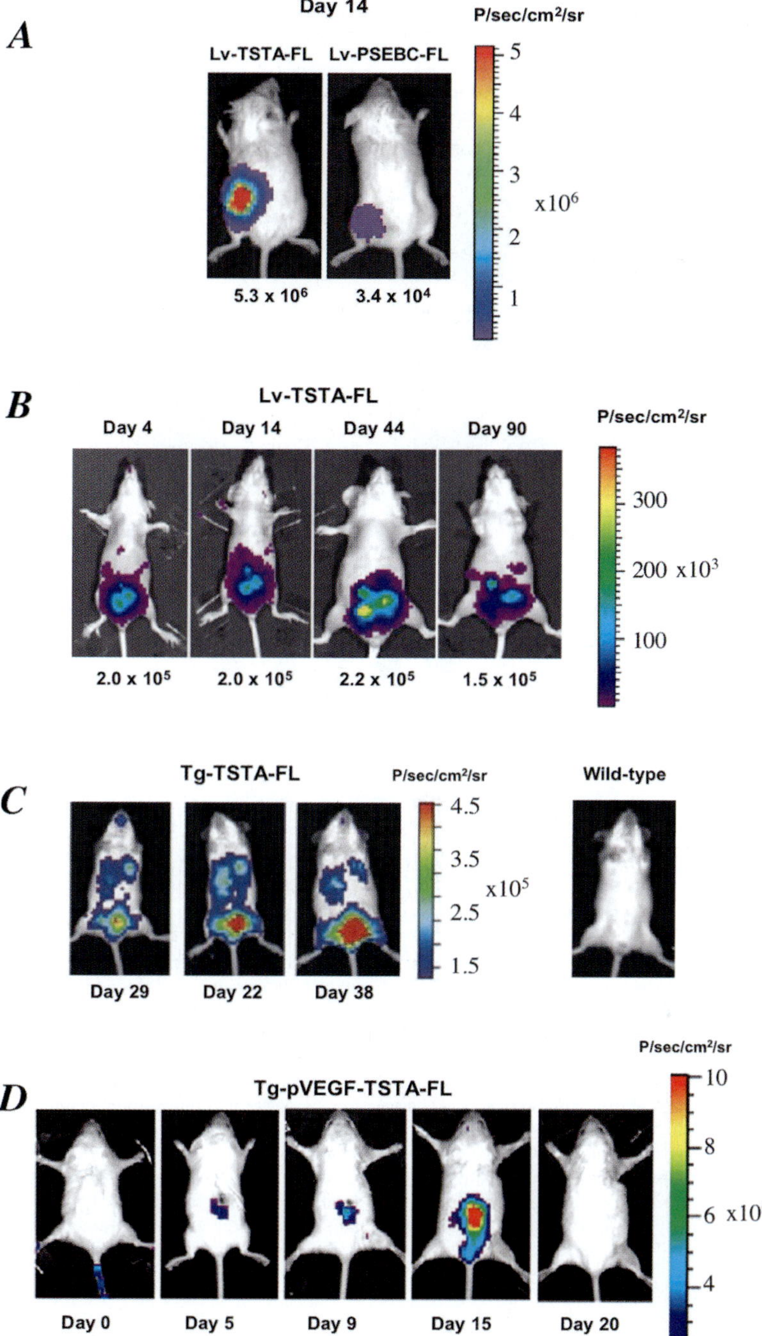

Figure 6.6. *In vivo* **stable, long-term optical imaging of TSTA-fl constructs. (A)** *Comparison of fl reporter gene expression using lentivirus carrying the one-step and two-step constructs.* Six-week-old male SCID mice were implanted with LAPC9 tumors. Equal titers of Lv-PBC-fl (Lv-PBC) and Lv-TSTA-fl (Lv-TSTA) were injected intratumorally (at ~0.6cm size) in two separate mice. The mice were imaged using the CCD camera on Day 14 following virus injection. The substrate, D-Luciferin was injected intraperitoneally 10 min prior to imaging. The *fl* gene expression was significantly greater in Lv-TSTA mice than in the Lv-PBC mice. **(B)** *In vivo optical imaging of mice following direct intraprostatic delivery of Lv-TSTA.* Six-week-old male nude mice were injected with Lv-TSTA in the dorsal lobe of the prostate. The mice were imaged using the CCD camera 2–4 d after virus injection. Subsequent imaging was performed up to Day 90. By Day 4, a high level of bioluminescence signal was observed in the prostate. The *fl* expression showed strong persistence with time. **(C)** *Optical imaging of TSTA-fl transgenic mice.* Transgenic positive male and female mice were imaged at 4 weeks of age. A significantly intense signal is observed only in males and it corresponds to the prostate area. In contrast, the female transgenic mouse shows no *fl* signal. The TSTA system retains long-term, androgen-responsive *fl* expression *in vivo* (12 weeks), and the prostatic *fl* signal is abolished following castration of male transgenic mice (23 d postcastration). *Ex vivo* examination of transgenic mouse organs for *fl* expression detected signals restricted to the prostate. **(D)** *Induction of fl expression during wound healing in a pVEGF-TSTA-fl transgenic mouse.* Mouse was imaged before wounding on the back skin (*day 0*) and imaged again every 4–5 d using D-luciferin (150 mg/kg ip). Color images of visible light are superimposed on photographic images of mice with a scale in photons per second per square centimeter per steradian (photons/sec/cm²/sr).

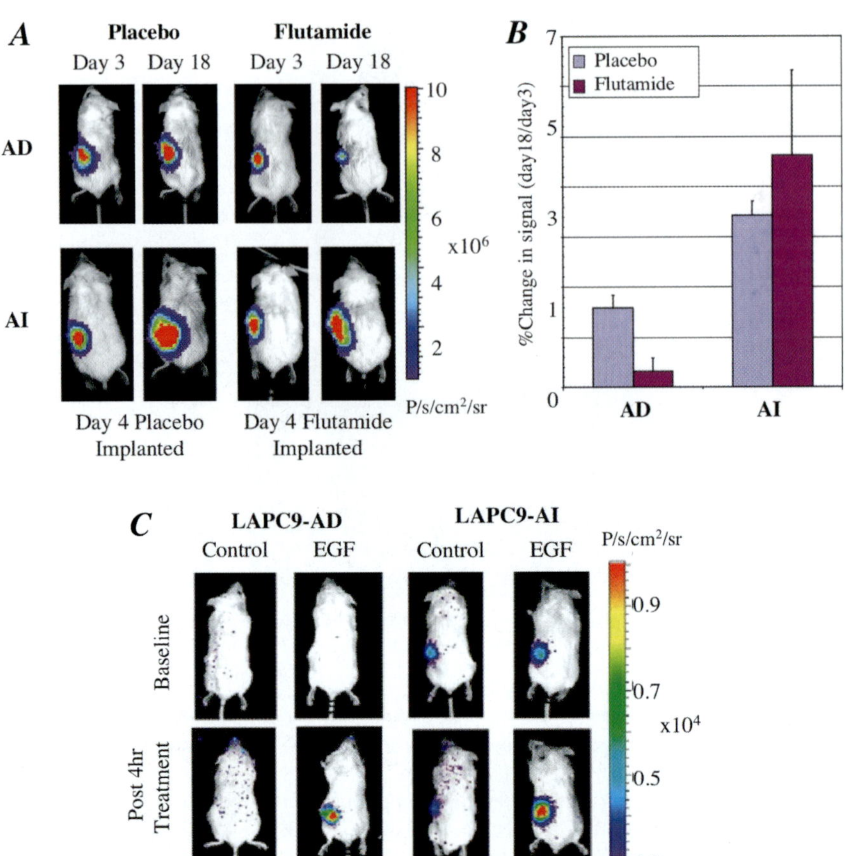

Figure 6.7. The TSTA system and imaging signaling pathway function *in vivo*. (A) *The TSTA system can be used to image the androgen-dependent to androgen-independent transition of prostate tumors in vivo.* Injection of AdTSTA results in strong expression and *fl* activity in AD LAPC9 xenograft tumors. Following castration, the signal decreases as the tumor is reduced in size. Development of a recurrent, androgen-independent tumor can be detected noninvasively by a rise in *fl* signals by Day 18. **(B)** *AdTSTA-mediated activity can be applied to examine functionality of the AR pathway in vivo.* AdTSTA were injected intratumorally into intact or castrated male mice bearing LAPC9 xenografts. By Day 3 a baseline image was acquired and flutamide or placebo pellets were implanted. The effect of treatment at Day 18 is shown. There was no detectable flutamide effect in the LAPC9 AI tumor model, suggesting resistance of AI to flutamide. The percentage change in signal versus Day 3 in placebo- and flutamide-treated animals bearing AD and AI tumors is shown. These results illustrate the application of imaging technology to provide mechanistic insight into antiandrogen targeting of androgen receptor during disease progression *in vivo*. **(C)** AdTSTA-ELK1 monitors both AR and MAPK activity *in vivo*. AdTSTA-ELK1 was constructed using a modified form of the AdTSTA system. Imaging AR and MAPK signaling in the LAPC9 xenograft model. Castrated AI or intact AD male SCID mice were implanted with LAPC9 tumors and were injected intratumorally with 2×10^7 pfu of AdTSTA-ELK1. After 3 d, mice were injected i.p. with EGF or vehicle control and then imaged 4 h later.

Examination of AR function *in vivo* in a progressive disease context is critically important because failure of androgen deprivation therapy tends to occur gradually over a period of time that can vary from weeks to years. The LAPC9 tumor is a good preclinical model for this transition because the AD-AI switch is gradual in castrated males and can be monitored using the TSTA system. The Ad-TSTA-*fl* allowed for visualization of the AD-AI transition *in vivo* and revealed that besides loss of AR activity upon androgen withdrawal, *fl* expression is reactivated as the tumor transitions into the recurrent state (Figure 6.7a) [15]. This example illustrates the ability to visualize reporter gene noninvasively over time in a highly sensitive manner. Androgen withdrawal by castration led to decreased serum PSA levels, decreased AR levels, and AR cytoplasmic localization within xenograft tumors. Furthermore, chromatin immunoprecipitation analysis revealed a loss of AR binding from the PSA promoter and enhancer. On transition of the tumor into the AI stage, AR levels increased, followed by AR translocation to the nucleus and binding to the PSA enhancer and promoter. Furthermore, RNA polymerase II also bound at both the promoter and the downstream exons. The data suggested that AR was fully active in AI tumors and that the TSTA system could sensitively measure AR activity in prostate tissues.

Much is currently known regarding the response of AR to antiandrogens in culture; however, little is known of how antiandrogen treatment affects AR activity in AD and AI tumors in animal models or how the effects might compare to androgen withdrawal. Because previous studies with the AdTSTA had shown that the amount of FL activity is proportional to AR activity both *in vitro* and in animals [9, 15], this system was used to examine the effect of flutamide, a nonsteroidal antiandrogen currently used for prostate cancer treatment in the context of AR function and activity. Importantly, the imaging measurements allowed for identification of time points at which AR expression declines within the AD tumor environment accompanying the onset of the therapeutic effect. These AR alterations suggested fundamental differences between androgen withdrawal and antiandrogen treatment, which might be related to the mechanism of hormone resistance. The effect of flutamide treatment on AR activity can be compared in AD and AI LAPC9 tumors. Whereas the AD tumors are grown in intact male mice, the AI tumors are propagated in castrated male SCID mice. Figure 6.7a shows that the imaging signal in AD tumors decreases over time in response to flutamide. In contrast, no inhibition of AR function is observed in AI tumors over the same period. In the AD group, the optical signals of placebo-treated animals were higher relative to flutamide-treated tumors. In contrast, the AI group had similar optical signals from both the flutamide-treated and the placebo cohort. Therefore, flutamide significantly inhibited AR function in AD prostate cancer xenografts, and the imaging trends accurately recapitulate the effect of flutamide on PSA levels, a clinical benchmark of prostate cancer (Figure 6.7b). Yet, imaging with AdTSTA proved more sensitive than serum PSA levels and reliably identified flutamide-mediated inhibition at earlier time points. Despite our previous observation that androgen receptor is fully active in AI cancer [15], AR signaling appears to be resistant to flutamide inhibition. Therefore, we concluded in this study that antiandrogens and androgen withdrawal have distinct mechanisms of inhibition within the context of a tumor.

Examining Signaling Pathway Output and Function Simultaneously with TSTA

Examination of signaling pathways *in vivo* has been explored in the context of the TSTA system for AR and of pharmaceuticals on AR, as discussed in the previous section. Another application of signaling pathway evaluation in a tissue-specific manner has consisted of detecting receptor tyrosine kinase activity *in vitro* and *in vivo*. Elevated mitogen-activated protein kinase (MAPK) activity is a hallmark of many malignancies. A current paradigm regarding the AD to AI transition of prostate cancer is that it is driven by receptor tyrosine kinases

and downstream MAPK pathway activity. One hypothesis is that the basal MAPK levels of AI tumors may greatly exceed that of AD tumors. To detect MAPK function in a prostate-specific manner during this AD-AI transition state, the binary design of TSTA was used to construct an imaging cassette able to simultaneously measure AR and MAPK function [30]. In the first step, a modified PSA regulatory region expresses GAL4-ELK1 rather than GAL4-VP16. GAL4-ELK1 contains the GAL4 DNA-binding domain fused to the activation domain of ELK1. Phosphorylation of the ELK1 activation domain by Erk1/2 kinase occurs at several sites and activates ELK1 to stimulate transcription from a GAL4-responsive reporter gene (*fl*).

Detection of EGF-activated MAPK in prostate cancer xenografts was achieved with the TSTA-ELK1 system. The amount of FL output was proportional to the activity of AR. The cell culture data showed evidence that the TSTA-ELK system responded synergistically to the combined action of AR and MAPK using either artificial or natural ligands for the Ras/MAPK pathway. This TSTA-ELK1 system, delivered via an Ad vector, can also measure the response to systemic EGF stimulation in both AD and AI tumor models *in vivo*. *Fl* expression was stimulated 4 h after stimulation in the LAPC9 model (Figure 6.7c). A consistent 2- to 3.5-fold induction of *fl* expression was detected in the EGF-treated animals compared to controls. AdTSTA-ELK1 could detect the systemic effect of EGF injection on MAPK-mediated activity in tumors *in vivo*. Therefore, data suggested that this imaging system could be used to screen for drugs that inhibit MAPK activity and thus facilitate analysis of signaling pathways during cancer progression and response of human cancer to therapy.

Coupling TSTA Imaging to Gene Therapy Applications

The molecular basis of imaging exploits specific molecular probes to achieve image contrast that differentiate target tissue from normal or background tissue in the subject [32]. The strengths of molecular imaging are particularly beneficial to cancer gene therapy [33], where it can aid in determining the efficacy of gene delivery and expression *in vivo*. Figure 6.8 illustrates the use of molecular imaging to assess the location, duration, and magnitude of gene transfer, as well as therapeutic activity.

The most common method to assess *in vivo* transgene expression is imaging reporter genes. Optical imaging technologies based on bioluminescent (e.g., luciferase genes) and fluorescent proteins produced in nature (e.g., GFP and RFP) have been applied widely as imaging reporters, allowing efficient, noninvasive, and rapid assessment of transgene expression in preclinical small-animal models (Figure 6.8a). However, the attenuation and scattering of visible light as it traverses through tissue

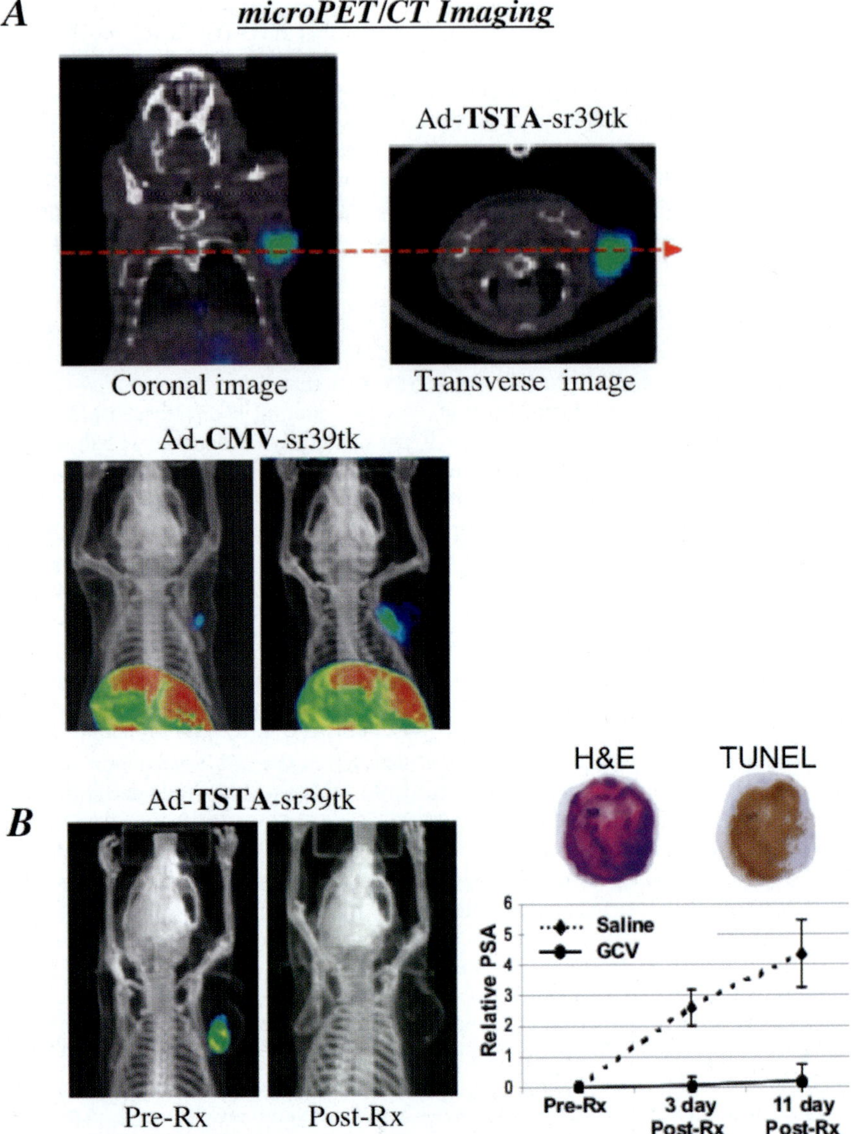

Figure 6.8. Utility of coupling *in vivo* imaging to prostate-targeted gene therapy. (A) *Magnitude and location of gene expression by vector mediated [^{18}F]FHBG-PET.* Robust HSV-sr39tk-dependent PET signal was detected in the LAPC-4 prostate tumor 8 d after intratumoral injection of 4×10^9 pfu of AdTSTA-sr39tk. The coronal and transverse images of a PET/CT study are represented. In the lower panel, both animals received the same 4×10^9 pfu of AdCMV-sr39tk injection directed into the tumor. However, their levels of transgene expression on Day 8, reflected by FHBG-PET signals, were different. Unexpectedly, strong PET signals were observed in the liver of both animals. This finding is attributed to spillage of AdCMV-sr39tk into systemic circulation despite the tumor-directed viral administration. **(B)** *Monitor suicide gene therapy by microPET/CT.* The paired images are the PET/CT of one representative animal injected with AdTSTA-sr39tk, pre- and posttreatment with ganciclovir (GCV). The tumor-localized PET signal declined post-GCV treatment. The graph on the right shows the serum PSA of the control group (saline treatment) continue to rise while the serum PSA of GCV-treated cohort showed no increase, indicating tumor growth was halted. Immunohistochemical analysis of the treated tumor showed substantial apoptosis, indicated by TUNEL-positive staining.

severely hinders optical imaging for human applications. The high-energy radionuclide imaging modality such as PET has the distinct advantage of producing 3-D localized signals, which can be translated from small animal to human subjects (Figure 6.8b). Three categories of PET reporter systems have been developed that either use intracellular enzymes (e.g., herpes simplex virus thymidine kinase (HSV1-tk)), cell membrane receptors (e.g., dopamine, somatostatin, and sodium iodide symporter (NIS) (refer to Chapter 3 for PET reporter details). We will focus our discussion on the *HSV1-tk* gene because it has the dual capacity to provide an imaging as well as tumoricidal function, and its expression has been targeted using TSTA.

Molecular imaging is a useful technology to validate the functionality of the TSTA vectors for *in vivo* applications. Hence, we used the TSTA system to amplify the expression of imaging reporter genes, *fl* and HSV1 mutant *tk* (*sr39tk*), which can be visualized by a cooled CCD optical camera and a positron emission tomography (PET) scanner, respectively. The engineered sr39TK enzyme exhibits significant higher binding affinity for Penciclovir (PCV) and ^{18}F-labeled PCV analogs, and thus, improved sensitivity for PET imaging (see also Chapter 3). A critical consideration in applying the prostate-targeted TSTA gene therapy to patients is whether it would be functional in hormone refractory prostate cancer (HRPC) [16]. The TSTA-driven vectors were tested on AD and AI prostate tumors. Real-time gene expression was monitored by optical and the combined PET and computed tomography (CT) modality [16]. Our results clearly illustrated that the activity of the TSTA vectors was AR dependent and recapitulated the functional status of endogenous AR in the tumors. In several cases, the TSTA-mediated expression was more robust in the AI tumors than in AD tumors, supporting the conclusion that AR function is activated in HRPC despite castrated levels of androgen.

Recently, we exploited the dual capacity of HSV1-tk as a toxic suicide and a PET reporter gene in an image-guided therapeutic study. In the study, we directly compared the therapeutic effects of Ad-TSTA- to Ad-CMV-driven sr39tk (Figure 6.8a). Vectors were delivered into human prostate tumors and serial optical and PET-CT imaging revealed gene expression restricted to tumors when the prostate-specific TSTA vector was employed [22]. In contrast, administration of a constitutively expressed Ad-CMV-*sr39tk* resulted in both strong tumor and liver signals. In fact, *sr39tk*-based PET enables the direct determination of the locations and magnitude of therapeutic gene expression prior to and post ganciclovir (GCV) prodrug treatment. The therapeutic outcome could be predicted based on the imaging findings. Overall, the augmented prostate-specific TSTA expression system was superior to the constitutive approach in safeguarding against systemic toxicity due to inadvertent leakage of vector outside of the tumor while achieving effective tumor killing (Figure 6.8b).

The integration of noninvasive imaging into prostate-specific cytotoxic gene therapy is an excellent example of how current treatment strategies can be improved. Several interesting and powerful applications of AdTSTA take advantage of its real-time production of imaging signals in living animals. Because both the PSA promoter and the TSTA system activities depend on the AR function *in vivo*, examination of AdTSTA-driven reporter

gene expression in a prostate tumor model under androgen ablation conditions enables the visualization of the tumor transitioning from the AD to the AI state. This transition was denoted by a recovery of luciferase optical signals despite the maintenance of an androgen-deprived environment [15]. AdTSTA was also applied to monitor the real-time treatment effects of antiandrogens, such as flutamide, in tumors [23]. In summary, the TSTA approach has shown to be an effective mechanism to boost many weak but cell-specific transcriptional regulatory elements. Besides the PSA promoter, many other promoters (e.g., PSMA, ARR2PB, and hK2) could be augmented by the TSTA strategy to target the primary tumor and its metastases (e.g., OC and PSA).

Imaging Cancer Metastasis

The robust and specific expression nature of TSTA imaging vectors, in particular for PET imaging, hold promise in that they can be developed to detect and treat metastatic lesions in preclinical and clinical settings. Another application for the TSTA system includes its use to detect occult metastases of prostate cancer at distant sites. The ability of AdTSTA to detect either lung or sentinel lymph node prostate cancer lesions has been confirmed by using both bioluminescence and PET imaging. First, AdTSTA-*fl* was used to detect metastases in a prostate tumor model that spontaneously metastasizes to lung and lymph nodes (intraprostatically-implanted LAPC9 cells overexpressing vascular endothelial growth factor C or VEGF-C) [34–35]. AdTSTA-*fl* (10^8 pfu) was administered via tail vein on Day 21 after tumor implantation and bioluminescence imaging revealed prominent signals in the chest region of the mouse (Figure 6.9a) at levels ~11-fold above background. Presence of metastatic tumor cells was confirmed by both *ex vivo* organ bioluminescence and histology in the same peripheral region of the lung. In Figure 6.9b, LAPC9/VEGF-C tumors also were used to determine feasibility of detecting lymph node micrometastases of prostate cancer. Fifteen days after intraprostatic tumor implantation, animals received AdTSTA-*fl* in each footpad to target lymph nodes possibly involved in prostate cancer metastases. FL imaging detected the presence of prostate cells in superficial lymph nodes *in vivo* and *ex vivo*. Histological examination of the node with the highest luminescence signal (periaortic) revealed a focus of subcapsular metastasis (indicated by a square). Using PET imaging, signals can be detected in both tumor and draining lymph nodes (Figure 6.9c). In this example, LAPC9/VEGF-C tumors were established on the right upper back and received 10^8 pfu of AdTSTA-*sr39tk* peritumorally. The viral particles were found to be lymphotropic, draining out of the tumor through the lymphatics and transducing any prostate tumor metastases located in the sentinel lymph nodes. [18]FHBG PET imaging detected the presence of

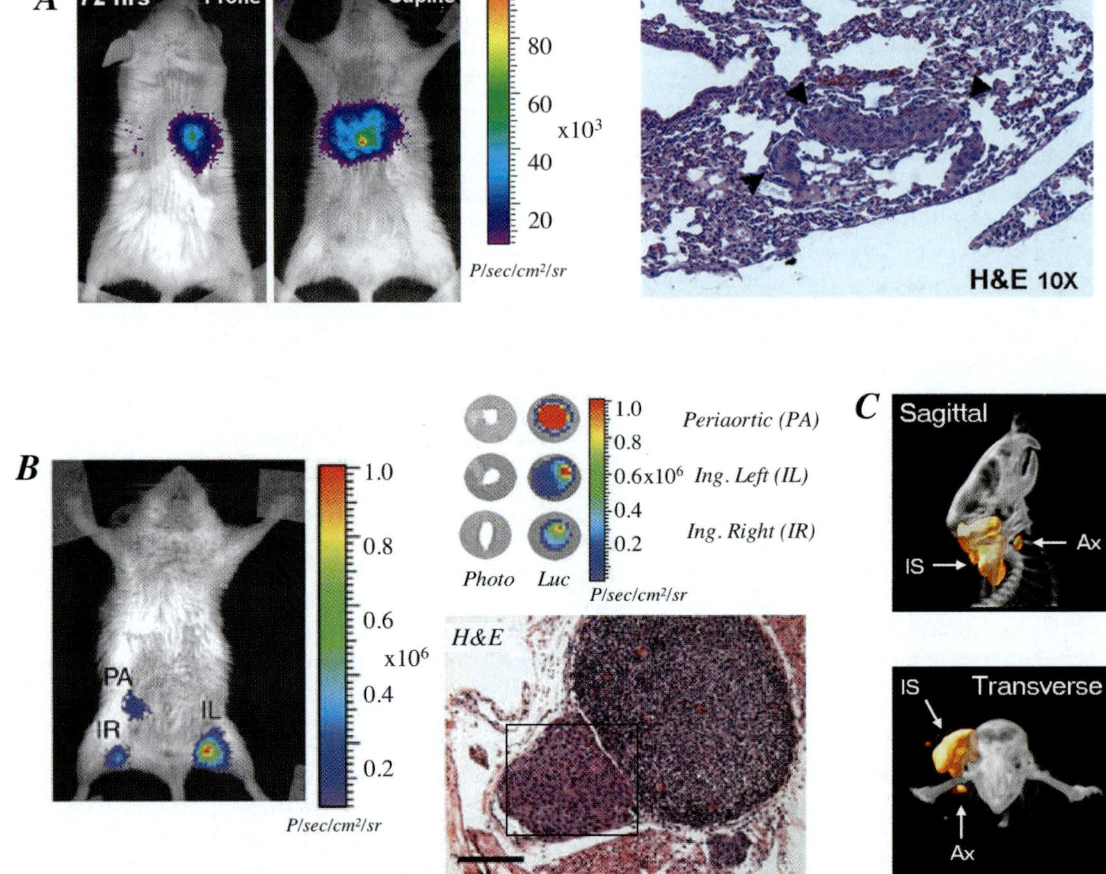

Figure 6.9. Application of TSTA to image occult lung and lymph node metastases. (A) *The prostate-specific AdTSTA-firefly luciferase (fl) vector can be used to detect lung metastases of prostate cancer.* 10^8 pfu of AdTSTA-fl were injected intravenously 21 d after orthotopic implantation of prostate tumors (LAPC9), and mice were imaged for FL expression 3 d later. This vector was generated in the new TSTA configuration diagrammed in Figure 6.5. **(B)** *AdTSTA-fl can be used to detect small lymphnode metastases of prostate cancer.* AdTSTA-fl was injected in both hind paws of an animal harboring an established orthotopic human prostate tumor typically able to metastasize to lymph nodes (LAPC9/VEGF-C)(left). *Ex vivo* imaging of the harvested pelvic nodes indicated that the largest volume of metastasis was located in the periaortic lymph node, as noted by its signal intensity. Histology (H&E) revealed a small subcapsular lesion (marked by a square) within this periaortic node. The number of cells detected within the node metastases was estimated to be ~4×10^3 cells, as determined by morphometry. **(C)** *Detection of occult lymph node metastasis by microPET/CT imaging.* A mouse with a tumor bearing tropism for metastasizing to lymph nodes received AdTSTA-tk via intratumoral injections. The hypothesis tested was that Ad could drain out of the tumor through peritumoral lymphatic vessels and into the regional sentinel lymph nodes. Signals were detected in both the tumor and the draining sentinel lymph node using ^{18}FHBG PET imaging. Representative sagittal (top) and transverse (bottom) microPET/CT images show that PET signal is emitted from the tumor injection site (IS) as well as the draining axillary sentinel lymph nodes (Ax). These results suggest that AdTSTA is a powerful and specific tool to reach distant prostate cancer metastases for their detection and potential eradication.

AdTSTA-*sr39tk* expression in both the tumor (injection site, IS) and the draining axillary lymph node (Ax), and these two signals could be distinguished in sagittal and transverse tomographic views. These findings support the potential of TSTA vectors as promising candidates to engineer future clinical tools to manage metastatic prostate cancer.

Translation of TSTA vector technology to image metastases in the clinics will require the use of high-energy radionuclide imaging approaches. TSTA strategies should enable safer and more efficient targeted diagnostic and therapies for prostate and other cancers. Based on current data, we anticipate that incorporating TSTA system in place of constitutive (CMV)-driven gene therapy strategies might improve the safety of the current status of localized and systemic prostate cancer therapies.

Other TSTA Systems

Other promoters have been optimized for use with the TSTA system and examples include promoters such as

VEGF, CEA, Tyrosinase, Muc-1, RPE95, and CCKAR [30, 36–41], to be discussed in the following. The vascular endothelial growth factor (VEGF) promoter activity could be augmented using the TSTA system in a transgenic mouse model [30], allowing examination of early angiogenesis in tumor progression as well as the wound healing process. The use of the VEGF promoter coupled to TSTA should allow for various applications in which it may be important to noninvasively examine VEGF gene expression *in vivo*. The carcinoembryonic antigen (CEA) promoter was used successfully to target expression of a GAL4-VP16 fusion protein to CEA-positive cells with an enhancement in expression 20–100-fold higher than CEA-negative colorectal and lung cancer cells [40]. The Mucin-1 (Muc1) promoter is expressed in a variety of colon carcinomas and also has been used to reduce systemic toxicity through specific gene expression of a GAL4-VP16 fusion protein [36]. Integration of the GAL4-VP16-binary system resulted in an up to 250-fold increase of Muc1-specific gene expression. In mucin-positive cell lines, expression was up to 590-fold higher as compared to the CMV promoter. Using a retinal pigment epithelial cell chimeric promoter (truncated human and mouse RPE95) in combination with GAL4-VP16 enhanced the transcriptional activity of the weak RPE promoter ~13-fold in human retinal pigment epithelial cells while retaining relative cell specificity following cotransfection [41]. The cholescystokinin type A receptor (CCKAR) pancreatic-specific promoter was used to drive expression of BikDD proapoptotic gene by the GAL4-VP16-WPRE integrated systemic amplifier (CCKAR-VISA) [37]. This adapted TSTA system was used to augment the CCKAR promoter activity 300- to 1000-fold and the C-VISA activity reached 60–300% of CMV promoter activity. The C-VISA system also targeted specific transgene expression to pancreatic cancer in animal models as examined by bioluminescence imaging. Most recently, a chimeric prostate-specific sequence was generated that combined PSA and PSMA enhancers and a T cell receptor gamma-chain alternate reading frame protein (TARP) promoter (PPT) [42]. The PPT sequence was shown to be transcriptionally active in human prostate cancer cells both in the presence and absence of testosterone. An adenovirus with TSTA-amplified firefly luciferase expression, Ad-PPT/TSTA, displayed up to 100-fold higher prostate-specific transcriptional activity than a nonamplified PPT-based adenovirus. The PPT-TSTA system yielded approximately 30-fold higher transgene expression than the nonamplified PPT-based adenovirus in both LNCaP tumor xenografts and murine TRAMP-C2 tumors. Overall, given the success of these cell- or tissue-specific TSTA systems, which achieved highly efficient and specific transgene expression, we anticipate they will likely contribute to the safety and efficacy of experimental approaches applied to cancer gene therapy.

A novel regulation system involves a positive feedback loop with prostate specificity (PFLPS) (Figure 6.10). This system incorporates a TRE (tet-responsive element) upstream of the prostate-specific ARR2PB promoter driving a tet-transactivator (tTA) to enhance its activity with Tet regulation *in vitro* [43]. The expression of both GFP and tTA were placed under the control of these TRE-ARR2PB promoters, and a positive feedback loop was demonstrated specifically in prostate cells. Hence, the innovation of this design resides in the combination of drug inducible and tissue-selective transcriptional control to achieve even greater regulation of expression levels in prostate cancer cells. A different conditional approach to enhancing tissue-specific promoter activity involves the bacteriophage P1-derived Cre-lox system (Figure 6.10), a powerful and versatile tool for *in vivo* DNA recombination. When Cre recombinase expression is targeted using a tissue-specific promoter, conditional "knockout" of a target gene can be achieved in a particular tissue or cell type. This system also can apply to enhance tissue- or tumor-specific promoter activities in cancer gene therapy. Target cells are transduced with a vector harboring the therapeutic transgene separated from a strong constitutive promoter by a translational stop cassette flanked by two loxP sites. Cotransduction with a vector expressing Cre from a tissue-specific promoter results in excision of the stop cassette and expression of the transgene in a highly tissue-specific manner [44]. The Cre-lox system has been applied to enhance the activity of prostate-specific promoters such as PSA and PSMA [45–47]. In this approach, the adenoviral vector-expressing Cre driven by the PSA promoter-enhancer activates expression of the secondary CMV-loxP-stop-luciferase vector. As expected, this PSA promoter-based system was regulated by androgen, and its use in a CD (cytosine deaminase) suicide gene therapy inhibited LNCaP tumor growth in nude mice [45]. The efficiency of this binary system might be limited by the need for two vectors to transduce the same cell for activation. As a result, the combined PSA promoter/enhancer and Cre-loxP system exhibited more modest increases in activity (3X stronger) than that of a direct PSA promoter/enhancer-driven vector. Using a liposome-mediated gene transfer, the combination of the Cre/lox system with the PSMA promoter/enhancer (PEPM) greatly enhanced the efficacy of HSV-tk suicide gene therapy as compared with that of the PEPM promoter alone in PSMA+ cells [46]. As the PSMA promoter is induced by androgen depletion, the PEPM-Cre/CMV-lox system also exhibited a stronger inhibitory effect on tumor growth in castrated compared to intact noncastrated mice. These findings indicate that the PEPM and Cre-lox-mediated cytotoxic therapy may apply to patients undergoing androgen deprivation therapy [47]. An interesting strategy that developed involved an altered form of Cre recombinase (CRE-M) that

Other gene expression amplification systems

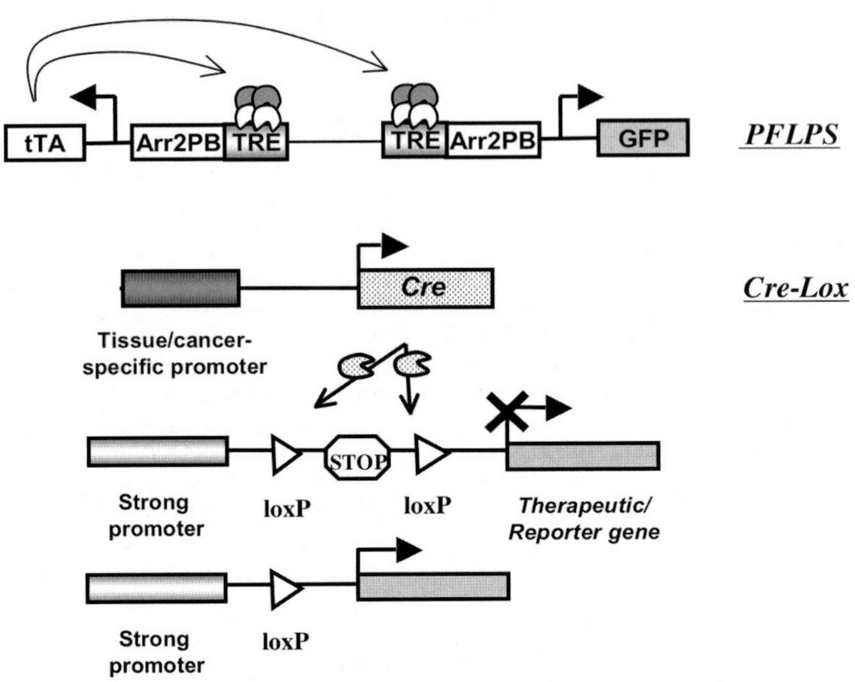

Figure 6.10. Other TSTA systems. Other amplification systems include a <u>novel positive feedback loop with prostate specificity</u> (PFLPS) [43], Cre-lox, and C-VISA. In the PFLPS, both tTA and GFP genes are expressed in a tissue-specific manner by ARR2PB promoters. Binding of tTA protein to the TRE sequences upstream of the ARR2PB promoters can induce GFP and tTA expression further, initiating a positive feedback loop. In nonprostate cells, where the ARR2PB promoter is inactive, gene expression would not be induced. In the <u>Cre-lox</u> system, mediated gene expression activation system involves Cre recombinase expression regulated by a tissue- or cancer-specific promoter. Activation of transgene expression is induced by removal of the translational inhibition sequence via a Cre-specific recombination between the two loxP sites.

displays an androgen-induced activity. CRE-M was generated by fusion of Cre to the AR ligand-binding domain (LBD), and three different forms of LBD (wild-type, nonligand binding, or Thr→Ala mutant) were examined [48]. The activity of CRE-M-LBDwt incorporated in Ad was stimulated tenfold by the addition of androgen. Interestingly, CRE-M-LBD mutant activity was inducible by both androgen and the antiandrogen flutamide, suggesting that the CRE-M-AR-LBD system can provide an additional level of regulation to the Cre/loxP-mediated prostate gene therapy applications [48].

Another application of the Cre-loxP system used PET imaging to assess Cre-loxP activity *in vivo*. An Ad containing a silent HSV1-*tk* gene was administered to transgenic mice expressing Cre in liver [49]. Activation of the HSV1-*tk* gene expression and tk activity in liver could be assessed noninvasively by PET, demonstrating a powerful tool for monitoring location, magnitude, and duration of conditional gene activation that may be applied for other animal tumor models or gene therapy applications.

Limitations of TSTA Systems and Future Directions

Several TSTA and Cre-lox systems were described that hold promise for therapeutic and gene expression applications *in vivo*. However, it remains critical that any reporter imaging or therapeutic approach does not significantly result in adverse effects. Several limitations may need to be overcome with each particular system to ensure success in safety, specificity, reversibility, low basal activity, and high inducibility. For both TSTA and Cre-lox, a potential limitation is induction of cellular toxicity. For example, in zebrafish, high levels of injected GAL4-VP16 can be deleterious to development [50], and in cultured mammalian cells, Cre expression/activity can result in reduced cell proliferation and accumulation of chromosomal aberrations [51]. The toxicity in these two examples appeared to correlate with higher levels and activity of effector proteins, suggesting that low-level or prolonged expression may allow for sufficient activity but without toxic effects. In this respect, the TSTA system has the advantage in that reporter/therapeutic gene levels can be modulated while the effects of Cre-lox (gene excision) may not be reversed once activated. The Cre-lox system, however, may hold an advantage over TSTA in that it requires a relatively small amount of Cre for effective conditional gene activation/amplification. Future studies will focus on modulating different levels of Cre expression to evaluate the sensitivity needed for imaging or therapeutic applications within an acceptable toxicity threshold.

Other strategies that might present some advantages over both TSTA and Cre-lox include regulatable expression systems using prokaryotic or eukaryotic transcriptional controls. A widely used prokaryotic inducible strategy is the "tetracycline (tet) transactivator system," which similar to TSTA consists of two-step transactivation but relies on tet binding to regulate activator protein function. The activator is a fusion between the tet repressor and HSV1-VP16 proteins (tetR-VP16 or tTA). When tTA associates with tet, its conformation is altered to render the DNA-binding domain inactive. In the absence of tet, tTA activates transcription by binding DNA at tetO sequences placed upstream of a minimal promoter and gene(s) of interest. Both uni- and bidirectional conformations have been shown to work well in this system. This system (and several mutant forms of tTA with improved tet regulatability) has been used successfully in transgenic animals and gene therapy vectors to achieve highly specific and inducible gene expression. Other inducible systems utilize eukaryotic regulatory elements such as fusions of GAL4-VP16 to the ligand binding domain of the estrogen receptor (ER) or progesterone receptor (PR) [52, 53]. The ER chimeric transcription factor responded to estradiol by binding to GAL4 sites and activating transcription ∼100-fold, although at relatively high doses, while the PR chimeric transcription factor allows tighter regulation by a progesterone antagonist (RU486) at low doses. One major drawback of using RU486 is its abortifacient action, which clearly compromises its use in gene therapy applications.

An apparently better class of inducible system uses an activation domain from the human transcription factor Nf-kB (p65) rather than that of VP16 [54]. The transcription factor in this case is a heterodimer, of which one monomer combines the DNA-binding domain of the nonmammalian protein ZFHD-1 and the human protein FKBP12. The second monomer is composed of a fusion between the human protein FRAP and p65. Because both FRAP and FKBP12 bind to the small molecule rapamycin, a heterodimeric transcription factor is formed that is able to bind at ZFHD-1 sites to activate transcription at responsive promoters. Advantages include induction at very low rapamycin doses and, theoretically, the use of p65 rather than VP16 should avoid toxic transcription "squelching" effects. Disadvantages include the fact that two proteins need to be expressed prior to heterodimer formation, and that the inducer is immunosuppressive, which may not be ideal for gene therapy use. Recent applications of this system have included the use of the rapamycin-based dimerizer system to regulate the expression of endogenous genes, the generation of new chemical dimerizers based on FK506, dexamethasone and methotrexate, and progress is being made toward the clinical use of adeno-associated virus and adenovirus vectors regulated by rapamycin analogs (reviewed in ref. [55]). Newer generation vectors include adeno-associated viruses (AAV) developed to induce multiple cycles of rapamycin-dependent antiangiogenic transgene (angiostatin) expression, and these may be promising for clinical applications [56].

In summary, the challenges encountered by these inducible systems have been high basal gene expression and low activity, which can be improved further with reducing minimal promoter sequences to reduce baseline expression, by refining the composition and/or number of transactivation domains within activator protein fusions, and by titrating the number of DNA-binding elements. The major advantage of these systems has been the ability to modulate and control spatial and temporal gene expression in a living organism or cells. This advantage can be incorporated into the TSTA system to enhance its promise for clinical applications. For example, TSTA might be further modified to encompass inducible regulation at the activator step (tet system or macrolides). Nevertheless, the challenges for translating these transcriptional regulation systems to a therapeutic setting lie in developing newer vectors that can achieve spatial and temporal gene expression control while maintaining expression at levels appropriate for effective therapy. The field will continue to improve upon current systems to evolve into strategies and vectors that can allow us to monitor noninvasively the magnitude and location of gene expression, the corresponding effectiveness of new therapies, and any toxic effects of vectors and expression systems within a living organism.

SUMMARY

Transcriptional targeting and gene expression amplification strategies are promising and feasible alternatives with which to improve the specificity and efficacy of current gene therapy modalities. However, significant obstacles remain in specifically seeking out and destroying cancer cells that may have spread to multiple sites within the organism (cancer metastases). To further enhance tumor selectivity and killing, approaches that augment the transcriptional targeting of the TSTA system could also target cell surface antigens or biochemical pathways unique to tumor cells. Combinations of multiple targeting strategies into tissue- or cancer-specific viral vectors should achieve synergistic selectivity and efficacy of therapy.

Although incorporation of transductional targeting will likely enhance the efficacy of gene expression-based approaches we have discussed, the field of prostate cancer transductional targeting is in the early stages of development. Some promising approaches have included redirecting Ad tropism to prostate cancer cells via alpha6beta1 integrins [57], epidermal growth factor receptor [58], and PSMA [59], either incorporated into the Ad vector or coupled to a polymer coating. Clearly, the efficacy of these newer targeted gene therapy approaches compared to traditional vectors in clinical settings will need to be assessed in clinical trials. Noninvasive imaging will be a critical tool to assess the performance of the targeted vectors *in vivo*. As these augmented tissue-specific regulatory systems (e.g., TSTA), more efficient viral vector configurations, and cell surface targeting approaches continue to evolve, it will remain critical to stringently test cancer-targeted gene therapies in preclinical settings and promote translation of the most promising strategies to the clinic.

REFERENCES

1 Ray, P., Bauer, E., Iyer, M., Barrio, J. R., Satyamurthy, N., Phelps, M. E. et al. (2001). Monitoring gene therapy with reporter gene imaging. *Semin Nucl Med* 31(4): 312–320.

2 Wu, L., Johnson, M., Sato, M. (2003). Transcriptionally targeted gene therapy to detect and treat cancer. *Trends Mol Med* 9(10): 421–429.

3 Wu, L., Matherly, J., Smallwood, A., Adams, J. Y., Billick, E., Belldegrun, A. et al. (2001). Chimeric PSA enhancers exhibit augmented activity in prostate cancer gene therapy vectors. *Gene Ther* 8(18): 1416–1426.

4 Lee, S. J., Kim, H. S., Yu, R., Lee, K., Gardner, T. A., Jung, C. et al. (2002). Novel prostate-specific promoter derived from PSA and PSMA enhancers. *Mol Ther* 6(3): 415–421.

5 Sadowski, I., Ma, J., Triezenberg, S., Ptashne, M. (1988). GAL4-VP16 is an unusually potent transcriptional activator. *Nature* 335(6190): 563–564.

6 Segawa, T., Takebayashi, H., Kakehi, Y., Yoshida, O., Narumiya, S., Kakizuka, A. (1998). Prostate-specific amplification of expanded polyglutamine expression: a novel approach for cancer gene therapy. *Cancer Res* 58(11): 2282–2287.

7 Figueiredo, M. L., Sato, M., Johnson, M., Wu, L. (2006). Specific targeting of gene therapy to prostate cancer using a two-step transcriptional amplification system. *Future Oncol* 2(3): 391–406.

8 Aumuller, G., Seitz, J., Lilja, H., Abrahamsson, P. A., von der Kammer, H., Scheit, K.H. (1990). Species- and organ-specificity of secretory proteins derived from human prostate and seminal vesicles. *Prostate* 17(1): 31–40.

9 Zhang, L., Adams, J. Y., Billick, E., Ilagan, R., Iyer, M., Le, K. et al. (2002). Molecular engineering of a two-step transcription amplification (TSTA) system for transgene delivery in prostate cancer. *Mol Ther* 5(3): 223–232.

10 Cleutjens, K. B., van der Korput, H. A., Ehren-van Eekelen, C. C., Sikes, R. A., Fasciana, C., Chung, L. W. et al. (1997). A 6-kb promoter fragment mimics in transgenic mice the prostate-specific and androgen-regulated expression of the endogenous prostate-specific antigen gene in humans. *Mol Endocrinol* 11(9): 1256–1265.

11 Cleutjens, K. B., van der Korput, H. A., van Eekelen, C. C., van Rooij, H. C., Faber, P. W., Trapman, J. (1997). An androgen response element in a far upstream enhancer region is essential for high, androgen-regulated activity of the prostate-specific antigen promoter. *Mol Endocrinol* 11(2): 148–161.

12 Schaffner, D. L., Barrios, R., Shaker, M. R., Rajagopalan, S., Huang, S. L., Tindall, D. J. et al. (1995). Transgenic mice carrying a PSArasT24 hybrid gene develop salivary gland and gastrointestinal tract neoplasms. *Lab Invest* 72(3): 283–290.

13 Yeung, F., Li, X., Ellett, J., Trapman, J., Kao, C., Chung, L.W. (2000). Regions of prostate-specific antigen (PSA) promoter confer androgen-independent expression of PSA in prostate cancer cells. *J Biol Chem* 275(52): 40846–40855.

14 Titus, M. A., Schell, M. J., Lih, F. B., Tomer, K. B., Mohler, J. L. (2005). Testosterone and dihydrotestosterone tissue levels in recurrent prostate cancer. *Clin Cancer Res* 11(13): 4653–4657.

15 Zhang, L., Johnson, M., Le, K. H., Sato, M., Ilagan, R., Iyer, M. et al. (2003). Interrogating androgen receptor function in recurrent prostate cancer. *Cancer Res* 63(15): 4552–4560.

16 Sato, M., Johnson, M., Zhang, L., Gambhir, M. Carey and L. Wu (2005). Functionality of androgen receptor-based gene expression imaging in hormone refractory prostate cancer. *Clin Cancer Res* 11(10): 3743–3749.

17 Pang, S., Dannull, J., Kaboo, R., Xie, Y., Tso, C. L., Michel, K. et al. (1997). Identification of a positive regulatory element responsible for tissue-specific expression of prostate-specific antigen. *Cancer Res* 57(3): 495–499.

18 Adams, J. Y., Johnson, M., Sato, M., Berger, F., Gambhir, S. S., Carey, M. et al. (2002). Visualization of advanced human prostate cancer lesions in living mice by a targeted gene transfer vector and optical imaging. *Nat Med* 8(8): 891–897.

19 Iyer, M., Wu, L., Carey, M., Wang, Y., Smallwood, A., Gambhir, S. S. (2001). Two-step transcriptional amplification as a method for imaging reporter gene expression using weak promoters. *Proc Natl Acad Sci U S A* 98(25): 14595–14600.

20 Qiao, J., Doubrovin, M., Sauter, B. V., Huang, Y., Guo, Z. S., Balatoni, J. et al. (2002). Tumor-specific transcriptional targeting of suicide gene therapy. *Gene Ther* 9(3): 168–175.

21 Sato, M., Johnson, M., Zhang, L., Zhang, B., Le, K., Gambhir, S. S. et al. (2003). Optimization of adenoviral vectors to direct highly amplified prostate-specific expression for imaging and gene therapy. *Mol Ther* 8(5): 726–737.

22 Johnson, M., Sato, M., Burton, J., Gambhir, S. S., Carey, M., Wu, L. (2005). Micro-PET/CT monitoring of herpes thymidine kinase suicide gene therapy in a prostate cancer xenograft: The advantage of a cell-specific transcriptional targeting approach. *Mol Imaging* 4(4): 463–472.

23 Ilagan, R., Zhang, L. J., Pottratz, J., Le, K., Salas, S., Iyer, M. et al. (2005). Imaging androgen receptor function during

flutamide treatment in the LAPC9 xenograft model. *Mol Cancer Ther* 4(11): 1662–1669.

24 Johnson, M., Huyn, S., Burton, J., Sato, M., Wu, L. (2006). Differential biodistribution of adenoviral vector in vivo as monitored by bioluminescence imaging and quantitative polymerase chain reaction. *Hum Gene Ther* 17(12): 1262–1269.

25 Sato, M., Figueiredo, M. L., Burton, J. B., Johnson, M., Chen, M., Powell, R. et al. (2008). Configurations of a two-tiered amplified gene expression system in adenoviral vectors designed to improve the specificity of in vivo prostate cancer imaging. *Gene Ther* 15(8): 583–593.

26 Ray, S., Paulmurugan, R., Hildebrandt, I., Iyer, M., Wu, L., Carey, M. et al. (2004). Novel bidirectional vector strategy for amplification of therapeutic and reporter gene expression. *Hum Gene Ther* 15(7): 681–690.

27 Ray, S., Paulmurugan, R., Patel, M. R., Ahn, B. C., Wu, L., Carey, M. et al. (2008). Noninvasive imaging of therapeutic gene expression using a bidirectional transcriptional amplification strategy. *Mol Ther.*

28 Iyer, M., Salazar, F. B., Lewis, X., Zhang, L., Carey, M., Wu, L. et al. (2004). Noninvasive imaging of enhanced prostate-specific gene expression using a two-step transcriptional amplification-based lentivirus vector. *Mol Ther* 10(3): 545–552.

29 Iyer, M., Salazar, F. B., Wu, L., Carey, M., Gambhir, S. S. (2006). Bioluminescence imaging of systemic tumor targeting using a prostate-specific lentiviral vector. *Hum Gene Ther* 17(1): 125–132.

30 Wang, Y., Iyer, M., Annala, A., Wu, L., Carey, M., Gambhir, S. S. (2006). Noninvasive indirect imaging of vascular endothelial growth factor gene expression using bioluminescence imaging in living transgenic mice. *Physiol Genomics* 24(2): 173–180.

31 Ilagan, R., Pottratz, J., Le, K., Zhang, L., Wong, S. G., Ayala, R. et al. (2006). Imaging mitogen-activated protein kinase function in xenograft models of prostate cancer. *Cancer Res* 66(22): 10778–10785.

32 Massoud, T. F., Gambhir, S. S. (2003). Molecular imaging in living subjects: seeing fundamental biological processes in a new light. *Genes Dev* 17(5): 545–580.

33 Iyer, M., Sato, M., Johnson, M., Gambhir, S. S., Wu, L. (2005). Applications of molecular imaging in cancer gene therapy. *Curr Gene Ther* 5(6): 607–618.

34 Brakenhielm, E., Burton, J. B., Johnson, M., Chavarria, N., Morizono, K., Chen, I. et al. (2007). Modulating metastasis by a lymphangiogenic switch in prostate cancer. *Int J Cancer* 121(10): 2153–2161.

35 Burton, J. B., Johnson, M., Sato, M., Koh, S. B., Mulholland, D. J., Stout, D. et al. (2008). Adenovirus-mediated gene expression imaging to directly detect sentinel lymph node metastasis of prostate cancer. *Nat Med* 14(8): 882–888.

36 Block, A., Milasinovic, D., Mueller, J., Schaefer, P., Schaefer, H., Greten, H. (2002). Amplified Muc1-specific gene expression in colon cancer cells utilizing a binary system in adenoviral vectors. *Anticancer Res* 22(6A): 3285–3292.

37 Xie, X., Xia, W., Li, Z., Kuo, H. P., Liu, Y., Ding, Q. et al. (2007). Targeted expression of BikDD eradicates pancreatic tumors in noninvasive imaging models. *Cancer Cell* 12(1): 52–65.

38 Richards, C. A., Austin, E. A., Huber, B. E. (1995). Transcriptional regulatory sequences of carcinoembryonic antigen: identification and use with cytosine deaminase for tumor-specific gene therapy. *Hum Gene Ther* 6(7): 881–893.

39 Siders, W. M., Halloran, P. J., Fenton, R. G. (1998). Melanoma-specific cytotoxicity induced by a tyrosinase promoter-enhancer/herpes simplex virus thymidine kinase adenovirus. *Cancer Gene Ther* 5(5): 281–291.

40 Koch, P. E., Guo, Z. S., Kagawa, S., Gu, J., Roth, J. A., Fang, B. (2001). Augmenting transgene expression from carcinoembryonic antigen (CEA) promoter via a GAL4 gene regulatory system. *Mol Ther* 3(3): 278–283.

41 Zhang, D., Sutanto, E. N., Rakoczy, P. E. (2004). Concurrent enhancement of transcriptional activity and specificity of a retinal pigment epithelial cell-preferential promoter. *Mol Vis* 10: 208–214.

42 Dzojic, H., Cheng, W. S., Essand, M. (2007). Two-step amplification of the human PPT sequence provides specific gene expression in an immunocompetent murine prostate cancer model. *Cancer Gene Ther* 14(3): 233–240.

43 Woraratanadharm, J., Rubinchik, S., Yu, H., Fan, F., Morrow, S. M., Dong, J. Y. (2004). Highly specific transgene expression mediated by a complex adenovirus vector incorporating a prostate-specific amplification feedback loop. *Gene Ther* 11(18): 1399–1407.

44 Sato, Y., Tanaka, K., Lee, G., Kanegae, Y., Sakai, Y., Kaneko, S. et al. (1998). Enhanced and specific gene expression via tissue-specific production of Cre recombinase using adenovirus vector. *Biochem Biophys Res Commun* 244(2): 455–462.

45 Yoshimura, I., Ikegami, S., Suzuki, S., Tadakuma, T., Hayakawa, M. (2002). Adenovirus mediated prostate specific enzyme prodrug gene therapy using prostate specific antigen promoter enhanced by the Cre-loxP system. *J Urol* 168(6): 2659–2664.

46 Ikegami, S., Tadakuma, T., Suzuki, S., Yoshimura, I., Asano, T., Hayakawa, M. (2002). Development of gene therapy using prostate-specific membrane antigen promoter/enhancer with Cre Recombinase/LoxP system for prostate cancer cells under androgen ablation condition. *Jpn J Cancer Res* 93(10): 1154–1163.

47 Ikegami, S., Tadakuma, T., Ono, T., Suzuki, S., Yoshimura, I., Asano, T. et al. (2004). Treatment efficiency of a suicide gene therapy using prostate-specific membrane antigen promoter/enhancer in a castrated mouse model of prostate cancer. *Cancer Sci* 95(4): 367–370.

48 Kaczmarczyk, S. J., Green, J. E. (2003). Induction of cre recombinase activity using modified androgen receptor ligand binding domains: a sensitive assay for ligand-receptor interactions. *Nucleic Acids Res* 31(15): e86.

49 Sundaresan, G., Paulmurugan, R., Berger, F., Stiles, B., Nagayama, Y., Wu, H. et al. (2004). MicroPET imaging of Cre-loxP-mediated conditional activation of a herpes simplex virus type 1 thymidine kinase reporter gene. *Gene Ther* 11(7): 609–618.

50 Koster, R. W., Fraser, S. E. (2001). Tracing transgene expression in living zebrafish embryos. *Dev Biol* 233(2): 329–346.

51 Loonstra, A., Vooijs, M., Beverloo, H. B., Allak, B. A., van Drunen, E., Kanaar, R. et al. (2001). Growth inhibition and

DNA damage induced by Cre recombinase in mammalian cells. *Proc Natl Acad Sci U S A* **98**(16): 9209–9214.

52　Wang, H. U., Chen, Z. F., Anderson, D. J. (1998). Molecular distinction and angiogenic interaction between embryonic arteries and veins revealed by ephrin-B2 and its receptor Eph-B4. *Cell* **93**(5): 741–753.

53　Braselmann, S., Graninger, P., Busslinger, M. (1993). A selective transcriptional induction system for mammalian cells based on Gal4-estrogen receptor fusion proteins. *Proc Natl Acad Sci U S A* **90**(5): 1657–1661.

54　Rivera, V. M., Clackson, T., Natesan, S., Pollock, R., Amara, J. F., Keenan, T. et al. (1996). A humanized system for pharmacologic control of gene expression. *Nat Med* **2**(9): 1028–1032.

55　Pollock, R., Clackson, T. (2002). Dimerizer-regulated gene expression. *Curr Opin Biotechnol* **13**(5): 459–467.

56　Nguyen, M., Huan-Tu, G., Gonzalez-Edick, M., Rivera, V. M., Clackson, T., Jooss, K.U. et al. (2007). Rapamycin-regulated control of antiangiogenic tumor therapy following rAAV-mediated gene transfer. *Mol Ther* **15**(5): 912–920.

57　Stevenson, M., Hale, A. B., Hale, S. J., Green, N. K., Black, G., Fisher, K. D. et al. (2007). Incorporation of a laminin-derived peptide (SIKVAV) on polymer-modified adenovirus permits tumor-specific targeting via alpha6-integrins. *Cancer Gene Ther* **14**(4): 335–345.

58　Bonsted, A., Engesaeter, B. O., Hogset, A., Maelandsmo, G. M., Prasmickaite, L., D'Oliveira, C. et al. (2006). Photochemically enhanced transduction of polymer-complexed adenovirus targeted to the epidermal growth factor receptor. *J Gene Med* **8**(3): 286–297.

59　Kraaij, R., van Rijswijk, A. L., Oomen, M. H., Haisma H. J., Bangma, C. H. (2005). Prostate specific membrane antigen (PSMA) is a tissue-specific target for adenoviral transduction of prostate cancer in vitro. *Prostate* **62**(3): 253–259.

Imaging Instrumentation

Physics, Instrumentation, and Methods for Imaging Reporter Gene Expression in Living Subjects

7

Craig S. Levin

INTRODUCTION

Achieving *in vivo* imaging of gene transfer provides molecular imaging with exciting opportunities in applications such as tracking stem or progenitor cells after transplantation or guiding efficacy studies of gene therapy [1, 2]. Monitoring of transgene expression is routinely accomplished by coexpressing marker genes with potential therapeutic transgenes in small laboratory animals. In this standard approach, the animals are sacrificed and methods such as histology or fluorescence microscopy (e.g., see [3]) are performed on appropriate tissue samples to analyze the presence of marker gene expression. As with other biological studies, it is desirable to perform these studies routinely using noninvasive imaging to increase efficiency and information content associated with monitoring transgene expression over time in the same research subject.

In this chapter we describe currently available imaging technologies utilized in the reporter gene imaging research described in this book. These technologies exploit energy emissions that span nearly the entire range of the electromagnetic spectrum. The imaging system's function is to collect these signals and form images that can be analyzed to monitor the spatiotemporal characteristics of certain cellular and molecular processes occurring in cells located within tissues of living subjects. Certain imaging systems actually excite the processes that produce the detected signal, as in the case of optical fluorescence or magnetic resonance imaging. The molecular imaging technologies described in this book for noninvasive reporter gene imaging include optical techniques that utilize fluorescence or bioluminescence light photon emissions, the radionuclide methods of positron emission tomography and single photon emission tomography that collect positron annihilation or gamma ray photons, respectively, and magnetic resonance methods that rely on emissions from the radiofrequency portion of the electromagnetic spectrum.

This chapter will give an overview of the principle of operation of each modality, state-of-the-art instrumentation available, methodologies, and important performance parameters. For a more comprehensive treatment of each modality, references will be provided. Discussions will also include comparisons and integration of multiple modalities. Although these systems can be used to image other biological phenomena besides reporter gene expression, we will discuss system design issues that result in improved ability to measure (e.g., visualize and quantify) a reporter gene signal, which we term *reporter gene sensitivity*, which essentially involves features that allow one to better extract a small reporter gene expression signal above the background signal inherent to that modality. The advantages and limitations of the different modalities will be described. Reporter gene probe optimization for better signal detection and detailed discussions of *in vivo* reporter gene imaging assays using these modalities are presented in the other chapters of this book.

OPTICAL IMAGING

Background

Of all reporter gene expression imaging assays used in biomedical research, optical imaging techniques are the most common. This fact is due to the strong history of optical reporter imaging assays developed in cell culture in the field of molecular biology using green fluorescent (FL) proteins (GFPs) [3–6], wide accessibility of optical imaging components, systems, and reagents, relatively low cost, and the capability to measure very low levels of signal (i.e., excellent *gene expression sensitivity*). To detect transgene expression with optical imaging, the coding region must be fused with either a FL protein such as GFP or red FL protein (RFP) (e.g., DsRed or HcRed) [6–9] or the bioluminescent (BL) protein luciferase [10, 11]. Another approach to measure the expression of a transgene is to detect the expressed protein indirectly through its activity; examples of this are the measurement of enzymatic conversion of a relevant substrate in a FL product, as in the case of a protease or beta-galactosidase [12, 13] or immunodetection techniques using quantum dots [14]. Detection of light emissions is possible with light source technology such as a laser and sensitive photon imaging devices such as a charge-coupled device (CCD), essentially a digital video camera residing in a dark box. Such an imaging system allows the measurement and relative quantification of gene expression. For BL, the light level is lower, so the camera must be sensitive to lower light levels.

Tissue Optical Properties

Overview: Optical properties of living tissues can cause substantial problems for *in vivo* imaging of gene expression using optical methods. Tissue is a *turbid* medium for light propagation, and optical photons, commonly referred to as *light*, are rapidly scattered and absorbed in tissue and a photon beam is quickly attenuated. However, for wavelengths in the range of ~650–900 nm, referred to as the *optical window* [15], absorption is relatively low (e.g., see Figure 7.1a), so it can penetrate roughly a centimeter through tissues even though it still undergoes substantial scatter and log orders of attenuation (see Figure 7.1b) [16]. Thus, light emitted from cells 1 cm deep within the tissues of a mouse has a reasonable chance of reaching the body surface for external detection, allowing measurements of intensity and spatial, spectral, and/or temporal characteristics of the emissions. However, due to scattering and the limited detection solid angle of the light sensor, most photons are lost.

Scattering results from refractive index mismatches between different cellular components and fluid in tissue and is especially high in skin. Microscopic inhomogeneities cause multiple scatter to occur, causing spreading of the light and resultant loss of directional information. As a result, imaging is relatively easy to perform when the signal is emitted close to the skin surface but becomes quite difficult or impossible for emission sources located deep (>1 cm) in tissue. Note that the limits in optical detection of gene expression caused by the skin can be solved by a surgical approach where a skin flap is opened, resulting in an increased signal and therefore higher detection sensitivity for gene expression in cells and increased depth at which these observations can be performed. Endoscopic methods allow microscopic inspection of the epithelial layers of accessible body cavities, such as the mucosal linings of sections of the gastrointestinal tract. However, the focus of this chapter will be noninvasive methods to image living subjects.

Absorption and scatter of light: Tissue comprises molecules that are natural chromophores and strongly absorb incident or emitted light. The extent of light absorption decreases with increasing wavelength, the cell and tissue type encountered, and their physical state [17]. Tissues with high hemoglobin content strongly absorb blue–green emissions (e.g., ~400–470 nm). Absorption effects are important up to ~580 nm. Tissues are relatively transparent for red and near infrared (NIR) wavelengths [18, 19]. Thus, depending upon the depth and wavelength of the light source(s), the emissions may not escape the subject's body to be imaged. Figure 7.1a depicts the dependence of the light absorption (μ_a) and reduced scatter (μ_s') attenuation coefficients on wavelength in highly vascular liver tissue [16]. It is often convenient to quantify the average scatter probability in terms of the reduced scattering coefficient $\mu_s' = (1 - g)\mu_s$, where g is the average cosine of the photon scatter angle over many scatters [18, 19]. In most tissues, the average scatter angle is typically small with g ~ 0.9. Because the lower wavelength light photons from the emission spectrum (<600 nm) tend to be absorbed in tissue >1 mm thick, and the higher wavelength photons (>600 nm) are much more likely to scatter ($\mu_s \gg \mu_a$), the remaining photons that do escape the subject's body are highly *diffuse* (have undergone many multiple scatters before radiating from the surface) and primarily reddish in color. This high degree of scatter causes the photons to take long and highly irregular *diffusive* paths through tissue. Thus, the result is that the emitted signal intensity varies strongly with wavelength and source depth. Red and yellow–green light sources 1 cm deep within soft tissue are attenuated by a factor of approximately 100 and 10^9, respectively [16]. Thus, *in vivo* applications of optical imaging are most useful for small, hairless (typically nude or shaved) mouse models of disease because most of the organs of interest are found at most 1–2 cm deep within the turbid tissue. It is also clear that for best depth sensitivity, the camera system should be very sensitive in the red and NIR portion of the optical emission spectrum (700–900 nm). At 900 nm light absorption in water molecules greatly increases, closing the optical window.

Optical Imaging Performance Issues

Photon sensitivity: A practical and useful definition of "sensitivity" for *in vivo* optical imaging approaches is the minimum light signal that may be detected, which is ultimately limited by the level of background signal present. Note that because there may be biological and physical background contributions present, the reporter gene expression sensitivity for optical imaging is not simply equivalent to its imaging system photon sensitivity. The imaging system photon sensitivity depends strongly upon the geometric light collection efficiency, which depends upon the light collection efficiency of the imaging sensor optics and the focal distance to the surface radiance of interest. Because FL and BL emissions comprise visible light, the imaging system photon sensitivity

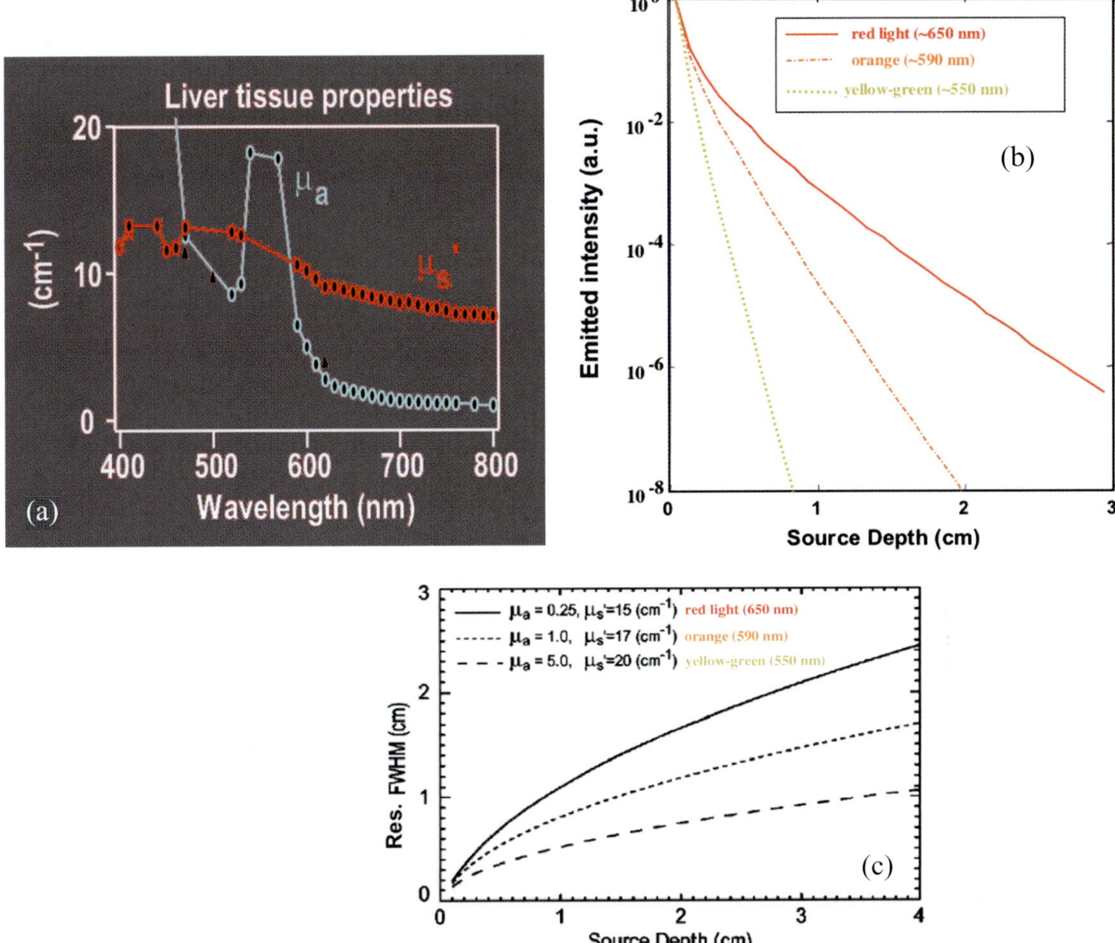

Figure 7.1. (A) Plot of light absorption and reduced scatter coefficients, μ_a and μ_s', versus wavelength in liver tissue. It is often convenient to characterize the scatter probability in terms of the reduced scattering coefficient $\mu_s' = (1 - g)\mu_s$, where g is the average cosine of the photon scatter angle over many scatters. In most tissues, the average scatter angle is typically small with $g \sim 0.9$ (i.e., $\mu_s \sim 10\ \mu_s'$). Light of wavelength below 600 nm is strongly absorbed in soft tissue. Above 600 nm light is strongly scattered rather than absorbed. The absorption of light is particularly high in liver tissue due to the high vascular content (presence of oxyhemoglobin and deoxyhemoglobin). **(B)** Intensity of transmitted light (arbitrary units) and **(C)** resolution (FWHM) of surface radiance spot versus the light source depth within tissue for yellow–green, orange, and red wavelength light. The absorption and reduced scatter coefficients assumed for **(B)** and **(C)** are in the legend of **(C)**. *Adapted from* [16].

also depends strongly upon the light sensor *quantum efficiency* or probability that light impinging upon the sensor is converted to electric charge through the photoelectric effect. As described, for emission sources >1 mm depth, mainly >600 nm light radiates from the body [20] and so the sensor quantum efficiency should be high for red and NIR light. The sensitivity to very low levels of light (from perhaps BL emission sources deep in tissue) also depends upon the level of the background *dark current* (the level of electronic current present in the sensor without a light source), which is mainly due to the inherent *thermoionic emission* level, which depends on the sensor temperature and electronic *readout noise* contributions of the sensor. These contributions should be

low for highest reporter gene expression sensitivity. The minimum detectable number of photons must exceed this inherent noise level of the light sensor system, which is easier for FL imaging (FLI) than for BL imaging (BLI) because the former uses a strong excitation light source. The minimum detectable signal of an optical imaging system is any radiance measurement just above the effective observed background radiance, which depends upon the total noise intensity measured for a given pixel size and exposure time. For BLI typical minimum photon sensitivity values are 20–100 photons/pixel/sec [20]. However, due to substantial light photon attenuation versus emission depth, and in case of FLI, the dependence of the signal on the excitation source intensity, it is difficult to convert an optical signal sensitivity into units of the molar concentration of and/or number of cells expressing the gene of interest. Finally, for FLI, due to strong autofluorescence background contamination from endogenous fluorochromes existing in tissue, the typical photon signal sensitivity is over an order of magnitude lower than for BLI [16].

Spatial resolution: The light photons that escape the body through a given area and direction from the surface of the animal are known as the surface *radiance*, typically expressed in units of watts/cm^2/steradians, where steradian is a unit of "solid" or volumetric angle. This surface radiance is used for noninvasive, *in vivo* optical imaging. Spatial resolution for *in vivo* optical imaging is mainly limited by the surface radiance spatial resolution (spot size), which is in turn a function of the emission source depth. The camera spatial resolution, which is a function of the camera lens magnification factor and the light sensor pixel size, is typically high (in the micron range), and so it contributes insignificantly to the overall measured surface radiance spot size compared to the emission source depth effect. Figure 7.1c plots the approximate surface radiance signal spot size (i.e., spatial resolution) versus emission source depth within tissue for red (~650 nm), orange (~590 nm), and yellow–green (~550 nm) light [16, 20]. Due to the highly diffusive nature of red and NIR light propagation, the measured spatial resolution is worse (surface radiance spot size is larger) for a light signal emitting in that portion of the spectrum. From Figure 7.1c we can induce the coarse rule of thumb that the measured spatial resolution (FWHM) of the surface radiance signal for a red light source is roughly equal to the depth below the surface in which it resides (e.g., ~1 cm FWHM for a 1 cm deep point source although somewhat less than 2 cm for a 2 cm deep 650 nm emission source) [16, 20]. Note that extrapolating from Figure 7.1c, if NIR wavelength emission sources are used, the FWHM resolution observed at the surface of the animal will be even greater than the source depth.

Thus, obtaining high resolution, quantitative optical detection, and imaging of gene expression *in vivo* and noninvasively is challenging due to the optical properties of intact biological tissues. Choosing excitation sources and reporters that emit in the red and NIR portions of the optical spectrum appear most promising with regard to penetration of tissues, with the drawback of producing and accepting a higher fraction of multiple-scattered photons into the detector that result in radiance resolution loss.

Fluorescence Imaging (FLI)

Overview

Optical fluorescence imaging (FLI) can image a variety of *in vivo* processes including gene expression occurring in cells located within tissues of live small laboratory animal subjects (mainly mice) by observing the body surface distribution of FL signal. Specific genes of interest can be linked with reporter genes in transgenic animals and their expression followed *in vivo* over the animal's lifetime. This approach has also been used in important *in vivo* applications such as monitoring therapeutic gene delivery strategies [21], tracking infectious diseases [22], and following the proliferation of cancer cells and their progeny in xenograft and transplant tumor models [23]. The FL proteins are isolated from living organisms and the gene that encodes for these proteins may be inserted into cells and used as a reporter gene [24]. GFP has been widely used in biological research for cell culture and *ex vivo* study of tissue sections [3, 4] as well as *in vivo* studies [25]. However, due to its low peak light absorption wavelength (~470–490 nm) a matched excitation source light cannot easily penetrate into tissue to excite GFP for *in vivo* FL imaging of live subjects. Similarly, due to the low peak emission wavelength (~510 nm) the GFP emission light cannot escape out of tissue. Furthermore, GFP's absorption and emission spectra also overlap with that of the background autofluorescence of surrounding tissue due to absorption by endogenous tissue fluorophores. This autofluoroescence signal is not relevant for characterization of the gene expression level and must be subtracted from the overall detected signal. Thus, GFP is poorly suited for a wide variety of *in vivo* imaging studies. For *in vivo* FLI it is clear that a high light yield fluorophore with longer wavelength absorption and emission spectra within the optical window are desirable. These properties can be achieved by exploiting mutations in the naturally occurring gene encoding the FL protein (e.g., GFP and other FL proteins) [7]. It is also desired to use a high intensity, matching long wavelength excitation source and appropriate optical filters.

FLI Instrumentation

The distribution of the reporter gene can be localized within the animal subject by illuminating the animal with an external light source that excites the fluorophore within the FL protein [25]. The light source can be an intense laser (e.g., a red-emitting Ti:sapphire laser) that emits a well-defined frequency of light, ideally well matched to the absorption spectrum of the fluorophore to maximize signal intensity, minimize autofluorescence, and minimize any damage to tissue. The light source can also be a broadband source such as a fluorescent lamp equipped with a bandpass, in particular, a *low pass* filter that preferentially passes the lower frequency (longer wavelength) portion of the spectrum and that will have better penetration into the subject tissues for excitation of the reporter molecule. The filters on the emission light reaching the detector should pass the emitted light and block the excitation light that happens to propagate

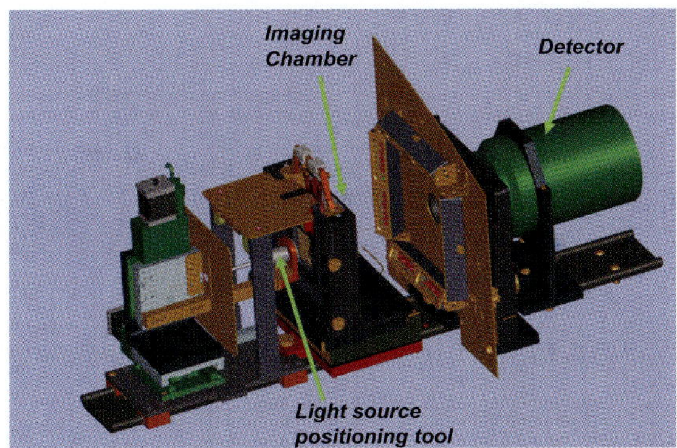

Figure 7.2. CAD drawing depicting the inside of the VisEn medical fluorescence molecular tomography (FMT) system. Tomographic images of fluorescent probes are available from a planar imaging geometry using a model of photon migration through tissue in the image reconstruction procedure. Transillumination and signal normalization reduce spatial variability and effects of tissue heterogeneity. Courtesy of Marion Kerr, VisEn Medical.

there through strong scattering. The slope of both the excitation and emission filters at the transition between blocking and transmission versus wavelength should be as sharp as possible, with high transmission probability at the desired wavelengths. But filtering always results in some light intensity loss and overall signal-to-noise ratio (SNR) degradation.

Basic FL imaging experiments require an excitation source (with an appropriate filter), a computer-controlled lens-coupled CCD imager, and a light-tight chamber to generate images of the surface light distribution (e.g., see Figure 7.2). Because reporter gene expression sensitivity in FL imaging is mainly limited by tissue autofluroescence rather than by the noise characteristics of the photodetector, conventional CCD cameras may be used, which helps control system costs. A typical CCD chip may have a 2.5×2.5 cm^2 sensitive area with a 1024×1024 array of 24 µm pixels. The optics facilitates collection of light from a relatively large field-of-view (FOV) (e.g., 10×10 cm^2) into the smaller (e.g., 2.5×2.5 cm^2) CCD sensitive area. Each pixel collects photons and converts them into photoelectrons for a selectable time (the frame duration or *integration time*) before it is read out. In a CCD, the pixels are connected in such a manner that during readout the charge passes through each pixel with the output of one pixel serving as the input to the next one. This yields a sequential pattern of the charge collected for every pixel from a given exposure, corresponding to the incoming light intensity. This sequence is read out into an output register, amplifier, and digitizer that converts the pattern to a corresponding digital image pixel intensity. The CCD display software in turn converts this intensity pattern into an image. Other details about CCD imagers will be discussed in

the BLI section of this chapter and can also be found in [26].

FLI Methodology

Photons from the excitation source are absorbed by the fluorophore within the FL protein, which brings the molecule to excited vibrational levels. A minute energy loss occurs as a result of an internal rearrangement between the different vibrational states. The remaining excitational energy can be converted and carried off by light photons, resulting in FL emission, with the resulting photons having lower energy (longer wavelength) than that absorbed by the excitation source. Thus, the FL intensity is proportional to the intensity of the incident light impinging upon the fluorophore, probability that the fluorophore absorbs the light, concentration of the fluorophore, and *quantum yield* of the fluorophore. The fluorophore concentration indicates the level of reporter gene expression. From an imaging system point of view, a main strength of FL imaging for detecting reporter gene signals is the relatively high quantum yield of available fluorophores yielding robust light signals with an appropriate excitation source [27]. Another advantage is that the reporters are available in the whole spectral range of visible light as well as into the NIR region, which is better for transmission out of tissue. See Chapter 1 for additional details about available fluorescent reporters.

The challenge with FL reporter gene imaging is the presence of an autofluorescence background signal from surrounding tissue [20], which limits gene expression signal detection sensitivity. The emission and absorption spectra of the reporter gene fluorophores are always broad and overlap with those of the tissue autofluorescence. However, autofluorescence is less of a problem if one uses NIR fluorochromes because hemoglobin and water, the major absorbers of visible and IR light, respectively, have their lowest absorption coefficients in the NIR. One may use an appropriate long wavelength (*low pass*) filter in front of the photodetector (e.g., CCD) to allow only the NIR light to pass. A second challenge is that the FL imaging approach requires excitation light to penetrate into the tissues to excite the fluorophores and the resulting FL emission light to escape for detection. The influence of tissue optical parameters is present during both the excitation and emission processes.

Because of the high degree of light scattering and strong depth dependence of both excitation and emission light intensity, it is not possible to irradiate a small volume of tissue and challenging to localize and quantify reporter gene distributions unless the region of interest is at or near the surface of the skin. Another reason it is challenging to absolutely quantify a gene expression signal with FL imaging is that the signal is always proportional to the excitation source intensity. A final challenge with FL imaging is that the incident excitation beam can be reflected by the skin of the subject, affecting the emission measurement and reducing the transmitted light intensity available for excitation of the fluorophore [28]. Optical filters are typically employed to select both excitation and emission wavelengths to avoid detection of the excitation light during the emission light measurement. Internal reflection of the emission light can also occur at the tissue–air boundaries due to the abrupt change to a lower refractive index.

Noninvasive *in vivo FLI System Technology Advances*

We have discussed the challenges confronting noninvasive optical imaging methods that make it difficult to obtain quantitatively accurate and spatially resolved detection of *in vivo* gene expression. However, the small size of typical animal models (usually mice) often permit the detection of enough light arising from relatively deep locations to allow capture of signals with acceptable SNR. As described, reporter gene expression sensitivity is also limited, especially in the visible spectral range, by the presence of ubiquitous autofluorescence signals (mostly arising from skin and gut), which need to be separated from those of fluorophores marking gene expression. In recent years there has been much research devoted to mitigating these problems. These methods attempt to exploit the optical properties of living, intact biological tissues, and some of the advances are applicable to BLI as well as FLI.

Multispectral imaging: Multispectral imaging is the ability to form images from certain selected regions of the optical emission spectrum. If the spectral selection is highly precise, multispectral imaging enables the ability to separate autofluorescence from the reporter gene expression fluorescence, effectively providing a substantial increase in signal sensitivity and quantification compared to conventional nonspectral approaches. In the NIR region, autofluorescence, although still significant, poses less of a problem to signal detection. However, the task of disentangling signals from multiple fluorophores remains. In biology there is also a need to perform multiplexed measurements of expression levels of several optical reporter genes simultaneously. If one had a series of optical reporters for different genes with distinct emission spectra peaks, using multispectral imaging techniques it is possible to spectrally unmix the measured signal to allow one to characterize and quantify expression levels of multiple optical reporters and separate the autofluorescence signal, simultaneously, in one imaging assay.

The spectral selection can be achieved with a high-quality set of filters that each selects a precise portion of the spectrum [29]. An FL imaging system innovation that facilitates *in vivo* multispectral imaging is the liquid crystal tunable filter (LCTF) (see Figure 7.3a). LCTFs [30] use electrically controlled liquid crystal elements to select a specific visible wavelength of light for transmission through the filter at the exclusion of all others (e.g., see Figures 7.3b–d). The LCTF provides rapid, vibrationless selection of any wavelength in the visible and near-infrared tuning ranges. Advantages of the LCTF technology over standard filters are that it allows extremely precise wavelength selection ($\pm$10–20 nm), it is a solid state crystal with no moving parts (in contrast to filter wheels), and it allows continuous tunability over hundreds of nanometers. Figures 7.3c,d show an example reporter gene expression imaging signal that is unmixed from background with the LCTF. Of course, if excitation and emission spectra of multiple FL proteins and/or endogenous fluorophores overlap too much, the classic bandpass filter approach does not work effectively. However, if reference spectra are available for each fluorophore present, a simple linear algorithm can be used to estimate the weights from each spectra. In the absence of such reference curves, more complex classification algorithms can be exploited for spectral separation [31].

Fluorescence tomography (FLT): FLT attempts to perform a volumetric reconstruction of the emission source distribution, using principles of tomographic image reconstruction of diffracting sources [32]. Until relatively recently, the field of FLI mainly used single-view, nontomographic, reflective, planar imaging to estimate the reporter gene expression distribution within a mouse [25, 33, 34]. Limitations of planar compared to tomographic imaging are: (1) planar images are a superposition of emissions from all depths, which limits image contrast resolution, especially if there is more than one FL focal source; (2) precise depth localization in specific tissues of interest is usually not possible; (3) there is strong depth-dependent resolution blurring; and (4) quantification in a given depth of interest is not possible due to significant photon attenuation. To address these issues, there has been significant research in the development of spatially resolved three-dimensional (3-D) FLT. In addition to addressing the problems with planar imaging, FLT is useful for applications in which the FL surface radiance signal is weak from certain views but stronger for others, and it is simply important to sample the signal from different projection angles. For the best 3-D projection imaging capabilities, the FL system includes a mechanism to acquire multiple planar views from several orientations about the animal. However, drawbacks to this multiview approach are system cost and/or imaging time, so there is great interest to solve the 3-D tomographic image reconstruction problem with a single-view, planar-chamber geometry [33] (e.g., see Figure 7.2).

NIR illumination may be achieved via multiple fiber optics located at different points on the animal subject [34–36] or by scanning the beam across the subject surface using a mirror [37, 38]. FL signal is collected from multiple points on the animal surface also using an array of fiber optics or a scanning mirror that reflects signal

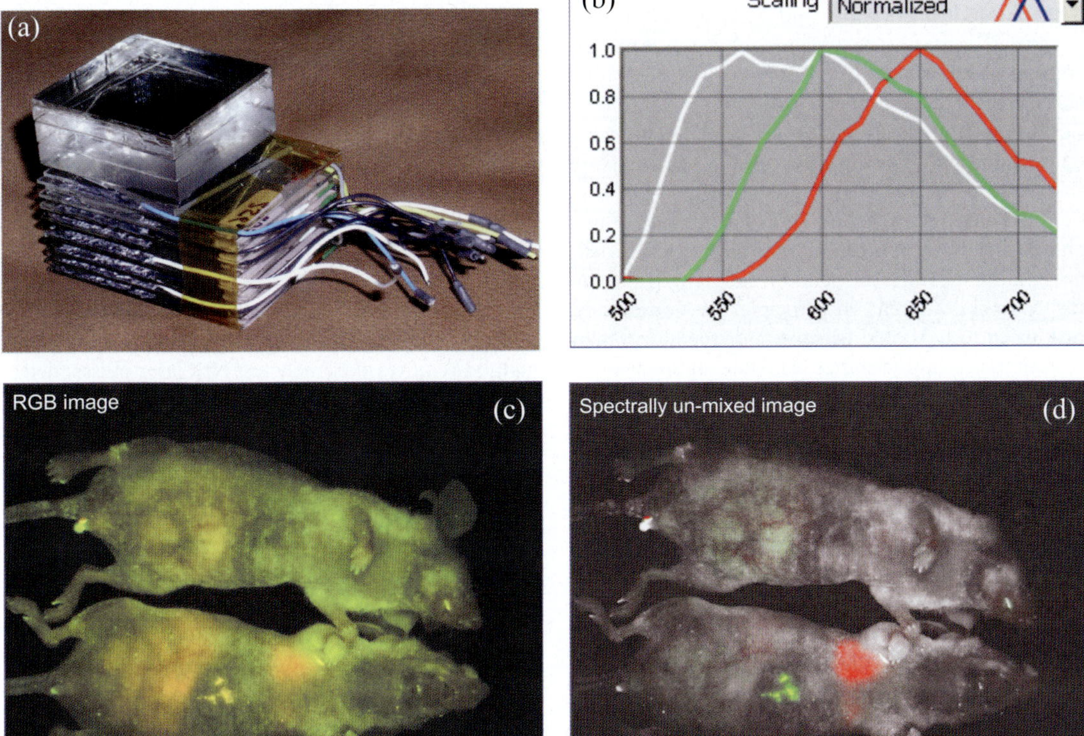

Figure 7.3. (A) Liquid crystal tunable filter (LCTF) used in the CRi Maestro *in vivo* spectral fluorescence imaging system. A LCTF's operation is based on a series of linear polarizers and birefringent retarders. Each "stage" (the combination of linear polarizer, retarder, linear polarizer) creates an interference pattern on the light passing through it. The total interference of all of the stages together yields the selected LCTF bandpass. **(B–D)** *In vivo* imaging of DDAOG in a lung model with Maestro. The lower mouse in **(C)**, which was infected with malaria parasites expressing a reporter gene that codes for beta galactosidase known as LacZ was given an intracardiac injection of DDAOG [13], a conjugate of ß-galactosidase and 7-hydroxy-9*H*-(1,3-dichloro-9,9-dimethylacridin-2-one) (DDAO), 30 min prior to imaging (the upper mouse in **(C)** is a negative control). On cleavage, the fluorophore shifts ~50 nm (see spectra in **(B)**). **(D)** shows a composite image indicating the locations of the fluors. Images courtesy James Mansfield, CRi, and Clemens Lowik, Ivo Que, Eric Kaijzel, Andy Waters, Blandine Franke-Fayard, and Chris Janse, Leiden University Medical Center.

into the detector. For a given view, the surface radiance is often very diffuse and the depth of the FL source is uncertain. FLT ideally provides the ability to reconstruct cross-sectional slices through the FL site(s) with the goals to recover depth, spatial resolution, contrast of FL markers *in vivo*, and quantitative accuracy of reconstructed FL distributions.

To pursue these goals with accuracy, FLT requires the incorporation of an accurate mathematical model of photon migration in the tissue toward the detector, called a *forward model*, to account for light attenuation and diffusion along any given path of the photons [35, 39, 40]. The model assumes point sources of excitation light and takes into account the dependence of the fluorophore emission intensity on the relative positions of the fluorophore and excitation source. Because the tissue is highly scattering, photon attenuation does not follow a simple exponential decay with depth rule, as for x-ray or gamma-ray photons, but it can be treated as particles

that elastically scatter through a random medium. Photon migration is governed by the general light transport equation, referred to as the Boltzmann transport equation [41], describing the photon flux within the subject tissues. This equation is in general difficult to solve, even numerically. Modeling may be performed numerically following the Boltzmann equation through Monte Carlo simulation techniques that treat light as a collection of discrete particles migrating through tissue or analytically using the diffusion approximation to this equation. If $\mu_s \gg \mu_a$ and tissue homogeneity is assumed, the *diffusion* approximation can be exploited to simplify the Boltzmann equation into a form relating the divergence and rate of change of the flux measured at the detector locations with the photon source density [42]:

$$\nabla \cdot (D(\mathbf{r})\nabla\Phi(\mathbf{r}, t)) - \frac{1}{v}\frac{\partial}{\partial t}\Phi(\mathbf{r}, t) - \mu_a(\mathbf{r})\Phi(\mathbf{r}, t)$$
$$= -s(\mathbf{r}, t), \tag{7.1}$$

where Φ is the measured photon flux at any point in 3-D space $\mathbf{r}$ and time t, $\mu_a(\mathbf{r})$ and $\mu_s'(\mathbf{r})$ the absorption and reduced scattering coefficients, $D(\mathbf{r})$ the diffusion coefficient given by $D = 1/3(\mu_a(\mathbf{r}) + \mu_s'(\mathbf{r}))$, $s(\mathbf{r}, t)$ the time varying photon source density, which is a sum of contributions from the excitation source and excited fluorophore, and $v = c/n$ the speed of light in the tissue with effective refractive index n. The situation is complex for FLT because the excitation source and excited fluorophore are both sources of the "photon source density" term. Because μ_a and μ_s' depend on wavelength and the two spectra are different, in principle, the fluorophore contribution to Φ should be solved for separately than that from the excitation source. The fluorophore source term is directly proportional to the excitation photon flux, the quantum yield of the fluorophore, and an exponential term representing the fluorophore decay [39, 40]. For FLT the goal is to determine the distribution and intensity of the FL source distribution $s(\mathbf{r}, t)$.

For tomographic imaging capabilities, because the surface radiance strongly depends upon the tissue thickness that the light traverses, a special light source should be available to measure the 3-D contours of the surface of the animal. This map of the animal's surface boundaries together with an accurate model of light propagation through the tissue along many different paths, and/or measured or estimated optical scatter and absorption parameters, may be incorporated into a 3-D image reconstruction algorithm to estimate the 3-D distribution of cells expressing the active gene of interest in the animal. The most rigorous inversion approach for FLT uses *iterative* reconstruction algorithms to solve an inverse equation to form a volumetric map of the fluorochrome distribution from the measurements [40, 43, 44]. The photon migration model is incorporated into the iterations to effectively deconvolve photon diffusion effects from the images. As stated, FLT may allow better resolution of structures deep within tissues, improve quantification of image data, and give a more faithful visual representation of the FL sources(s). If successful, the result is a 3-D quantitative data set with millimeter spatial resolutions throughout the live subject. These properties will provide more useful correlation to images from other modalities. However, the high scatter, absorption, and significant tissue heterogeneity make *in vivo* FLT an extremely challenging problem [35, 37, 39, 40, 42–44].

FLT and *diffuse optical tomography* (DOT) have appeared in research literature for years [45–47]. FLT images molecular processes such as reporter gene expression occurring within cells that are millimeters to $\sim$1 cm deep in tissue by reconstructing the 3-D distribution of probes tagged with FL proteins, preferably emitting in the NIR for better tissue transmission. In FLT typically an intense laser excitation source propagates light into the subject and the emitted FL signals are collected from multiple views. There are three approaches used to probe deep tissue volumes in FLT. All three approaches to FLT require an accurate photon transport model to achieve accurate quantification and spatially resolved detection of *in vivo* gene expression. Note that because the diffusion approximation (Equation (7.1)) breaks down at and near boundaries of heterogeneous tissue, so do these approaches.

The *time-domain* (TD) approach [48] uses the fact that those paths of light propagation that arrive first at the photodetector have undergone the least scatter and have therefore on average interacted with less diffusive tissue than photon tracks arriving at later times. The TD method requires extremely fast NIR laser pulses, detectors (e.g., photomultiplier tubes (PMTs), not CCDs), and electronics to measure the *time-of-flight* FL distribution over the scanned tissues, referred to as the temporal PSF (TPSF). This approach uses the time-dependent diffusion equation (7.1) and the assumed form for the FL source distribution $s(\mathbf{r}, t) = \Phi(\mathbf{r}, t)\varepsilon(\mathbf{r})e^{-t/\tau(\mathbf{r})}$, where $\Phi(\mathbf{r}, t)$ is the excitation source (laser) photon flux at $\mathbf{r}$, $\varepsilon(\mathbf{r})$ the effective quantum yield of the FL source after absorption of the excitation light it receives, and $\tau(\mathbf{r})$ the FL lifetime. Thus, in addition to the FL source intensity distribution, the temporal information available with the TD approach in theory gives the FL lifetime, independent determination of absorption and scatter coefficients, and fluorophore depth [49].

The *frequency domain* (FD) approach [50–53] uses an intensity-modulated excitation light source wave. Typically an NIR laser is modulated at a single frequency and a sensitive photodetector (e.g., a PMT, photodiode, or modulated, image-intensified CCD) is used to determine the intensity and phase from the modulation envelope of the measured photon flux. The wave is distorted in optically heterogeneous tissue, resulting in reductions in amplitude and phase shifts of the excitation light with respect to the emitted light wave. These changes are measured in the photodetectors placed at the surface of the body. The FL parameters are obtained by solving a version of Equation (7.1) that is reparameterized for frequency, and the resulting information is converted into maps of the tissue interior.

In the *continuous wave* (CW) approach [50, 54], the excitation and FL emission light are steady state light sources (time-invariant, thus $\partial/\partial t\Phi(\mathbf{r}, t) = 0$ in Equation 7.1), and the distribution of FL emitters throughout the subject is reconstructed from intensity measurements at the subject boundaries. Note that with the CW technique, the FL lifetime information is lost in the integration over time. Using the CW assumption, the problem of solving Equation (7.1) is reduced to a simple linear inverse problem, where the spatially varying, time-independent photon flux $\Phi(\mathbf{r})$ measured at the detector location $\mathbf{r}$ determines the unknown FL source intensity at $\mathbf{r}'$, $s(\mathbf{r}')$. It is also typically assumed that the animal subject may be treated as an infinite slab of tissue so that there are no reflected waves. A common assumption is that $\Phi(\mathbf{r})$ has the following form

$$\Phi(\mathbf{r}) = \int g(\mathbf{r}, \mathbf{r}')s(\mathbf{r}')d\mathbf{r}, \qquad (7.2)$$

where $g(\mathbf{r}, \mathbf{r}')$ is known as a Green's function, which is essentially a blurring kernel or point spread function (PSF) that may be determined from tomographic measurements [55]. Thus, the problem of estimating the source $s(\mathbf{r})$ is reduced to a simple deconvolution equation.

A number of groups (e.g., see [56–59]) are also investigating the use of multispectral acquisition to exploit the fact that different portions of the fluorochrome emission spectra have different optical properties. The information from this additional multispectral data is added to the system matrix and used in the iterative image reconstruction process. A challenge with this approach is obtaining high enough photon statistics in each spectral acquisition window to yield good SNR.

In summary, there has been significant progress in FLT research due to assumptions incorporated into the formalism, such as the diffusion, infinite slab, point source, homogeneous optical parameter approximations, and signal dependence upon the excitation source intensity. Studying the potential resolution, sensitivity, and quantitative accuracy of FLT is still an active research topic.

Invasive *in vivo* *FL Microscopy Methods*

As stated, a brute force approach around the tissue turbidity issue for imaging gene expression deeper than 1 cm in tissue is the invasive route. For completeness we include a brief discussion of these methods that have a goal to obtain microscopy-like resolution in live subjects.

Intravital microscopy of exposed organs requires anesthesia and surgery to open a skin flap or insert a chronic–transparent optical window to bring the objective lenses close to the tissues of interest [60, 61]. Endoscopy methods allow access to the epithelial layers in accessible portions of the gastrointestinal tract [62]. These invasive approaches can yield exquisite lateral spatial resolution of 1–15 µm. A popular approach for cellular imaging is *laser scanning confocal microcopy* [63]. Using a conventional wide-field optical microscope, out-of-focus plane FL emissions from the specimen interfere with features in the region of interest, reducing the contrast and/or spatial resolution of the perceived structures with respect to background. In a confocal microscope, using a pinhole aperture placed close to the light sensor, the undesired out-of-focal plane light is removed, resulting in the ability to generate various focal-plane sections through the top surface of the tissue. However, in confocal microscopy illumination of the specimen is

focused only on a small section, in contrast to conventional wide-field microscopy in which the entire specimen is illuminated. This greatly increases the scan time required to image a given specimen. Decreasing scan time requires fast scanning of the aperture, which may be accomplished with a multihole, rapidly spinning disk placed in between the light source and the specimen [64].

Multiphoton (usually two photon) microscopy [65, 67] can probe gene expression in cells at the skin surface or, with an intravital preparation, in organs inside the subject. Multiphoton excitation is based on the small but finite probability that multiple low-energy photons (e.g., two red wavelength photons from a Ti:sapphire laser source) arrive simultaneously at a fluorophore and induce an electronic transition associated with a high-energy (e.g., single blue) photon [65]. Because efficient two-photon excitation requires a high spatial and temporal concentration of photons, the result is a confocal effect at the focal spot without requiring complex confocal detection optics. The structures can be observed only at the microscope focus, which is scanned across one plane of the specimen at a time. Using NIR excitation this technique produces confocal-like sections up to ~500 µm deep without background contamination from out-of-focus planes [65]. Because the majority of this book focuses on noninvasive, nonmicroscopic methods of imaging reporter gene expression in living subjects, we leave a more extensive discussion of the invasive approaches to the references [60–67].

Bioluminescence Imaging (BLI)

Overview

Molecular imaging using BLI requires cellular expression of an enzyme known as *luciferase* that is responsible for making some insects, jellyfish, and bacteria glow [67]. The gene for this enzyme is incorporated into DNA of cells, microorganisms, or animal models of disease. If an appropriate substrate is available for the enzyme to act upon, the result is a reaction that emits a subtle glow of visible light called *bioluminescence* (BL) that can be used to monitor cellular and genetic activity of every cell that expresses the luciferase enzyme [10, 11, 68]. In the case of firefly luciferase, the substrate *D-luciferin* must be present (introduced into the subject) as well as oxygen, magnesium, and adenosine triphosphate (ATP) for the reaction to occur. Because BL emissions result from an ongoing chemical reaction with inherent kinetic variations, the resulting light is emitted in a continuous glow of visible light photons peaked at a most probable emission wavelength. Various forms of luciferase and associated substrates are available that produce light in the visible range (wavelengths of 400–700 nm or energies of 1.5–3 eV). The peak emission wavelength for naturally occurring firefly luciferase is at ~560 nm. For

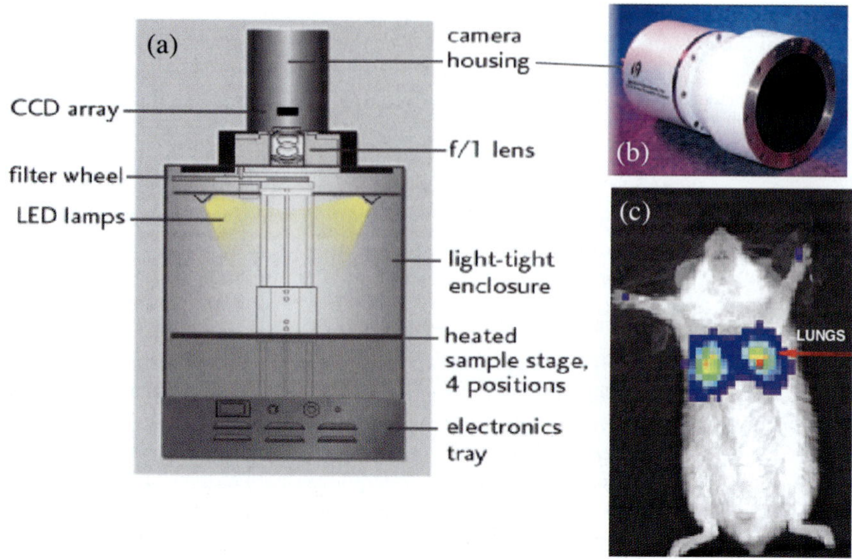

Figure 7.4. (**A**) Schematic of the basic configuration for the *in vivo* bioluminescence imaging systems (IVIS) available from Caliper-Xenogen Corporation. (**B**) Picture of the camera housing the low-noise cooled CCD array and adjustable lens components. (**C**) Bioluminescence imaging of cancer cells expression of a bioluminescent reporter gene. 7×10^5 A375 cancer cells stably expressing the *hrl-mtfp-ttk* fusion reporter gene [23] were injected via tail vein in an SCID mouse and 2 h later imaged for *Renilla* bioluminescence signal following tail vein injection of the substrate coelenterazine. Prominent bioluminescence signal was obtained from both lungs [$\sim 1.3 - 1.5 \times 10^5$ max (photons/sec/cm²/steradian)] (Image courtesy Andrew Loening, Stanford University).

bacterial luciferase, the substrate is produced endogenously and the peak emission wavelength is ~490 nm. Recently, luciferase genes isolated from insects and sea organisms were genetically modified to be efficiently expressed in mammalian cells [69]. The result of two of these mutations has yielded a shift of the BL emission peak to ~615 nm [16, 69]. BL is emitted from every cell throughout the body of the subject that expresses the luciferase enzyme and that reacts to the appropriate substrate. Thus, for most applications, the cells transfected with the luciferase reporter gene are introduced exogenously and tracked for various assay types in the animal model of interest. Please see Chapter 2 for further details on bioluminescent reporters.

BLI Instrumentation

As for FLI, in BLI light collimation and response line identification are not possible because the signal comprises many photons that are scattering multiple times before exiting the tissue. Furthermore, because tissue and air have different indices of refraction, the light reaching the tissue surface will undergo significant refraction before entering the camera. Therefore what remains for imaging at a particular single projection view is a diffuse "blob" of photons and, as for CW FLI, imaging systems used for BL are typically similar to ordinary digital cameras. In addition, considerations of how the light from BL surface radiance is collected and focused onto the light sensor are important. Most often a standard optical lens

system collects light from the animal (optically focusing on the surface of interest) and focuses the image into the relatively small light sensor that will be used to create a digital image. Finally, for highest sensitivity to low levels of BL light, there should be no contamination from ambient laboratory light.

Figure 7.4a depicts an *in vivo* imaging system (IVIS) configuration offered by Caliper Life Siences [Alameda, CA] for BLI. BLI requires a light-tight box that isolates the animal from ambient light. A highly sensitive digital camera system is carefully coupled to this box. The camera system comprises an optical lens system that collects light from the animal with a focus on the surface of interest. The animal stage can translate toward and away from the lens to vary the FOV. The lens transmits the collected light image and focuses it onto a highly sensitive, low noise CCD (Figure 7.4b) that converts the light image into a digital image (e.g., Figure 7.4c). In BLI, a continuous current of photons is collected (integrated) and processed in a single or multiple *exposures* of the optical camera sensor. Typically, extremely low levels of light (low quantum yield) are generated from BL reactions and significantly less light escapes the subject, so exposure times are required to be relatively long compared to FLI or ordinary photography. Thus the electronic noise level of the sensor should be very low.

The CCD chip in a typical BLI system is different from that used in an ordinary digital camera in a few ways. First, the CCD sits in a cryogenic chamber for cooling to very low temperatures (e.g., -105 to $-120°C$).

Cooling reduces the dark noise by significantly reducing the background rate of *thermoionic emission* (constant thermal release of electrons from the silicon crystal lattice) in the silicon CCD chip. The dark current falls by a factor of ten for every $20°C$ drop in temperature. Second, the quantum efficiency of the CCD can be significantly improved for low light detection by thinning the backside of the conventional CCD, turning the chip around so that light illuminates the backside, and using antireflective coating. The quantum efficiency (QE) of such a high-sensitivity, back-thinned, back-illuminated CCD camera is typically greater than 80% for red and NIR light. The SNR ratio for such a device is described by $(I \cdot t \cdot QE)/\sqrt{(I \cdot t \cdot QE + I_d \cdot t + N^2)}$, where I is the incoming light photon flux (photons/pixel/sec), t the light photon collection time (sec), I_d the dark current (electrons/pixel/sec), and N the read noise of the device (electrons) [20, 26]. In absence of the dark current and read noise, the SNR follows $\sqrt{(I \cdot t \cdot QE)}$, which is just the Poisson fluctuation of the collected light.

The SNR should be at least two to three for statistically significant signal detection. In FLI the SNR can be improved by increasing the photon flux reaching the detector, which can be accomplished by increasing the incident illumination intensity and using NIR-emitting FL reporter gene probes. However, these approaches are not available to BLI. Only cooling (to reduce I_d) and low-noise electronic components (to reduce N) can help. An effective increase in flux per pixel can be achieved by adding together detected photons from neighboring pixels at a loss of intrinsic spatial resolution and increase in noise.

A variable optical filter is employed to allow the user to select certain regions of the BL emission spectrum, if desired, although significant filtering will reduce the overall detected signal. A red light-emitting diode (LED) is also present to allow ordinary reflected light photographic images to be taken of the surface of the animal to have a natural anatomical framework on which to fuse the BL emission images. If the imaging system also includes a powerful excitation source that can excite an FL center inside the animal and a filter wheel, continuous wave optical FL emissions can also be imaged with a typical BLI system configuration.

BLI Methodology

If firefly luciferase is the BL-producing enzyme under study, the substrate luciferin is introduced into the subject an appropriate time before the study begins and the animal subject is positioned on a shelf within the dark box. A typical BLI acquisition consists of first activating the LEDs, acquiring a photographic image with the CCD camera system, and displaying the image with a gray-scale color scheme. The LEDs are turned off and then the camera is exposed to the BL surface radiance within a dark box. The exposure duration can take from seconds to minutes depending on the amount of active luciferase available within the subject, the depth of the active region of interest, the amount of light escaping the subject, and the number of exposures desired. Due to the very low light level emitted in a typical BL reaction, a typical single-frame exposure duration might be 30 sec to minutes. For very low light levels, the image SNR may be improved by summing the charge generated from a square group of multiple small pixels (e.g., 10×10 pixels each 24×24 μm) together into one big pixel [20]. This operation increases the photon signal per image pixel but reduces the image resolution along any direction by the same factor (e.g., a factor of ten). Because the spatial resolution is limited by the surface radiance spot size rather than the CCD pixel size, as long as the grouping is not too large, this resolution-sensitivity trade-off is typically acceptable. The acquired image results in a picture of the BL surface radiance in units of photons/sec/cm^2/steradian. The BL image intensity is often displayed using a red–green–blue color scale (with red indicating the highest and violet the lowest intensity), and the BL image is overlaid onto the gray-scale photographic image for anatomical correlation (e.g., Figure 7.4c).

For background correction, typically images without luminescent sources present and the camera shutter closed are acquired daily for various acquisition times, and all subsequent studies of a given duration are corrected using the appropriate background file. Reliability of system measurements is checked weekly with a calibrated light source of known intensity. Because a significant portion of the BL emission spectrum is strongly attenuated in tissue, the intensity of light detected at the surface from a given population of BL cells is strongly dependent upon its depth. Thus, as was the case for FLI, for planar BLI it is not possible for absolute quantification of signals *in vivo* from an unknown distribution of BL-emitting cells, especially if any of the cells are ≥ 1 mm deep into tissue below the surface from which a radiance measurement is taken. However, it is possible to perform an absolute intensity calibration of the BL imaging system to convert the resulting image intensity measured into units of photon radiance (photons/cm^2/steradian). To determine this conversion factor, typically a calibrated, very weak light source of known photon radiance is placed in front of the camera and imaged. The conversion factor of image intensity into radiance is determined as a function of the lens position, the imaging FOV, the selected imaging pixel bin size, and the wavelength of the calibration source light.

BL Tomography (BLT)

Most current users of BLI use single-view, non-tomographic, planar imaging to estimate the luciferase-producing cell distribution within a mouse. The very low BL quantum yield, lack of a temporally and spatially correlated excitation source, high scatter and absorption, and significant tissue heterogeneity make *in vivo* BLT a challenging problem [55, 70, 71]. However, there have been several attempts to use the diffusion equation (Equation (1)) and methods discussed for FLT to achieve BLT [55, 59, 70, 71]. There are a few special considerations for BLT. First, because there is no excitation source, $s(\mathbf{r})$ in Equation (7.1) comprises only a contribution from the BL emission source, and optical parameters for only one source spectrum are considered when solving the equation. Second, because after injection of the substrate the BL emission intensity does not vary much during the acquisition and signal modulation is not available, only CW approaches are possible for BLT. Third, multispectral BLT (e.g., [59]) is challenging due to the very low light level per band. Finally, aside from the same problems with depth-dependent quantification and resolution issues, due to the low quantum yield, acquiring adequate photon statistics for multiple-view BLT can be very time-consuming and therefore less attractive to users.

BLI versus FLI

BLI has an advantage over FLI in that an external excitation light source is not required. Thus, because external excitation light required for FLI will be significantly attenuated in tissue and produce significant amounts of autofluorescence, especially for FL sources deep within tissue that emit at wavelengths <600 nm, the background light levels are much lower for BLI compared to FLI. Another advantage with imaging BL-tagged cells in tissue is that the data are more easily quantified because the signal measured on the surface of the animal subject is simply proportional to the number of luminescent cells. For FLI, the signal level is proportional to both the number of FL cells and the intensity of the external excitation light, which is strongly attenuated in tissue present in front of the target fluorophore. This problem is especially evident if the target fluorophore absorption spectrum peaks at lower wavelengths. These factors make it more difficult to quantify the FL reporter gene probe concentration distribution. Nevertheless, photon migration models have been developed to account for these effects in an attempt to restore quantitative accuracy for FLI [52].

FLI has some major advantages over BLI. The FL quantum yield is orders of magnitude higher than for BL. Genetically modified luciferases that emit at higher wavelengths [69] are critical to improve tissue penetration of the weak BL light signal. Thus, high-sensitivity, low-noise imaging detectors such as cooled CCD imaging systems are required for BLI [20]. In contrast, there has been much work developing red and NIR-emitting FL reporter gene probes with high quantum yields that will

facilitate more robust signals emitted from deeper within tissue, which somewhat relaxes the noise requirements of the photodetector. FLI can be performed in both live and fixed cells and no substrate is required. Although not necessarily related to reporter gene imaging, fluorochromes can be coupled to peptides and antibodies and FL signals may be *activatable* or switched on and off by the presence or absence of specific molecules or molecular events [72], which, in addition to enabling reversible modulation of the signal, can help further reduce the background autofluorescence signal. In contrast, the generation of BL is specific to cells that contain the luciferase reporter gene and is thus limited to studying genetically manipulated cells, transgenic mice, or infectious agents such as bacteria or viruses tagged with the reporter gene and introduced into the subject.

Other distinct advantages of FLI over BLI are with regard to *in vivo* tomography. Because there is no control or *modulation* of BL intensity once the reaction begins, the image reconstruction problem for BLT is more *ill-posed* than for FL tomography (FLT) [69, 70]. As a consequence, most BL imaging studies are still acquired in nontomographic, 2-D planar projection mode. In contrast, FLT has appeared in research literature for years [45–47].

RADIONUCLIDE IMAGING
Background

There are key differences between radionuclide imaging and optical imaging methodologies. In contrast to the spectrum of light emission wavelengths for FLI or BLI, radionuclides of interest emit high-energy photons of a single energy. Radionuclide decay events are discrete in time. Monoenergetic high-energy photon events are emitted in a coincident pair for positron emission tomography (PET) or one at a time for single-photon emission computed tomography (SPECT), in both cases as a result of decay of a single atomic nucleus. Thus, whereas in FLI or BLI a continuous current of photons are collected, integrated over time, and processed in single or multiple exposures of the optical camera sensor with no individual photon information provided, in radionuclide imaging individual high-energy photons are detected and processed one at a time to measure their interaction location, absorbed energy, and in the case of PET, arrival time. Like optical photons, high-energy photon emissions are isotropic in 3-D and propagate through and interact with tissue molecules through scatter and absorption. The photon beam is also attenuated as a function of penetration depth. However, tissue is not a turbid medium for high-energy photon propagation, and the scatter or absorption attenuation coefficients (μ_s, μ_a) are typically a factor of fifty to one hundred smaller for radionuclide-generated photons compared to visible light (e.g., $\mu_s \sim 0.1$ cm^{-1} and $\mu_a \sim 0.0005$ cm^{-1} for 511 keV photons, whereas $\mu_s' \sim 10$–20 cm^{-1} ($\mu_s = 100$–200 cm^{-1}) and $\mu_a \sim 0.5$ cm^{-1} for red light propagating in tissue).

One key advantage of radionuclide photon imaging (PET and SPECT) methodology is that it is possible to utilize photon *collimation*, which precisely determines the direction of photon incidence into the system. Due to the relatively low tissue scatter probability there is a strong correlation between the photons collected along a collimated direction and a response line through the subject that encompasses the locations of the radionuclides emitting those photons. This facilitates spatially resolved, quantitative measurements of reporter gene expression deep (>30 cm) in subject tissues. In optical imaging due to the high probability of scatter, all but a very small percentage of the optical photons undergo numerous scatter interactions and thus photon collimation (spatially or temporally) is not considered practical for *in vivo* optical imaging. Thus, *in vivo*, noninvasive radionuclide imaging reporter gene imaging assays that inspect tissues greater than 1 cm deep can in principle be more easily translated from small-animal research to the clinic. Also, since the collimation methods are different for PET and SPECT, so are certain aspects of their instrumentation, methodology, and performance, which will be discussed in separate sections.

Optical versus Radionuclide Tomography

In nondiffractive, nondiffusive tomographic imaging techniques [32] such as PET, SPECT, and x-ray computed tomography (CT), the photons are assumed to travel along straight lines between the source and detector, and image reconstruction does not critically rely on incorporating an accurate model of photon transport through the tissue, although there are resolution recovery benefits associated with accurate system response models [73, 74]. In contrast, accurate BLT or FLT image reconstructions rely on including an accurate photon migration model (the *forward model*). In optical tomography methods (BLT and FLT) it is extremely unlikely that light photons will arrive at a photodetector without scattering multiple times. Due to strong attenuation of the light signals in tissue, BLT and FLT are most sensitive to shallow sources but can perhaps resolve light sources up to ~ 1 cm deep. Thus, *in vivo*, noninvasive BLT and FLT assays are confined to small laboratory animal imaging and are not easily translated to the clinic with the exception of imaging thin extremities such as fingers, toes, and possibly compressed breast tissue [52]. In contrast, PET and SPECT are clinically used to image structures deep within the human body. Quantitative radionuclide tomography is facilitated with accurate photon attenuation correction methods and various calibrations. Like BL, radionuclide imaging does not require an excita-

tion source, but it does require an exogenous molecular contrast agent to be introduced into the subject. Finally, whereas most optical tomography systems acquire data from one or just a few projection views (e.g., see the FLT system depicted in Figure 7.2), radionuclide tomography systems typically acquire many projection data sets at fine angular sampling all the way around the subject. PET photon sensors are typically configured in a fixed ring, whereas for SPECT they are configured as heads that are rotated (in angular step and shoot mode) around the patient to collect the full orbit of projection data sets.

Most reporter gene strategies using radionuclides exploit the expression of the gene producing the thymidine kinase (*tk*) enzyme. This gene can transfect a cell nucleus of interest using the herpes simplex virus (HSV) [75, 76], bacteria genetically modified to express the HSV1-*tk* reporter gene [77], or by infecting the cells with bacteria that possess an endogenous *tk* gene homologous to that expressed by HSV1-*tk* [78]. Upon administration of positron-emitting substrates such as ^{18}F labeled 9-[4-fluoro-3-(hydroxymethyl)butyl] guanine (FHBG) or ^{18}F and radioactive iodine (e.g. ^{125}I, ^{131}I)-labeled 1-(2-deoxy-2-fluoro-D-arabinofuranosyl)-5-iodouracil (FIAU), respectively, for PET and SPECT, the phosphorylation of the substrate by the *tk* enzyme results in trapping of the radioactive probe in the transfected cells. This strategy has now been widely used for imaging of stem cell [79] and T-cell trafficking [80, 81], imaging of tumors and metastases [82], imaging of intracellular pathways such as the induction of p53 gene expression, and T-cell activation [83].

Positron Emission Tomography (PET)

Overview

PET reporter gene imaging requires a reporter probe labeled with a radionuclide that emits *positrons*. A positron is a particle that has the same mass as an electron but has opposite charge. Proton-rich (or neutron-deficient) nuclei emit positrons. Common examples are ^{18}F, ^{15}O, ^{13}N, and ^{11}C. Positron-emitting radionuclides (e.g., ^{18}F) may be synthesized by accelerating protons using a particle accelerator such as a *cyclotron* and directing the resulting proton beam into an appropriate target (e.g., H$_2$O with isotopically enriched ^{18}O), where a nuclear reaction occurs. Proton-rich nuclei may also be created using an appropriate nuclear *generator*, which creates short-lived positron-emitting radionuclides (e.g., ^{82}Rb) from the decay of a long-lived parent (e.g., ^{82}Sr). In PET, a positron-emitting radionuclide is attached to atoms of the molecule of interest (e.g., reporter gene probe molecule) to track its biodistribution *in vivo* within tissues of the imaging subject. One positron is ejected, with a range of initial velocities, from each radionuclide that decays. The emitted positrons encounter and

Table 7.1. Properties of some of the most common inorganic scintillation crystals that have been used in PET (BGO, LSO, GSO, and LYSO) and SPECT (sodium iodide) systems. The scintillation crystal converts the absorbed high-energy photon energy into visible light, which is subsequently detected and converted into a robust electronic signal using a photodetector

Scintillator	Effective	Density (g/cc)	1/e Attenuation Length at 511 keV (cm)	Relative Light Yield (% NaI)	Refractive Index	Decay Time (ns)	Peak Emission Wavelength (nrn)	Rugged?
"BGO" $Bi_4(GeO_4)_3$	75	7.13	1.06	IS	2.15	300	480	Yes
"LSO" $Lu_2(SiO_4)O{:}Ce$	66	7.4	1.13	75	1.82	42	420	Yes
GSO $Gd_2(SiO_4)O{:}Ce$	59	6.71	1.4	20	1.85	60	440	Yes
"LYSO" $Lu_{1.8}Y_{0.2}(SiO4)O{:}Ce$	65	7.1	1.2	107	1.81	40	420	Yes
"Sodium Iodide" NaI(Tl)	51	3.67	2.94	100	1.85	230	410	No

interact with electrons and nuclei of nearby atoms of the tissue. During its trajectory the positron scatters off the atomic nuclei, loses energy, and slows down through excitation and ionization of the atoms it encounters [84]. Once a positron slows down enough it may combine with an atomic electron in the vicinity and subsequently the pair will combine and *annihilate*, whereby their mass is converted into electromagnetic energy in the form of high-energy photons. If the positron and electron are at rest when they annihilate the result is almost always two photons emitted simultaneously in opposite directions, each with energy of 511 keV, the rest-mass energy of both the positron and electron.

PET Instrumentation

The most common PET system configuration surrounds the subjects with a complete cylindrical shell comprising contiguous rings of many position-sensitive 511 keV photon detectors [84]. A PET acquisition consists of detecting and positioning millions of oppositely directed 511 keV coincident photon pairs emitted in the system detector gantry. A PET scan can require 5–60 min, depending on parameters such as the system photon sensitivity, the mode of acquisition, the size of the imaging subject region of interest, and the amount of injected activity.

The front-end photon sensors (a.k.a. *detectors*) are arguably the most important (and expensive) components of a PET system because their characteristics determine important system performance parameters such as photon sensitivity and spatial, energy, and temporal resolutions. The standard configuration for a PET detector utilizes inorganic *scintillation crystals*, which absorb the 511 keV photons and generate a flash of light. Most state-of-the-art PET systems use discrete scintillation crystal arrays of individual crystal rods optically isolated by reflectors (e.g., see [84–91]). The 511 keV photons are highly penetrating, and for them to be stopped efficiently to promote good photon sensitivity, the array crystals must have high atomic number (Z) and density and be relatively thick (long). For excellent spatial resolution,

the crystals must also be narrow for precise localization of the incoming photon interactions in the detector. Finally, for excellent spatial and temporal resolutions, the scintillation light yield should be bright and fast. Table 7.1 lists important properties of the most commonly used scintillation crystals used in PET. Typically the crystals are arranged into arrays (e.g., as in Figure 7.5a), coupled to photodetectors, and built into modules (e.g., as in Figure 7.5b). The modules are fixed together to form a ring as depicted in Figure 7.5c for a small-animal system. Figure 7.5a,b shows the scintillation crystal array/detector submodules used to build the state-of-the-art Concorde Microsystems/Siemens Inveon system, which is a high-resolution PET system dedicated to preclinical small-animal molecular imaging research [85]. The crystal pixel dimension and detector gantry diameter determine the main differences between human and small-animal imaging systems. High-resolution animal imaging systems use ≤2 mm crystal width and <20 cm detector diameter [85]. Human systems typically use ≥4 mm crystal width and 80 cm detector diameter.

In the conventional approach used in clinical PET systems (e.g., see [88–90]), the discrete crystal arrays are coupled directly to sensitive light detectors called *photomultiplier tubes* (PMTs), which collect the light from the crystals and convert it to a robust electronic signal that can be used for spatial, energy, and time information. However, to accurately position the light flashes from miniscule array crystals required for high-resolution PET, often specialized components such as position-sensitive PMTs (PSPMT; e.g., see [91]), and, to facilitate high packing fraction of PSPMT-based detector modules, fiber optics [86, 87] (e.g., Figure 7.5b), have been used. In general, fiber optic coupling should be avoided whenever possible because it introduces substantial light signal loss and energy and arrival time dispersion [85].

PET Methodology

The reporter gene probe or other molecular probe radiolabeled with a positron emitter is introduced into the body of the imaging subject in trace (e.g., picomolar

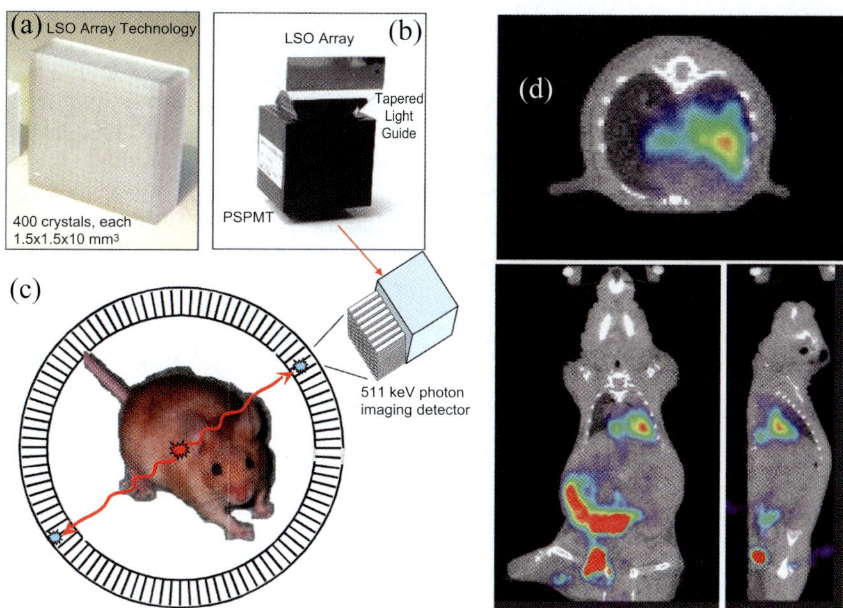

Figure 7.5. (**A**) High-resolution LSO (see Table 7.1) scintillation crystal array used in the Siemens Inveon small-animal PET system. (**B**) Scintillation imaging detector module in the Inveon system comprising the LSO crystal array, tapered light guide, and position sensitive photomultiplier tube (PSPMT) (Courtesy of Robert Nutt, Siemens Preclinical Solutions). (**C**) Depiction of the standard PET system comprising many of the imaging detector modules arranged in multiple rings. A "true" two-annihilation-photon event emitted from a mouse, producing two coincident light flashes in two modules of the PET system is shown. (**D**) Example image slices through a reconstructed volume [transverse (top), coronal (bottom left) and sagittal (bottom right)] of a PET reporter gene imaging study. Images show fused Siemens microPET R4/GE Explore Locus CT images of melanoma cancer cell animal model. The cancer cells are marked with HSV1-sr39tk [1, 76] and have metastasized to lungs allowing imaging following phosphorylation of the [18]FHBG substrate introduced into the body (Images courtesy of Andrew Loening, Stanford University).

to nanomolar) quantities. After an appropriate waiting period for the probe molecules to reach the desired target molecules, the subject is placed within a PET system that surrounds him/her with many position-sensitive high-energy photon detectors that record the position, time, and energy of the incoming photons from many (e.g., 180) two-dimensional (2-D) views called *projections*. Photons that are absorbed in the body or that are not directed at these detectors do not contribute to the data set. Photons that are absorbed in detectors create scintillation light pulses. The light pulses are converted into electronic signals in the four PMTs coupled to the crystal array. The electrical signals from the PMTs are used to localize the incoming photon event to a given crystal within a given detector module. A weighted mean of the pulse heights is used to identify which individual crystal element within an array was hit. The total scintillation pulse height created for that event represents the absorbed photon energy. This energy is compared to a predetermined photopeak *energy window setting* for that identified crystal, obtained during calibrations. If the total event pulse height is within the window, the event is accepted, and the location of that crystal within the module, the location of the module within the system, and a time stamp for that event are recorded. Coincident events are selected as pairs of single events with time stamps that

match to within the *coincidence time window setting* for the system. These latter two steps highlight the importance of excellent energy and coincidence time resolutions. The resulting event is assigned to the appropriate tomographic *line-of-response* (LOR) drawn between the two detector element pairs, and that LOR value is incremented by one count. The collected data set comprises the number of coincident photon pair events emitted from the subject and recorded along all system LORs (the response lines between any two detector elements on either side of the PET system). PET requires many such coincident photon pairs to be positioned (assigned to an LOR) in this manner to build up high statistical quality image data.

There are two modes in which the coincidence photon pairs can be acquired in PET. In 2-D PET acquisition, the coincident photon LORs are confined to essentially 2-D detection planes corresponding to each ring (or perhaps two adjacent rings) of scintillation crystals within the cylindrical detector system. These detection planes are oriented perpendicular to the system axis. This 2-D confinement is accomplished using lead "washers" called *interplane septa* inserted orthogonal to the system axis and in between every scintillation crystal ring. These septa prevent photons from entering at oblique angles with respect to the septa plane. At present, only one vendor (GE Healthcare) supplies a clinical PET system with 2-D acquisition capability, all others run only in *3-D acquisition mode*. In 3-D acquisition mode, the lead septa are removed and all LORs formed between any two crystals from any two detector rings are allowed. All small-animal PET systems currently available run only in 3-D acquisition mode. The advantage of the 3-D mode is a significant ($\sim$5–10 fold) photon sensitivity increase because of the large increase of collected photon flux. The drawback of the 3-D mode, especially for clinical PET systems, is a considerable increase in background scatter and random coincidence event rate and system dead time due to possible crystal saturation.

There are three different event types in which two photons are considered to arrive simultaneously and are thus recorded as coincidence events. The good events are called *true*, where the line drawn between the two hit detector elements for that event passes through the point of emission of both photons (e.g., as depicted in Figure 7.5c). In *scatter* events, one or both 511 keV photons undergo *Compton* scatter in tissue before they are detected and the line drawn between the hit elements does not pass through the point of emission. Compton scatter can also occur in the detector crystals, which is a significant source of photon positioning errors in small-animal PET imaging because the crystals are small and the subject scatter is less likely. *Random* events occur when two distinct radionuclei each contribute one detected photon within the time resolution of the system and the line drawn between the two hit elements does not pass through the point of emission of either photon.

Both scatter and random events are an undesirable source of *background* counts that will cause grossly mispositioned events and therefore a loss in contrast resolution and quantitative accuracy, analogous to the background autofluorescence in FLI. These undesired background events may be reduced by narrowing the energy and coincidence time window settings and limiting the FOV activity, but residual scatter and random counts can also be subtracted from the data set using other techniques [84, 92]. Note that Compton scatter events can account for >70% of all accepted events in a 3D-acquired whole-body clinical PET study [93], even after energy window discrimination is applied, so it is important that the residual scatter correction method is highly accurate in clinical PET.

The acquired data are often organized into sets of parallel LORs, called *projections*, that give 2-D representations of the probe distribution for all angular views about the subject [84]. Often a continuous sequence or *cine* view of the sequential projection view data is displayed to allow gross visualization of the 3-D radionuclide distribution rotating like a top about the system axis. The organization of data facilitates *tomographic image reconstruction*, which is a process that uses mathematical algorithms to estimate the 3-D probe distribution volume

from the 2-D projection data and yields cross-sectional slices through the probe distribution [84]. The image reconstruction algorithm is a key component that turns raw hits recorded into 3-D images. There are two basic classes of reconstruction schemes, *analytical* and *iterative*. Analytic approaches consider the acquisition process, the measurements, and the reconstructed image as continuous functions. The analytical image reconstruction algorithm (e.g., filtered back projection or FBP) is based on direct computation of an inverse transform formula that converts the recorded detector hits into an image [94, 95].

Iterative techniques consider the previously mentioned functions to be discrete quantities. The iterative process starts with a "guess" of the 3-D probe distribution and goes through iterative (successive) modifications of that estimate until a solution is reached [16, 96–100]. Iterative algorithms differ by the algorithm by which the measured and current estimated projections are compared for a given iteration and the algorithm for the correction that is applied to modify the current estimate for a given iteration. Iterative techniques may incorporate statistical methods as well as accurate system models to find the best solution. Iterative approaches may be appropriate for photon count limited data such as whole-body clinical studies and for PET systems with nonstandard geometry.

Analytical methods are linear, but due to statistical noise they require spatial frequency filtering that results in a compromise in spatial resolution. Iterative methods allow an improved trade-off between spatial resolution and noise and enable a mechanism to incorporate accurate system modeling, but they are more computationally intensive. The analytic methods are typically more computationally efficient. Typically both analytical and iterative image reconstruction options are available on most clinical and small-animal PET systems.

The most common image reconstruction algorithm currently employed in PET is ordered subsets expectation maximization (OSEM) [100], which is essentially an accelerated version of the statistics-based maximum likelihood estimation maximization (MLEM) iterative algorithm expressed by the following equation:

$$\bar{f}_j^{(k+1)} = \frac{\bar{f}_j^{(k)}}{\sum\limits_{i=1}^{n} a_{ij}} \left(\sum_{i=1}^{n} \frac{g_i}{\sum\limits_{j'=1}^{m} a_{ij'} \bar{f}_{j'}^{(k)}} a_{ij} \right), \qquad (7.3)$$

where, $f^{(k+1)}$ is the current image estimate descretized via pixel j, $f^{(k)}$ is the previous image estimate, g_i is the value of the measured projection bin i, and a_{ij} is the weighting factor representing the contribution of the pixel j to the number of counts detected in projection bin i, or equivalently, the probability that a photon emitted from image pixel j is detected in measured projection bin i. The iterative procedure comprises a set of successive projections along lines from the current image estimate into the measured detector bins, known as a *forward projection*, and the reverse process, known as *back projection*. For 3-D-acquired data the most common image reconstruction approach first appropriately rebins the 3-D-acquired data set into a pseudo-2-D data set, using a process known as Fourier rebinning [95], and then the resulting data set is reconstructed with OSEM. However, assuming the availability of more powerful computational resources, other iterative algorithmic preferences such as 3D-OSEM [101] and 3-D maximum a posteriori (MAP) [102] are also employed. Once the cross-sectional image volume is reconstructed, the user may slice through it at any orientation for localization of any molecular-based signature of interest (e.g., thymidine-kinase reporter gene expression as in Figure 7.5d).

PET Performance Issues

As for FLI and BLI systems, performance parameters dictate a PET and SPECT system's reporter gene expression signal sensitivity. However, because radionuclide-generated decays comprise discrete monoenergetic photon events rather than a continuous glow of a spectrum of light, the performance parameters for the former are quite distinct from optical imaging systems. There are several important parameters of PET system performance such as *photon sensitivity, spatial resolution, energy resolution, coincident time resolution,* and *count rate performance*. The energy and temporal resolutions as well as count rate performance work together to define the available system *contrast resolution*, which is the ability to differentiate two slightly different concentration levels of probe in adjacent targets. The photon sensitivity, spatial resolution, and contrast resolution work together to define the reporter gene expression sensitivity of a PET instrument.

Photon sensitivity: Because in radionuclide imaging the photon events are collected and processed one at time rather than running in photon integration mode, the photon sensitivity is defined differently. The system *photon sensitivity* is the fraction of all coincident 511 keV photon pairs emitted from the imaging subject that are recorded by the system, and is also referred to as the *coincidence photon detection efficiency*. This parameter determines the statistical quality of image data for a given acquisition time. Photon sensitivity impacts image quality because it influences the noise level of images reconstructed at a desired *spatial resolution*. Photon sensitivity in PET is improved by (1) increasing the probability that emitted photons will traverse detector material, which is known as the *geometric efficiency*, and by (2) increasing the likelihood that photons traversing detector material will be stopped and an acceptable signal created, termed the *intrinsic detection efficiency*. The geometric efficiency is enhanced by tightly packing the detector elements together with little or no spaces, bringing the

detectors as close as possible to the body, and covering the subject with as much detector area as possible; these factors decrease the chance that photons will escape without traversing detector material. However, bringing the detectors closer to the subject can lead to position-dependent PET parallax positioning errors (hence loss of spatial resolution uniformity) due to annihilation photon penetration into the detector elements.

The intrinsic detection efficiency is improved by using denser, higher atomic number (Z), and thicker (longer) detector elements to improve the 511 keV stopping power. Annihilation photons interact with the medium they traverse through two processes. In *Compton scatter*, the photon scatters off a single electron in the outer shell of a traversed atom. The scattered photon changes its energy (frequency) and the outer shell electron is ejected from the atom. In the *photoelectric effect* the entire photon energy is absorbed by an inner shell atomic electron that is ejected from that atom. These two interaction mechanisms work together to *attenuate* (reduce) the number of photons traveling along a given direction, with a given photon attenuation factor $e^{-\mu x}$, where μ is the *linear attenuation coefficient*, which is related to the interaction probability of a photon with a medium and is a function of the atomic number Z, the attenuating material density, and the incoming photon energy; x is the material thickness traversed by the photon beam. Ideally one would like minimal attenuation in the subject tissues and maximum attenuation in the sensitive detector materials. Typical PET detector system photon sensitivities range from <1% (one coincidence photon pair collected for every one hundred emitted) for clinical systems to a few percent for small-animal systems.

Spatial resolution: The *spatial resolution* describes a system's ability to distinguish two closely spaced molecular probe concentrations and is important to detect and visualize subtle molecular signals from miniscule structures. PET spatial resolution is limited by the fact that one is trying to precisely determine the location of a positron-emitting nucleus attached to the probe molecule indirectly using the line drawn between the two annihilation photon hits in the detectors. Because this line results from two electronically determined interaction positions, this process is called *electronic collimation*. The spatial resolution is typically measured by imaging a point-like positron radioactive source and measuring its observed spread in the reconstructed images. The fundamental spatial resolution limit is dictated by (1) the *positron range* effect, which is due to variations in direction and path length of all the possible positron trajectories (see Figure 7.6a) created from a given point positron source; the extent of this resolution-degrading effect depends upon the range of energies of the emitted positrons and the medium traversed by the positrons before they annihilate; (2) the *photon acollinearity* effect, which is caused by the fact that since the positron and electron are not always at rest when they combine, the two annihilation

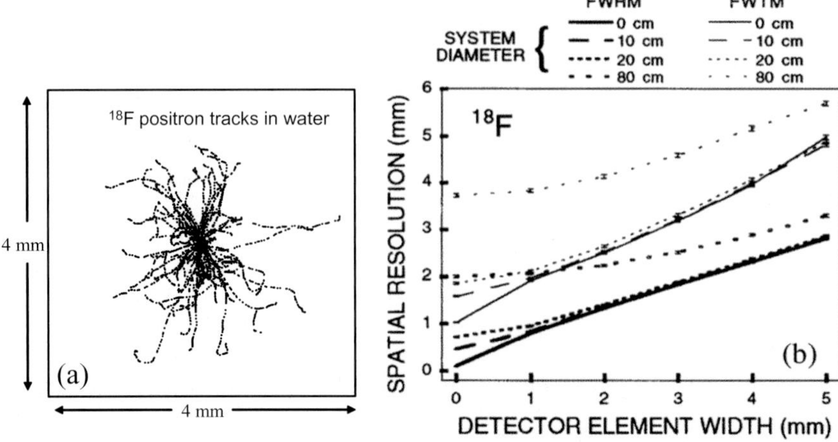

Figure 7.6. (**A**) Plot of computer-simulated energy deposition tracks for positrons emitted from an ^{18}F point source in water. (**B**) PET spatial resolution limit (FWHM and FWTM) for a point source of ^{18}F determined by a convolution of positron range, photon acollinearity, and detector element width contributions for different system diameters plotted versus detector element width. In theory, for system diameter <20 cm and 1 mm detector pixel size, submillimeter reconstructed spatial resolution is possible. Adapted from [103].

photons are not always emitted 180° apart, and hence the line defined by the two detectors hit will not always pass through the point of the positron–electron annihilation; This acollinearity affect on spatial resolution is worse for larger system diameters; (3) the size of the photon detector element, which determines how precisely a system can localize the photon interactions. The size of the detector element used in PET has been gradually decreasing over time to improve spatial resolution. Typical clinical systems use 4–6 mm detector pixels and small-animal systems use 1.5–2 mm detector pixels [85].

Figure 7.6b plots the combined spatial resolution limit from these three effects as a function of detector pixel size for various system diameters ranging from small-animal to clinical PET systems for an ^{18}F point source [103]. We see that in principle spatial resolution may be improved significantly by reducing the 511 keV photon detector pixel size. The element size dominates spatial resolution for small-diameter (<20 cm) animal PET systems, because the acollinearity effect on spatial resolution is minor for small-detector diameters. However, developing 511 keV photon-detector arrays with miniscule detector elements is challenging and typically results in performance compromises in other important system parameters. For example, using a point ^{18}F positron source, a 15-cm detector system diameter for small animal, and 1 mm scintillation crystal pixels, Figure 7.6b indicates that it is possible in principle to achieve submillimeter full-width-at-half-maximum (FWHM) spatial resolution at the center of the system, provided there are enough counts in the acquired data (adequate photon sensitivity) to reconstruct images at that desired spatial resolution without requiring significant smoothing. However, it is difficult to collect a high fraction of the available light out of narrow (1 mm width) and long (>2 cm) scintillation crystals [104]. Furthermore, this *light collection efficiency* varies as a function of interaction location within the crystal, so energy and time resolutions suffer as a result. Typically, to achieve acceptable light collection with 1 mm crystal pixels, their length is limited to ∼10 mm [86, 87, 105, 106], but this significantly compromises the probability of crystals to absorb incoming 511 keV photons (e.g., ∼58% intrinsic detection efficiency for single photons and ∼34% for paired coincident photons in LSO), known as the *intrinsic detection efficiency*, and hence limits the overall photon sensitivity performance.

For standard human systems, the detector pixel dimensions are typically ≥4 mm. Other resolution-blurring terms enter the equation, the photon sensitivity is much lower, and so the reconstructed ^{18}F point source resolution is typically 7–10 mm FWHM at the system center, depending upon image reconstruction parameters. Going to narrower detector pixel dimension (<4 mm) is not practical for standard whole-body human systems because the diameter is large (typically ∼80 cm), which increases the chance that photons will escape undetected. Hence photon sensitivity is too low (<1%) to provide adequate counts for higher-resolution image reconstructions that match the higher detector resolution because significant smoothing is required.

Energy and coincidence time resolution: Energy resolution is the precision to which one can measure the incoming photon energy. Because photons that scatter lose energy, good energy resolution means one may use a narrow energy window setting during data acquisition to reduce scatter photon contamination in image data without significantly compromising photon sensitivity.

A narrow energy window setting also helps reduce the rate of random photon contamination because many of these photons also undergo scatter. The *coincidence time resolution* determines how well one can decide whether two coincident photons truly arrive simultaneously. Analogous to benefits of good energy resolution, good coincidence time resolution means one may use a narrow time window setting to reduce random events without compromising photon sensitivity. Good energy and coincidence time resolution are enabled by using scintillation crystals that generate brighter and faster light pulses, low noise photodetectors, and by collecting a higher fraction of the scintillation light into the photodetector to create larger, more robust electronic pulses. A typical value for PET energy resolution is 20–25% FWHM at 511 keV and 2–3 ns FWHM for coincidence time resolution.

Count rate performance: Each detector signal recorded in a PET system has a finite processing time. If too many photons hit the detectors in a given time, the front-end photon detectors or subsequent acquisition electronics in the PET system can saturate due to piling up of more than one electronic detector pulse within the required signal processing duration. Typically the degree of pile up is limited by the photon detector signal processing time, which depends upon the decay time of the scintillation crystal, the effective integration time of the electronics, and the photon event rate seen by the detector. For example, suppose a 10 mCi (370 MBq) point source is placed at the center of a PET system with ten detector modules providing 5% coincidence photon detection efficiency. Then the average photon event rate per detector module is roughly 3.7×10^{8}(radionuclide decays per second)$\times 2$(photons per event)$\times 0.05$(photon sensitivity)$\div 10$(photon detector modules) $= 3.7 \times 10^{6}$ counts per second. If each system detector module required 1 μs of processing time per event, there could be significant pile up of events. For a given system photon sensitivity, for the best count rate performance the system should use scintillation crystals with fast decay time (see Table 7.1), detectors with excellent time resolution, fast processing electronics, and limited activity within the sensitive FOV.

Quantification of PET data: Unlike *in vivo*, noninvasive optical imaging, there is a clearer path to quantification of PET data. There are several undesired physical effects inherent to the process of detecting annihilation photons in PET, which must be compensated for either before or during image reconstruction to facilitate spatially resolved, quantitatively accurate data [84]. Because acquired PET data is often organized in terms of sets of LORs, to understand how these correction factors are applied to the 3-D volume of data acquired, it is simplest to visualize these correction factors being applied to one LOR at a time. The undesired physical factors include (1) 511 keV photon attenuation within the tissue, which causes the probe distribution to appear less intense for deeper structures; (2) detector response *nonuniformity*, which causes the probe distribution to appear artificially nonuniform due to nonuniform photon detector response throughout the system; (3) detector saturation or dead time, which can cause artificial loss in spatial and contrast resolutions; (4) random coincidence background, which can cause loss in quantitative accuracy and contrast resolution; (5) scatter coincidence background, which can also cause quantitative accuracy and contrast resolution degradation; (6) *isotope decay*, which leads to an artificially lower measured probe concentration as a function of time; (7) *partial volume effect*, which can artificially reduce intensity for structures that are on the order of the system spatial resolution or smaller. Finally, quantitative accuracy of PET image data further relies upon proper calibration of image counts to true isotope activity.

PET System Technology Advances

There have been several recent advances in PET system technology that will yield improvements in reporter gene expression sensitivity [85, 107]. Instrumentation and algorithmic innovations yield enhancements to parameters such as spatial resolution, contrast resolution, and quantification of data [107]. Preclinical system improvement trends have been described in a recent article [85]. Spatial resolution of small-animal PET has improved substantially in recent years, owing to technologies that facilitate the development of arrays of smaller scintillation crystals [85, 105–107] (e.g., Figure 7.5a) and new detector arrangements [e.g., 107–109]. Significant photon sensitivity improvements have been achieved by configuring detectors closer to the subject [107–109], increasing axial extent of the detectors [85], and reducing interarray gaps [85, 107]. Innovative statistical iterative image reconstruction algorithm implementation has enabled the processing of sparse data from systems with an enormous number of system response lines formed between all miniscule crystal elements [85, 107, 110].

In the clinical arena, precise measurements of the system response function have yielded improvements in reconstructed spatial resolution owing to resolution recovery within the OSEM reconstruction algorithm [111]. There has also been renewed interest in the concept of time-of-flight (ToF)-PET, owing to the development of high Z, high-density scintillation crystal systems that also achieve superior coincidence time resolution (≤ 1 ns FWHM) [112, 113]. This excellent coincidence time resolution constrains each event and all line projections performed during image reconstruction to a smaller region along each LOR. Because essentially more counts are placed in a smaller region along an LOR, the ToF information improves SNR and signal-to-background ratio (SBR) in the image reconstruction process [114–115]. The FWHM of the constrained line projection,

$\Delta l = c \times T_{\rm r}/2$ [116], where c is the speed of light and $T_{\rm r}$ is the FWHM coincidence time resolution. For example, if $T_{\rm r} = 500$ ps FWHM, then the line projection is spread in a Gaussian distribution with FWHM = 7.5 cm centered on the estimated emission point. Note that such a ToF resolution does not itself provide adequate spatial resolution to avoid the need for the forward and backward line projection reconstruction methods, but SBR improvements can enhance contrast resolution. The SNR improvement factor, f, provided by ToF information along any LoR is roughly equal to the square root of the ratio of the subject thickness, L, along that line to the FWHM of the constrained line projection, Δl; that is, $f = \sqrt{(L/\Delta l)}$ [116]. Thus, there is potential for significant SBR improvement for imaging the human trunk (e.g., if $L = 60$ cm, and $\Delta l = 7.5$ cm, then $f = 2.8$).

Single Photon Emission Computed Tomography (SPECT)

Overview

Molecular imaging using SPECT requires a molecular probe (e.g., reporter gene probe) labeled with a radionuclide that results in the emission of *gamma-ray photons or characteristic x-ray photons*. In contrast to PET only a single photon is detected per event, and that photon is emitted directly from the radioactive atom. Gamma-ray photons are emitted from the nucleus as a result of relaxation of neutrons and protons that are in an excited energy state of the nucleus. Common examples of gamma-ray emitters are ^{123}I, ^{125}I, and ^{99m}Tc. X-ray photons may result from alternate nuclear relaxation or decay processes that involve the removal of an inner-shell atomic electron. When an outer-shell electron fills this inner-shell vacancy, x-rays may be emitted from the atom. Common examples of radionuclides that result in x-ray emissions used in single-photon imaging are ^{111}In and ^{201}Tl, which undergo *electron capture decay*, whereby an inner-shell electron is "captured" (absorbed) by a proton within the nucleus. This leaves an inner electron shell vacancy of the atom that results in the emission of a characteristic x-ray. Single photon-emitting radionuclides can be created using a nuclear reactor by bombarding reactor-generated neutrons onto high Z targets or as a product of the fission process itself. Single-photon emitters can also be produced using a nuclear *generator*, which creates short-lived positron-emitting radionuclides (e.g., ^{99m}Tc) from the decay of a long-lived parent isotope (e.g., ^{99}Mo).

Single photons are ejected at the speed of light from every radioactive atom attached to the molecules of the SPECT probe distributed throughout the body of the subject. The emitted photons interact with electrons and nuclei of nearby atoms of the tissue or external detectors through Compton scatter or photoelectric absorption.

Unlike a beam of positrons traversing matter, the energetic photons do not "slow down" from interactions in body tissues or external detector materials, but rather the photon beam is attenuated (the number of photons traveling along a given line is reduced). The photons that escape from the body can be used for SPECT imaging.

SPECT Instrumentation

In the most common SPECT system configuration, the subject is surrounded with one or more position-sensitive gamma-ray photon detector panels, which typically are large scintillation detectors that rotate around the patient to collect the projection data. A SPECT acquisition consists of detecting and positioning many single photons traversing the detectors and can take 20 min to an hour depending upon parameters such as the collimator used, the number of detector projection angles, the acquisition time per projection, the size of the imaging subject region of interest, and the amount of activity available.

For single-photon imaging, in order to determine the correct LOR, just as was required for PET, it is crucial to be able to precisely fix the photon direction of incidence into the system and determine its interaction location in the position-sensitive detectors. In PET, the two-photon hits on either side of the system give the estimation of the LOR assignment through electronic collimation. In SPECT, because there is only one photon per event, the electronic collimation used for PET is not possible, and *physical collimation* must be used to determine the photon's incident direction. Physical collimation is accomplished using a structure made of high-density, high-Z material such as lead or tungsten with a well-defined configuration of holes for the photons to enter. Photons that hit the holes at the wrong angle do not make it through the collimator, these are absorbed in the collimator material, and thus do not contribute to the image. Photons that make it through the collimator holes have a well-defined direction of entrance into the camera. The LOR is determined by the hole the photon entered and the interaction location within the position-sensitive detector.

There are two common types of gamma-ray collimators employed in molecular imaging with SPECT, the *parallel-hole* and *pin-hole* collimators. Figure 7.7a depicts single photons entering a *pin-hole* collimator. For a pin-hole collimator the possible photon response lines pass through the point-like pin-hole, and hit the surface of a given position-sensitive scintillation detector panel at a range of incident angles, yielding a 2-D projection of the single-photon radionuclide distribution onto the face of the camera. Due to this geometry the projected object appears inverted with a projected magnification factor that depends upon the ratio of the pin-hole to detector distance (i.e., the height of the pin-hole collimator cone) to the object to pin-hole distance. This magnification

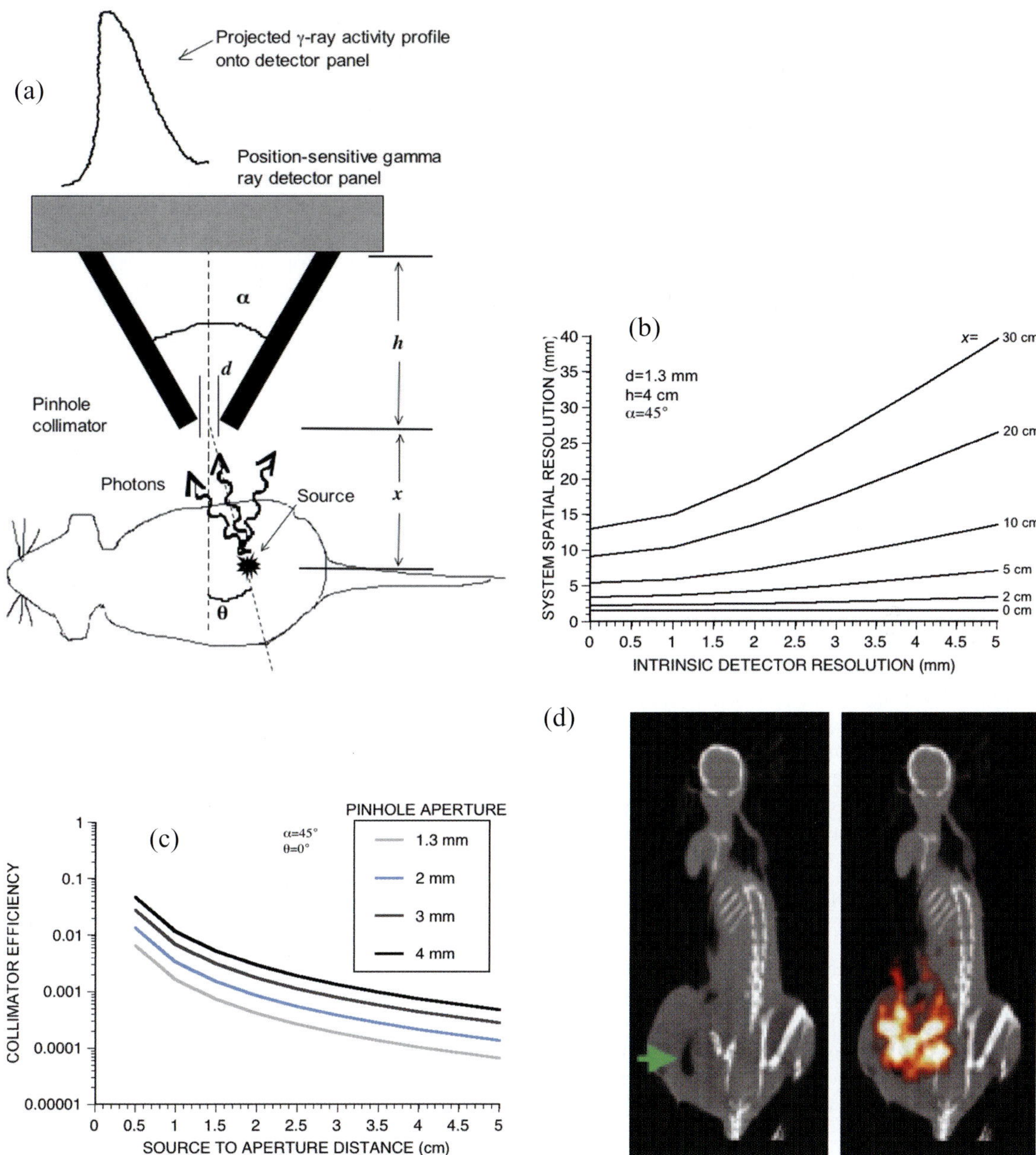

Figure 7.7. (**A**) Depiction of a point source projecting gamma rays through a single pin-hole collimator onto a detector panel. The important parameters determining the collimator efficiency and spatial resolution are defined in the drawings. (**B**) Calculated total system spatial resolution (FWHM) (convolution of collimator and detector contribution) for different source to pin-hole distances, x, using a single pin-hole collimator (assuming $d = 1.3$ mm, $h = 4$ cm, $\alpha = 45°$ in Equation 7.4) versus intrinsic detector panel resolution. (**C**) Calculated efficiency for a single lead pin-hole collimator versus source to pin-hole distance, x for various pin-hole diameters (assuming $\alpha = 45°$ and $\theta = 0°$). (**D**) Tracking of bacterial infection using a SPECT reporter gene strategy [78]. A BALB/C mouse harboring a CT-26 mouse tumor was treated with Clostridium novyi-NT (bacteriolytic therapy), a bacteria possessing an endogenous thymidine kinase (*tk*) gene homologous to that expressed by the herpes simplex virus (HSV) *tk* gene. After 24 h of treatment, the animal was injected with ^{125}I labeled 1-(2-deoxy-2-fluoro-D-arabinofuranosyl)-5-iodouracil (FIAU), a substrate that is phosphoralated by the *tk* enzyme. The animal was scanned 24 h later after the [125]IFIAU injection for 40 min with a Gamma Medica X-SPECT/CT system using a low-energy, high-resolution parallel-hole collimator. The green arrow in the bottom left CT image indicates gas in the tumor caused by the infection and the hot orange signal in the bottom right fused SPECT/CT image represents [125]IFIAU uptake/phosphorylation by the C. novyi–NT bacteria infecting the tumor cells. Images courtesy of Martin Pomper, Johns Hopkins University, and reproduced from [78] with permission from Proceedings of the National Academy of Sciences.

factor together with <1 mm pin-hole diameter can yield submillimeter resolution (see Figure 7.7b), which makes pin-hole-based SPECT attractive for small-animal imaging. The drawback of a small pin-hole size is that the photon sensitivity is quite low (see Figure 7.7c).

Many instrumentation requirements for PET apply for SPECT, with a few exceptions: (1) Because collimators are required for SPECT, the detectors are not configured in rings but rather in flat panels called *heads* that must rotate around the patient to view the photon activity from all angles. Only one rotating head is required for SPECT, but most systems have two heads to improve system photon sensitivity. (2) Because photon energies are lower in SPECT, crystals do not have to be as thick (e.g., <1 cm vs. >2 cm), dense, and high Z in order to have high intrinsic detection efficiency (see NaI(Tl) entry in Table 7.1 compared to the others listed); most of the world's SPECT systems use NaI(Tl) scintillation crystals. (3) Because the photon collimator determines the geometric efficiency and spatial resolution performance of SPECT, the detector crystal design requirements are somewhat relaxed. For example, most clinical systems use a continuous sheet of NaI(Tl) scintillation crystal rather than discrete crystal arrays. Another example is that high-spatial-resolution small-animal SPECT can be accomplished with standard clinical scintillation detector heads by substituting a special high-resolution multipinhole insert for a standard clinical collimator [117, 118]. However, special miniscule discrete crystal element arrays, similar to those used for PET (e.g., Figure 7.5a), are most common for small-animal SPECT system detector designs [84, 119, 120].

SPECT Methodology

There are several possible single-photon events that can occur [16]. The good events occur when the line drawn between the interaction location within the scintillation crystal and photon emission location within the subject passes through a collimator hole. For a *scatter* event, the photon undergoes a Compton scatter in the tissues before it enters the collimator and the line drawn between the crystal interaction point and the collimator hole does not pass through the point of photon emission. As was the case for PET, scatter events are background events that can cause photon positioning errors and a background "haze" in the image, and therefore reduce contrast resolution. Because scatter photons lose energy when they change direction, these undesired events are reduced without significant loss in photon sensitivity by having good energy resolution and using a narrow energy window setting around the photopeak. Photons may be absorbed in the patient or collimator or may not be directed toward a detector at all. Photons may interact in the crystal but lose some of their energy due to escape of a resulting characteristic x-ray that results from

photoelectric absorption, they may scatter first and then leave the crystal, they may back scatter off external materials and reenter the crystal, or they may pass through the crystal undetected. Similar detection principles hold for 511 keV photons interacting in scintillation crystals.

In principle, SPECT data from a distribution of a single photon-emitting probe is acquired and organized into 2-D projections in a very similar manner to that described previously for PET, with a few exceptions [84]. First, the camera heads are not rings of open crystal, but typically one or more rectangular plates covered with collimator that do not completely cover the subject and must be rotated around the patient to acquire the full angular sampling/projection data sets required for tomographic reconstruction. Second, if the detector plates use continuous scintillation crystal slabs rather than discrete pixels, as is the case for most clinical systems, each interaction in the detector is positioned using the light recorded from several PMTs (not just four), and the positioned event typically assigned to fictitious *bins* in order to have well-defined LOR assignment of events. Data from discrete crystal designs of SPECT cameras are processed much differently. If a discrete crystal design is used [84, 119, 120] the event position is assigned to the individual crystal that was hit, similar to the process used in PET. Third, only single photons are recorded per positioned event and so time resolution performance is not critical, and there is no coincidence event processing step. Similar to PET, the total pulse height is measured for each event and is compared to the energy window setting for the individual crystal element. However, if continuous sheet crystals are used, there is just one energy window for the entire crystal. If the measured energy is within the window for the crystal, the event x–y position is recorded and assigned to the appropriate position bin. The collected data set comprises the number of photons emitted from the subject and recorded along all system LORs, which are the response lines formed by the collimator and the assigned positioning bin.

As was the case for PET, both iterative and analytic approaches are used for SPECT image reconstruction. Due to the relatively low statistical quality of data and the need for incorporating models of collimator effects such as resolution blurring, typically iterative reconstruction algorithms such as OSEM are widely utilized for both clinical and preclinical SPECT imaging. The iterative image reconstruction process for SPECT is analogous to that for PET. An example reconstructed image of a SPECT reporter gene expression assay is presented in Figure 7.7d.

SPECT System Performance Issues

Similar performance parameters such as photon sensitivity (efficiency), spatial resolution, and energy resolution used to characterize a PET system are also used for

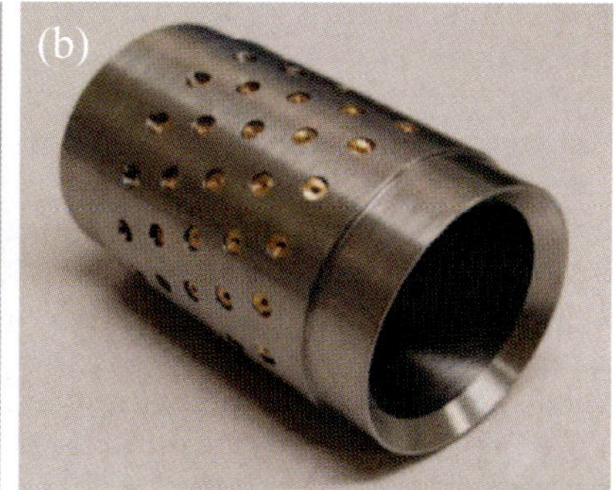

Figure 7.8. Examples of gamma-ray collimators for SPECT. (**A**) Section from a low-energy, high-resolution lead collimator commonly used in clinical SPECT imaging. The parallel holes are each ∼1.2 mm diameter with 0.2 mm lead septa separating the holes. (for 140 keV photons from ^{99m}Tc, $\mu_{lead} = 30$ cm^{-1}). (**B**) Multipinhole structure for ultra-high resolution small-animal SPECT imaging. The cylinder body is made of tungsten and has seventy-five gold pin-hole apertures inserted into holes placed in a hexagonal pattern throughout the cylinder. All pin-holes are tilted toward the center such that they are precisely directed toward and focused on the center point of the cylinder. The peak efficiency (photon sensitivity) at the center, for 0.6 mm diameter pin-hole apertures, is 0.22% at the cylinder center point and 0.1% 1 cm away from the center (multipinhole collimator picture is courtesy of Freek Beekman, Delft University of Technology). Within 1 cm from the center, the insert can clearly resolve 0.6 mm diameter hot rods separated by 1.2 mm [125].

SPECT. The combined effects of these performance parameters dictate a SPECT system's reporter gene expression sensitivity.

Photon sensitivity: SPECT system photon sensitivity is a product of the *collimator geometric efficiency* and the *intrinsic detector efficiency* [84]. The collimator efficiency is the probability that an emitted photon passes through the collimator and depends upon the collimator type and material and collimator properties such as hole size, thickness of the septa between holes, length of the holes, and distance of the activity source(s) from the collimator (for pin-hole only). The intrinsic detection efficiency quantifies how well the scintillation detector absorbs incoming photons, which depends upon the photon energy and the crystal material effective density, Z, and thickness along the photon's path. Typically, the collimator efficiency determines the overall system efficiency. The photon sensitivity (a.k.a. "efficiency") E and spatial resolution R for a pin-hole collimator are given by the following equations [121, 122]:

$$E \approx d[d + 2\mu^{-1}\tan(\alpha/2)] \cdot \cos^3\theta/16x^2,$$

$$R \approx \sqrt{d[d + 2\mu^{-1}\tan(\alpha/2)]} \cdot (h + x)/h, \qquad (7.4)$$

where the parameter definitions are provided in Figure 7.7a, aside from μ, which is the gamma-ray attenuation coefficient for the collimator material at the incident energy (e.g., $\mu = 27$ cm^{-1} for 140 keV photons in lead). We see that parameter choices that increase efficiency E will also broaden R, therefore decreasing spatial resolution.

Figure 7.7c plots efficiency for a single pin-hole collimator versus source-to-collimator distance for various aperture sizes. Unlike the pin-hole collimator, parallel-

hole collimator efficiency is insensitive to source–collimator distance [84, 121, 122]. Typical collimator efficiencies for SPECT imaging conditions range from 10^{-3} (one gamma-ray photon collected per 10^3 emitted) for small animal systems to 10^{-5} for clinical systems, and thus photon sensitivity for standard SPECT is relatively low compared to PET, limiting statistical quality of data for the same study duration. For small-animal pin-hole SPECT, efficiency can be improved by increasing the number of pin-holes [117, 118].

Spatial resolution: Because the single photons in SPECT are emitted directly from the radioactive atom of interest (unlike in PET), spatial resolution is mainly limited by how well the incoming photons can be collimated. Spatial resolution in SPECT is a convolution of the collimator resolution and the intrinsic detector resolution but typically limited by the collimator hole size. The collimator spatial resolution depends upon the collimator type (e.g., parallel-hole or pin-hole) and collimator properties such as hole size(s), thickness of the septa between holes, length of the holes, and distance of the activity source(s) from collimator. Typical high-resolution collimator hole sizes are ∼1.3 mm diameter for a low-energy, parallel-hole version for human systems (see Figure 7.8a), and 0.5–1.0 mm (pin-holes) for small-animal SPECT systems. The intrinsic detector resolution depends on the shape of the light distribution created from each interaction, which mainly depends on the crystal design and properties and how that distribution is sampled by the photodetector array. Typical intrinsic (detector only) spatial resolutions are ∼3.5 mm FWHM for clinical SPECT systems that use continuous sheet crystals [84, 123] and ∼1.5 mm FWHM for small-animal systems that use arrays of discrete crystals [16, 84, 119, 120, 122].

Spatial resolution in SPECT is typically measured by imaging fine line or point sources placed on top of the collimator and measuring the observed spread in the reconstructed images. Because typically the SPECT photon sensitivity is significantly lower than for PET, the statistical quality of the data is lower and more smoothing is required during the image reconstruction process, further reducing spatial resolution. In human systems using low-energy, high-resolution (LEHR) parallel-hole collimator, a typical value for the best point source resolution achievable is 7 cm FWHM at 10 cm source–collimator distance. In addition to collimator properties, the SPECT spatial resolution depends strongly upon the statistical quality of the data, the reconstruction algorithm used, and the degree of smoothing required. As a result, clinical image spatial resolution can be well over 1 cm FWHM.

Figure 7.7b plots total spatial resolution for a single, 1.3 mm diameter pin-hole collimator versus intrinsic detector resolution for various source to aperture distances. We see that SPECT spatial resolution can be significantly improved by using smaller hole size (e.g., smaller pin-holes). For pin-hole collimation there is also a *magnification factor* given by the ratio of the pin-hole-to-detector crystal distance divided by the activity-source-to-pin-hole distance that allows further improvement of spatial resolution. For close proximity SPECT imaging, where the source is <5 cm from the collimator, the intrinsic spatial resolution of the detector plays a significant role in determining the spatial resolution. For a 1 mm pin-hole in a small-animal SPECT system, 1 mm detector crystal pixels, and 5 mm source to collimator distance, <2 mm FWHM spatial resolution may be easily achieved. Submillimeter resolution is possible using <1 mm diameter pin-hole that is held very close to the object of interest.

Energy resolution: Typical energy resolutions are superior in SPECT versus PET because SPECT systems use NaI(Tl) as a scintillation crystal (see Table 7.1), which produces brighter scintillation light flashes and creates larger, more robust electronic pulses. Clinical systems use a continuous sheet crystal geometry, which provides a high aspect ratio for light collection into the PMTs. A typical energy resolution for a SPECT system that uses a NaI(Tl) crystal scintillation is 10% FWHM at 140 keV, which allows excellent rejection of scatter photons. However, Compton scatter is a more significant problem for PET due to the fact that two photons must be detected per decay, at 511 keV essentially all interactions in tissue are due to scatter, and the system has relatively wide acceptance angles for scatter photons.

Timing and count rate performance: Note that unlike for PET, because SPECT requires only one single photon detected per event, timing resolution is not important in SPECT. Due to the presence of a highly inefficient collimator, count rate performance is not critical either for a SPECT camera unless an enormous amount of high-energy photon (e.g., ^{131}I) activity is injected into the patient, one is performing first pass blood pool imaging, or there is some other application where a large amount of activity is present in a relatively focal region in front of the camera. Of course, count rate performance is critical if the collimator is removed and coincidence imaging is to be performed with a multiheaded SPECT system [124].

Quantification of SPECT data: SPECT also has several corrections that must be implemented for best qualitative and quantitative accuracy of image data. The method of corrections is in general different for continuous sheet versus discrete crystal designs. The most common corrections are the following: (1) *Uniformity correction* is used to account for the variations in intrinsic detection efficiency or PMT response that can cause hot or cold spot artifacts to appear. (2) *Energy (a.k.a. pulse height or gain) correction* is used in continuous sheet crystal detector head designs to account for the different total pulse magnitude created across a given detector head due to variations in properties such as crystal light yield and PMT *quantum efficiency* (probability of turning light into electrons through the photoelectric effect) as a function of the position across the detector head. A good energy correction will allow the camera to achieve the best energy resolution, which is useful to reject scatter photons. (3) *Spatial linearity correction* is implemented to correct for spatial distortions that misrepresent the true radionuclide distribution.

SPECT System Technology Advances

For SPECT systems that use pin-hole collimation, efficiency may be significantly improved by increasing the number of pin-hole apertures used [117, 118] and by bringing the tissue(s) of interest in close proximity to the pin-hole apertures. This multipinhole SPECT approach and associated image reconstruction algorithms that model the pin-hole geometry and physics have been applied successfully for small-animal imaging [19–20]. For multipinhole SPECT with many submillimeter pin-holes (e.g., see Figure 7.8b), submillimeter reconstructed resolution with improved photon sensitivity can be achieved [117, 125].

A new detector material that has become popular in recent years to replace the scintillation crystal in SPECT is cadmium zinc telluride (CZT) [16, 84, 107, 122] (see Figure 7.9a). CZT is a semiconductor crystal, not a scintillation crystal. Photons absorbed in CZT create electron-hole pairs, as is the case for scintillation crystals, but instead of producing light, these electrons and holes drift in opposite directions in a strong electric field applied across the material, and the electronic pulse is directly extracted from the crystal (see Figure 7.9b), rather than having to go through the intermediate steps of scintillation light creation, collection, and

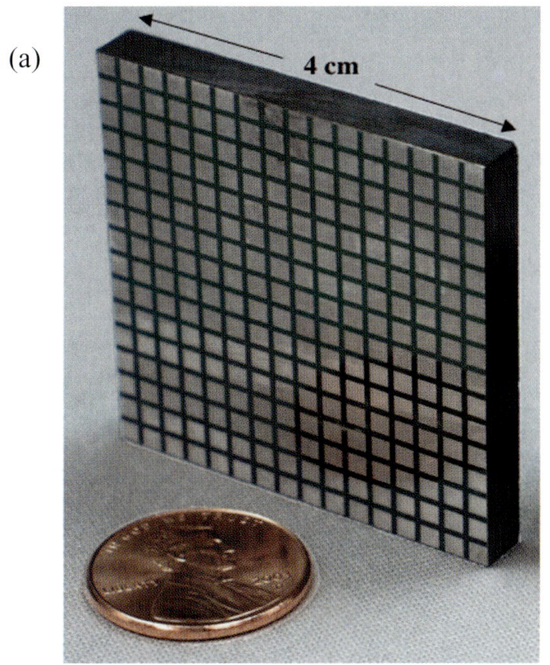

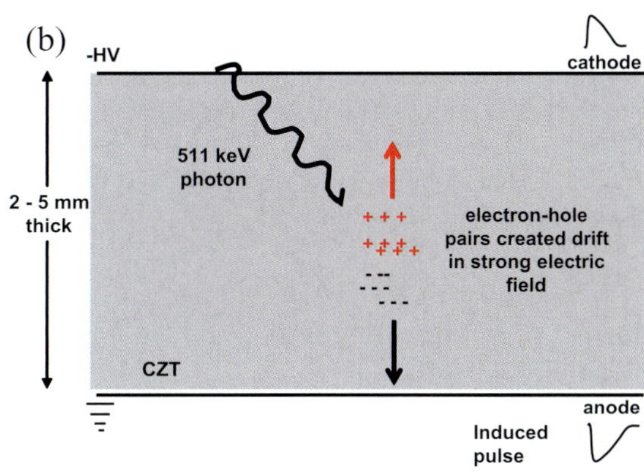

Figure 7.9. **(A)** Picture of a $40 \times 40 \times 5$ mm^3 cadmium-zinc-telluride (Cd$_{0.8}$Zn$_{0.2}$Te) array. The 256-pixel device shown has 2.25 mm anode pixels (indium) deposited on a 2.5-mm pitch. The backside has a continuous cathode. Adapted from [107]. **(B)** Depiction of a basic planar electrode detector. A high-energy photon interacts in the CdZnTe and creates electron-hole pairs that drift in a strong electric field ($\sim$1 kV/cm) established across the detector. Opposite polarity signals are induced on the anode and cathode planes with pulse height directly proportional to the amount of charge liberated in the photon interaction. Because the electrons drift over an order of magnitude faster than holes, the signals induced are dominated by electron motion only. Adapted from [16].

photoelectric conversion in a photodetector. The main advantages of this direct conversion of the incoming photon energy to an electronic signal are that (1) the energy resolution ($\sim$5–6% FWHM at 140 keV) and therefore scatter rejection capability is typically much better than for scintillation detectors [84, 122], and (2) high intrinsic spatial resolution may be achieved using fine "pixels" defined by the electrode pattern deposited on the detector faces (Figure 7.9a), rather than requiring arrays of miniscule discrete crystals. Because semiconductor crystals cannot be formed to be as large as scintillation crystals such as NaI(Tl), a CZT-SPECT detector panel typically comprises many submodules tiled together. Perhaps a disadvantage of the pure semiconductor approach is that the detector signals are relatively small in comparison to scintillation detectors and special low-noise electronics are required for electronic readout of each electrode of each module, which significantly increases electronic readout complexity.

MAGNETIC RESONANCE IMAGING (MRI)

Overview

In recent years there has been considerable effort to achieve MR imaging of gene expression [126, 127].

Although at the moment MR methods are roughly 10^4– 10^6 less sensitive for gene expression probe detection than radionuclide and optical methods [128], the former can achieve much higher spatial resolution. MRI is based on the physics principle of nuclear magnetic resonance (NMR), a technique that depends upon the quantum mechanical property of atomic nuclei called *spin* [129– 132]. Spin is a fundamental property of nature like electrical charge or mass. NMR essentially detects the results of externally manipulating the quantum states of nuclear spin within biologically relevant molecules. Individual unpaired protons, electrons, and neutrons possess spin, which comes in multiples of 1/2 and can be in "up" or "down," positive and negative states, respectively. Nuclei having an odd number of neutrons or protons will have an unpaired spin and a net magnetic moment, causing them to behave like a tiny magnet. In NMR, it is unpaired nuclear spins that are important. The most common examples of such atomic nuclei are ^{1}H, ^{13}C, ^{19}F, ^{23}Na, and ^{31}P. The most commonly used NMR technique in biomedical research studies is ^{1}H (proton) NMR, due to the high proton concentration in the body tissues (from water), robust NMR signals, and relatively high signal sensitivity [129].

To create an MR signal that can be detected, a resonance condition must be established. In other words, there must exist a situation of alternating absorption and dissipation of energy. The process of alternating absorption and emission of radiofrequency (RF) energy by the nuclear spins of material is the origin of the term *nuclear magnetic resonance*. The signal in NMR results from the difference between the energy absorbed by the spins that make a transition from a lower energy state to the higher energy state, and the energy emitted by the spins that simultaneously make a transition from the higher energy state to the lower energy state. The signal is thus proportional to the population difference between the states. NMR is a rather sensitive spectroscopy method because it is capable of detecting these very small population differences. It is the resonance, or exchange of energy at a specific frequency between the spins and the spectrometer, that gives NMR its sensitivity.

Basics of Nuclear Magnetic Resonance (NMR)

Consider the imaging subject as an ensemble of hydrogen nuclei, each with a nuclear spin. In absence of any external forces, the spins are oriented randomly. For protons there are two states of the spin, positive and negative. If one places the subject in a strong homogeneous, static magnetic field, B_0, some of the magnetic moments associated with the two spin states tend to align themselves parallel (low-energy state) or antiparallel (high-energy state) to B_0. In nature, a slightly larger fraction of spins align themselves in a lower energy state, parallel to B_0, establishing a net longitudinal magnetization vector, $\mathbf{M}$, that is also parallel to B_0. With a static field strength (magnitude of B_0) of 1.5 Tesla (T), only about 1–10 of every 10^6 individual proton spin magnetic moments align with B_0 and contribute to $\mathbf{M}$ and the observed MR signal. Thus, molar concentrations of water protons are required to provide sufficient signal intensities in MR.

If this net magnetic moment $\mathbf{M}$ is perturbed by an orthogonal external electromagnetic (EM) field, it will be temporarily directed at some nonzero angle with respect to B_0, and as a result will experience a torque, which causes it to precess about B_0. This situation is somewhat analogous to a spinning top precessing about the gravitational field after it tips. The nuclei precess at a frequency that is directly proportional to the static field strength. The precession frequency is known as the Larmor frequency, f given by $f = \gamma B_0$, where γ is a proportionality constant known as the gyromagnetic ratio, which relates the magnetic moment to the angular momentum and is specific to the nuclear species of interest and strength of its individual magnetic moment; B_0 is the static magnetic field strength. For protons, $\gamma \approx 43$ MHz/B_0. For $B_0 = 1.5$ Tesla, $f \approx 64$ MHz. In NMR spectroscopy, f is between 60 and 800 MHz for hydrogen nuclei. In clinical MRI, f is typically between 15 and 128 MHz for hydrogen (proton) imaging [130].

In an NMR system, the net magnetization $\mathbf{M}$ is tilted by an external oscillating magnetic field perpendicular to B_0, with oscillations in the RF range of the EM spectrum [131]. If the external field oscillation frequency matches the Larmor precession frequency the ensemble demonstrates a resonant absorption whereby the spins will absorb energy and become excited. The proton spin can undergo a transition between the two energy states by the absorption of an RF photon. A proton in the lower energy state absorbs a photon and ends up in the upper energy state. The energy of this photon must exactly match the energy difference between the two states. The energy of the photon needed to cause a transition between the two spin states is $E = hf = h\gamma B_0$, where h is known as Planck's constant. When the energy of the photon matches the energy difference between the two spin states, an absorption of energy occurs, but any other external frequency has no effect.

After altering the direction of the net magnetization vector $\mathbf{M}$ by exposing the nuclear spin system to an EM field energy equal to the energy difference between the spin states, $\mathbf{M}$ will be tipped at a "flip" angle between 0 and $90°$ with respect to B_0 and precess around the latter at the Larmor frequency. The flip angle is proportional to the amplitude and duration of the RF pulse. If enough energy is put into the system, it is possible to saturate the spin system and make the component of $\mathbf{M}$ along the z-direction, M_Z, equal to 0. If the system is pumped with enough energy so that $\mathbf{M}$ lies in the XY plane it will rotate about the z-axis at a frequency equal to that which would cause a transition between the two energy levels of the proton spin, which is the Larmor frequency.

In an NMR system, a time-varying perpendicular RF pulse produced by a current through the system transmit coil, tuned to the same frequency as the nuclei precession, produces transitions into a higher energy state [132]. At the end of the applied RF pulse, the RF signal emitted by the material is at its maximum intensity. Once the excitation field is switched off, the signal intensity diminishes rapidly (within a few hundred milliseconds) as the higher energy state (the antiparallel state) is depopulated and the nuclei return to their original energy state. $\mathbf{M}$ returns to its equilibrium state according to the exponential decay process that is characteristic of the nuclear species of interest and the energy is emitted in the form of a weak RF signal. The time constant that describes how M_Z returns to its equilibrium value is called the spin lattice relaxation time (*T1*). The frequency of the emitted signal depends on the strength of the applied static magnetic field as well as the type of nuclei producing the signal. If a receiver coil is placed nearby, the precessing $\mathbf{M}$ will induce a small RF current on the coil. That is, this decaying RF signal is picked up by the NMR system's receiver coil. Detection and analysis of this signal provide insight into the chemical composition of the material. The waveform of this signal is an exponentially damped sine wave and is called the *free induction decay* (FID) [131].

M, a vector, comprises two components, *longitudinal* (parallel to $\mathbf{B}_0$) and *transverse* (in the plane orthogonal to $\mathbf{B}_0$). At the end of the applied RF transmit pulse, the higher energy state decays through two deexcitation processes known as *spin–lattice* (*T1*) and *spin–spin* (*T2*) relaxation depending upon whether the decay of **M** is along the longitudinal or transverse directions, respectively [132]. That is, after *T1* relaxation, **M** realigns with the $\mathbf{B}_0$, and after *T2* relaxation, the precession of **M** in the transverse plane dephases. The particular decay process governing *T2* relaxation is a result of phase dispersal of the signal [129]. The physical mechanisms responsible for phase dispersal are the incoherent interactions between individual nuclear spin constituents of the net magnetization, a property that is specific to the nuclear species and environment. In the presence of an inhomogeneous magnetic field, the spins will have variations in precession rates that mirror the inhomogeneity. These different precession rates cause further phase dispersal and more rapid signal decay with a net decay constant referred to as *T2** [130].

MRI Instrumentation

An MRI system comprises three basic subsystems [133], the *static magnetic field*, the RF *transmit* and *receive coils*, and three orthogonal magnetic field *gradient coils*. These subsystems are configured concentrically as depicted in Figure 7.10a.

The static (a.k.a. "main") magnet is the largest and most costly scanner component around which the other subsystems of the scanner are built. Both the strength and precision of the main magnet are important. The magnet field lines near the magnet center (a.k.a. the "iso-center") should be extremely straight and highly uniform, or *homogeneous*. Slight inhomogeneities in the field strength within the scan region should be less than three parts per million (3 ppm) [133]. State-of-the-art MR magnet systems are based upon a cylindrically shaped superconducting electromagnet. When a niobium–titanium alloy is cooled by liquid helium to 4 K ($-269°$ C, $-452°$ F) it becomes a superconductor, losing resistance to flow of electrical current. An electromagnet constructed with superconductors can have extremely high field strengths, with very high stability. The construction of such superconducting magnets is very complex and expensive, and the cryogenic helium is costly and difficult to handle.

Most superconducting magnets comprise tightly wound coils of superconductive wire immersed in liquid helium, inside a vessel called a *cryostat*. Despite thermal insulation, ambient heat causes the helium to slowly boil off, and so a constant supply of liquid helium is necessary. A cryo-cooler (a.k.a. coldhead) is used to recondense a fraction of the helium vapor back into the liquid helium bath. Several manufacturers now offer "cryogenless" scanners, where instead of being immersed in liquid helium the magnet wire is cooled directly by a cryo-cooler [133].

Magnetic field strength is an important factor in determining image quality. Higher magnetic fields increase SNR, permitting higher resolution or faster scanning. However, higher field strengths mean more expensive magnets, higher maintenance costs, and increased safety concerns. A 1.5 Tesla (T) field strength is a good compromise between cost and performance for general biomedical imaging use, including gene expression imaging in mice [134, 135]. However, for certain specialist uses (e.g., high-resolution small-animal imaging), field strengths up to 11.7 T or more may be desirable [136–137].

Using conventional body surface coils or custommade high-resolution coils, clinical 1.5–3.0 Tesla MR imaging systems can acquire high-quality images in small animals with good SNR and volumetric resolutions of ≤ 0.1 mm^3 in 5–10 min [134–135]. Field strengths greater than 3.0 T are not commonly employed for routine clinical imaging. High-field (e.g., up to 11.7 T) MR systems, with special transmit, receive (see Figure 7.10b), and gradient coil design and reconstruction algorithms can achieve microscopic resolutions at high SNR [136–137].

The RF transmission system comprises an RF synthesizer, power amplifier, and transmitting coil, usually built into the scanner body [133]. The power of the transmitter is variable. Current high-end scanners can have a peak output power of up to 35 kW and sustain an average power of 1 kW. The receiver comprises the coil, preamplifier, and signal processing system. Although one may acquire MR data using an integrated coil for both transmitting and receiving, if a small region is under study then better image quality is obtained by using a close-fitting smaller receiver coil. For example, for *in vivo* small animal imaging of gene expression, specialized high-resolution coils (e.g., Figure 7.10b) are available or can be custom manufactured to suit the user's special needs. Due to the fact that the signal is inversely dependent on the square of the distance to the receiver coil, these specialized coils fit closely to the contours of the subject.

To turn the basic NMR signal into images, MRI requires spatially varying magnetic fields known as *gradients* [133]. Magnetic field gradients are generated by three coils that produce orthogonal gradients in the x, y, and z directions of the scanner (z is along the system axis). The gradient coils are typically resistive electromagnets powered by sophisticated amplifiers, which permit rapid and precise field strength and duration

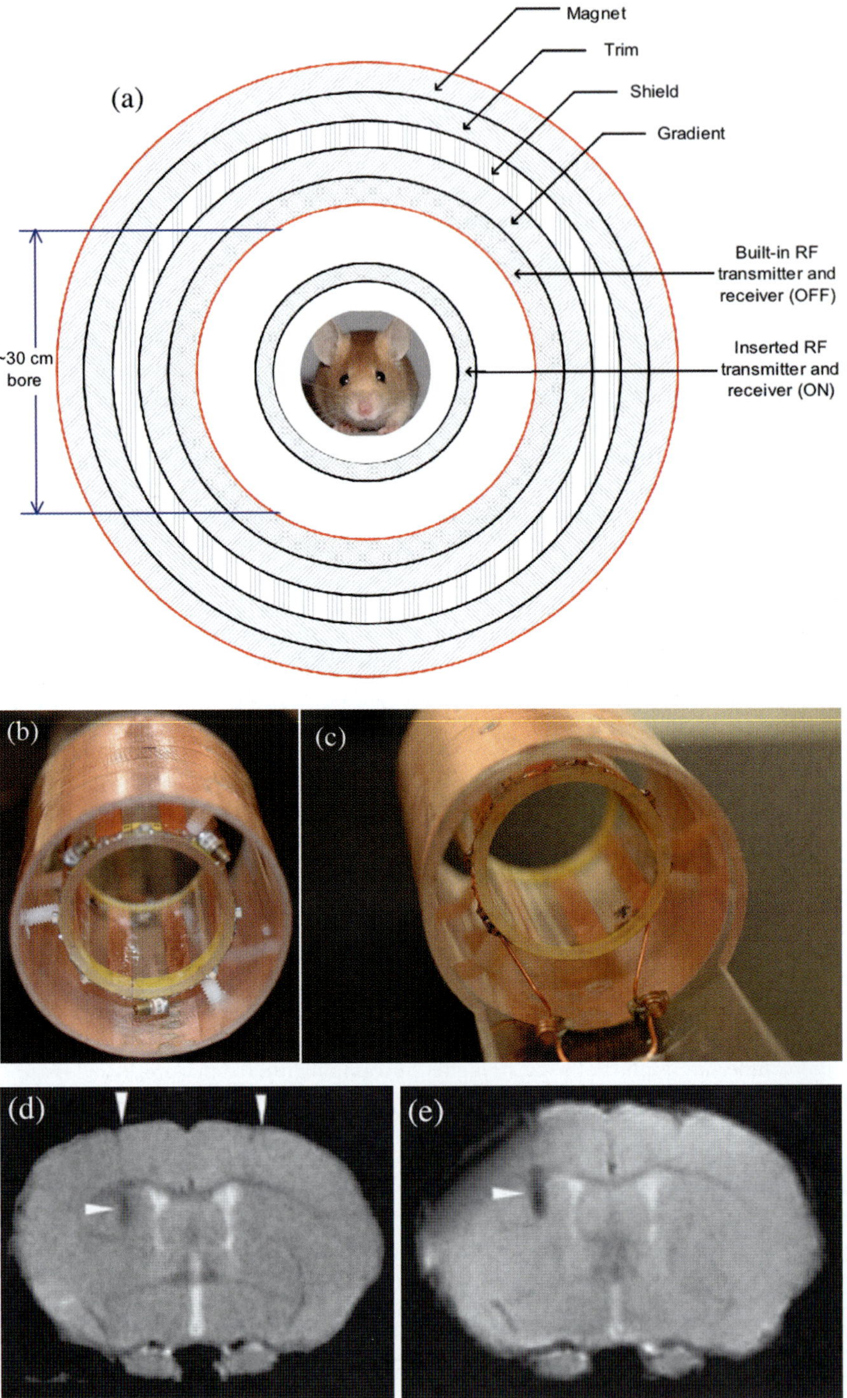

Figure 7.10. (**A**) Depiction of an end-on view of basic concentric subsystems of a 7T MR scanner. (**B**) and (**C**) Two views of the quadrapole field "bird-cage" RF transmitter/receiver coil for a 7T small-animal MR system. (Coil picture courtesy of Laura Pisani and coil design/construction by Ron Watkins, Stanford University). (**D,E**) *In vivo* results ferritin transgene expression in the mouse brain using MRI. An adenovirus encoding the MRI reporter gene (ferritin) was inoculated into the striatum. (**D**) T2-weighted image 5 d after injection shows the inoculated sites (left arrow, MRI reporter gene; right arrow, adenovirus-lacZ control) with a contrast-to-noise enhancement of 6.6 in the left region indicating robust transgene expression. (**E**) T2*-weighted image at Day 5 after injection shows a contrast-to-noise enhancement of 16.2. Images were acquired using an 11.7 T MR 89 mm vertical-bore Bruker microimaging system. Coronal slices were acquired at the injection site using T2-weighted spin echo and T2*-weighted gradient echo sequences. The parameters were TR/TE = 1200/35 ms and TR/TE = 1200/6.7 ms for the spin echo and gradient echo sequences, respectively. Image data were acquired at 256 × 256 points with 2.6 cm field-of-view, 0.75 mm slice thickness, and an in-plane resolution of 98 μm. Adapted from reference [143] with permission from *Nature Medicine*.

adjustments. Typical gradient systems generate gradients from 20 mT/meter to 100 mT/meter. For example, in a 1.5 T magnet, when a maximal z-axis gradient is applied, the field strength may be 1.45 T at one end of a 1 m long bore and 1.55 T at the other). The magnetic field gradients determine the plane (i.e., cross-sectional slice) of imaging. By carefully combining the orthogonal gradients, any plane can be selected for imaging. Scan speed strongly depends upon performance of the gradient system. Stronger gradients enable faster imaging or higher resolution. Faster switching gradient systems can also permit faster scanning. However, in the clinic gradient performance is limited by safety concerns associated with the well-known occurrence of nerve stimulation in an MRI system.

A recent advance in MRI technology has been the development of sophisticated multielement phased array coils capable of acquiring multiple channels of data in parallel [138]. This *parallel-imaging* technique uses unique acquisition schemes that enable accelerated imaging by replacing some of the spatial coding requirements of the magnetic gradients with the increased spatial sensitivity of the multiple coil elements. However, the increased acceleration also reduces the SNR and can cause image reconstruction artifacts. A detailed review of parallel imaging (acquisition and reconstruction) techniques is discussed in reference [138].

MRI Methodology

Overview: To produce an image for MRI, each MR signal must be referenced to a specific region of tissue. This is accomplished by applying a *gradient magnetic field* in which the field strength varies linearly with position. The gradient gradually varies the magnetic field strength resulting in a corresponding shift in the RF frequency needed to stimulate the tissue. Because emitted RF signals will also demonstrate a shift in frequency, the excited tissue from which the signals originated can be localized. Using a computer-aided reconstruction program, somewhat analogous to that used in computed tomography (e.g., PET or SPECT), except with precise depth encoding information along any line enabled by the gradients, the signals attributed to individual volume elements of tissue can be resolved and reconstructed into an image. The most common method of image reconstruction is the two-dimensional Fourier transform. Image slice selection is achieved by applying a magnetic gradient in addition to the external magnetic field during the radio frequency pulse. Only one plane within the object will have protons that are on–resonance and contribute to the signal.

The simplest method in MRI is to build up the data for an image slice from a series of discrete signal samples [131]. For example, a typical $T2$-weighted imaging series, used to form an image whose contrast depends predom-

inantly on the intrinsic tissue magnetization parameter, $T2$, requires that the time between excitation pulses, known as "TR," be two to three times longer than the intrinsic tissue magnetization parameter, $T1$. The $T1$ of biological samples is typically on the order of a second; TR must therefore be 3 sec or more (note that cerebrospinal fluid (CSF) can have much longer $T1$s of several seconds) [139]. A typical MR image is formed from 128 repeated samples so that the imaging time for a standard $T2$-weighted scan is roughly 400 sec, or nearly 7 min.

Spatial encoding: Tomographic image formation requires spatial encoding in three dimensions. Data from one dimension is typically determined by *slice selective excitation* [132], where a narrow-band RF excitation pulse is transmitted to the subject in the presence of a magnetic field gradient. Because the NMR phenomenon depends on an exact match between the RF excitation pulse frequency and the proton spin frequency, which depends in turn on the local magnetic field, this pulse will excite the MR signal over a correspondingly narrow range of locations, defining an imaging slice. The other two gradient fields determine the "in-plane" spatial encoding.

A magnetic field gradient applied across the excited slice causes the spin frequency to be position dependent. There are three axes used for spatial encoding of MR images. One dimension of spatial encoding is achieved by slice selective excitation (the "slice selection" axis). The other two are encoded by phase and frequency. Some texts refer to the slice selection axis as the "z" axis. The "readout axis" is often labeled the "frequency" or "x" axis; the "phase encoding" axis is typically labeled the "y" axis.

k-space: An image can be considered to be composed of a number of spatial frequencies at different orientations. A two-dimensional Fourier transformation of an image will express these components as a matrix of spatial frequencies known as k-space. That is, k-space represents the raw MRI data before it has been Fourier transformed in order to make an image [129]. Low spatial frequencies are represented at the center of k-space and high spatial frequencies at the periphery. Frequency and phase encoding are used to measure the amplitudes of a range of spatial frequencies within the object being imaged.

As described, the frequency-encoding gradient is applied during readout of the signal and is orthogonal to the slice selection gradient [130]. During application of the gradient the frequency differences in the readout direction progressively change. At the midpoint of the readout these differences are small, and the low spatial frequencies in the image are sampled, filling the center of k-space. Higher spatial frequencies will be sampled toward the beginning and end of the readout filling the periphery of k-space. Phase encoding is applied in the remaining orthogonal plane and uses the same principle of sampling the object for different spatial frequencies. However, phase encoding is applied for a brief period before the readout and the strength of the gradient is changed incrementally between each RF pulse. For each phase encoding step a line of k-space is filled.

In standard MRI, k-space is covered line by line, and "k" is considered a vector with both direction and magnitude. Following each RF excitation, a single line of raw data is collected along k_x (the readout axis), with sequential lines acquired at different displacement along the k_y axis. Because a separate excitation step is required prior to the collection of each data line, the total imaging time depends on the time between excitations, "TR," as well as on the total number of lines of data collected. The latter depends on the desired spatial resolution and FOV in the final images.

The k-space formalism simplifies several complex concepts. For example, it becomes easy to understand the role of phase encoding. In a standard spin echo or gradient echo scan (described in the following), where the readout gradient (e.g., G_x) is constant, a single line of k-space is scanned per RF excitation pulse. When the phase-encoding gradient (e.g., G_y) is zero, the line scanned is the k_x axis. When a nonzero phase-encoding pulse is added in between the RF excitation and the start of the readout gradient, the line scanned is $k_y =$ constant.

The center of k-space determines image contrast, a fact often exploited in advanced MR imaging techniques. One such advanced technique is known as *spiral acquisition*, where a rotating magnetic field gradient is applied, causing the trajectory in k-space to trace a spiral out from the center to the edge. Due to $T2$ and $T2^*$ decay the signal is greatest at the start of the acquisition. Thus, acquiring the center of k-space first improves contrast-to-noise ratio (CNR) when compared to conventional zig-zag acquisitions, especially in the presence of motion.

The concept of k-space also simplifies the comparison of different scanning techniques. For example, in single-shot *echo planar imaging* (EPI) [131] used in functional MRI (fMRI), all of k-space is scanned in a single shot, following either a sinusoidal or zig-zag trajectory. Because alternating lines of k-space are scanned in opposite directions, this must be taken into account in the image reconstruction. Multishot EPI and fast spin echo techniques [132] acquire only part of k-space per excitation. In each shot, a different interleaved segment is acquired, and the shots are repeated until k-space is sufficiently filled. Because the data at the center of k-space represent lower spatial frequencies than the data at the edges of k-space, the time-to-echo (TE) value for the center of k-space determines the image's $T2$ contrast.

3-D acquisition: A 3-D volume MR acquisition simultaneously excites a set of contiguous slices, known as a *slab* during each TR interval. The slices in the 3-D slab are acquired and reconstructed differently than for 2-D acquisitions. The most common acquisition strategy for 3-D MR imaging is to use rectilinear sampling [140–141], where the 3-D volume is spatially encoded with phase encoding along two perpendicular spatial directions and frequency encoding along the third. The resulting raw data fills a 3-D k-space matrix, which is reconstructed by a 3-D Fourier transform. The main advantage of 3-D acquisitions is their ability to acquire thin contiguous slices ideal for flexible retrospective visualization techniques such as volume rendering, maximum intensity projection, or multiplanar reformatting. Because phase encoding is performed along two spatial directions, the acquisition time for 3-D scans is longer. However, for short $T1$ it is the time spent on this extra phase-encoding step that provides the SNR advantage for 3-D over 2-D acquisitions [141].

Pulse sequences: The local environment determines the MR signals created. Different physical characteristics, such as proton density, $T1$ and $T2$ relaxation, flow, diffusion, perfusion, or temperature, can be probed with different acquisition parameters, called *pulse sequences*, which select features such as RF pulse strength, shape, and repetition rate, and duration between transmit and receive, that perturb **M** in different ways [140–141]. These different perturbations can be converted to high-resolution images of a wide range of processes that can probe anatomic, physiological, or functional pathologies in living subjects. The most common pulse sequences employed in MR for gene expression imaging are *spin–echo* and *gradient–echo* sequences, which will be described in the following.

To understand mechanisms of MRI contrast for imaging gene expression, it is important to have a basic understanding of the NMR time constants involved in relaxation processes that establish equilibrium following RF excitation. As the excited nuclei relax and realign they emit energy at rates recorded to provide information about the material they are in. We have learned that the process of realignment of nuclear spins with the main magnetic field is called the longitudinal relaxation, and the time required for $1/e$ of the tissue's nuclei to realign is $T1$, which is typically about 1 sec at 1.5 T main field strength. $T2$-weighted imaging relies upon local dephasing of spins following the application of the transverse energy pulse, which has a transverse relaxation time of $T2$. Typically $T2 < 100$ ms for tissue at 1.5 Tesla main field strength. A subtle but important variant of the $T2$ technique is called $T2^*$ imaging. $T2$ imaging utilizes a *spin echo* technique (see the following), in which spins are refocused to compensate for local magnetic field inhomogeneities, whereas $T2^*$ imaging is performed without this refocusing. As a result, $T2^*$ imaging sacrifices some image integrity (spatial resolution loss) but provides additional sensitivity to relaxation processes that cause incoherence of transverse magnetization.

Echoes: In MRI, an *echo* is the emission of energy from a proton in the form of an electromagnetic resonance signal at a certain well-defined time after its excitation [140]. At the time of the echo, spins are back in phase again and the signal is measured. In MRI, the desired number of echoes is selectable. Often up to eight echoes are permissible for 2-D or 3-D scans using a *spin echo* sequence. An echo signal is generated from an FID by means of a bipolar, switched magnetic gradient. The echo is produced by reversing the direction of a magnetic field gradient or by applying balanced pulses of magnetic field gradient before and after a refocusing RF pulse to cancel out the position-dependent phase shifts that have accumulated due to the gradient. In the latter case, the gradient echo is generally adjusted to be coincident with the RF spin echo. When the RF and gradient echoes are not coincident, the time of the gradient echo is denoted echo time (TE) and the difference in time between the echoes is denoted time difference (TD). The resulting image slice $I(x, y)$ can be expressed in terms of the magnitude of the magnetization vector $M(x, y)$, by the following equation [141]:

$$I(x, y) \alpha M(x, y)[1 - e^{-TR/T1(x,y)}]e^{-TE/T2^*(x,y)}. \quad (7.5)$$

Spin echo: In NMR, *spin echo* (SE) refers to the refocusing of precessing nuclear spin magnetization by applying a *180° pulse* of resonant RF energy. The SE pulse sequence was devised in the early days of NMR by Carr and Purcell [140]. SE is the most common pulse sequence used in MR imaging [141]. The technique uses 90° RF pulses to excite the magnetization and one or more 180° pulses to refocus the spins to generate signal echoes (hence the name) detected by the RF receiver coil. The 90° excitation pulse rotates the longitudinal magnetization (M_z) into the xy-plane, and at that point in time the dephasing of the transverse magnetization (M_{xy}) begins. A subsequent application of a 180° refocusing pulse rotates the magnetization in the x-plane and generates the signal echoes. The purpose of the 180° pulse is to rephase the spins, causing them to regain coherence and thereby recover transverse magnetization, producing a spin echo. Because typically *T1* is in the range of 100–2000 ms, which is greater than *T2* for living tissues [139], the recovery of the z-magnetization occurs with the *T1* relaxation time and typically at a much slower rate than for *T2* decay. If the inversion pulse is applied after a period *T* of dephasing, the inhomogeneous evolution will rephase to form an *echo* at time 2*T*. The intensity of the echo relative to the initial signal is given by $\exp(-2T/T2)$ where *T2* is the time constant for spin–spin relaxation. With this type of spin echo imaging no *T2** decay occurs due to the 180° refocusing pulse. For this reason, spin echo sequences are more robust against problems such as susceptibility artifacts than *gradient echo* sequences. In the simplest form of SE imaging,

the pulse sequence has to be repeated as many times as the image has lines. The SE pulse sequence has many forms such as [141]: (1) the *multi echo* pulse sequence using single or multislice acquisition; (2) the *fast spin echo* (FSE/TSE) pulse sequence; (3) echo planar imaging (EPI) pulse sequence; and (4) the gradient and spin echo (GRASE) pulse sequence.

Examples of different contrast mechanisms and pulse sequence parameters are: (1) proton density (PD) weighted: short *TE* (20 ms) and long TR; (2) *T1* weighted: short *TE* (10–20 ms) and short *TR* (300–600 ms); and (3) *T2* weighted: long *TE* (greater than 60 ms) and long *TR* (greater than 1600 ms).

Another technique for generating spin echoes is to apply three successive 90° RF pulses [140]. After the first 90° pulse, the magnetization vector exchanges energy through dipole–dipole interactions and, in a time τ, forms what is often referred to as a "pancake" in the x′–y′ plane. A further 90° pulse is then applied such that the "pancake" reorients in the x′–z′ plane. When considering the two types of relaxation, spin–lattice and spin–spin (*T1* and *T2*, respectively), we assume the first to take an infinite amount of time, causing the spin vectors to precess about the z axis. At this point, the angle each spin makes with the z′ axis is equal to the angle it previously made about the y′ axis. Any change in angle that subsequently takes place will require a change in energy, thus implying a spin–lattice interaction is required. This effectively yields a recording of the state of the system as it was at time τ. After a further time $\tau 2$ a third pulse is applied, bringing **M** back in the x′–y′ plane oriented in the same direction as for the (90–τ–180) spin echo sequence. Next, after final delay of τ one arrives at what is commonly referred to as a *stimulated echo*. This technique is commonly used when studying *T1* relaxation times because by measuring the magnitude of the correct echo and its decay with pulse width separation we can determine *T1*. The resulting echo magnitude will depend on the factor $\exp(-\tau 2/T1)$.

Gradient echo: The gradient echo (GRE) pulse sequence is also one of the most often used in modern MR imaging. It was the first fast MR pulse sequence available, and its introduction in the 1980s permitted for the first time MR image acquisition times in the range of seconds rather than minutes. Several variations of the basic GRE sequence have been devised and are known by many names and acronyms [140–141]. Basically, a GRE is generated by applying a pair of bipolar gradient pulses. In the basic gradient echo sequence there is no refocusing 180° pulse and the data are sampled during a gradient echo, which is achieved by dephasing the spins with a negatively pulsed gradient before they are rephased by an opposite gradient with opposite polarity to generate the echo. The gradient echo does not refocus the effects of main field inhomogeneity and therefore is generally used with a short echo time [140].

The pulse sequence begins with an excitation pulse, termed the *alpha* (α) *pulse*. This pulse tilts the magnetization by a flip angle α, which is typically between 0 and $90°$. In the special case where $\alpha = 90°$ the sequence is identical to the so-called *partial saturation* or *saturation recovery* pulse sequence [141]. The flip angle can also be slowly increased during data acquisition. The variable flip angle approach is called *tilted optimized nonsaturation excitation* or TONE [141]. Then, instead of acquiring data in a steady state where z-magnetization recovery and destruction by pulses are balanced, the data are acquired such that all of the z-magnetization is expended during imaging by tilting a little more of the remaining z-magnetization into the xy-plane for each acquired imaging response line. The readout or acquisition stage occurs during an FID once the read gradient pulse is turned on to enable localization of the signal in the readout direction. To achieve this goal, the data are sampled during the "gradient echo," which is achieved by properly dephasing the spins before they are rephased by an equal but opposite gradient to generate the echo, when the areas under the negative and positive gradients are equal. The final portion of the sequence may involve introducing additional gradients, or RF pulses with the aim to "spoil" any remaining xy-magnetization or to refocus the xy-magnetization at the moment when the spin system encounters the next pulse.

In conventional GRE imaging, the described basic pulse sequence is repeated for each image line that is acquired. As a result of the short repetition time, the z-magnetization cannot fully recover, and after a few initial α-pulses an equilibrium is established between z-magnetization recovery and reduction. Ultrafast GRE sequences are obtained by reducing the repetition time (*TR*), which results in image acquisitions of less than 1 sec and typically less than 500 ms. Such sequences are often labeled with the prefix "turbo" (e.g., turboFLASH, turboFFE, turboGRASS [141]). Most 3-D pulse sequences use gradient echoes because their short minimum TR allows the acquisition to be completed in several minutes or less. Disadvantages of gradient echo imaging are compromised anatomic details and artifacts in regions with varying susceptibility (e.g., between the air-containing sinuses and brain and especially between fluid from trauma and normal tissue).

Contrast enhancement: Both *T1*-weighted and *T2*-weighted images are acquired for most MRI studies. However, they do not always adequately show the desired anatomy or pathology. One option to improve contrast is to utilize advanced image acquisition techniques such as fat suppression or chemical-shift imaging [139]. Another option is to administer a contrast agent, typically a substance with specific magnetic properties, to delineate areas of interest. In clinical imaging, most commonly a paramagnetic contrast agent, usually a gadolinium compound [139] is given. Gadolinium-enhanced tissues and fluids appear extremely bright on *T1*-weighted images. This provides high sensitivity for detection of vascular tissues (e.g., tumors) and permits assessment of brain perfusion (e.g., in stroke). More recently, superparamagnetic contrast agents (e.g., iron oxide nanoparticles [142]) have become available. These agents appear very dark on $T2^*$-weighted images.

Contrast agent mechanisms: MR contrast agents are employed to alter the biochemical environment inside the subject in a manner that results in shortening the relaxation times of water protons in a desired local region. Different contrast agents affect *T1* and *T2* differently. Paramagnetic ions such as Gd^{3+}, in chelated form, mainly shorten *T1*. Just as high molar concentrations of water protons are required to provide sufficient signal intensities, a high concentration of an MR contrast agent is needed to induce sufficient signal modulation. For example, compared to radionuclide or optical methods, MR requires 10^4–10^6 times higher concentration of contrast agents, which complicates the detection of small molecules or subtle processes such as gene expression. Thus, MR contrast agents are especially useful where high local concentration is achieved, such as to differentiate vessels and interstitial spaces of diseased and normal tissues. The use of polychelates of paramagnetic ions or superparamagnetic compounds such as iron oxide nanoparticles containing a crystalline core of oxygen and ferrous (Fe^{2+}) and ferric (Fe^{3+}) ions coated with dextran can further increase the resulting signal and generate an MR gene expression signal at relatively low concentrations (<100 µM Fe) in tissue.

Targeted contrast agents: There has been some work to develop targeted MR contrast agents for molecular imaging of certain receptor systems at relatively high molar concentrations. To be able to detect less abundant targets associated with transgene expression *in vivo* requires efficient signal amplification strategies that depend on the specific gene product to be imaged [127], described in more detail in another chapter of this book. These schemes ultimately involve MR contrast agents involving paramagnetic Gd^{3+} [136] and superparamagnetic Fe^{2+} and Fe^{3+} [142] ions, the latter in oxide forms FeO ("iron oxide") and Fe_2O_3 ("magnetite"), respectively. Several strategies have been used to bring these ions onto or within cells. One robust signal amplification approach exploits the overexpression of an iron-internalizing receptor transgene and accumulation of targeted iron oxide nanoparticles in endosomes within cells [142]. Another scheme overexpresses a target protein within the cell that stores iron [143] (e.g., see Figure 10d,e). These mechanisms change the magnetic characteristics of the cells in such a way that *T2*-weighted MRI signal at longer echo times shows a significant reduction in signal compared with cells not overexpressing that receptor, resulting in marked hypointensity. Activatable gene expression strategies are possible by exploiting enzyme–substrate mechanisms and the need for Gd^{3+} to interact with water to generate a signal change based upon a relaxation enhancement resulting in increased contrast [136]. Another potential contrast mechanism to delineate gene expression is based upon chemical exchange saturation transfer (CEST) [144].

Pulse sequences for gene expression detection: The pulse sequences used for *in vivo* MRI of transgene expression of appropriate targeted contrast agents typically involve a mixture between *T1*- and *T2*-weighted spin echo and gradient echo sequences at different time points after intravenous administration of the contrast agent [126, 127, 134–137, 142–143]. These studies often employ a special surface coil that covers the length of the mouse (~5 inches) (e.g., Figure 7.10b). Important parameters to vary are the repetition time TR, typically on the order of seconds, and time to echo (*TE*), typically on the order of milliseconds to minutes.

Image formation: The demodulated MR signal $S(t)$ generated by freely precessing nuclear spins in the presence of an applied linear magnetic field gradient G equals the Fourier transform of the *effective spin density* [131]. In other words, as time progresses the signal traces out a trajectory in k-space with the velocity vector of the trajectory proportional to (aligned with) G. Effective spin density is the true proton spin density corrected for the effects of *T1* preparation, *T2* decay, dephasing due to field inhomogeneity, flow, diffusion, and any other phenomena that affect the magnitude of transverse magnetization available to induce signal in the RF receiver coil. From the basic k-space formalism, one can reconstruct an image I simply by taking the inverse Fourier transform of the sampled data [131]:

$$I(\mathbf{x}) = \int S(\mathbf{k}(t)) \cdot e^{-i2\pi \mathbf{k} \cdot \mathbf{x}} d^3 k, \qquad (7.6)$$

where $\mathbf{k}$ is a vector with components in each of the three gradient directions (x, y, z) represented by the vector $\mathbf{x}$, and t is the gradient on-time. The signal location in k-space is the integral of the gradient amplitude over time:

$$\mathbf{k} \equiv \gamma \int_0^t \mathbf{G}(t') dt'. \qquad (7.7)$$

As the gradient-time product increases, that is, as the signal is encoded to higher k values, the image resolution increases. Thus, to form an MR image of any desired final spatial resolution requires collecting MR data over an appropriate corresponding area of k-space.

Because $\mathbf{x}$ and $\mathbf{k}$ are conjugate variables (with respect to the Fourier transform) one can use the Nyquist theorem [145] to show that the step size Δk in k-space determines maximum frequency that is correctly sampled, and thus the FOV of the image (i.e., FOV $\propto 1/\Delta k$, where

k is the magnitude of $\mathbf{k}$). Likewise, k_{max}, the maximum value of k sampled, determines the spatial resolution R (i.e., $R \propto 1/k_{max}$) [132]. These relationships apply to each axis (x, y, and z) independently.

MRI Performance Issues

Spatial resolution: Spatial resolution, or reconstructed image voxel size of an MR image depends on the integral product of the imaging gradient amplitudes and the duration of time they are switched on, referred to as the "on-time"; the highest resolution available is given by the maximum gradient amplitude–time product in the raw data. Specifically, the pixel size is roughly equal to $1/\gamma Gt$, where γ is the Larmor constant (4258 Hz/gauss), G is the gradient amplitude, usually expressed in gauss/cm (1 Gauss $= 10^{-4}$ T), and t is the gradient "on-time." For example, a gradient of 0.5 gauss/cm, left on for 10 msec, yields a spatial resolution of 0.47 mm along one in-plane dimension only – the "readout" direction (e.g., x). In standard 2-D Fourier transform MRI, the spatial encoding for the second in-plane dimension (e.g., y) is created by applying a brief gradient pulse in the y-direction before each readout line. For 128 resolution lines along this axis, 128 separate lines must be acquired, each for 10 msec. The total readout duration is therefore 128 × 10 msec, or 1.28 sec. Unfortunately, the MR signal lasts for only roughly 100 msec (limited by *T2*), and over the course of a 1.28-sec readout duration (the spatial encoding period) the signal will have completely decayed. As another example, for rapid "one-shot" EPI acquisitions used for fMRI, gradient sets reaching amplitudes of up to ~4 gauss/cm with a rise time of ~180 μsec using a sinusoidal waveform yield roughly 3 mm pixel size in the readout axis.

Increases in spatial resolution require either increases in gradient amplitude, duration, or both, but neither is easy to obtain. Switching gradients rapidly to very high amplitudes can cause tissue heating. Increasing the duration of the gradient pulses lowers the effective image bandwidth and increases image susceptibility to non-linearities/shape distortion and other artifacts.

There are k-space-encoding schemes that can be exploited to improve spatial resolution [129–132]. Increases in resolution along the phase-encoding axis are achieved simply by extending the total duration of the readout. This increases the total displacement along k_y (phase-encoding k axis) at the cost of a decrease in bandwidth and an increase in minimum echo time. For example, doubling the encoding period reduces the pixel size and the bandwidth, per pixel, by a factor of two. Shape distortions from field inhomogeneity remain constant, although they will cover twice as many pixels.

Another way to increase spatial resolution along the readout axis results from a symmetry property of k-space. This symmetry property implies that it is necessary to acquire only half of the entire MR raw data space to form a complete image. A very efficient way to achieve high resolution in a single-shot EPI experiment is to use a long readout duration along k_y and to acquire only the positive (or negative) values in k_x. It is then a relatively simple matter to calculate the missing data corresponding to the uncollected portion of the image and then to Fourier transform the entire raw data set to form a complete image.

SNR: SNR in MRI ultimately determines the sensitivity for gene expression detection. SNR is a function of [131, 141] (1) available transverse magnetization; this factor is ultimately limited by the population difference between the proton spin states but is strongly affected by the pulse sequence used and inherent contrast; (2) field strength; (3) imaging time, or more precisely, time spent receiving the signal; (4) bandwidth, or, essentially, the signal sampling rate; (5) RF coil loading, coupling, and sensitivity; and (6) voxel volume. The SNR of MR acquisition is often simply stated in the form of a scaling relationship that depends only on the voxel volume and the total acquisition time:

$$SNR \propto \Delta x \cdot \Delta y \cdot \Delta z \sqrt{T_{\text{acq, total}}}.$$

For 3-D imaging, because the entire 3-D slab of slices is reconstructed with a single Fourier transform, the total acquisition time includes the number of slice encoding steps, which for nuclei with short $T1$ is much greater for 3-D compared to 2-D acquisitions [141]. For $T1$ that is much longer than the sequence time period, there is no SNR advantage of 3-D over 2-D [141].

CNR: CNR $\equiv (S_A - S_B)/noise$, where S_A and S_B are signal intensities of tissues A and B. The *noise* is inherent fluctuations in the image data and can arise from imaging hardware as well as from the tissue itself. The image *noise* is typically defined as the measured standard deviation of an appropriate region of interest (ROI) in the background of the image [132].

MULTIPLE IMAGING MODALITY APPROACH TO REPORTER GENE EXPRESSION IMAGING

We have described five imaging modalities that have shown strong promise for *in vivo* reporter gene expression imaging. There is no best modality for *in vivo* gene expression imaging, and one may have to use a combination of more than one imaging modality and contrast strategy to answer the questions of interest [128]. Combining data from two or more *in vivo* imaging modalities can for example add anatomic and/or physiologic information to reporter gene imaging studies, enable time correlation of two or more distinct gene expression imaging strategies, or allow simultaneous imaging of gene expression with other complementary biological parameters by exploiting multiple molecular targets and/or probes. Software fusion of data from two separate

imaging modalities is possible with the help of anatomical or fiducial markers that allow spatial registration of the two image volumes, but such efforts are most successful for studies of organs and tissues that do not move with time, such as the brain [146].

The other approach for multimodality imaging is to develop a system that integrates more than one modality into a single instrument. Such a hybrid system allows either simultaneous (temporally and spatially registered) or sequential acquisitions with the different modalities. A clinical example of the power of multimodality imaging is the combining of PET and CT to localize primary, recurrent, and metastatic cancer throughout the body [147]. PET/CT is an ideal combination because the result is a tool that provides information that cannot be obtained as easily using the two modalities separately. PET is used to measure the increased metabolic or cellular activity of the cancer and CT is used to provide high-resolution visualization of the corresponding anatomy where the cancer resides. Adding CT to PET has the additional benefit of enhancing PET's accuracy and throughput by facilitating a rapid, low-noise, accurate estimate of photon attenuation coefficients [148–149]. Furthermore, the integrated PET/CT system does not compromise performance of either system. Clinical SPECT/CT systems that have become available recently will likely play important roles as well in characterizing diseases for which SPECT has desirable characteristics [150–151].

Although the only commercially available multimodality hybrid clinical systems are either PET/CT or SPECT/CT scanners, the greater system flexibility allowed in small-animal imaging research has resulted in the development of several high-resolution dual- and tri-modality systems such as PET/CT [152–154], SPECT/CT [155–157], PET/SPECT/CT [158–160], PET/MRI [161–172], PET/optical [173–175], and SPECT/optical [176–177]. Such multimodality systems facilitate a range of *in vivo* strategies to obtain rich, correlative information about the molecular basis of disease, such as reporter gene expression, and enhance interpretation and quantification capabilities of data from the individual modalities involved.

FUTURE DIRECTIONS FOR REPORTER GENE IMAGING SYSTEM TECHNOLOGIES

Although great progress has been made in the endeavor to image reporter gene expression in living subjects, we emphasize there are still substantial improvements needed to advance signal detection sensitivity and quantification accuracy for all the modalities described in this chapter. For preclinical imaging, improving signal detection capabilities sharpens the ability to study subtle biological signatures of disease and guide the discovery and development of new gene- and cell-based treatments. For clinical imaging, detecting and quantifying a smaller number of cells expressing a desired reporter gene could in the future impact early disease detection, tracking of cell proliferation and propagation, and monitoring the efficacy of novel therapies. Improving imaging system capabilities is also important because it can in some cases relax the challenging requirements on new reagents, molecular targets, and assays under study for more accurate reporter gene expression imaging. We leave discussion of future directions for reporter and assay development that enhance transgene expression imaging to other chapters in this book. Here we summarize a general list of challenges for reporter gene expression imaging system capabilities from the perspective of imaging system technology only.

Higher contrast: Better signal and lower background. If the level of background events that mimic the desired signal is high, the ability to detect a subtle transgene expression signal above background is diminished. It is thus critical to mitigate physical system-related background. For FLI instrumentation this means advancing excitation light source and fluorescent light source as well as filter technologies for more precise selection of specific bands of interest or development of techniques with insignificant or absent autofluorescence mechanisms. One of the reasons BLI imaging is so sensitive is that there is no excitation source and thus no background autofluorescence signal. However, advancing BL tomographic capabilities as well as resolution recovery may further improve the signal-to-background ratio. For radionuclide imaging (PET and SPECT), increasing photon sensitivity enables collection of more signal within a given study time. Improving detector energy resolution can mitigate measured background from photon Compton scatter in tissue as it enables one to use a narrower pulse height window to reduce scatter contamination without significant effect on the event acceptance statistics. For PET, improving detector coincidence time resolution down to a few hundred picoseconds or less can further reduce the effects of contamination from background random and multiple (>2) photon coincidences, enhance contrast-to-noise ratio improvements available with ToF, and mitigate the effects of detector count rate saturation on image contrast. In MRI, better signal-to-background ratios and thus higher reporter gene signal sensitivities require further research in coil design, pulse sequences, and methods to achieve higher reporter gene signal intensities and lower noise. The discovery of probes/targets that result in increased accumulation of paramagnetic and especially superparamagnetic particles will likely play a role in signal amplification for gene expression imaging using MRI. Of course, advancing reconstructed spatial resolution for all these imaging modalities can also improve the resulting contrast between transgene expression in a miniscule cluster of cells and the background, when the cluster size approaches the system spatial resolution to mitigate partial volume effects.

Higher (e.g., $\leq 1\,mm$) spatial resolution. Higher system spatial resolution means an enhanced ability to resolve a subtle reporter gene expression signal emanating from a smaller localized cluster of cells (a smaller region of interest in an image) at any given background level. It is thus also critical to improve reconstructed spatial resolution of reporter gene expression for sources deep within heterogeneous tissue. The high spatial resolution should also be uniform throughout the subject. For the optical methods, to achieve ≤ 1 mm resolution in mouse models, continued efforts are needed to advance accuracy of photon migration models and other dispersion effects incorporated into the image reconstruction process as well as the reconstruction algorithm itself to recover ultrahigh, uniform spatial resolution throughout heterogeneous subject tissues. Reconstructing images at higher resolution might also require more efficient optical photon collection instrumentation, especially in the case of BL tomography. Due to physical barriers of light penetration in tissue, translation of optical imaging assays into humans requires increased efforts on the development of endoscopic methods that can probe accessible portions of certain internal organs with ultra-high resolution. For the radionuclide methods, continued efforts are required to develop higher-resolution photon collimation methods (electronic for PET, physical for SPECT) while at the same time increasing photon collection efficiency to acquire high statistics data in a reasonable study duration. Efforts to improve resolution modeling and image reconstruction strategies to reconstruct high spatial resolution images with excellent SNR are important. For PET, meeting these needs requires the development of systems with smaller photon detection elements while bringing the detectors closer to the subjects and/or covering more of the body with detectors. For SPECT, continued efforts are needed to advance physical collimator design to enable higher spatial resolution with adequate photon collection efficiency. Currently, only MRI can reach ≤ 1 mm resolution for gene expression imaging in rodents, although any further advancements in coil design and other features that enhance spatial resolution without reducing SNR could mean the possibility of visualizing fewer cells expressing the reporter gene of interest.

Better quantification of signal. For the role of *in vivo* imaging in biological studies that exploit transgene expression to gain importance and acceptance, continued evaluation of and enhancements to image data quantification are critical. Accurate quantification means that the reconstructed signal accurately reflects the real biological target-to-background ratio achieved by the reporter gene probe inside the subject. There are two basic sources of quantification error for *in vivo* imaging. The first is from systematic errors and statistical noise in the image acquisition process. These include effects from attenuated energy signal propagation through tissue, flaws in the imaging system signal detection process, and the limited amount of signal present. Equipment developers should continue to improve the spatial resolution and signal-to-background capabilities of hardware as described previously. The imaging equipment industry should also continue to develop practical, easy to use data correction and calibration methods for sources of physical artifacts such as photon attenuation, scatter, and nonuniform signal sensitivity. The field also needs continued development of practical image reconstruction algorithms that incorporate these corrections as well as accurate models of system resolution-blurring kernels to generate accurate, high-resolution images in practical reconstruction times with standard computational resources. These algorithms are key to contrast recovery and accurate quantification of small signals. The second systematic effect is from the analysis software used to quantify ROIs in the image data. The image analysis packages used to generate results that quantify the reporter gene signal from the images should be evaluated thoroughly for accuracy by equipment vendors as well as the sites using them. This includes studying the effects of different methods to draw ROIs and quantify ROIs and understanding effects of interuser variations. More robust software should be developed to reduce these systematic errors introduced in the image analysis procedure.

REFERENCES

1 Gambhir, S. S., Herschman, H. R., Cherry, S. R. et al. (2000). Imaging transgene expression with radionuclide imaging technologies. *Neoplasia* **2**: 118–138.

2 Weissleder, R., Mahmood, U. (2001). Molecular imaging. *Radiology* **219**: 316–333.

3 Luers, G. H., Jess, N., Franz, T. (2000). Reporter-linked monitoring of transgene expression in living cells using the ecdysone-inducible promoter system. *Eur. J. Cell Bio* **79**(9): 653–657.

4 Kain, S. R., Adams, M., Kondepudi, A., Yang, T. T. et al. (1995). Green fluorescent protein as a reporter of gene expression and protein localization. *BioTechniques* **19**: 650–655.

5 Prasher, D. C., Eckenrode, V. K., Ward, W. W. et al. (1992). Primary structure of the Aequorea Victoria green fluorescent protein. *Gene* **111**: 229–233.

6 Tsien, R. Y. (1998). The green fluorescent protein. *Annu Rev Biochem* **67**: 509–544.

7 Baird, G. S., Zacharias, D. A., Tsien, R. Y. (2000). Biochemistry, mutagenesis, and oligomerization of DsRed, a red fluorescent-protein from coral. *Proc Natl Acad Sci* **22**: 11984–11989.

8 Matz, M. V., Fradkov, A. F., Labas, Y. A. et al. (1999). Fluorescent proteins from nonbioluminescent Anthozoa species. *Nat Biotechnol* **10**: 969–973.

9 Zhang, J., Cambell, R. E., Ting, A. Y., Tsien, R. Y. (2002). Creating new fluorescent probes for cell biology. *Nat Rev Mol Cell Biol* **12**: 906–918.

10 Contag, C. H., Contag, P. R., Mullins, J. I., et al. (1995). Photonic detection of bacterial pathogens in living hosts. *Mol Microbiol* **18**: 593–603.

11 Contag, C. H., Spilman, S., Contag, P. (1997). Visualizing gene expression in living animals using a bioluminescent reporter. *Photochem Photobiol* **66**: 523–531.

12 Mahmood, U., Weissleder, R. (2003). Near-infrared optical imaging of proteases in cancer. *Mol Cancer Ther* **2**: 489–496.

13 Tung, C. H. et al. (2004). In vivo imaging of beta-galactosidase activity using far red fluorescent switch. *Cancer Res* **64**: 1579–1583.

14 West, J. L., Hlas, N. J. (2003). Engineered nanomaterials for biophotonics applications: improving sensing, imaging, and therapeutics. *Annu Rev Biomed Eng* **5**: 285–292.

15 Chance, B. (1998). Near-infrared images using continuous, phase-modulated, and pulsed light with quantitation of blood and blood oxygenation. *Ann New York Acad Sci* **838**: 29–45.

16 Levin, C. S. (2005). Primer on molecular imaging technology. *Eur J Nucl Med Mol Imaging* **32**: S325–S345.

17 Natasha, S. et al. (2001). Noninvasive functional optical spectroscopy of human breast tissue. *Proc Natl Acad Sci USA* **98**: 4420–4425.

18 Ishimaru, A. (1978). *Wave Propagation and Scattering in Random Media*. Academic Press: New York.

19 Cheong, W. F., Prahl, S. A., Welch, A. J. (1990). A review of the optical properties of biological tissues. *IEEE J. Quantum Electronics* **26**: 2166–2185.

20 Rice, B. W., Cable, M. D., Nelson, M. B. (2001). In vivo imaging of light-emitting probes. *J Biomed Opt* **6**(4), 432–440.

21 Bogdanov, A., Wiessledder, R. (2002). In vivo imaging of gene delivery and expression. *Trends in Biotech* **20**(8): S11–S18.

22 Ruthel, G., Ribot, W. J., Bavari, S., Hoover, T. A. (2004). Time-lapse confocal imaging of development of *Bacillus anthracis* in macrophages. *J Infect Dis* **189**: 1313–1316.

23 Ray, P., De, A., Min, J. J., Tsien, R. Y., Gambhir, S. S. (2004). Imaging tri-fusion multimodality reporter gene expression in living subjects. *Cancer Research* **64**: 1323–1330.

24 Campbell, R. E., Tour, O., Palmer, A. E. et al. (2002). A monomeric red fluorescent protein. *Proc Natl Acad Sci USA* **99**: 7877–7882.

25 Hoffman, R. M. (2002). *In vivo* imaging of metastatic cancer with fluorescent proteins. *Cell Death & Diff* **9**(8): 786–789.

26 Holst, G. C. (1998). *CCD Arrays, Cameras, and Displays*. SPIE: Bellingham, WA.

27 Wessleder, R., Ntziachristos, V. (2003). Shedding light onto live molecular targets. *Technol Trends Nat Med* **2**:123–128.

28 Nickell, S. et al. (2000). Anisotropy of light propagation in human skin. *Phys Med Biol* **45**: 2873–2886.

29 Roblyer, D., Richards-Kortum, R., Sokolov, K. et al. (2008). Multispectral optical imaging device for *in vivo* detection of oral neoplasia. *J Biomed Opt* **13**(2): 024019.

30 Miller, P. J., Hoyt, C. C. (1995). Multispectral imaging with a liquid crystal tunable filter. *Proc SPIE* **2345**: 354.

31 Duda, R. O., Hart, P. E., Stork, D. G. (2001). *Pattern Classification*. 2nd ed. John Wiley and Sons: New York.

32 Kak, A. C., Slaney, M. (1998). *Principles of Computerized Tomographic Imaging.* IEEE Press: New York.

33 Graves, E. E., Ripoll, J., Weissleder, R., Ntziachristos, V. (2003). A submillimeter resoution fluorescence molecular imaging system for small animal imaging. *Med Phys* 5: 901–911.

34 Ntziachristos, V., Bremer, C., Weissleder, R. (2003). Fluorescence imaging with near-infrared light: New technological advances that enable in vivo molecular imaging. *Eur Radiol* 13: 195–208.

35 Ntziachristos, V., Bremer, C., Graves, E. E. et al. (2002). In vivo tomographic imaging of near-infrared fluorescent probes. *Mol Imaging* 2: 82–88.

36 Ntziachristos, V., Tung, C. H., Bremer, C., Weissleder, R. (2002). Fluorescence molecular tomography resolves protease activity *in vivo. Nat Med* 7: 757–760.

37 Patwardhan, S. V., Bloch, S., Achilefu, S., Culver, J. P. (2006). Quantitative small animal fluorescence tomography using an ultrafast gated image intensifier. *Proc 28th IEEE EMBS Ann Intl Conf*, New York City, Aug. 30–Sept 3, 2006, 2675–2678.

38 Domañski, A. W. (2002). Optical sensors and microsystems. In: Martellucci, S., Chester, A. N., Mignani, A. G. (Eds.), *Optical Tomography: Techniques and Applications.* Springer: New York.

39 Hutchinson, C. L., Troy, T. L., Sevick-Muraca, E. M. (1996). Fluorescence lifetime determination in tissues and other random media from measurement of excitation and emission kinetics. *Applied Optics* 35: 2325–2332.

40 Milstein, A. B., Oh, S., Webb, K. J. et al. (2003). Flurorescence optical diffusion tomography. *Appl Opt* 42(16): 3081–3094.

41 Cai, W., Lax, W., Alfano, R. R. (2000). Cumulant solution of the elastic Boltzmann transport equation in an infinite uniform medium. *Phys Rev E Stat Phys Plasmas Fluids Relat Interdiscip Topics* 61(4A): 3871–3876.

42 Cai, W., Das, B. B., Liu, F. et al. (1996). Time-resolved optical diffusion tomographic image reconstruction in highly scattering turbid media. *Proc Nat Acad Sci USA* 93(24): 13561–13564.

43 Ye, J. C., Bouman, C. A., Webb, K. J., Millane, R. P. (2001). Nonlinear multigrid algorithms for Bayesian optical diffusion tomography. *IEEE Trans Imag Proc* 10(5): 909–922.

44 Oh, S., Milstein, A. B., Bouman, C. A. Webb, K. J. (2005). A general framework for nonlinear multigrid inversion. *IEEE Trans Imag Proc* 1: 125–140.

45 Ntziachristos, V., Weissleder, R. (2001). Experimental three-dimensional fluorescence reconstruction of diffuse media by use of a normalized born approximation. *Optics Letters* 26: 893–895.

46 Arridge, S. R. (1993). Diffuse optical tomography. In: Muller, G. (Ed.), *Medical Optical Tomography: Functional Imaging and Monitoring.* for Advanced /Optical/ Technologies Series, Vol. IS11, SPIE /Optical/ Engineering Press: 31–64.

47 Pogue, B. W., McBride, T. O. et al. (1999). Comparison of imaging geometries for diffuse optical tomography of tissue. *Opt Express* 4: 270–286.

48 Schmidt, F. E. W., Fry, M. E., Hillman, E. M. C. et al. (2000). A 32-channel time-resolved instrument for medical optical tomography. *Rev Sci Instr* 71(1): 256–265.

49 ART Advanced Research Technologies Inc. (2003). *Preclinical optical molecular imager.* (white paper). September 2003. (http://www.art.ca).

50 Culver, J. P., Choe, R., Holboke, M. J. et al. (2003). 3d diffuse optical tomography in the plane parallel transmission geometry; Evaluation of a hybrid frequency domain/continuous wave clinical system for breast imaging. *Medical Physics* 30(2): 235–247.

51 Yu, G., Durduran, T., Furuya, D. et al. (2003). Frequency-domain multiplexing system for in vivo diffuse light measurements of rapid cerebral hemodynamics. *Appl Opt* 42(16): 2931–2939.

52 Hawrysz, D. J., Sevick-Muraca, E. M. (2000). Developments towards breast cancer imaging using near-infrared optical measurements and fluorescence contrast agents. *Neoplasia* 2: 388–417.

53 Sevick-Muraca, E. M., Houston, J. P., Gurfinkel, M. (2002). Fluorescence-enhanced, near infrared diagnostic imaging with contrast agents. *Curr Opin Chem Biol* 6: 642–650.

54 Corlu, A., Durduran, T., Choe, R. et al. (2003). Uniqueness and wavelength optimization in continous-wave multispectral diffuse optical tomography. *Opt Lett* 28: 2339–2341.

55 Wang, G., Hoffman, E. A. et al. (2003). Development of the first bioluminescent CT scanner. *Radiology* 229: 566–572.

56 Corlu, A. (2007). *Multi-spectral and fluorescence diffuse optical tomography of breast cancer.* Ph.D. Dissertation, University of Pennsylvania, 2007.

57 Chaudhari, A. J. (2006). *Hyperspectral and multispectral optical bioluminescence and fluorescence tomography in small animal imaging.* Ph.D. Dissertation, University of Southern California, 2006.

58 Psycharakis, S., Zacharakis, G., Garofalakis, A. et al. (2007). Autofluorescence removal from fluorescence tomography data using multispectral imaging. *Proc SPIE* 6626: 662601–662607.

59 Chaudhari, A. J., Darvas, F., Bading, J. R. et al. (2005). Hyperspectral and multispectral bioluminescence optical tomography for small animal imaging. *Phys Med Biol* 50: 5421–5441.

60 Jain, R. K., Munn, L. I., Fukumura, D. (2002). Dissecting tumour pathophysiology using intravital microscopy. *Nat Rev Cancer* 4: 266–276.

61 Dobschuetz, E., Pahernik, S., Hoffmann, T. et al. (2004). Dynamic intravital fluorescence microscopy – A novel method for the assessment of microvascular permeability in acute pancreatitis. *Microvasc Res* 67(1): 55–63.

62 Kimura, T., Muguruma, N., Ito, S. et al. (2007). Infrared fluorescence endoscopy for the diagnosis of superficial gastric tumors. *Gastrointest Endosc* 66(1): 37–43.

63 Pawley, J. B. (1995). *Handbook of Biological Confocal Microscopy.* (2nd ed.) Plenum Press: New York.

64 Nakano, A. (2002). Spinning-disk confocal microscopy – a cutting-edge tool for imaging of membrane traffic. *Cell Struct Funct* 27(5): 349–355.

65 Williams, R. M., Zipfel, W. R., Webb, W. W. (2001). Multi-photon microscopy in biological research. *Curr Opin Chem Biol* 5: 603–608.

66 Oheim, M. et al. (2001). Two-photon microscopy in brain tissue: parameters influencing the imaging depth. *J Neurosci Meth* 111: 29–37.

67 Mason, W. T. (ed.) (1999). *Fluorescent and Luminescent Probes for Biological Activity*. Academic Press: London.

68 Contag, C. H., Jenkins, D., Contag, P. R., Negrin, R. S. (2000). Use of reporter genes for optical measurements of neoplastic disease *in vivo*. *Neoplasia* 2, 41–52.

69 Zhao, H., Doyle, T. C., Coquoz, O. et al. (2005). Emission spectra of bioluminescent reporters and interaction with mammalian tissue determine the sensitivity of detection in vivo. *Journal of Biomedical Optics* 10(4).

70 Wang, G., Li, Y., Jiang, M. (2004). Uniqueness theorems in bioluminescence tomography. *Med Phys* 31(8): 2289–2299.

71 Gu, X., Zhang, Q., Larcom, L., Jiang, H. (2004). Three-dimensional bioluminescence tomography with model-based reconstruction. *Opt Express* 12(17): 3996–4000.

72 Weissleder, R., Tung, C. H., Mahmood, U., Bogdanov, A. (1999). *In vivo* imaging of tumors with protease-activated near-infrared fluorescent probes. *Nat Biotechnol* 17: 375–378.

73 Qi, J., Leahy, R. M., Cherry, S. R. et al. (1998). High resolution 3D Bayesian image reconstruction using the microPET small animal scanner. *Phys Med Biol* 43: 1001–1013.

74 Alession, A. M., Kinahan, P. E. (2006). Improved quantitation for PET/CT image reconstruction with system modeling and anatomical priors. *Med Phys* 33(11): 4095–4103.

75 Tjuvajev, J. G., Stockhammer, G., Desai, R. et al. (1995). Imaging the expression of transfected genes in vivo. *Cancer Res* 55: 6126–6132.

76 Gambhir, S. S., Barrio, J. R., Herschman, H. R., Phelps, M. E. (1999). Assays for noninvasive imaging of reporter gene expression. *Nucl Med Biol* 26: 481–490.

77 Soghomonyan, S. A., Doubrovin, M., Pike, J. et al. (2005). Positron emission tomography (PET) imaging of tumor-localized Salmonella expressing HSV1-TK. *Cancer Gene Ther* 12: 101–108.

78 Bettegowda, C., Foss, C. A., Chong, I. et al. (2005). Imaging bacterial infections with radiolabeled 1-(2′deoxy-2′-fluoro-beta-D-arabinofuranosyl)-5-iodouracil. *Proc Natl Acad Sci USA* 102: 1145–1150.

79 Ivanova, A., Ponomarev, V., Ageyeva, L. et al. (2002). Imaging adoptive stem cell therapy with HSV-tk/GFP reporter gene. *Mol Imaging* 1: 208–209.

80 Koehne, G., Doubrovin, M., Doubrovina, E. et al. (2003). Seriel in vivo imaging of the target migration of human HSV-TK-transduced antigen-specific lymphocytes. *Nat Biotechnol* 21: 405–413.

81 Dubey, P., Su, H., Adonai, N. et al. (2003). Quantitative imaging of the T cell anti-tumor response by positron emission tomography. *Proc Natl Acad Sci USA* 100: 1232–1237.

82 Minn, A. J., Kang, Y., Serganova, I. et al. (2005). Distinct organ-specific metastatic potential of individual breast cancer cells and primary tumors. *J Clin Invest* 115: 44–55.

83 Ponomarev, V., Doubrovin, M., Lyddane, C. et al. (2001). Imaging TCR-dependent NFAT-mediated T-cell activation with positron emission tomography *in vivo*. *Neoplasia* 3: 480–488.

84 Wernick, M., Aarsvold, J. (eds.). (2004). *Emission Tomography: The Fundamentals of PET and SPECT*. Elsevier Academic Press: San Diego.

85 Levin, C. S., Zaidi, H. (2007). Current trends in preclinical PET system design. In Zaidi, Alavi, A. (Eds.), *PET Instrumentation and Quantification* 2(2): 125–160.

86 Tai, Y. C., Chatziioannou, A. et al. (2001). Performance evaluation of the microPET P4: a PET system dedicated to animal imaging. *Phys Med Biology* 46(7): 1845–1862.

87 Tai, Y. C., Ruangma, A., Laforest, R., Siegel, S., Newport, D. F. (2003). Performance evaluation of the microPET (R) Focus: A second generation small animal PET system. *J Nucl Med* 44(5): 159–160.

88 Karp, J. S., Surti, S., Daube-Witherspoon, M. E. et al. (2003). Performance of a brain PET camera based on anger-logic gadolinium oxyorthosilicate detectors. *J Nucl Med* 44(8): 1340–1349.

89 Bettinardi, V., Danna, M., Savi, A. (2004). Performance evaluation of the new whole-body PET/CT scanner: Discovery ST. *European Journal of Nucl Med and Mol Im* 31(6): 867–881.

90 Schmand, M., Eriksson, L., Casey, M. E., Wienhard, K., Flugge, G., Nutt, R. (1999). Advantages using pulse shape discrimination to assign the depth of interaction information (DOI) from a multi layer phoswich detector. *IEEE Trans Nucl Sci* 46(4): 985–990.

91 Seidel, J., Vaquero, J. J., Green, M. V. (2003). Resolution uniformity and sensitivity of the NIH ATLAS small animal PET scanner: Comparison to simulated LSO scanners without depth-of-interaction capability. *IEEE Trans Nucl Sci* 50(5): 1347–1350.

92 Levin, C. S., Dahlbom, M., Hoffman, E. J. (1995). A Monte Carlo correction for the effect of Compton scattering in 3-D PET brain imaging. *IEEE Trans Nucl Sci* 42: 1181–1188.

93 Levin, C. S., Tai, Y.-C., Hoffman, E. J., Dahlbom, M. et al. (1995). Removal of the effect of Compton scattering in 3-D whole body positron emission tomography by Monte Carlo. *1995 IEEE MIC Conf Rec* II: 1050–1054.

94 Kinahan, P. E., Rogers, J. G. (1989). Analytic 3D image reconstruction using all detected events. *IEEE Trans Nucl Sci* 36: 964–968.

95 Defrise, M., Kinahan, P. E., Townsend, D. W., Michel, C., Sibomana, M., Newport, D. F. (1997). Exact and approximate rebinning algorithms for 3-D PET data. *IEEE Trans Med Imaging* 16(2): 145–158.

96 Lange, K., Carson, R. (1984). EM reconstruction algorithms for emission and transmission tomography. *J Comput Assist Tomography* 8: 306–316.

97 Vardi, Y., Shepp, L. A., Kaufman, L. (1985). A statistical model for positron emission tomography. *J Amer Stat Assoc* 80: 8–37.

98 Hebert, T., Leahy, R. (1989). A generalized EM algorithm for 3-D Bayesian reconstruction from Poisson data using Gibbs priors. *IEEE Trans Med Imaging* 8: 194–202.

99 Green, P. J. (1982). Bayesian reconstructions from emission tomography data using a modified EM algorithm. *IEEE Trans Med Imaging* 1: 113–122.

100 Hudson, H. M., Larkin, R. S. (1994). Accelerated image reconstruction using ordered subsets of projection data. *IEEE Trans Med Imaging* 13: 601–609.

101 Liu, X., Comtat, C., Michel, C. et al. (2001). Comparison of 3-D reconstruction with 3D-OSEM and FORE+OSEM for PET. *IEEE Trans Med Imag* 20: 804–814.

102 Qi, J., Leahy, R. M. (2006). Iterative reconstruction techniques in emission computed tomography. *Phys Med Biol* 51: R541–578.

103 Levin, C. S., Hoffman, E. J. (1999). Calculation of positron range and its effect on the fundamental limit of positron emission tomography system spatial resolution. *Phys Med Biol* 44: 781–799.

104 Levin, C. S. (2002). Design of a high resolution and high sensitivity scintillation crystal array for PET with nearly complete light collection. *IEEE Trans Nucl Sci* 45(5): 2236–2243.

105 Tai, Y. C., Chatziioannou, A. F., Yang, Y. F., Silverman, R. W., Meadors, K., Siegel, S., Newport, D. F., Stickel, J. R., Cherry, S. R. (2003). The MicroPET II: design, development and initial performance of an improved microPET scanner for small-animal imaging. *Phys Medicine Biol* 48(11): 1519–1537.

106 Miyaoka, R. S., Kohlmyer, S. G., Lewellen, T. K. (2001). Performance characteristics of micro crystal element (MiCE) detectors. *IEEE Trans Nucl Sci* 48(4;2): 1403–1407.

107 Levin, C. S. (2008). New imaging technologies to enhance the molecular sensitivity of positron emission tomography. *Proc IEEE* 2008(96): 439–467.

108 Wu, H., Pal, D., O'Sullivan, J. A., Tai, Y. C. (2008). A feasibility study of a prototype PET insert device to convert a general-purpose animal PET scanner to higher resolution. *J Nucl Med* 49: 79–87.

109 Tai, Y. C., Wu, H., Pal, D., O'Sullivan, J. A. (2008). Virtual-pinhole PET. *J Nucl Med* 49: 471–479.

110 Pratx, P. D., Olcott, G., Chinn, G., and Levin, C. S. (2009). Accelerated Fully 3-D List-Mode OSEM for High-Resolution PET using Graphics Processing Units. *IEEE Transactions in Medical Imaging* 28(3): 435–445.

111 Anderson, J., Oz, O., Brandon, D. et al. (2008). Initial evaluation of a new spatial resolution-recovery method for PET reconstruction. *J Nucl Med* 49(Suppl 1): 391.

112 Moses, W. W. (2003). Time of flight PET revisited. *IEEE Trans Nucl Sci* 50: 1325–1330.

113 Karp, J. S., Surti, S., Daube-Witherspoon, M. E., Muehllehner, G. (2008). Benefit of time-of-flight in PET: Experimental and clinical results. *J Nucl Med* 49(3): 462–470.

114 Surti, S., Karp, J. S., Popescu, L. M. et al. (2006). Investigation of time-of-flight benefit for fully 3-D PET. *IEEE Trans Med Imag* 25: 529–538.

115 Conti, M. (2006). Effect of randoms on signal-to-noise ratio in TOF PET. *IEEE Trans Nucl Sci* 53: 1183–1193.

116 Budinger, T. F. (1983). Time-of-flight positron emission tomography; status relative to conventional PET. *J Nucl Med* 24: 73–78.

117 Schramm, N. U., Ebel, G., Engeland, U. et al. (2003). High resolution SPECT using multi-pinhole collimation. *IEEE Trans Nucl Sci* 51: 757–763.

118 Beekman, F. J., Vastenbouw, B. (2004). Design and simulation of a high-resolution stationary SPECT system for small animals. *Phys Med Biol* 49: 4579–4592.

119 Weisenberger, A. G., Wojcik, R., Bradley, E. L., Brewer, P., Majewski, S., Qian, J., Ranck, A., Saha, M. S., Smith, M. F., Welsh, R. E. (2003). SPECT-CT system for small animal imaging. *IEEE Trans Nucl Sci* 50(1): 74–79.

120 MacDonald, L. R., Patt, B. E., Iwanczyk, J. S. et al. (2001). Pinhole SPECT of mice using the LumaGEM Gamma Camera. *IEEE Trans Nucl Sci* 48(3).

121 Cherry, S. R., Sorenson, J. A., Phelps, M. E. (Eds.). (2003). *Physics in Nuclear Medicine*. (3rd. ed.). Elsevier Science: Philadelphia.

122 Levin, C. S. (2003). Detector design issues for compact nuclear emission cameras dedicated to breast imaging. *Nucl Inst Meth A* 497(1): 60–74.

123 Graham, L. S., Levin, C. S., Muehllehner, G. (2003). Anger scintillation camera. In: Sandler, M. P., Coleman, R. E., Patton, J. A., Wackers, F. J., Gottschalk, A. (Eds.), *Diagnostic Nuclear Medicine*. 4th ed. Lippincott Williams & Wilkins: Philadelphia.

124 Lewellen, T. K., Miyaoka, R. S., Jansen, F., Kaplan, M. S. (1997). A data acquisition system for coincidence imaging using a conventional dual-headed gamma camera. *IEEE Trans Nucl Sci* 44(3): 1214–1218.

125 Vastenhouw, B., Beekman, F. (2007). Submillimeter total-body murine imaging with U-SPECT-I. *J Nucl Med* 48: 487–493.

126 Gilad, A. A., Winnard, P. T., van Zijl, P. C. M., Bulte, J. W. M. (2007). Developing MR reporter genes: promises and pitfalls. *NMR Biomed* 20: 275–290.

127 Hogemann, D., Basilion, J. P. (2002). Seeing inside the body: MR imaging of gene expression. *Eur J Nucl Med* 29(3): 400–408.

128 Massoud, T. F., Gambhir, S. S. (2003). Molecular imaging in living subjects: Seeing fundamental biological processes in a new light. *Genes & Develop* 17: 545–580.

129 Smith, R. C., Lange, R. C. (1997). *Understanding Magnetic Resonance Imaging*. CRC Press, Taylor & Francis Group, New York, NY.

130 Mitchell, D. G. (1999). *MRI Principles*. W.B. Saunders Company Philadelphia, PA.

131 Liang, Z.-P., Lauterbur, P. C. (2000). Principles of Magnetic Resonance Imaging: A Signal Processing //Perspective/. IEEE Press, New York, NY.

132 Vlaardingerbroek, M. T., den Boer, J. A. (2004). *Magnetic Resonance Imaging*. Springer Verlag Berlin, Germany.

133 Jin, J. (1998). *Electromagnetic Analysis and Design in Magnetic Resonance Imaging*. Taylor & Francis New York, NY.

134 Weissleder, R., Moore, A., Mahmood, U. et al. (2000). *In vivo* magnetic resonance imaging of transgene expression. *Nat Med* 6(3): 351–354.

135 Hogemann, D., Josephson, L., Weissleder, R., Basilion, J. P. (2000). Improvement of MRI probes to allow efficient detection of gene expression. *Bioconj Chem* 11: 941–946.

136 Louie, A. Y., Huber, M. M., Ahrens, E. T. et al. (2000). *In vivo* visualization of gene expression using magnetic resonance imaging. *Nat Biotech* 18: 321–325.

137 Gilad, A. A., McMahon, M. T., Walczak, P. et al. (2007). Artificial reporter gene providing MRI contrast based on proton exchange. *Nat Biotech* 25(2): 217–219.

138 Hoge, W. S., Brooks, D. H., Madore, B., Kyriakos, W. E. (2005). A tour of accelerated parallel MR imaging from a linear systems perspective. *Conc Magn Reson* 27A: 17–37.

139 Brown, M. A., Semelka, R. C. (2003). *MRI: Basic Principles and Applications*. 3rd ed. Wiley-Liss. New York, NY.

140 Haacke, E. M., Brown, R. W., Thomson, M. R., Venkatesan, R. (1999). *Magnetic Resonance Imaging. Physical Principles and Sequence Design*. Wiley-Liss (John Wiley & Sons): New York.

141 Bernstein, M. A., King, K. F., Zhou, X. J. (2004). *Handbook of MRI Pulse Sequences*. Academic Press Amsterdam, Netherlands

142 Shen, T., Weissleder, R., Papisov, M. et al. (1993). Monocrystalline iron oxide nanocompounds (MION): physicochemical properties. *Magn Reson Med* 29: 599–604.

143 Genove, G., DeMarco, U., Xu, H. et al. (2005). A new transgene reporter for in vivo magnetic resonance imaging. *Nat Med* 11(4): 450–454.

144 Gilad, A. A., McMahon, M. T., Walczak, P. et al. (2007). Artificial reporter gene providing MRI contrast based on proton exchange. *Nat Biotech* 25(2): 217–219.

145 Roberts, R. A., Barton, B. F. (1965). *Theory of Signal Detectability: Composite Deferred Decision Theory*. R. A. Roberts and B. F. Barton, Theory of Signal Detectability: Composite Deferred Decision Theory. Ann Arbor, MI: Univ. of Michigan, 1965.

146 Hill, D. L. G., Batchelor, P. G., Holden, M., Hawkes, D. (2001). Medical image registration. *Phys Med Biol* 46: R1–45.

147 Schoder, H., Erdi, Y. E., Larson, S. M., Yeung, H. W. (2003). PET/CT: a new imaging technology in nuclear medicine. *European J Nucl Med and Mol Imaging* 30 (10): 1419–1437.

148 Kinahan, P. E., Townsend, D. W., Beyer, T., Sashin, D. (1998). Attenuation correction for a combined 3D PET/CT scanner. *Medical Physics* 25(10): 2046–2053.

149 Townsend, D. W., Beyer, T. (2002). A combined PET/CT scanner: the path to true image fusion. *British Journal of Radiology* 75: S24–30.

150 Forster, G. J., Laumann, C., Nickel, O. et al. (2003). SPECT/CT image co-registration in the abdomen with a simple and cost-effective tool. *Eur J Nucl Med and Mol Im* 30(1): 32–39.

151 Keidar, Z., Israel, O., Krausz, Y. (2003). SPECT/CT in tumor imaging: Technical aspects and clinical applications. *Seminars in Nuclear Medicine* 33(3): 205–218.

152 Goertzen, A. L., Meadors, A. K., Silverman, R. W., Cherry, S. R. (2002). Simultaneous molecular and anatomical imaging of the mouse in vivo. *Phys Med Biol* 47: 4315–4328.

153 Liang, H., Yang, Y., Yang, K., Wu, Y., Boone, J. M., Cherry, S. R. (2007). A microPET/CT system for in vivo small animal imaging. *Phys Med Biol* 52: 3881–3894.

154 Bérard, P., Riendeau, J., Pepin, C. et al. (2007). Investigation of the LabPET^{TM} detector and electronics for photon-counting CT imaging. *Nucl Instr Meth A* 571: 114–117.

155 Goertzen, A. L., Jones, D. W., Seidel, J. et al. (2005). First results from the high-resolution mouseSPECT annular scintillation camera. *IEEE Trans Med Imaging* 24: 863–867.

156 Kim, H., Furenlid, L. R., Crawford, M. J. et al. (2006). Semi-SPECT: a small-animal single-photon emission computed tomography (SPECT) imager based on eight cadmium zinc telluride (CZT) detector arrays. *Med Phys* 33: 465–474.

157 Madsen, M. T. (2007). Recent advances in SPECT imaging. *J Nucl Med* 48(4): 661–673.

158 Parnham, K. B., Chowdhury, S., Li, J., Wagenaar, D. J., Patt, B. E. (2006). Second-generation, tri-modality preclinical imaging system. *IEEE Nuclear Science Symposium Conference Record*, Oct. 29–Nov. 4, 2006, San Diego: 1802–1805.

159 Patt, B., Parnham, K., Li, J., Iwata, K., Vandehei, T. (2005). FLEX: Tri-modality small animal tomography combining PET, SPECT and CT in a single modular gantry [abstract]. *J Nucl Med* 46: 207.

160 Saoudi, A., Lecomte, R. (1999). A novel APD-based detector module for multi-modality PET/SPECT/CT scanners. *IEEE Trans Nucl Sci* 46: 479–484.

161 Farahani, K., Slates, R., Shao, Y., Silverman, R., Cherry, S. (1999). Contemporaneous positron emission tomography and MR imaging at 1.5 T. *J Magn Reson Imaging* 9: 497–500.

162 Shao, Y., Cherry, S. R., Farahani, K., Meadors, K. (1997). Simultaneous PET and MR imaging. *Phys Med Biol* 42: 1965–1970.

163 Marsden, P. K., Strul, D., Keevil, S. F., Williams, S. C., Cash, D. Simultaneous PET and NMR. *Br J Radiol* 75: S53–59.

164 Mackewn, J. E., Strul, D., Hallett, W. A. et al. (2005). Design and development of an MR-compatible PET scanner for imaging small animals. *IEEE Trans Nucl Sci* 52: 1376–1380.

165 Pichler, B. J., Judenhofer, M. S., Catana, C. et al. (2006). Performance test of an LSO-APD detector in a 7-T MRI scanner for simultaneous PET/MRI. *J Nucl Med* 47: 639–647.

166 Judenhofer, M. S., Catana, C., Swann, B. K. et al. (2007). Simultaneous PET/MR images, acquired with a compact MRI compatible PET detector in a 7 Tesla magnet. *Radiology* 244: 807–814.

167 Catana, C., Wu, Y., Judenhofer, M. S., Qi, J., Pichler, B. J., Cherry, S. R. (2006). Simultaneous acquisition of multislice PET and MR images: Initial results with a MR-compatible PET scanner. *J Nucl Med* 47: 1968–1976.

168 Raylman, R. R., Majewski, S., Velan, S. S. et al. (2007). Simultaneous acquisition of magnetic resonance spectroscopy (MRS) data and positron emission tomography (PET) images with a prototype MR-compatible, small animal PET imager. *J Magn Reson* 186: 305–310.

169 Woody, C., Schlyer, D., Vaska, P. et al. (2007). Preliminary studies of a simultaneous PET/MRI scanner based on the RatCAP small animal tomograph. *Nucl Instr Meth A* 571: 102–105.

170 Handler, W. B., Gilbert, K. M., Peng, H., Chronik, B. A. (2006). Simulation of scattering and attenuation of 511 keV photons in a combined PET/field-cycled MRI system. *Phys Med Biol* 51: 2479–2491.

171 Lucas, A. J., Hawkes, R. C., Ansorge, R. E. et al. (2006). Development of a combined microPET-MR system. *Technol Cancer Res Treat* 5: 337–341.

172 Yamamoto, S., Takamatsu, S., Murayama, H., Minato, K. (2005). A block detector for a multislice, depth-of-interaction MR-compatible PET. *IEEE Trans Nucl Sci* 52: 33–37.

173 Rannou, F. R., Kohli, V., Prout, D. L., Chatziioannou, A. F. (2004). Investigation of OPET performance using GATE, a Geant4-based simulation software. *IEEE Trans Nucl Sci* 51: 2713–2717.

174 Alexandrakis, G., Rannou, F. R., Chatziioannou, A. F. (2005). Tomographic bioluminescence imaging by use of a

combined optical-PET (OPET) system: a computer simulation feasibility study. *Phys Med Biol* **50**: 4225–4241.

175 Vu, N. T., Silverman, R. W., Chatziioannou, A. F. (2006). Preliminary performance of optical PET (OPET) detectors for the detection of visible light photons. *Nucl Instr Meth Phys Res A* **569**: 563–566.

176 Peter, J., Ruehle, H., Stamm, V. et al. (2005). Development and initial results of a dual-modality SPECT/Optical small animal imager. *IEEE Nucl Sci Symp Conf Rec* **4**.

177 Peter, J., Semmler, W. (2007). A modular design triple-modality SPECT-CT-ODT small animal imager [abstract]. *Eur J Nuc Med Mol Imaging* **34**: S158.

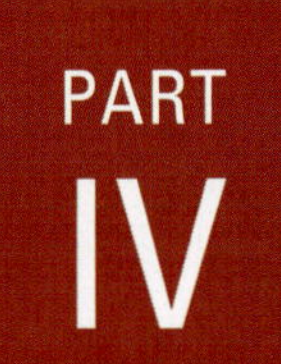

Current Applications of Imaging Reporter Genes

Reporter Gene Imaging of Cell Signal Transduction

Tarik F. Massoud, Ramasamy Paulmurugan, Carmel Chan, Hua Fan-Minogue, and Sanjiv Sam Gambhir

INTRODUCTION

Signal transduction pathways enable cells to act in response to the perception of stimuli in their environment and the integration of external and internal signals by changes in transcriptional activity, metabolism, or other regulatory measures. The proper functioning of these pathways is vital for cell adaptation and survival under varying conditions, as well as for cell differentiation, fate, and death [1].

Cell signal transduction can be simplistically recognized as three steps: reception, transduction, and induction. Reception entails the binding of a signal molecule (e.g., a hormone) to its specific receptor. Transduction is the process by which, for example, a second messenger is formed in or released into the cytosol, thus amplifying the stimulus and initiating the cell's response to the signal. Induction results in activation of the cellular process.

The study of individual cells and cell lines allows us to identify features of signal transduction, but work on physiological models defines what is relevant in the physiology of a given cell type at a given stage [2]. Indeed, the simplified experimental paradigms of conventional cell biology systems used for the study of signal transduction are a double-edged sword. At each level from the *in vivo* microscopic study of protein–protein interactions (PPIs) in single cells to the precise molecular structural definition of proteins by X-ray crystallography, what we gain in precision, rigor, and definition we may lose in relevant biology [2].

Molecular imaging is a rapidly emerging biomedical research discipline that can extend such biological observations to a more meaningful dimension in living subjects. It may be defined as the visual representation, characterization, and quantification of biological processes at the cellular and subcellular levels within intact living organisms. It is a novel multidisciplinary field where the images produced reflect cellular and molecular pathways and *in vivo* mechanisms of disease present within the context of physiologically authentic environments [3]. Although *in vitro* and cell culture studies in basic signal transduction research have been and remain a mainstay for defining biochemical and gene expression pathways, these approaches have been less successful in deciphering physiological whole-body contributions of proteins where redundancies and differences in regulation can alter the outcome from that initially predicted [4]. In contrast, *in vivo* animal models allow the assessment of phenomena such as tolerances, complementation, and redundancy in biological pathways [5]. Molecular imaging allows both the temporal and the spatial biodistribution of a molecular probe and related biological processes to be determined in a more meaningful manner throughout an intact living subject. Visualization of functions and interactions of a particular gene or protein pertaining to cell signal transduction (and drugs that target faulty pathways) becomes easier in a more realistic manner that respects the dynamics of complex biological networks and of complete and holistic biological systems in the entire living subject [3].

REPORTER GENE IMAGING

We previously reviewed in detail the many advantages afforded by molecular imaging in living subjects, such as assessment of whole-body phenomena, longitudinal monitoring of individual variation, functionality, and quantification [3]. One subset of molecular imaging techniques comprises reporter gene expression imaging. This represents an "indirect" imaging method involving multiple components and entailing use of a pretargeting molecule (reporter protein(s) expressed from a reporter gene(s)) subsequently activated upon occurrence of a specific molecular event. Following this, a molecular probe (a substrate or a ligand) specific for the activated pretargeting molecule (an enzyme or receptor) is used to image its activation [6]. In the last few years there has been a veritable explosion in the field of reporter gene imaging with the aim of determining location(s), time variation, and magnitude of gene expression within living subjects. Interesting variations on standard reporter gene assays have been adapted recently for imaging of complex molecular interactions in living subjects. Elsewhere in this book, we review the general principles of reporter gene expression imaging in living subjects, the characteristics of an ideal imaging reporter gene/probe, and the various categories of reporter gene imaging systems. There are four broad but overlapping applications of reporter gene imaging: (1) gene marking of cells, (2) imaging gene therapies, (3) imaging transgenic animals, and (4) imaging molecular interactions [3, 7]. With continued and rapid technological advancements in this field, reporter gene expression imaging has the potential to greatly refine our preclinical animal models of disease and will ultimately contribute to better understanding of disease development and the discovery, validation, and fine-tuning of new and improved clinical diagnosis, treatment, and prevention regimens.

SIGNAL TRANSDUCTION OVERVIEW

It is essential that cells communicate with each other for optimal functioning of a multicellular organism. Cell growth, migration, shape, differentiation, and apoptosis are examples of cellular events regulated by information (signals) that cells receive from their physical environment (e.g., light on retinal cells [8], odorants on nasal epithelium [9], and taste receptor stimulation in taste buds [10]) or through contact with other cells (direct contact, juxtacrine signaling [11]; or over short distances, paracrine signaling [12]), or interaction with matrix molecules (e.g., fibronectin), or commonly through stimulation by soluble chemicals acting as signaling factors (e.g., growth factors, hormones, cytokines, chemokines, neurotransmitters, and neurotrophins). Moreover, within each cell is a highly complex spatiotemporal organization necessary to carry out these wide varieties of different functions. Successfully executing these complex processes requires highly dynamic information transfer between different regions and compartments within cells [13]. Cell signal transduction encompasses all the biochemical and biophysical phenomena that lead from the perception of a signal by a cell to the biological response of the cell [2]. The functioning of cells and tissues therefore relies on the fidelity of signaling pathways and the overall molecular basis of intercellular and intracellular communication [14].

In 1979 Korman et al. reported on the relationship between the degree of chemotactic activity in *Escherichia coli* and the amount of galactose-, glucose-, maltose-, and ribose-binding proteins in its cells [15]. They concluded that the response of the bacterial flagella was proportional to the amount of a specific receptor on its cell surface. Since then a large number of different cell surface receptors for signaling molecules have been identified, as well as intracellular signaling pathways that connect the receptors with effector molecules in the cytoplasm or nucleus [16] (Figure 8.1). Such cell signaling pathways frequently consist of ubiquitous building blocks, and their design seems to be conserved through evolution and between species [1].

Many diseases result from the malfunction of signaling pathways, and there is an increasing trend in modern molecular therapeutics to seek ways to manage disease through the manipulation of signal transduction pathways [14]. For example, many cancer cells expressing overactive signaling pathway are dependent for their survival on these pathways, and unlike normal cells, they seem to have lost the redundancy in signal transduction pathways. This feature probably accounts for the preferential vulnerability of cancer cells to drugs acting as signal interceptors. Although universal signal transduction inhibitors would be expected to inhibit the proliferation of diseased cells, they also could affect normal healthy cells. Selective blockers, on the other hand, are expected to be targeted exclusively toward a single hyperactive signaling pathway in a particular disease [14]. However, a significant downside to the principle of selective signaling blockade is the unfortunate occurrence of drug resistance [17]. This may be exemplified in the case of the drug trastuzumab, a monoclonal antibody targeted against the human epidermal growth factor receptor (HER) 2 tyrosine kinase receptor, which is overexpressed in approximately 25% of invasive breast cancers. The majority of patients with metastatic breast cancer who initially respond to trastuzumab demonstrate disease progression within 1 year of starting treatment. Preclinical studies have indicated several molecular mechanisms that could contribute to the development of trastuzumab resistance. Increased signaling via the phosphatidylinositol 3-kinase/Akt pathway (see the following) could contribute to trastuzumab resistance because of activation of multiple receptor pathways that include HER2-related receptors or non-HER receptors such as the insulin-like growth factor 1 receptor, which appears to be involved in a cross-talk (see the following) with HER2 in resistant cells. Moreover, loss of function of the tumor suppressor PTEN gene, the negative regulator of Akt (see the following), results in heightened Akt signaling that leads to decreased sensitivity to trastuzumab. Decreased interaction between trastuzumab and its target receptor HER2 owing to steric hindrance of HER2 by cell surface proteins such as mucin-4 (MUC4) may block the inhibitory actions of trastuzumab. Novel therapies targeted against these aberrant molecular pathways offer hope that the effectiveness and duration of response to trastuzumab can be greatly improved [17] (Figure 8.2).

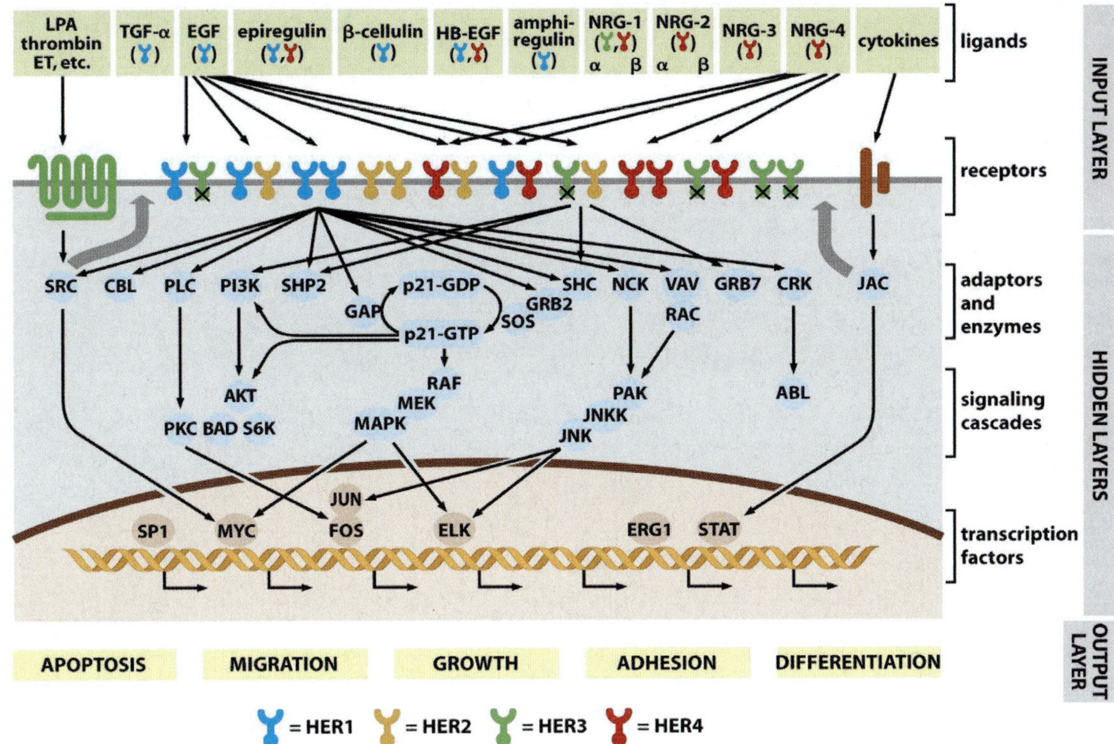

Figure 8.1. The EGFR signaling network. Schematic representation of how cells can communicate with their surroundings. A variety of proteins messengers (ligands, green squares, top) interact with a complex array of cell surface receptors, which transduce signals across the plasma membrane (gray) into the cytoplasm, where a complex network of signal-transducing proteins processes these signals, funnels signals into the nucleus (bottom), and ultimately evokes a variety of biological responses ("output layer," yellow rectangles, bottom). Reproduced with permission from Weinberg (2007).

MECHANISMS OF CELL SIGNAL TRANSDUCTION

A cell signal is whatever is recognized as such by a plasma membrane-bound receptor that itself initiates an intracellular response to this signal. A receptor is the structure that recognizes and reacts to the signal and interprets the specificity of the signal [2]. Cell surface receptors recognize various signaling molecules with their extracellular domains (see the following for definition of a protein domain), resulting in activation of their intracellular domains. Cellular physiology depends on thousands of signals. The number of receptors is a multiple of the number of signals, hence the tremendous complexity and specificity of signals and their receptors [2].

The entire repertoire of human plasma membrane receptors is listed at the Human Plasma Membrane Receptome (http://www.receptome.org). Several types of receptors may be identified depending on their functions, ligands, and relative positioning to the plasma membrane:

(1) Some receptor proteins are "peripheral membrane proteins." These proteins adhere only temporarily to the biological membrane with which they are associated. These molecules attach to integral membrane proteins or penetrate the peripheral regions of the lipid bilayer. The regulatory protein subunits of many ion channels and transmembrane receptors, for example, may be defined as peripheral membrane proteins. In contrast to integral membrane proteins, peripheral membrane proteins tend to collect in the water-soluble component, or fraction, of all proteins extracted during a protein purification procedure. Proteins with Glycosylphosphatidylinisotol (GPI) anchors are an exception to this rule and can have purification properties similar to those of integral membrane proteins. The reversible attachment of proteins to biological membranes has been shown to regulate cell signaling and many other important cellular events through a variety of mechanisms [18].

(2) Transmembrane receptors are embedded in the phospholipid bilayer of cell membranes and allow the activation of signal transduction pathways in response to the activation by the ligand. These receptors are integral membrane proteins, which reside and operate typically within a cell's plasma membrane, but also in the membranes of some subcellular compartments

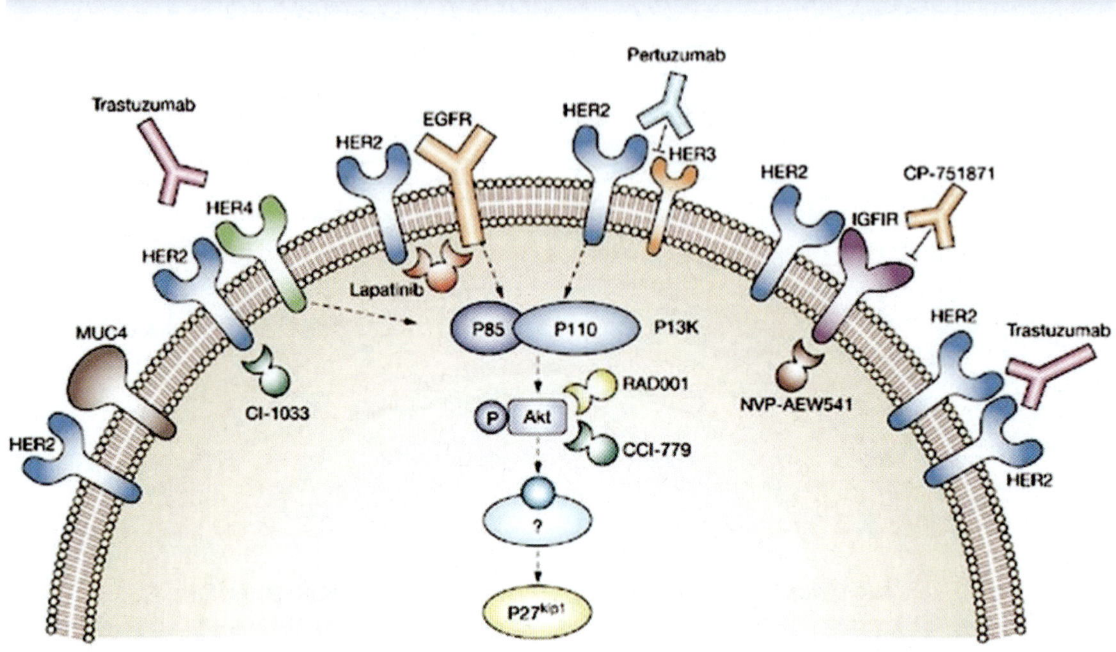

Figure 8.2. Proposed mechanisms of trastuzumab resistance. The EGFR family consists of four different members: EGFR; human epidermal growth factor receptor (HER) 2; HER3; and HER4. Upon ligand binding, these receptors can homodimerize or heterodimerize with each other to form several receptor combinations. This ligand–receptor binding in turn induces tyrosine kinase activity within the cytoplasmic domains of these receptors, which leads to activation of downstream signaling pathways, in particular the PI3K/Akt pathway. Activation of components of this pathway leads to elevated levels of the p27^{Kip1} protein. The membrane-associated glycoprotein mucin-4 (MUC4) could block inhibitory actions of trastuzumab by directly binding with HER2, preventing interaction between the drug and its molecular target. Furthermore, in trastuzumab-resistant cells HER2 and IGF1R physically interact and are involved in a crosstalk that results in p27^{Kip1} downregulation. Potential novel agents that could prove beneficial in trastuzumab-resistant tumors include the following: the pan-HER tyrosine kinase inhibitor CI-1033; lapatinib (LAP), an EGFR/HER2 tyrosine kinase inhibitor; pertuzumab, an HER2-targeted antibody that disrupts receptor heterodimerization between EGFR/HER2 and HER2/HER3; CP-751871, an IGF1R-targeted antibody; the IGF1R kinase inhibitor NVP-AEW541; and the mammalian target of rapamycin inhibitors RAD001, CCI-779, and AP23573. HER, human epidermal growth factor receptor; IGF1R, insulin-like growth factor 1 receptor; MUC4, membrane-associated glycoprotein mucin-4; P, phosphorylated; PI3K, phosphatidylinositol 3-kinase. Reproduced with permission from Nahta et al. (2006).

and organelles. Binding to a signaling molecule or sometimes to a pair of such molecules on one side of the membrane, transmembrane receptors initiate a response on the other side. In this way they play a unique and important role in cellular communications and signal transduction. Many transmembrane receptors are composed of two or more protein subunits that operate collectively and may dissociate when ligands bind, fall off, or at another stage of their "activation" cycles. They are often classified based on their molecular structure or because the structure is unknown in any detail for all but a few receptors, based on their hypothesized (and sometimes experimentally verified) membrane topology. The polypep-tide chains of the simplest are predicted to cross the lipid bilayer only once, while others cross as many as seven times (the so-called G-protein-coupled receptors [19]). Transmembrane receptors may be subdivided into three parts or domains: extracellular, transmembrane, and intracellular. These receptors may be (a) "metabotropic" receptors, for example, when coupled to G proteins and affect the cell indirectly through enzymes that control ion channels. All metabotropic receptors have seven hydrophobic transmembrane domains (hence they are also known as "seven transmembrane receptors" or 7TM receptors [20] because they pass through the membrane seven times, and the protein's N-terminus is located

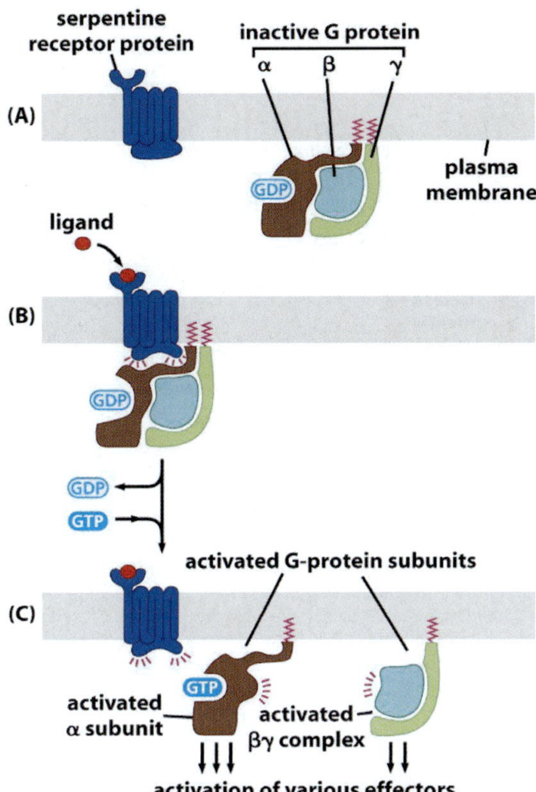

Figure 8.3. Signaling by "seven transmembrane receptors." **(A)** The 7TM receptors are associated with heterodimeric G proteins, which are composed of α, β, and γ subunits. **(B)** Once a receptor has bound its cognate ligand (red), it simulates the α subunit of the G protein to release its GDP (guanosine triphosphate) and bind GTP (guanosine triphosphate) instead (light blue). **(C)** As part of the response, the α subunit dissociates from the β and γ pair. Now the α subunit (brown) and, independently, the β and γ complex (blue and light green) can each proceed to regulate enzymes that evoke a variety of downstream responses. The signaling is halted when the α subunit hydrolyzes its GTP, causing it to reassociate with the β and γ complex (not shown). Reproduced with permission from Weinberg (2007).

on the extracellular side of the membrane and its C-terminus is on the intracellular side), and most are monomeric proteins, although GABA$_B$ receptors require heterodimerization to function properly. Examples include G-protein-coupled receptors (Figure 8.3), receptor tyrosine kinases (Figure 8.4), and guanylyl cyclase receptors. The 7TM receptors are coded by approximately 700 genes, that is, approximately 2% of all genes in the human genome. (b) "Ionotropic" receptors (also known as ligand-gated ion channels) contain a central pore that opens in response to the binding of ligand [21]. The direct link between ligand binding and opening or closing of the ion channel, which is characteristic of

ligand-gated ion channels, contrasts with the indirect function of metabotropic receptors, which use second messengers. Ligand-gated ion channels are also different from voltage-gated ion channels (which open and close depending on membrane potential) and stretch-activated ion channels (which open and close depending on mechanical deformation of the cell membrane). Examples include Cys-loop receptors (such as the nicotinic acetylcholine receptor) [22], ionotropic glutamate receptors [23], and adenosine triphosphate (ATP)-gated channels [24].

(3) Another major class of receptors includes intracellular proteins such as those for steroid and intracrine peptide hormone receptors. These receptors often enter the cell nucleus and modulate gene expression in response to the activation by the ligand. Examples are the class of nuclear receptors located in the cell nucleus [25] and the inositol trisphosphate (IP$_3$) receptor located on the endoplasmic reticulum [26]. The ligands that bind to them are usually intracellular second messengers, for example, IP$_3$ and extracellular lipophilic hormones, for example, steroid hormones. Some intracrine peptide hormones also have intracellular receptors.

Soluble signaling molecules regulating cell growth often bind to receptors with an intrinsic tyrosine kinase domain. Such receptors are activated by ligand-induced receptor dimerization or oligomerization, whereby the intracellular kinase domains are juxtaposed and thus may activate each other through mutual phosphorylation (Figure 8.5). Such autophosphorylation serves two important functions: it alters its conformation so that the active site becomes accessible to substrate and establishes docking sites for intracellular signaling molecules with SH2 or PTP domains (see the following) [16].

Activation of a cell surface receptor can trigger several intracellular signaling pathways, resulting in an information transfer between the membrane and other cellular locations and compartments. This involves the physical movement of signaling molecules through the cell, either by diffusion or with the aid of local regeneration of the messenger or active transport [13]. There is a limited repertoire of known signal transduction pathways modulated by cell surface receptors, with only a few ubiquitous intracellular signal molecules (cAMP, cGMP, Ca^{2+}, etc.), phosphorylation cascades, and other pathways [2]. There are two different ways by which Ca^{2+} acts as a second messenger and can regulate proteins [27] – by direct recognition of Ca^{2+} by the protein, and by binding of Ca^{2+} in the active site of an enzyme. One of the best-studied interactions of Ca^{2+} with a protein is the regulation of calmodulin by Ca^{2+} [28]. Calmodulin itself can regulate other proteins or be part of a larger protein (e.g., phosphorylase kinase). Other types of second messengers include lipophilic molecules (e.g., diacylglycerol,

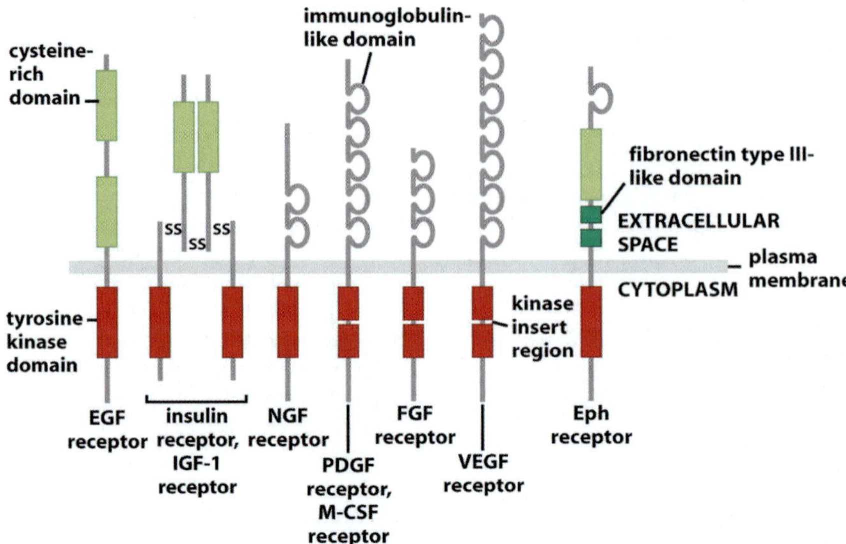

Figure 8.4. Structure of tyrosine kinase receptors. The EGF receptor is only one of a large number of similarly structured receptors encoded by the human genome. These tyrosine kinase receptors (TKRs) can be placed into distinct families, depending on the details of their structure. Representatives of most of these families are shown here. All of these share in common similar cytoplasmic tyrosine kinase domains (red). The ectodomains of these receptors (which protrude into the extracellular space, green, gray) have highly variable structures, reflecting the fact that they recognize and bind a wide variety of extracellular ligands. Reproduced with permission from Weinberg (2007).

required for the activation of protein kinase C [29]), and the free radical nitric oxide [30].

There are four categories of intracellular signaling pathways. The most common and best studied are where signals are transduced from cell surface receptors to the nucleus through relay systems (Figure 8.6). Such a pathway involves the occurrence of protein–protein interactions inside the cell, induced by an external signal. For instance, many growth factors bind to receptors at the cell surface and stimulate cells to progress through the cell cycle and divide. Several of these receptors are kinases that start to phosphorylate themselves and other proteins

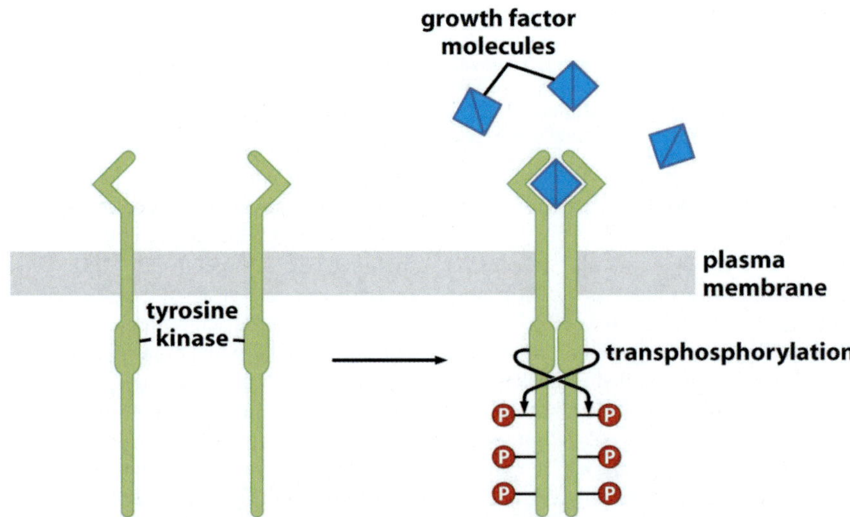

Figure 8.5. Receptor dimerization following ligand binding. In the absence of ligand, receptor molecules (green) are free to move laterally in the plane of the plasma membrane. In the presence of growth factor ligand, two receptor molecules are brought together to form a dimer. Once the receptor is dimerized, the tyrosine kinase domain of each receptor subunit is able to phosphorylate the C-terminal cytoplasmic tail of the other subunit; the process of transphosphorylation (arrows). Reproduced with permission from Weinberg (2007).

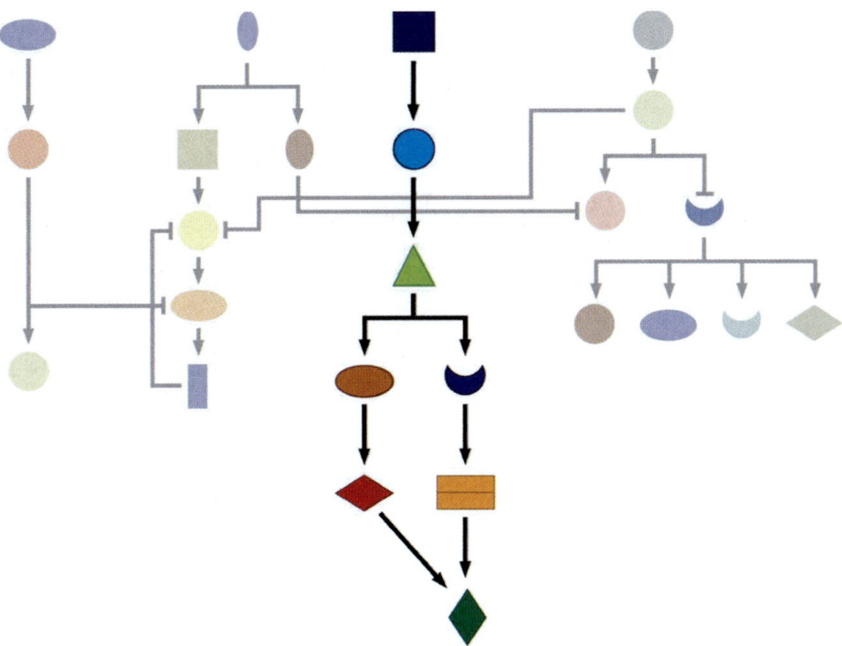

Figure 8.6. Signaling cascades. A diagram of hypothetical signaling circuitry shows how a signal transduction cascade (a series of signaling proteins that operate much like a molecular bucket brigade) passes signals from an upstream source (purple square) to its intended downstream target (dark green diamond) and, at the same time, avoids inadvertent activation of dozens of other signaling proteins in the cell (faintly drawn symbols). Reproduced with permission from Weinberg (2007).

when binding to a ligand. This phosphorylation can generate a binding site for a different protein and thus induce protein–protein interaction(s). For example, the ligand epidermal growth factor (EGF) binds to its receptor EGFR, activating it and resulting in autophosphorylation [31]. The phosphorylated receptor binds to an adaptor protein (GRB2), which couples the signal to further downstream signaling processes. One of the signal transduction pathways that can be activated is called the mitogen-activated protein kinase (MAPK) pathway. The signal transduction component labeled as "MAPK" in the pathway was originally called "ERK," so the pathway is called the MAPK/ERK pathway (Figure 8.7). The MAPK protein is an enzyme, a protein kinase that can attach phosphate to target proteins such as the transcription factor MYC and, thus, alter gene transcription

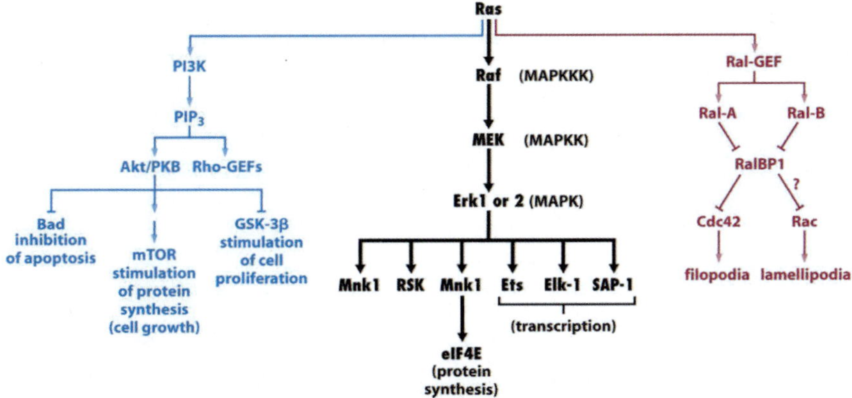

Figure 8.7. The Ras → Raf → MAP kinase pathway. This signaling cascade (black) is one of a series of similarly organized pathways in mammalian cells that have the overall plan MAPKKK → MAPKK → MAPK. In this particular case, Ras activates the Raf kinase (a MAPKKK), the latter proceeds to phosphorylate and activate MEK (a MAPKK), and MEK then phosphorylates and activates ErK1 and Erk2 (MAPKs). The latter can then phosphorylate kinases in the cytoplasm that regulate translation as well as transcription factors in the nucleus. Reproduced with permission from Weinberg (2007).

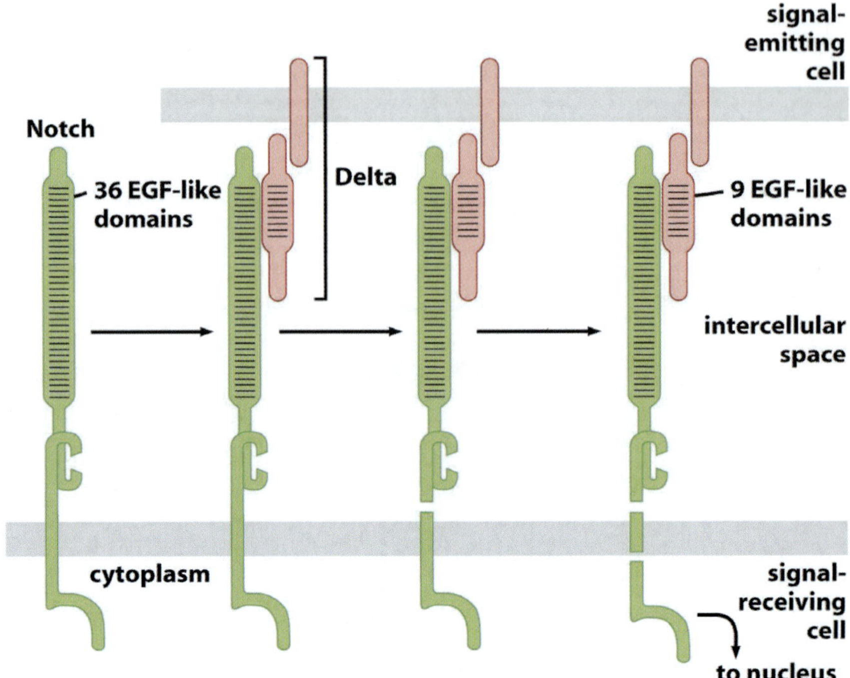

Figure 8.8. The structure of the Notch receptor. Notch receptor (green) appears to embody a very primitive type of signaling. After it has bound a ligand (e.g., NotchL or, as shown here, Delta, pink), which is being displayed on the surface of an adjacent cell, Notch undergoes two successive proteolytic cleavage events, and the resulting C-terminal cytoplasmic fragment is thereby freed, allowing it to migrate to the cell nucleus, where it alters the expression of certain genes. Reproduced with permission from Weinberg (2007).

and, ultimately, cell cycle progression. Many cellular proteins are activated downstream of the growth factor receptors (such as EGFR) that initiate this signal transduction pathway. Some signaling transduction pathways respond differently depending on the amount of signaling received by the cell. For example, the hedgehog protein activates different genes, depending on the amount of hedgehog protein present. Complex multicomponent signal transduction pathways provide opportunities for feedback, signal amplification, and interactions inside one cell between multiple signals and signaling pathways.

The other three types of signaling pathways involve: (a) Intramembrane cleavage of activated receptor and release of the intracellular fragment that translocates to the nucleus. In the case of Notch-mediated signaling, the activation of Notch can cause the Notch protein to be altered by a protease [32]. Part of the Notch protein is released from the cell surface membrane and can act to change the pattern of gene transcription in the cell nucleus (Figure 8.8). (b) Ligand passage through the plasma membrane, followed by ligand binding to receptor to result in its activation either in the cytoplasm or nucleus. Intracellular receptors include nuclear receptors and cytoplasmic receptors and are soluble pro-

teins localized within the nucleoplasm or the cytoplasm, respectively. The typical ligands for nuclear receptors are lipophilic hormones, with steroid hormones (e.g., testosterone, progesterone, and cortisol) and derivatives of vitamin A and D among them [33]. To reach its receptor and initiate signal transduction, the hormone must pass through the plasma membrane, usually by passive diffusion. The nuclear receptors are ligand-activated transcription activators; on binding with the ligand (the hormone), the ligands will pass through the nuclear membrane into the nucleus and enable gene transcription. (c) Direct signaling to the nucleus. For example, emerging evidence suggests the existence of a "direct" mode of the EGFR pathway distinct from the traditional transduction pathway. This new mode of EGFR signaling involves cellular transport of EGFR from the cell surface to the cell nucleus, association of nuclear EGFR complex with gene promoters, and transcriptional regulation of the target genes. Although the nature and pathological consequences of the nuclear EGFR pathway remain elusive, accumulating evidence suggests its association with increased tumor cell proliferation and poor survival rate in breast cancer patients. Although several anti-EGFR agents are being tested in breast cancer patients clinically and others under preclinical development, a better

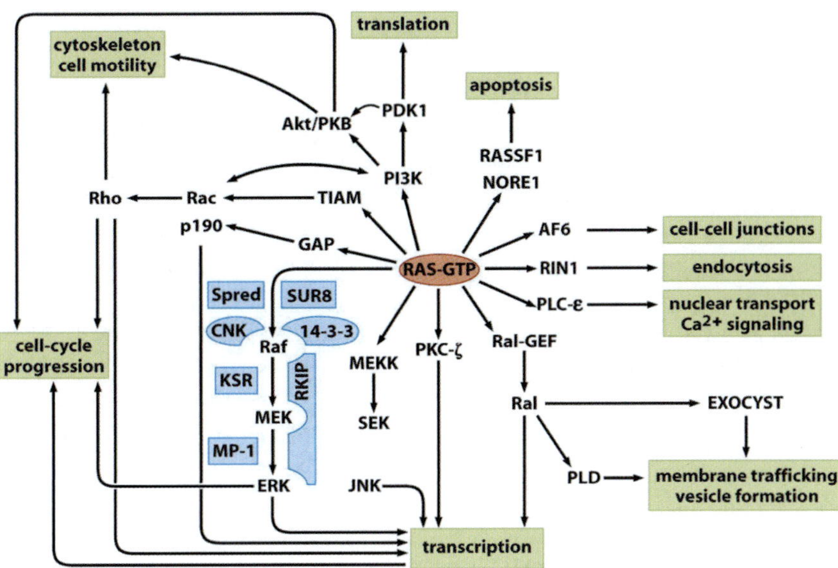

Figure 8.9. Ras effector pathways. Detailed biochemical analyses of the GTP-bound, activated form of the Ras proteins has indicated that these proteins bind to far more than the three major effectors of Ras signaling: Raf, Ral-GEF, and PI3 kinase. At least eight additional Ras-interacting proteins have been uncovered, most of which are known or suspected to play key roles in relaying Ras signals to specialized downstream signaling circuits involved in diverse functions. Reproduced with permission from Weinberg (2007).

understanding of the traditional and nuclear EGFR pathways will facilitate the identification of patients likely to respond to these agents as well as future development of more effective anti-EGFR therapeutic interventions [34].

Space constraints in this chapter limit detailed scrutiny of the large variety of signaling pathways studied to date. The following nine pathways are the main ones relevant to development and maintenance of cancer: Ras signaling (three subtypes: Ras-Raf-MAPK (mentioned previously) [35], Ras-PI3K-Akt/PKB [36], and Ras-Ral-GEF [37]), Jak-STAT signaling [38], integrin signaling [39], Wnt-β-catenin signaling [40], GPCR signaling [41], nuclear factor-κB signaling [42], Notch signaling [32], TGF-β signaling [43], and hedgehog signaling [44].

At first glance, signaling is a linear connection between input elements (receptors) and output elements (regulators of gene expression). A closer inspection reveals that signaling pathways interact with each other, forming a network or map [1]. Signaling maps, therefore, can be considered schematically as boxes indicating molecules connected by arrows that delineate the possible flow of information (signals) between them to result in specific cellular actions such as gene expression, movement, cell division, etc. [13] (Figure 8.9). The generation of networks arises because intracellular signal transduction does not simply occur through activation of a number of parallel pathways; rather, the components of the different pathways show extensive crosstalk between each other [45]. Schwartz [46] introduced the notion

of crosstalk, referring to the case that two inputs work through distinct signaling pathways but cooperate to regulate an output (Figure 8.10). There are also examples of inactivating crosstalk between signaling pathways [47]. The complexity created by the crosstalk in the signal transduction network makes it virtually impossible to infer by hand all the consequences that follow after the modification of one part of the network. Fortunately, a number of databases have been constructed to bring signal transduction knowledge into a well-organized format, providing simple and fast access to the signal transduction system. The Signal Transduction Classification Database (STCDB) is a database of information relative to the classification of signal transduction [48]. It describes each type of characterized signal transduction (ST) for which a unique ST number has been provided. Approved classifications are available for Web browsing at http://www.techfak.uni-bielefeld.de/~mchen/STCDB

The strengths of intracellular signals are carefully controlled; such control mechanisms occur at a number of different levels [16]. One important aspect of signal transduction control is the control of the quantity of signaling molecules by regulated proteolysis through ubiquitin-mediated degradation. Ubiquitination and, in particular, phosphorylation, are well-established posttranslational modifications that regulate signal transduction events. It is likely that other types of posttranslational modifications occur, for example, acetylation.

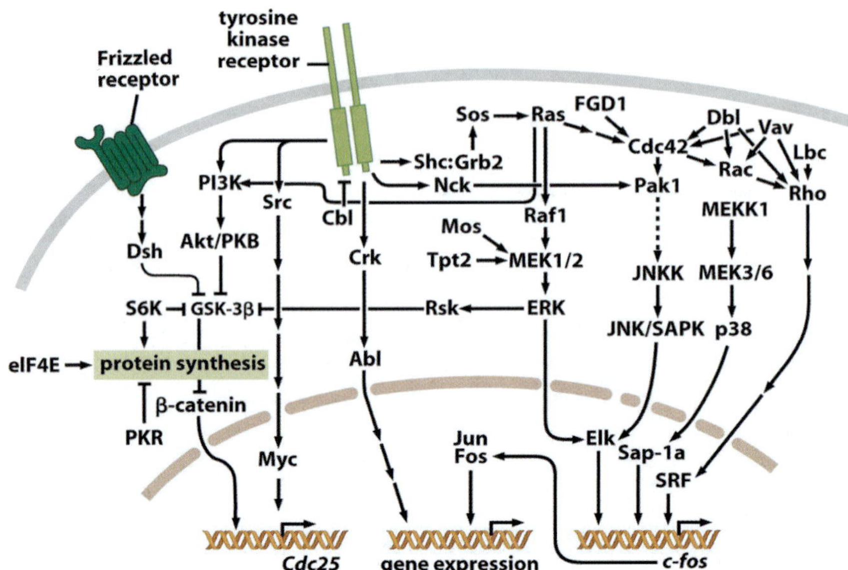

Figure 8.10. Two-dimensional signaling maps. While signaling pathways can be depicted as linear cascades reaching from the plasma membrane to the nucleus, with occasional branch points, recent research reveals that these pathways interact via numerous cross-connections and points of convergence. This diagram, itself a simplification, provides some hints of these cross-connections. Note, for example, that the presence of GSK-3β, which plays a central role in the Wnt-β-catenin pathway, is also regulated by signaling proteins lying downstream of tyrosine kinase receptors. Reproduced with permission from Weinberg (2007).

Another mechanism for control of signaling involves induction of inhibitory signals in parallel to stimulatory ones. Another question governing control of this flow of information is how the translocation to the nucleus of components in these signaling pathways is regulated. Further research will be important to elucidate the mechanisms for nuclear import and export. In addition to spatial aspects, temporal aspects of signal transduction are important determining factors of control. There is evidence that the duration of a signal is crucial for cellular response [16].

Many signaling reactions taking place in cells involve complex positive and negative nonlinear feedback as well as transport, and their dynamics can give rise to a wide variety of nonintuitive behaviors. To interpret and understand these data it is becoming increasingly necessary to model and analyze them using qualitative and quantitative mathematical models, often within a systems biology analytical framework. This may become increasingly necessary in the study of any human diseases caused by the malfunctioning of signaling components within pathways that form complex networks. The combined effects of multiple malfunctioning complexes can be substantial. An understanding of how individual components function within the context of an entire system under a variety of situations should help to further understand why multiple interactions between aberrant signaling pathways often result in pathophysiology [49]. An emerging concept in medicine, therefore, is that

disease itself may perturb the normal network structures of an intracellular system through genetic effects and/or pathological microenvironmental cues [50]. Systems biology is a scientific discipline aiming to identify and quantify all the molecular elements of a biological system, to measure their interactions, and to integrate that information into graphical network models that can be used to generate predictive hypotheses in explaining emergent behaviors [50].

ROLE OF PROTEIN–PROTEIN INTERACTIONS IN SIGNAL TRANSDUCTION AND TRANSCRIPTIONAL REGULATION

The initiation of downstream signaling pathways leads to various cellular responses. The mechanism whereby components in such signaling pathways interact with each other came through the discovery that signaling molecules consist of different types of conserved domains arranged in a modular fashion. Through these modules, signaling molecules form physical contacts with upstream and downstream components in signaling pathways.

Owing to their biological importance and growing interest in the mechanism of their function, the protein–protein recognition surfaces of some of the proteins involved in intracellular signal transduction merit a separate brief mention. These are of great interest because of their role in the control of a myriad of cellular activities. Please see Royer for a more detailed account [51].

A protein domain is part of a protein sequence and structure that can evolve, function, and exist independently of the rest of the protein chain. Each domain forms a compact three-dimensional structure and often can be independently stable and folded. Many proteins consist of several structural domains. One domain may appear in a variety of evolutionarily related proteins. Protein domains involved in signal transduction include Src homology domains 2 and 3, commonly referred to as SH2 and SH3 domains [52]. The SH2 domains recognize tyrosine phosphorylated proteins, especially autophosphorylated growth factor receptors. They are found in growth factor receptor binding proteins present in signal transduction downstream of the receptors but upstream of Ras; they have also been found in docking proteins. On the other hand, SH3 domains bind to proline-rich sequences in their target protein partners. SH3 domains are also found in proteins involved in signal transduction, such as the protein tyrosine kinases. They recognize polyproline type II helical structures (PXXP motifs) in cell signaling proteins.

Unlike the SH2 domains, the phosphopeptide-interacting domains, or PI/PTB (phosphotyrosine-binding) domains, recognize the sequence that is N-terminal, rather than C-terminal, to the phosphotyrosine of the receptor [53]. These adaptor proteins, once tyrosine phosphorlyated by the receptor, become capable of binding to the Grb2/Sos complex, thus coupling the receptor to the Ras signaling pathway.

PDZ domains (also termed GLFG repeats or DHR domains) were first identified as ninety amino acid segments in three proteins, PSD-95, DlgA, and ZO-1 (hence the acronym PDZ domain), which are all guanylate kinases [54]. Proteins containing PDZ domains have been implicated in ion channel receptor clustering, receptor/enzyme coupling, and a variety of other protein associations.

LIM domains are zinc finger cysteine-rich domains found in many homeodomain proteins involved in development and in nonhomeodomain proteins involved in differentiation, associated with the cytoskeleton or cellular senescence [55]. These domains bind to PDZ motifs, bHLH (helix-loop-helix) transcription factors, as well as other LIM domains, and LIM binding proteins, and thus mediate PPIs implicated in important cell functions.

An incomplete list of other protein interaction motifs includes the pleckstrin homology domain, which binds to acidic domains in signal transduction proteins as well as to phosphoinositides [56], the WW domain, a semiconserved region of thirty-eight to forty amino acids termed WW because of the two conserved tryptophan residues spaced twenty amino acids apart and which interacts with proteins involved in cell signaling [57], the WSXWS amino acid motif in cytokine recep-

tors [58], and the WD repeat (WD-repeat-containing proteins contain four or more copies of the WD-repeat [tryptophan-aspartate repeat], a sequence motif approximately thirty-one amino acids long that encodes a structural repeat) [59].

Transcriptional regulation is mediated by a number of protein–protein and protein–DNA complexes in addition to that between the RNA polymerase and the gene and transcript. Protein factors that modulate gene transcription have been termed transcription factors. They fall into three large groups, general transcription factors involved in general multiprotein transcriptional machinery, transcription-associated or bridging factors, and specific factors that recognize control sequences in the noncoding regions of the DNA. Modulating the expression of genes by these factors occurs through their ability to enhance (activation) or disfavor (repression) recruitment of the proteins involved in the transcription. This process therefore involves specific, high-affinity PPIs, and understanding the structural and energetic basis for the affinity and specificity of these interactions is paramount to the understanding of transcriptional control.

One important family is the bHLH transcription factors, so named because of a structural motif that includes a basic region that binds to DNA and a helix-loop-helix region involved in dimerization [60]. Members of this family include the MyoD family of transcription factors that regulate the expression of muscle-specific structural genes and thus control the differentiation of muscle cells. Other members of the bHLH family of transcription factors are the ubiquitous products of the E2A gene, E12 and E47. Another class of transcription factors is represented by the leucine zipper transcription factors of which c-jun and c-fos (constituents of the AP1 transcription factor) are the most well known [61]. AP1 is involved in regulating gene transcription linked to cell proliferation. Another large family of ligand-responsive transcription factors is embodied by the superfamily of steroid/nuclear receptors. These proteins include the steroid hormone receptors glucocorticoid, estrogen, mineralocorticoid, and androgen receptors, lipophilic hormone receptors such as the vitamin D, retinoic acid, and thyroid hormone receptors, and a number of orphan receptors and transcription factors involved in Drosophila development. These receptors exhibit a homologous modular architecture with an N-terminal constitutive and tissue-specific transcriptional activation domain (TAF1), which is quite variable in length and is responsible for interacting with tissue-specific additional transcription factors, a central DNA binding and dimerization domain (DBD), which also may bear determinants for interaction with other transcriptional regulators, a linker region involved in nuclear localization, and a C-terminal ligand binding domain (LBD) that also bears ligand-dependent transcriptional activation activity (TAF2) and dimerization determinants. The protein–protein interactions operative in these systems are numerous and key to receptor function.

IMAGING CELL SIGNAL TRANSDUCTION

The development of reliable and sensitive indicators to measure signaling at the single cell level has been an active area of research in cell biology for many years [13]. Because many signaling reactions involve proteins that undergo reversible conformational changes and/or formation of complexes with other proteins, it is to be expected that FRET (fluorescence resonance energy transfer)-based techniques and measurements will become increasingly important in the analysis of signaling pathways in cell culture [62].

Undoubtedly, FRET microscopic techniques are generally well suited to quantitative study of protein–protein interactions and other formation of complexes in single cells in culture. Interestingly, they are less well adapted to measuring homodimerization of proteins, as discussed in Massoud et al. [63]. Furthermore, any extrapolation of FRET techniques to potential whole-body imaging in living subjects would likely encounter the ever-present constraint of normal tissue autofluorescence common to whole-body fluorescence imaging in general. With this in mind, our aim as well as that of others in the field of molecular imaging technique development and innovation has been to establish separate assays based in particular on bioluminescence imaging. These would complement the highly quantifiable FRET microscopic techniques in single cells or the β-galactosidase complementation assay in cell culture and at the same time allow meaningful evaluation of protein–protein interactions in two different dimensions, that is, *in vitro* using benchtop luminometry, and by charge coupled device (CCD) camera imaging of intact living subjects. The previously exemplified fluorescence- and bioluminescence-based *in vivo* imaging techniques possess distinct advantages and disadvantages, and some are also shared between them. For example, the monitoring of weak or low levels of protein dimerization may be comparatively and especially more difficult for FRET microscopy using different fluorophores because the incremental changes in fluorescence are already small [64] and any unpaired fluorophores add noise to the system [65]. FRET requires generally that overexpression of the protein chimeras take place to allow quantification of small changes in fluorescence emissions [66], and the same may also hold true for endogenous proteins evaluated with bioluminescence imaging, although this is yet to be ascertained. Conversely, the systems we describe using bioluminescence reporters have the important advantage of enzymatic amplification [3], rendering them as more physiologically relevant assays. The inherent high sensitivity

of bioluminescence imaging (possibly able to detect $10^{-15}-10^{-17}$ mole/L of molecular probe) in living subjects is also likely to be a benefit when monitoring low levels of signaling protein interactions in living subjects [67]. Objective evaluation of the relative merits of complementation-based bioluminescence assays in living subjects and FRET for microscopic imaging of protein interactions would be difficult to achieve in absolute terms. Indeed, this may not be necessary if their predicted complementary roles prove well founded in practice.

Noninvasive methods for molecular imaging of the activity of different signal transduction pathways and the expression of different genes in living subjects would be of considerable value [68]. They would help understand the role specific genes and signal transduction pathways have in various diseases and could elucidate temporal dynamics and regulation at different stages of disease and during various therapeutic interventions [69]. There are two general strategies when attempting molecular imaging of cell signal transduction in living subjects. The first is to target the various biochemical and biophysical components of the signaling pathway and to tailor the imaging strategy to each component separately and in isolation. This may be particularly appropriate, for example, when attempting to study the effect of an inhibitor or enhancer drug on a specific protein–protein interaction at a single node along a signal transduction pathway. This strategy may be considered a direct one to image components of cell signal transduction. Conversely, an indirect strategy to image any given pathway does away with focusing on individual nodes and instead establishes the product of gene expression downstream of the pathway as the target for imaging, that is, it attempts to image transcriptional activation of endogenous genes downstream of a given pathway. Imaging reveals the presence, quantity, duration, and location of this product and therefore indirectly signifies the general occurrence of successful signal transduction further upstream along the particular pathway under scrutiny.

Imaging the effects of modulator drugs on signal transduction pathways follows the same principles adopted for similar assessments in single cells. The best argument in favor of the hypothesis that a biological event is necessary for a signal transduction pathway is to show that its suppression inhibits more downstream events [2]. Suppression is possible by pharmacological inhibitors, antibodies, dominant negative or competing peptides or proteins, deletion of the protein by inhibition of its synthesis (e.g., antisense and RNA interference), or gene knockout. For inhibitors, the postulated suppression must take place in the system studied under the conditions used and it must be specific. Such controls are often missing in the artificial confines of single cells in culture. Molecular imaging of signaling pathways in intact living subjects can provide more authentic physiological environments to provide such controls.

Imaging cell signal transduction is in its infancy, and therefore, what follows is an outline of the generic methods and approaches possible, rather than specific examples, for direct and indirect molecular imaging of cell signal transduction. Direct tactics are presented sequentially as they might be used along a hypothetical prototypical signal transduction pathway from cell surface receptors to the nucleus via relay nodes.

Imaging of Receptor Presence, Density, Occupancy, and Activation

The high-affinity binding of a radiopeptide tracer to receptors expressed on the surface of specific cells is a characteristic that can be exploited for imaging. Receptor scintigraphy using radiolabeled peptide ligands has proven its effectiveness in laboratory animals and in clinical practice. Over the last few years, substantial documentation with receptor-mediated imaging has been presented [70]. However, among thousands of possible candidates, only a few radioligands are useful on a daily basis in clinical practice. Over the past 3 decades, such radioligands have evolved from monoclonal antibodies [71] through F(ab')$_2$ and Fab fragments [72] to engineered antibody fragments [73] and small biologically active (synthetic) peptides [74]. Receptor/ligand imaging has been most useful in the brain and to some extent in assessing body tumors.

The mechanism of action of neurotransmitters depends on the features of the two receptor subfamilies. Ligand-gated receptors contain an intrinsic channel that is rapidly opened in response to transmitter binding, whereas G-protein-coupled receptors activate G proteins in the membrane, which then stimulate various membrane effector proteins. Membrane proteins act on the synthesis of second messengers (e.g., cAMP, cGMP, and calcium ions), which in turn act on intracellular protein kinases.

Imaging of neurotransmitters and neuroreceptors has become critical to understanding the intrinsic neurochemical basis of neurologic and psychiatric disease. The dopaminergic system has been extensively investigated in terms of both presynaptic and postsynaptic processes by means of selective PET tracers. ^{18}F-Fluoro-DOPA has been used as a probe of the presynaptic dopaminergic system, as a probe of amino acid decarboxylase activity (the rate-limiting enzyme in the synthesis of dopamine) and thus, nigrostrial neuron density and presynaptic function [75]. The dopaminergic system [76] has also been studied with tracers binding to the presynaptic dopamine reuptake system, such as ^{11}C-nomifensine, ^{18}F-GBR13,119, ^{11}C-cocaine, ^{11}C-CFT, ^{11}C-WIN35,428, ^{123}I-β-CIT.WIN35,428, and ^{123}I-β-CIT. Dopamine receptors can be grouped into two major families: one including D1 and D5 receptors and the other including the D2, D3, and D4 receptors. PET

tracers to measure D1 and D2 have been developed. Wagner et al., using 11C-N-methyl-spiperone, a D2 receptor antagonist, reported the first visualization of dopamine receptors in live human subjects with PET [77]. Subsequently, several other D2-receptor tracers have been synthesized including ^{11}C-raclopride, ^{18}F-haloperidol, ^{76}Br-bromospirone, and ^{18}F-fluoro-ethyl-spiperone [78]. For SPECT studies of the D2 receptors ^{123}I-Iodobenzamide has been used [79]. Specific D1 ligands SCH 23,390 and SCH 39,166 labeled with ^{11}C have also allowed investigation of D1-receptor subtypes in human subjects with PET [80, 81].

Only a few tracers are available for assessing the serotoninergic system [82], including ^{11}C-ketanserin, ^{18}F-setoperone, and ^{18}F-altanserin. Tracers for central-type benzodiazepine binding sites, which are postsynaptic membrane receptor ionophore complexes with a GABA$_A$ receptor, have also been developed for studies in humans, including ^{11}C-flumazenil and ^{123}I-iomazenil.

Tracers have been developed to assess cholinergic presynaptic function including acetylcholinesterase activity, by ^{11}C-N-methyl-piperydinil-proprionate [83] and the vesicular acetylcholine transporter, by vesamicol and benzovesamicol labeled with either ^{11}C or ^{18}F or ^{123}I. Nicotinic receptor function assessment has been pursued with ^{11}C-labeled nicotine [84], whereas muscarinic receptor function assessment [85] has been evaluated with ^{123}I-quinuclinidylbenzilate, ^{11}C-scopolamine, ^{11}C-tropanylbenzilate, and ^{11}C-Nmethyl-piperydil-benzilate. A major problem with muscarinic receptors is still the lack of radiolabeled tracers for differentiating cerebral M1 postsynaptic receptors from presynaptic M2 subtype.

Opiate receptors have been studied with two ligands: ^{11}C-carfentanil, a potent opiate agonist highly selective for mu receptors [86], and ^{11}C-diprenorphine, a partial agonist of the same system but with no specificity for the opiate receptor subtypes mu, delta, or kappa. Delta receptors, on the other hand, can be imaged using ^{11}C-methyl-naltrindole [87]. The histamine receptors of the H1 subtype are involved with arousal, locomotor activity, appetite, cardiovascular regulation, and thermoregulation. They can be imaged and saturated with ^{11}C-pyrilamine [88]. Finally, the activity of the mitochondrial enzyme monoamine oxidase B can be investigated by using ^{11}C-L-deprenyl [89].

Somatostatin (SST) and vasoactive intestinal peptide (VIP) analogs have been used with success in imaging tumor surface receptors. SST and VIP receptors are widely distributed throughout the human body but are overexpressed on tumor cells. SST acts as a neurotransmitter or hormone depending on the site of action and target cell type. VIP is a neuroendocrine mediator with a broad range of biological activities in various cells and tissues. Because SST is readily attacked by aminopeptidases and endopeptidases and has a short half-life,

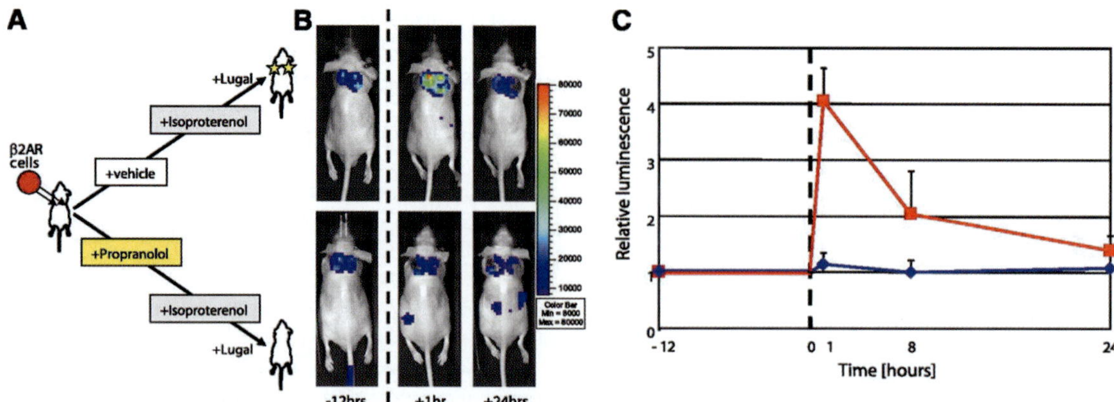

Figure 8.11. β2-adrenergic receptor blocker pharmacodynamics monitored *in vivo* using bioluminescence imaging. Cells expressing the β2 adrenergic receptor construct as well as Fluc and Ras were injected into the backs of BALB/c nude mice. **(A)** At 7 to 14 days after cell implantation, when cells had grown into small tumors, baseline luminescence was acquired by i.p. injection of lugal. Mice were pretreated with the β-adrenergic receptor antagonist propranolol (3 mg/kg i.p.) or vehicle. One hour later, all mice were injected with the agonist isoproterenol (6 mg/kg i.p.) or vehicle. **(B,C)** Isoproterenol injection induced a robust, ~fourfold luminescence increase (red line). In contrast, propranolol pretreatment completely abrogated isoproterenol-induced GPCR activation (blue line), showing the specificity of the system. Thus, this technology could be used to test activators as well as inhibitors of GPCRs (mean ± SE; $n = 6$/group). Reproduced with permission from von Degenfeld et al.

synthetic analogs have been developed for imaging. Three commercially available SST analogs are octreotide, lanreotide, and vapreotide. These receptor radiopharmaceuticals are effective in diagnosing and staging tumors and their metastases due to binding to SST receptors [90]. ^{123}I-labeled VIP and ^{99m}Tc-labeled VIP analog are also promising tumor tracers [91].

G-protein-coupled receptors (GPCRs) are a versatile and ubiquitous family of membrane receptors that transmit extracellular signals to mammalian cells and constitute the most important class of drug targets. Yet sensitive and specific methods are lacking that would allow quantitative comparisons of pharmacologic properties of these receptors in physiological or pathological settings in live animals. Von Degenfeld et al. [92] sought to overcome these limitations by employing low-affinity, reversible beta-galactosidase complementation (i.e., a sensing system not based on split reporter technology but instead based on the principle of lacZ intracistronic complementation, previously described by the same authors [93]) to quantify GPCR activation via interaction with beta-arrestin. A panel of cell lines was engineered expressing different GPCRs together with the reporter system. *In vitro* evaluation revealed highly sensitive, dynamic, and specific assessment of GPCR agonists and antagonists. Following implantation of the cells into mice, it was possible for the first time to monitor pharmacological GPCR activation and inhibition in their physiological context by noninvasive Fluc bioluminescence imaging in living animals (Figure 8.11). This technology has unique advantages that may enable novel applications in the functional investigation of GPCR modulation in live animals in biological research and drug discovery.

Imaging of Receptor Conformational Changes

Paulmurugan et al. have used the estrogen receptor (ER) as an example of a receptor targeted for imaging its conformational changes upon ligand interaction [94]. Indeed, strategies for high-throughput analysis using reporter gene imaging of interactions between various hormones and drugs with the ER are crucial for accelerating the understanding of ER biology and pharmacology. Through careful analyses of the crystal structures of the human ER (hER) ligand-binding domain (hER-LBD) in complex with different ligands, it was hypothesized that the hER-LBD intramolecular folding pattern could be used to distinguish ER agonists from selective ER modulators and pure antiestrogens. They therefore constructed and validated intramolecular folding sensors encoding various hER-LBD fusion proteins that could lead to split Renilla/firefly luciferase reporter complementation in the presence of appropriate ligands (Figure 8.12a). A mutant hER-LBD with low affinity for circulating estradiol was also identified for imaging in living subjects. Cells stably expressing intramolecular folding sensors expressing wild-type and mutant hER-LBD were used to image ligand-induced intramolecular folding in living mice (Figure 8.12b). This was the first hER-LBD intramolecular folding sensor suited for high-throughput quantitative analysis of interactions between hER with hormones and drugs using cell lysates,

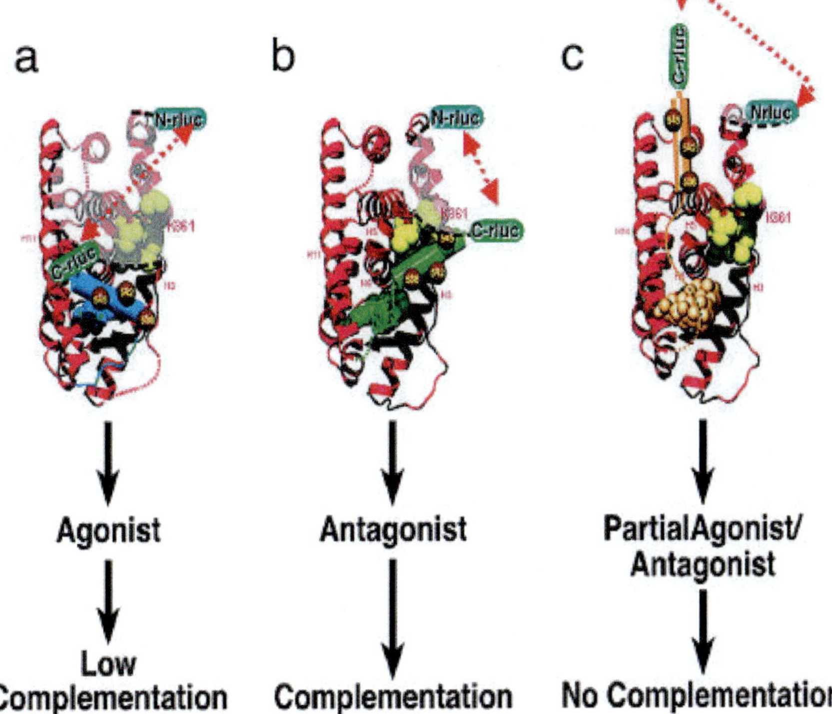

Figure 8.12a. Schematic representation of the hypothetical model of ligand-induced intramolecular folding of ER that leads to split RLUC complementation. The N- and C-terminal fragments of split RLUC were fused to the N and C terminus, respectively, of the hERα of various lengths (amino acids 281–549 and 281–595). Binding of ER ligands to the intramolecular folding sensor (N-RLUC-hER-C-RLUC) induces different potential folding patterns in the LBD based on the type of ligand. This folding leads to split RLUC complementation for ER antagonist **(A)** (H12 and ligands are colored green), low complementation for ER agonist **(B)** (H12 and ligands are colored blue), and no complementation for partial ER agonist/antagonist **(C)** (H12 and ligands are colored gold) with the selective folding sensor. Even though the distance between the N- and C-RLUC fragments after binding with partial agonist **(C)** is smaller than that of agonists **(B)**, this model depicts the importance of the orientations of the split RLUC fragments in complementation. The yellow spheres are hydrophobic amino acids located between helix 3 and helix 5 of LBD. Reproduced with permission from Paulmurugan and Gambhir (2006).

intact cells, and molecular imaging of small living subjects. Importantly, the strategies developed can also be extended to study and image other important protein intramolecular folding systems.

Imaging of Receptor Dimerization

BRET technology involves the nonradioactive transfer of energy between donor and acceptor molecules by the Förster mechanism [95] (Figure 8.13). The energy transfer primarily depends on (i) an overlap between the emission and excitation spectra of the donor and acceptor molecules, respectively, and (ii) the proximity of <100 Å between the donor and acceptor entities. As F(fluorescence)RET/BRET-based technologies assume more prominent roles in the field of studying protein–protein interactions, manufacturers are continually developing new instrumentations for measuring FRET/BRET ratios, which are, in general, low-intensity signals. BRET measurements are usually obtained with a microplate reader equipped with specific filter sets for detecting donor and acceptor emission peaks. This cellular assay has been applied to real-time imaging of cells, high-throughput screening of drugs, and small-animal and plant models. We [96] used the BRET[2] system (Biosignal Packard), which involves Rluc as a bioluminescent donor, and mutant green fluorescent protein (GFP[2]) as a fluorescent acceptor, adapted for expression in mammalian cells and characterized by a significantly red-shifted Stokes shift that emits transferred energy at 508 nm. The resonance energy transfer from the reaction of the reconstructed Rluc protein with its substrate Deep Blue Coelenterazine (DBC) excites the GFP[2] protein when interaction of the two fused proteins Id and MyoD or FKBP12 and FRB in the presence of a small molecule mediator (rapamycin) occurs. We demonstrated the ability to detect signal from PPIs in cultured cells, as well as from the surface and deeper tissues of small living

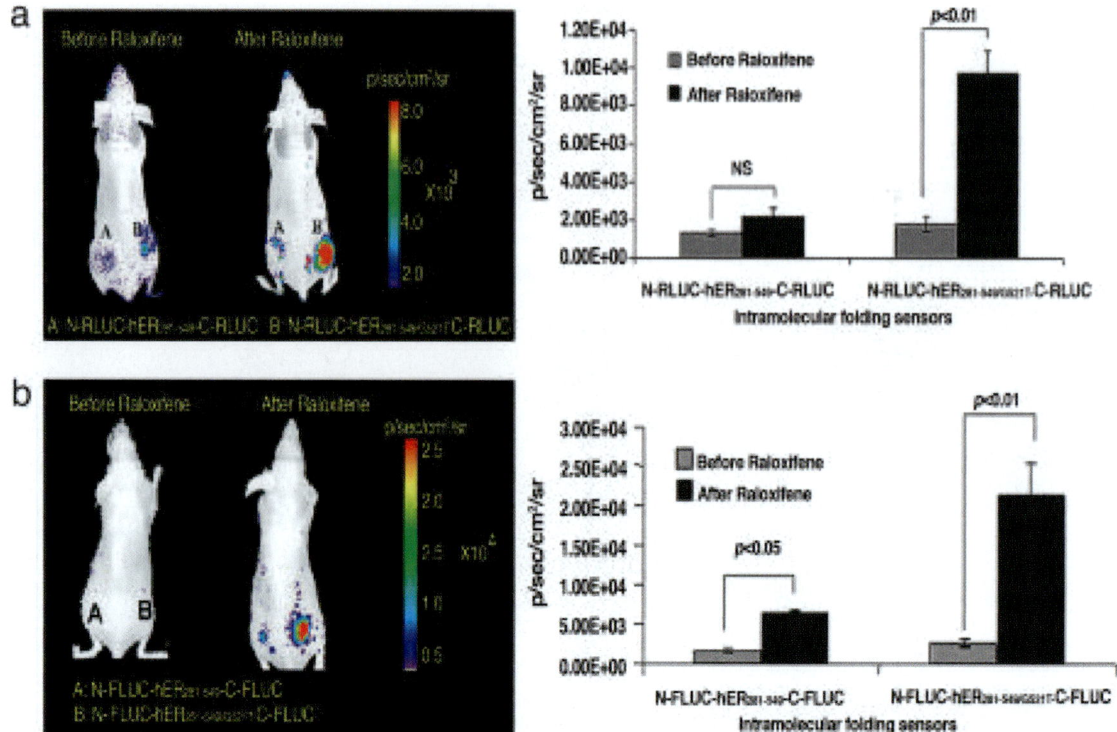

Figure 8.12b. Bioluminescence imaging of ER antagonist-induced intramolecular folding in a mouse model. **(A)** Shown is optical CCD camera imaging of 293T cells stably expressing intramolecular folding sensors N-RLUC-hER$_{281-549}$-C-RLUC and N-RLUC-mutant-hER$_{281-549/G521T}$-C-RLUC in living female nude mice before and after treatment with antagonist raloxifene (0.5 mg per mouse) and the corresponding quantitative graph. **(B)** Similar imaging conducted by using the same sensors with the split FLUC fragment system (N-FLUC-hER$_{281-549}$-C-FLUC and N-FLUC-mutant-hER$_{281-549/G521T}$-C-FLUC). The site implanted with the cells expressing the intramolecular folding sensor with the mutant hER (G521T) shows a higher RLUC complementation signal after raloxifene treatment compared with that of the wild-type hER. Reproduced with permission from Paulmurugan and Gambhir (2006).

animals with implanted cells overexpressing the fusion constructs (Figure 8.13).

We recently showed that the BRET^2 assay sensitivity can be significantly improved by using Rluc mutants with improved quantum efficiency and/or stability (e.g., Rluc8 and RlucM) as a donor [97, 98]. To extend the time of light measurement, we also developed CLZ400 (also known as bisdeoxycoelenterazine) analogs, showing that signal from our improved BRET^2 vector can be monitored for up to 6 h. This approach, currently undergoing continued validation, should have important implications for the study of protein–protein interactions in cells maintained in their natural environment, particularly if it can be effectively applied to evaluate new pharmaceuticals.

Further advances in this field have led us to develop a highly photon-efficient, self-illuminating fusion protein combining a mutant red fluorescent protein (mOrange) and a mutant Rluc (Rluc8) [99]. This new BRET fusion protein (BRET^3) exhibits severalfold improvement in light intensity in comparison to existing BRET fusion proteins. BRET^3 also exhibits the most red-shifted

light output (564 nm peak wavelength) of any reported bioluminescence protein that uses its natural coelenterazine substrate, a benefit of which can be demonstrated at various tissue depths in small animals. The imaging utility of BRET^3 at the single-cell level was demonstrated using an intramolecular sensor incorporating two mTOR pathway proteins (FKBP12 and FRB) that dimerize only in the presence of rapamycin. With increased photon intensity, red-shifted light output, and good spectral resolution (~85 nm), BRET^3 showed improved spatial and temporal resolution for measuring intracellular events in single cells and when using living small-animal models. The development of further BRET^3-based assays will allow imaging of PPIs using a single assay directly scalable from intact living cells to small living subjects, allowing for potential accelerated drug discovery.

The main advantages of BRET, when compared to FRET, arise through its high sensitivity for measurement of interactions avoiding the consequences of the required excitation of the donor with an external light source. BRET assays show no photo bleaching or photoisomerization of the donor protein, no photo damage to cells,

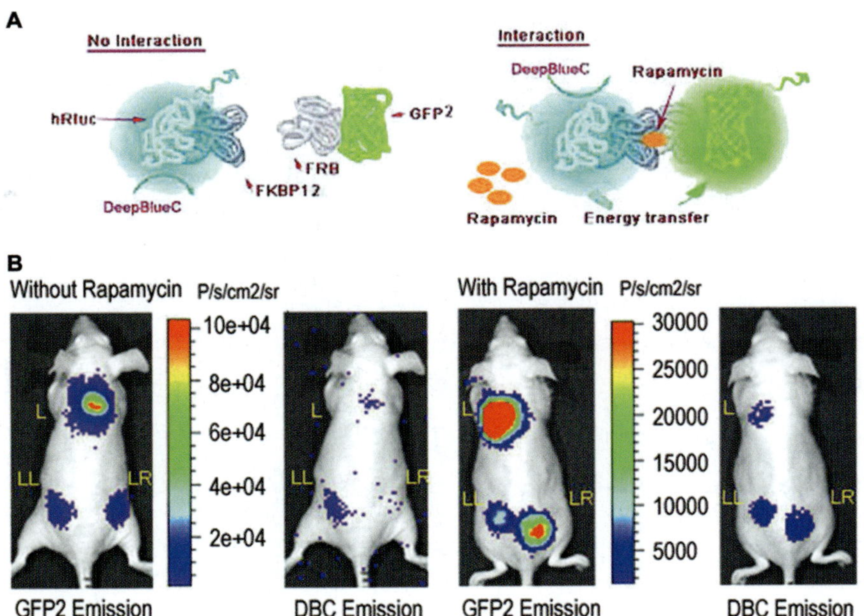

Figure 8.13. (A) Schematic showing small molecule-mediated protein–protein interaction leading to bioluminescence resonance energy transfer (BRET). FKBP12 is fused to the N-terminus of RLUC donor protein, and a FRB is fused to the C-terminus of GFP^2 acceptor protein. When these genes encoding for both of these two fusion proteins are expressed inside cells and rapamycin is present to mediate FRB-FKBP12 interaction, then resonance energy transfer occurs. This BRET signal can be detected by using the Deep Blue Coelenterazine (DBC) substrate for RLUC. **(B)** Detection of *in vivo* BRET^2 signal from specific protein–protein interaction. Dorsal view of a nude mouse implanted s.c. with 5×10^6 293T cells either transiently transfected with p*BRET^2* (L) or with p*FKBP12-hRluc* (LL) alone or cotransfected with p*FKBP12-hRluc* and p*GFP^2-FRB* (LR) in presence (right panel) or absence (left panel) of rapamycin. Mice that received the small molecule mediator drug, rapamycin, (5 mg/kg) were injected i.p. immediately after cell implantation. The scan was performed 7 h after drug administration. Mice were scanned for 5 min integration time using either GFP^2 or DBC filters in succession by injecting with 25 μg DBC intravenously. Reproduced with permission from De and Gambhir (2005).

and no light scattering or autofluorescence from cells or microplates (when used *in vitro*), which can be caused by incident excitation light. In addition, one main advantage of BRET over FRET is the lack of emission arising from direct excitation of the acceptor. This reduction in background should permit detection of interacting proteins at much lower concentrations than possible for FRET. Adapting BRET technology for imaging protein–protein interactions in living subjects is currently being validated with a view to effective application in evaluating new pharmaceuticals.

Imaging of Protein Phosphorylation

Protein phosphorylation mediated by protein kinases has been studied recently using a genetically encoded, generalizable, split Fluc-assisted complementation system [100]. This was developed for noninvasive monitoring of phosphorylation events and efficacies of kinase inhibitors in cell culture and in small living subjects by optical bioluminescence imaging. The serine/threonine kinase Akt mediates mitogenic and antiapoptotic responses that result from activation of multiple signaling cascades.

It is considered a key determinant of tumor aggressiveness and is a major target for anticancer drug development. An Akt sensor (AST) was constructed to monitor Akt phosphorylation and the effect of different PI-3K and Akt inhibitors (Figure 8.14a). Specificity of AST was determined using a nonphosphorylable mutant sensor containing an alanine substitution (ASA). It was found that the PI-3K inhibitor LY294002 and Akt kinase inhibitor perifosine led to temporal- and dose-dependent increases in complemented Fluc activities in 293T human kidney cancer cells stably expressing AST (293T/AST) but not in 293T/ASA cells. Inhibition of endogenous Akt phosphorylation and kinase activities by perifosine also correlated with an increase in complemented Fluc activities in 293T/AST cells but not in 293T/ASA cells. Treatment of nude mice bearing 293T/AST xenografts with perifosine led to a twofold increase in complemented Fluc activities compared to that of 293T/ASA xenografts (Figure 8.14b). Our system was used to screen a small chemical library for novel modulators of Akt kinase activity. We foresee that this generalizable approach for noninvasive monitoring of phosphorylation events will accelerate the discovery and

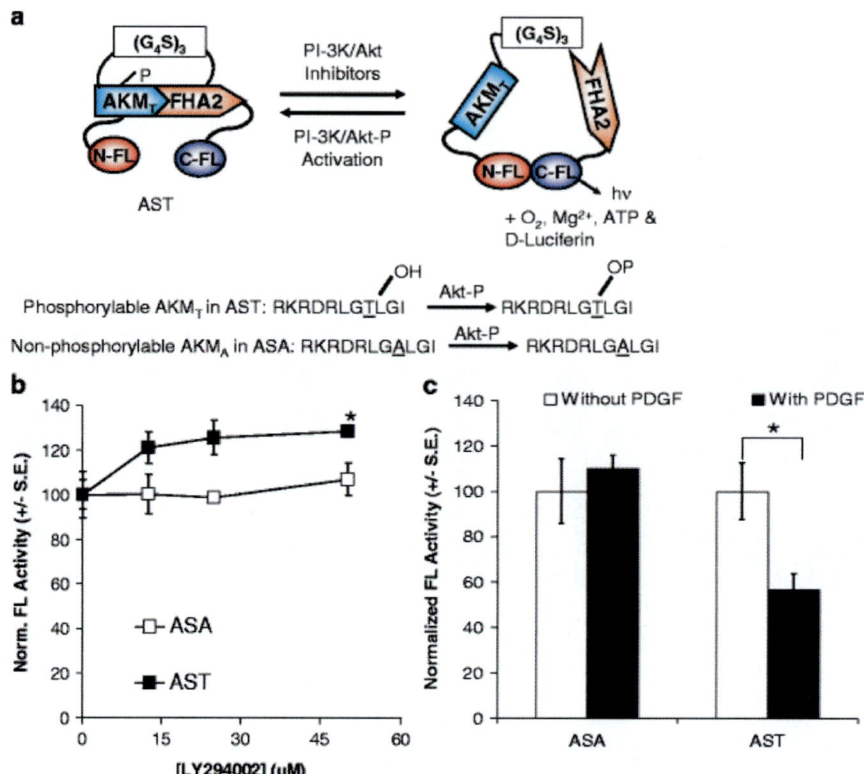

Figure 8.14a. A schematic diagram for Akt kinase sensor (*AST*). Inhibition of Akt kinase activity by PI-3K/Akt inhibitors leads to decreased phosphorylation of AST at the Akt kinase motif (*AKM_T*) and interaction with the phosphothreonine binding domain (*FHA2*). This leads to increased complementation of split FL fragments (*NFL* and *CFL*) and light production in the presence of the FL substrate D-luciferin. On the other hand, activation of Akt kinase activity (by *PI-3K/AKT-P*) leads to phosphorylation of AST at the AKM_T and increased interaction with FHA2, thus hindering complementation between NFL/CFL. The ASA sensor with a nonphosphorylable AKM_A motif served as a negative control. **(B)** Inhibition of Akt kinase activity led to increase in complemented FL activity in BT474 cells transiently transfected with AST. BT474 cells were transiently transfected with ASA or AST for 24 h in the presence of the PI-3K inhibitor LY294002 or carrier control prior to bioluminescence imaging of intact cells upon addition of D-luciferin. Total flux (complemented FL activities) was normalized for transfection efficiency using RL activities, protein content, and carrier control treated cells (100%). LY led to a dose-dependent increase in complemented FL activities in BT474 cells transiently transfected with AST compared to that of carrier control treated cells. *$p < 0.05$ relative to carrier control treated cells. **(C)** Activation of Akt kinase activity by platelet-derived growth factor led to decrease in complemented FL activity. 293T cells transiently transfected with ASA or AST for 24 h were treated with PDGF or carrier control for 30 min prior to analysis of complemented FL activity as described in 1B. In 293T cells transiently transfected with AST, PDGF led to a decrease in complemented FL activity. On the other hand, in 293T cells transiently transfected with ASA, PDGF did not lead to significant decrease in complemented FL activity. *$p < 0.05$ relative to carrier control treated cells. Reproduced with permission from Chan et al. (2008a).

validation of novel kinase inhibitors and modulators of phosphorylation events.

Zhang et al. [101] have also described a new reporter molecule whose bioluminescence activity within live cells and in mice can be used to measure Akt activity. Akt activity in cultured cells and tumor xenografts was monitored quantitatively and dynamically in response to activation or inhibition of receptor tyrosine kinase, inhibition of phosphoinositide 3-kinase, or direct inhibition of Akt (Figure 8.15a, b). The results provided unique insight into the pharmacokinetics and pharmacodynamics of agents that modulate Akt activity, revealing the usefulness of this reporter for rapid dose and schedule optimization in the drug development process.

Having constructed a genetically engineered hybrid bioluminescent Akt reporter (BAR) molecule that reports on Akt serine/threonine kinase activity (containing an Akt consensus substrate peptide consisting of a domain that binds phosphorylated amino acid residues (FHA2) flanked by nFluc and cFluc reporter domains), the same authors described a modified version of this reporter molecule (myristoylated and palmitoylated bioluminescent Akt reporter (MyrPalm-BAR)), which is membrane bound and whose bioluminescence activity can be used to monitor Akt activity at the cell membrane [102]. This was based on the fact that Akt is recruited to the plasma membrane upon activation. Using changes in Akt activation status with small-molecule inhibitors of

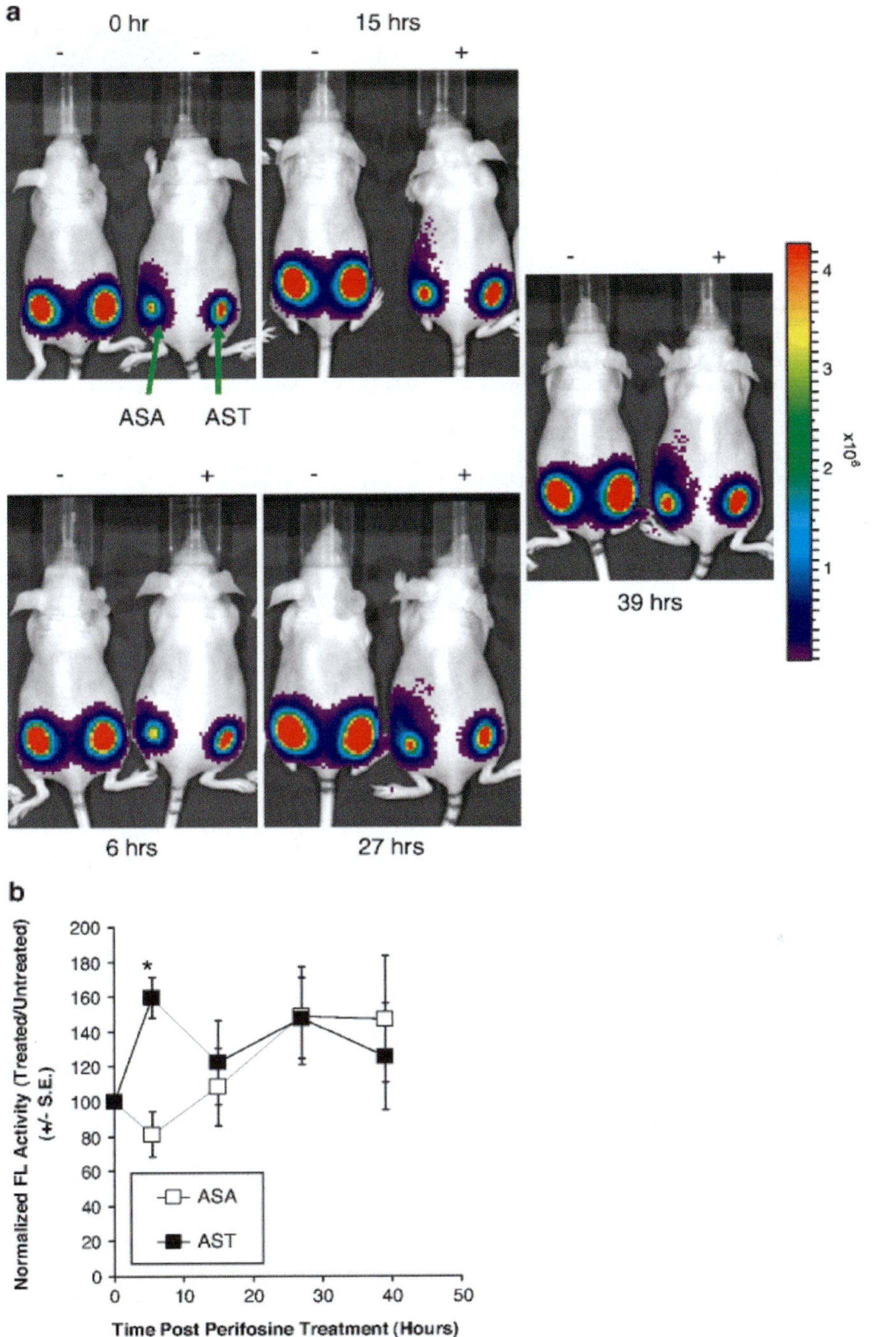

Figure 8.14b. Inhibition of Akt kinase activity by perifosine in living mice. **(A)** The efficacy of perifosine in inhibition of Akt kinase activity in living mice was determined by bioluminescence imaging of nude mice bearing 293T/ASA (*left*) and 293T/AST (*right*) xenografts. Upon determination of baseline signals at time 0 h, mice were treated with 30 mg/kg of perifosine or carrier control and reimaged at 6, 15, 27, and 39 h upon i.p. injection of D-luciferin. **(B)** Max photons at each time point were normalized to that of carrier control treated mice and expressed as average normalized max photons $\pm$ SEM for each xenograft. *$p < 0.05$ relative to mice bearing 293T/ASA xenografts. Reproduced with permission from Chan et al. (2008a).

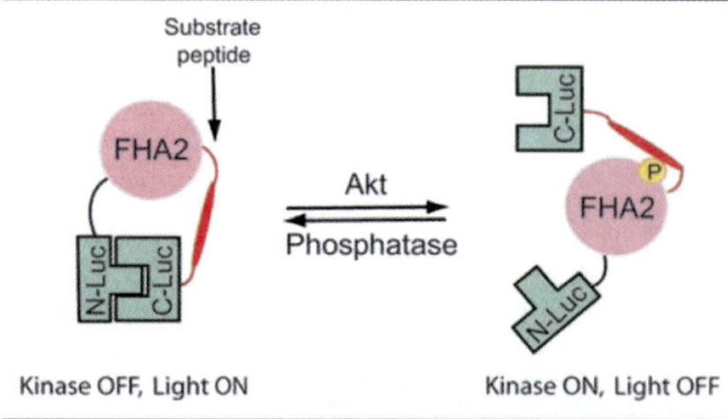

Figure 8.15a. The proposed mechanism of action for the BAR reporter involves Akt-dependent phosphorylation of the Aktpep domain (thick line), which results in its interaction with the FHA2 domain (right). In this form, the reporter has minimal bioluminescence activity. In the absence of Akt activity, the N-Luc and C-Luc domains reassociate, restoring bioluminescence activity (left). Reproduced with permission from Zhang et al. (2007).

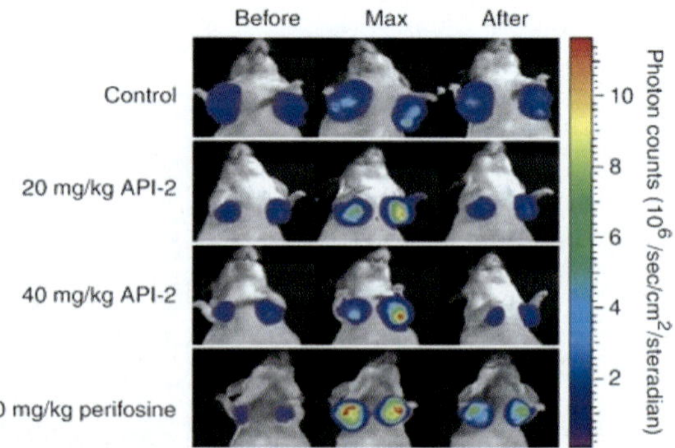

Figure 8.15b. Tumor-bearing mice were treated with vehicle control (20% DMSO in PBS), API-2 (20 mg/kg or 40 mg/kg), or perifosine (30 mg/kg). Images of representative mice are shown before treatment, during maximal luciferase signal upon treatment (Max), and after treatment. Reproduced with permission from Zhang et al. (2007).

Akt, they demonstrated that the membrane-targeted Akt reporter was more sensitive and quantitative. In addition, inhibition of upstream signaling kinases such as epidermal growth factor receptor and phosphatidylinositol 3-kinase activity resulted in changes in Akt activity that were quantitatively monitored by bioluminescence imaging. Based on these results, the authors proposed that the membrane-associated Akt reporter may be better suited for high-throughput screening and identification of novel compounds that modulate the Akt pathway.

Imaging of Cytoplasmic Protein–Protein Interactions

An important application of *in vivo* molecular imaging focuses on characterization of interacting protein partners along signal transduction pathways. The imaging of such intracellular protein–protein interactions is an area of current intensive research. The use of protein–protein interaction imaging in various combinations should help monitor different components of intracellular pathways and networks, including their application to logical circuitry analysis within cells to provide protein-based and transcription-based biological "computation" [103], with potential for future extrapolation to imaging in living subjects. Imaging interacting protein partners in living subjects could also pave the way to functional proteomics in whole animals, the assessment of dysfunctional signaling networks in diseased cells, and the ability to provide a tool to evaluate new pharmaceuticals targeted to modulate protein–protein interactions along signaling pathways.

Functional proteins can be assembled from one or more noncovalently attached polypeptides with the occurrence and efficiency of assembly used for measuring real-time protein–protein interactions. This process suggests that synthetically separated fragments of a single

polypeptide might be able to complement and give rise to an enzymatically active protein, particularly if the interaction is helped by fusion of the halves to strongly interacting moieties. In the "split protein" strategy, a single reporter protein/enzyme is cleaved into N-terminal and C-terminal segments; each segment is fused to one of two interacting proteins (X and Y). Physical interactions between the two proteins X and Y reconstitute the functional reporter protein leading to signal generation that can be measured. This split protein strategy can work either through protein-fragment complementation assays (PCA), or intein-mediated reconstitution assays. In the latter case reconstitution of the full reporter protein occurs, unlike in the former where no actual chemical reconstitution occurs.

The PCA strategy for detecting protein–protein interactions was first demonstrated by Pelletier et al. using the enzyme dihydrofolate reductase (DHFR) [104] following inspiration from a 1994 paper by Johnsson and Varshavsky describing what they called the "ubiquitin split protein sensor" [105]. In all PCAs, splitting a specific reporter protein into two distinct fragments abolishes its function. Bringing the two fragments back together in a controlled manner then restores functional activity [106] (Figure 8.16). Selected fragments of many proteins can associate to produce functional bimolecular complexes [107]; the PCA system can therefore be generalized for a number of enzymes for PPI detection, examples including DHFR, glycinamide ribonucleotide (GAR) transformylase, aminoglycoside and hygromycin B phosphotransferases, all reviewed by Michnick et al. [106], *E. coli* TEM-1 β-lactamase [108, 109], green fluorescent protein and its variants [107], and the molecular imaging reporters Fluc [110] and Rluc [111].

Ozawa et al. [112] initially demonstrated that firefly luciferase can be split between amino acid positions 437 and 438 and used with inteins (DnaE) in a reconstitution

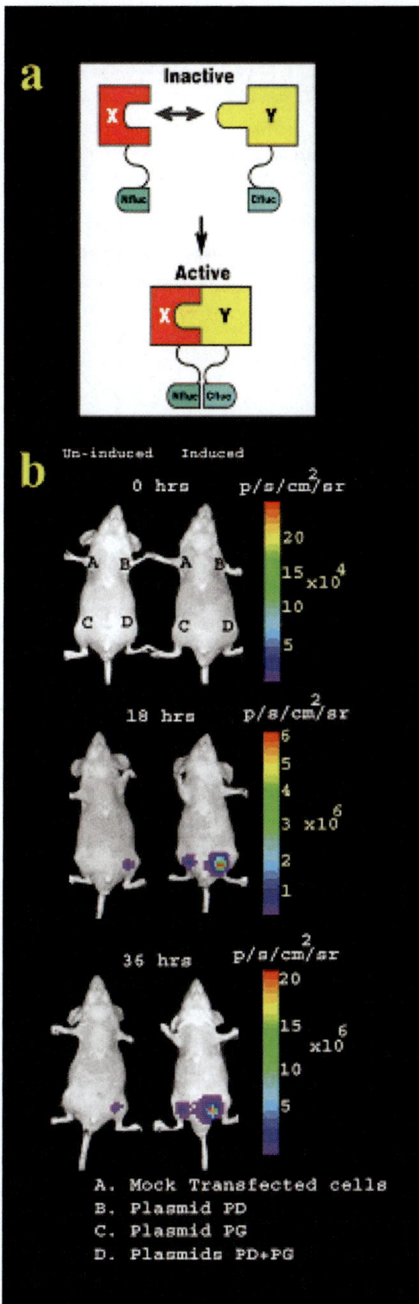

strategy to detect insulin-induced interaction of phosphorylated insulin receptor substrate 1 (IRS-1) and its target N-terminal SH2 domain of PI 3-kinase in a cell culture assay. We subsequently reasoned that it may be possible to split firefly luciferase and use split reporter complementation *without* inteins. We therefore studied PCA and intein-mediated reconstitution of firefly luciferase fragments and found that a complementation strategy was as sensitive as the intein-mediated reconstitution strategy under the conditions tested [110].

We demonstrated for the first time the feasibility of imaging protein–protein interactions using split reporters in small living animals. We studied a PCA based on split firefly luciferase (cleaved into two fragments nFluc: residues 1–437; and cFluc: residues 438–550), using the interaction of Id and MyoD as test proteins [110] (Figure 8.16). Separately, Luker et al. [113] described a systematic truncation library yielding alternative complementary N and C fragments of firefly luciferase (nFluc: residues 2–416; and cFluc: residues 398–550). These fragments were used to monitor rapamycin-mediated interaction of rapamycin-binding proteins FRB and FKBP12. We similarly used the firefly luciferase fragments previously tested with Id and MyoD to study rapamycin-mediated interactions and found the complementation to be too weak for optical imaging in living animals using the CCD camera (unpublished data). Further studies on reporter complementation assays for imaging of PPIs in living subjects led us to use a combinatorial strategy to identify a novel split site for Fluc with improved imaging characteristics over previously published split sites [114]. This newly developed split Fluc system was recently used to study the crucial role of tumor hypoxia in tumorigenesis [115] as well as the application to image protein phosphorylation mediated by protein kinases described previously [100].

The enzyme Renilla luciferase (or the synthetically mutated humanized version, hRluc), is a 36-kDa monomeric bioluminescence imaging reporter protein, the smallest optical reporter protein identified to date for studying protein–protein interactions in a PCA strategy [111, 116]. This PCA strategy, using N- and C-terminal halves of split Renilla luciferase functions in both cell

Figure 8.16. Schematic diagram of the split reporter-based complementation strategy used to optically image protein–protein interactions in living mice. The N-terminal half of firefly luciferase is attached to protein X through a short peptide FFAGYC, and the C-terminal half of firefly luciferase is connected to protein Y through the peptide CLKS. Interaction of proteins X and Y recovers Fluc activity through protein complementation **(A**, top panel). *In vivo* optical cooled CCD imaging of mice carrying transiently transfected 293T cells for the induction of the complementation-based split luciferase system. All images shown are the visible light image superimposed on the optical CCD bioluminescence image with a scale in photons/second/cm²/steradian (sr). Mice were imaged in a supine position after IP injection of D-luciferin. **(B**, lower panel) A set of nude mice was repetitively imaged after s.c. implantation of 293T cells transiently transfected with various plasmids as described in reference (Paulmurugan et al. (2002)). One group of mice was induced with TNFα, and the other group was not induced. The images are from one representative mouse from each group immediately after implanting cells (0 h) and 18 and 36 h after TNFα induction. The induced mouse showed higher Fluc signal at site D (where interacting proteins result in reporter protein complementation) when compared with the mouse not receiving TNFα. The Fluc signal significantly increases after receiving TNFα. Reproduced with permission from Paulmurugan et al. (2002).

culture and living animals, has been demonstrated with several different protein partners. Because of its smaller fragment size than Fluc, Rluc has less hindrance with interacting protein partners, and works more efficiently with different imaging assays. Another clear advantage of using the split Rluc system when compared to the split Fluc is that the former's enzymatic reaction is ATP independent and therefore could be used in specific situations where the PPI under study itself requires ATP. We used fragments generated by splitting between residues 229 and 230 to study rapamycin-induced interaction of human proteins FRB and FKBP12 [117] (Figure 8.17). Moreover, protein interaction between IRS1 and the SH2 domain of PI3K in the insulin-signaling pathway was located in living mammalian cells using Rluc split between residues 91 and 92 [111]. One limitation associated with the use of Rluc is its relatively rapid reaction kinetics requiring early time-point measurements [118]. Nevertheless, this split reporter system appears highly suitable for studying protein–protein interactions in cells and in living animals owing to its optical bioluminescence nature and amplifiable signal through an enzymatic process. Further, the complementation strategies based on Rluc fragments, with smaller fragment size than firefly luciferase, have less hindrance with interacting protein partners and work more efficiently with different imaging assays including protein–protein interactions, small molecule-induced protein–protein interactions, small molecule-mediated inhibition of protein–protein interactions, and protein homodimerization [116, 117, 119, 100, 63].

Imaging of Nucleocytoplasmic Trafficking

Nucleocytoplasmic trafficking of functional proteins plays a vital role in regulating gene expressions in response to extracellular signals. Kim et al. developed a genetically encoded bioluminescence indicator for monitoring the nuclear trafficking of target proteins *in vitro* and *in vivo* [120]. The principle is based on reconstitution of split fragments of Renilla luciferase (Rluc) by protein splicing with a DnaE intein (a catalytic subunit of DNA polymerase III). A target cytosolic protein fused to the N-terminal half of Rluc is expressed in mammalian cells. If the protein translocates into the nucleus, the Rluc moiety meets the C-terminal half of Rluc, and full-length Rluc is reconstituted by protein splicing (Figure 8.18a). They demonstrated quantitative cell-based *in vitro* sensing of ligand-induced translocation of the androgen receptor, which allowed high-throughput screening of exogenous and endogenous agonists and antagonists. Furthermore, the indicator enabled noninvasive *in vivo* bioluminescence imaging of the androgen receptor translocation in the brains of living mice with the CCD imaging system (Figure 8.18b). These rapid and quantitative analyses *in vitro* and *in vivo* provide a wide variety of potential

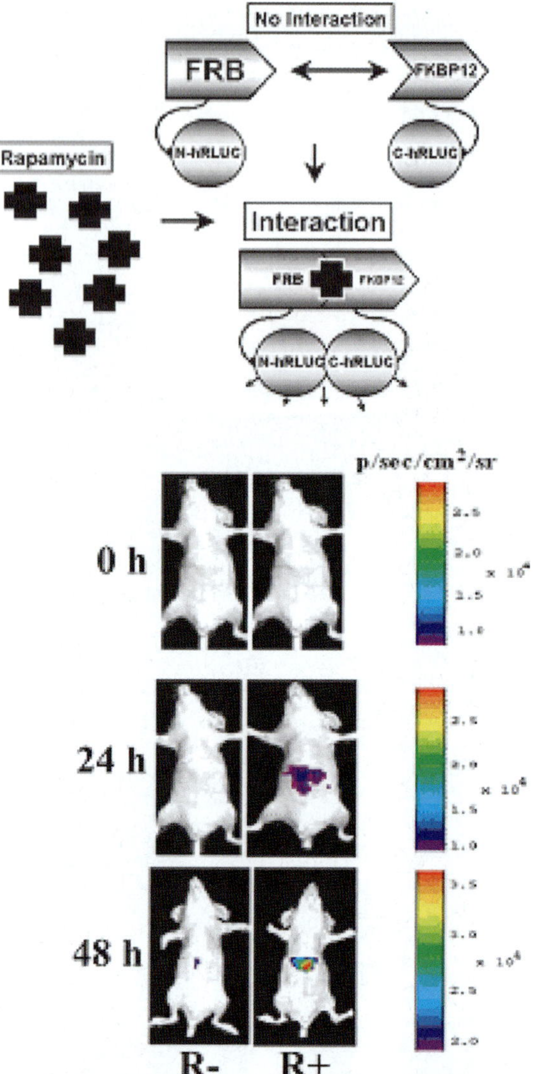

Figure 8.17. Schematic diagram of rapamycin-mediated synthetic Renilla luciferase (*hRLUC*) protein fragment-assisted complementation strategy. In this strategy, N-terminal and COOH-terminal portions of hRLUC fragments are attached to proteins X and Y, respectively, through a short peptide linker GGGGSGGGGS. The N and C portions of hRLUC fragments are closely approximated by the dimerization of proteins FRB and FKBP12 only in the presence of the small molecule rapamycin, and this, in turn, leads to recovered activity of the hRLUC protein. Optical charged coupled device imaging of living mice carrying i.v. injected 293T cells transiently cotransfected with Nhrluc-FRB and FKBP12-Chrluc. The animals not receiving rapamycin showed only a mean background signal of $4 \pm 1 \times 10^3$ p/s/cm²/sr at all of the time points studied. The animals receiving repeated injections of rapamycin emitted signals, originating from the region of the liver, that were threefold (mean, 1.6×10^4 p/s/cm²/sr) and fivefold (mean, 3.0×10^4 p/s/cm²/sr) higher than background ($P < 0.05$) at 24 h and 48 h after the injection of rapamycin, respectively. ($R-$, animals not receiving rapamycin; $R+$, animals receiving rapamycin). Reproduced with permission from Paulmurugan et al. (2004).

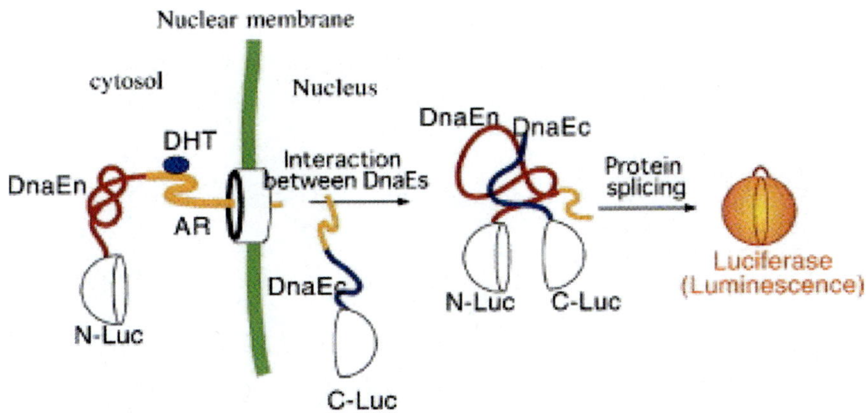

Figure 8.18a. Imaging of receptor trafficking across nuclear membrane. When the androgen receptor (AR) is bound to 5α-dihydrotestosterone (DHT), it translocates into the nucleus and brings the N- and C-terminal halves of DnaEs close enough to fold correctly, thereby initiating protein splicing to link the concomitant Rluc halves with a peptide bond. The C-terminal half of split Rluc was located beforehand in the nucleus by a fused nuclear localization signal. The cells containing this reconstituted Rluc allow one to monitor nuclear translocation of AR with its luminescence by coelenterazine as the substrate. Reproduced with permission from Umezawa et al. (2005).

applications for screening pharmacological or toxicological compounds affecting nucleocytoplasmic translocation of signaling molecules and testing them in living animals.

Imaging of Nuclear Protein–Protein Interactions

It has been estimated that more than 50% of all protein interactions described in the literature have been detected using the yeast two-hybrid system [121, 122]. In a two-hybrid assay, two proteins are expressed in yeast with one fused to a DNA-binding domain (BD) and the other fused to a transcription activation domain (AD). If the two proteins interact, they activate transcription of one or more reporter genes that are distal to binding sites for the BD. Studied first in yeast, this classical two-hybrid system was later adapted for mammalian cells using different expression plasmids but similar assay principles. Our group has adapted this system further (Figure 8.19) using pBIND-Id (which contains the yeast GAL4 DNA-binding domain fused to the Id protein) and pACT-MyoD (which contains the herpes simplex virus VP16 activation domain fused to a segment of the murine MyoD protein) as the two-hybrid proteins [123]. Id and MyoD are the members of the helix-loop-helix family of nuclear proteins and are known to strongly interact *in vivo* during myogenic differentiation [124]. To inducibly modulate the expression of these two-hybrid proteins, we replaced the CMV promoter of pBIND-Id and pACT-MyoD with TNFα inducible NFκB response elements. TNFα is a pleiotropic cytokine secreted by lipopolysaccharide (LPS)-stimulated macrophages that induces a variety of cell-specific events, including NFκB activation, and causes tumor necrosis *in vivo* when injected

in tumor-bearing mice [123]. Finally, we used pG5-luc (which contains five GAL4 binding sites upstream of a minimal TATA box, followed by the firefly luciferase gene) as a two-hybrid reporter. In cells treated with TNFα, activated endogenous NFκB moves to the nucleus and activates transcription of the two hybrid proteins: they in turn interact and activate transcription of the firefly luciferase reporter. We studied this system in detail in cell culture and in 293T cells implanted in mice using cooled CCD camera bioluminescence imaging while using TNFα to modulate the system [123].

Separately, but following a similar design, Luker et al. [125] developed a tetracycline (or doxycycline)-inducible, bidirectional vector carrying in one direction the tumor suppressor p53 gene fused with Gal4, and in the other direction, the T antigen (TAg) of SV 40 fused with VP16. Expression of the p53 and TAg hybrid proteins is induced by doxycycline, resulting in their interaction and formation of a VP16-Gal4 transactivator complex. This complex binds to the Gal4 binding sequences in the promoter of a HSV1-sr39tk-GFP reporter fusion protein. Tumor xenografts of HeLa cells stably expressing both the reporter plasmid and the bidirectional two-hybrid expression plasmid were implanted in living mice, and doxycycline-induced protein–protein interactions were imaged by microPET at different time points. MicroPET imaging was performed using a Fluorine-18 positron labeled analog of penciclovir, which is trapped in cells expressing the reporter fusion (through action of the sr39TK protein). The fusion protein showed correlated increase in expression as detected by fluorescence microscopy (for the GFP component) and by microPET (for the TK component) when induced with increasing doses of doxycycline both in cells and in tumor xenografts of living mice. A subsequent study defined quantitative

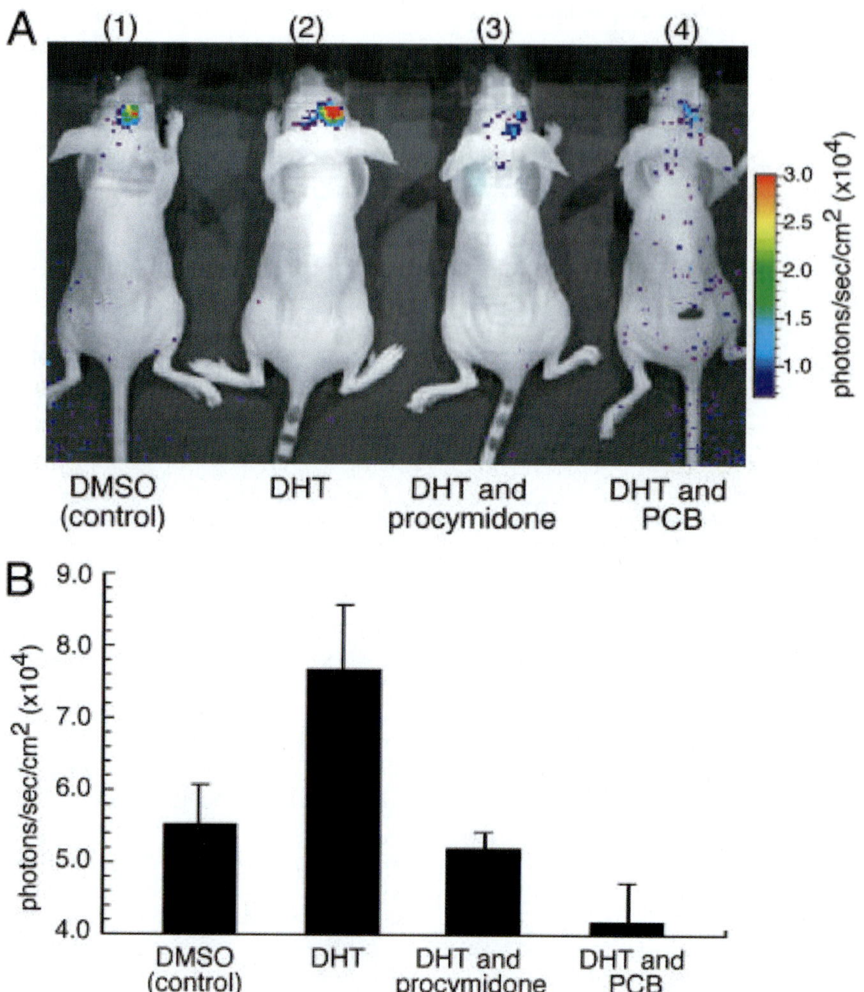

Figure 8.18b. An inhibitory effect of procymidone or PCB on the bioluminescence developed by DHT (10 μg/kg of body weight) in living mice. **(A)** The inhibitory effect of chemicals on AR translocation into the nucleus in the mouse brain. The COS-7 cells transiently cotransfected with pcRDn-NLS and pcDRc-AR were implanted in the forebrain of the nude mice at a depth of 3 mm through a 1-mm burrhole. Of mouse groups 1–4, groups 1 and 2 were stimulated with 1% DMSO, whereas groups 3 and 4 were stimulated with procymidone (10 mg/kg body weight) and PCB (10 mg/kg of body weight), respectively. Two hours after the stimulation, mouse groups 2–4 were then stimulated with DHT (10 μg/kg of body weight). Two hours after DHT stimulation, the mice were imaged in 2-min intervals until reaching the maximum photon counts after intracerebral injection of coelenterazine (1.4 mg/kg of body weight). **(B)** The average of photon counts from each implanted site in *A* ($n = 3$). The averages of three mice were $(5.53 \pm 0.53) \times 10^4$ (group 1), $(7.68 \pm 0.91) \times 10^4$ (group 2), $(5.07 \pm 0.23) \times 10^4$ (group 3), and $(4.18 \pm 0.55) \times 10^4$ (group 4) (photons per sec per cm^2). Reproduced with permission from Kano et al. (2004).

(relative differences in amounts of interacting proteins and expression of reporter gene) and kinetic (time interval between induction of interacting proteins and detection of reporter activity *in vivo*) parameters pertaining to this system [126].

Imaging of Transcriptional Activation of Endogenous Genes

This imaging approach is presented in greater detail elsewhere in this book. Only a summary of some examples is presented here to provide a useful perspective to the more direct strategies of imaging signal transduction presented previously.

The first example of this imaging approach was presented by Green et al., who performed repetitive imaging with microPET of endogenous albumin gene expression [127]. Transgenic mice in which the herpes simplex virus type 1 thymidine kinase (HSV1-tk) reporter gene was driven by the albumin promoter (AL-HSV1-tk) were used for this. These mice were imaged repeatedly in a microPET scanner after administering an imaging probe

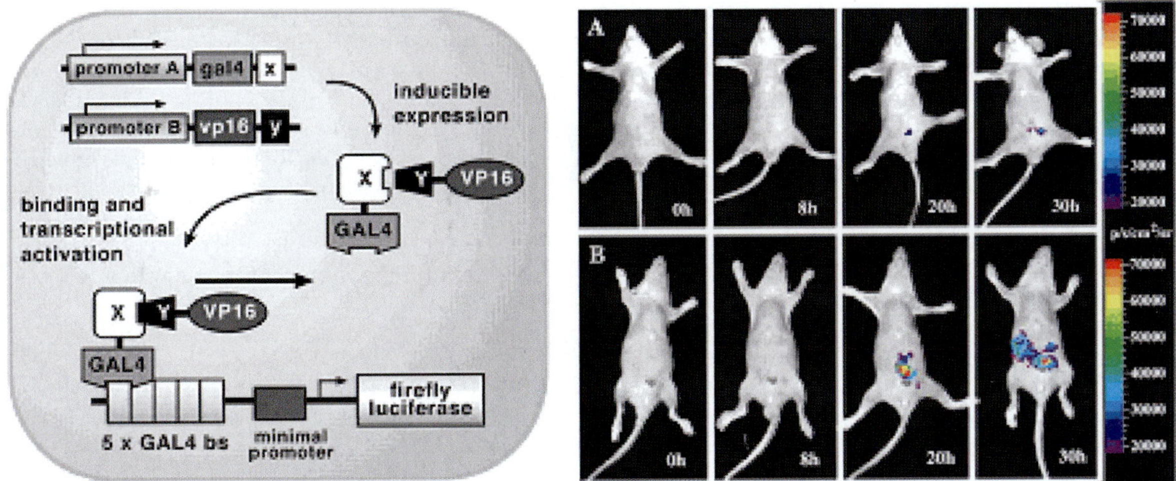

Figure 8.19. Imaging protein–protein interaction in living mice with a modified yeast two-hybrid strategy. Schematic diagram of the system for imaging the interaction of proteins X and Y. The first step involves the vectors pA-gal4-x and pB-vp16-y, which are used to drive transcription of *gal4-x* and *vp16-y* through use of promoters A and B. In the second step, the two fusion proteins GAL4-X and VP16-Y interact because of the specificity of protein X for protein Y. Subsequently, the GAL4-X-Y-VP16 binds to GAL4-binding sites (five GAL4-binding sites (bs) are available) on a reporter template. This leads to VP16-mediated transactivation of firefly luciferase reporter gene expression under the control of GAL4 response elements in a minimal promoter. Transcription of the firefly luciferase reporter gene leads to firefly luciferase protein, which, in turn, leads to a detectable visible light signal in the presence of the appropriate substrate (D-Luciferin), ATP, Mg^{2+}, and oxygen. The NFκB promoter was used for either pA or pB and TNFα-mediated induction. *In vivo* optical CCD imaging of mice carrying transiently transfected 293T cells for induction of the yeast two-hybrid system. All images shown are the visible light image superimposed on the optical CCD bioluminescence image with a scale in photons/sec/cm^2/steridian (sr). Mice in top row were imaged after injection of D-Luciferin but with no TNFα-mediated induction. Mice in bottom row were imaged after injection of D-Luciferin after TNFα-mediated induction, showing marked gain in signal from the peritoneum over 30 h. Reproduced with permission from Ray et al. (2002).

of approximately 200 microCi of 9-[4-[18F]fluoro-3-(hydroxymethyl)butyl]guanine (FHBG) (a substrate for HSV1-TK enzyme). Four transgenic mice were monitored for body weight, serum albumin, and imaged at the end of each of three dietary phases (17%, 0%, and 25% protein diet). Each phase lasted 14–21 d. The 0% protein diet had been reported previously to reduce albumin gene expression in rats. Twenty nontransgenic mice of the same strain followed a similar feeding schedule and were monitored for serum albumin, body weight, and sacrificed at various time points for determination of their GAPDH normalized albumin mRNA levels. As expected, transgenic mice showed a relatively high FHBG signal from the liver region. Variation of the mean FHBG signal in two mice with a fixed 17% protein diet over a 4-month period was <19% s.d. The mean +/− s.e. FHBG liver standardized uptake value (SUV) in four transgenics went from 4.49 +/− 0.32 to 2.17 +/− 0.52 to 6.21 +/− 0.72 as the mice went through the three diets of 17%, 0%, and 25% sequentially. Nontransgenic mice showed GAPDH normalized albumin mRNA that went from 37.68 +/− 6.04 to 26.41 +/− 4.29 to 52.42 +/− 4.09. The FHBG SUV from transgenics was well correlated with GAPDH normalized albumin mRNA from non-transgenics (r(2) = 0.97), supporting the notion that endogenous gene expression of albumin can be indirectly

imaged with FHBG. This research demonstrated that HSV1-tk gene can be used to noninvasively image the modulated expression of transgenes by visualizing and quantifying correlated changes in albumin expression in wild-type mice, as well as HSV1-TK expression using microPET in transgenic mice in which the reporter gene is driven by the albumin promoter.

Doubrovin et al. also developed and assessed a method for monitoring the transcriptional activation of endogenous genes by positron-emission tomography (PET) imaging [69]. The HSV1-tk/GFP (TKGFP) dual-reporter gene was used to monitor transcriptional activation of p53-dependent genes. A retrovirus bearing the Cis-p53/TKGFP reporter system was constructed in which the TKGFP reporter gene was placed under control of an artificial cis-acting p53-specific enhancer. U87 glioma and SaOS-2 osteosarcoma cells were transduced with this retrovirus and used to establish xenografts in rats. It was demonstrated that DNA damage-induced upregulation of p53 transcriptional activity correlated with the expression of p53-dependent downstream genes, such as p21, in U87 (wild-type p53) but not in SaOS-2 osteosarcoma (p53 −/−) cells. PET, with [(124)I]FIAU (2′-fluoro-2′-deoxy-1-beta-d-arabinofuranosyl-5-[(124)I]iodouracil) and the Cis-p53TKGFP reporter system, was sufficiently sensitive to image the transcriptional regulation of genes in the p53 signal transduction pathway. These imaging results were confirmed by independent measurements of p53 activity and the expression levels of downstream genes (e.g., p21) by using conventional molecular–biological assays. PET imaging of p53 transcriptional activity in tumor xenografts by using this Cis-p53TKGFP reporter system was deemed useful in assessing novel therapeutic approaches.

The same research group adopted similar strategies to study tumor hypoxia, a spatially and temporally heterogeneous phenomenon consequent to several tumor and host tissue-specific processes [128]. To study the dynamics and spatial heterogeneity of hypoxia-inducible factor-1 (HIF-1)-specific transcriptional activity in tumors, Serganova et al. used repetitive noninvasive PET imaging of hypoxia-induced HIF-1 transcriptional activity in tumors in living mice. This approach used a novel retroviral vector bearing an HIF-1-inducible "sensor" reporter gene (HSV1-tk/GFP fusion) and a constitutively expressed "beacon" reporter gene (DsRed2/XPRT). C6 glioma cells transduced with this multireporter system revealed dose-dependent patterns in temporal dynamics of HIF-1 transcriptional activity induced by either $CoCl_2$ or decreased atmospheric oxygen concentration. Multicellular spheroids of C6 reporter cells developed a hypoxic core when >350 μm in diameter. 18F-2′-fluoro-2′deoxy-1beta-D-arabionofuranosyl-5-ethyl-uracil (FEAU) PET revealed spatial heterogeneity of HIF-1 transcriptional activity in reporter xenografts in mice as a function of size or ischemia-reperfusion injury. With increasing tumor diameter (>3 mm), a marked increase in HIF-1 transcriptional activity was observed in the core regions of tumors. Even a moderate ischemia-reperfusion injury in small C6 tumors caused a rapid induction of HIF-1 transcriptional activity, which persisted for a long time because of the inability of C6 tumors to rapidly compensate acute changes in tumor microcirculation.

More recently, while recognizing that vascular endothelial growth factor (VEGF) plays a critical role in the early activation of stromal tissues during wound healing and tumor growth, Wang et al. reported the use of a two-step transcriptional amplification (TSTA) approach to augment the transcriptional activity of the relatively weak VEGF promoter (pVEGF) using firefly luciferase (*fl*) reporter gene and bioluminescence imaging [129]. In cell culture, they demonstrate that TSTA-based *fl* gene expression can be significantly enhanced over the direct one-step system. Using a transgenic mouse model (pVEGF-TSTA-fl), they demonstrated the induction of VEGF gene expression using a wound-healing model and a subcutaneous mammary tumor model. In skin-wounding experiments, pVEGF-induced *fl* expression in the wound lesion was detected on Days 4 and 5 and peaked on Days 15–22. Furthermore, the

bioluminescence signal showed good correlation with the endogenous VEGF protein levels in the wound tissue ($r^2 = 0.70$). In the mammary tumor model, *fl* expression was detected on Day 3, peaked at Day 17, and declined thereafter. These results support the use of noninvasive bioluminescence imaging for the longitudinal monitoring of VEGF induction during wound healing and tumor progression, and this mouse model should find use in various applications in which it is important to noninvasively study signaling pathways leading to VEGF gene expression.

CONCLUDING REMARKS AND FUTURE OUTLOOK

Signal transduction is one of the fundamental processes of living cells, being second in importance only to cell division [130]. In this chapter we have presented a brief overview of the fundamental biological principles of cell signaling along with an outline of ongoing early attempts at development and feasibility studies within the emerging field of molecular imaging of cell signal transduction in intact living animals.

In simple terms, there are two general strategies to noninvasively image these processes in living subjects. The first is a "direct" strategy to target the various biochemical and biophysical components of a signaling pathway under scrutiny and to tailor the imaging strategy to each component separately and in isolation. This may be particularly appropriate, for example, when attempting to study the effect of an inhibitor or enhancer drug on a specific protein–protein interaction occurring at a single node along a signaling cascade. Examples of various possible approaches currently adopted within this particular strategy are presented in some detail previously. Conversely, an "indirect" strategy to image any given pathway does away with focusing on its individual nodes and instead establishes the product of gene expression downstream of the pathway as the target for imaging, that is, it attempts to image transcriptional activation of endogenous genes downstream of a given cascade. Imaging reveals the presence, quantity, duration, and location of this product and therefore indirectly signifies the general occurrence of successful signal transduction further upstream along the pathway under scrutiny. Further details of this strategy can be found elsewhere in this book.

We foresee that current simple approaches to molecular imaging of signal transduction in living subjects will undergo considerable evolution in the years to come, driven by two main forces: greater understanding of the biology of cellular signaling and its relevance to disease, and continued technical iterations and novel developments within molecular imaging, particularly through use of reporter gene technology. Innovative

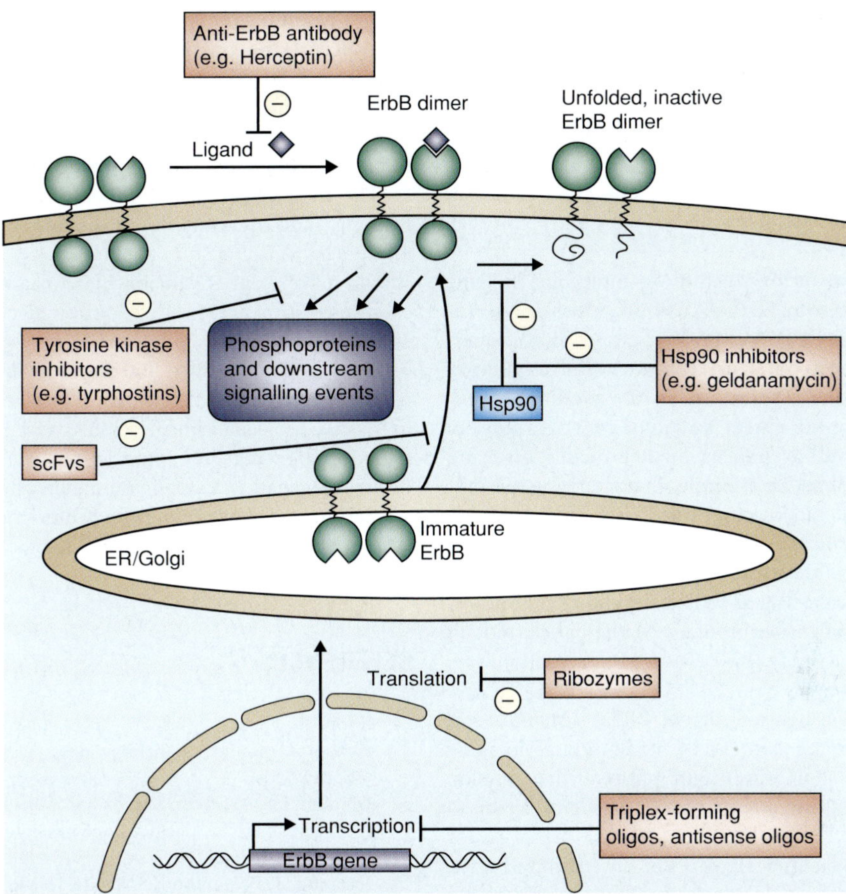

Figure 8.20. Anti-ErbB antibodies (such as Herceptin®, which binds ErbB2) block ligand binding and stimulate receptor internalization. Tyrosine kinase inhibitors such as tyrphostins block downstream signaling of the receptor–ligand complex, and Hsp90 inhibitors (for example, geldanamycin) prevent stabilization of ErbBs at the membrane. The active conformation of ErbB2 is maintained through interactions with a chaperone (Hsp90), and therefore chaperone antagonists inactivate the oncoprotein. It might also be possible to prevent ErbBs from reaching the cell surface by blocking their transcription with triplex-forming oligonucleotides, their translation with antisense oligonucleotides or ribozymes, or their trafficking to the cell surface with intracellular single-chain Fv fragments of antibodies (scFvs). (ER, endoplasmic reticulum.) (Nature reviews Cell & Molecular Biology). Reproduced with Permission from Yarden et al. (2001).

synergy between these two fields would be inevitable, especially in the clinical context of developing and validating drugs that act as signal interceptors, whether universal signal transduction inhibitors or selective blockers targeted exclusively toward a single hyperactive signaling pathway in a particular disease (Figure 8.20).

There are many aspects of cell signaling where it is safe to predict that rapid research progress will be made in the future. However, because of its preeminence as a mechanism of signal transduction, most attention will be paid to protein phosphorylation and its role in intracellular signaling [131]. On a different scale, the scene is dominated primarily by the continued and increasing realization of the complexity of cell signaling, where the presence of functional and molecular redundancy along with apparent coincidental signaling, signaling crosstalk, and signal overlapping all demand a greater understanding of

mechanisms involved in signal integration between different pathways (see reference to "systems biology" in the previous section) [130]. These pathways interact with one another to form complex networks [132]. The need to study such networks has recently prompted the development of The Alliance for Cellular Signaling (AfCS; http://www.signaling-gateway.org): a multidisciplinary, multi-institutional consortium to study cellular signaling systems and to provide large-scale research collaborations designed to answer global questions about signaling networks. Their same goal, stated from a slightly different perspective, is to understand fully how cells interpret signals in a context-dependent manner. How does a cell respond appropriately to one voice when it must listen simultaneously to many, and how does it alter this response in the context of other concurrent or recent signaling events? Answers to these questions will require identification of all the molecules that comprise the various signaling systems, assessment of time-dependent information flow through the systems in both normal and pathological states, and reduction of the mass of detailed data into a set of interacting models that describe cellular signaling. Further details regarding the central questions of the AfCS and an overview of their experimental strategy can be found in Gilman et al. [133].

Several areas of investigation in molecular imaging are also required to further customize these noninvasive approaches to answer biological and clinical questions pertaining to cell signal transduction. For example, further refinement in the use of the modified mammalian two-hybrid system for noninvasive imaging of PPIs is required, as well as further quantitative and kinetic evaluations, for example, characterizing the ability to follow interactions over time based on the half-life of the reporter protein(s) that are transactivated. Relevant studies are therefore also needed to optimize the choice of transactivator as well as the choice of promoters and levels of fusion proteins. Mathematical models to correlate and study these effects are currently under development.

The emission light of current BRET systems is still suboptimal in the context of small-animal imaging because of the short wavelength nature of the emission light and the strong attenuation of biological tissue to photons of wavelength less than 600 nm. Therefore, for *in vivo* molecular imaging of PPIs, an optimized technique using resonance transfer methods would utilize a single-chain biosensor (to fix molar ratios of donor and acceptor) with a bioluminescence donor in conjunction with a red-shifted fluorophore. In our laboratory, mutation studies on Rluc have led to the development of several Rluc variants that exhibit significantly enhanced light output and stability compared to the native enzyme, as well as red-shifted light output, which yields a green-peaked emission spectrum. The combination of such luciferases with red fluorescent proteins is currently being investigated.

Unlike fluorescence microscopy-based techniques, studies of the kinetics of PPIs, including analysis of complementation reversibility, are in their initial phases at present, although this will be an area of active investigation. These future experiments will also require assessment in several cell lines, as well as with a greater variety of protein partners of different sizes and interaction affinities (weak transient to strong obligate) to establish the general widespread applicability of this technique.

The opportunity to measure two different protein interactions at the same time by spectrally unmixing output colors will be useful in attempts to multiplex image protein interaction networks [134]. Recent advances in processing two-color imaging now allow for the total spectral deconvolution of multicolored bioluminescent images, assuming the spectra are different enough to reliably calculate the contribution of each individual emitter within each detection window, based on their published spectra. Simultaneous imaging of multiple interactions should allow deconvolution of complex protein interactions and, eventually, protein interactomes.

The high sensitivity of these assays for detecting, locating, and quantifying PPIs, combined with the advantages of doing so in a living subject environment, should make them of valuable in many areas of biological investigation and future clinical molecular medicine applications. Indeed, endpoints in molecular imaging of PPIs can be quantified and therefore are particularly useful for translational research. Ultimately, we foresee innovative molecular imaging tools, such as the one presented, enhancing our appreciation of entire biological signaling pathway systems and their pharmacological regulation and accelerating the achievement of a "systems biology" understanding of biological complexity [135].

REFERENCES

1 Klipp, E., Liebermeister, W. (2006). Mathematical modeling of intracellular signaling pathways. *BMC Neurosci* 7 (Suppl 1): S10.

2 Dumont, J. E., Dremier, S., Pirson, I., Maenhaut, C. (2002). Cross signaling, cell specificity, and physiology. *Am J Physiol Cell Physiol* 283: C2–28.

3 Massoud, T. F., Gambhir, S. S. (2003). Molecular imaging in living subjects: seeing fundamental biological processes in a new light. *Genes Dev* 17: 545–580.

4 Livingston, J. N. (1999). Genetically engineered mice in drug development. *J Intern Med* 245: 627–635.

5 Gassmann, M., Hennet, T. (1998). From genetically altered mice to integrative physiology. *News Physiol Sci* 13: 53–57.

6 Gambhir, S. S., Barrio, J. R., Herschman, H. R., Phelps, M. E. (1999). Assays for noninvasive imaging of reporter gene expression. *Nucl Med Biol* 26: 481–490.

7 Massoud, T. F., Singh, A., Gambhir, S. S. (2008). Noninvasive molecular neuroimaging using reporter genes: part II, experimental, current, and future applications. *AJNR Am J Neuroradiol* 29: 409–418.

8 Burns, M. E., Arshavsky, V. Y. (2005). Beyond counting photons: trials and trends in vertebrate visual transduction. *Neuron* 48: 387–401.

9 Ronnett, G. V., Moon, C. G. (2002). Proteins and olfactory signal transduction. *Annu Rev Physiol* 64: 189–222.

10 Wong, G. T., Gannon, K. S., Margolskee, R. F. (1996). Transduction of bitter and sweet taste by gustducin. *Nature* 381: 796–800.

11 Singh, A. B., Harris, R. C. (2005). Autocrine, paracrine and juxtacrine signaling by EGFR ligands. *Cell Signal* 17: 1183–1193.

12 Gnecchi, M., Zhang, Z., Ni, A., Dzau, V. J. (2008). Paracrine mechanisms in adult stem cell signaling and therapy. *Circ Res* 103: 1204–1219.

13 Weijer, C. J. (2003). Visualizing signals moving in cells. *Science* **300**: 96–100.

14 Levitzki, A. (2004). Introduction: signal transduction therapy-10 years later. *Semin Cancer Biol* **14**: 219–221.

15 Koman, A., Harayama, S., Hazelbauer, G. L. (1979). Relation of chemotactic response to the amount of receptor: evidence for different efficiencies of signal transduction. *J Bacteriol* **138**: 739–747.

16 Heldin, C. H. (2001). Signal transduction: multiple pathways, multiple options for therapy. *Stem Cells* **19**: 295–303.

17 Nahta, R., Yu, D., Hung, M. C., Hortobagyi, G. N., Esteva, F. J. (2006). Mechanisms of disease: understanding resistance to HER2-targeted therapy in human breast cancer. *Nat Clin Pract Oncol* **3**: 269–280.

18 Malmberg, N. J., Falke, J. J. (2005). Use of EPR power saturation to analyze the membrane-docking geometries of peripheral proteins: applications to C2 domains. *Annu Rev Biophys Biomol Struct* **34**: 71–90.

19 Hanson, M. A., Stevens, R. C. (2009). Discovery of new GPCR biology: one receptor structure at a time. *Structure* **17**: 8–14.

20 Lefkowitz, R. J. (2004). Historical review: a brief history and personal retrospective of seven-transmembrane receptors. *Trends Pharmacol Sci* **25**: 413–422.

21 Collingridge, G. L., Olsen, R., Peters, J. A., Spedding, M. (2009). Ligand gated ion channels. *Neuropharmacology* **56**: 1.

22 Pless, S. A., Lynch, J. W. (2008). Illuminating the structure and function of Cys-loop receptors. *Clin Exp Pharmacol Physiol* **35**: 1137–1142.

23 Swanson, G. T., Sakai, R. (2009). Ligands for ionotropic glutamate receptors. *Prog Mol Subcell Biol* **46**: 123–157.

24 Jarvis, M. F., Khakh, B. S. (2009). ATP-gated P2X cation-channels. *Neuropharmacology* **56**: 208–215.

25 Kucharova, S., Farkas, R. (2002). Hormone nuclear receptors and their ligands: role in programmed cell death (review). *Endocr Regul* **36**: 37–60.

26 Chalmers, M., Schell, M. J., Thorn, P. (2006). Agonist-evoked inositol trisphosphate receptor (IP3R) clustering is not dependent on changes in the structure of the endoplasmic reticulum. *Biochem J* **394** (Pt 1): 57–66.

27 Hardingham, G. E., Bading, H. (1999). Calcium as a versatile second messenger in the control of gene expression. *Microsc Res Tech* **46**: 348–355.

28 Takai, N., Ueda, T., Nasu, K., Yamashita, S., Toyofuku, M., Narahara, H. (2009). Targeting calcium/calmodulin-dependence kinase I and II as a potential anti-proliferation remedy for endometrial carcinomas. *Cancer Lett* **277**: 235–243.

29 Tu, P., Kunert-Keil, C., Lucke, S., Brinkmeier, H., Bouron, A. (2009). Diacylglycerol analogues activate second messenger-operated calcium channels exhibiting TRPC-like properties in cortical neurons. *J Neurochem* **108**: 126–138.

30 Kalinowski, L., Dobrucki, L. W., Malinski, T. (2001). Nitric oxide as a second messenger in parathyroid hormone-related protein signaling. *J Endocrinol* **170**: 433–440.

31 Normanno, N., De Luca, A., Bianco, C., Strizzi, L., Mancino, M., Maiello, M. R., Carotenuto, A., De Feo, G., Caponigro, F., Salomon, D. S. (2006). Epidermal growth factor receptor (EGFR) signaling in cancer. *Gene* **366**: 2–16.

32 Wang, Z., Li, Y., Banerjee, S., Sarkar, F. H. (2008). Exploitation of the Notch signaling pathway as a novel target for cancer therapy. *Anticancer Res* **28**: 3621–3630.

33 McEwan, I. J. (2009). Nuclear receptors: one big family. *Methods Mol Biol* **505**: 3–18.

34 Lo, H. W., Hsu, S. C., Hung, M. C. (2006). EGFR signaling pathway in breast cancers: from traditional signal transduction to direct nuclear translocalization. *Breast Cancer Res Treat* **95**: 211–218.

35 Downward, J. (2003). Targeting RAS signalling pathways in cancer therapy. *Nat Rev Cancer* **3**: 11–22.

36 LoPiccolo, J., Blumenthal, G. M., Bernstein, W. B., Dennis, P. A. (2008). Targeting the PI3K/Akt/mTOR pathway: effective combinations and clinical considerations. *Drug Resist Updat* **11**: 32–50.

37 Kfir, S., Ehrlich, M., Goldshmid, A., Liu, X., Kloog, Y., Henis, Y. I. (2005). Pathway- and expression level-dependent effects of oncogenic N-Ras: p27(Kip1) mislocalization by the Ral-GEF pathway and Erk-mediated interference with Smad signalling. *Mol Cell Biol* **25**: 8239–8250.

38 Li, W. X. (2008). Canonical and non-canonical JAK-STAT signalling. *Trends Cell Biol* **18**: 545–551.

39 Harburger, D. S., Calderwood, D. A. (2009). Integrin signalling at a glance. *J Cell Sci* **122**: 159–163.

40 Takigawa, Y., Brown, A. M. (2008). Wnt signaling in liver cancer. *Curr Drug Targets* **9**: 1013–1024.

41 Bhola, N. E., Grandis, J. R. (2008). Crosstalk between G-protein-coupled receptors and epidermal growth factor receptor in cancer. *Front Biosci* **13**: 1857–1865.

42 Baud, V., Karin, M. (2009). Is NF-kappaB a good target for cancer therapy? Hopes and pitfalls. *Nat Rev Drug Discov* **8**: 33–40.

43 Tian, M., Schiemann, W. P. (2009). The TGF-beta paradox in human cancer: an update. *Future Oncol* **5**: 259–271.

44 Jiang, J., Hui, C. C. (2008). Hedgehog signaling in development and cancer. *Dev Cell* **15**: 801–812.

45 Smith, N. J., Luttrell, L. M. (2006). Signal switching, crosstalk, and arrestin scaffolds: novel G protein-coupled receptor signaling in cardiovascular disease. *Hypertension* **48**: 173–179.

46 Schwartz, M. A., Baron, V. (1999). Interactions between mitogenic stimuli, or, a thousand and one connections. *Curr Opin Cell Biol* **11**: 197–202.

47 O'Neill, L. A. (2008). When signaling pathways collide: positive and negative regulation of toll-like receptor signal transduction. *Immunity* **29**: 12–20.

48 Chen, M., Lin, S., Hofestaedt, R. (2004). STCDB: Signal Transduction Classification Database. *Nucleic Acids Res* **32**: D456–458.

49 Weng, G., Bhalla, U. S., Iyengar, R. (1999). Complexity in biological signaling systems. *Science* **284**: 92–96.

50 Hood, L., Heath, J. R., Phelps, M. E., Lin, B. (2004). Systems biology and new technologies enable predictive and preventative medicine. *Science* **306**: 640–643.

51 Royer, C. (2004). *Protein-protein interactions.* Available from: http://www.biophysics.org/education/croyer.pdf

52 Engen, J. R., Wales, T. E., Hochrein, J. M., Meyn, M. A. 3rd, Banu Ozkan, S., Bahar, I., Smithgall, T. E. (2008). Structure and dynamic regulation of Src-family kinases. *Cell Mol Life Sci* **65**: 3058–3073.

53 Uhlik, M. T., Temple, B., Bencharit, S., Kimple, A. J., Siderovski, D. P., Johnson, G. L. (2005). Structural and evolutionary division of phosphotyrosine binding (PTB) domains. *J Mol Biol* **345**: 1–20.

54 Jemth, P., Gianni, S. (2007). PDZ domains: folding and binding. *Biochemistry* **46**: 8701–8708.

55 Grunewald, T. G., Butt, E. (2008). The LIM and SH3 domain protein family: structural proteins or signal transducers or both? *Mol Cancer* **7**: 31.

56 Lemmon, M. A. (2007). Pleckstrin homology (PH) domains and phosphoinositides. *Biochem Soc Symp* **74**: 81–93.

57 Ilsley, J. L., Sudol, M., Winder, S. J. (2002). The WW domain: linking cell signalling to the membrane cytoskeleton. *Cell Signal* **14**: 183–189.

58 Weidemann, T., Höfinger, S., Müller, K., Auer, M. (2007). Beyond dimerization: a membrane-dependent activation model for interleukin-4 receptor-mediated signaling. *J Mol Biol* **366**: 1365–1373.

59 Smith, T. F. (2008). Diversity of WD-repeat proteins. *Subcell Biochem* **48**: 20–30.

60 Yamada, K., Miyamoto, K. (2005). Basic helix-loop-helix transcription factors, BHLHB2 and BHLHB3; their gene expressions are regulated by multiple extracellular stimuli. *Front Biosci* **10**: 3151–3171.

61 Manna, P. R., Dyson, M. T., Stocco, D. M. (2009). Role of basic leucine zipper proteins in transcriptional regulation of the steroidogenic acute regulatory protein gene. *Mol Cell Endocrinol* **302**: 1–11.

62 Zhang, J., Allen, M. D. (2007). FRET-based biosensors for protein kinases: illuminating the kinome. *Mol Biosyst* **3**: 759–765.

63 Massoud, T. F., Paulmurugan, R., Gambhir, S. S. (2004). Molecular imaging of homodimeric protein-protein interactions in living subjects. *FASEB J* **18**: 1105–1107.

64 Wehrman, T., Kleaveland, B., Her, J. H., Balint, R. F., Blau, H. M. (2002). Protein-protein interactions monitored in mammalian cells via complementation of beta-lactamase enzyme fragments. *Proc Natl Acad Sci USA* **99**: 3469–3474.

65 Truong, K., Ikura, M. (2001). The use of FRET imaging microscopy to detect protein-protein interactions and protein conformational changes in vivo. *Curr Opin Struct Biol* **11**: 573–578.

66 Hu, C. D., Chinenov, Y., Kerppola, T. K. (2002). Visualization of interactions among bZIP and Rel family proteins in living cells using bimolecular fluorescence complementation. *Mol Cell* **9**: 789–798.

67 Paulmurugan, R., Gambhir, S. S. (2003). Monitoring protein-protein interactions using split synthetic renilla luciferase protein-fragment-assisted complementation. *Anal Chem* **75**: 1584–1589.

68 Rudin, M. (2008). Noninvasive imaging of receptor function: signal transduction pathways and physiological readouts. *Curr Opin Drug Discov Devel* **11**: 606–615.

69 Doubrovin, M., Ponomarev, V., Beresten, T., Balatoni, J., Bornmann, W., Finn, R., Humm, J., Larson, S., Sadelain, M., Blasberg, R., Gelovani Tjuvajev, J. (2001). Imaging transcriptional regulation of p53-dependent genes with positron emission tomography in vivo. *Proc Natl Acad Sci USA* **98**: 9300–9305.

70 van de Wiele, C., Boersma, H., Dierckx, R. A., De Spiegeleer, B., van Waarde, A., Elsinga, P. H. (2008). Growth factor/peptide receptor imaging for the development of targeted therapy in oncology. *Curr Pharm Des* **14**: 3340–3347.

71 Elsässer-Beile, U., Reischl, G., Wiehr, S., Bühler, P., Wolf, P., Alt, K., Shively, J., Judenhofer, M. S., Machulla, H. J., Pichler, B. J. (2009). PET imaging of prostate cancer xenografts with a highly specific antibody against the prostate-specific membrane antigen. *J Nucl Med* **50**: 606–611.

72 Macfarlane, D., Socrates, A., Eisenberg, P., Larcos, G., Roach, P., Gerometta, M., Smart, R., Tsui, W., Scott, A. M. (2009). Imaging of deep venous thrombosis in patients using a radiolabelled anti-D-dimer Fab' fragment (99mTc-DI-DD3B6/22–80B3): results of a phase I trial. *Eur J Nucl Med Mol Imaging* **36**: 250–259.

73 Leyton, J. V., Olafsen, T., Lepin, E. J., Hahm, S., Bauer, K. B., Reiter, R. E., Wu, A. M. (2008). Humanized radioiodinated minibody for imaging of prostate stem cell antigen-expressing tumors. *Clin Cancer Res* **14**: 7488–7496.

74 Zaccaro, L., Del Gatto, A., Pedone, C., Saviano, M. (2009). Peptides for tumour therapy and diagnosis: current status and future directions. *Curr Med Chem* **16**: 780–795.

75 Bose, S. K., Turkheimer, F. E., Howes, O. D., Mehta, M. A., Cunliffe, R., Stokes, P. R., Grasby, P. M. (2008). Classification of schizophrenic patients and healthy controls using [18F] fluorodopa PET imaging. *Schizophr Res* **106**: 148–155.

76 Marek, K., Jennings, D. (2009). Can we image premotor Parkinson disease? *Neurology* **72**: S21–26.

77 Wagner, H. N. Jr., Dannals, R. F., Frost, J. J., Wong, D. F., Ravert, H. T., Wilson, A. A., Links, J. M., Burns, H. D., Kuhar, M. J., Snyder, S. H. (1985). Imaging neuroreceptors in the human brain in health and disease. *Radioisotopes* **34**:103–107.

78 Wong, D. F., Singer, H. S., Brandt, J., Shaya, E., Chen, C., Brown, J., Kimball, A. W., Gjedde, A., Dannals, R. F., Ravert, H. T., Wilson, P. D., Wagner, H. N. Jr. (1997). D2-like dopamine receptor density in Tourette syndrome measured by PET. *J Nucl Med* **38**: 1243–1247.

79 Vlaar, A. M., de Nijs, T., Kessels, A. G., Vreeling, F. W., Winogrodzka, A., Mess, W. H., Tromp, S. C., van Kroonenburgh, M. J., Weber, W. E. (2008). Diagnostic value of 123I-ioflupane and 123I-iodobenzamide SPECT scans in 248 patients with parkinsonian syndromes. *Eur Neurol* **59**: 258–266.

80 Nikolaus, S., Antke, C., Kley, K., Poeppel, T. D., Hautzel, H., Schmidt, D., Müller, H. W. (2007a). Investigating the dopaminergic synapse in vivo. I. Molecular imaging studies in humans. *Rev Neurosci* **18**: 439–472.

81 Nikolaus, S., Larisch, R., Beu, M., Antke, C., Kley, K., Forutan, F., Wirrwar, A., Müller, H. W. (2007b). Investigating the dopaminergic synapse in vivo. II. Molecular imaging studies in small laboratory animals. *Rev Neurosci* **18**: 473–504.

82 Costes, N., Merlet, I., Ostrowsky, K., Faillenot, I., Lavenne, F., Zimmer, L., Ryvlin, P., Le Bars, D. (2005). A 18F-MPPF PET normative database of 5-HT1A receptor binding in men and women over aging. *J Nucl Med* **46**: 1980–1989.

83 Hirano, S., Shinotoh, H., Arai, K., Aotsuka, A., Yasuno, F., Tanaka, N., Ota, T., Sato, K., Fukushi, K., Tanada, S.,

Hattori, T., Irie, T. (2008). PET study of brain acetylcholinesterase in cerebellar degenerative disorders. *Mov Disord* 23: 1154–1160.

84 Sihver, W., Fasth, K. J., Ogren, M., Lundqvist, H., Bergström, M., Watanabe, Y., Långström, B., Nordberg, A. (1999). In vivo positron emission tomography studies on the novel nicotinic receptor agonist [11C]MPA compared with [11C]ABT-418 and (S)(-)[11C]nicotine in rhesus monkeys. *Nucl Med Biol* 26: 633–640.

85 van Oosten, E. M., Wilson, A. A., Stephenson, K. A., Mamo, D. C., Pollock, B. G., Mulsant, B. H., Yudin, A. K., Houle, S., Vasdev, N. (2009). An improved radiosynthesis of the muscarinic M2 radiopharmaceutical, [18F]FP-TZTP. *Appl Radiat Isot* 67: 611–616.

86 Newberg, A. B., Ray, R., Scheuermann, J., Wintering, N., Saffer, J., Schmitz, A., Freifelder, R., Karp, J., Lerman, C., Divgi, C. (2009). Dosimetry of 11C-carfentanil, a mu-opioid receptor imaging agent. *Nucl Med Commun* 30: 314–318.

87 Weerts, E. M., Kim, Y. K., Wand, G. S., Dannals, R. F., Lee, J. S., Frost, J. J., McCaul, M. E. (2008). Differences in delta- and mu-opioid receptor blockade measured by positron emission tomography in naltrexone-treated recently abstinent alcohol-dependent subjects. *Neuropsychopharmacology* 33: 653–665.

88 Kim, S. E., Szabo, Z., Seki, C., Ravert, H. T., Scheffel, U., Dannals, R. F., Wagner, H. N. Jr. (1999). Effect of tracer metabolism on PET measurement of [11C]pyrilamine binding to histamine H1 receptors. *Ann Nucl Med* 13: 101–107.

89 Johansson, A., Engler, H., Blomquist, G., Scott, B., Wall, A., Aquilonius, S. M., Långström, B., Askmark, H. (2007). Evidence for astrocytosis in ALS demonstrated by [11C](L)-deprenyl-D2 PET. *J Neurol Sci* 255: 17–22.

90 Reubi, J. C., Maecke, H. R. (2008). Peptide-based probes for cancer imaging. *J Nucl Med* 49: 1735–1738.

91 Moody, T. W., Gozes, I. (2007). Vasoactive intestinal peptide receptors: a molecular target in breast and lung cancer. *Curr Pharm Des* 13: 1099–1104.

92 von Degenfeld, G., Wehrman, T. S., Hammer, M. M., Blau, H. M. (2007). A universal technology for monitoring G-protein-coupled receptor activation in vitro and noninvasively in live animals. *FASEB J* 21: 3819–3826.

93 Rossi, F., Charlton, C. A., Blau, H. M. (1997). Monitoring protein-protein interactions in intact eukaryotic cells by beta-galactosidase complementation. *Proc Natl Acad Sci USA* 94: 8405–8410.

94 Paulmurugan, R., Gambhir, S. S. (2006). An intramolecular folding sensor for imaging estrogen receptor-ligand interactions. *Proc Natl Acad Sci USA* 103: 15883–15888.

95 Förster, T. (1948). Intermolecular energy transference and fluorescence. *Ann Phys* 2: 54–75.

96 De, A., Gambhir, S. S. (2005). Noninvasive imaging of protein-protein interactions from live cells and living subjects using bioluminescence resonance energy transfer. *FASEB J* 19: 2017–2019.

97 De, A., Loening, A. M., Gambhir, S. S. (2007). An improved bioluminescence resonance energy transfer strategy for imaging intracellular events in single cells and living subjects. *Cancer Res* 67: 7175–7183.

98 Levi, J., De, A., Cheng, Z., Gambhir, S. S. (2007). Bisdeoxycoelenterazine derivatives for improvement of bioluminescence resonance energy transfer assays. *J Am Chem Soc* 129: 11900–11901.

99 De, A., Ray, P., Loening, A. M., Gambhir, S. S. (2009). BRET3: A red-shifted bioluminescence resonance energy transfer (BRET) based integrated platform for imaging protein-protein interactions from single live cell and living animals. *FASEB J* 23: 2702-2709.

100 Chan, C. T., Paulmurugan, R., Reeves, R. E., Solow-Cordero, D., Gambhir, S. S. (2008). Molecular imaging of phosphorylation events for drug development. *Mol Imaging Biol* 11: 144-158.

101 Zhang, L., Lee, K. C., Bhojani, M. S., Khan, A. P., Shilman, A., Holland, E. C., Ross, B. D., Rehemtulla A. (2007). Molecular imaging of Akt kinase activity. *Nat Med* 13: 1114–1119.

102 Zhang, L., Bhojani, M. S., Ross, B. D., Rehemtulla, A. (2008). Enhancing Akt imaging through targeted reporter expression. *Mol Imaging* 7: 168–174.

103 Xu, C. W., Mendelsohn, A. R., Brent, R. (1997). Cells that register logical relationships among proteins. *Proc Natl Acad Sci USA* 94: 12473–12478.

104 Pelletier, J. N., Campbell-Valois, F., Michnick, S. W. (1998). Oligomerization domain-directed reassembly of active dihydrofolate reductase from rationally designed fragments. *Proc Natl Acad Sci USA* 95: 12141–12146.

105 Johnsson, N., Varshavsky, A. (1994). Split ubiquitin as a sensor of protein interactions in vivo. *Proc Natl Acad Sci USA* 91: 10340–10344.

106 Michnick, S. W., Remy, I., Campbell-Valois, F. X., Vallee-Balisle, A., Pelletier, J. N. (2000). Detection of protein-protein interactions by protein fragment complementation strategies. *Methods Enzymol* 328: 208–230.

107 Hu, C. D., Kerppola, T. K. (2003). Simultaneous visualization of multiple protein interactions in living cells using multicolor fluorescence complementation analysis. *Nat Biotech* 21: 539–545.

108 Galarneau, A., Primeau, M., Trudeau, L. E., Michnick, S. W. (2002). Beta-lactamase protein fragment complementation assays as in vivo and in vitro sensors of protein-protein interactions. *Nat Biotechnol* 20: 619–622.

109 Spotts, J. M., Dolmetsch, R. E., Greenberg, M. E. (2002). Time-lapse imaging of a dynamic phosphorylation-dependent protein-protein interaction in mammalian cells. *Proc Natl Acad Sci USA* 99: 15142–15147.

110 Paulmurugan, R., Umezawa, Y., Gambhir, S. S. (2002). Noninvasive imaging of protein-protein interactions in living subjects by using reporter protein complementation and reconstitution strategies. *Proc Natl Acad Sci USA* 99: 15608–15613.

111 Kaihara, A., Kawai, Y., Sato, M., Ozawa, T., Umezawa, Y. (2003). Locating a protein-protein interaction in living cells via split Renilla luciferase complementation. *Anal Chem* 75: 4176–4181.

112 Ozawa, T., Umezawa, Y. (2001). Detection of protein-protein interactions in vivo based on protein splicing. *Curr Opin Chem Biol* 5: 578–583.

113 Luker, K. E., Smith, M. C., Luker, G. D., Gammon, S. T., Piwnica-Worms, H., Piwnica-Worms, D. (2004). Kinetics of regulated protein-protein interactions revealed with

firefly luciferase complementation imaging in cells and living animals. *Proc Natl Acad Sci USA* **101**: 12288–12293.

114 Paulmurugan, R., Gambhir, S. S. (2007). Combinatorial library screening for developing an improved split-firefly luciferase fragment-assisted complementation system for studying protein-protein interactions. *Anal Chem* **79**: 2346–2353.

115 Choi, C. Y., Chan, D. A., Paulmurugan, R., Sutphin, P. D., Le, Q. T., Koong, A. C., Zundel, W., Gambhir, S. S., Giaccia, A. J. (2008). Molecular imaging of hypoxia-inducible factor 1 alpha and von Hippel-Lindau interaction in mice. *Mol Imaging* **7**: 139–146.

116 Paulmurugan, R., Gambhir, S. S. (2003). Monitoring protein-protein interactions using split synthetic renilla luciferase protein-fragment-assisted complementation. *Anal Chem* **75**: 1584–1589.

117 Paulmurugan, R., Massoud, T. F., Huang, J., Gambhir, S. S. (2004). Molecular imaging of drug-modulated protein-protein interactions in living subjects. *Cancer Res* **64**: 2113–2119.

118 Bhaumik, S., Gambhir, S. S. (2002). Optical imaging of Renilla luciferase reporter gene expression in living mice. *Proc Natl Acad Sci USA* **99**: 377–382.

119 Paulmurugan, R., Gambhir, S. S. (2005). Novel fusion protein approach for efficient high-throughput screening of small molecule-mediating protein-protein interactions in cells and living animals. *Cancer Res* **65**: 7413–7420.

120 Kim, S. B., Ozawa, T., Watanabe, S., Umezawa, Y. (2004). High-throughput sensing and noninvasive imaging of protein nuclear transport by using reconstitution of split Renilla luciferase. *Proc Natl Acad Sci USA* **101**: 11542–11547.

121 Fields, S., Song, O. (1989). A novel genetic system to detect protein-protein interactions. *Nature* **340**: 245–246.

122 Fields, S., Sternglanz, R. (1994). The two-hybrid system: an assay for protein-protein interactions. *Trends Genet* **10**: 286–292.

123 Ray, P., Pimenta, H., Paulmurugan, R., Berger, F., Phelps, M. E., Iyer, M., Gambhir, S. S. (2002). Noninvasive quantitative imaging of protein-protein interactions in living subjects. *Proc Natl Acad Sci USA* **99**: 3105–3110.

124 Benezra, R., Davis, R. L., Lockshon, D., Turner, D. L., Weintraub, H. (1990). The protein Id: a negative regulator of helix-loop-helix DNA binding proteins. *Cell* **61**: 49–59.

125 Luker, G. D., Sharma, V., Pica, C. M., Dahlheimer, J. L., Li, W., Ochesky, J., Ryan, C. E., Piwnica-Worms, H., Piwnica-Worms, D. (2002). Noninvasive imaging of protein-protein interactions in living animals. *Proc Natl Acad Sci USA* **99**: 6961–6966.

126 Luker, G. D., Sharma, V., Pica, C. M., Prior, J. L., Li, W., Piwnica-Worms, D. (2003). Molecular imaging of protein-protein interactions: controlled expression of p53 and large T-antigen fusion proteins in vivo. *Cancer Res* **63**: 1780–1788.

127 Green, L. A., Yap, C. S., Nguyen, K., Barrio, J. R., Namavari, M., Satyamurthy, N., Phelps, M. E., Sandgren, E. P., Herschman, H. R., Gambhir, S. S. (2002). Indirect monitoring of endogenous gene expression by positron emission tomography (PET) imaging of reporter gene expression in transgenic mice. *Mol Imaging Biol* **4**: 71–81.

128 Serganova, I., Doubrovin, M., Vider, J., Ponomarev, V., Soghomonyan, S., Beresten, T., Ageyeva, L., Serganov, A., Cai, S., Balatoni, J., Blasberg, R., Gelovani, J. (2004). Molecular imaging of temporal dynamics and spatial heterogeneity of hypoxia-inducible factor-1 signal transduction activity in tumors in living mice. *Cancer Res* **64**: 6101–6108.

129 Wang, Y., Iyer, M., Annala, A., Wu, L., Carey, M., Gambhir, S. S. (2006). Noninvasive indirect imaging of vascular endothelial growth factor gene expression using bioluminescence imaging in living transgenic mice. *Physiol Genomics* **24**: 173–180.

130 Dhanasekaran, N. (1998). Cell signaling: an overview. *Oncogene* **17**: 1329–1330.

131 Hunter, T. (2000). Signaling – 2000 and beyond. *Cell* **100**: 113–127.

132 Iyengar, R. (2005). Teaching resources. Introduction: Overview of pathways and networks and GPCR signaling. *Sci STKE* 2005: tr4.

133 Gilman, A. G. et al. (2002). Overview of the alliance for cellular signaling. *Nature* **420**: 703–706.

134 Gammon, S. T., Leevy, W. M., Gross, S., Gokel, G. W., Piwnica-Worms, D. (2006). Spectral unmixing of multicolored bioluminescence emitted from heterogeneous biological sources. *Anal Chem* **78**: 1520–1527.

135 Massoud, T. F., Gambhir, S. S. (2007b). Integrating noninvasive molecular imaging into molecular medicine: an evolving paradigm. *Trends Mol Med* **13**: 183–191.

136 Yarden, Y., Sliwkowski, M. X. (2001). Untangling the ErbB signalling network. *Nat Rev Mol Cell Biol* **2**: 127–137.

137 Weinberg, R. A. (2007). *The biology of cancer.* Garland Science, New York.

9

Gene Therapy and Imaging of Transgene Expression in Living Subjects

Shahriar S. Yaghoubi and Sanjiv Sam Gambhir

GENE THERAPY

Broadly defined, gene therapy is the process of delivering genetic materials into a patient or cells from the patient to modify the expression of one or more genes to achieve a therapeutic outcome for the patient. Hence, a transgene delivery procedure may be intended for the following reasons: (1) to initiate the expression of the transgene in specific tissues or the entire body of the patient, (2) to reduce the expression of specific endogenous genes at a target site or the entire body of the patient, and (3) to enhance the expression of specific endogenous genes in target tissues or the entire body of the patient. The success of a gene therapy procedure depends on several factors including (1) specific cell targeting when it is necessary to cause changes in gene expression only in specific cells of the patient's body and avoid modification of gene expression in all other cells. This may be necessary to avoid side effects and deliver sufficient gene expression modifying genetic material to the target cells; (2) efficiency of transgene delivery to achieve sufficient modification of gene expression in target cells; (3) modified gene expression for sufficient time duration; and (4) lack of immune reaction to the product of the therapeutic transgene, when immune reaction will reduce efficacy or potency of the therapeutic procedure. Investigators in the field of gene therapy are working hard to solve problems associated with all of these factors. Optimization of gene therapy techniques is essential to fulfill the promise of gene therapy as a curative strategy in medicine. Many of these problems will be solved by optimizing techniques for delivering genetic materials.

How Does One Deliver Therapeutic Transgenes?

Gene therapy is accomplished through two different general delivery mechanisms, *ex vivo* or *in vivo*. *Ex vivo* delivery involves harvesting target cells from the patient or a human donor, transferring the transgene(s) into the cells while they are being cultured outside of the patient, and then administering the genetically engineered autologous or allogeneic (when it is potentially feasible) cells back into the patient. *In vivo* delivery is direct administration of the transgene into the patient. Both of these delivery mechanisms often require the use of transgene delivery vectors, which are either viral or nonviral vectors [1, 2]. Electronic, ultrasound-facilitated, gene gun, and hydrodynamic transfer are among nonviral methods of gene transfer, which can be used to deliver naked DNA plasmid and in some cases RNA into target cells or tissues of a patient [1]. In most cases, delivery methods are designed to achieve specific targeting. Despite significant advances, efficient tissue-specific delivery of transgenes without adverse effects remains a major challenge of gene therapy. Monitoring therapeutic transgene (TG) expression in living subjects is an essential part of the gene therapy development process.

MOLECULAR IMAGING AND GENE THERAPY

Imaging living subjects can be incorporated into preclinical and clinical gene therapy protocols to accomplish several objectives (Figure 9.1). Gene therapy vectors can be tagged such that one can image their biodistribution following administration into the preclinical living research subjects or in human patients. The molecules tagging gene therapy vectors can be radionuclides or fluorescent molecules that emit a signal continuously or upon exposure to a range of optical wavelengths [3]. Alternatively, vectors can be tagged by molecules, which can be detected by specific molecular probes. Molecular imaging can also provide more direct information by monitoring the expression of therapeutic transgenes. Potentially, imaging probes can specifically detect the mRNA or proteins encoded by the therapeutic transgene. For example,

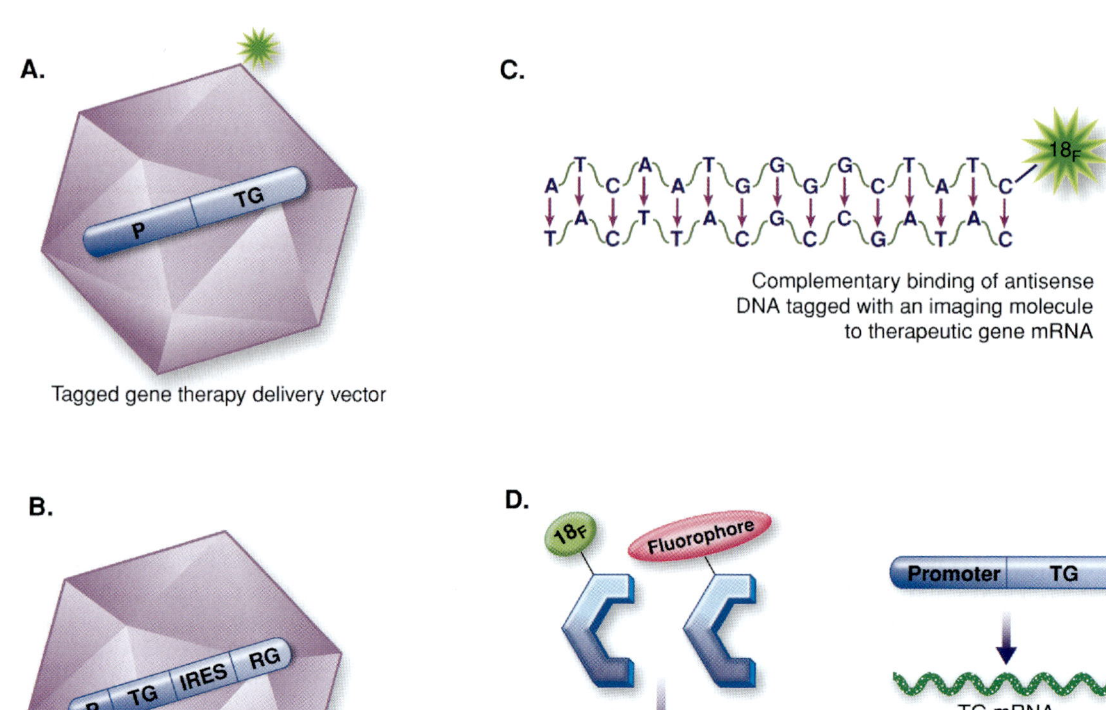

Figure 9.1. Applications of molecular imaging in gene therapy. **(A)** Imaging of a tagged gene therapy vector. **(B)** Indirect imaging of therapeutic gene expression by genetic linkage to the expression of an imaging reporter gene. **(C)** Imaging therapeutic gene expression with a labeled antisense oligonucleotide probe. **(D)** Imaging the protein encoded by a therapeutic gene directly with a specific imaging probe.

we may one day be able to design radiolabeled antisense oligonucleotides that can specifically detect mRNAs in living subjects. Meanwhile, other strategies have been studied for RNA imaging [4, 5]. Imaging probes are currently available that can detect the protein products of some therapeutic transgenes directly in living subjects, and more specific probes for detection of other TG will likely be developed in the future. Meanwhile, as discussed in the next section, TG expression can be imaged indirectly using imaging reporter transgenes (RGs) through linked genetic constructs. Beyond imaging TG expression or the biodistribution of the delivery vectors, molecular imaging is also useful for assessing response to gene therapy. For example, TG expression within target cells may induce expression of certain endogenous genes, which one can potentially image with imaging probes [e.g., p53 induction]. Also, in cancer gene therapy, the effect of TG expression on the tumor cells can be evaluated soon after treatment with many different imaging probes. For

example, one can image effects on metabolism, proliferation, tumor vasculature, and oxygen availability [6].

WHAT CAN REPORTER GENE IMAGING DO FOR THE FIELD OF GENE THERAPY?

Reporter gene imaging can provide information at almost all of the key stages of gene therapy. First, it can be used to monitor the biodistribution of the delivery vehicle itself (e.g., adenovirus by labeling virus) and targeting of vector to intended/unintended sites. Second, it can be used to image the biodistribution, magnitude, and time variation of TG expression. Third, it can be used to monitor molecular events that are initiated within cells following expression of TG. For example, the protein encoded by a TG may activate transcription of certain endogenous genes or induce binding of other proteins to each other. As discussed in Chapters 8 and 10, both of these molecular events may be imageable using reporter genes.

Imaging Biodistribution of Gene Therapy Vectors Using Imaging Reporter Genes

Biodistribution of gene therapy vectors can be imaged either by attaching detectable molecules directly to the vector or by incorporating a reporter gene into the vector. There are reasons to pursue either or both approaches. When the vector is directly labeled, one can monitor its distribution even if the transgene(s) it is carrying are not being expressed and regardless of whether the vector is within cells or not. This may allow prediction of adverse effects due to vector administration. Reporter transgene imaging will only reveal vector biodistribution when the reporter gene is expressed within cells. Also, when the vector is labeled with radionuclide probes and then injected, imaging sensitivity should be higher, due to the absence of background radioactivity. However, direct vector labeling also has shortcomings or problems. When the vector is labeled with a radionuclide probe the duration of image monitoring is dependent on the half-life of the radioisotope. Possible detachment of the probe molecule from the vector can result in misleading images. Moreover, the attachment of a bulky probe molecule may alter the biodistribution of the original vector. For example, it can influence the infectivity of the vector so that it can't transduce target cells as well as was intended. Reducing transduction efficiency would interfere with therapy and adversely affect efficacy. An important principle of molecular imaging is that the imaging procedure should not interfere with the therapeutic procedure.

Usually imaging is intended not only for detection of vector pharmacokinetics to predict adverse effects but also to predict efficacy and potency. Despite specific vector targeting, if the delivered therapeutic transgene is not being expressed or the expression is too low, the therapy will not be efficacious. Therefore, it will be necessary to determine in which tissues the therapeutic transgene that will be delivered by a specific vector will be expressed, at what magnitude, and for how long. In Section 3.3 we describe methods for linking the expression of a TG to an imaging reporter gene when they are delivered together. However, by inserting an imaging RG into the exact vector carrying the exact gene construct that will be used to deliver the TG, one may be able to predict the biodistribution, magnitude, and time variation of TG. Imaging RGs have been used to monitor the biodistribution of different viral [7–14] and nonviral vectors [15, 16]. Transgene delivery with lentiviral vectors into the brains of rodents and mouse neuroblastoma cells and tumors has been studied using firefly luciferase (fluc)[12, 17] and a mutant of the herpes simplex 1 virus thymidine kinase (HSV1-sr39tk)[17]. Although the Fluc bioluminescence

signal declined by about 20%, it remained constant thereafter for more than 10 months [12]. Adenoviral transgene delivery into the heart [10] and lungs [11] of rats has been studied using the HSV1-sr39tk/9-(4-[^{18}F]fluoro-3-(hydroxymethyl)butyl)guanine ([^{18}F]FHBG) PET RG (PRG)/PET reporter probe (PRP) system. HSV1-sr39tk and [^{18}F]FHBG have also been used to image targeted delivery of modified adenoviruses to tumor cells expressing integrin $\alpha_v\beta_3$[9] and intracoronary TG delivery into rabbit hearts using cationic liposomes [15]. Other nuclear imaging RG systems have also been used to study adenoviral transgene delivery. The human type 2 somatostatin receptor gene (hSSTr2), which can be imaged with radiolabeled peptide ligands such as ^{99m}Tc-P829, ^{188}Re-P829, ^{99m}Tc-P2045, and ^{111}In-Octreotide, was investigated as a SPECT RG in a study of adenoviral transgene delivery to ovarian cancer xenografts [8]. Furthermore, the sodium iodide symporter, which can serve as a PET or a SPECT RG, was used to demonstrate that all-*trans*-retinoic acid can enhance the expression of an adenovirally transduced transgene in MCF-7 breast cancer cells [13]. Finally, our laboratory has demonstrated imaging of transgene delivery with the nonviral vector polyethylenimine conjugated to transferrin to tumors in mice, using the Fluc bioluminescent RG [16]. Basically, a variety of RGs are available to image vector delivery of transgenes in living subjects using most available molecular imaging modalities.

A special case of imaging viral transgene delivery with RGs is that of imaging the expression of RGs delivered by oncolytic viruses. The reason is that delivered RGs need to be expressed within living cells and oncolytic viruses kill the cells they transduce. Hence, it is possible that the expression of RGs delivered by lytic or replication-competent viruses can only be imaged for a short period following transduction. The transduction of oncolytic herpes simplex viruses has been monitored by imaging their viral thymidine kinase enzymes with [^{18}F]FHBG [14] or ^{124}I-5-iodo-2′-fluoro-1-β-D-arabinofuranosyluracil ([^{124}I]FIAU) [18, 19]. 2-[^{14}C]FIAU uptake studies demonstrated increased accumulation in remaining viable cultured cancer cells as a function of time posttransduction and increasing multiplicities of infection [18, 19]. However, [^{124}I]FIAU tumor uptakes were relatively low, when [^{124}I]FIAU had been injected 8 h [18, 19], 48 h [19], or 4 days [18] after infection of the tumors with oncolytic viruses, probably due to the lysis of most infected cancer cells by that time. (Note: Actual imaging of mice was either 48 h or 3 days after [^{124}I]FIAU injection.) A more dynamic follow-up of oncolytic HSV-1 tumor infection was performed with [^{18}F]FHBG PET at 2, 6, 24, 48, and 72 h postinfection [14]. (Note: actual images of mice were acquired 1 h after [18F]FHBG injections.) This study demonstrated that [^{18}F]FHBG accumulation was highest at 6 h and was

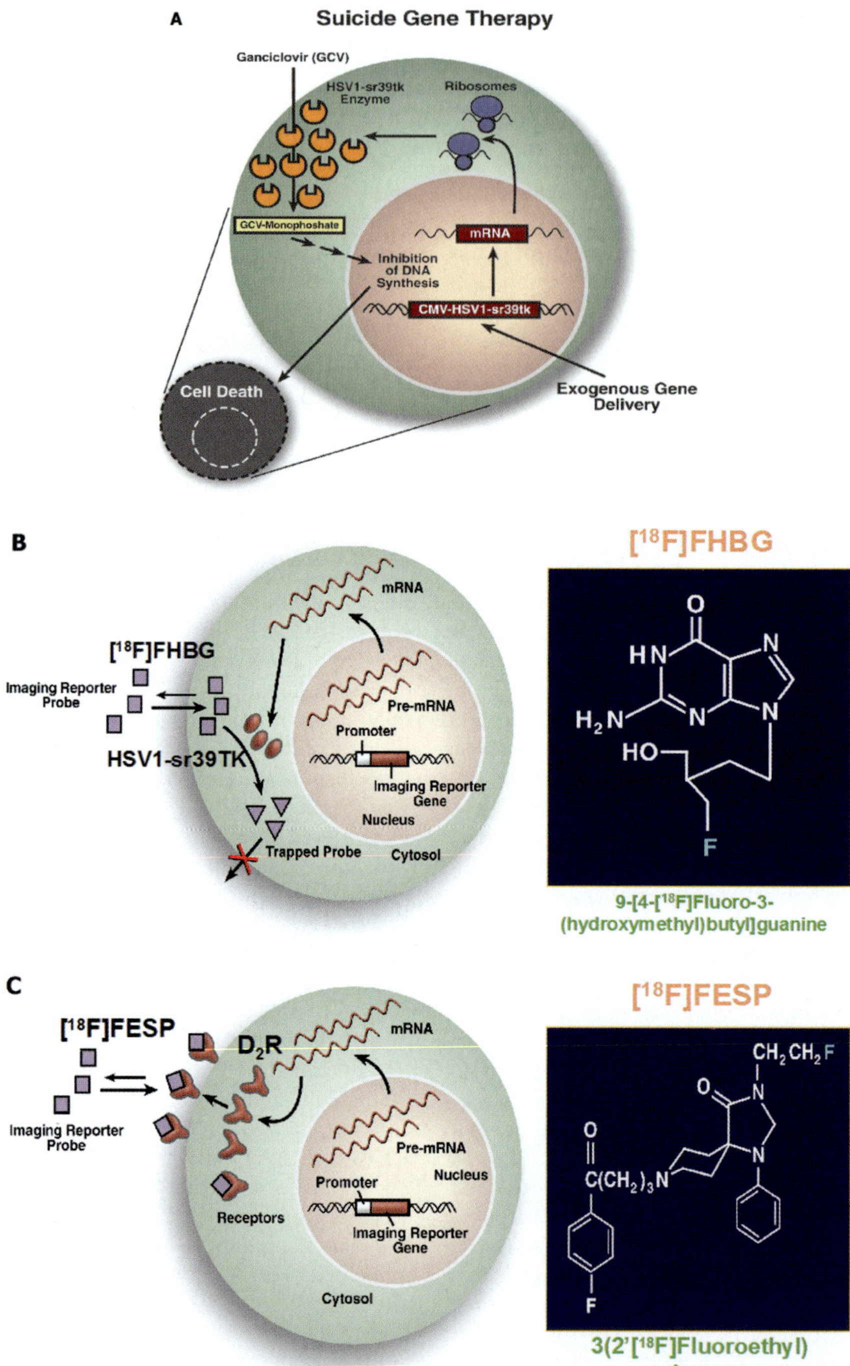

Figure 9.2. Diagrams of mechanisms of **(A)** suicide gene therapy and **(B–C)** PET reporter gene (PRG)/probe (PRP) systems. **(A)** Herpes simplex virus type 1 thymidine kinase (HSV1-tk) and its mutant HSV1-sr39tk can serve as suicide genes, causing programmed cell death when cells expressing them are exposed to pharmacological doses of ganciclovir (GCV) or penciclovir (PCV). This is the result of monophosphorylation of these prodrugs by HSV1-TK or HSV1-sr39TK enzymes, which could lead to DNA chain termination or inhibition of DNA polymerases. **(B)** HSV1-tk or HSV1-sr39tk can also serve as imaging reporter genes, because their enzyme products can phosphorylate certain radiolabeled nucleoside analogs, such as [18F]FHBG, which are administered at trace doses, and trap them within cells expressing them. The HSV1-sr39tk PRG is more efficient in catalyzing monophosphorylation of GCV, PCV, or [18F]FHBG. **(C)** The dopamine type 2 receptor PRG is a receptor-based imaging reporter gene, the protein product of which specifically binds the PRP [18F]FESP. D2R is also an endogenous gene and should not be immunogenic. Although D2R is not itself a therapeutic transgene (TG), it can be used for indirect imaging of TGs.

then lower at 24, 48, and 72 h [14]. Therefore, imaging of oncolytic viruses carrying the HSV1-tk PRG should start at earlier time points after virus administration.

Molecular Imaging: Imaging the Expression of Therapeutic Transgenes That are Also Imaging Reporter Genes, Directly

Molecular imaging probes are currently available that can specifically detect the proteins of some TG. The herpes simplex virus 1 thymidine kinase (HSV1-tk) or the mutant HSV1-sr39tk can be used as suicide genes (Figure 9.2a) as well as imaging reporter genes (Figure 9.2b). As discussed in Chapter 3, there are a variety of acycloguanosine and uracil nucleoside PET and SPECT analogs that can detect HSV1-TK or HSV1-sr39TK enzyme activity in living subjects. We originally reported PET imaging of 9-(4-[18F]fluoro-3-(hydroxymethyl)butyl)guanine ([18F]FHBG) accumulation in stable HSV1-sr39tk expressing C6 glioma tumors (C6sr39) in nu/nu mice in response to ganciclovir (GCV) treatment [20]. We observed a significant decrease of [18F]FHBG accumulation in the shrinking C6sr39 tumors following GCV injections; whereas [18F]FHBG accumulation did not increase significantly in the regrown tumors (Figure 9.3). Decrease of tumor cell viability following GCV injection and increase of viable cells in regrown tumors was confirmed by coimaging all the mice with [18F]FDG PET. [18F]FHBG has also been used to image delivery of the HSV1-sr39tk transgene with adenoviral vectors for gene therapy of prostate cancer tumors [21, 22]. A clever application of adenoviruses carrying the HSV1-sr39tk transgene regulated by the prostate-specific antigen promoter, amplified using the two-step transcriptional activation system (see Chapter 6) is using them as diagnostic agents to image metastasis of prostate cancer cells to sentinel lymph nodes [23]. Deng et al. originally reported imaging the regression of tumors, composed of murine NG4TL4 cancer cells transduced *ex vivo* with HSV1-tk carrying retroviruses, with [131I]-5-iodo-2′-fluoro-1-β-D-arabinofuranosyluracil ([131I]FIAU) SPECT in mice treated with ganciclovir (GCV) [24]. Then in a follow-up study Wang et al. incorporated [18F]Fluorodeoxyglucose ([18F]FDG), 2–18F-fluoroethyl-L-tyrosine ([18F]FET) and 5–18F-fluoro-2′-deoxyuridine ([18F]FUdR) PET imaging into their protocol along with [131I]FIAU SPECT imaging and observed significant declines in [131I]FIAU and [18F]FUdR 4 days after GCV treatments; whereas lesser declines in [18F]FDG and [18F]FET accumulation occurred 7 days after GCV treatment [25]. However, the fact that both GCV and [131I]FIAU are HSV1-TK enzyme substrates and may compete with each other was not taken into consideration in this study. Finally, Deng et al. have also demonstrated that [131I]FIAU imaging would be useful in monitoring GCV

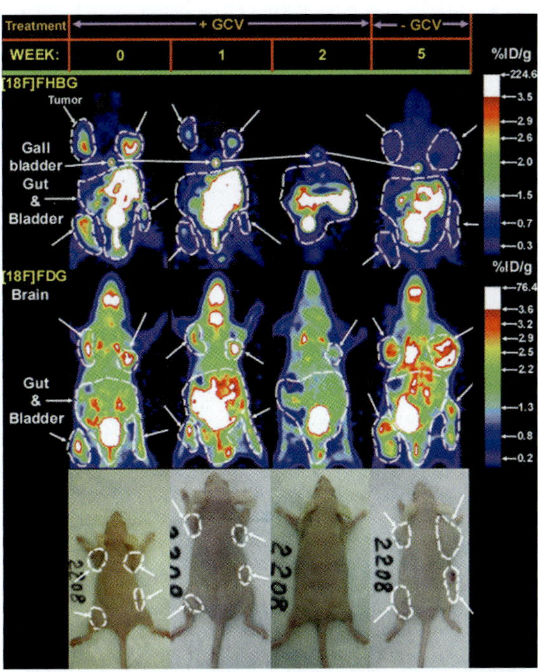

Figure 9.3. Four C6sr39 tumor xenografts were implanted subcutaneously on four sites of the nude mouse shown. All four tumors highly accumulated [18F]FHBG and [18F]FDG prior to starting GCV treatment (week 0). The mouse was administered daily IP injections of GCV (100 mg/kg) for 2 weeks, during which time the tumors regressed (three of them visually eradicated) and [18F]FHBG and [18F]FDG accumulation declined to background levels. The mouse was monitored up to 3 weeks after halting GCV treatment. The tumors regrew but only accumulated [18F]FHBG at background levels, despite robust ability to accumulate [18F]FDG.

therapy of metastatic lung tumor cells expressing the HSV1-tk gene [26].

Imaging of HSV1-tk has also been incorporated into gene therapy clinical trials [27]. The first HSV1-tk/HSV1-sr39tk imaging reporter probe studied in humans was [18F]FHBG [28] (Figure 9.4), which is now approved by the United States Food and Drug Administration (FDA) as an investigational new drug (IND# 61,880). Intravenous injection of up to 7 mCi (2 µg) of [18F]FHBG has not caused any adverse effects in all human subjects thus far studied [28] (unpublished data), and a dose of 14 µg/kg of nonradioactive FHBG was safe in rats and rabbits [29]. Later, [18F]FHBG was used to image direct adenoviral delivery of HSV1-tk into the tumors of hepatocarcinoma patients [27, 30]. In this study, the only patients who responded to suicide gene therapy with HSV1-tk were the ones in whose livers adenovirus injection sites specific above background accumulation of [18F]FHBG had been detected. Prior to this adenoviral gene therapy imaging trial, Jacobs et al. for the first time reported imaging liposome-delivered HSV1-tk in the glioma tumor of one of five patients with

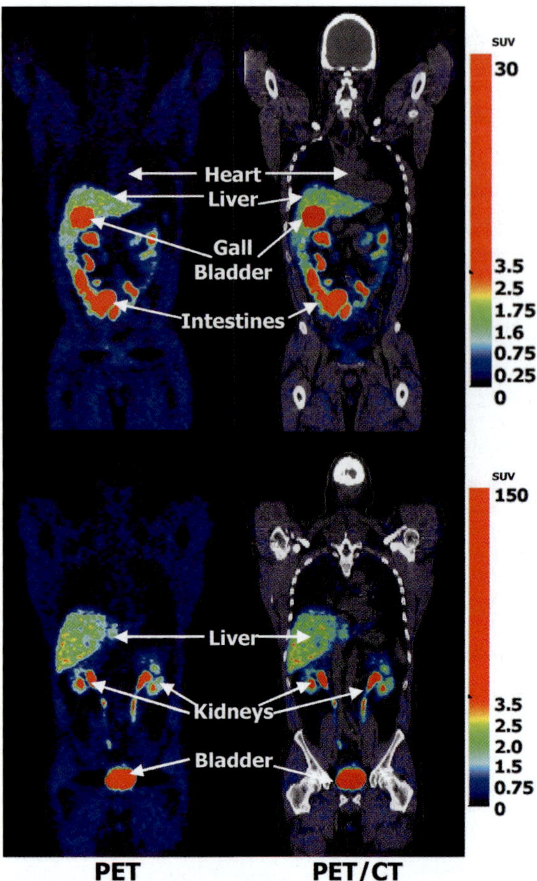

Figure 9.4. PET and PET/CT image views of two different coronal planes of a human volunteer who had been injected with approximately 7 mCi of [18F]FHBG intravenously. Images were acquired approximately 2 h after tracer injection. This figure illustrates the normal biodistribution of [18F]FHBG into all of the tissues on its clearance pathway. [18F]FHBG clears rapidly from background tissues outside of its clearance pathway and does not cross the blood–brain barrier.

[124I]FIAU [31]. Finally, [123I]FIAU SPECT imaging was not able to detect conditionally replicating oncolytic herpes simplex virus 1716 in a single study on glioma patients [32]. Clinical application of reporter genes (discussed in greater detail in Chapter 13) is not only limited to gene therapy. Radionuclide-based reporter gene/probe systems are now being used to image adoptively transferred genetically engineered therapeutic cells [33], and we expect accelerating use of clinical reporter gene–based imaging in future cellular gene therapy trials.

Molecular Imaging: Indirect Imaging of Therapeutic Transgene Expression Using Imaging Reporter Genes

It is often ideal to image the activity of the protein encoded by the therapeutic transgene using a highly specific probe. However, imaging probe development is not an easy task as it requires demonstrating appropriate chemistry, stability, suitable pharmacokinetics, safety, and high detection sensitivity in addition to specificity for the target protein [28, 29, 34–36]. Many probes will fail to meet these requirements, making the development process expensive and economically undesirable. Currently, specific imaging probes are not available for many of the therapeutic transgenes being investigated. A general method for indirect imaging of a therapeutic transgene's expression is linking its expression to that of an imaging reporter gene [37–40]. This may be accomplished through any of the methods illustrated in Figure 9.5 and described next.

A relatively simple approach is simultaneous coadministration of two vectors, identical in all respects with the single exception that one vector carries the therapeutic transgene but in the other vector the therapeutic transgene is replaced by an imaging RG. This approach was first validated using two replication-deficient adenoviral vectors, one of which carried the HSV1-sr39tk PRG and the other the D_2R PRG; otherwise they were identical (Figure 9.6A) [41]. Coinfection of two different cancer cell lines and coinjection of the adenoviruses into the tail vein of immunodeficient mice resulted in correlated expression of both PRGs by the infected cultured cells and in the livers of immunodeficient mice (Figures 9.6B–D). This strategy was also validated in implanted tumors, and the correlated expression of both PRGs was shown for up to 1 month in the livers of both immunodeficient and immunocompetent mice [41]. Despite its relative technical simplicity and potentially rapid generalizability, the dual vector method may have some drawbacks, including a relatively higher chance that correlation between the RG and TG will not be as good as with the techniques described in the following, because not every cell will receive the same number of vectors carrying each transgene. Though this may not matter, as

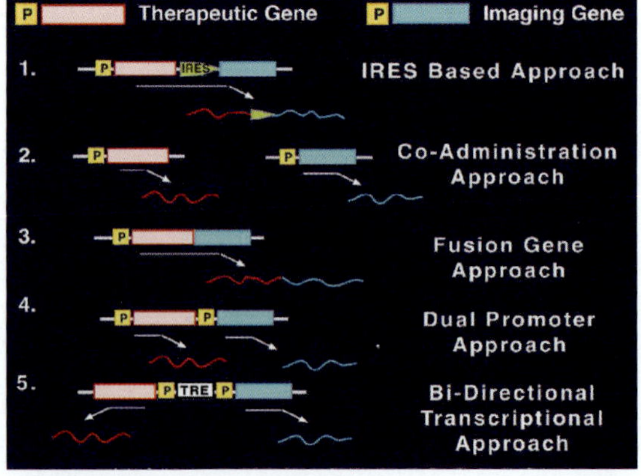

Figure 9.5. Methods for linking the expression of a therapeutic gene to an imaging reporter gene.

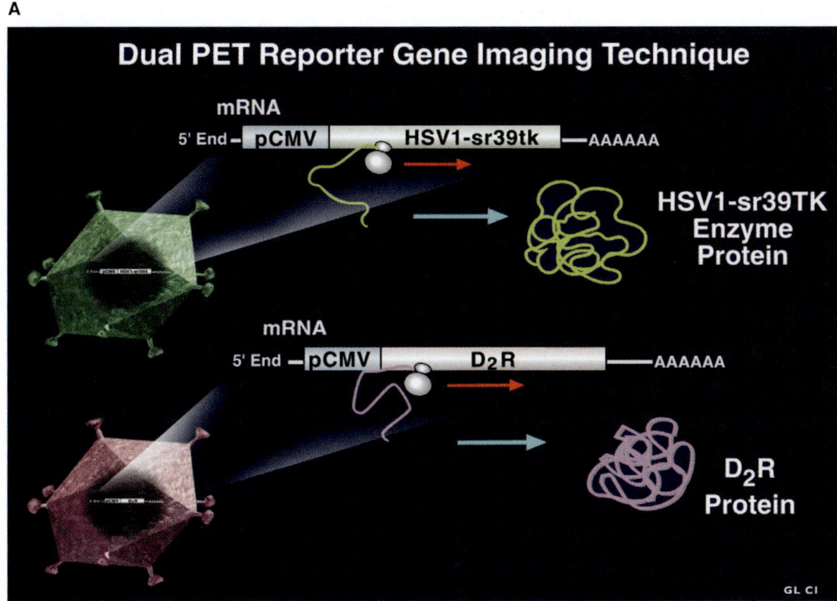

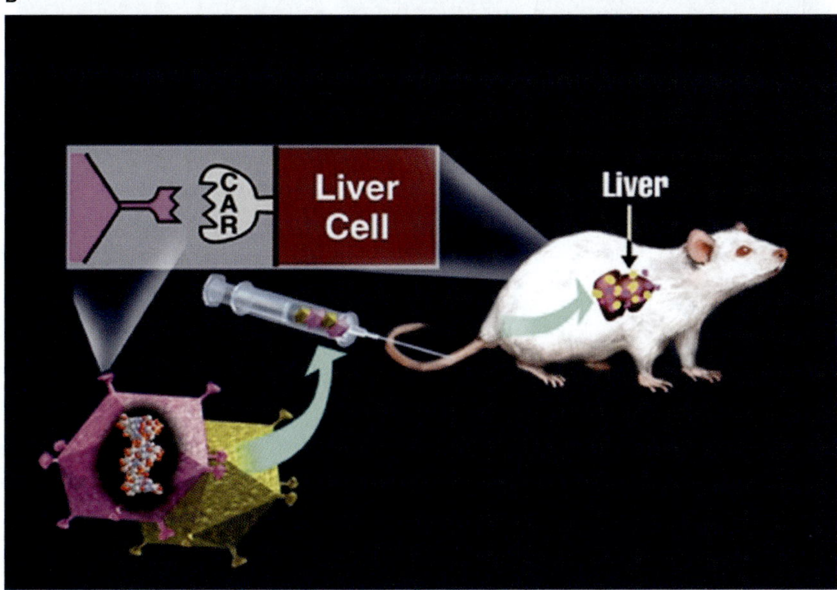

Figure 9.6. (A) Two identical replication-deficient adenoviral vectors, with the exception that each carries a different PRG. **(B)** Tail vein injection of adenoviruses mainly leads to infection of the liver due to the presence of the Coxsackie adenovirus receptors highly expressed on hepatocytes. **(C)** MicroPET whole body [18F]FESP and [18F]FHBG images of nude mice coinjected Ad-CMV-D$_2$R and Ad-CMV-HSV1-sr39tk intravenously. Equal viral titers of Ad-CMV-D$_2$R and Ad-CMV-HSV1-sr39tk were dissolved in 200 ul 0.9% NaCl and injected into the tail vein of nude mice. After 3 d, the mice were whole body scanned for 28 min (4 min per each of 7 beds) with MicroPET for signals from either [18F]FHBG or [18F]FESP, injected through the tail vein. The next day, the same mice were scanned identically with the alternate PRP. Images illustrate degree of accumulation and biodistribution of [18F]FHBG and [18F]FESP in individual nude mice coinjected with different but equivalent titers of Ad-CMV-HSV1-sr39tk and Ad-CMV-D$_2$R. From left to right, the first mouse received 0 pfu, the second mouse received 0.4 X 10^8 pfu, the third mouse received 0.5 X 10^8 pfu, and the fourth mouse received 3.6 X 10^8 pfu. The upper panel illustrates biodistribution of [18F]FESP and the lower panel that of [18F]FHBG. Images are the average of several whole-body slices containing the liver. The images are scaled based on the injected dose. Images shown are based on MAP reconstruction of the microPET data.

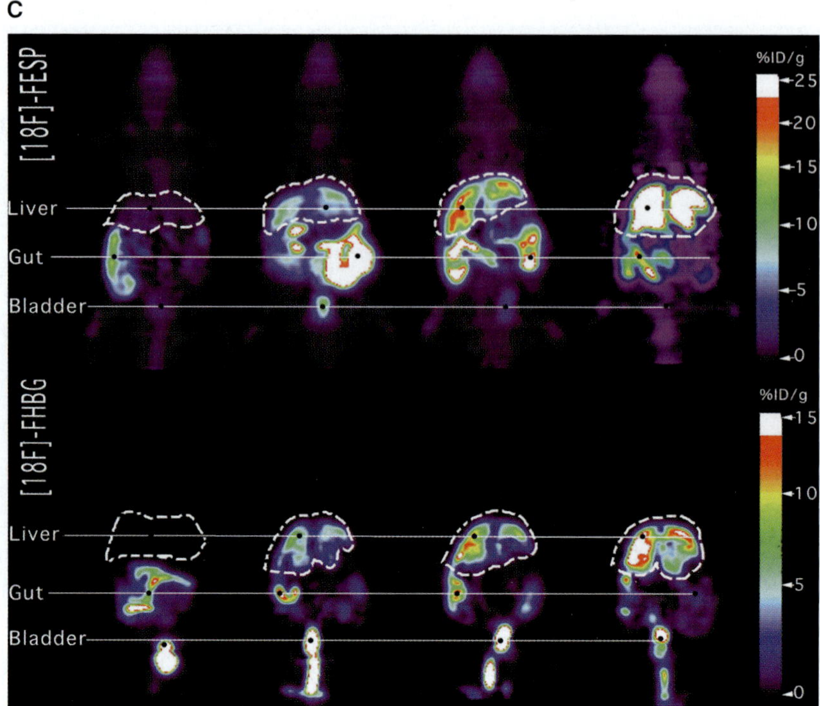

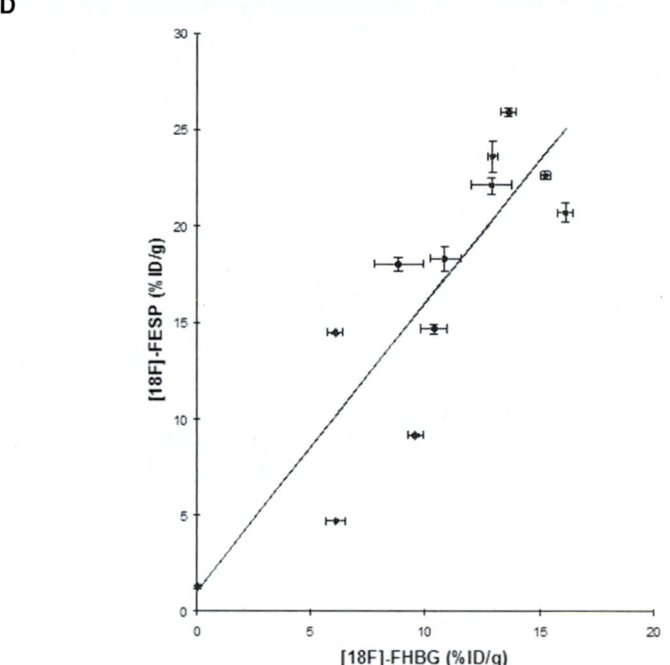

Figure 9.6 (*cont.*) (D) Quantitative determination of correlation between [18F]FHBG and [18F]FESP activity in the livers of nude mice injected with equivalent doses of Ad-CMV-HSV1-sr39tk and Ad-CMV-D$_2$R. Twelve nude mice (four of whom are shown in Figure 9.5C) were coinjected intravenously with various equivalent titers of Ad-CMV-D$_2$R and Ad-CMV-HSV1-sr39tk. The mice were imaged with microPET for the detection of [18F]FHBG and [18F]FESP (which had been injected through tail veins) on 2 consecutive days. Whole-body images were reconstructed by filtered back projection. The whole-body slices containing the liver were averaged. Three different regions of interest (ROI) were drawn over the livers and average counts/second/pixel in each ROI was determined. The graph illustrates percentage ID/g retention of [18F]FHBG versus percentage ID/g retention of [18F]FESP in the livers of the nude mice ($r^2 = 0.76$; $y = 1.49x + 1.06$). The error bars are standard deviations of the three average percentage ID/g. Figures 9.6C and D are reprinted with permission from Figure 4A and B of Yaghoubi et al. *Gene Therapy* 8:1072–1080 (2001).

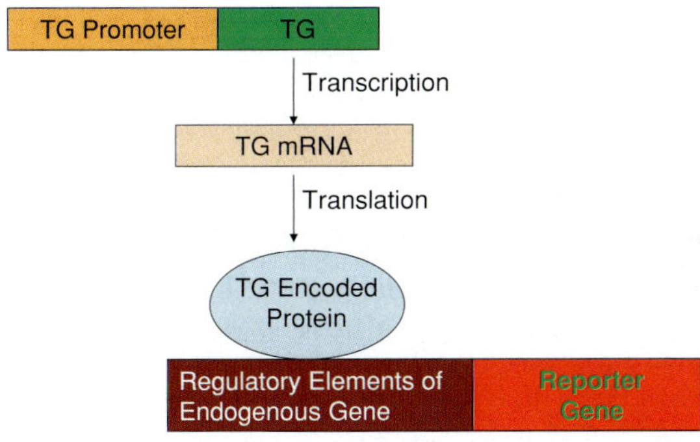

Figure 9.7. A method for imaging the effect of proteins encoded by a therapeutic gene on the expression of an endogenous gene.

the resolution of most imaging modalities is not at single cell level and all cells in a tissue will have equal chance of taking up both vectors. Another disadvantage when using viral vectors is the double viral burden, which may increase chances of immune reaction when using certain viruses. More genetic cloning-involved strategies include fusion gene, bicistronic, double identical promoter, and bidirectional transcriptional approaches. A bicistronic gene construct would consist of a single promoter driving transcription of the TG and RG, resulting in a single mRNA. However, the TG and RG are separated by an internal ribosomal entry site (IRES); hence the RG and TG are translated into two separate proteins. This approach has been validated for linking the expression of HSV1-sr39k and D_2R PRGs [42–45], HSV1-tk and LacZ RGs [46], and firefly luciferase (fluc) and Renilla luciferase (Rluc) bioluminescent RGs [44]. Although the transcription of a single mRNA increases the probability of obtaining significantly correlated expression of the linked TGs and RGs using the bicistronic system, the expression of the transgene downstream of the originally used encephalomyocarditis IRES is usually attenuated. This problem of attenuation was addressed using a 196-nucleotide super IRES sequence (from the 5$'$ UTR of the Gtx homeodomain mRNA), which increased the expression of the downstream RG [44, 47]. Unlike the bicistronic vector approach, the double identical promoter constructs and bidirectional promoter systems result in the transcription of two separate mRNAs. Nonetheless, double CMV promoter systems have thus far been the only constructs used to successfully link the expression of two actual TGs, the vascular endothelial growth factor and p53, to the HSV1-sr39tk PRG, albeit only in cell culture [48, 49]. However, we have also performed a preliminary study linking p53 TG expression to HSV1-sr39tk PRG expression using the covector infection of C6 glioma cells and demonstrated correlated expression [49]. The bidirectional promoter approach is useful as an inducible system allowing regulation of both RG and TG expression in a correlated manner [50]. This

technique has been validated for linking the expression of HSV1-sr39tk and D_2R PRGs [50], and Fluc and Rluc bioluminescent RGs [51]. The fusion gene approach has the highest probability of correlated expression because both TG and RG are transcribed in a single mRNA as well as a single protein. However, it is likely that the quarternary structure of the single fusion protein can hinder the function of the therapeutic protein or the detectability of the reporter protein. The fusion gene system has now been extensively used for multimodality RG imaging [52–58], which is described extensively in Chapter 5.

IMAGING THE EFFECT OF THERAPEUTIC TRANSGENE EXPRESSION USING IMAGING REPORTER GENES

Even though the immediate aim of gene therapy is to cause a change in the expression of specific gene(s) in target cells, like all other therapeutic approaches, the ultimate aim is to treat or cure a disease. Molecular imaging is increasingly being used to monitor the effect of therapeutic regimens in preclinical research and in patients. These same approaches can be used to monitor the effect of a gene therapy procedure. For example, when patients are treated with a suicide gene and its prodrug, [^{18}F]FDG imaging can be used to analyze the effect of suicide gene therapy on tumor metabolism [20, 25].

RG imaging can provide a greater level of information, before symptomatic changes are observed. It can be used to monitor the regulation of endogenous gene expression following introduction of a therapeutic transgene, the expression of which may induce activation or inhibit expression of specific endogenous genes. One can make genetic constructs such that an RG is under the control of the regulatory elements of an endogenous gene. If these regulatory elements are controlled by the protein encoded by the TG, then the expression levels of endogenous genes regulated by the TG can be monitored with the RG. Figure 9.7 describes the general strategy. This strategy has been used to image transcriptional regulation of a p53-dependent downstream gene [59]. The HSV1-tk/GFP fusion gene was cloned downstream of a p53 enhancer element. Investigators used Etoposide, BCNU, and UV irradiation to induce the p53 pathway and assay the cells for upregulated GFP and HSV1-tk expression. Induction of p53 not only upregulated the expression of the RGs but also the expression of p21, which is a p53-dependent downstream gene [59]. This system may also be used to image the effect of administering the p53 TG into cancer cells. Finally, one can use transgenic mice with endogenous tissue-specific promoters to follow downstream gene expression effects of therapeutic agents, including TG in living animal models [60].

FUTURE OF REPORTER GENE IMAGING IN GENE THERAPY

As it pertains to gene therapy, thus far imaging reporter genes have mainly been used to image the pharmacokinetics of therapeutic transgenes. That is, as discussed in this chapter, the magnitude, time variation, and location of some therapeutic transgenes have been either directly or indirectly imaged using different modalities of imaging reporter genes. The application of imaging reporter genes in gene therapy can be extended to noninvasive monitoring of therapeutic transgene pharmacodynamics. That is, one can use reporter genes to monitor up- or downregulation of endogenous genes following introduction of a TG into cells. For example, if the protein product of a TG is a transcription factor, one can monitor the upregulation of promoters that are affected by that transcription factor (Figure 9.7). A transgene construct can be introduced into cells containing an RG under the control of the suspected promoter. These studies can also be done in transgenic animals, who have an imaging RG under the control of a promoter that may be regulated by the protein product of the TG. In that case, the activity of the downstream promoter can be monitored within different tissues at the same time.

Among the most underdeveloped techniques in RG-based TG imaging is quantification. Until now, most quantitative analysis has been relative measurements of signal intensities. Basically, investigators report image signal intensities in TG target tissues to describe and compare TG delivery efficiencies. However, signal intensities do not directly translate into the number of protein molecules encoded by a TG. This type of quantification analysis can most accurately be performed using PET reporter genes and through pharmacokinetic modeling of the PET reporter probe [61]. Absolute quantification of gene expression and quantitative analysis of TG pharmacokinetics will allow more accurate prediction of TG efficacy and potency through statistical analysis.

Ultimately, the most needed application of RG imaging of TG is in gene and adoptive cellular gene therapy patient clinical trials and perhaps eventually incorporation into routine TG and genetically engineered therapeutic cell treatments. The reason is that many of the invasive TG pharmacokinetic studies that can be performed in animal disease models are simply impractical when dealing with human patients, yet it is as critical to study TG pharmacokinetics in humans to prevent long- and short-term adverse effects. However, so far RG imaging in clinical trials has been rare. One reason is that to incorporate RG imaging it is often necessary to plan it during the early stages of preparing for the gene and cellular gene therapy trial for both technical and regulatory reasons. This has not occurred much due to a relative disconnect between many of the translational gene/cell therapy investigators and companies and the translational molecular imaging investigators. Additionally, because regulatory agencies do not currently require whole-body imaging of TG in patients enrolled in gene therapy trials, many would avoid the cost of RG imaging, despite the potentially critical information this procedure may provide.

Technical factors are an additional hinderance to incorporating RG imaging in many clinical trials. Thus far, RG imaging has only been applied in the low-hanging fruit trials, where the TG is itself an RG (For example, when HSV1-tk is used as a suicide transgene in cancer gene therapy trials or a safety gene in adoptive cellular gene therapy trials). Many of these trials do not use the HSV1-sr39tk PET RG (which can also serve as a better suicide or safety transgene in addition to being a more sensitive RG for [^{18}F]FHBG), because many of the trials were planned years ago with HSV1-tk, and using HSV1-sr39tk would require obtaining new regulatory approvals. It gets much more complicated when the TG is not itself an RG, because of the requirement for the novel linking techniques we discussed in previous sections. The safety and feasibility of the genetic TG/RG linking constructs still need to be evaluated before being used in humans, and then these novel techniques will need to obtain regulatory approval. Finally, radionuclide and MRI-based RG systems need to be further optimized for more widespread application. Immune reaction against some reporter proteins are a potential problem for use in some patients whose immune systems have not been suppressed. For that reason, a variety of human-derived RGs, discussed in Chapter 3, have been developed.

Despite these difficulties, we believe that the most exciting future application of RG imaging will be in patients. Clinical imaging can hasten development of gene or cellular gene therapy strategies that were originally promised to be curative and low-side-effect procedures. We are also excited about RG imaging fulfilling its promise of being a generalized TG imaging technique as it is used to image TGs other than HSV1-tk. Finally, we expect RG imaging will expand beyond pharmacokinetics to pharmacodynamics, helping elucidate the effect of TGs on molecular events within tissues of living subjects, noninvasively.

REFERENCES

1   Gao, X., Kim, K., Liu, D. (2007). Nonviral gene delivery: What we know and what is next. *AAPS Journal* 9(1): E92–E104.

2   De, A., Gambhir, S. S. (2003). PET in imaging gene expression and therapy. In: Bailey, D. et al. (Eds.), *Positron Emission Tomography: Basic Science and Clinical Practice*, New York: Springer-Verlag: 845–868.

3   Min, J. J., Gambhir, S. S. (2004). Gene therapy progress and prospects: noninvasive imaging of gene therapy in living subjects. *Gene Therapy* 11: 115–125.

4 Walls, Z., Gambhir, S. S. (2007). BRET-based method for detection of specific RNA species. *Bioconjug Chem* **19**(1): 178–184.

5 Walls, Z. et al. (2008). A generalizable strategy for imaging pre-mRNA levels in living subjects using spliceosome-mediated RNA trans-splicing. *J Nucl Med* **49**(7): 1146–1154.

6 Yaghoubi, S. S., Gambhir, S. S. (2008). Imaging and cancer. In: Mendelsohn, J. et al. (Eds.), *The Molecular Basis of Cancer*. Saunders Elsevier: Philadelphia: 309–323.

7 Tarantal, A. F. et al. (2006). Fetal gene transfer using lentiviral vectors: In vivo detection of gene expression by microPET and optical imaging in fetal and infant monkeys. *Hum Gene Ther* **17**: 1254–1261.

8 Chaudhuri, T. R. et al. (2001). A noninvasive reporter system to image adenoviral-mediated gene transfer to ovarian cancer xenografts. *Gynecol Oncol* **83**: 432–438.

9 Xiong, Z. et al. (2006). Imaging chemically modified adenovirus for targeting tumors expressing integrin avb3 in living mice with mutant herpes simplex virus type 1 thymidine kinase PET reporter gene. *J Nucl Med* **47**(1): 130–139.

10 Wu, J. C. et al. (2002). Positron emission tomography imaging of cardiac reporter gene expression in living rats. *Circulation* **106**: 180–183.

11 Richard, J.-C. et al. (2003). Imaging the spatial distribution of transgene expression in the lungs with positron emission tomography. *Gene Therapy* **10**: 2074–2080.

12 Deroose, C. M. et al. (2006). Noninvasive monitoring of long-term lentiviral vector-mediated gene expression in rodent brain with bioluminescence imaging. *Mol Ther* **14**(3): 423–431.

13 Lim, S. J. et al. (2007). Enhanced expression of adenovirus-mediated sodium iodide symporter gene in MCF-7 breast cancer cells with retinoic acid treatment. *J Nucl Med* **48**(3): 398–404.

14 Kuruppu, D. et al. (2007). Positron emission tomography of herpes simplex virus 1 oncolysis. *Cancer Research* **67**(7): 3295–3300.

15 Sen, L. et al. (2005). Noninvasive imaging of ex vivo intra-coronarily delivered nonviral therapeutic transgene expression in heart. *Mol Ther* **12**(1): 49–57.

16 Hildebrandt, I. J. et al. (2003). Optical imaging of transferrin targeted PEI/DNA complex in living subjects. *Gene Therapy* **10**: 758–764.

17 De, A., Lewis, X. Z., Gambhir, S. S. (2003). Noninvasive imaging of lentiviral-mediated reporter gene expression in living mice. *Mol Ther* **7**(5): 681–691.

18 Jacobs, A. et al. (2001). Positron emission tomography-based imaging of transgene expression mediated by replication-conditional, oncolytic herpes simplex virus type 1 mutant vectors in vivo. *Cancer Research* **61**: 2983–2995.

19 Bennett, J. J. et al. (2001). Positron emission tomography imaging for herpes virus infection: Implications for oncolytic viral treatments of cancer. *Nat Med* **7**(7): 859–863.

20 Yaghoubi, S. S. et al. (2005). Imaging progress of herpes simplex virus type 1 thymidine kinase suicide gene therapy in living subjects with positron emission tomography. *Cancer Gene Therapy* **12**: 329–339.

21 Pantuck, A. J. et al. (2002). CL1-SR39: A noninvasive molecular imaging model of prostate cancer suicide gene therapy using positron emission tomography. *J Urol* **168**: 1193–1198.

22 Pantuck, A. J. et al. (2002). Optimizing prostate cancer suicide gene therapy using herpes simplex virus thymidine kinase active site variants. *Hum Gene Ther* **13**: 777–789.

23 Burton, J. B. et al. (2008). Adenovirus-mediated gene expression imaging to directly detect sentinel lymph node metastasis of prostate cancer. *Nat Med* **14**(8): 882–888.

24 Deng, W. et al. (2004). Non-invasive in vivo imaging with radiolabelled FIAU for monitoring cancer gene therapy using herpes simplex virus type 1 thymidine kinase and ganciclovir. *Eur J Nucl Med Mol Imaging* **31**(1): 99–109.

25 Wang, H. et al. (2006). Molecular imaging with 123I-FIAU, 18F-FUdR, 18F-FET, and 18F-FDG for monitoring herpes simplex virus type 1 thymidine kinase and Ganciclovir prodrug activation gene therapy of cancer. *J Nucl Med* **47**: 1161–1171.

26 Deng, W. et al. (2006). Serial in vivo imaging of lung metastases model and gene therapy using HSV1-tk and Ganciclovir. *J Nucl Med* **47**: 877–884.

27 Penuelas, I. et al. (2005). Gene therapy imaging in patients for oncological applications. *Eur J Nucl Med Mol Imaging* **32**(14): S384–S403.

28 Yaghoubi, S. S. et al. (2001). Human pharmacokinetic and dosimetry studies of [18F]FHBG: a reporter probe for imaging herpes simplex virus type-1 thymidine kinase reporter gene expression. *J Nucl Med* **42**: 1225–1234.

29 Yaghoubi, S. S. et al. (2006). Preclinical safety evaluation of 18F-FHBG: A PET reporter probe for imaging herpes simplex virus type 1 thymidine kinase (HSV1-tk) or mutant HSV1-sr39tk's expression. *J Nucl Med* **47**(4): 706–715.

30 Penuelas, I. et al. (2005). Positron emission tomography imaging of adenoviral-mediated transgene expression in liver cancer patients. *Gastroenterology* **128**(7): 1787–1795.

31 Jacobs, A. et al. (2001). Positron-emission tomography of vector-mediated gene expression in gene therapy for gliomas. *Lancet* **358**: 727–729.

32 Dempsey, M. F. et al. (2006). Assessment of 123I-FIAU imaging of herpes simplex viral gene expression in the treatment of glioma. *Nucl Med Commun* **27**: 611–617.

33 Yaghoubi, S. S. et al. Noninvasive detection of therapeutic cytolytic T cells with ^{18}F-FHBG in a patient with glioma. *Nat Clin Pract Oncol* **6**: 53–58.

34 Chin, F. T. et al. (2008). Semiautomated radiosynthesis and biological evaluation of [18F]FEAU: a novel PET imaging agent for HSV1-tk/sr39tk reporter gene expression. *Mol Imaging Biol* **10**(2): 82–91.

35 Alauddin, M. M., Conti, P. S. (1998). Synthesis and preliminary evaluation of 9-(4-[18F]-Fluoro-3-Hydroxymethylbutyl)Guanine ([18F]FHBG): a new potential imaging agent for viral infection and gene therapy using PET. *Nucl Med Biol* **25**: 175–180.

36 Alauddin, M. A. et al. (2001). Preclinical evaluation of the penciclovir analog 9-(4-[18F]fluoro-3-hydroxymethylbutyl)guanine for in vivo measurement of suicide gene expression with PET. *J Nucl Med* **42**(11): 1682–1690.

37 Ray, P. et al. (2001). Monitoring gene therapy with reporter gene imaging. *Semin Nucl Med* **31**(4): 312–320.

38 Iyer, M. et al. (2005). Applications of molecular imaging in cancer gene therapy. *Curr Gene Ther* 5: 607–618.

39 Gambhir, S. S. et al. (2000). Imaging transgene expression with radionuclide imaging technologies. *Neoplasia* 2(1–2): 118–138.

40 Herschman, H. R. et al. (2002). Monitoring gene therapy by positron emission tomography. In: Curiel, D. T., and Douglas, J. T. (Eds.), *Vector Targeting for Therapeutic Gene Delivery* 661–685. Wiley-Liss: New York.

41 Yaghoubi, S. S. et al. (2001). Direct correlation between positron emission tomographic images of two reporter genes delivered by two distinct adenoviral vectors. *Gene Therapy* 8: 1072–1080.

42 Yu, Y. et al. (2000). Quantification of target gene expression by imaging reporter gene expression in living animals. *Nat Med* 6(8): 933–937.

43 Liang, Q. et al. (2002). Noninvasive, repetitive, quantitative measurement of gene expression from a bicistronic message by positron emission tomography, following gene transfer with adenovirus. *Mol Ther* 6(1): 73–82.

44 Wang, Y. et al. (2005). Noninvasive monitoring of target gene expression by imaging reporter gene expression in living animals using improved bicistronic vectors. *J Nucl Med* 46: 667–674.

45 Chen, I. Y. et al. (2004). Micro–positron emission tomography imaging of cardiac gene expression in rats using bicistronic adenoviral vector-mediated gene delivery. *Circulation* 109: 1415–1420.

46 Tjuvajev, J. G. et al. (1999). A general approach to the non-invasive imaging of transgenes using cis-linked herpes simplex virus thymidine kinase. *Neoplasia* 1: 315–320.

47 Chappell, S. A., Edelman, G. M., Mauro, V. P. (2000). A 9-nt segment of a cellular mRNA can function as an Internal Ribosome Entry Site (IRES) and when present in linked multiple copies greatly enhances IRES activity. *PNAS* 97(4): 1536–1541.

48 Anton, M. et al. (2004). Coexpression of herpesviral thymidine kinase reporter gene and VEGF gene for noninvasive monitoring of therapeutic gene transfer: an in vitro evaluation. *J Nucl Med* 45(10): 1743–1746.

49 Yaghoubi, S. S., Gambhir, S. S. (2003). Monitoring p53 therapeutic transgene expression in mice by imaging the expression of a linked HSV1-sr39tk PET reporter transgene. *J Nucl Med* 44(5): 30P [Abstract#95].

50 Sun, X. et al. (2001). Quantitative imaging of gene induction in living animals. *Gene Therapy* 8: 1572–1579.

51 Ray, S. et al. (2004). Novel bidirectional vector strategy for amplification of therapeutic and reporter gene expression. *Hum Gene Ther* 15: 681–690.

52 Ray, P. et al. (2004). Imaging tri-fusion multimodality reporter gene expression in living subjects. *Cancer Research* 64: 1323–1330.

53 Ray, P., Tsien, R., Gambhir, S. S. (2007). Construction and validation of improved triple fusion reporter gene vectors for molecular imaging of living subjects. *Cancer Research* 67(7): 3085–3093.

54 Ray, P., Wu, A. M., Gambhir, S. S. (2003). Optical bioluminescence and positron emission tomography imaging of a novel fusion reporter gene in tumor xenografts of living mice. *Cancer Research* 63: 1160–1165.

55 Doubrovin, M. et al., Development of a new reporter gene system-dsRed/Xanthine Phosphoribosyltransferase-Xanthine for molecular imaging of processes behind the intact blood–brain barrier. *Mol Imaging* 2(2): 93–112.

56 Ponomarev, V. et al. (2004). A novel triple-modality reporter gene for whole-body fluorescent, bioluminescent, and nuclear noninvasive imaging. *Eur J Nucl Med Mol Imaging* 31: 740–751.

57 Jacobs, A. et al. (1999). Functional coexpression of HSV-1 thymidine kinase and green fluorescent protein: implications for noninvasive imaging of transgene expression. *Neoplasia* 1(2): 154–161.

58 Hwang, D. W. et al., Development of a dual membrane protein reporter system using sodium iodide symporter and mutant dopamine D2 receptor transgenes. *J Nucl Med* 48(4): 588–595.

59 Doubrovin, M. et al. (2001). Imaging transcriptional regulation of p53-dependent genes with positron emission tomography in vivo. *PNAS* 98(16): 9300–9305.

60 Green, L. A. et al. (2002). Indirect monitoring of endogenous gene expression by positron emission tomography (PET) imaging of reporter gene expression in transgenic mice. *Mol Imaging Biol* 4(1): 71–81.

61 Green, L. A. et al. (2004). A tracer kinetic model for ^{18}F-FHBG for quantitating herpes simplex virus type 1 thymidine kinase reporter gene expression in living animals using PET. *J Nucl Med* 45(9): 1560–1570.

Imaging Regulation of Endogenous Gene Expression in Living Subjects

10

Vladimir Ponomarev

The past decade has witnessed a remarkable increase in our knowledge and understanding of the genetics and molecular biology of human diseases. Significant progress in the understanding of the molecular–genetic mechanisms of many diseases has been achieved with the advent of the modern molecular–biological assays. Analytical methods used in molecular biology have been introduced into mainstream research and clinical practice. In parallel, imaging methods are transforming into widely available diagnostic procedures for molecular imaging showing excellent spatial and temporal characteristics. However, there remains a tremendous gap between the advances made in molecular biology and their application to preclinical and clinical studies. Molecular reporter gene imaging can bridge contemporary molecular biology and imaging by noninvasive monitoring biochemical processes at cellular and subcellular levels *in vivo*.

The development of transgenic animal models of human diseases, which allows the molecular basis of the disease to be studied in a living organism, has provided new insight into disease development, progression, and treatment [1, 2]. Established methods for noninvasive imaging of reporter gene expression can be introduced into the existing reporter gene-based molecular–biological assay systems. In such assay systems, reporter gene expression is linked to an endogenous molecular genetic process of interest. These molecular genetic processes of interest include regulation of endogenous gene expression at the transcriptional (*cis*-reporter systems) and posttranscriptional (RNA-dependent reporter systems) levels, activation of specific signal transduction pathways, and specific transcription factors.

CIS-REPORTER SYSTEMS

Extracellular signals trigger the sequential activation of a series of intracellular signaling molecules, such as protein kinases and phosphatases. The final step in the activation of many pathways is the binding of an activated transcription factor to specific enhancer elements found in the promoters of various cellular genes [3]. This binding modulates transcription of these genes. The transcription level of cellular genes reflects the activation status of the involved signaling events. Gene transcription and mRNA translation, hence a gene level of expression, are typically monitored by Northern and Western blotting. To determine the activation of the promoter, and hence signaling events, different vectors have been designed that use a reporter gene in place of the endogenous cellular gene [4]. The reporter gene usually encodes an enzyme, receptor, transporter, or fluorescent protein that can be monitored easily, inexpensively, and quantitatively. Reporter assays replace tedious blotting procedures and are widely used due to their convenience, sensitivity, and ability to generate reliable *in vivo* data. The *cis*-reporting systems feature a *cis*-reporting plasmid or viral vector that contains a reporter gene driven by a basic promoter element (TATA box) joined to direct repeats of *cis*-acting DNA elements, also known as enhancers (Figure 10.1)[4]. The *cis*-reporting systems can be used to study the effects of a new gene, growth factor, or drug candidate. After a vector expressing a *cis*-reporter system of interest is transfected into mammalian cells, the cells are stimulated with an exogenous stimulus, treated with a candidate drug, or transfected to overexpress a new gene. The increased reporter enzyme activity indicates transcriptional activation of the reporter gene and involvement of the signaling pathways converging at these *cis*-acting enhancer elements in cellular responses to the exogenous or endogenous stimuli, drug action, or functions of the gene of interest.

Cis-reporter systems have been used to study the regulation of expression of different genes and signal transduction pathways. Transgenic models using different response elements or complete promoters may serve as a general approach to monitor gene expression *in vivo*

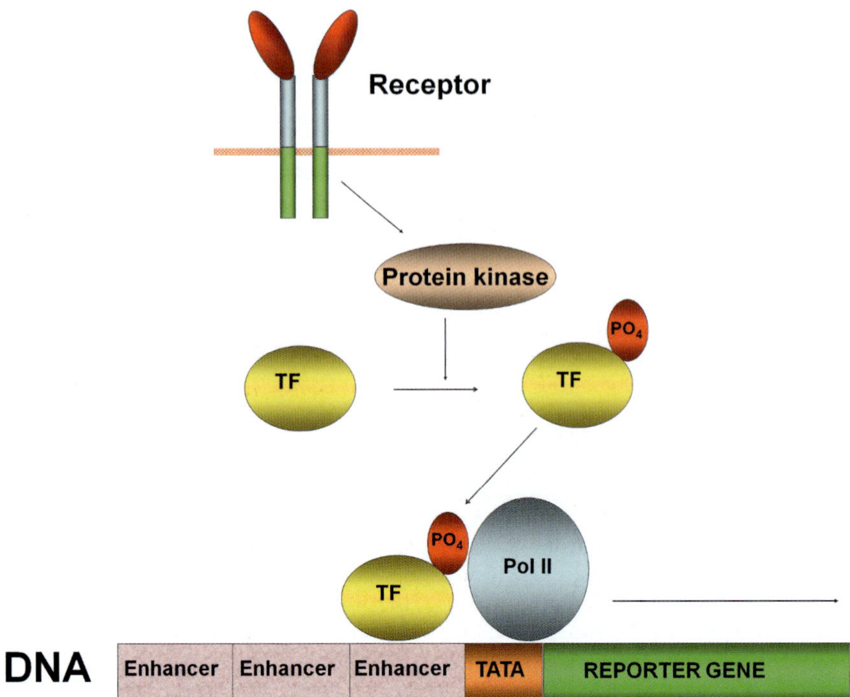

Figure 10.1. The paradigm of the *Cis*-reporter system. The transcription factor (TF) is activated (phosphorylated, PO₄) during the activation of a particular signal transduction pathway. The activated transcription factor binds to the specific *cis*-acting DNA element, enhancer, and initiates transcription of the reporter gene. TATA – minimal promoter; Pol II – RNA polymerase type II.

and to screen exogenous factors, such as nutrients and pharmaceuticals, for modulation of gene expression [5]. Several examples of *cis*-reporter systems are described next.

Hormone Receptors

Steroid hormone receptors are ligand-dependent transcription factors whose activity is modulated by estrogens and androgens. They control the growth, differentiation, and function of reproductive tissues and other systems through intracellular signaling and activation of corresponding target genes [6]. In a study by Lemmen et al. [7] transgenic mice carrying a luciferase reporter gene under the control of three consensus estrogen-responsive elements coupled to a minimal TATA box were generated for direct detection of activated estrogen receptors (ERs). Transgenic animals were exposed to 17β-estradiol, 17β-estradiol-dipropionate, or diethylstilbestrol to compare and characterize their ability to activate the reporter construct via endogenous ERs. Background levels of luciferase activity varied extensively between different organs studied due to a difference in the level of endogenous estrogens in the tissues. The highest (>10,000-fold) induction of luciferase was measured in bone and kidney 24 h after exposure to 17β-estradiol-dipropionate. Other highly responsive organs included

liver, testis, pituitary, brain, prostate, and colon, which show different activity profiles. In all tissues except the testis, luciferase activity was blocked after treatment with the ER antagonist.

The application of imaging technology to study animal models of cancer provides mechanistic insight into antiandrogen targeting of androgen receptors during disease progression. Ilagan et al. [8] tested a novel androgen receptor-specific luciferase-based reporter system driven by the prostate specific antigen (PSA) promoter to detect the action of the antiandrogen flutamide on androgen receptor function in xenograft models of prostate cancer (Figure 10.2). Flutamide inhibited the androgen-signaling pathway in androgen-dependent but not refractory tumors. Bioluminescent imaging of tumors revealed that the response of the androgen-dependent luciferase reporter system to flutamide was more sensitive and robust than serum PSA measurements.

Ciana et al. [9] addressed the importance of hormone-independent activation of the estrogen receptor using a transgenic mouse that expresses a luciferase reporter gene under the control of activated estrogen receptors. This group showed that ERs might participate in the development or function of nonreproductive organs such as bone and brain, independent of gonadal activity.

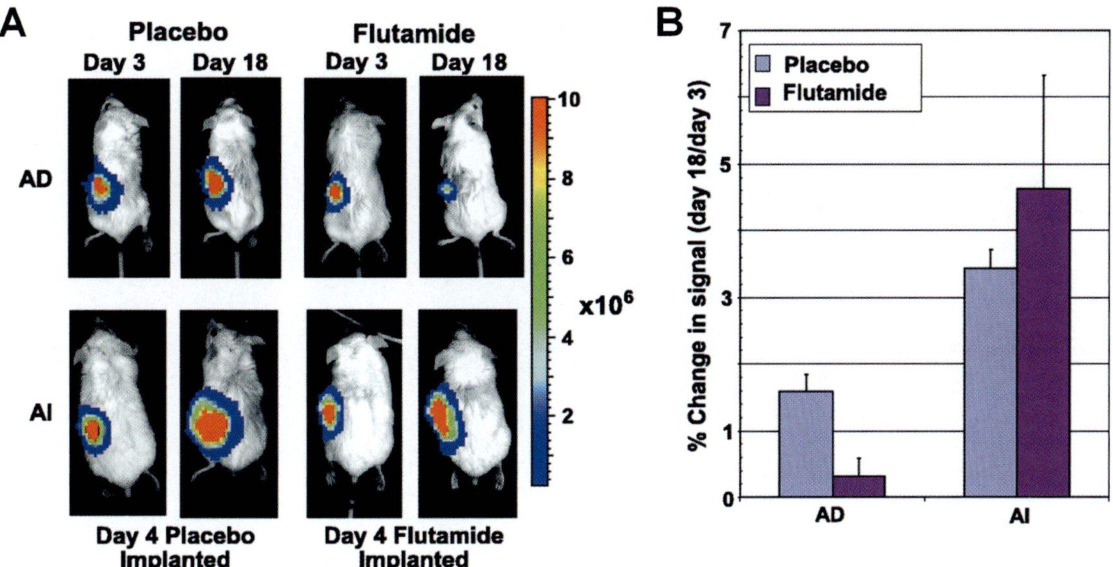

Figure 10.2. Bioluminescence imaging (BLI) of prostate cancer tumors response to flutamide. Androgen-dependent (AD) and -independent (AI) flank LAPC9 tumors were transduced with adenoviral vector encoding luciferase under control of androgen-dependent promoter. (**A**) On Day 3 a baseline image was acquired and flutamide or placebo pellets were implanted. The effect of treatment at Day 18 is shown. (**B**) Cohorts of androgen-dependent ($n = 7$) and androgen-independent ($n = 6$) animals were studied. Percentage change in signal on Day 18 versus Day 3 in placebo- and flutamide-treated animals bearing androgen-dependent and androgen-independent tumors is shown. (Adapted from (8) with permission).

The previously mentioned studies have implications for estrogens, androgens, and their antagonists in therapies such as hormone replacement therapy or hormone-dependent cancer treatment.

Proliferation

Initially bioluminescence imaging (BLI) of xenografts was studied in which firefly luciferase was driven by a constitutive promoter (e.g., cytomegalovirus promoter, pCMV) and the light output was a measure of cell number [10, 11]. Uhrbom et al. [12] developed a new transgenic mouse for BLI of the proliferative activity of glioma cells. In this mouse line, the firefly luciferase is controlled by the human E2F1 promoter, which exhibits proliferation-specific activity *in vivo* [13]. Unlike the mice with the constitutively expressed reporter gene, the E2F1-luc mouse can be used in functional studies, as it generates light proportional to both cell number and proliferation index. Authors have shown that the proliferation of platelet-derived growth factor (PDGF)-induced gliomas is dependent on both PDGF receptor activation and its downstream signaling, the mammalian target of rapamycin (mTOR), a serine/threonine protein kinase that regulates cell growth and proliferation. The reduction in light production after treatment with either PDGF or mTOR inhibitors is predominantly an effect of cell proliferation inhibition (Figure 10.3). Therefore, the E2F1-luc model provides a means to investigate downstream signaling pathways in tumor maintenance.

NF-κB is the collective name for a family of transcription factors that regulate cellular proliferative and survival responses [14]. Aberrant and increased activity of NF-κB is characteristic of some lymphoid tumors. In particular, the activation of NF-κB by viral oncogenes is a mechanism used by lymphomagenic viruses [15]. In a study by Keller et al. [16] the inhibition of NF-κB was shown to be a viable therapeutic approach for treating EBV- and KSHV-associated lymphomas. A human lymphoma cell line was engineered to express luciferase under the control of NF-κB-specific promoter, with decreased activity on inhibition of NF-κB by treatment with specific NF-κB inhibitor Bay 11–7082. Bioluminescence was decreased in all mice with visible tumors treated with Bay 11–7082 by 24 h, indicating effective *in vivo* NF-κB inhibition by this drug. The decrease of luciferase activity *in vivo* correlated with tumor growth delays and increased disease-free survival.

Human telomerase reverse transcriptase (hTERT) is an attractive target for anticancer therapeutics because of its requirement for cellular immortalization and expression in greater than 85% of human neoplasms [17]. Padmanabhan et al. [18] attempted to develop noninvasive methods of measuring hTERT transcription using a multimodality approach. A fusion reporter construct containing humanized Renilla luciferase, monomeric red fluorescence protein, and a truncated mutant herpes simplex virus 1 thymidine kinase (HSV1-tsr39tk) was placed under the control of the hTERT promoter and introduced into tumor cell lines with and without

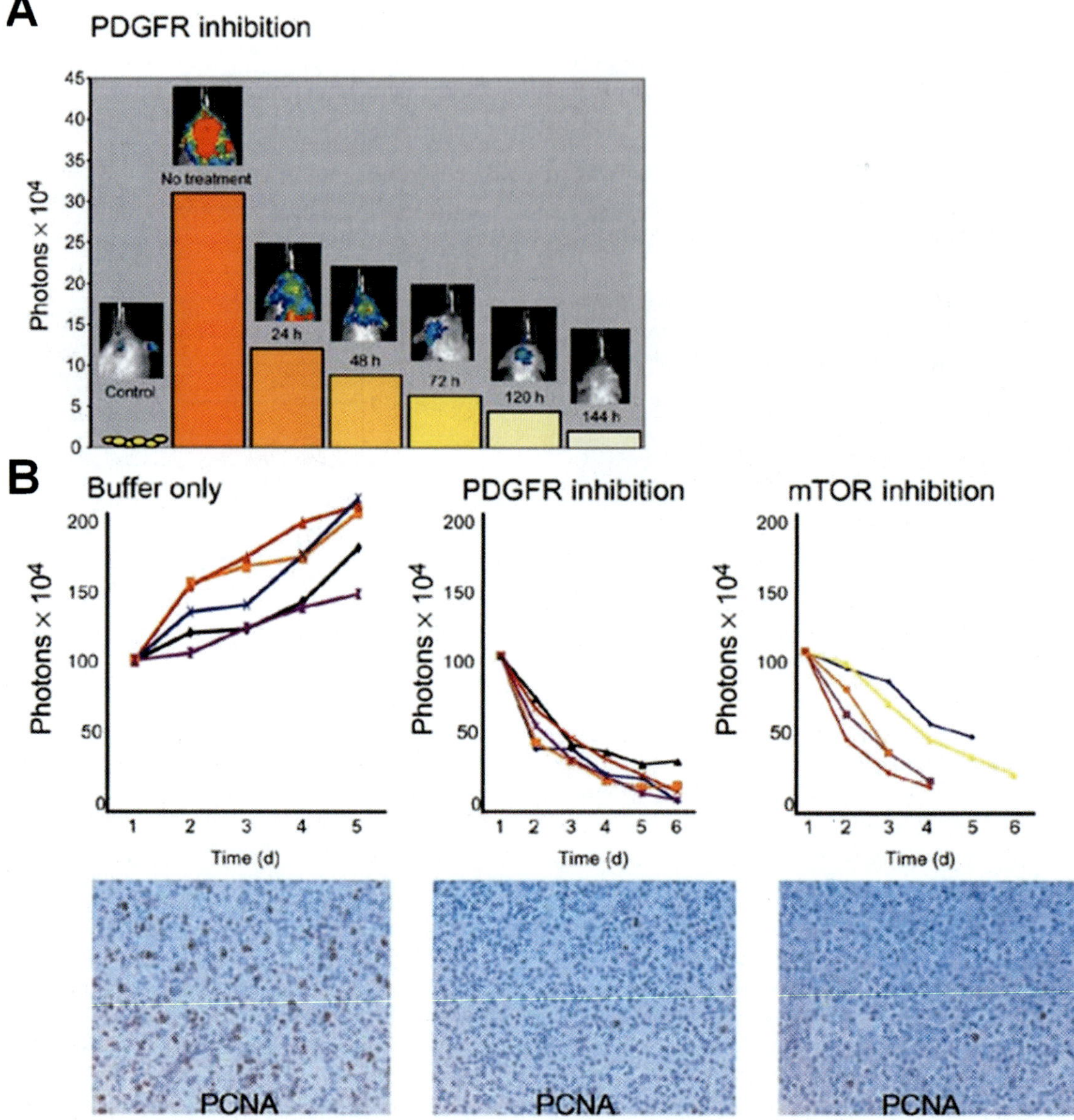

Figure 10.3. BLI of platelet-derived growth factor (PDGF)-induced glioma-bearing *cis*-EF-luciferase mice engineered to express luciferase reporter gene in spontaneous brain tumors. (**A**) Longitudinal imaging of one tumor-bearing mouse treated with PDGFR-specific inhibitor PTK787/ZK222584 daily for 6 d. (**B**) Longitudinal study with five *cis*-EF-luciferase tumor-bearing mice in each cohort: untreated (left panel) or treated daily with PTK787/ZK222584 (middle panel) or mTOR-specific inhibitor CCI-779 (right panel). Upper panels show photon counts of emitted light and lower panels show immunohistochemical staining for PCNA as a measure of cell proliferation. (Adapted from (12) with permission).

hTERT expression. Upon treatment with 5-fluorouracil, a chemotherapeutic agent that decreases hTERT gene expression, fluorescent microscopy and bioluminescent imaging of living cells demonstrated downregulation of hTERT promoter activity. This study provided evidence that the hTERT-sensitive reporter construct can be used to study the expression of a critical tumor growth gene in living subjects for *in vivo* assessment of cell proliferation, aging, and hTERT-specific drug screening.

Metastasis-Promoting Stimuli

Tumor cells that evade antiproliferative control may display an altered sensitivity to tumor growth factor-β (TGF-β) and undergo tumorigenic progression in response to this cytokine [19]. In mouse models of breast cancer, TGF-β signaling promotes lung [20] and bone metastasis [21]. Kang et al. [22] created a retroviral reporter vector in which a fusion protein containing

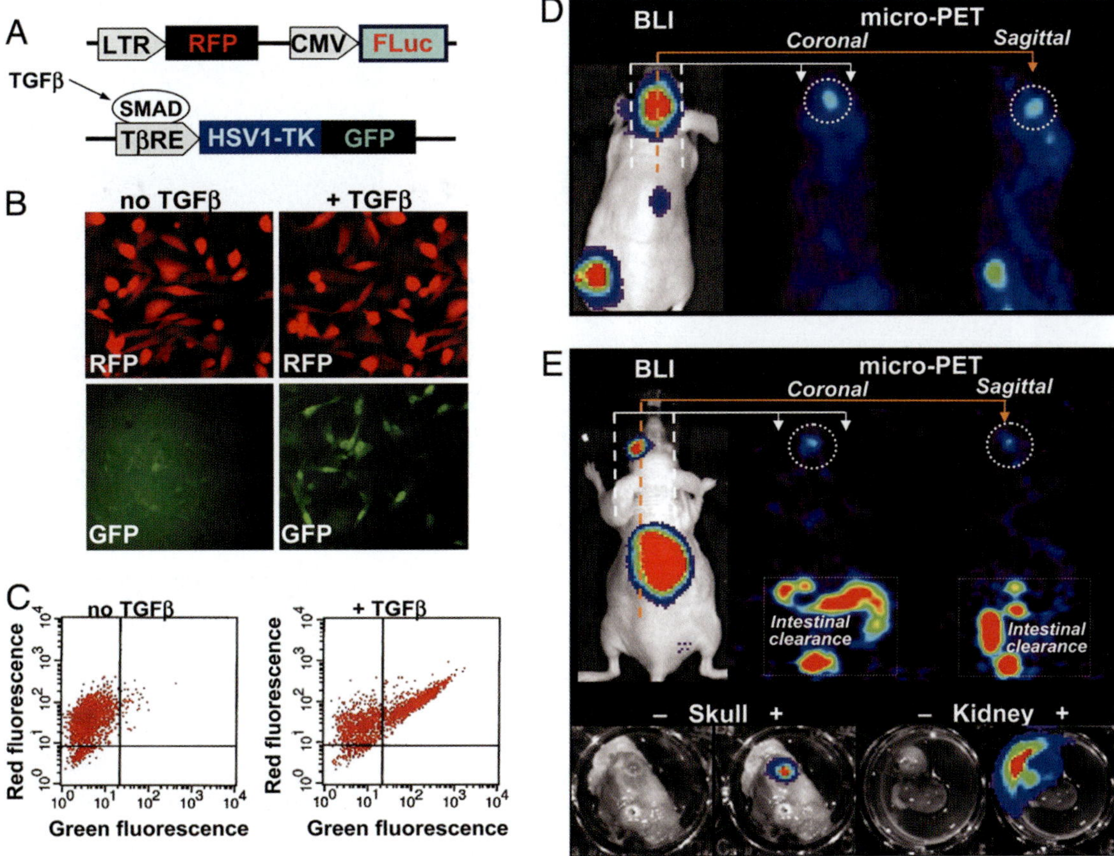

Figure 10.4. Functional imaging of TGF-β – Smad signaling in breast cancer bone metastasis model. (**A**) Schematic representation of the retroviral vectors: SFG-RFP-cmvFLuc constitutively expressing RFP and firefly luciferase and *cis*-TGF-β1–Smads–HSV1tk/GFP expressing HSV-tk/GFP fusion reporter protein in response to TGF-β. (B and C) SCP3 transduced with these two vectors were treated with TGF-β or no additions for 24 h and analyzed by (**B**) fluorescence microscopy or (**C**) two-color FACS. The constitutive RFP fluorescence is shown on the ordinate, and the HSV1tk/GFP fusion fluorescence, inducible by TGF-β, is shown on the abscissa. (**D** and **E**) In vivo bioluminescence and microPET imaging of metastases in mice. SCP2 (**D**) and SCP3 (**E**) cells bearing the SFG-RFP-cmvFLuc and *cis*-TGF-β 1–Smads–HSV1tk/GFP vectors were injected into the left cardiac ventricle and analyzed after 4 weeks (SCP2) or 18 weeks (SCP3). Bioluminescence imaging shows sites of metastases in the skull (**D** and **E**) and adrenal gland (**E**). ^{18}F-2′-fluoro-2′deoxy-1-D-arabionofuranosyl-5-ethyl-uracil (^{18}F-FEAU) microPET images of HSV1tk/GFP reporter activation shows localization of radioactivity to the skull in the coronal and sagittal image planes. No visualization of the adrenal metastasis was seen on microPET imaging. (**F**) At necroscopy, the head showing the skull and the adrenal metastasis plus kidney were removed and imaged *ex vivo* for photographic (–) and bioluminescence (+) imaging. (Adapted from (22) with permission).

HSV1-tk and GFP was placed under the transcriptional control of a TGF-β-responsive promoter element. This vector was introduced into breast cancer cells expressing firefly luciferase under pCMV constitutive promoter. When inoculated into the arterial circulation of immunodeficient mice, transduced cells formed aggressive bone metastases, as visualized by luciferase bioluminescence imaging (Figure 10.4). TGF-β signaling activity could be readily detected in bone metastases by microPET but not in adrenal gland metastases, although all of the metastases were formed by the same cell line. This study showed that breast cancer cells undergo TGF-β-dependent Smad-mediated transcriptional activation in the bone microenvironment and the possibility of

therapeutic targeting of this pathway in TGF-β-rich metastatic sites.

Differentiation

Retinoic acids are natural derivatives of vitamin A and play important roles in modulating tumor cell growth by regulating differentiation, thus suggesting the potential for using these derivatives in cancer therapy and prevention [23]. To visualize the intranuclear responses of functional retinoic acid receptors, So et al. [24] developed a dual-imaging reporter gene system based on the use of the sodium/iodide symporter and luciferase under the control of an artificial *cis*-acting retinoic acid-responsive

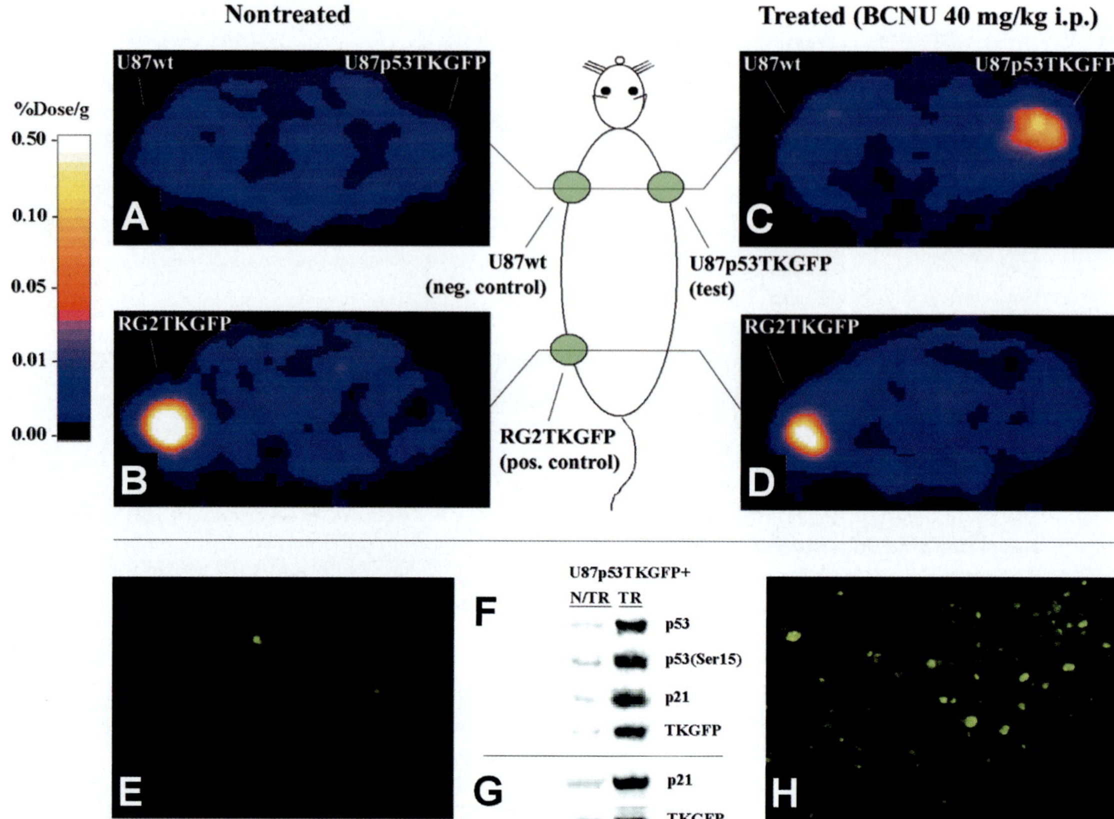

Figure 10.5. PET imaging of endogenous p53 activation in U87 p53+/+ xenografts using *cis*-p53-HSV1tk/GFP reporter system. (**B–E**) Transaxial PET images through (**A** and **C**) the shoulder and (**B** and **D**) pelvis of two rats are color-coded to the same radioactivity scale (% dose/g). A nontreated animal is shown on the left, and a BCNU-treated animal is shown on the right. Both animals had three s.c. tumor xenografts: U87/*cis*-p53-HSV1tk/GFP (test) in the right shoulder, U87 wild-type (negative control) in the left shoulder, and RG2/HSV1tk/GFP (positive control) in the left thigh. The nontreated animal on the left shows localization of radioactivity only in the positive control tumor (RG2/HSV1tk/GFP); the test (U87/*cis*-p53-HSV1tk/GFP) and negative control (U87wt) tumors are at background levels. The BCNU-treated animal on the right shows significant radioactivity localization in the test tumor (right shoulder) and in the positive control (left thigh) but no radioactivity above background in the negative control (left shoulder). Fluorescence microscopy images of U87/*cis*-p53-HSV1tk/GFP s.c. tumor samples obtained from (**E**) nontreated rats and (**H**) rats treated with 40 mg/kg BCNU i.p. (**F**) The same U87/*cis*-p53-HSV1tk/GFP s.c. tumor samples obtained from nontreated (N/TR) and BCNU-treated (TR) animals were assessed for the levels of activated p53 (Ser-15 phosphorylated), total p53 protein, p21, and HSV1tk/GFP proteins by immunoblot analysis; and (**G**) for the levels of p21 and HSV1tk/GFP mRNAs by RT-PCR. (Adapted from (28) with permission).

element (RARE). The *cis*-RARE-NIS/luciferase imaging reporter system was sufficiently sensitive to allow the visualization of intranuclear retinoic acid receptor activity. *In vivo* luciferase and ^{99m}Tc scintigraphic images showed higher intensity in tumors after retinoic acid treatment. This cis-RARE-sensitive imaging reporter system will be useful *in vitro* and *in vivo* for evaluating retinoic acid responses in such areas as cellular differentiation and chemoprevention.

Apoptosis and Stress Response

In recent years, cancer-related genes have been analyzed as predictive indicators for cancer therapies. Among those genes, the gene product of a tumor suppressor gene

p53 plays an important role in control of tumor progression and response to therapy [25]. p53 induces cell-cycle arrest, apoptosis, and depression of DNA repair after cancer therapies such as radiation, hyperthermia, and anticancer agents [26]. Mutations of p53 are observed at high frequency in human tumors and are recognized in about half of all malignant tumors in humans [27]. An original article by Doubrovin et al. [28] was the first to show that p53-dependent gene expression can be imaged *in vivo* with PET and [124]I-FIAU and by *in situ* fluorescence. In this study, a fusion dual-reporter gene HSV1tk/GFP was expressed under control of a p53-specific response element (*cis*-p53-HSV1tk/GFP reporter system). The PET images corresponded with upregulation of genes in the p53 signal transduction pathway (e.g., p21WAF downstream gene) in response to DNA damage induced by chemotherapy (Figure 10.5). These findings were observed in U87 (p53 +/+) cells and xenografts but not in SaOS (p53−/−) cells. PET imaging of p53 transcriptional activity in tumors using the *cis*-p53-HSV1tk/GFP reporter system can assess the effects of new drugs or other novel therapeutic paradigms mediated through p53-dependent pathways. For example, specific p53 gene therapy strategies based on p53 overexpression [29, 30] could be monitored by noninvasive imaging.

In the study by Che et al. [31] a reporter system was created to study stress and drug responses in transduced cells and tissues. A retroviral vector containing the hNIS and GFP reporter genes under the control of an inducible human heat shock protein (HSP)70 promoter was constructed and introduced into rat glioma cells. It was demonstrated that the local application of heat on transduced xenografts can effectively induce hNIS and eGFP gene expression *in vivo* and that this expression can be efficiently visualized by fluorescence, scintigraphic, and micropositron emission tomography imaging.

Hypoxia and Neoangiogenesis

Tissue hypoxia has been recognized as a significant component of normal and pathological neoangiogenesis. In tumors it is associated with malignant progression and resistance to radiation therapy [32]. Cellular responses to hypoxia are mediated largely through a transcriptional activator, hypoxia-inducible transcription factor 1-alfa (HIF-1α)[33]. A heterodimer HIF-1α is formed in response to hypoxia, which binds to the hypoxia response element (HRE) in the enhancer region of many target genes including erythropoietin [34], vascular endothelial growth factor (VEGF) [35], glycolytic enzymes [36], and enhances gene transcription. Serganova et al. [37] developed a noninvasive imaging paradigm to monitor and assess the level of HIF-1α activity during tumor development and growth. A dual-reporter vector containing HSV1tk/GFP fusion reporter gene under control of a *cis*-HRE-sensor element ("sensor" part) and constitutively expressing the red fluorescent protein (RFP) ("beacon" part) was used to transduce C6 glioma cells. Transcriptional activation of the HIF-1α signaling pathway was observed *in vitro* after 16 h of hypoxia induced with low oxygen conditions or CoCl₂, a known inducer of HIF-1α activity. When studied *in vivo* with increasing tumor diameter, a marked increase in HIF-1α transcriptional activity was observed by PET (^{18}F-FEAU accumulation) and GFP fluorescence in the core regions of tumors (Figure 10.6). In addition, using the same vector this group showed that a moderate ischemia-reperfusion injury in small C6 tumors caused a rapid induction of HIF-1α transcriptional activity *in vivo*, which persisted for several hours. Repetitive ^{18}F-FEAU PET imaging

of HIF-1α-dependent HSV1tk reporter gene expression revealed that the constitutive oncogenic signaling rather than hypoxia-induced HIF-1α transcriptional activity plays a leading role in the induction of neoangiogenesis during the early phases of tumor development. With the increase in tumor size, a decompensation of blood flow and oxygen supply by the existing tumor vessels triggers a marked increase in HIF-1α transcriptional activity in the tumor core regions, which further promotes tumor neoangiogenesis. PET imaging showed that even a moderate ischemia-reperfusion injury in tumors causes a rapid induction of HIF-1α transcriptional activity, which persists for a longer time than in normal tissues because of the inability of tumors to rapidly compensate for changes in tumor microcirculation.

VEGF is a well-known angiogenic factor important for vascular development and maintenance in all mammalian organs [38]. The development of molecular tools and pharmacological agents to selectively inhibit VEGF function and block angiogenesis and/or vascular permeability has led to great promise in the treatment of various cancers [39], macular degeneration [40], and wound healing [41]. The ability to monitor upregulated VEGF levels *in vivo* in a noninvasive and quantitative manner will allow monitoring of early events during angiogenesis. Wang et al. [42] noninvasively monitored the endogenous VEGF gene expression. Transgenic mice carrying full-length human VEGF promoter were used to study the expression of firefly luciferase reporter gene during wound healing and subcutaneous mammary tumor development. The bioluminescence signal showed good correlation with the endogenous VEGF protein levels in the wounded tissues (Figure 10.7) and growing tumors. This noninvasive imaging approach will be useful for repetitive assessment of normal tissue and tumor hypoxia during preclinical development of novel radiation therapies and radiation sensitizers, antiangiogenic drugs, and different hypoxia-specific small molecular imaging agents.

Inflammation and Activation of Immune Cells

Inflammation and activation of immune cells play an important role in autoimmune diseases and host defense against pathogens and cancer. Many inflammatory responses are mediated by the activation of NF-κB, a family of transcription factors present in virtually every cell type [43]. Animal models with targeted disruption of NF-κB subunits demonstrate the importance of NF-κB in cellular immunity, inflammation, and lymphoid organ development [44]. Carlsen et al. [45] have developed transgenic mice that express luciferase under the control of NF-κB, enabling real-time *in vivo* imaging of NF-κB activity in intact animals. In the absence of extrinsic stimulation, strong luminescence was observed in lymph nodes in the neck region, thymus, and Peyer's patches

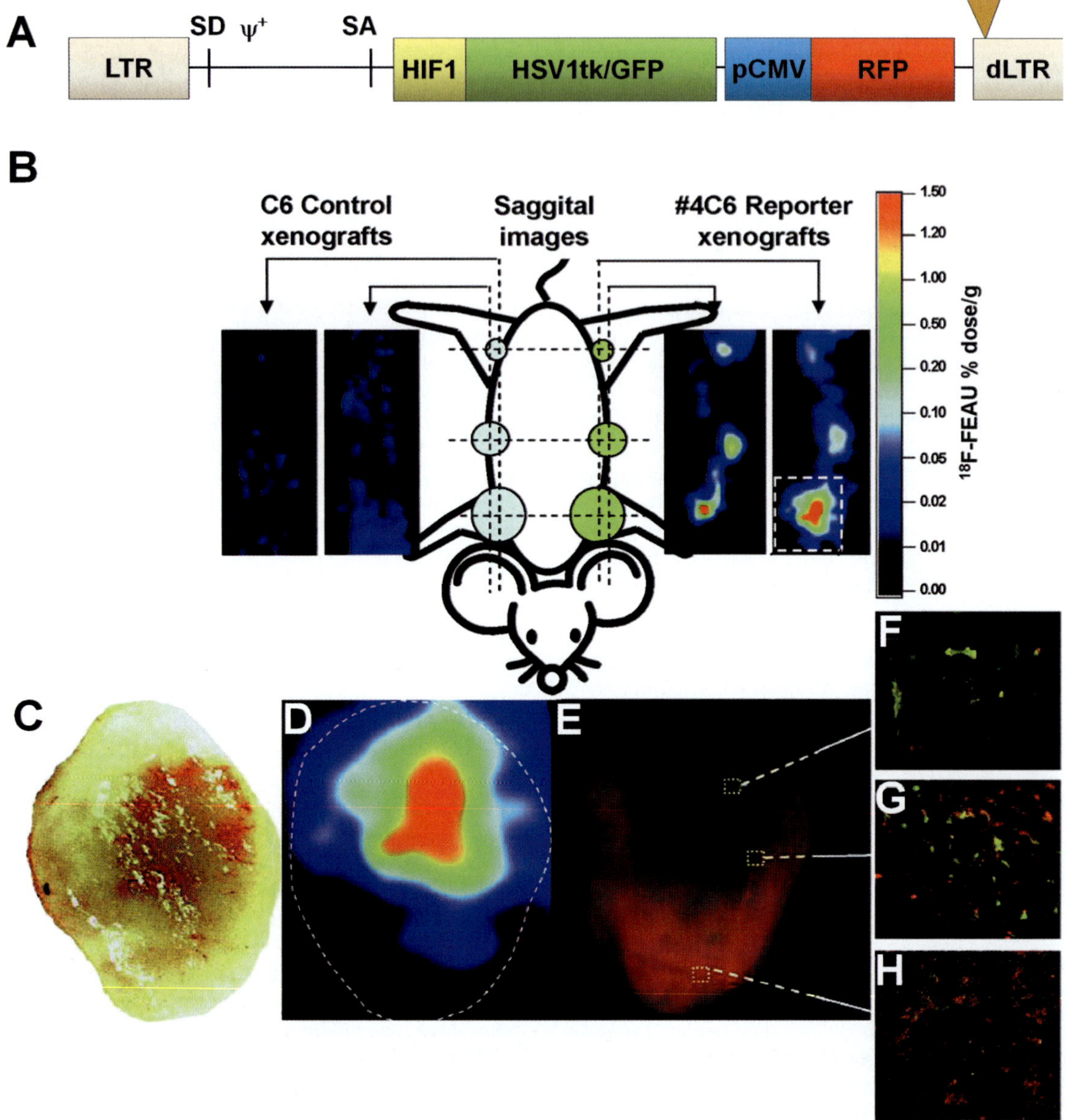

Figure 10.6. *In vivo* microPET imaging of HIF-1α transcriptional activity using *cis*-HRE-HSV1tk/GFP reporter system: tumor size dependency. (**A**) Structure of the self-inactivating retroviral vector bearing the *cis*-HIF1α-HSV1tk/GFP reporter system. The expression of the HSV1tk/GFP reporter gene is regulated by an artificial promoter containing multiple tandem repeats of an HIF1α-specific DNA-binding motif. Constitutive expression of the red fluorescent protein (RFP) is driven by the cytomegalovirus promoter. (**B**) ^{18}F-FEAU microPET images of HIF-1α transcriptional activity of a mouse bearing multiple wild-type C6 (control, blue circles) and 4C6 (reporter, green circles) tumor xenografts of different size (image planes are indicated by dotted lines continuing into solid arrows). Two sagittal image planes are shown through the 4C6-reporter xenografts (left side of animal + through wild-type C6 xenografts right side). Differently sized xenografts in the same animal provide simultaneous visualization of the central regions in each tumor and show a size-dependent magnitude and heterogeneity of hypoxia-induced HIF-1α transcriptional activity. The dotted white square around the large 4C6 tumor defines the portion of the image used for *in vivo-in situ* comparison presented in **C–E**. Macrophotographic image (**C**) of the cut surface through the large 4C6 tumor is shown; the cut surface was in a plane similar to the corresponding sagittal PET image (**D**); the tumor borders are outlined (white dotted line) based on the corresponding macro fluorescent image of the tumor (**E**). Fluorescence microscopic examination of these different areas at cellular resolution (x40 magnification) shows the hypervascular hypoxic core region (**F**), hypoxic–normoxic "border-zone" (**G**), and a normoxic region (**H**). (Adapted from (37) with permission).

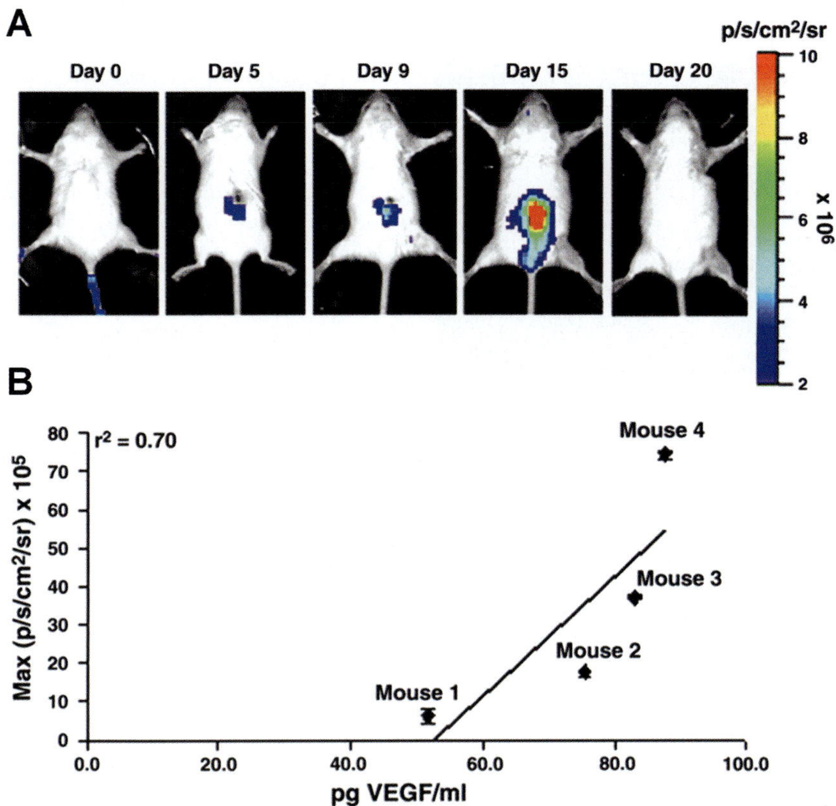

Figure 10.7. Imaging VEGF expression *in vivo*. Induction of firefly luciferase expression during wound healing in a transgenic mouse expressing firefly luciferase under control of VEGF promoter. (**A**) Mouse was imaged before wound creation in the CCD camera (Day 0) and imaged again subsequent to wound creation every 4–5 d. (**B**) Correlation plot of maximum bioluminescence signal ($p \cdot s^{-1} \cdot cm^{-2} \cdot sr^{-1}$) versus endogenous VEGF levels (pg/ml) in the wound tissue ($r^2 = 0.70$). (Adapted from (42) with permission).

(Figure 10.8). Treating mice with TNF-α, IL-1, or LPS increased the luminescence in a tissue-specific manner, with the strongest activity observed in skin, lungs, spleen, Peyer's patches, and the wall of the small intestine. This transgenic mouse model can be used to screen potential candidate drugs for the treatment of inflammatory conditions associated with aberrant NF-κB activation.

The elevated production of prostaglandins in inflammatory responses is largely dependent on induced cyclooxygenase-2 (COX-2) expression [46]. Elevated levels of COX-2 gene expression have been implicated in the initiation and progression of a variety of cancers, including colon, breast, and lung tumors [47]. The induction of luciferase gene expression following systemic interferon gamma/endotoxin administration was robustly demonstrated by Ishikawa et al. [48] using a dose–response relationship and a time course for luciferase expression from the COX-2 promoter in transgenic mice. Others have shown upregulation of luciferase expression under control of COX-2 promoter in tumor xenografts [49].

Systemic administration of antigen-specific T-lymphocytes is one of the most studied and clinically applicable methods of adoptive anticancer cell therapy

[50, 51]. Methods for noninvasive and repetitive evaluation of trafficking, homing, tumor targeting, differentiation, and persistence of adoptively transferred cells have recently been described [52, 53]. Long-term trafficking and localization of T-lymphocytes is an important component of anticancer immune response and in the elimination of abnormal cells and infectious agents from the body. An essential component of the immune response in many normal and disease states is T-cell activation [54]. T-cell receptor (TCR) interactions with MHC–peptide complexes expressed on antigen-presenting cells initiate T-cell activation, resulting in transcription mediated by a number of factors [55]. These factors contribute to the regulation of a number of target genes including IL-2 and other cytokines [56] through several activating pathways and involve several transcription factors such as the nuclear factor of activated T cells (NFAT) [57]. Ponomarev et al. [58] have monitored and assessed TCR-dependent NFAT-mediated activation of T cells *in vivo* using noninvasive optical and PET imaging. The cis-NFAT-HSV1-tk/GFP reporter system was introduced into the human T-cell leukemia cell line, Jurkat, that expresses a functional TCR complex. In mice with

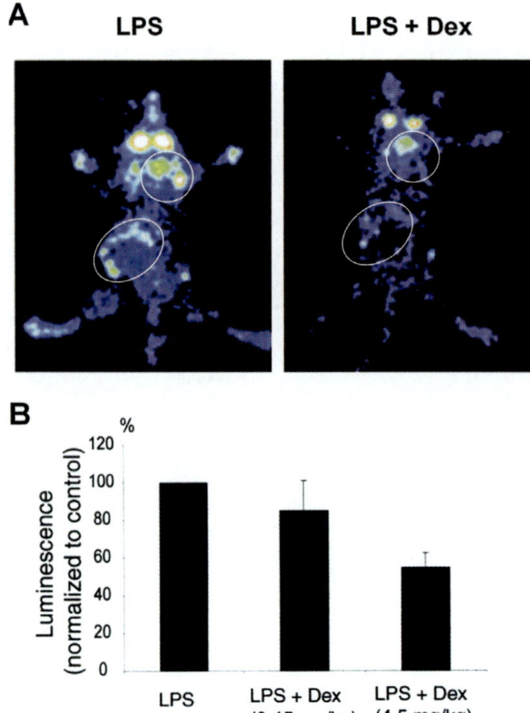

Figure 10.8. Imaging of lipopolysaccharide (LPS)-induced NF-κB activity. (**A**) Transgenic mice expressing firefly luciferase under control of NF-κB promoter were pretreated with either dexamethasone (0.45 or 4.5 mg/kg i.v.) or saline (control). One hour later LPS (2 mg/kg) was given to animals by IV injection. Images of one representative mouse from the control group (LPS alone) and one from the group pretreated with the highest concentration of dexamethasone followed by LPS are presented. Areas of lymph nodes in the neck region, thymus, and Peyer's patches are marked. (**B**) Five to 6 h after the LPS injection skin was removed from the abdominal part of the animal and the photons from the whole animal were counted. Values are expressed as the mean luminescence normalized to the control value from three mice. (Adapted from (45) with permission).

focal Jurkat infiltrates, known activators of T cells, anti-CD3, and anti-CD28 antibodies produced significantly higher levels of HSV1-tk/GFP reporter gene expression in cis-NFAT-HSV1-tk/GFP+ Jurkat cells than in non-treated or nontransduced cells as assessed by PET imaging with ^{124}I-FIAU (Figure 10.9). A strong correlation between HSV1-tk/GFP expression and upregulation of T-cell activation markers (CD69 and IL-2 production) was demonstrated both *in vitro* and *in vivo*. Furthermore, T-cell activation can be arrested pharmacologically by the administration of clinically used calcineurin inhibitors such as cyclosporin A and FK506 [59]. This circumstance makes this imaging approach clinically valuable, as it can be used in patients receiving immunosuppressive drugs after bone marrow and organ transplantation. When combined with imaging of lymphocyte activation, noninvasive PET imaging should allow monitoring of the

trafficking, proliferation, and antigen-specific activation of T cells in antitumor clinical trials.

In a recent study Radu et al. [60] visualized the process of immune cell activation using direct radiotracer approach. 1-(2′-deoxy-2′-[^{18}F]fluoroarabinofuranosyl) cytosine (^{18}F-FAC) was shown to be specific for imaging activation-specific upregulation of the deoxyribonucleotide salvage pathway in lymphoid cells (Figure 10.3). ^{18}F-FAC enabled visualization of lymphoid organs and was sensitive to localized immune activation in a mouse model of antitumor immunity. PET imaging detected early changes in lymphoid mass in systemic autoimmunity and allowed evaluation of immunosuppressive therapy. Such an approach can be combined with an appropriate reporter gene imaging system (e.g., NFAT-responsive reporter system) for noninvasive PET monitoring of immune cell activation during the antitumor immunological response.

Tissue-Specific Expression

Imaging of tissue- or organ-specific reporter gene expression can be achieved using a reporter construct driven by a tissue-specific promoter. Kang et al. [61] developed a human sodium/iodide symporter (hNIS)-transgenic mice with hNIS expression driven by alpha-myosin heavy-chain (α-MHC) promoter. hNIS expression in transgenic mice was specific to cardiomyocytes, as shown on γ-camera and microPET images and confirmed by immunohistochemistry using an hNIS-specific antibody (Figure 10.10). These transgenic mice could provide a valuable source to study the differentiation of cardiomyocytes and contribute to the development of optimal cell-based therapy in various heart diseases.

Iyer et al. [62] studied androgen deprivation in a prostate-specific-luciferase transgenic mouse model using firefly luciferase driven by the prostate-specific antigen (PSA) promoter. The transgenic mice showed prostate-specific expression of luciferase in male mice as early as 3 weeks of age. Furthermore, this study demonstrated that blocking androgen availability can downregulate luciferase expression in the mouse prostate.

Green et al. [63] attempted to noninvasively image the process of albumin synthesis in the liver using a genetic-reporter system and PET. Albumin is the most abundant endogenous protein in the serum and accounts for approximately 80% of the secretory protein synthesized and expressed in the liver. The synthesis rate of albumin is known to correlate with its mRNA level and is regulated by dietary protein. In this study the synthesis of albumin was assessed in a transgenic animal model noninvasively by examining the level of HSV1-tk reporter gene transcription from the promoter of the albumin gene after manipulation of the dietary protein content (Figure 10.11). ^{18}F-FHBG was used as a probe for PET imaging of HSV1-tk reporter gene expression. Mice were placed

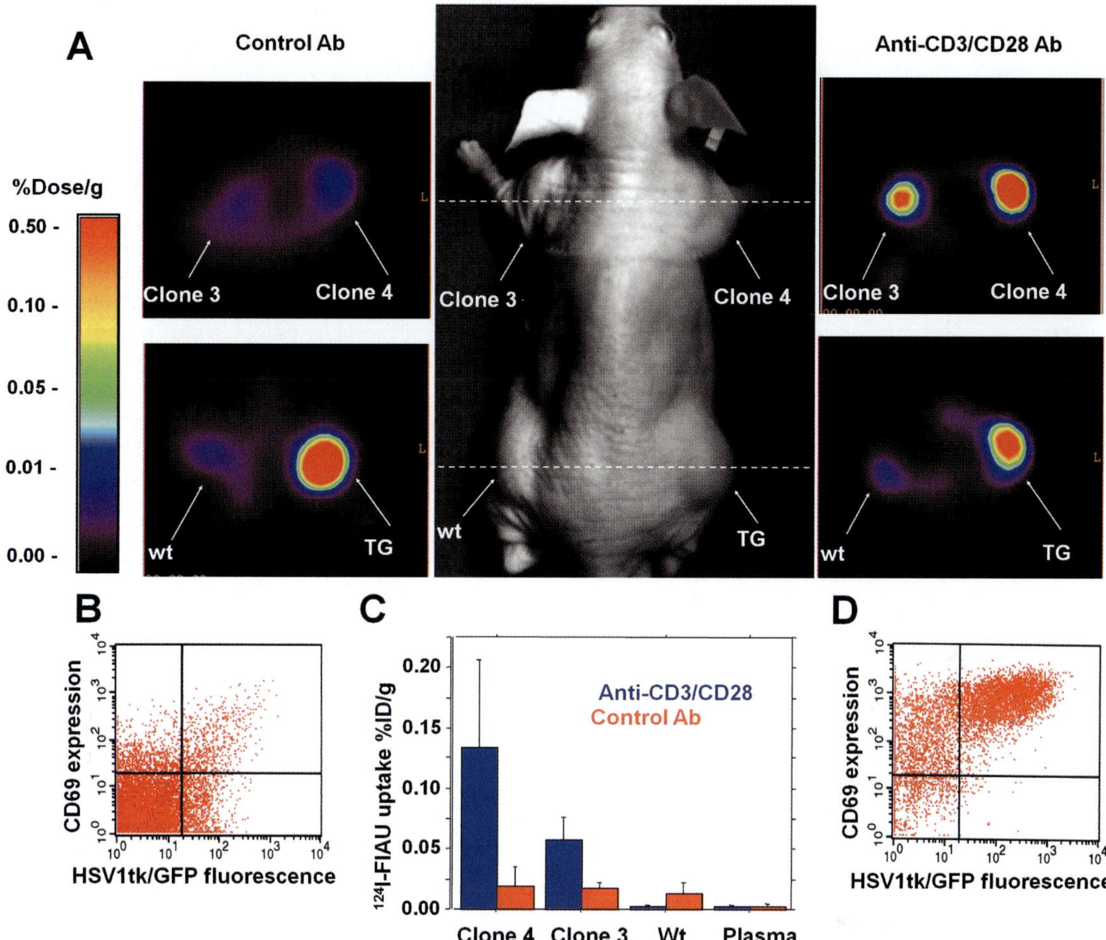

Figure 10.9. Imaging T-cells activation using *cis*-NFAT-HSV1tk/GFP reporter system *in vivo*. (**A**) Middle panel, a photographic image of a typical mouse bearing different s.c. Jurkat cell infiltrates: wild-type Jurkat cells (wt) or Jurkat cells transduced either with *cis*-NFAT-HSV1tk/GFP reporter vector (clones 3 and 4) or constitutively expressed HSV1tk/GFP (TG) reporter vector. Transaxial ^{124}I-FIAU PET images of HSV1tk/GFP expression in a mouse treated with control antibody (left panel) and anti-CD3/CD28 antibodies (right panel) were obtained at the levels indicated by the dashed lines. Note specific radiotracer accumulation in TG infiltrates (positive control, left and right panels) and in clones 3 and 4 infiltrates in the animal injected with anti-CD3/CD28 antibodies (right panel). (**C**) ^{124}I-FIAU accumulation (%dose/g) in tissue samples of the Jurkat/*cis*-NFAT-HSV1tk/GFP clone 3 and 4 infiltrates, wild-type Jurkat infiltrates, and blood plasma, obtained after PET imaging. (**B, D**) FACS profiles of HSV1tk/GFP and T-cell activation marker CD69 expression in a tissue sample from the same Jurkat/*cis*-NFAT-HSV1tk/GFP clone 4 infiltrate that was imaged with PET treated with control antibody (**B**) and anti-CD3/CD28 antibodies (**D**). (Adapted from (58) with permission).

on a low-protein diet for 14 d, scanned at the end of diet phase, and then returned to a normal nutrition regimen. The ^{18}F-FHBG accumulation in livers of the transgenic mice decreased at the end of low-protein diet and correlated with albumin mRNA, indicating that endogenous gene expression of albumin can be indirectly imaged with PET. Serum albumin levels were reduced during protein deprivation and then returned to normal levels when the protein diet was restored.

Chen et al. explored a similar approach [64]. This group showed the feasibility of monitoring tissue-specific expression of human sodium iodide symporter (hNIS) reporter gene driven by the murine albumin

enhancer/promoter with a gamma camera and ^{99m}Tc-pertechnetate in hepatoma cells.

In the study by Qiao et al. [65] colorectal liver metastases were successfully targeted by adenoviral vector-expressing HSV1tk under control of CEA-specific promoter. As imaged by gamma camera, ^{131}I-FIAU accumulated after intratumoral injection of HSV1tk-encoding adenovirus only in the areas of CEA-positive tumors with significantly less spread to the adjacent liver tissue. The increased imaging and therapeutic specificity of this novel CEA-driven suicide gene therapy vector is a proof of principle for effective tumor restricted gene expression using a tissue-specific promoter.

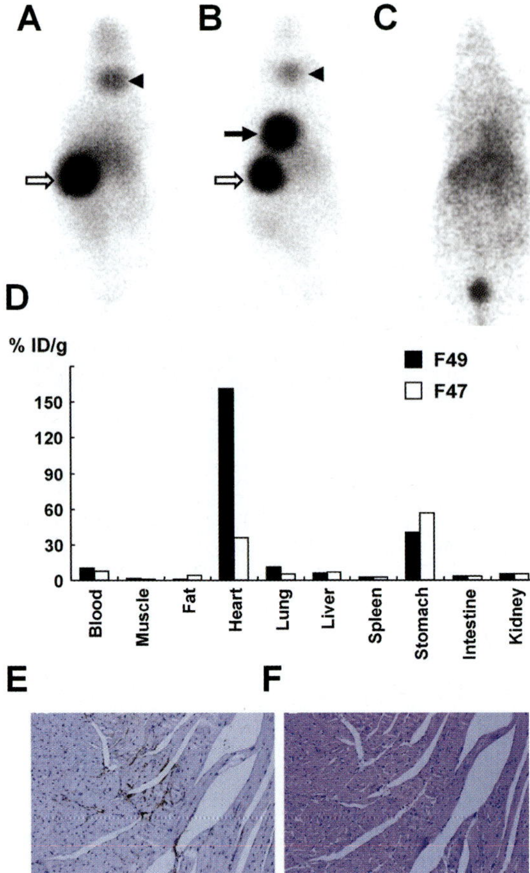

Figure 10.10. Imaging of cardiospecific reporter gene expression. Whole-body ^{99m}Tc-pertechnetate gamma camera scans of (**A**) control wild-type mouse and (**B**) transgenic mouse expressing human sodium/iodide symporter (hNIS) under control of alpha-myosin heavy chain (α-MHC) promoter. Transgenic mouse shows higher uptake of ^{99m}Tc-pertechnetate in heart (black arrow) than in thyroid (arrowhead) or stomach (outlined arrow). (**C**) Radiotracer uptake was completely blocked by KClO$_4$, an inhibitor for hNIS. (**D**) Biodistribution data of two transgenic mice that showed highest (mouse F49) and lowest (mouse F47) cardiospecific radiotracer uptake after injection of ^{99m}Tc-pertechnetate. (**E**) Immunohistochemical staining for (**E**) hNIS expression and (**F**) hematoxylin–eosin staining (x 400). Brown staining along cytoplasmic membrane indicates extensive expression of NIS. Normal histology of myocardium is confirmed. (Adapted from (61) with permission).

Certain applications require imaging transcriptional activity from weak promoters (e.g., CEA promoter and PSA promoter). *Cis*-reporter gene systems do not provide sufficient transcriptional activity for reliable visualization of reporter gene expression driven by weak promoter elements *in vivo*. The two-step transcriptional amplification (TSTA) strategies are used to image signaling processes that involve protein–protein interactions and/or signal amplification from weak promoter elements and are discussed in details in corresponding chapters ("Reporter gene expression amplification strategies" and

"Reporter gene strategies for imaging protein–protein interactions, protein phosphorylation, intramolecular folding, and protein metabolism").

IMAGING OF RNA

An alternative approach to imaging gene expression at a transcriptional level involves radiolabeled antisense oligonucleotides (RASONs). RASONs are complementary to a small segment of target mRNA and could potentially target any specific mRNA sequence [66]. In this context, imaging specific mRNAs with RASONs can be considered a more "direct" approach for imaging molecular genetic events. Nevertheless, RASON imaging has several serious limitations, including: (a) low specificity of localization (low number of target mRNA molecules per cell, poor stability of binding, degradation by H-RNAse); (b) high background activity (high nonspecific extra- and intracellular distribution); (c) limited delivery of reporter probe (poor vascular and cell membrane permeability; inability to penetrate blood–brain barrier); and (d) slow clearance (slow washout of nonbound oligonucleotides from intracellular space).

In contrast, "indirect" reporter gene imaging of RNA has several major advantages: (a) highly specific localization; (b) signal amplification (through a unique enzymatic activity of the reporter gene product that results in local trapping and accumulation of the substrate); (c) low background activity (rapid clearance and low nonspecific activity); and (d) simplified reporter probe delivery (higher vascular and cell membrane permeability).

Indirect Imaging of Pre-mRNA Splicing

Indirect quantitation of transcribed pre-mRNA molecules can be achieved by using a *trans*-splicing approach. *Trans*-splicing refers to a process whereby an intron of one pre-mRNA interacts with an intron of a second pre-mRNA, enhancing the recombination of splice sites between two conventional pre-mRNAs. Spliceosomes mediate *trans*-splicing (SMaRT) between 5' and 3' splice sites present on two individual pre-mRNAs resulting in a new, chimeric product. *Trans*-splicing has broad applications, depending on the nature of the sequences inserted or *trans*-spliced to the defined target, including repair of disease-causing mutant genes at the level of RNA splicing for several disorders including hemophilia [67] and cystic fibrosis [68]. The specificity of the *trans*-splicing reaction is conferred primarily by the binding domain of the pre-*trans*-splicing molecules (PTMs), which are designed to be complementary to intronic sequences in the target of interest [69].

Numerous studies to develop ribozyme- or spliceosome-mediated RNA *trans*-splicing have used the repair of lacZ [70] to quantify the efficiency of *trans*-splicing

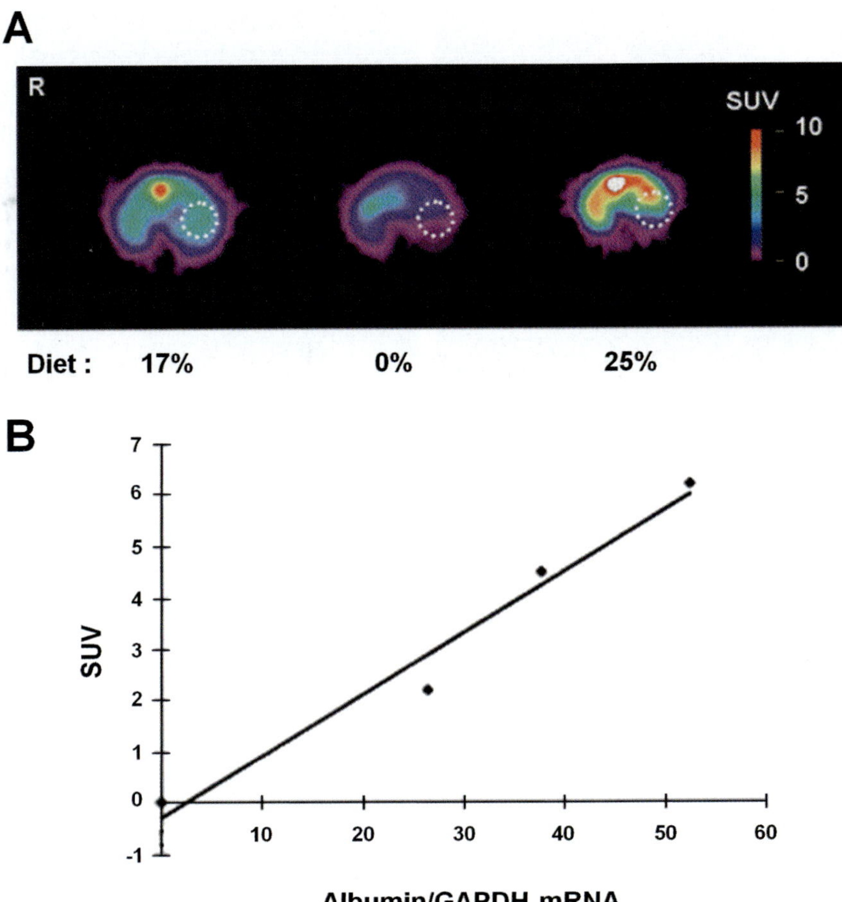

Figure 10.11. Imaging albumin synthesis *in vivo*. (**A**) Three transverse microPET images of a single transgenic mouse expressing HSV1tk under control of albumin promoter. Images were taken at the level of the liver 1 h after ^{18}F-FHBG tail-vein injection at the end of each of the three dietary phases: before dietary protein manipulation (Phase I), the liver demonstrates high uptake of ^{18}F-FHBG (left panel); at the end of Phase II, tracer uptake in the liver is greatly diminished (middle panel), but 18 days after the restoration of protein to the diet (Phase III), a marked increase in ^{18}F-FHBG retention in the liver was observed (right panel). The color scale represents standardized uptake value (SUV); the circular dotted region of interest represents the region chosen for quantitative analysis. (**B**) Correlation between SUV (transgenic mice) and GAPDH normalized albumin mRNA (control mice) across all three protein diets ($r^2 = 0.97$). (Adapted from (92) with permission).

in cells. Bhaumik et al. [71] recently demonstrated *in vivo* proof of concept of SMaRT(tm) molecular imaging to visualize endogenous gene expression using a split-reporter gene system. Mouse neuroblastoma cells were transduced with a target plasmid expressing the 5′ portion of Renilla luciferase followed by the E6–E7 region of the human papillomavirus (HPV) gene. A PTM was constructed to *trans*-splice to the E6–E7 region of HPV and insert the 3′ portion of Renilla luciferase into the target pre-mRNA, generating the full-length coding sequence of luciferase in target neuroblastoma cells. Following injection of the coelenterazine, the luciferase activity was detected in subcutaneous tumors and deeper tissues, affected by the tumor cells (Figure 10.12). This was the first study to visualize real-time gene expression in live animals by RNA *trans*-splicing. It represents

an important step toward the universal imaging of an endogenous pre-mRNA target, with potential applications in research and diagnostics. The coding domain of the PTM can be easily changed to encode a reporter suitable for imaging with other modalities, including clinical PET and MRI, or to produce a protein that acts therapeutically. These studies have been recently expanded by Walls et al. [72] using a generalized strategy to visualize RNA *trans*-splicing of virtually any RNA of interest.

Hasegawa et al. [73] explored another method to visualize RNA *trans*-splicing based on the *trans*-splicing Tetrahymena ribozyme derived from the self-splicing group I introns of Tetrahymena thermophila. This ribozyme recognizes a target mRNA by its complementarity to its internal guide sequence (IGS) and cleaves

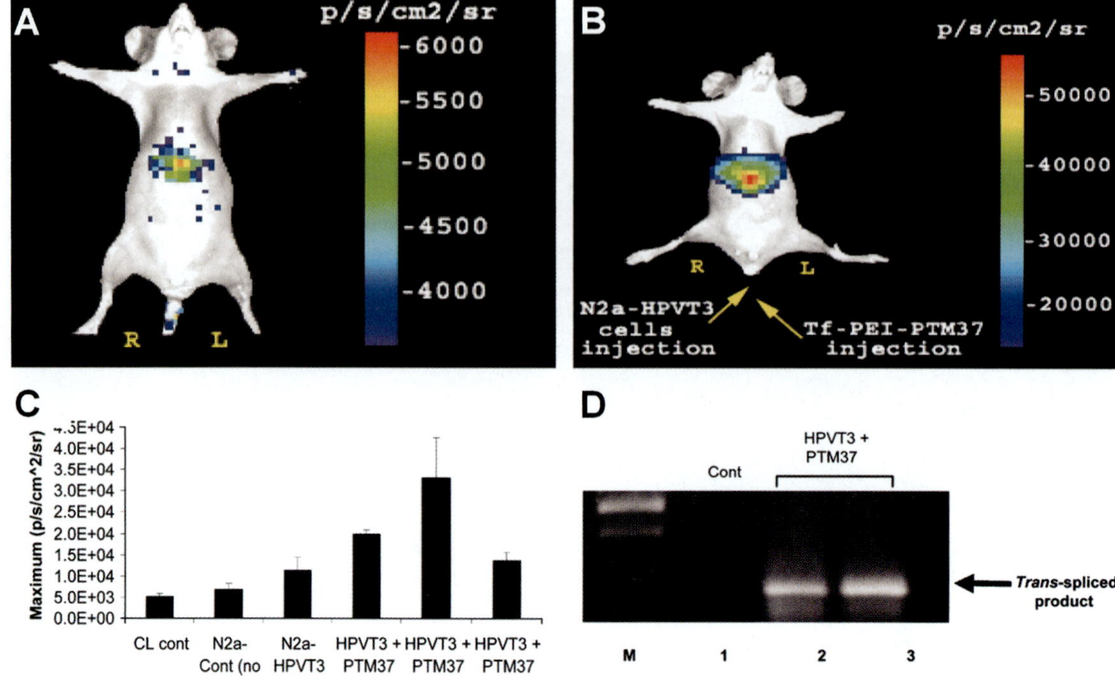

Figure 10.12. Imaging Renilla luciferase reporter gene expression using spliceosomes mediate *trans*-splicing (SMaRT). (**A**) Mice were injected with mouse neuroblastoma cells that contained a target plasmid expressing the 5′ portion of Renilla luciferase followed by the E6–E7 region of the human papillomavirus (HPV) gene (N2a-HPVT3 cells). No significant bioluminescent activity was detected in these mice following intravenous injection of the luciferase substrate. (**B**) PTM expression plasmids containing the E6–E7 region of HPV and the 3′ portion of Renilla luciferase (Tf-Pei-PTM37) were injected into the tail vein of test animals 48 h after target cell injection. Following injection of the coelenterazine, the luciferase activity was detected in subcutaneous tumors and deeper tissues affected by the tumor cells. (**C**) ROI analysis of bioluminescent signals emitted as a function of time after PTM injection. The signal peaks at 2 d after injection and then subsides by Day 3. (**D**) RT-PCR results using total RNA isolated from the liver of mice injected with either zero (lane 1), one (lane 2), or two (lane 3) doses of PTMs and using primers specific for opposite sides of the junction created by transsplicing. (Adapted from (71) with permission).

the target mRNA into a 5′ and 3′ fragment. The 5′ fragment is subsequently ligated to an exon attached to its 3′ end. *Trans*-splicing ribozyme is able to repair defective mRNA targets in a sequence-specific manner and is a promising therapeutic and imaging agent. This strategy was applied to visualize the activity of a bacterial enzyme β-lactamase in single living mammalian cells by fluorescence imaging.

RNA *Stab*-Reporter Systems

Gene expression is controlled at the transcriptional and posttranscriptional levels. Posttranscriptional regulation of gene expression in eukaryotic cells includes mRNA processing, turnover, and translation. The control of gene transcription by extracellular stimuli through DNA-binding proteins (transcription factors) has been widely studied. In contrast, relatively little is known about regulation of mRNA turnover. Stability of mRNA is determined by *cis*-acting elements within the mRNA molecule, believed to be recognized by regulatory proteins [74]. Such *cis*-elements positively or negatively modulate mRNA stability and are present throughout the mRNA, including the coding region and 3′UTR (3′-untranslated region). The 3′UTRs of rapidly decaying mRNAs usually contain an adenosine- or uridine-rich element (AURE), characterized by multiple copies of the pentanucleotide AUUUA [75]. In chimeric constructs, the AURE element can destabilize normally stable transcripts. Many genes including cytokine and regulatory genes are regulated transcriptionally, but their mRNAs also contain multiple AUUUA motifs, and their mRNA stability is regulated in response to extracellular stimuli. When produced in nonstimulated cells, such mRNA transcripts are unstable, but their half-lives are prolonged after cell activation. The RNA *Stab*-reporter system (Figure 10.13) is based on a fusion of a constitutively expressed reporter gene fused to the 3′UTR of the gene of interest. The resulting reporter gene mRNA contains the 3′UTR mRNA sequence of the gene of interest and is degraded at a similar rate as the mRNA of the endogenous gene of interest. Therefore, the steady-state level

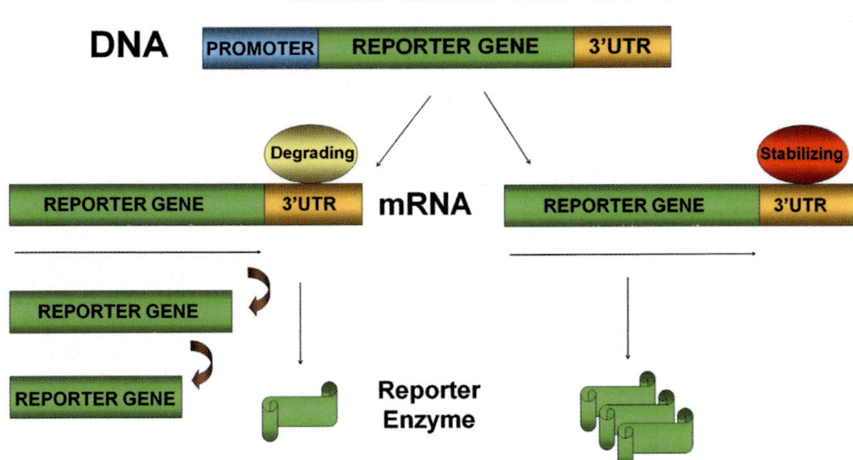

Figure 10.13. The paradigm of the RNAStab-reporter system. The system involves constitutive expression of a reporter gene fused to the 3'-untranslated region (UTR) of the gene of interest. In nonstimulated cells, the reporter gene/3'UTR fusion mRNA is produced and degraded with the same rate as the mRNA of endogenous gene of interest. In stimulated cells, the mRNA of endogenous gene of interest is stabilized by stabilizing proteins that bind to the 3'UTR. As a result, increased amounts of encoded protein are produced (translated). Also, the reporter gene/3'UTR fusion mRNA is also stabilized by the same stabilizing proteins that bind to its 3'UTR. This results in the increased production (translation) of the reporter gene product. Thus, up- or downregulation of a reporter gene expression reflects the posttranscriptional regulation of endogenous gene expression through mRNA stabilization.

of a reporter gene mRNA (its half-life) determines the level of reporter gene expression, translation, and activity and reflects the steady-state level of mRNA of the endogenous gene of interest. The reporter-based mRNA stabilization reporter systems have been widely used to study regulation of expression of different cytokines (e.g., IL2 [76], GM-CSF[77], growth factors (e.g., VEGF[78]), and receptors (e.g., Glut1[79]).

Lee et al. [80] constructed a new hypoxia-inducible RNA *Stab*-luciferase plasmid with the Epo 3'-untranslated region, which is known to stabilize the Epo mRNA under hypoxia [81]. Smooth-muscle A7R5 cells were transfected with the reporter plasmid and incubated under hypoxia conditions. Using a luciferase detection assay it was shown that the Epo UTR increased reporter gene expression in RNA*Stab* transduced cells specifically under hypoxia that correlated with increased expression of VEGF. This hypoxia-driven RNA*Stab* gene expression reporter system will be useful for the development of gene therapy for various ischemic diseases.

Posttranscriptional Modulation

Gene expression levels are also regulated by posttranscriptional modulation, including the translation of mRNA. For example, studies on the enzyme dihydrofolate reductase (DHFR) indicated that DHFR binds to its own mRNA in the coding region and that inhibition of DHFR by MTX releases the DHFR enzyme from the mRNA [82]. Consequently, this release results in an increase in translation of DHFR protein. A study by Mayer-Kuckuk et al. [83] demonstrated that it is feasible

to image posttranscriptional regulation of DHFR gene expression. This was shown by exposing 8LT-CT colorectal cancer cells to antifolates and inducing a rapid increase in the levels of DHFR. This effect was visualized by ^{124}I-FIAU and PET imaging studies performed on nude rats bearing fusion DHFR-HSV1-TK-transduced 8LT-CT xenografts (Figure 10.14). The results of this study indicated that the increase in reporter protein and enzyme (DHFR-HSV1tk) activity was occurring at a translational level rather than at the transcriptional level. This study demonstrated that this adaptive cellular response mechanism could determine whether posttranscriptional regulation of gene expression could be monitored by noninvasive imaging.

CHOICE OF REPORTER GENE AND IMAGING MODALITY FOR INDUCIBLE GENETIC-REPORTER SYSTEMS

The dynamic nature of the regulation of endogenous gene expression demands the use of genetic-reporter systems with reporter genes that have relatively short half-lives. Luciferase is well suited for monitoring transcription due to its relatively fast induction and to the relatively short biological half-lives of both luciferin and luciferase [10]. These features explain the predominant use of luciferases in inducible genetic-reporter systems suitable for whole-body optical imaging in animal models [84]. In addition, *in vivo* bioluminescence reporter imaging currently remains more sensitive than *in vivo* fluorescence reporter imaging due to the absence of

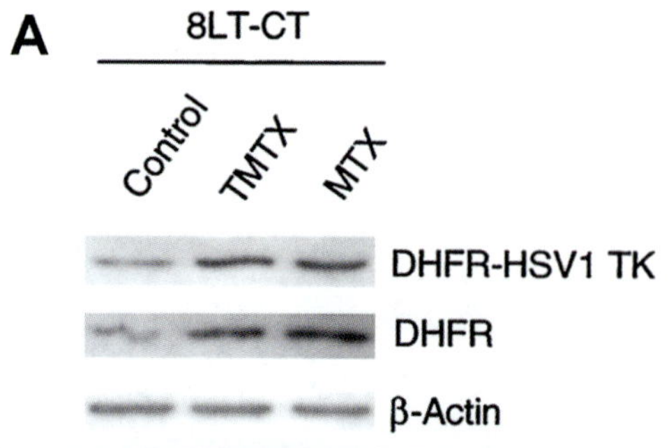

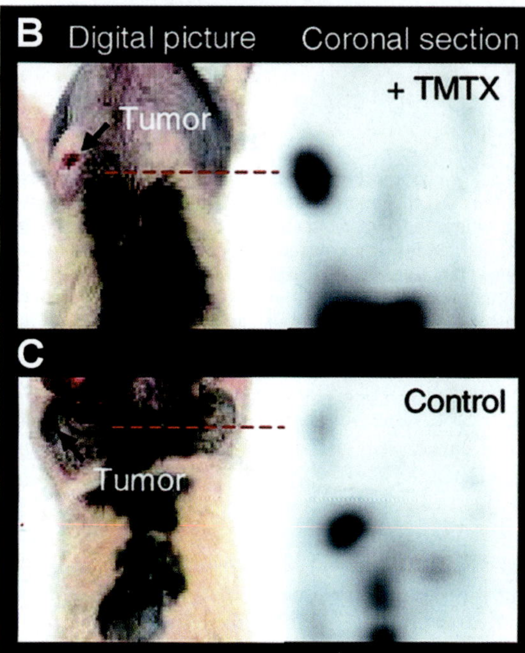

Figure 10.14. Imaging posttranscriptional regulation of gene expression *in vivo*. (**A**) Increased dihydrofolate reductase (DHFR) and HSV1tk fusion protein levels in 8LT-CT colon cancer cells exposed to antifolate trimetrexate (TMTX) for 24 h. Western blot analysis using an anti-DHFR antibody. (**B, C**) Rats bearing flank 8LT-CT tumors transduced with fusion DHFR-HSV1tk were treated with 10 mg/kg antifolate for 3 d. Digital pictures and coronal PET scans obtained from (**B**) TMTX-treated and (**C**) water-treated (control) rats. TMTX treatment resulted in increased ^{124}I-FIAU accumulation in DHFR-HSV1tk transduced tumor xenografts. (Adapted from (83) with permission).

autofluorescence and low background activity [10, 85]. However, this advantage over the longer-lived eGFP has recently been mitigated by the development of short-lived, rapidly degradable variants of eGFP [86] and HSV1-tk [87]. These short-lived variants have been used for higher temporal resolution imaging of chimeric proteins within cells. Combining these reporter genes into a single fusion gene could provide an additional tool for the analysis of endogenous molecular genetic processes both *in vivo* and *ex vivo*. Such multifunction reporter genes were created and the single encoded

proteins were shown to be functional with all modalities [88–90].

Reporter gene technology has been applied toward studying the physiologic phenomena associated with different signaling pathways, including hypoxia. The potential adverse effects of hypoxic conditions on the effectiveness of commonly used reporter genes have been investigated. Studies by Cecic et al. [91] found that bioluminescent and fluorescent reporter activities are decreased under hypoxia despite minimal variations in protein production, whereas beta-galactosidase reporter activity per unit protein was unchanged. Therefore, researchers should be careful in designing a genetic reporter system for certain applications and include appropriate (constitutively expressed) controls for normalization of pathway-specific differences during reporter gene expression.

In summary, imaging regulation of endogenous genes in living animals and potentially in human subjects using noninvasive imaging techniques can provide a better understanding of normal and cancer-related biologic processes. The merger of the existing molecular–biological assay systems that utilize different reporter genes and established techniques for noninvasive reporter gene imaging is a logical step in understanding an endogenous gene expression *in vivo*. Such a merger allows for whole-body noninvasive imaging of different genetic and molecular biological events in transgenic animals and transduced tissues. In the future, with the advent of tissue-targeted delivery vectors that could be administered systemically, the noninvasive imaging of different genetic and molecular biological events could potentially be extended into human subjects. For example, the intratumoral localization of adoptively administered antitumor-specific lymphocytes and their tumor-specific activation at the tumor site could be monitored using an inducible, activation-specific, promoter-driven "sensor" reporter gene. Such a reporter system has a potential to be used as a powerful tool for tracking cells and imaging inducible upregulation of reporter gene expression upon tumor-specific activation at the same time. This imaging paradigm could be easily translated into clinical protocols for monitoring the efficacy of adoptive anticancer immunotherapies.

But even today, a tremendous need exists for the genetic and molecular characterization of normal tissues as well as primary and metastatic tumor lesions by means of noninvasive imaging in preclinical models. The latter could potentially be achieved using reporter gene expressing cells or xenografts or by the direct transduction of target tissue with a reporter system followed by noninvasive imaging of the molecular genetic process of interest. Both academia and industry provide a number of preestablished reporter vectors, reporter cell lines, and reporter transgenic animals suitable for targeted noninvasive visualization of the molecular–biological process of interest *in vivo*. The development of databases of currently available reporter vectors, cells, and animals will facilitate the assessment of new therapeutic strategies and drugs and will ensure the standardization of the *in vivo* evaluation of regulation of endogenous gene expression in living subjects.

REFERENCES

1   Hanahan, D., Wagner, E. F., Palmiter, R. D. (2007). The origins of oncomice: a history of the first transgenic mice genetically engineered to develop cancer. *Genes Dev* 21(18): 2258–2270.

2   Frese, K. K., Tuveson, D. A. (2007). Maximizing mouse cancer models. *Nat Rev Cancer* 7(9): 645–658.

3   Blackwell, T. K., Walker, A. K. (2006). Transcription mechanisms. *WormBook*: 1–16.

4   Arnone, M. I., Dmochowski, I. J., Gache, C. (2004). Using reporter genes to study cis-regulatory elements. *Methods Cell Biol* 74: 621–652.

5   Albanese, C., Hulit, J., Sakamaki, T. et al. (2002). Recent advances in inducible expression in transgenic mice. *Semin Cell Dev Biol* 13(2): 129–141.

6   Kato, S., Sato, T., Watanabe, T. et al. (2005). Function of nuclear sex hormone receptors in gene regulation. *Cancer Chemother Pharmacol* 56(Suppl 1): 4–9.

7   Lemmen, J. G., Arends, R. J., van Boxtel, A. L. et al. (2004). Tissue- and time-dependent estrogen receptor activation in estrogen reporter mice. *J Mol Endocrinol* 32(3): 689–701.

8   Ilagan, R., Zhang, L. J., Pottratz, J. et al. (2005). Imaging androgen receptor function during flutamide treatment in the LAPC9 xenograft model. *Mol Cancer Ther* 4(11): 1662–1669.

9   Ciana, P., Raviscioni, M., Mussi, P. et al. (2003). In vivo imaging of transcriptionally active estrogen receptors. *Nat Med* 9(1): 82–86.

10   Contag, C. H., Jenkins, D., Contag, P. R. et al. (2000). Use of reporter genes for optical measurements of neoplastic disease in vivo. *Neoplasia* 2(1–2): 41–52.

11   Contag, C. H., Bachmann, M. H. (2002). Advances in in vivo bioluminescence imaging of gene expression. *Annu Rev Biomed Eng* 4: 235–260.

12   Uhrbom, L., Nerio, E., Holland, E. C. (2004). Dissecting tumor maintenance requirements using bioluminescence imaging of cell proliferation in a mouse glioma model. *Nat Med* 10(11): 1257–1260.

13   Johnson, D. G., Ohtani, K., Nevins, J. R. (1994). Autoregulatory control of E2F1 expression in response to positive and negative regulators of cell cycle progression. *Genes Dev* 8(13): 1514–1525.

14   Karin, M., Lin, A. (2002). NF-kappaB at the crossroads of life and death. *Nat Immunol* 3(3): 221–227.

15   Cahir-McFarland, E. D., Davidson, D. M., Schauer, S. L. et al. (2000). NF-kappa B inhibition causes spontaneous apoptosis in Epstein–Barr virus-transformed lymphoblastoid cells. *Proc Natl Acad Sci U S A* 97(11): 6055–6060.

16   Keller, S. A., Hernandez-Hopkins, D., Vider, J. et al. (2006). NF-kappaB is essential for the progression of KSHV- and EBV-infected lymphomas in vivo. *Blood* 107(8): 3295–3302.

17   Fakhoury, J., Nimmo, G. A., Autexier, C. (2007). Harnessing telomerase in cancer therapeutics. *Anticancer Agents Med Chem* 7(4): 475–483.

18   Padmanabhan, P., Otero, J., Ray, P. et al. (2006). Visualization of telomerase reverse transcriptase (hTERT) promoter activity using a trimodality fusion reporter construct. *J Nucl Med* 47(2): 270–277.

19   Siegel, P. M., Massague, J. (2003). Cytostatic and apoptotic actions of TGF-beta in homeostasis and cancer. *Nat Rev Cancer* 3(11): 807–821.

20   Siegel, P. M., Shu, W., Cardiff, R. D. et al. (2003). Transforming growth factor beta signaling impairs Neu-induced mammary tumorigenesis while promoting pulmonary metastasis. *Proc Natl Acad Sci U S A* 100(14): 8430–8435.

21   Yin, J. J., Selander, K., Chirgwin, J. M. et al. (1999). TGF-beta signaling blockade inhibits PTHrP secretion by breast cancer cells and bone metastases development. *J Clin Invest* 103(2): 197–206.

22   Kang, Y., He, W., Tulley, S. et al. (2005). Breast cancer bone metastasis mediated by the Smad tumor suppressor pathway. *Proc Natl Acad Sci U S A* 102(39): 13909–13914.

23   Garattini, E., Gianni, M., Terao, M. (2007). Retinoids as differentiating agents in oncology: a network of interactions with intracellular pathways as the basis for rational therapeutic combinations. *Curr Pharm Des* 13(13): 1375–1400.

24   So, M. K., Kang, J. H., Chung, J. K. et al. (2004). In vivo imaging of retinoic acid receptor activity using a sodium/iodide symporter and luciferase dual imaging reporter gene. *Mol Imaging* 3(3): 163–171.

25   Haupt, S., Haupt, Y. (2006). Importance of p53 for cancer onset and therapy. *Anticancer Drugs* 17(7): 725–732.

26   Ohnishi, T. (2005). The role of the p53 molecule in cancer therapies with radiation and/or hyperthermia. *J Cancer Res Ther* 1(3): 147–150.

27   Soussi, T., Wiman, K. G. (2007). Shaping genetic alterations in human cancer: the p53 mutation paradigm. *Cancer Cell* 12(4): 303–312.

28   Doubrovin, M., Ponomarev, V., Beresten, T. et al. (2001). Imaging transcriptional regulation of p53-dependent genes with positron emission tomography in vivo. *Proc Natl Acad Sci U S A* 98(16): 9300–9305.

29   Merritt, J. A., Roth, J. A., Logothetis, C. J. (2001). Clinical evaluation of adenoviral-mediated p53 gene transfer: review of INGN 201 studies. *Semin Oncol* 28(5 Suppl. 16): 105–114.

30   Roth, J. A., Grammer, S. F., Swisher, S. G. et al. (2001). P53 gene replacement for cancer–interactions with DNA damaging agents. *Acta Oncol* 40(6): 739–744.

31   Che, J., Doubrovin, M., Serganova, I. et al. (2007). HSP70-inducible hNIS-IRES-eGFP reporter imaging: response to heat shock. *Mol Imaging* 6(6): 404–416.

32   Riesterer, O., Milas, L., Ang, K. K. (2007). Use of molecular biomarkers for predicting the response to radiotherapy with or without chemotherapy. *J Clin Oncol* 25(26): 4075–4083.

33   Liao, D., Johnson, R. S. (2007). Hypoxia: a key regulator of angiogenesis in cancer. *Cancer Metastasis Rev* 26(2): 281–290.

34 Semenza, G. L. (1994). Regulation of erythropoietin production. New insights into molecular mechanisms of oxygen homeostasis. *Hematol Oncol Clin North Am* **8**(5): 863–884.

35 Forsythe, J. A., Jiang, B. H., Iyer, N. V. et al. (1996). Activation of vascular endothelial growth factor gene transcription by hypoxia-inducible factor 1. *Mol Cell Biol* **16**(9): 4604–4613.

36 Semenza, G. L., Roth, P. H., Fang, H. M. et al. (1994). Transcriptional regulation of genes encoding glycolytic enzymes by hypoxia-inducible factor 1. *J Biol Chem* **269**(38): 23757–23763.

37 Serganova, I., Doubrovin, M., Vider, J. et al. (2004). Molecular imaging of temporal dynamics and spatial heterogeneity of hypoxia-inducible factor-1 signal transduction activity in tumors in living mice. *Cancer Res* **64**(17): 6101–6108.

38 Thurston, G., Gale, N. W. (2004). Vascular endothelial growth factor and other signaling pathways in developmental and pathologic angiogenesis. *Int J Hematol* **80**(1): 7–20.

39 Ruegg, C., Mutter, N. (2007). Anti-angiogenic therapies in cancer: achievements and open questions. *Bull Cancer* **94**(9): 753–762.

40 Waisbourd, M., Loewenstein, A., Goldstein, M. et al. (2007). Targeting vascular endothelial growth factor: a promising strategy for treating age-related macular degeneration. *Drugs Aging* **24**(8): 643–662.

41 Yla-Herttuala, S. (2006). An update on angiogenic gene therapy: vascular endothelial growth factor and other directions. *Curr Opin Mol Ther* **8**(4): 295–300.

42 Wang, Y., Iyer, M., Annala, A. et al. (2006). Noninvasive indirect imaging of vascular endothelial growth factor gene expression using bioluminescence imaging in living transgenic mice. *Physiol Genomics* **24**(2): 173–180.

43 Baldwin, A. S., Jr. (1996). The NF-kappa B and I kappa B proteins: new discoveries and insights. *Annu Rev Immunol* **14**: 649–683.

44 Doi, T. S., Takahashi, T., Taguchi, O. et al. (1997). NF kappa B RelA-deficient lymphocytes: normal development of T cells and B cells, impaired production of IgA and IgG1 and reduced proliferative responses. *J Exp Med* **185**(5): 953–961.

45 Carlsen, H., Moskaug, J. O., Fromm, S.H. et al. (2002). In vivo imaging of NF-kappa B activity. *J Immunol* **168**(3): 1441–1446.

46 Rajakariar, R., Yaqoob, M. M., Gilroy, D. W. (2006). COX-2 in inflammation and resolution. *Mol Interv* **6**(4): 199–207.

47 Liao, Z., Mason, K. A., Milas, L. (2007). Cyclo-oxygenase-2 and its inhibition in cancer: is there a role? *Drugs* **67**(6): 821–845.

48 Ishikawa, T. O., Jain, N. K., Taketo, M. M. et al. (2006). Imaging cyclooxygenase-2 (Cox-2) gene expression in living animals with a luciferase knock-in reporter gene. *Mol Imaging Biol* **8**(3): 171–187.

49 Nguyen, J. T., Machado, H., Herschman, H. R. (2003). Repetitive, noninvasive imaging of cyclooxygenase-2 gene expression in living mice. *Mol Imaging Biol* **5**(4): 248–256.

50 Rooney, C., Smith, C. A., Ng, C. Y., Loftin, S., Li, C., Krance, R. A., Brenner, M. K., Heslop, H. E. (1995). Use of gene-modified virus-specific T lymphocytes to control Epstein–Barr-virus-related lymphoproliferation. *Lancet* **345**(8941): 9–13.

51 Rosenberg, S. A., Yang, J. C., Robbins, P. F., Wunderlich, J. R., Hwu, P., Sherry, R. M., Schwartzentruber, D. J., Topalian, S. L., Restifo, N. P., Filie, A., Chang, R., Dudley, M. E. (2003). Cell transfer therapy for cancer: lessons from sequential treatments of a patient with metastatic melanoma. *J Immunother* **26**(5): 385–393.

52 Dubey, P., Su, H., Adonai, N., Du, S., Rosato, A., Braun, J., Gambhir, S. S., Witte, O. N. (2003). Quantitative imaging of the T cell antitumor response by positron-emission tomography. *Proc Natl Acad Sci U S A* **100**(3): 1232–1237.

53 Koehne, G., Doubrovin, M., Doubrovina, E., Zanzonico, P., Gallardo, H. F., Ivanova, A., Balatoni, J., Teruya-Feldstein, J., Heller, G., May, C., Ponomarev, V., Ruan, S., Finn, R., Blasberg, R. G., Bornmann, W., Riviere, I., Sadelain, M., O'Reilly, R. J., Larson, S. M., Tjuvajev, J. G. (2003). Serial in vivo imaging of the targeted migration of human HSV-TK-transduced antigen-specific lymphocytes. *Nat Biotechnol* **21**(4): 405–413.

54 Soldevila, G., Castellanos, C., Malissen, M., Berg, L. J. (2001). Analysis of the individual role of the TCRzeta chain in transgenic mice after conditional activation with chemical inducers of dimerization. *Cell Immunol* **214**(2): 123–138.

55 Macian, F., Lopez-Rodriguez, C., Rao, A. (2001). Partners in transcription: NFAT and AP-1. *Oncogene* **20**(19): 2476–2489.

56 Kalli, K., Huntoon, C., Bell, M., McKean, D. J. (1998). Mechanism responsible for T-cell antigen receptor- and CD28- or interleukin 1 (IL-1) receptor-initiated regulation of IL-2 gene expression by NF-kappaB. *Mol Cell Biol* **18**(6): 3140–3148.

57 Li, W., Handschumacher, R. E. (1996). Regulation of the nuclear factor of activated T cells in stably transfected Jurkat cell clones. *Biochem Biophys Res Commun* **219**: 96–99.

58 Ponomarev, V., Doubrovin, M., Lyddane, C. et al. (2001). Imaging TCR-dependent NFAT-mediated T-cell activation with positron emission tomography in vivo. *Neoplasia* **3**(6): 480–488.

59 Kiani, A., Rao, A., Aramburu, J. (2000). Manipulating immune responses with immunosuppressive agents that target NFAT. *Immunity* **12**: 359–372.

60 Radu, C. G., Shu, C. J., Nair-Gill, E. et al. (2008). Molecular imaging of lymphoid organs and immune activation by positron emission tomography with a new [18F]-labeled 2'-deoxycytidine analog. *Nat Med* **14**(7): 783–788.

61 Kang, J. H., Lee, D. S., Paeng, J. C. et al. (2005). Development of a sodium/iodide symporter (NIS)-transgenic mouse for imaging of cardiomyocyte-specific reporter gene expression. *J Nucl Med* **46**(3): 479–483.

62 Iyer, M., Salazar, F. B., Lewis, X. et al. (2005). Non-invasive imaging of a transgenic mouse model using a prostate-specific two-step transcriptional amplification strategy. *Transgenic Res* **14**(1): 47–55.

63 Green, L. A., Yap, C. S., Nguyen, K. et al. (2002). Indirect monitoring of endogenous gene expression by positron emission tomography (PET) imaging of reporter gene expression in transgenic mice. *Mol Imaging Biol* **4**(1): 71–81.

64 Chen, L., Altman, A., Mier, W. et al. (2006). 99mTc-pertechnetate uptake in hepatoma cells due to tissue-specific human sodium iodide symporter gene expression. *Nucl Med Biol* **33**(4): 575–580.

65 Qiao, J., Doubrovin, M., Sauter, B. V. et al. (2002). Tumor-specific transcriptional targeting of suicide gene therapy. *Gene Ther* **9**(3): 168–175.

66 Gauchez, A. S., Du Moulinet D'Hardemare, A., Lunardi, J. et al. (1999). Potential use of radiolabeled antisense oligonucleotides in oncology. *Anticancer Res* **19**(6B): 4989–4997.

67 Chao, H., Mansfield, S. G., Bartel, R. C. et al. (2003). Phenotype correction of hemophilia A mice by spliceosome-mediated RNA trans-splicing. *Nat Med* **9**(8): 1015–1019.

68 Liu, X., Jiang, Q., Mansfield, S. G. et al. (2002). Partial correction of endogenous DeltaF508 CFTR in human cystic fibrosis airway epithelia by spliceosome-mediated RNA trans-splicing. *Nat Biotechnol* **20**(1): 47–52.

69 Puttaraju, M., DiPasquale, J., Baker, C. C. et al. (2001). Messenger RNA repair and restoration of protein function by spliceosome-mediated RNA trans-splicing. *Mol Ther* **4**(2): 105–114.

70 Dallinger, G., Puttaraju, M., Mitchell, L. G. et al. (2003). Development of spliceosome-mediated RNA trans-splicing (SMaRT) for the correction of inherited skin diseases. *Exp Dermatol* **12**(1): 37–46.

71 Bhaumik, S., Walls, Z., Puttaraju, M. et al. (2004). Molecular imaging of gene expression in living subjects by spliceosome-mediated RNA trans-splicing. *Proc Natl Acad Sci U S A* **101**(23): 8693–8698.

72 Walls, Z. F., Puttaraju, M., Temple, G. F. et al. (2008). A generalizable strategy for imaging pre-mRNA levels in living subjects using spliceosome-mediated RNA trans-splicing. *J Nucl Med* **49**(7): 1146–1154.

73 Hasegawa, S., Jackson, W. C., Tsien, R. Y. et al. (2003). Imaging Tetrahymena ribozyme splicing activity in single live mammalian cells. *Proc Natl Acad Sci U S A* **100**(25): 14892–14896.

74 Beelman, C. A., Parker, R. (1995). Degradation of mRNA in eukaryotes. *Cell* **81**(2): 179–183.

75 Chen, C. Y., Shyu, A. B. (1995). AU-rich elements: characterization and importance in mRNA degradation. *Trends Biochem Sci* **20**(11): 465–470.

76 Ragheb, J. A., Deen, M., Schwartz, R. H. (1999). CD28-Mediated regulation of mRNA stability requires sequences within the coding region of the IL-2 mRNA. *J Immunol* **163**(1): 120–129.

77 Shaw, G., Kamen, R. (1986). A conserved AU sequence from the 3′ untranslated region of GM-CSF mRNA mediates selective mRNA degradation. *Cell* **46**(5): 659–667.

78 Dibbens, J. A., Miller, D. L., Damert, A. et al. (1999). Hypoxic regulation of vascular endothelial growth factor mRNA stability requires the cooperation of multiple RNA elements. *Mol Biol Cell* **10**(4): 907–919.

79 Boado, R. J., Pardridge, W. M. (1998). Ten nucleotide cis element in the 3′-untranslated region of the GLUT1 glucose transporter mRNA increases gene expression via mRNA stabilization. *Brain Res Mol Brain Res* **59**(1): 109–113.

80 Lee, M., Choi, D., Choi, M. J. et al. (2006). Hypoxia-inducible gene expression system using the erythropoietin enhancer and 3′-untranslated region for the VEGF gene therapy. *J Control Release* **115**(1): 113–119.

81 McGary, E. C., Rondon, I. J., Beckman, B. S. (1997). Post-transcriptional regulation of erythropoietin mRNA stability by erythropoietin mRNA-binding protein. *J Biol Chem* **272**(13): 8628–8634.

82 Ercikan-Abali, E. A., Banerjee, D., Waltham, M. C. et al. (1997). Dihydrofolate reductase protein inhibits its own translation by binding to dihydrofolate reductase mRNA sequences within the coding region. *Biochemistry* **36**(40): 12317–12322.

83 Mayer-Kuckuk, P., Banerjee, D., Malhotra, S. et al. (2002). Cells exposed to antifolates show increased cellular levels of proteins fused to dihydrofolate reductase: a method to modulate gene expression. *Proc Natl Acad Sci U S A* **99**(6): 3400–3405.

84 Southern, M. M., Brown, P. E., Hall, A. (2006). Luciferases as reporter genes. *Methods Mol Biol* **323**: 293–305.

85 Choy, G., Choyke, P., Libutti, S. K. (2003). Current advances in molecular imaging: noninvasive in vivo bioluminescent and fluorescent optical imaging in cancer research. *Mol Imaging* **2**(4): 303–312.

86 Li, X., Zhao, X., Fang, Y. et al. (1998). Generation of destabilized green fluorescent protein as a transcription reporter. *J Biol Chem* **273**(52): 34970–34975.

87 Hsieh, C. H., Chen, F. D., Wang, H. E. et al. (2008). Generation of destabilized herpes simplex virus type 1 thymidine kinase as transcription reporter for PET reporter systems in molecular genetic imaging. *J Nucl Med* **49**(1): 142–150.

88 Ponomarev, V., Doubrovin, M., Serganova, I., Vider, J., Shavrin, A., Beresten, T., Ivanova, A., Ageyeva, L., Tourkova, V., Balatoni, J., Bornmann, W., Blasberg, R., Gelovani Tjuvajev, J. G. (2004). A novel triple modality reporter gene for whole body fluorescent, bioluminescent and nuclear non-invasive imaging. *Eur J Nucl Med* **31**(5): 740–751.

89 Ray, P. D. A., Min, J. J., Tsien, R. Y., Gambhir, S. S. (2004). Imaging tri-fusion multimodality reporter gene expression in living subjects. *Cancer Res* **64**(4): 1323–1330.

90 Ray, P., Tsien, R., Gambhir, S. S. (2007). Construction and validation of improved triple fusion reporter gene vectors for molecular imaging of living subjects. *Cancer Res* **67**(7): 3085–3093.

91 Cecic, I., Chan, D. A., Sutphin, P. D. et al. (2007). Oxygen sensitivity of reporter genes: implications for preclinical imaging of tumor hypoxia. *Mol Imaging* **6**(4): 219–228.

92 Green, L., Yap, C., Nguyen, N., Barrio, J., Namavari, M., Satyamurthy, N., Phelps, M., Sandgren, E., Herschman, H., Gambhir, S. (2002). Indirect monitoring of endogenous gene expression by positron emission tomography (PET) imaging of reporter gene expression in transgenic mice. *Mol Imaging & Biol* **4**(1): 71–81.

11

Imaging Cell Trafficking and Immune Cell Activation Using PET Reporter Genes

Evan D. Nair-Gill, Chengyi J. Shu, Isabel J. Hildebrandt, Dean O. Campbell, Owen N. Witte, and Caius G. Radu

MAJOR IMMUNE CELL TYPES AND THEIR FUNCTIONS

Pathogen invasion of an immunocompetent host induces a coordinated response from a network of diverse immune cell types. The interactions between these various immune cell types are spatially and temporally regulated to facilitate the acquisition of effector mechanisms that ensure pathogen clearance. This section briefly summarizes the major components of the immune network and their actions during an immune response. A more detailed description of the development and function of specific immune cell types can be found in [1].

The immune network has two major components: the innate and adaptive immune systems. Cells of the innate immune system such as macrophages and dendritic cells generally make first contact with pathogens. Pathogen-derived molecules activate specific receptors on innate immune cells leading to the release of chemoattractant molecules and recruitment of other inflammatory cells such as neutrophils. Furthermore, macrophages and dendritic cells ingest foreign proteins (or antigens) and migrate to nearby lymph nodes where they serve as antigen-presenting cells (APCs) in the initiation of the adaptive immune response.

T and B lymphocytes are the key cell types of the adaptive immune system. Both B and T cells express dedicated and highly variable cell surface receptors for antigen. Exposure to antigen together with help from T cells activates B cells to proliferate and differentiate into plasma cells that secrete antigen-specific antibodies that, by various effector mechanisms participate in antigen clearance. Once the invading pathogen has been cleared, the expanded antigen-specific B cell population contracts through apoptosis. B cells that remain are memory cells capable of mounting a rapid antibody response against the same antigen independent of T-cell help.

T cells can recognize an enormous diversity of antigens through their highly variable, heterodimeric T-cell receptors (TCRs). The initiation of a T-cell immune response is triggered by the interaction of the TCR with processed antigenic peptides bound to major histocompatibility complex (p/MHC) molecules expressed on the surface of APCs. Antigen-induced activation of T cells triggers a series of changes in their number, function, and location. $CD4^+$ T cells proliferate in response to antigen and can differentiate into T helper 1 (Th1) or T helper 2 (Th2) cells. These T-cell subsets differ in their cytokine secretion profile and their effects on other immune cell types. Th1 cells mediate cellular immunity and are crucial to the function of $CD8^+$ T cells and macrophages during infections with viruses and intracellular bacteria. Th2 cells predominate during antibody-mediated responses, aiding B-cell activation and differentiation into antibody-secreting plasma cells that are especially important in extracellular bacterial infections and parasitic infestations.

Antigen-induced activation of $CD8^+$ T cells leads to marked proliferation and differentiation into effector cells that express an arsenal of cytotoxic proteins such as granzymes and perforin. $CD8^+$ T cells home to sites of inflammation and kill pathogen-infected or transformed host cells. Similar to B cells, upon resolution of the immune response, both $CD4^+$ and $CD8^+$ T-cell populations expanded by the antigen contract to ensure the maintenance of immune homeostasis. The remaining antigen-specific memory cells are capable of mounting a rapid response against the same antigen (reviewed in [2]).

The ability to quantitatively assess the dynamic perturbations of the immune network in response to antigen challenge is generally applicable to both the study and clinical management of infections, autoimmune and inflammatory disorders, as well as tumor immunity. Sophisticated *ex vivo* analytical tools can readily identify the phenotype and function of immune cells isolated from peripheral blood and tissue biopsies (reviewed in [3]). Data from such *ex vivo* assays have greatly advanced the understanding of essential immune cell types and their role in different forms of immunity. However, these assays provide only a "static image" of a highly dynamic process. Their accuracy can be affected by sampling errors and integrating *ex vivo* measurements to provide a more complete picture of the spatial and temporal kinetics of immune cells *in vivo* can be challenging.

In Vivo Imaging of Immune cells

Immunologists now have access to a toolbox of imaging modalities that, in principle, should be applicable to a wide range of immunological studies. Diverse platforms are included in this toolbox: optical imaging, magnetic resonance imaging, and radionuclide-based imaging modalities such as single photon emission computed tomography (SPECT) and positron emission tomography (PET) (reviewed in [4]). Chapter 7 contains a detailed review of these imaging modalities. However, multiple challenges need to be addressed before molecular imaging techniques gain universal acceptance in the immunology community. This chapter summarizes some of these challenges along with representative examples of recent studies that have successfully used molecular imaging to noninvasively visualize immune responses in preclinical and clinical settings.

OPTICAL IMAGING APPROACHES TO VISUALIZE IMMUNE RESPONSES

Most of the initial imaging studies of immune cells *in vivo* have relied on optical imaging reporters that are either bioluminescent (BLI) or fluorescent (reviewed in [5]). In BLI, light is emitted as a consequence of intracellular enzymatic reactions. This offers the advantage of high sensitivity due to a high signal to noise ratio – only cells that express the reporter are labeled with virtually zero background signal. Two frequently used enzymes are firefly and renilla luciferases. Firefly luciferase (Fluc) oxidizes the substrate D-Luciferin in the presence of ATP and oxygen. The yellow–green light emitted by this reaction enables the detection of cells engineered to express the FLuc gene. Renilla luciferase (RLuc) uses a different substrate (Coelenterazine) than FLuc. Whole-body BLI studies of immune trafficking patterns using these luciferases in various immunological models have been reported [6, 7]. Substantial effort has been made to enhance the sensitivity of bioluminescence reporters. Rabinovich et al. (8) reported genetic modifications to Fluc, including codon optimization and removal of cryptic splice sites that increased the sensitivity of this enzyme more than 100-fold and allowed visualization of as few as three T cells *in vivo*. A new luciferase [9] derived from the marine organism Gaussia princeps and optimized for

human codon usage (hGluc) can also be used to monitor tumor growth or to genetically mark specific populations of immune cells. This luciferase utilizes Coelenterazine as its substrate and is 200-fold more sensitive *in vivo* than RLuc [9]. Genetic modification of Gluc fusing the enzyme to the transmembrane portion of the CD4 cell surface antigen results in a membrane bound reporter (extGluc) [10] that has enhanced sensitivity over both Gluc and Rluc. In this study, tumor-targeted T cells were labeled with extGluc and tumor cells were labeled with Fluc and the antitumor immune response was visualized with concomitant imaging of these distinct cell populations in the same organism.

Fluorescence imaging using fluorescent proteins and dyes in conjunction with *in vivo* imaging devices, especially intravital microscopy (IVM), has enabled detailed studies of immune cell trafficking and the molecular interactions between immune cells in tissues (reviewed in [11]). The development of highly sensitive fluorescent imaging systems has enabled *in vivo* imaging of cells labeled with a variety of red-shifted fluorescent proteins [12]. Although the limits of this technology are still being explored, it offers the possibility of "multicolor" *in vivo* imaging of diverse immune cell populations in action. Further advantages of optical imaging approaches include high sensitivity, excellent specificity, and relatively low cost. There are substantial drawbacks to this technology, however, including limited penetration of light through tissue which, together with photon scattering, rapidly degrades the sensitivity of detection for deeper-seated organs or lesions.

MAGNETIC RESONANCE IMAGING (MRI) OF IMMUNE CELLS

MRI is based on detecting the perturbation of magnetic nuclei that are pulsed with radiofrequency irradiation within a magnetic field. The nucleus that is most frequently measured in MRI is the hydrogen nucleus (proton MRI). Proton MRI signal is dependent on the amount of water contained in the tissues being imaged as well as the chemical environment of these tissues. Compounds that influence these parameters change the proton MRI signal and provide contrast. Contrast agents like gadolinium lead to increased signal intensity whereas iron oxide-based contrast agents darken MRI images (reviewed in [13–15]). Because of its exquisite resolution (25–100 μm) MRI has been used primarily in the clinic to image soft tissues.

MRI studies to image immune cells have been centered primarily on *ex vivo* labeling of cultured immune cells. Phagocytic cells accumulate high levels of magnetic contrast agent and several studies have demonstrated the use of MRI to track these cells *in vivo* (reviewed in [16], [17] and [18]). In a clinical study, de Vries et al. used MRI to monitor a cell-based vaccine utilizing dendritic cells labeled with iron particles that were injected directly into regional lymph nodes of melanoma patients [19]. The investigators were able to track the migration of labeled dendritic cells from the site of injection to nearby lymph nodes and could detect as few as 1.5×10^5 cells in these sites.

Compared to phagocytes, *ex vivo* labeling of lymphocytes is more difficult due to the relatively low level of uptake of magnetic contrast agent. In a proof-of-concept study, Yeh et al. [20] isolated rat T cells and labeled them with superparamagnetic iron oxide particles (SPIOs) before infusion into recipient animals. Injecting a calcium ionophore into the testicles induced inflammation and led to a decrease in MRI signal that correlated with recruitment of the iron-oxide labeled T cells [20]. Kircher et al. [21] optimized uptake of paramagnetic particles into T cells *ex vivo* by coupling cross-linked iron-oxide to a polycationic peptide derived from the cell membrane-penetrating HIV1-Tat protein. Imaging antigen-specific T cells labeled with this contrast agent demonstrated preferential homing of these cells to antigen-expressing tumors.

In vivo labeling of immune cells has centered on using ultra-small paramagnetic iron oxide particles (USPIO) to visualize macrophages during graft rejection, atherosclerosis, and neuroinflammation [22], [23], [24]. A method to broaden the use of *in vivo* labeling for MRI was reported by Pirko et al. [25] who coupled superparamagnetic particles to T cell-specific antibodies. This novel contrast agent visualized regions in the central nervous system of mice during both viral and autoimmune encephalitis that corresponded to areas of lymphocyte infiltration [25]. These images, however, showed a low signal-to-noise ratio and the limit of detection of this technology was not measured.

Reporter genes provide a way to potentially amplify MRI signal in immune cells *in vivo*. Several MRI reporter gene strategies have been developed (reviewed in [26]). One of the more promising reporter gene candidates for immune cell imaging purposes is ferritin, a storage protein for excess iron in cells under normal conditions. Genove et al. [27] exploited this property by infecting neurons *in vivo* with a human ferritin-expressing adenovirus. They were able to visualize the expression of the reporter gene using MRI and confirmed the signal with histology. Recently, this approach has been extended to transgenic mice expressing the ferritin gene under the control of an inducible promoter [28]. These studies demonstrate the potential of targeting ferritin reporter expression to specific cell types, including immune cells.

MRI labeling approaches have certain disadvantages. First, magnetic contrast agents that are loaded into cells *ex vivo* can be diluted out after repeated cell divisions *in vivo* leading to diminished signal over time. Second, many types of iron-based particles are metabolized *in vivo* leading to potentially toxic levels of iron in cells. Third, cell death *in vivo* can lead to a release of contrast agent leading to uptake into surrounding cells and false signal during imaging.

A promising approach using chemical exchange saturation transfer (CEST) contrast agents is being developed. CEST contrast relies on the chemical exchange of protons between the contrast agent and water *in vivo*. CEST MRI imaging involves pulsing with a specific radiofrequency (termed saturation pulse) triggering proton exchange between the contrast agent and water that manifests as loss of MRI signal. Importantly, when pulsing at other frequencies, the contrast agent is silent and there is no change in image signal (reviewed in [29]). Nonmetallic chemical compounds such as poly-amide polypeptides have been shown to contain enough protons with suitable chemical exchange rates that can be excited at specific MR frequencies and detected by imaging. These polypeptides can be genetically encoded to enable cell-type specific expression of the MR reporter gene [30]. Because each contrast agent has a unique frequency required for proton exchange, it is conceivable to visualize multiple contrast agents within a single organism [31]. The possibility of multiplexing with different MRI probes could have immense impact on immune imaging where simultaneous visualization of diverse cell types *in vivo* is desired.

POSITRON EMISSION TOMOGRAPHY (PET) IMAGING OF IMMUNE CELL POPULATIONS

In recent years PET has emerged as a powerful method for *in vivo* imaging of the immune system. PET provides quantitative measurements of the 3-D distribution of molecular probes labeled with positron-emitting radionuclides [32, 33]. PET scanners routinely measure radioactivity concentrations in the 10^{-12} M range, making PET one of the most sensitive imaging technologies applicable to both preclinical and clinical studies. MicroPET scanners for imaging small animals approach a spatial resolution of $\sim$1 mm^3 [34–36], whereas clinical scanners have an intrinsic resolution of $\sim$4–5 mm^3 [37]. Although PET was initially developed for clinical use, the invention of the microPET has enabled imaging studies in small animal models (reviewed in [38]).

The following sections summarize the advantages and limitations of PET assays to visualize immune responses *in vivo*. We will highlight various cell-labeling techniques including those using transgenic reporters to monitor and quantify adaptive immune responses and proceed to address efforts that may improve PET imaging for basic science and clinical applications.

LABELING TECHNIQUES FOR RADIONUCLIDE-BASED IMAGING OF THE IMMUNE RESPONSES

Techniques to label immune cells so that they can be visualized with PET or SPECT can be subdivided into direct and indirect labeling approaches.

Direct Labeling Approaches

First attempts to use microPET for immune monitoring studies in small animals used direct labeling techniques. A typical direct labeling application involves collecting various immune cell types from a donor animal and incubating then *ex vivo* with labeled probes trapped within the cells by various mechanisms. The free, unincorporated label is eliminated by successive washes and the cells are infused into a recipient animal. Direct labeling has enabled studies of the *in vivo* trafficking patterns of monocytes [39], dendritic cells [40], and lymphocytes [41]. Adonai et al. [42] used ^{64}Copper (Cu)-pyruvaldehyde-bis(N^4-mehtylthiosemicabazone) (^{64}Cu-PTSM), a lipophilic copper chelator that facilitates the passive diffusion of copper into cells to monitor lymphocyte trafficking *in vivo*. Lymphocyte migration from the lungs (the first pass organ of intravenous injection) to the spleen was visualized by microPET within the first 4 h after cell transfer. SPECT imaging of ^{111}Indium (^{111}In) oxine-labeled cytotoxic T lymphocytes (CTLs) specific for a model antigen was used to determine how homeostatic proliferation following myeloablation impacts antitumor immunity [43]. Although direct labeling has several obvious advantages, its drawbacks include potential toxic effects on lymphocytes, trapping of the probe in nonfunctional lymphocytes, limited potential for longitudinal studies due to the short isotope half-life, and dilution of the probe due to cell division, as well as release of probe by dying cells [44–46].

In addition to *ex vivo* labeling of immune cells, other direct imaging approaches for detecting sites of inflammation use radiolabeled cytokines and antibodies that, following systemic delivery, bind to cell surface targets. Examples include the use of interleukines 12 and 2 labeled with various isotopes to visualize T cells in murine inflammatory models [47, 48]. Cao et al. labeled an engineered tumor necrosis factor alpha (TNF-α) receptor (etanercept) with ^{64}Cu and successfully detected localized acute inflammation in mice that correlated with elevated secretion of TNF-α [49]. Moreover, monoclonal radiolabeled antibodies to immune cell surface markers and secreted factors may allow visualization-specific populations *in vivo* [50, 51]. A humanized anti-CD3 antibody labeled with ^{99m}Tc demonstrated the ability to detect 20×10^6 human T cells xenografted into mice.

^{111}In-anti-CD4$^+$ antibodies have also been used to visualize CD4$^+$ T cells during inflammation in a colitis model [52]. Signal in the colon during the study correlated with both histological score and total number of CD4 cells in the tissue.

Indirect Labeling Approaches

A PET or SPECT reporter gene encodes a protein that is the target of a specific radiolabeled probe. Imaging PET or SPECT reporter gene expression *in vivo* by injecting the cognate probe (called indirect labeling) enables detection of immune cells over long periods of time. We refer you to Chapter 3 for a more detailed discussion of PET and SPECT reporter genes.

TYPES OF PET REPORTER GENES

Reporter gene imaging is an evolving field. New reporter strategies are constantly being developed every year and multiple criteria should be observed when designing and evaluating them with regard to immune monitoring studies (reviewed in [4]).

Reporter approaches, especially PET, amenable to visualizing the immune system can be divided into three categories: (i) enzymatic reporters that biochemically modify probes and trap them within cells; (ii) transporter-based reporters that funnel probes across cell membranes; and (iii) receptor-based reporters that bind cognate radiolabeled ligands in a stoichiometric manner. Figure 11.1 depicts these different reporter strategies, giving examples of each kind [53].

Enzymatic reporters are most often kinases that specifically phosphorylate small molecule probes. Addition of the electronegative phosphate group to the probe inhibits its efflux across the cell membrane, effectively trapping the probe within the cells. The major advantage of enzymatic reporters is the signal amplification that results from enzyme-driven intracellular trapping. Signal amplification allows for greater sensitivity *in vivo* especially when imaging small populations of cells. It is difficult, however, to correlate enzymatic reporter activity to cell number. This is due in part to differences in probe metabolism in various tissues *in vivo* that are dependent on numerous factors.

Receptor-based reporter approaches, in theory, allow better quantification of the number reporter because of one-to-one binding between probe and receptor. The lack of signal amplification, however, limits the sensitivity of this approach. Membrane-bound transporters that translocate the PET probe from the extracellular environment into the intracellular space may allow for more sensitive imaging, but in the absence of a dedicated intracellular trapping mechanism probe may diffuse out

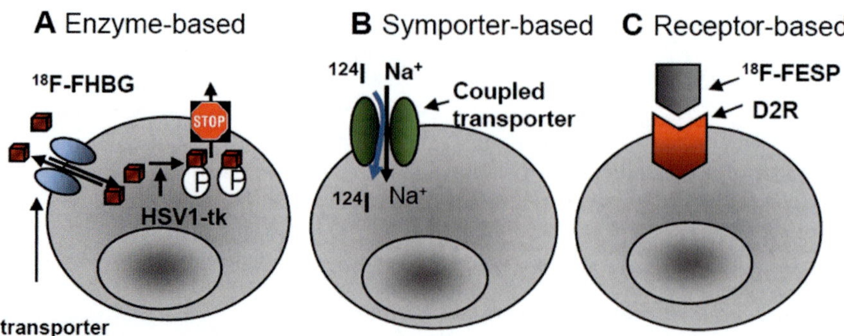

Figure 11.1. Common PET reporter gene/PET reporter probe systems. (A) Enzyme-based approaches. The PET probe is trapped in the cell following phosphorylation by a transgenic kinase PET reporter gene ("metabolic trapping"). Example: the acycloguanosine analog ^{18}F-FHBG and HSV1-tk. **(B)** Symporter-based approaches. The PET probe is a substrate for an active transmembrane transporter. Example: ^{124}I and the human sodium iodide symporter (NIS). **(C)**, Receptor-based approaches. The PET probe is a ligand for a cell surface receptor. Example: ^{18}F-fluoro-ethyl-spiperone (^{18}F-FESP) and the dopamine-2 receptor (DR2).

of the cell over time leading to attenuation of the imaging signal. Moreover, *in vivo* transporter kinetics is difficult to measure because of its dependence on a variety of factors in the intra- and extracellular environments (e.g., the local concentration of specific ions or small molecule metabolites) that impact probe distribution across the cell membrane. These confounding factors may increase the difficulty of rigorously quantifying PET signals resulting from transporter activity to extrapolate the number of reporter-expressing cells.

APPROACHES TO INTRODUCE REPORTER GENES INTO CELLS

Reporter gene imaging requires the delivery of the reporter gene to specific cell populations. This can be accomplished with three different approaches: (i) *ex vivo* transduction of cells with a reporter gene construct; (ii) introduction of the reporter gene into the germline to generate transgenic animals; (iii) *in vivo* delivery of reporter genes to immune cells using viral or nonviral vectors.

EX VIVO DELIVERY OF REPORTER GENES

Ex vivo transduction of cells with viral vectors containing reporter constructs can efficiently label immune cells with a PET reporter gene. Addition of a genetically encoded fluorescence reporter to the PET reporter construct using multicistronic vector systems further enables isolation and analysis of individual reporter-expressing cells with fluorescence-activated cell sorting (FACS). *Ex vivo* transduction methods often involve *in vitro* cell handling and culture steps that may alter viability or function of specific cell populations *in vivo*. This obstacle is germane to the immune system. Retroviral transduction of

T cells requires that the cells are stimulated *in vitro* so that they are in active cell cycle upon infection with the viral vector. Murine T cells activated *in vitro* and infected with a Fluc-encoding retrovirus have been shown to have a greatly diminished capacity to cause graft-versus-host disease in allogeneic recipients when compared to naïve T cells [54]. This is because *in vitro* stimulation of T cells results in the downregulation of critical cell surface proteins expressed on naïve cells that enable homing to specific lymphoid organs required for disease initiation *in vivo*. Labeling immune cells independent of *in vitro* manipulations can be accomplished by transducing hematopoietic or lymphoid progenitors with the reporter construct followed by transplant into myeloablated recipients. With good transduction efficacy, the immune system in the recipients should be reconstituted so that cells from diverse lineages will express the reporter gene and can be subsequently transplanted to study a variety of immune response models [55].

GERMLINE TRANSDUCTION OF REPORTER GENES

Insertion of PET reporter genes in the germline allows for continuous noninvasive imaging of key immune cell populations throughout the lifetime of an animal as well as transmission of the reporter gene to progeny. Two types of genetic manipulation can be used to generate transgenic strains of mice carrying a reporter gene: (i) homologous recombination-based approaches allowing location-specific integration of the reporter into the genome, and (ii) random integration of the reporter gene construct using conventional transgenic approaches, virus-mediated delivery of the reporter to stem cells [56], or bacterial artificial chromosome (BAC) techniques [57]. IVM studies using transgenic mice carrying fluorescent reporters knocked into specific cytokine loci have revealed some of the functional dynamics of adaptive and innate immunity *in vivo* [58]. The combination of IVM with two-photon microscopy (2P-IVM) allows cell imaging in solid organs at depths of up to 500 μm below the surface (reviewed in [11]). Several immune applications of 2P-IVM have been reported [59, 60]. Particularly relevant to this review, the combination of 2P-IVM and immunofluorescence on ordered sequential sections allowed Boissonnas et al. [61] to analyze the infiltration and destruction of solid tumors by cytotoxic T lymphocytes. Transgenic animals expressing a PET reporter in addition to a fluorescent marker would enable similar analyses on longer time scales, noninvasively and at the whole-body level. Immune lineage-specific PET reporter gene expression could help track the origin, proliferation, and movements of these cells during the evolution of diverse immune responses without *ex vivo* manipulations or invasive procedures.

IN VIVO DELIVERY OF PET REPORTER GENES

In vivo gene delivery strategies have been of great interest for cancer gene therapy applications and serve as a starting point for the introduction of reporter genes into specific cell types. Examples include the retargeting of adenoviral vectors using antibodies or other ligands that bind tumor-specific markers. This approach has been used to image the PET reporter mutant Herpes Simplex virus type 1 thymidine kinase(HSV1-sr39tk) delivered to tumors *in vivo* [62, 63] and could be modified to retarget adenovirus carrying a PET reporter to cell surface antigens that define particular immune cell subsets. Xiong et al. [64] used chemically modified adenoviral particles carrying the HSV1-sr39tk PET reporter with polyethylene glycol (PEG) coupled to a cyclic peptide motif (RGD). RGD cyclic peptides specifically bind α_v-integrins that are expressed on a variety of tumors. "PEGylation" of adenoviruses increases viral persistence in the circulation by reducing immune-mediated clearance. In this study, targeting the PET reporter-carrying, PEGylated adenovirus with the integrin-binding peptide motif diminished reporter gene expression in the liver (the tropism of nontargeted adenovirus) evidenced by reduced signal in this organ during imaging with 9-(4-^{18}F-Fluoro-3-[hydroxymethyl]butyl)guanine ([18]F- FHBG).

Yang et al. showed that it is possible to induce protein expression, in particular immune cell types *in vivo* by injecting mice with a lentiviral vector targeted specifically to dendritic cells through the dendritic cell-specific surface protein DC-SIGN. The lentivirus encoded the ovalbumin transgene and 2 weeks after innoculation, mice were able to mount cellular and humoral responses against tumors that expressed ovalbumin [60]. Ziegler

et al. further demonstrated the ability to target lentiviral vectors to antigen-specific immunoglobulins *in vivo* by generating virus particles that express the antibody's cognate antigen [65, 66].

Systemic administration of viral vectors raises concerns about viral toxicity, immune-mediated clearance of the virus and, in the case of lentiviral vectors, insertional mutagensis. Introducing reporter genes into cells utilizing nanoparticles, liposomes, and cell membrane-penetrating peptides may allow targeted labeling of immune cells without using viral vectors (reviewed in [67]).

PRECLINICAL IMMUNE MONITORING STUDIES USING PET AND MULTIMODALITY REPORTER IMAGING WITH HSV1-TK

Herpes Simplex virus type 1 thymidine kinase (HSV1-tk) is the most commonly used PET reporter gene in immune studies. HSV1-tk encodes a viral protein that is required for replication of the viral genome during infection. This enzyme phosphorylates acycloguanosine and thymidine analogs but, importantly, has a low affinity for thymidine itself. This reduces the competition of thymidine with acycloguanosine or thymidine analogs *in vivo*. Radiolabeling of various acycloguanosine and thymidine analogs with the positron emitters [18]Fluorine (^{18}F) or [124]Iodine ([124]I) has resulted in a series of nuclear imaging probes for HSV1-tk expression (reviewed in [53]). Mutational analysis of HSV1-tk resulted in HSV1-sr39tk, a novel kinase containing five distinct point mutations that confer an increased affinity for acycloguanosine analogs [68]. Gambhir et al. [69] compared HSV1-tk to HSV1-sr39tk in mice and found a twofold increase in the uptake of ^{18}F-pencyclovir in cells expressing the mutant kinase. Recently, Likar et al. [70] made further mutations to HSV1-tk and developed two kinase variants specific either for pyrimidine or acycloguanosine probes. This study showed that it was possible to image distinct cell populations in the same animal using these orthogonal PET reporter-probe sets [70].

PROOF-OF-CONCEPT STUDIES DEMONSTRATING THE FEASIBILITY OF USING HSV1-TK BASED PET REPORTER GENES TO IMAGE T-CELL RESPONSES

Koehne et al. infected cytotoxic T lymphocytes specific for the Epstein-Barr virus (EBV) with a retroviral vector encoding the HSV1-tk PET reporter gene [71]. Transduced cells were infused into mice bearing major histocompatibility (MHC)-matched, EBV-positive tumors. MicroPET analysis with the pyrimidine analog

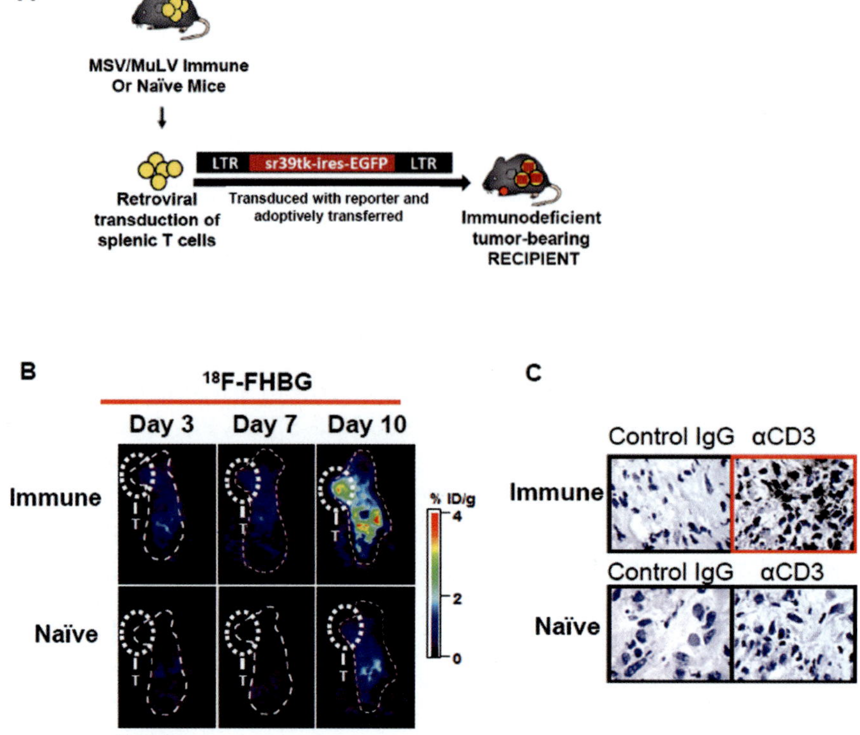

Figure 11.2. PET reporter gene imaging of a secondary immune response against a virally induced tumor. (A) Schematic for *ex vivo* transduction and adoptive transfer of naïve or memory lymphocytes into an animal bearing an oncoretroviral-induced tumor (MSV/MuLV). LTR – Retroviral Long Terminal Repeat, is the promoter that drives reporter expression. **(B)** Longitudinal PET imaging shows a strong signal increase in the tumor in animals that received memory cells on day 10 after adoptive transfer. Tumors in animals that received naïve cells do not show increased signal. **(C)** Immunohistochemistry analysis indicates that tumors in animals receiving memory cells are infiltrated with T cells. No T cells, however, homed to the tumors in animals that received naïve cells.

2′-fluoro-2′-deoxy-1ß-D-arabinofuranosyl-5–^{124}I-iodouracil (^{124}I-FIAU) showed that lymphocyte trafficking and subsequent tumor infiltration required the correct MHC molecules as well as EBV antigens. Importantly, this study found a correlation between the amount of ^{124}I-FIAU signal in the tumor and the number of reporter-expressing T cells in that tissue.

Dubey et al. [72] used a tumor model based on the Moloney murine sarcoma and leukemia virus complex (MSV/MuLV). Injecting this virus complex into immunocompetent mice triggers rapidly growing rhabdomyosarcomas that are strongly immunogenic and induce a potent CD4 and CD8 T cell-mediated immune response, leading to their rejection [73]. To noninvasively visualize immune rejection of the MSV-MuLV tumors, T cells were isolated from donor mice that had previously rejected the rhabdomyosarcoma. This heterogeneous population contained memory T cells for MSV-MuLV and were transduced *in vitro* with a retrovirus encoding HSV1-sr39tk. Reporter-expressing T cells were adoptively transferred into MHC-matched, sarcoma-bearing,

immunodeficient mice [72] (Figure 11.2A). Recipient animals were serially imaged by ^{18}F- FHBG microPET. Control animals received HSV1-sr39tk marked T cells from animals that had not been innoculated with the MSV-MuLV virus complex and did not harbor memory cells for the virus. ^{18}F- FHBG microPET imaging of mice that received memory T cells revealed a gradual increase in the ^{18}F-FHBG signal at the tumor site in contrast to control animals that maintained a low level of ^{18}F-FHBG signal in the tumor (Figure 11.2B). Reporter-labeled T cell from naïve donors were not found in the tumor, demonstrating that immunization was required for lymphocyte homing (Figure 11.2C). Mice bearing antigenically distinct tumors also did not show appreciable ^{18}F-FHBG signals in the tumor area, indicating that ^{18}F-FHBG signals were dependent on specific tumor antigen recognition by CD8^{+} T cells. These studies were an important step in demonstrating that PET reporter gene imaging can be used to serially visualize homeostatic expansion and trafficking of adoptively transferred T cells to tumors in immunodeficient hosts.

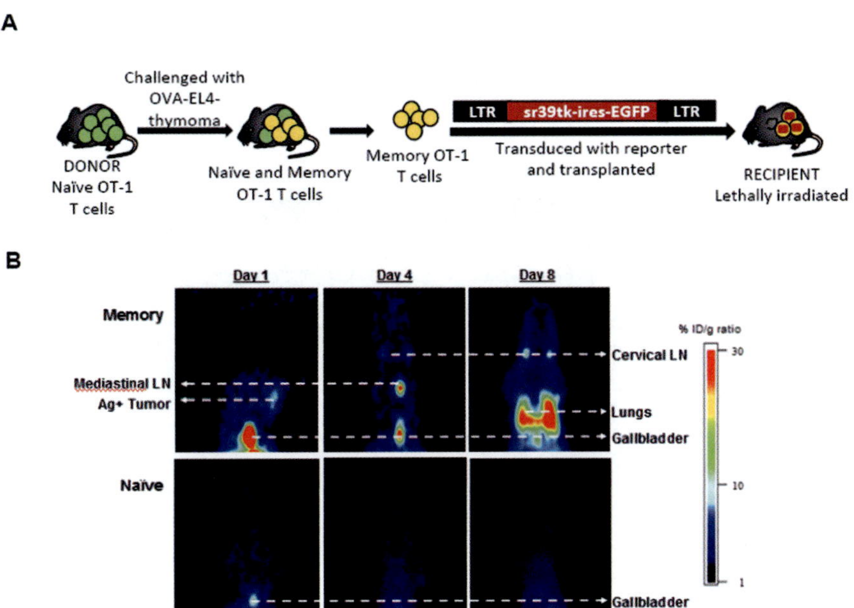

Figure 11.3. Kinetics of memory immune response against a model antigen. (**A**) Schematic showing the harvesting of memory and naïve T cells from animals challenged with the OVA-EL4 tumor followed by transduction with the HSV1-sr39tk PET reporter gene and adoptive transfer into immuno-ablated recipients. (**B**) An ^{18}F-FHBG microPET scan shows accumulation of radioactivity in the tumor and draining lymph nodes of animals receiving memory cells. Animals receiving naïve cells show no signal in the tumor or lymph nodes.

A follow-up study used the HSV1-sr39tk/^{18}F-FHBG reporter-probe system to measure the differential response kinetics of adoptively transferred memory and naïve CD8[+] T cells during an antitumor response [74]. The T-cell repertoire of the OT-1 TCR transgenic mice is limited to only CD8[+] T cells specific for a peptide antigen derived from the chicken ovalbumin protein. Su et al. harvested OT-1 transgenic T cells from mice that had been immunized with ovalbumin (memory cells) or vehicle control (naïve cells) and transduced them with a retrovirus encoding HSV1-sr39tk. Reporter-labeled OT-1 CD8[+] T cells were adoptively transferred into mice bearing tumors engineered to express the ovalbumin antigen (Figure 11.3A). Confirming the results from Dubey et al., longitudinal microPET imaging showed that clonal expansion and trafficking to the tumor occurred more rapidly in animals that received CD8[+] T cells from immunized animals than those that received naïve cells. This difference was apparent as early as 1 day after T cell transplant (Figure 11.3B). Both memory and naïve antigen-specific CD8[+] T cells could induce tumor rejection, but memory cell-mediated rejection occurred with faster kinetics. In addition, memory populations persisted for a longer period of time in secondary lymphoid organs after tumor eradication.

PET reporter gene imaging studies described so far have been focused mainly on imaging immune responses after adoptive transfer of immune cells into tumor-bearing recipients. Visualization and quantification of the whole-body kinetics of endogenous antitumor responses, however, is important to further define parameters that impact rejection of tumors by the immune system. To investigate the progression of an endogenous antitumor response, Shu et al. [55] used a lentiviral vector carrying HSV1-sr39tk coupled to a fluorescent marker and a BLI reporter gene [75] to transduce hematopoietic stem cells. These transduced multipotent cells were engrafted into lethally irradiated recipient mice. The expression of the multimodality reporter construct could be detected *ex vivo* by FACS in all major hematopoietic lineages including T lymphocytes (Figure 11.4A). Furthermore, the luciferase reporter provided a fast, low-cost way to evaluate hematopoietic reconstitution in recipient mice. After reconstitution, bone marrow chimeric animals were challenged with the MSV/MuLV retroviral complex and imaged over time with ^{18}F-FHBG microPET. MicroPET imaging revealed that transduced immune cells localized and expanded in the tumor and in the draining lymph nodes at the height of the antitumor immune response from day 10 to 14 after viral challenge (Figure 11.4B). BLI of the same set of animals did not show the presence of reporter-expressing cells in draining lymph nodes emphasizing the superior ability of PET to visualize deeper seated anatomical structures. Treatment of tumor-bearing chimeric animals with the immunosuppressive drug dexamethasone

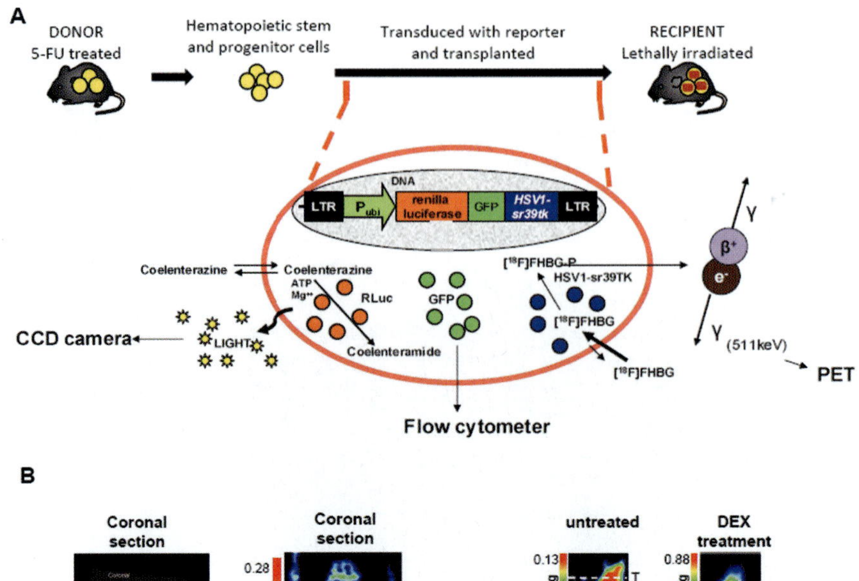

Figure 11.4. Imaging primary immune responses against virally induced tumors. (A) Schematic for the generation of bone-marrow chimeric animals in which the hematopoietic stem and progenitor cells express HSV1-sr39tk, Renilla luciferase (Rluc), and EGFP. The multimodality reporter construct allows for BLI, fluorescence imaging, and flow cytometry in addition to microPET. **(B)** Coronal section microPET image of infiltration of tumor and draining lymph nodes with transduced cells during a primary antitumor response. Chronic administration of the immunosuppressive drug dexamethasone (DEX) reduces the signal in both the tumor and the lymph nodes relative to untreated animals. T: tumor. DLN: draining lymph nodes.

diminished the ^{18}F-FHBG signals in draining lymph nodes. This result showed it is possible to use PET reporter gene imaging to evaluate the effects of drug treatment on select immune populations (Figure 11.4B).

QUANTIFYING CELLULAR IMMUNE RESPONSES WITH PET REPORTER GENE IMAGING

An advantage of PET reporter imaging is its potential ability to quantify the number of reporter-expressing cells in an anatomical region of interest, such as lymphoid organs and tumors. Further studies went on to refine the correlation of microPET signal to cell numbers *in vivo* and determine the limit of detection for HSV1-sr39tk PET reporter imaging. Su et al. [76] injected different numbers of HSV1-sr39tk$^+$-transduced CD8$^+$ T cells directly into established vascularized tumors and used microPET to measure the amount of ^{18}F-FHBG

that accumulated in these sites. To do this, the investigator drew a three-dimensional region of interest (ROI) around the tumors and normalized the radioactivity within the ROI both to the total amount of radiolabeled probe injected into the animal and to the mass of the tissue analyzed. The ^{18}F-FHBG signal in these tumors was dependent on the number of HSV1-sr39tk$^+$-transduced CD8$^+$ T cells injected and the limit of detection during these experiments was approximately 10^6 reporter-expressing T cells in a volume of 100 μL.

Using optical and microPET imaging, Yaghoubi et al. [7] have recently developed methods for determining the number of cells accumulating in arthritic mouse paws. To quantify the number of T-cell hybridomas that homed to the arthritic mouse paws, the cells were transduced with a triple fusion reporter gene that expressed Rluc, HSV1-sr39tk$^+$, and EGFP. Using nonarthritic animals the authors developed equations that related the number of Rluc-expressing cells in the paws to the signal intensity detected by optical imaging. Then, they quantitatively monitored the trafficking of two different collagen type II-specific T-cell hybridomas (A2-TFR and AG6-TFR). In arthritic animals injected systemically with the transduced hybridomas they observed that ~500–5000 A2-TFR cells and ~100–1200 AG6-TFR cells migrated to the inflamed paws. The number of cells that accumulated in the paws did not depend on the number of cells injected or the degree of inflammation in the paws. Above background ^{18}F-FHBG accumulation was also observed in some inflamed paws of T-cell hybridoma-injected arthritic mice [7].

PET REPORTER IMAGING OF CLINICALLY RELEVANT MODELS OF CANCER IMMUNOTHERAPY

The tumor rejection studies described so far rely on strong viral or model antigens to stimulate the immune system. However, human cancers generally express non-mutated or subtly mutated self-antigens that do not induce potent immune responses. Tumor rejection in these instances depends on breaking tolerance against such tumor-expressed self-antigens. Using PET to visualize and quantify immune responses against weak self-antigens is significantly more challenging: there are generally much fewer antigen-specific T cells in these cases than in immune responses against viral antigens.

A clinically relevant model of melanoma rejection was used to assess the utility of microPET to monitor tumor immunity dependent on T-cell recognition of self-antigens [77]. These studies used the Pmel-1 cancer immunotherapy model developed by Overwijk et al. [78]. In this model, rejection of established murine B16 melanoma tumors can be achieved by using a combined immunotherapy protocol consisting of adoptive transfer (AT) of CD8^{+} T cells specific for a self/tumor antigen (gp100$_{25-33}$) in conjunction with lymphodepletion (XRT, 500 cGy), dendritic cell (DC) vaccination and high dose (HD) systemic IL-2 [78]. Prior to adoptive transfer, CD8^{+} Pmel-1 T cells were transduced with a multicistronic retroviral vector encoding the PET reporter gene HSV1-sr39tk and the enhanced yellow fluorescent protein (EYFP) to allow *ex vivo* detection of these cells by microPET and flow cytometry, respectively. Antigen-specific T cells were adoptively transferred into lymphodepleted BL/6 mice bearing B16 melanoma tumors. Recipient mice were serially imaged by ^{18}F-FHBG microPET/CT (Figure 11.5).

Key findings from these experiments include: (i) ^{18}F-FHBG microPET imaging allowed the visualization of homing, expansion, contraction, and movement to secondary lymphoid organs of melanoma-specific T cells; (ii) these parameters varied considerably from animal to animal, indicating that longitudinal immune monitoring

studies might be crucial for generating surrogate markers of treatment efficacy; (iii) ROI-based quantification of ^{18}F-FHBG microPET data showed that the limit of detection in a lymph node was 1×10^4 reporter gene-expressing cells, and that a minimal threshold of ~3% of the total cells in the lymph node had to be HSV1-sr39tk^{+} T cells. The limit of detection in the spleen was 7×10^5 cells with a minimal threshold of ~10% of the total cells being HSV1-sr39tk^{+} T cells. The discrepancy in the limit of detection between these lymphoid organs may be due to differences in the density of HSV1-sr39tk^{+} T cells in the spleen and lymph node. Therefore, because T cells are more densely packed in lymph nodes than in the spleen, PET reporter imaging is more sensitive for detecting T cells localized in lymph nodes than in the spleen; (iv) varying the immunotherapeutic regime by complementing adoptive transfer and lymphodepletion with additional administration of IL-2 and DC vaccination resulted in a dramatic increase in HSV1-sr39tk^{+} T cell-dependent ^{18}F-FHBG accumulation at multiple lymphoid sites in the body; (v) the quantification of T cells at the tumor site can sometimes be hampered by the high nonspecific trapping of the ^{18}F-FHBG probe by the B16 tumors. Alternative HSV1-tk specific substrates such as ^{18}F-FEAU [7, 79–83] may result in lower tumor background, enabling quantitative measurements of T cell homing at the tumor site. Importantly, it was possible to detect the presence of HSV1-sr39tk^{+} Pmel-1 cells for extended periods of time (3–4 weeks following adoptive transfer) implying that reporter gene expression and repetitive microPET/CT scanning does not affect the lifespan of adoptively transferred T cells. These studies demonstrate that microPET/CT imaging can be used to quantitatively and longitudinally monitor the persistence of adoptively transferred T cells specific for a weak self/tumor antigen.

In a groundbreaking clinical study, Yaghoubi et al. monitored the trafficking of cytolytic T cells labeled with HSV1-tk to tumors in a glioma patient [84]. This strategy allowed visualization of T cells as they infiltrated a previously diagnosed tumor then migrated to a new tumor that had not yet been characterized in that patient (Figure 11.6). PET imaging was greatly enhanced by coregistration with MRI, which provided high-resolution anatomical imaging. This is the first study of its kind to use PET reporter gene imaging to evaluate cellular anticancer therapy.

FURTHER DEVELOPMENT OF PET REPORTER GENE IMAGING TECHNOLOGY FOR CLINICAL APPLICATIONS

As the studies previously described have shown, PET reporter imaging can greatly enhance the evaluation of

Evan D. Nair-Gill, Chengyi J. Shu, Isabel J. Hildebrandt, Dean O. Campbell, Owen N. Witte, and Caius G. Radu

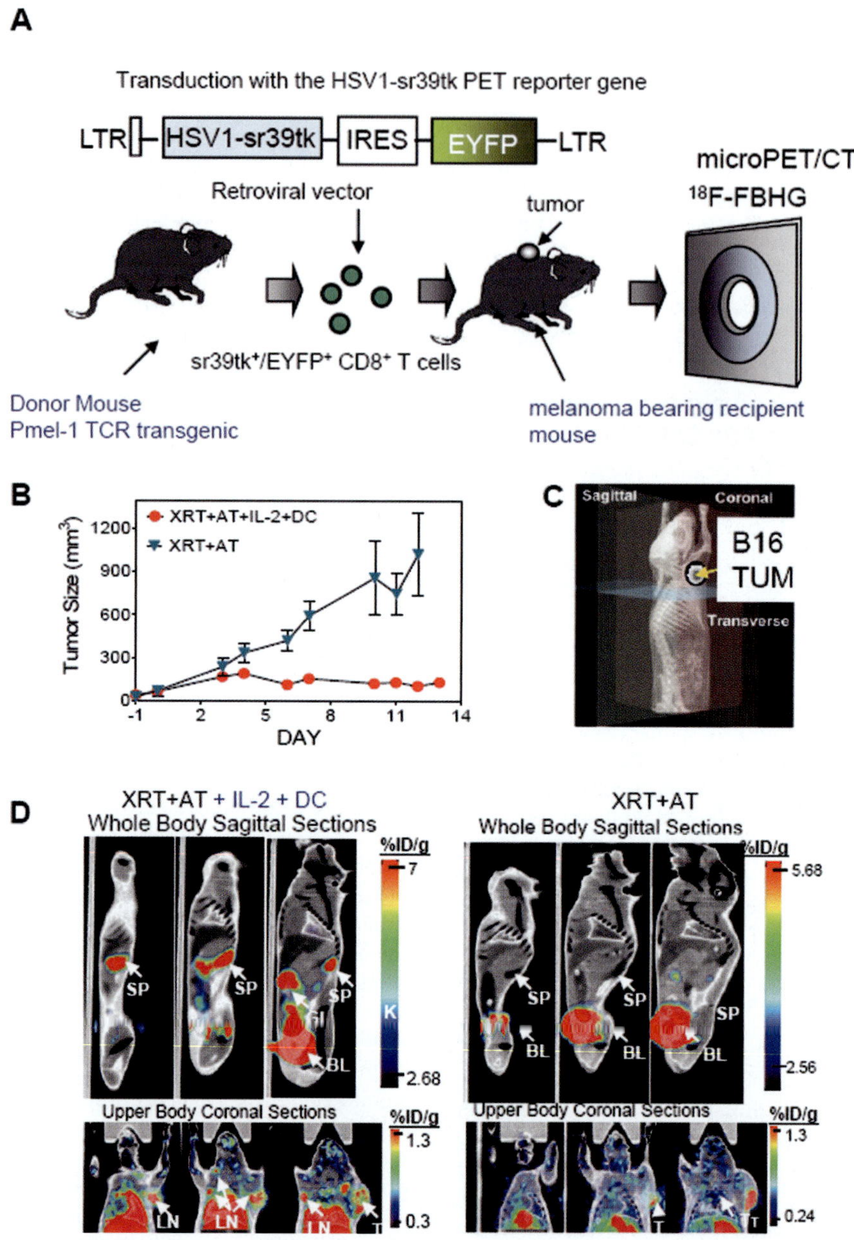

Figure 11.5. MicroPET/CT imaging of T-cell trafficking in a mouse model of adoptive cell transfer therapy against melanoma. (**A**) Experimental design. (**B**) Only mice receiving combined immunotherapy – whole body irradiation (XRT, −500 cGy) + adoptive transfer (AT) + IL-2 + DC- respond to treatment. (**C**) Whole-body microCT image indicating the tumor site. (**D**) On day 3 postadoptive transfer, recipient mice were injected i.v. with ^{18}F-FHBG and Fenestra LC contrast agent and imaged by microPET/CT. Slices of whole-body sagittal sections and coronal section (1 mm thickness) of the upper body are shown. SP, spleen; GI, gastrointestinal tract; BL, bladder; T, tumor; C, cervical LN; A, axillary LN; B, brachial LN. Adapted from Shu *et al.* ref. [77].

preclinical and clinical cancer immunotherapy protocols. However, to allow routine use of this technology in translational studies additional issues must be considered. The most significant issue is the immunogenicity of commonly used PET reporter genes such as those based on viral thymidine kinases.

THE IMMUNOGENICITY OF HSV1-TK-BASED PET REPORTERS

A major concern with the HSV1-tk reporter system in clinical applications is its known immunogenicity arising from the fact that it is a viral protein. In theory, the

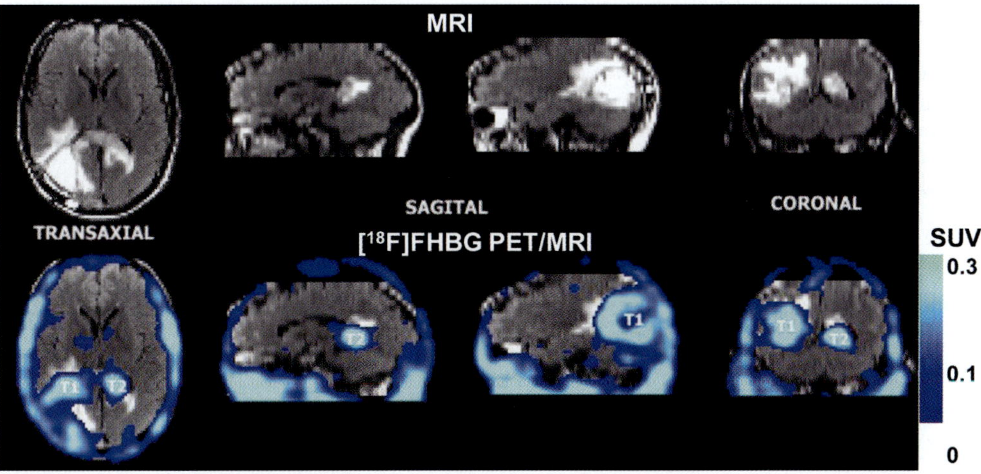

Figure 11.6. MRI and PET over MRI superimposed brain images of the patient who had been infused autologous CTLs expressing the HSV1-tk PET reporter gene. Images were acquired approximately 2 h after ^{18}F-FHBG injection. The patient had a surgically resected tumor (T1) in the left corner and a new nonresected tumor in the center (T2), near corpus callosum of his brain. Infused cells had localized at the site of T1 and also trafficked to T2. ^{18}F-FHBG activity is higher than the brain background at both sites. Background ^{18}F-FHBG activity is low within the central nervous system due to its inability to cross the blood–brain barrier. Background activity is relatively higher in all other tissues. Activity can also be observed in the meninges. Reprinted by permission from Macmillan Publishers Ltd: Nature Clinical Practice Oncology (84), 2009.

host immune system will recognize this protein as foreign and mount an immune response against reporter-labeled transplanted cells. Importantly, clinical studies have shown that the immunogenicity of HSV1-tk does not limit the half-life of reporter-labeled therapeutic cells in the blood if the patient had myeloablation prior to transplant [85]. As expected, without complete myelodepletion the recipient is able to clear the infused lymphocytes, effectively wiping out the therapeutic agent [86]. Thus, it is critical to monitor the peripheral blood for the presence of circulating host capable of lysing cells that express the PET reporter. It is also important to measure host humoral responses to the reporter protein. Analogous to graft rejection after organ transplantation, the occurrence of either of these events suggests rejection of transplanted cells leading to a failure in therapy.

To combat this problem, several humanized reporters have been investigated in various applications including the dopamine type II receptor [87], the somatostatin receptor [88], and the sodium-iodide symporter [89], but these have not demonstrated the same sensitivity as kinase-based reporter systems.

Soghomonyan et al. [80] demonstrated that human thymidine kinase 2 (hTK2) could be engineered to be a PET reporter. Although hTK2 is expressed primarily in the mitochondria, removing the N-terminus sorting signal leads to cytosolic expression of the kinase. In this study, the investigators imaged tumor xenografts expressing the modified hTK2 with the thymidine-based probes ^{18}F-FEAU and [124]I-FIAU. Importantly, like HSV1-tk-based reporters, the hTK2 PET reporter gene

can also function as a suicide gene if transduced cells are treated with a pharmacologic dose d-arabinofuranosyl-cytosine. This strategy is an initial step in the direction of clinically applicable, nonimmunogenic PET reporter genes.

The human norepinephrine transporter (hNET) has been recently evaluated as a potentially nonimmunogenic PET reporter gene. Using the norepinephrine analogue, [124]I-MIBG, Doubrovin et al. [90] showed that as few as 10^4 T lymphocytes transduced *ex vivo* with the hNET construct could be visualized by microPET. The investigators used this approach to monitor tumor infiltration by antigen-specific T lymphocytes. Development of ^{18}F-labeled hNET substrates could widen the applicability of this approach for monitoring cell-based therapies.

Wei et al. [91] demonstrated the feasibility of using an antibody fragment engineered to be expressed on the cell surface as a reporter gene. The authors showed that the expression of the DOTA antibody reporter 1 (DAbR1) can be detected using the PET reporter probe [86]Yttrium-acrylamidobenzyl-DOTA ([[86]Y]-AABD). This complex binds irreversibly to a cysteine residue in the DAbR1 reporter gene. U-87 glioma cells transduced with the reporter gene and transplanted into mice showed intense uptake of the probe. The abdominal background and nonspecific tumor uptake of the probe were very low. Whereas this study used a murine reporter gene construct, the concept can be applied to human antibody fragments. As an "inverse" of this concept, Kenanova et al. transfected a human T-cell line with recombinant human carcinoembryonic antigen (CEA), a tumor antigen with limited expression in normal human tissues. PET imaging could detect these cells *in vivo* when a high-affinity human antibody fragment labeled with [124]I was used as a probe (92). The advantages of using antibody fragments as either reporter genes or probes are: (i) the ability to first identify or design a target molecule and then generate an antibody against it; (ii) the high affinity and high specificity binding of antibodies to target molecules; (iii) a potentially high number of available antibodies available as reporter genes or probe candidates.

DIRECT PET IMAGING APPROACHES TO VISUALIZE IMMUNE RESPONSES

PET measurements of biochemical and cell surface biomarkers could also be used to assess cellular processes during an immune response independent of introducing a genetically encoded reporter. PET probes for biochemical pathways are of particular interest given their potential to detect and measure functional changes induced by lymphocyte activation. These types of functional measurements should complement reporter gene imaging, which provides information primarily on the location and number of specific cell populations within the body. Two PET probes of cellular biochemistry are used regularly for cancer detection: (i) ^{18}F-fluorodeoxyglucose (^{18}F-FDG), which visualizes cellular glucose metabolism through phosphorylation and trapping by hexokinase (reviewed in [93]); (ii) ^{18}F-fluorothymidine (^{18}F-FLT), which reflects cellular proliferation through phosphorylation and trapping by thymidine kinase 1 (TK1), a rate-limiting enzyme in the thymidine salvage pathway (reviewed in [94]). Although these probes have strong clinical applications in detecting metabolically active tumors and monitoring treatment responses, they could also be useful for evaluating the immune system. ^{18}F-FDG was used to measure increased glucose metabolism in activated immune cells in the context of neuroinflammation during experimental autoimmune encephalitis (EAE), a mouse model of multiple sclerosis [95]. ^{18}F-FDG microPET/CT could detect EAE disease progression and its response to immunosuppressive therapy. This study provided proof-of-concept for exploiting biochemical pathways as "endogenous reporters" of immune activation. An increase in glucose metabolism is not a specific marker for activated immune cells. Several tissues in the body, including brain, heart, and a variety of tumor and immune cells, will accumulate FDG.

To identify PET probes with enhanced immune specificity, our group used a differential screening strategy that measured the preferential retention of various small

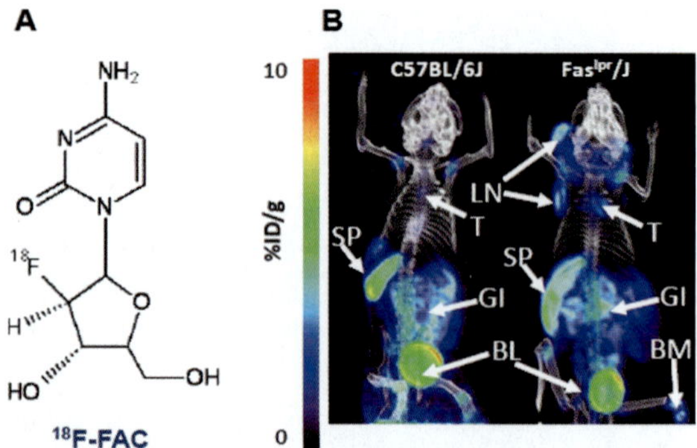

Figure 11.7. ^{18}F-FAC-PET/CT imaging of increased lymphoid mass in systemic autoimmunity imaging. (**A**) Chemical structure of ^{18}F-FAC. (**B**) Representative scans of ^{18}F-FAC in a C57BL/6J wild-type mouse (**left**) and a B6.MRL-Fasᴵᵖʳ/J mouse (**right**), which has an autoimmune phenotype. The increased size of the lymphoid organs in the B6.MRL-Fasᴵᵖʳ/J mouse is reflected in the increased uptake of ^{18}F-FAC in the lymph nodes and spleen. BM – bone narrow. LN – lymph node. T – thymus. SP – spleen. GI – gastrointestinal tract. BL – bladder. microPET/CT images correspond to 10-min scans 1 h after injection of the tracer.

molecules in activated T cells. This approach led to the development of 1-(2′-deoxy-2′-^{18}F-fluoroarabinofuranosyl) cytosine ([18]F- FAC) [96], a deoxycytidine analog that can be synthesized as a PET tracer. ^{18}F-FAC PET enables visualization of lymphoid organs such as thymus, spleen, and bone marrow in naïve mice. This probe detects localized immune activation during an antitumor immune response and also visualizes increased lymphoid mass in a murine model of systemic autoimmunity (Figure 11.7). A direct comparison of FAC and FDG showed that these probes preferentially labeled different immune cell types during an antitumor immune response (Nair-Gill et al., unpublished data). Whereas FDG accumulated predominantly in innate immune cells, FAC accumulated in both T cells and innate cells directly correlated to cell proliferation. This result demonstrates that imaging probes for different metabolic pathways can give distinct information about immune cell function during a response. Further studies utilizing PET reporter gene imaging together with metabolic PET could identify site-specific variations in immune cell function during a whole-body immune response in an organism.

Greater degrees of cell subtype and functional specificity in PET imaging may also be possible with the development of PET probes specific for immune cell surface markers. In this context, the recent demonstration of the ability to ^{18}F-label antibody fragments specific for cell surface proteins [49] could enable future imaging applications to monitor immune responses based on the expression of functionally relevant T-cell activation markers.

CONCLUSIONS

The first PET reporter studies were performed more than a decade ago [97], [98]. Ever since, we have witnessed an exciting expansion of this technology and its applications to preclinical studies of immunity. The ability of PET to visualize cell populations in a quantitative and longitudinal manner independent of anatomical location makes this imaging modality a unique and valuable tool for immune biology. Studies utilizing PET reporters to follow transduced lymphocytes *in vivo* in tumor rejection models have laid the foundation for imaging the progression and resolution of immune responses. In addition to ongoing efforts to develop improved reporter genes and probes for immune monitoring studies summarized in this chapter, a key direction for the future concerns translating animal studies to the clinic. Future development of reporter genes will enable tracking the movement, proliferation, and viability of transplanted cells repeatedly during treatment. Such studies will help more clearly define the parameters that determine therapeutic success and provide noninvasive surrogate biomarkers for clinical endpoints.

REFERENCES

1 Murphy, K. P., Travers, P., Walport, M., Janeway, C. (2008). *Janeway's Immunobiology*. New York: Garland Science.

2 Goldrath, A. W., Bevan, M. J. (1999). Selecting and maintaining a diverse T-cell repertoire. *Nature* 402: 255–262.

3 Clay, T. M., Hobeika, A. C., Mosca, P. J., Lyerly, H. K., Morse, M. A. (2001). Assays for monitoring cellular immune responses to active immunotherapy of cancer. *Clin Cancer Res* 7: 1127–1135.

4 Massoud, T. F., Gambhir, S. S. (2003). Molecular imaging in living subjects: seeing fundamental biological processes in a new light. *Genes Dev* 17: 545–580.

5 Hildebrandt, I. J., Gambhir, S. S. (2004). Molecular imaging applications for immunology. *Clin Immunol (Orlando, Fla)* 111: 210–224.

6 Costa, G. L., Sandora, M. R., Nakajima, A., Nguyen, E. V., Taylor-Edwards, C., Slavin, A. J., Contag, C. H., Fathman, C. G., Benson, J. M. (2001). Adoptive immunotherapy of experimental autoimmune encephalomyelitis via T cell delivery of the IL-12 p40 subunit. *J Immunol* 167: 2379–2387.

7 Yaghoubi, S. S., Creusot, R. J., Ray, P., Fathman, C. G., Gambhir, S. S. (2007). Multimodality imaging of T-cell hybridoma trafficking in collagen-induced arthritic mice: image-based estimation of the number of cells accumulating in mouse paws. *J Biomed Opt* 12: 064025.

8 Rabinovich, B. A., Ye, Y., Etto, T., Chen, J. Q., Levitsky, H. I., Overwijk, W. W., Cooper, L. J., Gelovani, J., Hwu, P. (2008). Visualizing fewer than 10 mouse T cells with an enhanced firefly luciferase in immunocompetent mouse models of cancer. *Proc Natl Acad Sci U S A* 105: 14342–14346.

9 Tannous, B. A., Kim, D. E., Fernandez, J. L., Weissleder, R., Breakefield, X. O. (2005). Codon-optimized Gaussia luciferase cDNA for mammalian gene expression in culture and in vivo. *Mol Ther* 11: 435–443.

10 Santos, E. B., Yeh, R., Lee, J., Nikhamin, Y., Punzalan, B., La Perle, K., Larson, S. M., Sadelain, M., Brentjens, R. J. (2009). Sensitive in vivo imaging of T cells using a membrane-bound Gaussia princeps luciferase. *Nat Med* 15: 338–344.

11 Sumen, C., Mempel, T. R., Mazo, I. B., von Andrian, U. H. (2004). Intravital microscopy: visualizing immunity in context. *Immunity* 21: 315–329.

12 Shu, X., Royant, A., Lin, M. Z., Aguilera, T. A., Lev-Ram, V., Steinbach, P. A., Tsien, R. Y. (2009). Mammalian expression of infrared fluorescent proteins engineered from a bacterial phytochrome. *Science (New York, N.Y)* 324: 804–807.

13 Strijkers, G. J., Mulder, W. J., van Tilborg, G. A., Nicolay, K. (2007). MRI contrast agents: current status and future perspectives. *Anticancer Agents Med Chem* 7: 291–305.

14 Bellin, M. F. (2006). MR contrast agents, the old and the new. *Anticancer Agents Med Chem* 60: 314–323.

15 Kirsch, J. E. (1991). Basic principles of magnetic resonance contrast agents. *Top Magn Reson Imaging* 3: 1–18.

16 Hoehn, M., Wiedermann, D., Justicia, C., Ramos-Cabrer, P., Kruttwig, K., Farr, T., Himmelreich, U. (2007). Cell tracking using magnetic resonance imaging. *J Physiol* 584: 25–30.

17 Rogers, W. J., Meyer, C. H., Kramer, C. M. (2006). Technology insight: in vivo cell tracking by use of MRI. *Nature Clinical Practice* 3: 554–562.

18 Arbab, A. S., Liu, W., Frank, J. A. (2006). Cellular magnetic resonance imaging: current status and future prospects. *Expert Rev Med Devices* 3: 427–439.

19 de Vries, I. J., Lesterhuis, W. J., Barentsz, J. O., Verdijk, P., van Krieken, J. H., Boerman, O. C., Oyen, W. J., Bonenkamp, J. J., Boezeman, J. B., Adema, G. J., *et al.* (2005). Magnetic resonance tracking of dendritic cells in melanoma patients for monitoring of cellular therapy. *Nat Biotechnol* 23: 1407–1413.

20 Yeh, T. C., Zhang, W., Ildstad, S. T., Ho, C. (1995). In vivo dynamic MRI tracking of rat T-cells labeled with super-paramagnetic iron-oxide particles. *Magn Reson Med* 33: 200–208.

21 Kircher, M. F., Allport, J. R., Graves, E. E., Love, V., Josephson, L., Lichtman, A. H., Weissleder, R. (2003). In vivo high resolution three-dimensional imaging of antigen-specific cytotoxic T-lymphocyte trafficking to tumors. *Cancer Res* 63: 6838–6846.

22 Mulder, W. J., Strijkers, G. J., Vucic, E., Cormode, D. P., Nicolay, K., Fayad, Z. A. (2007). Magnetic resonance molecular imaging contrast agents and their application in atherosclerosis. *Top Magn Reson Imaging* 18: 409–417.

23 Ye, Q., Yang, D., Williams, M., Williams, D. S., Pluempitiwiriyawej, C., Moura, J. M., Ho, C. (2002). In vivo detection of acute rat renal allograft rejection by MRI with USPIO particles. *Kidney Int* 61: 1124–1135.

24 Dousset, V., Delalande, C., Ballarino, L., Quesson, B., Seilhan, D., Coussemacq, M., Thiaudiere, E., Brochet, B., Canioni, P., & Caille, J. M. (1999). In vivo macrophage activity imaging in the central nervous system detected by magnetic resonance. *Magn Reson Med* 41: 329–333.

25   Pirko, I., Johnson, A., Ciric, B., Gamez, J., Macura, S. I., Pease, L. R., Rodriguez, M. (2004). In vivo magnetic resonance imaging of immune cells in the central nervous system with superparamagnetic antibodies. *Faseb J* 18: 179–182.

26   Gilad, A. A., Winnard, P. T., Jr., van Zijl, P. C., Bulte, J. W. (2007). Developing MR reporter genes: promises and pitfalls. *NMR in Biomed* 20: 275–290.

27   Genove, G., DeMarco, U., Xu, H., Goins, W. F., Ahrens, E. T. (2005). A new transgene reporter for in vivo magnetic resonance imaging. *Nat Med* 11: 450–454.

28   Cohen, B., Ziv, K., Plaks, V., Israely, T., Kalchenko, V., Harmelin, A., Benjamin, L. E., Neeman, M. (2007). MRI detection of transcriptional regulation of gene expression in transgenic mice. *Nat Med* 13: 498–503.

29   Sherry, A. D., Woods, M. (2008). Chemical exchange saturation transfer contrast agents for magnetic resonance imaging. *Ann Rev Biomed Eng* 10: 391–411.

30   Gilad, A. A., McMahon, M. T., Walczak, P., Winnard, P. T., Jr., Raman, V., van Laarhoven, H. W., Skoglund, C. M., Bulte, J. W., van Zijl, P. C. (2007). Artificial reporter gene providing MRI contrast based on proton exchange. *Nat Biotechnol* 25: 217–219.

31   McMahon, M. T., Gilad, A. A., DeLiso, M. A., Berman, S. M., Bulte, J. W., van Zijl, P. C. (2008). New "multicolor" polypeptide diamagnetic chemical exchange saturation transfer (DIACEST) contrast agents for MRI. *Magn Reson Med* 60: 803–812.

32   Phelps, M. (2000). Inaugural article: positron emission tomography provides molecular imaging of biological processes. *PNAS* 97: 9226–9233.

33   Phelps, M. E., Hoffman, E. J., Mullani, N. A., Ter-Pogossian, M. M. (1975). Application of annihilation coincidence detection to transaxial reconstruction tomography. *J Nucl Med* 16: 210–224.

34   Cherry, S., Gambhir, S. (2001). Use of positron emission tomography in animal research. *ILAR J* 42: 219–232.

35   Chatziioannou, A., Tai, Y., Doshi, N., Cherry, S. (2001). Detector development for microPET II. a 1 microl resolution PET scanner for small animal imaging. *Phys Med Biol* 46: 2899–2910.

36   Chatziioannou, A. F., Cherry, S. R., Shao, Y., Silverman, R. W., Meadors, K., Farquhar, T. H., Pedarsani, M., Phelps, M. E. (1999). Performance evaluation of microPET: a high-resolution lutetium oxyorthosilicate PET scanner for animal imaging. *J Nucl Med* 40: 1164–1175.

37   Dahlbom, M., Hoffman, E., Hoh, C., Schiepers, C., Rosenqvist, G. (1992). Whole-body positron emission tomography: Part I. Methods and performance characteristics. *J Nucl Med* 33: 1191–1199.

38   Cherry, S. R., Gambhir, S. S. (2001). Use of positron emission tomography in animal research. *ILAR Journal/National Research Council, Institute of Laboratory Animal Resources* 42: 219–232.

39   Ridolfi, R., Riccobon, A., Galassi, R., Giorgetti, G., Petrini, M., Fiammenghi, L., Stefanelli, M., Ridolfi, L., Moretti, A., Migliori, G., *et al.* (2004). Evaluation of in vivo labelled dendritic cell migration in cancer patients. *J Transl Med* 2: 27.

40   Paik, J. Y., Lee, K. H., Byun, S. S., Choe, Y. S., Kim, B. T. (2002). Use of insulin to improve [18 F]fluorodeoxyglucose labelling and retention for in vivo positron emission tomography imaging of monocyte trafficking. *Nucl Med Commun* 23: 551–557.

41   Matsui, K., Wang, Z., McCarthy, T. J., Allen, P. M., Reichert, D. E. (2004). Quantitation and visualization of tumor-specific T cells in the secondary lymphoid organs during and after tumor elimination by PET. *Nucl Med Biol* 31: 1021–1031.

42   Adonai, N., Nguyen, K. N., Walsh, J., Iyer, M., Toyokuni, T., Phelps, M. E., McCarthy, T., McCarthy, D. W., Gambhir, S. S. (2002). Ex vivo cell labeling with 64Cu-pyruvaldehyde-bis(N4-methylthiosemicarbazone) for imaging cell trafficking in mice with positron-emission tomography. *Proc Natl Acad Sci U S A* 99: 3030–3035.

43   Pittet, M. J., Grimm, J., Berger, C. R., Tamura, T., Wojtkiewicz, G., Nahrendorf, M., Romero, P., Swirski, F. K., Weissleder, R. (2007). In vivo imaging of T cell delivery to tumors after adoptive transfer therapy. *Proc Natl Acad Sci U S A* 104: 12457–12461.

44   Botti, C., Negri, D. R. M., Seregni, E., Ramakrishna, V., Arienti, F., Maffioli, L., Lombardo, C., Bogni, A., Pascali, C., Crippa, F., *et al.* (1997). Comparison of three different methods for radiolabelling humanactivated T lymphocytes. *Eur J Nucl Med Mol Imaging* 24: 497–504.

45   Balaban, E. P., Simon, T. R., Sheehan, R. G., Frenkel, E. P. (1986). Effect of the radiolabel mediator tropolone on lymphocyte structure and function. *J Lab Clin Med* 107: 306–314.

46   Balaban, E. P., Simon, T. R., Frenkel, E. P. (1987). Toxicity of indium-111 on the radiolabeled lymphocyte. *J Nucl Med* 28: 229–233.

47   Signore, A., Picarelli, A., Annovazzi, A., Britton, K. E., Grossman, A. B., Bonanno, E., Maras, B., Barra, D., Pozzilli, P. (2003). 123I-Interleukin-2: biochemical characterization and in vivo use for imaging autoimmune diseases. *Nucl Med Commun* 24: 305–316.

48   Annovazzi, A., D'Alessandria, C., Bonanno, E., Mather, S. J., Cornelissen, B., van de Wiele, C., Dierckx, R. A., Mattei, M., Palmieri, G., Scopinaro F. *et al.* (2006). Synthesis of 99mTc-HYNIC-interleukin-12, a new specific radiopharmaceutical for imaging T lymphocytes. *Eur J Nucl Med Mol Imaging*: 1–9.

49   Cao, Q., Cai, W., Li, Z. B., Chen, K., He, L., Li, H. C., Hui, M., Chen, X. (2007). PET imaging of acute and chronic inflammation in living mice. *Eur J Nucl Med Mol Imaging* 34: 1832–1842.

50   Malviya, G., Conti, F., Chianelli, M., Scopinaro, F., Dierckx, R. A., Signore, A. (2009). Molecular imaging of rheumatoid arthritis by radiolabelled monoclonal antibodies: new imaging strategies to guide molecular therapies. *Eur J Nucl Med*.

51   Malviya, G., D'Alessandria, C., Bonanno, E., Vexler, V., Massari, R., Trotta, C., Scopinaro, F., Dierckx, R., Signore, A. (2009). Radiolabeled humanized anti-CD3 monoclonal antibody visilizumab for imaging human T-lymphocytes. *J Nucl Med* 50: 1683–1691.

52   Kanwar, B., Gao, D. W., Hwang, A. B., Grenert, J. P., Williams, S. P., Franc, B., McCune, J. M. (2008). In vivo imaging of mucosal CD4+ T cells using single photon emission computed tomography in a murine model of coli. *J Immunol Methods* 329: 21–30.

53 Gambhir, S. S., Barrio, J. R., Herschman, H. R., Phelps, M. E. (1999). Assays for noninvasive imaging of reporter gene expression. *Nucl Med Biol* 26: 481–490.

54 Beilhack, A., Schulz, S., Baker, J., Beilhack, G. F., Wieland, C. B., Herman, E. I., Baker, E. M., Cao, Y. A., Contag, C. H., Negrin, R. S. (2005). In vivo analyses of early events in acute graft-versus-host disease reveal sequential infiltration of T-cell subsets. *Blood* 106: 1113–1122.

55 Shu, C. J., Guo, S., Kim, Y. J., Shelly, S. M., Nijagal, A., Ray, P., Gambhir, S. S., Radu, C. G., Witte, O. N. (2005). Visualization of a primary anti-tumor immune response by positron emission tomography. *Proc Natl Acad Sci U S A* 102: 17412–17417.

56 Pfeifer, A., Kessler, T., Yang, M., Baranov, E., Kootstra, N., Cheresh, D., Hoffman, R., Verma, I. (2001). Transduction of liver cells by lentiviral vectors: analysis in living animals by fluorescence imaging. *Mol Ther* 3: 319–322.

57 Lee, G. R., Fields, P. E., Griffin, T. J., Flavell, R. A. (2003). Regulation of the Th2 cytokine locus by a locus control region. *Immunity* 19: 145–153.

58 Reinhardt, R. L., Hong, S., Kang, S. J., Wang, Z. E., Locksley, R. M. (2006). Visualization of IL-12/23p40 in vivo reveals immunostimulatory dendritic cell migrants that promote Th1 differentiation. *J Immunol* 177: 1618–1627.

59 Halin, C., Rodrigo Mora, J., Sumen, C., von Andrian, U. H. (2005). In vivo imaging of lymphocyte trafficking. *Ann Rev Cell Dev Biol* 21: 581–603.

60 Germain, R. N., Bajenoff, M., Castellino, F., Chieppa, M., Egen, J. G., Huang, A. Y., Ishii, M., Koo, L. Y., Qi, H. (2008). Making friends in out-of-the-way places: how cells of the immune system get together and how they conduct their business as revealed by intravital imaging. *Immunol Rev* 221: 163–181.

61 Boissonnas, A., Fetler, L., Zeelenberg, I. S., Hugues, S., Amigorena, S. (2007). In vivo imaging of cytotoxic T cell infiltration and elimination of a solid tumor. *J Exp Med* 204: 345–356.

62 Liang, Q., Gotts, J., Satyamurthy, N., Barrio, J., Phelps, M. E., Gambhir, S. S., Herschman, H. R. (2002). Noninvasive, repetitive, quantitative measurement of gene expression from a bicistronic message by positron emission tomography, following gene transfer with adenovirus. *Mol Ther* 6: 73–82.

63 Liang, Q., Nguyen, K., Satyamurthy, N., Barrio, J. R., Phelps, M. E., Gambhir, S. S., Herschman, H. R. (2002). Monitoring adenoviral DNA delivery, using a mutant herpes simplex virus type 1 thymidine kinase gene as a PET reporter gene. *Gene Ther* 9: 1659–1666.

64 Xiong, Z., Cheng, Z., Zhang, X., Patel, M., Wu, J. C., Gambhir, S. S., Chen, X. (2006). Imaging chemically modified adenovirus for targeting tumors expressing integrin alphavbeta3 in living mice with mutant herpes simplex virus type 1 thymidine kinase PET reporter gene. *J Nucl Med* 47: 130–139.

65 Yang, L., Yang, H., Rideout, K., Cho, T., Joo, K. I., Ziegler, L., Elliot, A., Walls, A., Yu, D., Baltimore, D., *et al.* (2008). Engineered lentivector targeting of dendritic cells for in vivo immunization. *Nat Biotechnol* 26: 326–334.

66 Ziegler, L., Yang, L., Joo, K., Yang, H., Baltimore, D., Wang, P. (2008). Targeting lentiviral vectors to antigen-specific immunoglobulins. *Hum Gene Ther* 19: 861–872.

67 Rettig, G. R., Rice, K. G. (2007). Non-viral gene delivery: from the needle to the nucleus. *Expert Opin Biol Ther* 7: 799–808.

68 Black, M. E., Newcomb, T. G., Wilson, H.-M. P., Loeb, L. A. (1996). Creation of drug-specific herpes simplex virus type 1 thymidine kinase mutants for gene therapy. *Proc Natl Acad Sci U S A* 93: 3525–3529.

69 Gambhir, S. S., Bauer, E., Black, M. E., Liang, Q., Kokoris, M. S., Barrio, J. R., Iyer, M., Namavari, M., Phelps, M. E., Herschman, H. R. (2000). A mutant herpes simplex virus type 1 thymidine kinase reporter gene shows improved sensitivity for imaging reporter gene expression with positron emission tomography. *Proc Natl Acad Sci U S A* 97: 2785–2790.

70 Likar, Y., Dobrenkov, K., Olszewska, M., Shenker, L., Cai, S., Hricak, H., Ponomarev, V. (2009). PET imaging of HSV1-tk mutants with acquired specificity toward pyrimidine- and acycloguanosine-based radiotracers. *Eur J Nucl Med* 36: 1273–1282.

71 Koehne, G., Doubrovin, M., Doubrovina, E., Zanzonico, P., Gallardo, H. F., Ivanova, A., Balatoni, J., Teruya-Feldstein, J., Heller, G., May, C., *et al.* (2003). Serial in vivo imaging of the targeted migration of human HSV-TK-transduced antigen-specific lymphocytes. *Nat Biotechnol* 21: 405–413.

72 Dubey, P., Su, H., Adonai, N., Du, S., Rosato, A., Braun, J., Gambhir, S. S., Witte, O. N. (2003). Quantitative imaging of the T cell antitumor response by positron-emission tomography. *Proc Natl Acad Sci U S A* 100: 1232–1237.

73 Milan, G., Zambon, A., Cavinato, M., Zanovello, P., Rosato, A., Collavo, D. (1999). Dissecting the immune response to moloney murine sarcoma/leukemia virus-induced tumors by means of a DNA vaccination approach. *J Virol* 73: 2280–2287.

74 Su, H., Chang, D. S., Gambhir, S. S., Braun, J. (2006). Monitoring the antitumor response of naive and memory CD8 T cells in RAG1-/- mice by positron-emission tomography. *J Immunol* 176: 4459–4467.

75 Ray, P., De, A., Min, J. J., Tsien, R. Y., Gambhir, S. S. (2004). Imaging tri-fusion multimodality reporter gene expression in living subjects. *Cancer Res* 64: 1323–1330.

76 Su, H., Forbes, A., Gambhir, S. S., Braun, J. (2004). Quantitation of cell number by a positron emission tomography reporter gene strategy. *Mol Imaging Biol* 6: 139–148.

77 Shu, C. J., Radu, C. G., Shelly, S. M., Vo, D. D., Prins, R., Ribas, A., Phelps, M. E., Witte, O. N. (2009). Quantitative PET reporter gene imaging of CD8+ T cells specific for a melanoma-expressed self-antigen. *Int Immunol* 21: 155–165.

78 Overwijk, W. W., Theoret, M. R., Finkelstein, S. E., Surman, D. R., de Jong, L. A., Vyth-Dreese, F. A., Dellemijn, T. A., Antony, P. A., Spiess, P. J., Palmer, D. C., *et al.* (2003). Tumor regression and autoimmunity after reversal of a functionally tolerant state of self-reactive CD8+ T cells. *J Exp Med* 198: 569–580.

79 Ponomarev, V., Doubrovin, M., Shavrin, A., Serganova, I., Beresten, T., Ageyeva, L., Cai, C., Balatoni, J., Alauddin, M., Gelovani, J. (2007). A human-derived reporter gene for noninvasive imaging in humans: mitochondrial thymidine kinase type 2. *J Nucl Med* 48: 819–826.

80 Serganova, I., Doubrovin, M., Vider, J., Ponomarev, V., Soghomonyan, S., Beresten, T., Ageyeva, L., Serganov, A.,

Cai, S., Balatoni, J., *et al.* (2004). Molecular imaging of temporal dynamics and spatial heterogeneity of hypoxia-inducible factor-1 signal transduction activity in tumors in living mice. *Cancer Res* 64: 6101–6108.

81 Alauddin, M. M., Shahinian, A., Park, R., Tohme, M., Fissekis, J. D., Conti, P. S. (2007). In vivo evaluation of 2′-deoxy-2′-[(18)F]fluoro-5-iodo-1-beta-D-arabinofuranosyluracil ([18F]FIAU) and 2′-deoxy-2′-[18F]fluoro-5-ethyl-1-beta-D-arabinofuranosyluracil ([18F]FEAU) as markers for suicide gene expression. *Eur J Nucl Med* 34: 822–829.

82 Tseng, J. C., Zanzonico, P. B., Levin, B., Finn, R., Larson, S. M., Meruelo, D. (2006). Tumor-specific in vivo transfection with HSV-1 thymidine kinase gene using a Sindbis viral vector as a basis for prodrug ganciclovir activation and PET. *J Nucl Med* 47: 1136–1143.

83 Buursma, A. R., Rutgers, V., Hospers, G. A., Mulder, N. H., Vaalburg, W., de Vries, E. F. (2006). 18F-FEAU as a radiotracer for herpes simplex virus thymidine kinase gene expression: in-vitro comparison with other PET tracers. *Nucl Med Commun* 27: 25–30.

84 Yaghoubi, S. S., Jensen, M. C., Satyamurthy, N., Budhiraja, S., Paik, D., Czernin, J., Gambhir, S. S. (2009). Noninvasive detection of therapeutic cytolytic T cells with 18F-FHBG PET in a patient with glioma. *Nat Clin Pract Oncol* 6: 53–58.

85 Bonini, C., Ferrari, G., Verzeletti, S., Servida, P., Zappone, E., Ruggieri, L., Ponzoni, M., Rossini, S., Mavilio, F., Traversari, C., *et al.* (1997). HSV-TK gene transfer into donor lymphocytes for control of allogeneic graft-versus-leukemia. Science (New York, N.Y.) 276: 1719–1724.

86 Berger, C., Flowers, M. E., Warren, E. H., Riddell, S. R. (2006). Analysis of transgene-specific immune responses that limit the in vivo persistence of adoptively transferred HSV-TK-modified donor T cells after allogeneic hematopoietic cell transplantation. *Blood* 107: 2294–2302.

87 Liang, Q., Satyamurthy, N., Barrio, J. R., Toyokuni, T., Phelps, M. P., Gambhir, S. S., Herschman, H. R. (2001). Noninvasive, quantitative imaging in living animals of a mutant dopamine D2 receptor reporter gene in which ligand binding is uncoupled from signal transduction. *Gene Ther* 8: 1490–1498.

88 Rogers, B. E., Parry, J. J., Andrews, R., Cordopatis, P., Nock, B. A., Maina, T. (2005). MicroPET imaging of gene transfer with a somatostatin receptor-based reporter gene and (94m)Tc-Demotate 1. *J Nucl Med* 46: 1889–1897.

89 Che, J., Doubrovin, M., Serganova, I., Ageyeva, L., Zanzonico, P., Blasberg, R. (2005). hNIS-IRES-eGFP dual reporter gene imaging. *Mol Imaging* 4: 128–136.

90 Doubrovin, M. M., Doubrovina, E. S., Zanzonico, P., Sadelain, M., Larson, S. M., O'Reilly, R. J. (2007). In vivo imaging and quantitation of adoptively transferred human antigen-specific T cells transduced to express a human norepinephrine transporter gene. *Cancer Res* 67: 11959–11969.

91 Wei, L. H., Olafsen, T., Radu, C., Hildebrandt, I. J., McCoy, M. R., Phelps, M. E., Meares, C., Wu, A. M., Czernin, J., Weber, W. A. (2008). Engineered antibody fragments with infinite affinity as reporter genes for PET imaging. *J Nucl Med* 49: 1828–1835.

92 Kenanova, V., Barat, B., Olafsen, T., Chatziioannou, A., Herschman, H. R., Braun, J., Wu, A. M. (2009). Recombinant carcinoembryonic antigen as a reporter gene for molecular imaging. *Eur J Nucl Med* 36: 104–114.

93 Phelps, M. E. (2000). PET: the merging of biology and imaging into molecular imaging. *J Nucl Med* 41: 661–681.

94 Mankoff, D. A., Shields, A. F., Krohn, K. A. (2005). PET imaging of cellular proliferation. *Radiol Clin North Am* 43: 153–167.

95 Radu, C. G., Shu, C. J., Shelly, S. M., Phelps, M. E., Witte, O. N. (2007). Positron emission tomography with computed tomography imaging of neuroinflammation in experimental autoimmune encephalomyelitis. *Proc Natl Acad Sci U S A* 104: 1937–1942.

96 Radu, C. G., Shu, C. J., Nair-Gill, E., Shelly, S. M., Barrio, J. R., Satyamurthy, N., Phelps, M. E., Witte, O. N. (2008). Molecular imaging of lymphoid organs and immune activation by positron emission tomography with a new [18F]-labeled 2′-deoxycytidine analog. *Nat Med* 14: 783–788.

97 Gambhir, S. S., Barrio, J. R., Wu, L., Iyer, M., Namavari, M., Satyamurthy, N., Bauer, E., Parrish, C., MacLaren, D. C., Borghei, A. R., *et al.* (1998). Imaging of adenoviral-directed herpes simplex virus type 1 thymidine kinase reporter gene expression in mice with radiolabeled ganciclovir. *J Nucl Med* 39: 2003–2011.

98 Tjuvajev, J. G., Avril, N., Oku, T., Sasajima, T., Miyagawa, T., Joshi, R., Safer, M., Beattie, B., DiResta, G., Daghighian, F., *et al.* (1998). Imaging herpes virus thymidine kinase gene transfer and expression by positron emission tomography. *Cancer Res* 58: 4333–4341.

12

Imaging of Reporter Genes and Stem Cells

Martin G. Rodriguez-Porcel and Sanjiv Sam Gambhir

BACKGROUND: WHY DO WE NEED TO IMAGE STEM CELLS?

Over the last few years, cell-based regenerative medicine has appeared as a therapeutic option for many different disease states, such as musculoskeletal [1, 2], endocrine [3, 4], neurodegenerative [5], coronary artery disease [6–9], autoimmune diseases [10–13], and malignancies [14] and has been a focus of significant attention of the scientific community. The main objective of cell-based therapies is to repopulate the damaged tissue with functional cells, with the final goal that these cells will integrate with the remaining functional native cells and contribute to the recuperation of lost function. Use of stem cells has been shown to regenerate different organs and systems, such as endocrine [3, 15] (e.g., pancreas), joints [16–18] (i.e., cartilage), musculoskeletal [19] (i.e., bone), and cardiovascular system [20–22] (i.e., myocardium), and has also been used as adjuvant treatment for malignancies [23]. A wide range of cell types have been used, including mesenchymal stem cells [16, 19, 24–26], embryonic stem cells [27–29], bone marrow-derived hematopoietic stem cells (HSCs) [30, 31], bone marrow-derived endothelial progenitor cells (EPCs) [32–34], and neural stem cells [35, 36]. A significant body of knowledge has been acquired in regard to the biology of stem cells and the potential benefit of their use in tissue and organ regeneration. However, several questions remain regarding the biology of stem cells in living subjects and how they integrate with native tissue. Thus, the need to further study the biology and behavior of stem cells after transplantation in the living subject (both in the preclinical and clinical setting) becomes of paramount importance and is critical for the advancement of the field of cell-based regenerative medicine.

Until recently, researchers have been limited in the capacity to study cell biology and cell survival, both in cell culture and in living subjects, in part due to limitations in imaging technology. Most studies that investigated the biology of stem cells relied on traditional *ex vivo* assays and molecular techniques (e.g., histology and Western blotting), with their limitations of being invasive and limited in the number of time points that can be studied in any given subject. In addition, the application of these techniques for clinical use is limited to biopsy specimens. Thus, there is a pressing need to develop novel modalities to study stem cell biology noninvasively. A noninvasive approach will permit a longitudinal study (in the same subject) of the biology of stem cells while diminishing the interference of biological variables (the same subject can be used for multiple studies) [37, 38]. In addition, a noninvasive approach has the potential to minimally disturb the microenvironment, allowing a more physiologic study. Last, such strategies allow investigators to translate these studies to patients; something not possible until recently. In previous chapters of this book we have learned of the different modalities used to image gene expression and proteins. Some developments in noninvasive imaging have been adapted to image stem cells in living subjects. The following section describes some of the variables needed to successfully perform stem cell imaging.

CELL IMAGING WISH LIST

Focus will be placed on the different characteristics that should be sought in an ideal imaging strategy for cell imaging (Figure 12.1). Stem cells can be delivered through many routes of administration (e.g., direct injection into tissues of interest, intravenous and intracoronary). However, regardless of the route used, the first step is to ensure that cells get to where one wants them to be. Thus, it is important to image stem cells at the time of transplantation. Whether stem cells are delivered directly to the area of interest (e.g., myocardium, liver, pancreas and muscle) or systemically (for some diseases stem cells may be delivered intravenously) it is critical to be able to

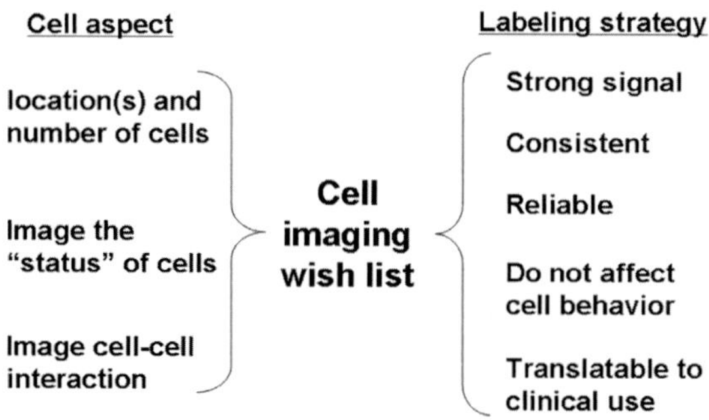

Figure 12.1. Cell imaging wish list. This list comprises the most important aspects an ideal strategy for stem cell imaging should have. They are separated in the aspects related to the cell information they should provide and those relevant to the labeling strategy used.

image the trafficking of cells throughout the body and see where they reside. Stem cell imaging at time of transplantation will provide information regarding the success or failure of cell delivery and will provide a basis from where one can begin the assessment of the effect of stem cell therapy. To successfully image cell transplantation/ delivery one needs a modality with sufficient sensitivity to detect the number of cells being delivered, and at the same time it should have enough spatial resolution to allow one to spatially localize where the cells have been delivered. This last point becomes even more important in diseases in which one wants the stem cells to be delivered to a specific part of an organ (e.g., cardiovascular and neurologic applications).

Subsequently, and after confirmation of the successful delivery of stem cells to the organ/tissue of interest it becomes critical to monitor the viability of stem cells over time. Being able to image stem cells repetitively will provide critical information regarding the survival of these cells whether they remain in the area where they were delivered or whether they have trafficked to other areas. To address the points mentioned, we need to use a labeling methodology, the emitted signal of which (output signal) accurately and linearly correlates with the number of viable stem cells.

Noninvasive imaging will also help study the biology of stem cells after transplantation. As opposed to what occurs in cell culture, when cells are transplanted into a living subject they are in contact with a new (and in general noxious) microenvironment and they have to adapt to it, which may be critical to stem cell survival and modulation. In addition, transplanted cells will come in contact with native cells, which may result in changes in stem cell functionality. Thus, it is critical that one is able to, in addition to monitor cell viability, monitor and assess the functionality of the transplanted cells. Furthermore, in the case of transplanted stem cells it is important to understand whether cells after transplantation

continue behaving like stem cells or differentiate into lineage-committed mature cells.

A similarly important variable to be contemplated is that the chosen labeling modality should not interact with the normal functions of the stem cell. Cell labeling should not impair *per se* the functionality and survival of the cells being transplanted. Otherwise, it will not be possible to accurately study the biology of these cells over time. Thus, it is imperative to test that the labeling approach does not impair cell biology.

Similar to cell-based regenerative medicine, one of the main objectives of noninvasive cell imaging is its clinical application. It is generally agreed upon that before its clinical use, therapeutic strategies should be tested in clinical models of disease. Only then one will accurately know the effect these therapies may or may not have on the diseases being treated. Imaging of these strategies should follow the same preclinical steps on their way to clinical applications. Use of these strategies in large-animal models is critical to accurately understand their biological behavior as it gets ready for use in the clinics. In addition, they will provide invaluable information for the therapeutics strategies under study. With that goal in mind, imaging in preclinical models is a necessary and critical step toward clinical application, and provides invaluable information that will ultimately be applied to clinical imaging. Furthermore, preclinical imaging is fundamental for the study of many aspects of the biology of transplanted stem cells and how they interact *in vivo*. Thus, an ideal imaging modality should be flexible across different imaging modalities, both in terms of spatial resolution and system sensitivity (the lowest amount of activity – amount of cells – that can be detected by that specific modality). In other words, it should allow us to the transfer of knowledge obtained preclinically (cell culture, small animals) and apply it (albeit with minimal modifications) to large-animal models and patients.

For stem cell imaging, an ideal imaging modality should provide integrated information related to the disease process (including functional assessment), status of the therapy delivered (stem cells in this case), and effect of this therapy on the organ/system function. Because the final goal is the restitution of structure and most importantly recovery of function of the organ of interest, it is of great interest that the imaging modality used provide an assessment of the functionality not only of the transplanted cells but also of the organ where the cells are transplanted to.

METHODOLOGIES FOR LABELING STEM CELLS

Direct Labeling

Among the most commonly used strategies for labeling stem cells for imaging in living subjects is that of

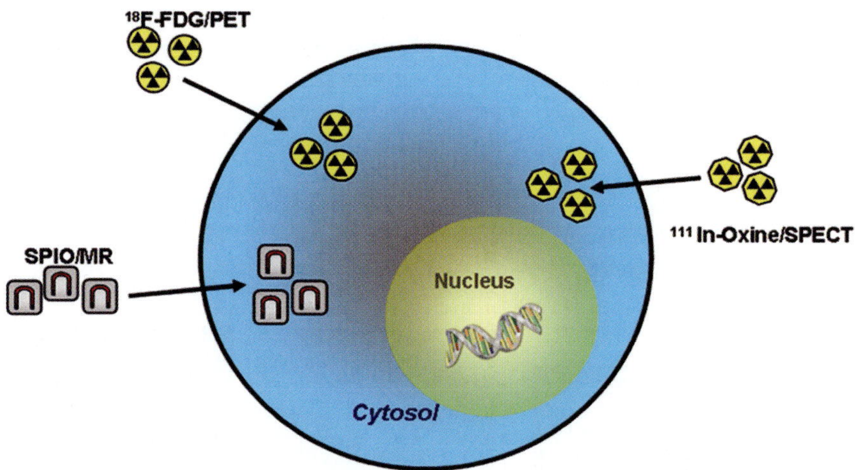

Figure 12.2. Direct cell labeling approach. In this strategy, labeling agents (for either MR, fluorescence, or nuclear imaging) are introduced *ex vivo*, stem cells are then transplanted to the organ of interest, and then noninvasive imaging is performed. SPIO superparamagnetic iron oxide particles, IFP iron fluorescent particles, FDG fluorodeoxyglucose, HMPAO exametazime, PET positron emission tomography, SPECT single-photon emission computed tomography, MRI magnetic resonance imaging.

direct labeling [39–43]. In a direct labeling strategy, labeling agents are introduced into the cells prior to transplantation, stem cells are transplanted and then followed in the living subject (Figure 12.2). Depending on the imaging modality used, cells can be labeled with quantum dots [44, 45] or fluorophores [46] (for optical fluorescence imaging), superparamagnetic iron oxide particles (SPIO, for magnetic resonance imaging (MRI)) [47, 48], radioisotopes (for single-photon emission computed tomography (SPECT), or positron emission tomography (PET)) [49–52].

Fluorescent Labeling

Fluorescent semiconductor nanocrystals (known as quantum dots, QDs [53]) or different fluorophores [54] can be used for imaging transplanted cells (Figure 12.3). They are incorporated into the cells *ex vivo*, and then the cells are transplanted. At each imaging time point, fluorophores are excited at predetermined wavelengths and emit a fluorescent signal. QDs have different composition and sizes that determine the excitation and emission of each QD. Thus, one can use QDs of different sizes to detect different types of stem cells, and in fact this approach has been used for different applications [55–58]. Similarly, different fluorophores (e.g., Cy5.5, Cy 7) can be used to image different cell types. However, fluorescence imaging has limited tissue penetration (around 2 mm) limiting the use of these techniques to superficial tissues in small animals (e.g., mice). Furthermore, fluorescence imaging devices commonly used do not have tomographic capabilities, which further limits the identification of nonsuperficial organs. Several efforts are under way to provide tomographic fluorescence

imaging [59–61]; however, its final applicability will only be known in the coming years.

MR Labeling

Another direct labeling approach uses MRI as the imaging modality (Figure 12.3). As mentioned in prior chapters of this book, SPIO are highly magnetic particles that can elicit changes in T_2 relaxivity (effect known as $T_2{}^*$), allowing their detection *in vivo* [62–64]. In a direct labeling strategy SPIOs are incorporated into the cells prior to transplantation to the living subject, and subsequently, MR imaging is performed. Because stem cells have been prelabeled with SPIOs, the signal originating from the SPIOs is used as a surrogate for number of cells. MRI offers the advantage of high spatial resolution, resulting in detailed organ morphologic and functional information, and thus appears to be a good candidate for an integrated stem cell imaging–functional assessment approach. This strategy has been used to label stem cells for the visualization of transplanted cells to organs like the myocardium [65, 66] or brain [42, 67]. However, the sensitivity of SPIO-MR imaging is in the micromolar range (10^{-9} mol/L) [68] and may not be sensitive enough to detect low signal levels (transplanted cells) over time. Sensitivity of the system can be increased using high-field magnets (11 T); however, for now, the use of these magnets is limited to small-animal preclinical models. Many efforts are under way to increase the sensitivity of this strategy, such as off-resonance MR imaging [69], which shows promise to be useful when low number of cells are delivered. SPIO-based imaging continues to be a good imaging strategy for initial localization of cells after transplantation and for the coregistration of cell

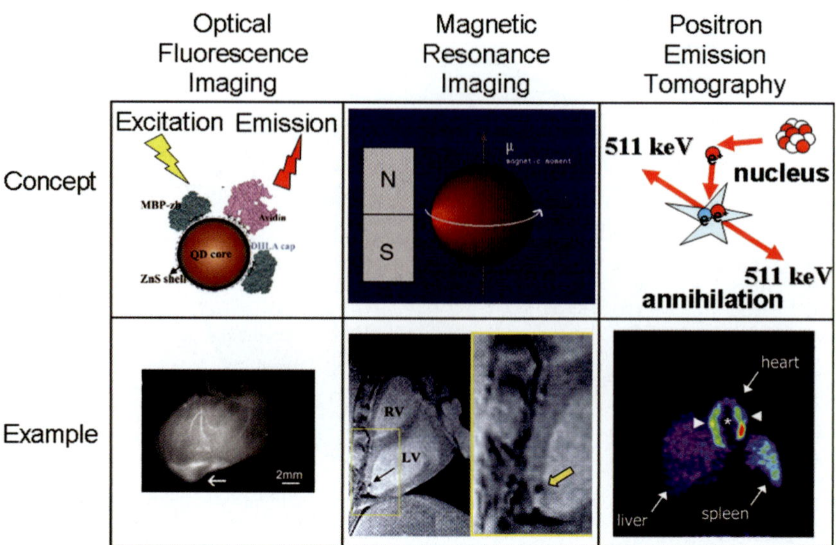

Figure 12.3. Direct cell labeling using different imaging modalities. Top panel illustrates the concept behind different imaging modalities, such as fluorescence (**left**), magnetic resonance imaging (MRI, **middle**), and nuclear medicine imaging (positron emission tomography (PET), **right**). Bottom panel presents examples of cell imaging using a direct labeling approach using the previously mentioned modalities. In the bottom left, cells have been labeled with quantum dots, transplanted, and then imaged *in vivo* (white arrow). In the center, cells have been labeled with paramagnetic particles delivered to the myocardium and then imaged using MRI. The black signal (yellow arrows) represents the paramagnetic signal, what has been used as surrogate of the amount of presence of stem cells. In the left panel, cells have been labeled with ^{18}F-Fluorodeoxyglucose (^{18}F-FDG), transplanted to the myocardium, and then imaged using PET. The white arrows point to the transplanted cells in the heart, as well as the liver and spleen uptake (route of tracer elimination). Adapted from Gao X et al. *Curr Opin Biotech* 16(1) 63–72, Kraitchman DL et al. *Circulation*. 2003 May 13; 107(18): 2290–3, and Hofmann et al. *Circulation* 2005 111: 2198–2202 with permission.

transplantation with areas of functional loss. In the case of the myocardium, it has been previously shown that MR imaging can be used to identify areas of infarction, and the same imaging modality (MR in this case) can be used to label stem cells for their localization immediately after delivery [66]. However, SPIO-based imaging appears to not be well suited for long-term stem cell monitoring [68]. Several reports suggested that SPIOs may not stay in the transplanted cells over time but rather be incorporated by macrophages and other cell types, like macrophages. In fact, SPIO-MR imaging has been used as a "macrophage detector" in many pathophysiological states [70, 71]. And because the effect of iron on the magnetic field continues (regardless of the location and status of the transplanted cells), there is an uncoupling between the MR signal and the viability of stem cells [72]. This effect may not be critical for the initial localization of cells but will likely preclude its use in the long-term monitoring of transplanted stem cells.

Radionuclide Labeling

Radioisotope direct labeling of cells has also been used for cell imaging using a strategy similar to SPIO-based techniques, which is to introduce a labeling agent to the cell prior to transplantation (Figure 12.2). Radioisotopes used for this purpose have different physical half-lives (e.g., [99m]Tc: 6 hours, [18]F: 109 minutes, [111]In:

2.8 days), signifying that cells can only be followed for a specified amount of time after cell labeling. For example, ^{111}In-labeled cells have been used for many years to clinically track the homing of inflammatory cells to localize inflammatory processes [40, 73]. More recently, this methodology has been applied to the labeling of stem cells, using different isotopes (e.g., ^{111}In for SPECT, ^{18}F-Fluoro-Deoxyglucose (^{18}F-FDG) for PET)[52, 74–76]. Isotopes like ^{18}F-FDG (physical half-life = 109 minutes) may allow tracking of cells for a few hours (after correction for isotope physical decay) after transplantation [49, 77]. If an isotope with a longer half-life is used (e.g., ^{111}In), cells may be followed for slightly longer time (up to a few days). Applying this concept, our laboratory has used an isotope with relatively long physical half-life (^{64}Cu) as the labeling agent for the monitoring of C6 rat glioma cells using PET as the imaging modality [51]. These strategies may be useful for the initial localization of transplanted cells (Figure 12.3) but will not be sufficient to monitor stem cells for extended periods of time. In addition to its physical half-life, each radioisotope also has a biological half-life (e.g., they may go in and out of the cell), which should be taken into consideration when performing these studies. Thus, when using this imaging strategy, physical as well biological properties of the labeling agent should be considered together with the properties of the cells to accurately determine the appropriate cell imaging modality. A major advantage of SPECT and

PET imaging is their high sensitivity (nano- and femto-molar, respectively), which permits the detection of relatively low amounts of signal [78–80]. However, SPECT and PET have relatively lower spatial resolution compared to other modalities (such as MRI), which may be a relative disadvantage for signal localization. The recent development of integrated PET-CT and SPECT-CT has overcome much of the limitations in spatial resolution previously seen with stand-alone PET or SPECT, allowing nuclear medicine modalities to also be used for the initial localization of transplanted cells after delivery.

Drawbacks and Unresolved Issues of Direct Labeling

The goal of imaging is to assess cells noninvasively without disturbing the biology of the transplanted cells (or their microenvironment). To our knowledge, quantum dots (QDs) have not been associated with alterations in cell biology but may have the potential to do so. On the other hand, SPIOs may have an effect on the gross morphology and doubling capacity of stem cells [81], which may preclude their widespread use as a labeling strategy. In the case of radioisotope labeling, cell toxicity will likely vary depending on the radioisotope and doses used. In summary, regardless of the labeling strategy used, it is recommended to assay for cell toxicity and determining cell dosimetry to make sure the strategy used does not cause toxicity neither to the transplanted cell nor to the recipient.

In addition, a direct labeling strategy does not account for cell division (i.e., cell numbers increase after cell division, but the number of radioisotope molecules stays the same), which results in "dilution" of the signal over time. Because the extent of this variable cannot be accounted for, it may represent a limitation because it does not allow accurate monitoring of cells over longer periods of time.

In summary, direct labeling strategies appear to be a good imaging strategy for detection of cells shortly after transplantation (e.g., to ensure that cells were delivered to the intended organ or region of an organ), providing a good signal-to-noise ratio. On the other hand, they may not be well suited for long-term monitoring of cell viability. Furthermore, in certain situations this approach can lead to false positive signal (i.e., MR), providing confounding information on cell viability.

Indirect Imaging: Reporter Gene Imaging

To overcome limitations of the strategies mentioned in prior sections, there was a need for a labeling strategy that would allow the short- and long-term monitoring of cells after transplantation to the living subject. In addition, to detect changes in cell survival and how they interact with the microenvironment, cells will need to be imaged repetitively and longitudinally. Over the

Table 12.1. Reporter genes

Fluorescence
Green Fluorescent Protein
Red Fluorescent Protein
Bioluminescence
Firefly Luciferase
Renilla Luciferase
Gaussia Luciferase
PET/SPECT
HSV Type 1 Thymidine Kinase
Dopamine Type 2 Receptor (D2R)
Sodium-Iodide Symporter (NIS)
Somatostatin Type 2 Receptor
MR
Transferrin/Ferritin Receptor

last decade, advances in molecular biology and imaging modalities have allowed repetitive and longitudinal imaging of transgene expression in many disease states (Figure 12.4) [37, 82]. In prior chapters of this book, the application of reporter gene technology for assessing several biological phenomena (e.g., transgene expression, protein–protein interaction, cellular protein folding) has been discussed. Recently, this technology has also been adapted for stem cell imaging.

Typically, reporter gene constructs consist of gene regulatory elements (promoters and enhancers), the reporter gene DNA sequence, and a polyA sequence. The reporter gene is first incorporated into the cell *ex vivo*, and then the stem cell is transplanted to the living subject. If the stem cells are viable, the reporter gene will be expressed and the protein (e.g., enzyme and cell surface receptor) will be encoded. At the specified imaging time point, an exogenously given substrate or light excitation (in the case of fluorescent reporter genes) is provided. The interaction between the substrate and the encoded protein will result in a signal that can be detected noninvasively. Depending on the promoter used (e.g., constitutive vs. protein-regulated expression) one will be able to obtain information on the viability of the cell (if we use a constitutive promoter) or the expression of certain protein (if one is interested in a specific activity of the stem cell, a protein-specific promoter can be used – this approach will be covered later in this chapter). On the other hand, if the reporter gene is not expressed (due to cell death or lack of expression of a protein in certain types of stem cells), there will be no production of the reporter protein and no interaction with the administered substrate; thus no signal will be emitted.

There are several types of reporter genes available for stem cell imaging (Table 12.1 lists the reporter genes most commonly used for *in vivo* imaging). For the most part,

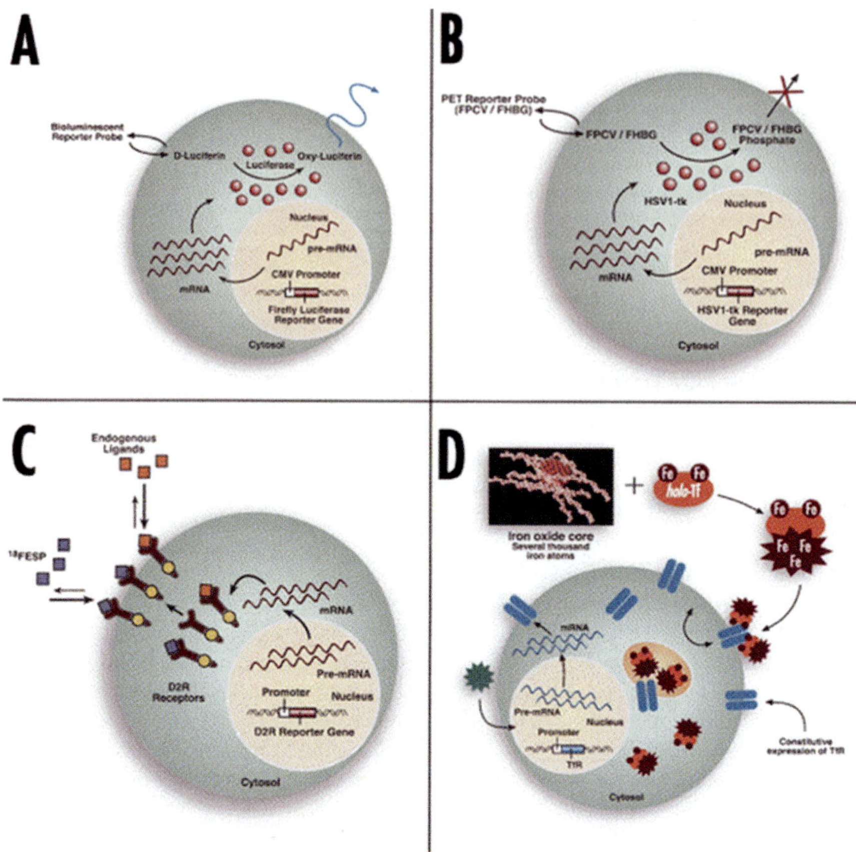

Figure 12.4. Four different strategies of imaging reporter gene/reporter probe. (**A**) Enzyme-based bioluminescence imaging. Expression of the firefly luciferase (fluc) reporter gene leads to the firefly luciferase reporter enzyme (FL), which catalyzes the reporter probe (D-luciferin) that results in a photochemical reaction. This yields low levels of photons that can be detected, collected, and quantified by a CCD camera. (**B**) Enzyme-based PET imaging. Expression of the herpes simplex virus type 1 thymidine kinase (HSV1-tk) reporter gene leads to the thymidine kinase reporter enzyme (HSV1-TK), which phosphorylates and traps the reporter probe (F-18 FHBG) intracellularly. Radioactive decay of F-18 isotopes can be detected via PET. (**C**) Receptor-based PET imaging. F-18 FESP is a reporter probe that interacts with D2R to result in probe trapping on or in cells expressing the D2R gene. (**D**) Receptor-based MRI imaging. Overexpression of engineered transferrin receptors (*TfRs*) results in increased cell uptake of the transferrin–monocrystalline iron oxide nanoparticles (Tf-MIONs). These changes result in a detectable contrast change on MRI. *CMV*, Cytomegalovirus. From Wu JC et al. *J Nuc Cardiol* 2004 Jul-Aug 11(4): 491–505 with permission.

reporter genes have been used to assess stem cell viability in a longitudinal manner. In addition, reporter genes can be used for the initial localization of stem cells after transplantation, albeit not with the spatial resolution as with SPIO–MR or the cell detection sensitivity of direct labeling with radionuclides. Different reporter genes also have unique qualities and applications for stem cell imaging. For the purposes of this chapter, we will classify reporter genes according to the imaging modality used to image their expression.

Fluorescent Reporter Genes

In previous chapters of this book the details of fluorescence reporter genes have been discussed. As a group, fluorescence reporter genes are sensitive and result in the emission of a strong signal (Figure 12.5). A main advantage of fluorescent reporter genes is that different fluorophores (e.g., red fluorescent protein and green fluorescent protein) can be used to label different cell populations permitting concomitant imaging of different cell populations [83]. However, there are some drawbacks to the use of fluorescent reporter genes. One main drawback of fluorescent reporter genes (when imaging in the living subject) is that emitted fluorescent light undergoes significant tissue attenuation (due to absorption) as well as tissue refraction. Due to this issue, the use of fluorescent reporter gene imaging is restricted to the imaging of superficial tissues (up to 2 mm) [37], such as subcutaneous tissue. In addition, and similar to what occurs to other imaging modalities, the detected signal constitutes only a small fraction of the emitted signal. Significant

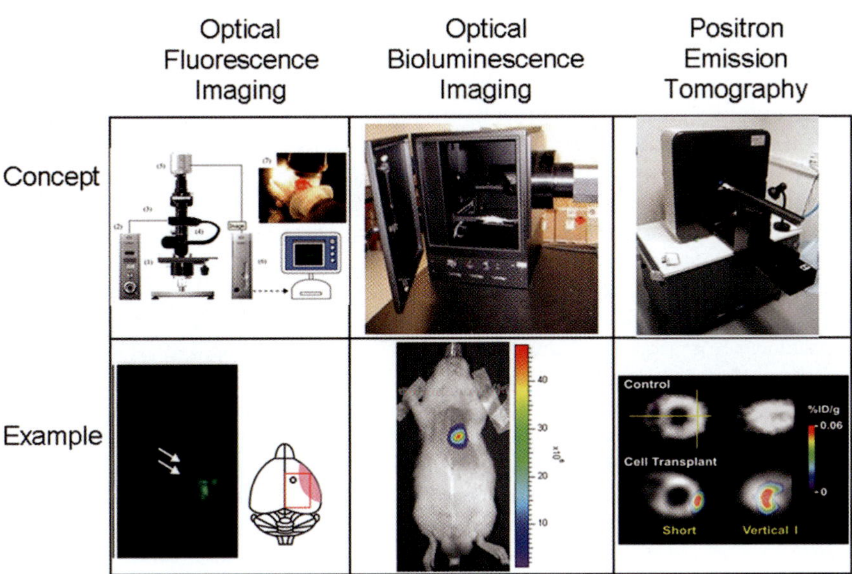

Figure 12.5. Examples of reporter gene imaging. Top panel shows the different imaging modalities used in reporter gene imaging, such as fluorescence imaging (**left**), BLI (**middle**), and nuclear medicine imaging (PET, **right**). In the fluorescent system, the animal is placed under the microscope and exposed skull is illuminated. In the bioluminescent and PET, animals are placed in the CCD camera system or scanner, respectively, and the substrate administered. Bottom panel presents examples of cell imaging using a reporter gene approach using the previously mentioned modalities. On the left panel, neural cells have been labeled with green fluorescence protein, transplanted to the brain and imaged in the living subject (white arrow). A diagram of the location of the scanning area is presented for orientation. On the center panel, cells that carry the optical reporter gene firefly luciferase were delivered to the myocardium and imaged using a CCD camera after administration of the substrate D-luciferin. Color images of visible light are superimposed on photographic images of mice with a scale in photons per second per square centimeter per steradian (sr). On the right panel, cells carrying the PET reporter gene HSV1-tk have been transplanted to the myocardium. The figure depicts tomographic images of the myocardium after the administration of the PET reporter probe ^{18}F-labeled 9-[4-fluoro-3-(hydroxymethyl)butyl]guanine. The color scale (%ID/g) indicates the percentage of injected dose that accumulates per gram of tumor. Red arrows indicate the area where cells are located. Adapted from Shichinohe H et al. *Brain Res Protoc* Aug 2004 13(3): 166–175 and Wu JC et al. *Circulation* 2003 018: 1302–1305 with permission.

efforts are under way to use multiple cameras, which in principle has the advantage of detecting a higher percentage of the emitted signal. Another issue to be considered in fluorescence imaging is that it is mainly a planar imaging technique, with no tomographic capabilities. Novel developments, such as time-domain imaging [84–86], incorporate the time domain in the analysis and have the potential to provide in-depth information of the fluorescent signal. However, such strategies are under development and are not ready to be routinely applied by mainstream imagers. In addition, fluorescence techniques have "autofluorescence" (an effect that can also be referred to as "fluorescent background"), which will lead to the need of stronger signal to obtain a good signal-to-noise ratio. The presence of autofluorescence could be a significant disadvantage, especially when the emitted signal is not high, leading to a low signal-to-noise ratio.

Perhaps the most used application of fluorescent reporter genes in cell imaging is for *ex vivo* analysis, where they can be used for cell sorting or histology (immunofluorescence). In Chapter 5 of this book, Ray et al. discussed bi- or trifusion reporter genes that can be easily used to identify and isolate groups of cells (using fluorescent activated cell sorting techniques) [87, 88]. For example, if stem cells are transduced with a bioluminescent or PET reporter gene, not all transduced stem cells will incorporate the gene of interest into their genome. Thus, a common way to sort the cells that actually have the gene of interest (for BLI or PET) is to use a fluorescent reporter gene cell marker for selection (e.g., Gfp and Rfp) [89]. Another routine use of fluorescent reporter genes is constituted by histology. Using a similar approach to the one previously described (bi- or trifusion reporter genes) one can "mark" cells with a fluorescent reporter genes, and then when the tissue is excised, the remaining cells can be easily identified using histological methods [89]. Similar to what happens *in vivo*, different reporter genes emit light in different spectra, permitting the detection of different fluorophores in any given sample. These applications play an important role in the study of stem cell biology at the cell culture and *ex vivo* level. Furthermore, due to their light emission and detection characteristics, they are commonly used as a starting strategy when designing novel reporter gene constructs.

Bioluminescent Reporter Genes

The basis of reporter gene-bioluminescence imaging (BLI) and the different types of BLI-reporter genes have already been covered in previous chapters of this book. In this section the application of BLI for the imaging of stem cells will be discussed. As mentioned in this and other chapters of this book, reporter genes (after introduction to the cell of interest) will encode for a specific protein (e.g., luciferase for optical imaging). However, they will only encode for that protein if they have the necessary cellular machinery (e.g., ribosomes, polymerase, and cofactors) to do so. If the cell is death, or apoptotic, transcription and translation will not take place, and thus no reporter protein will be produced. Furthermore, in some cases, the interaction between substrate (D-luciferin) and reporter protein (firefly luciferase) needs cofactors like oxygen, magnesium, and ATP (only available in viable cells). In other words, light emission will take place only if the cell is viable and healthy enough to carry out the previously mentioned processes. Thus, because of these characteristics, one can use these imaging strategies to longitudinally monitor stem cell survival and study cell status (Figure 12.5). Because BLI is also based on light emission and detection, absorption of emitted light by tissue remains a factor that limits the use of BLI to small animals and relatively superficial tissues and/or organs [37, 82].

For these reasons reporter gene-BLI strategies are commonly used for the tracking of stem cells delivered to organs located at different depths within the living subject [90–92] and are not restricted to the subcutaneous layer (as in fluorescence optical imaging). firefly luciferase (fluc) and Renilla luciferase (Rluc) are the two most common reporter genes used for imaging. In Chapter 2 of this book, Contag and Doyle cover the different characteristics of each luciferase: basis of light production, cofactors needed for light production (oxygen, magnesium, ATP, etc.). Firefly luciferase activity results in a more red-shifted light emission (wavelength: 500–700 nm) compared to Renilla luciferase (wavelength: 450–550 nm), which results in higher signal-to-background ratio. Hence it is usually a more "attractive" reporter gene for imaging cells in living subjects. In addition, Fluc's substrate (D-luciferin) is delivered intraperitoneally, compared to intravenous delivery of the Renilla luciferase substrate (coelenterazine), making the firefly luciferase easier to use. Our laboratory has developed variations (mutants) of preexisting Renilla luciferases [93] that result in increased light emission and a more red-shifted spectrum, which improves their sensitivity and detection capabilities. The use of two BLI reporter genes for some applications may be useful (e.g., one reporter to measure viability and another to measure differentiation status – this approach will be discussed later in this chapter). Because of its ease of use and relatively low expense compared to other modalities, BLI has been extensively used for high throughput small-animal studies and is one of the leading imaging modalities used for drug development and assessment of novel therapeutic strategies (at the small-animal level).

Using a reporter gene strategy, BLI has been successfully used for the *in vivo* study of cell delivery and monitoring of stem cell viability in small animals. BLI has been used for tracking and monitoring different types of stem cells after transplantation, such as cardiomyoblasts [90–92], HSCs [94], and embryonic stem cells [89, 95]. BLI has also been used to evaluate the response of stem cells transplanted to organs like the pancreas [96, 97].

As with other imaging modalities, BLI has drawbacks and unsolved issues. As it stands right now, it is a planar imaging modality, not able to provide depth information and signal localization within the living subject. Also, because of the lack of tomographic information, although BLI can provide information on the trafficking of stem cells to different organs, it may not be an ideal modality to accurately assess stem cell viability, as changes in signal depth (therefore higher light absorption by the tissues) can be confused with less cell survival. Significant efforts are being devoted to the development of tomographic BLI, either by rotating the subject under study or detecting light using two or more cameras, or by spectral imaging [98, 99]. However, those systems are under development and not available for high-throughput use of BLI for cell imaging in living subjects. As mentioned before, and due to the limited tissue depth studied, BLI is for the most part restricted to use in small animals (rats, mice) or very superficial tissues in larger living subjects (even patients). However, because it can be adapted to clinically used imaging modalities (e.g., PET and SPECT), BLI is routinely used as a starting step in the development of novel imaging strategies, and once the efficacy of these reporter gene strategies is proven they can be adapted for its use in the clinic.

In summary, BLI is a useful imaging modality for the monitoring of stem cell trafficking and survival in small animals and for the study of cell biology and its interaction with the microenvironment. Furthermore, it will play an important role in the translation of many of these novel molecular imaging strategies to the clinics.

Positron Emission Tomography/Single Photon Emission Computed Tomography Reporter Genes

The principles of PET/SPECT reporter gene-based cell imaging are basically similar to those previously described for BLI. A main advantage of the PET/SPECT imaging is that it can provide tomographic and volumetric information, allowing us to better localize and quantify the detected signal within the subject under study (Figure 12.5). In addition, the sensitivity of PET is in the femtomolar range (10^{-12} mol/L), higher than SPIO-MR (10^{-9} mol/L), but not as sensitive as optical imaging (10^{-15} mol/L). As previously mentioned, there are mainly three reporter gene systems (for PET or SPECT) used for cell imaging. The mostly used system is based on the production of a herpes simplex virus type 1 thymidine kinase enzyme (HSV1-TK) that phosphorylates an exogenously given substrate (e.g., ^{18}F-9-[4-fluoro-3-(hydroxymethyl)butyl]guanine or ^{18}F-FHBG, Figure 12.4 panel B) and retains it inside the cell. This strategy is powerful because the enzyme can phosphorylate many molecules of the radioactive substrate, increasing the signal retained in the tissue of interest and improving the signal-to-background ratio. PET imaging can also be performed using a receptor-based approach: the most used is a mutant version of the dopamine receptor (D_2R) [82] (Figure 12.4 panel C). Using a cell membrane receptor-based strategy has the advantage that the substrate does not have to cross the cell membrane to interact with the reporter protein, eliminating the biological variable of the cell membrane crossing (needed in the intracellular enzyme approach described earlier in this section), but at the same level of reporter protein expression this approach may be limited by the amount of signal it can produce, as one receptor conceivably interacts with only one molecule of the substrate. In addition, because it is an endogenous naturally occurring protein, this approach should not lead to an immune reaction by the host. Our laboratory and others have extensively used reporter gene-PET for the visualization of transplanted cells to the myocardium and showed that a PET-based reporter gene strategy can be used to monitor the fate of stem cells after transplantation to the living subject [89, 91]. PET imaging has also been used to monitor stem cells in other organs, like the pancreas [100, 101]. Another reporter gene system used for both PET and SPECT is that of the sodium-iodine symporter (NIS) [102–104]. The thyroid gland uses the sodium-iodine symporter system to transport iodine into the cells in exchange for sodium. Successful sequencing and cloning of the NIS gene sequence allowed its use as a reporter gene in the thyroid and other organs. This system has been used not only for PET (with ^{124}I as the substrate) but also for imaging with SPECT (using ^{123}I or ^{99}Tc-perthechnetate as substrates). Furthermore, the NIS transgene has also been used as a therapeutic gene with ^{131}I.

The most common use of reporter gene imaging for cell monitoring consists of transplanting cells (that carry the reporter gene) to the living subject and then administering a substrate (e.g., intravenously and intraperitoneally) that will interact with the reporter gene. However, immediately after transplantation, cells may not interact well with the administered substrate (lack of an appropriate blood supply may be one of the factors), which may limit the accuracy of this strategy in short-term engraftment of transplanted cells. To overcome this limitation, cells can be exposed to the substrate

(tracer) *ex vivo*, and then transplanted into the living subject, where noninvasive monitoring is performed. This approach has the benefit of a better interaction between cell and substrate, providing better and larger trapping of the probe and a more accurate monitoring of cell homing in the tissue of interest. Long-term monitoring of transplanted stem cells would then be performed in the usual way (after intravenous administration of substrate, as by that time the interaction between transplanted cells and vascular network would have improved and stabilized). Caution would need to be taken when comparing short- and long-term monitoring, as the administration of substrate in the two strategies (short- and long-term) would be different.

The major advantage of PET and SPECT is their potential for clinical use. Furthermore, several PET and SPECT probes are routinely used radionuclides in the clinics, and several lessons have been learned from that. So when these imaging strategies are ready to be tested in the clinics, many of the lessons from previously tested compounds will undoubtedly come in handy.

These imaging modalities have issues that need to be considered from the operational standpoint. On one side, PET has significant flexibility for the production of specific probes for the detection of different processes in the living subject, which is a significant advantage because it allows the researcher to first identify the molecule that he needs to image and then design a specific probe that will target that molecule. However, the production of reporter probes is complex, needing advanced chemistry and tight quality control. In addition, depending on the half-life of the radioisotope used, it requires an on site (or at least nearby) cyclotron, which limits this strategy to medium to large research centers. From an imaging standpoint, in PET imaging all electron–positron annihilations result in the production of photons of 511 KeV, and as such we cannot detect differences in registered signals. SPECT, on the other hand, can detect simultaneous signals (by varying the detection window/s, as it is routinely done with ^{201}Tl and ^{99}Tc), and at the same time tracer labeling is less complex (compared to PET) and, for the most part, it can be done in a nuclear pharmacy. However, the spatial resolution of SPECT is less than that of PET, and this variable may be of importance when we try to spatially localize a low number of cells. From the tracer perspective, SPECT has less flexibility on the tracer production compared to PET. What this means is that although fewer options are available for SPECT, if one is identified and answers the research questions posed by the investigator, its production and availability is more accessible to many academic and research centers.

Drawbacks and Unresolved Issues of Reporter Gene Imaging

The goal of imaging is to assess the status of cells non-invasively, without disturbing the biology of the

MR

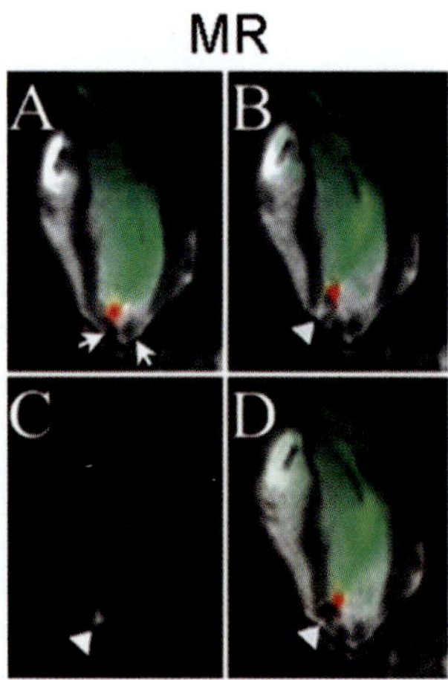

Ultrasound

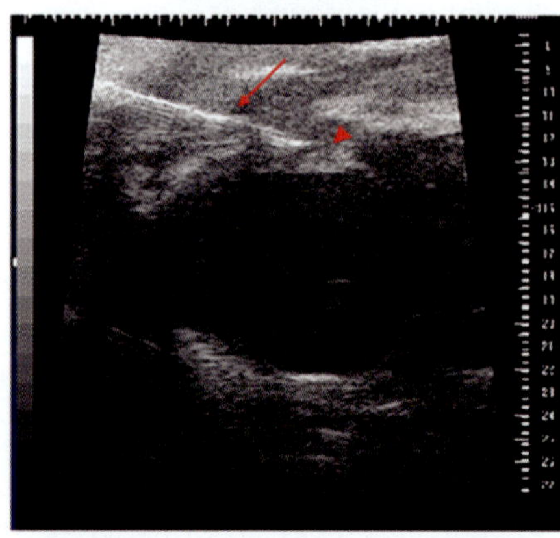

Figure 12.6. Right, MR fluoroscopy injection sequence. (**A**) Stiletto is engaged at apical septal border of anterior myocardial infarction. MRI signal from needle tip is red and that from guiding catheter green. Arrows indicate previous injections of iron-labeled stem cells, which show as dark signal voids. (**B**) An injection of Gd-DTPA is indicated by arrowhead and shows as white. (**C**) Saturation-preparation enhances appearance of test injectate compared with black myocardium and blood. (**D**) Iron-labeled MSCs (1×10^6) are injected into same spot, extinguishing local signal, and appear dark. Left, ultrasound guided delivery of stem cells. The image depicts a parasternal long axis view of the myocardium, obtained with a dedicated small animal high-resolution ultrasound (at a frequency of 30 MHz) with visualization of the injection needle (28G) as it enters the myocardium. Red arrow points to the needle, and red arrowhead marks the needle point as it enters the myocardium. From Dick AJ et al. *Circulation*. 2003 Dec 9; 108(23): 2899–904 with permission.

transplanted cells (or their microenvironment). As mentioned in other sections of this chapter, reporter genes are introduced in the DNA genome, what can cause disruption of the DNA sequence, with potential toxicity to the transplanted cells. Furthermore, the process behind the transgene expression is complex, involving many molecular steps, what may limit the amount of protein produced, potentially a limiting factor in the resultant signal. In addition, reporter genes do account for cell division (i.e., copies of DNA are integrated in the genome), but the strength of the signal may be not as sufficient for the monitoring of a low number of cells. In that respect, different imaging modalities have different sensitivities (lowest amount of cells that can be detected).

In what relates to radioisotopes, as is the case of direct labeling, radioisotopes could potentially be toxic not only to the transplanted cells but also to the recipient. Thus, we recommend assaying for cell toxicity and determining dosimetry in the living subject.

In the next section we present a discussion of the different applications of reporter gene imaging to address biological research questions in cell-based therapies.

STUDY OF STEM CELL BIOLOGY USING REPORTER GENES

In the previous sections of this chapter we discussed different imaging modalities that can be used to image stem cells in the living subject. Furthermore, the advantages and disadvantages of each imaging modality, in

what relates specifically to stem cell imaging, including reporter gene imaging, have been discussed. The next section will focus on some of the different research questions addressed using molecular imaging.

Cell Delivery

Stem cells can be delivered to the organ of interest through many routes of administration (e.g., intravenous, intracoronary, and direct injection in the tissue) [105]. Regardless of the route of administration, the first step is to make sure the stem cells actually got where we want them to be. In that respect, molecular imaging provides significant advantages as it permits us to "see" where cells are being injected. Stem cells can be previously loaded with different markers (e.g., SPIO for MRI, ^{18}F-FDG for PET imaging) and their delivery monitored using imaging (Figure 12.6) [66, 106]. In addition, our laboratory and others have performed ultrasound-guided injection of stem cells into the myocardium of small animals (Figure 12.6) [92, 107]. A more detailed discussion on the technical specifications of each imaging strategy exceeds the scope of this chapter and can be found elsewhere [66, 106, 107].

Cell Homing and Survival

The main goal of stem cell therapy (for most applications) is the recovery of lost organ function. In many cases, after stem cell transplantation, organ function can improve, sometimes transiently, or in other cases

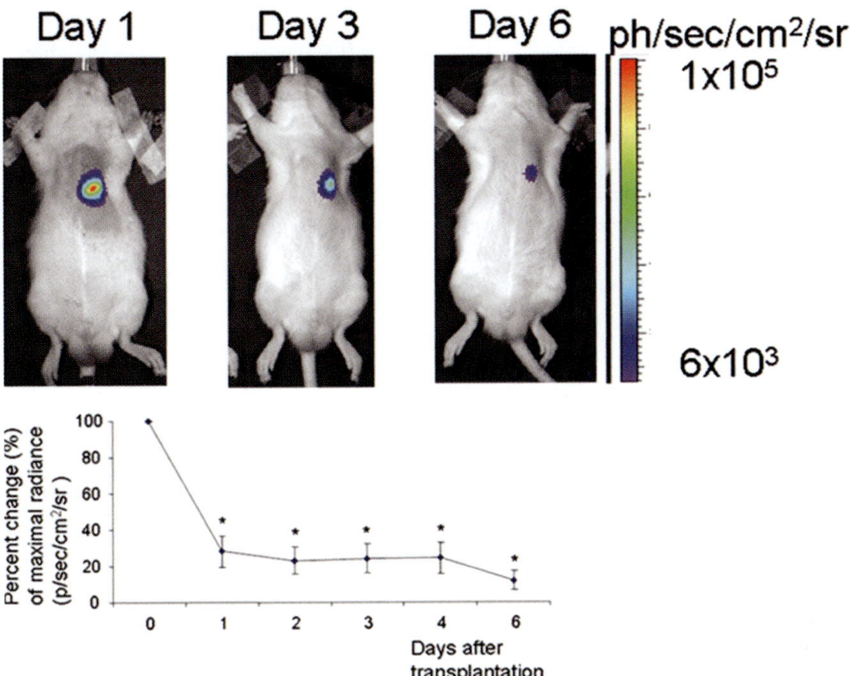

Figure 12.7. Top, Images of an optical bioluminescence scan at Days 1, 3, and 6 after cell transplantation to the myocardium, clearly demonstrating the capabilities of monitoring cell survival using reporter genes. Images were obtained after intraperitoneal injection of D-luciferin (substrate, 375 mg/kg of body weight) and represent 5-min acquisition scans. Bottom, longitudinal quantification of cell survival using optical imaging after transplantation in living subjects. Data are expressed as percentage change (%) of maximal radiance (in photons/sec/cm²/steradian) observed at Day 0. Error bars represent SEM. * represents $p < 0.05$ compared to Day 0.

result in no improvement. In any case, it is important to know what happened with the transplanted cells, whether they are viable, and if so, whether they are still in the organ where they have been transplanted to. As previously discussed, reporter gene imaging technology has been adapted to monitor stem cell viability over time.

Reporter gene imaging also permits the assessment of the viability and functionality of more than one population of cells. In this strategy, different cell populations can be transfected with different reporter genes (e.g., Rluc, Fluc for BLI), each one of which will provide a distinctive signal and thus can be imaged at the same time. This strategy provides a unique opportunity to monitor different populations of cells simultaneously.

Because of cell division and natural aging of cellular components, cells are constantly repairing and producing proteins for their cytoplasm, and for that function cells need to have their protein transcription/translation machinery constantly working for the reinstitution of structural proteins. Therefore, this system (protein production machinery) can be used to assess whether a cell is viable or not. The most common approach to study stem cell viability with reporter genes is to use a promoter that is constantly "ON" and does not depend on specific signals from the cell. Several promoters of viral origin (e.g., simian virus 40, Rous

sarcoma virus, Moloney murine leukemia virus, human cytomegalovirus (CMV)) have been used for this purpose and linked to DNA sequences (reporter genes in this case). The encoding signal (promoter) will always be "ON"; however, only in a viable cell (with a working transcription/translation protein machinery) would the protein (e.g., reporter gene) be synthesized. Our laboratory and others have extensively used this approach for the imaging of cell viability after transplantation to different organs such as the myocardium (Figure 12.7) [91, 92] and peripheral muscle [108]. Furthermore, Kutshcka et al. have used reporter genes to show that modification of the local cellular microenvironment results in improved survival of transplanted cardiomyoblasts [109, 110]. Neural cell engraftment has also been monitored using reporter genes. However, use of viral promoters has some drawbacks: it has been shown that the CMV promoter can undergo "gene silencing" (the gene is turned OFF) over time [111] (proposed to be due to a large number of Cpg repeats in the CMV promoter) that will result in reduced signal. In a reporter gene strategy, a decrease in the activity of the CMV promoter will result in decreased production of the reporter gene, which could be mistaken for decreased stem cell survival. Although there are ways to attenuate gene silencing (use of 5-azacytidine), these therapies by themselves can be toxic to stem cells and living subjects and are not commonly used. Thus, the use of alternate promoters should be sought.

Another approach that has emerged in the last few years is the use of promoters of constitutive mammalian cell proteins (e.g., α-actin, ubiquitin, and albumin), which are proteins that are constantly produced and replaced in a cell. Most of these promoters do not undergo gene silencing (low number of Cpg repeats) and are always turned "ON" as long as the cell is viable and may represent a better alternative as a promoter choice for stem cell imaging.

The final choice of promoter to be used for the assessment of cell viability will depend on many factors: size of the promoter, final size of the vector (large vectors may not penetrate and integrate easily into the cell of interest). In addition, and probably most importantly, it will depend on the strength of the signal produced by the reporter gene construct (combination of promoter plus reporter gene), which will be a critical factor in its translation from the bench to the living subject.

Interaction Between Transplanted Cells and Their Microenvironment

After stem cells are transplanted into a living subject they come in contact with a new environment. Furthermore, we know that the extracellular environment is active and likely determines cell fate, making the interaction of transplanted cells with their environment critical for their viability and function [112–115]. Stem cells are being transplanted to regain lost function in an already damaged organ, so it is reasonable to assume that in most cases the damaged tissue microenvironment will be noxious and may be detrimental to cell survival and function. A better understanding of this interaction (cell–microenvironment) will be critical to improve and develop novel therapeutic approaches. Modulation of the microenvironment has been shown to increase cell survival and organ function in models of hindlimb ischemia [116] and myocardial infarction [117]. However, most of these studies could not follow cell viability over time but rather depend on organ function or *ex vivo* analysis of stem cell survival. Reporter gene imaging provides a unique opportunity to study this interaction in living subjects over time. Using approaches like the one described previously (constitutive promoter driving the expression of a reporter gene) it will be possible to study cell viability before and after intervention (e.g., modulations of the microenvironment) with clinically accepted medical therapies (e.g., in cardiology with statins or angiotensin enzyme inhibitors).

Combined Gene and Cell Therapy

Cell-based therapies have great potential for treating many disease states. However, it is possible that cells may need to be genetically modified to design a more efficacious therapeutic approach. In fact, there is a large body of evidence suggesting that "preconditioned" stem cells, either by genetic modification [118, 119] or exogenous preconditioning [120], lead to a larger beneficial effect and improved cell survival after transplantation. Furthermore, evidence suggests that gene and cell therapy may result in greater benefit when administered concomitantly [121]. The rationale behind the concomitant use of gene and cell therapy lies on the concept that although cell therapy targets the tissue repopulation, one of the goals of gene therapy is to prevent further cell death and to provide a more "welcoming" microenvironment for the engraftment of transplanted stem cells. One of the great benefits of reporter gene imaging is that it can be used to concomitantly monitor cell and gene therapy. For that purpose, a reasonable strategy would be to use different reporter genes: for example, cells can be monitored using firefly luciferase while gene therapy can be monitored using Renilla luciferase. A similar strategy can be used if the goal is to monitor the expression of a therapeutic gene inside the stem cell that was transplanted. Another potential approach is the use of 2 PET reporter genes (e.g., herpes simplex virus 1 thymidine kinase-HSV1tk-, mutant dopamine receptor-D2R-) or even one PET reporter gene and one optical imaging gene. However, care should be exercised when using two different imaging modalities (PET and optical imaging) as they differ in sensitivity, making the results somewhat more difficult to interpret and correlate.

Stem Cell Differentiation

Since the birth of the field of stem cell therapy, there has been significant interest in the differentiation capacity of stem cells [122–125]. A large body of knowledge has been acquired regarding the biological mechanisms that determine when stem cells commit to a specific lineage of differentiation. Several groups have been able to monitor stem cell differentiation in cell culture, using different histological or noninvasive imaging techniques (some of those have been explained in other chapters of this book). Significant efforts in our laboratory and others are on their way to image the differentiation of stem cells in the living subject. Reporter gene imaging provides a unique opportunity to monitor stem cell differentiation in living subjects.

One of the most common strategies for monitoring stem cell differentiation is using protein-specific promoters (PSPs) driving the expression of reporter genes. In this imaging strategy, only when the specific promoters are turned ON by the cell, the reporter gene will be encoded, and thus the signal can be emitted. What this means for imaging of stem cell differentiation is that when stem cells differentiate into mature cells they turn ON specific signaling to encode proteins characteristic of "mature" cells, and one can take advantage of this intracellular signaling activation and use specific promoters to drive reporter genes (using PSPs). Concomitantly, stem cell viability can be monitored using a separate reporter gene (driven by constitutive promoters, like ubiquitin or alpha actin), as previously described. For example, stem cells can be labeled with two reporter genes: one reporter gene to assess cell viability = Rluc; and one reporter gene to assess cell differentiation = Fluc. After cells are transplanted to the living subject, three imaging scenarios are possible:

1. Cells are viable but have not differentiated: signal will be obtained from Rluc (viability) but not from FLuc (differentiation).
2. Cells are viable and have differentiated: signal will be obtained from both RLuc (viability) and from RLuc (differentiation).
3. Cells are not viable, thus no differentiation will take place: in this case, no signal will be detected from either reporter gene.

Reporter genes for monitoring stem cell viability have been extensively used by our and other laboratories, and limitations and disadvantages have been mentioned in other sections of this chapter. However, use of PSPs for reporter gene imaging is an area in development, and several issues need to be considered: first, expression of PSPs almost invariably results in weak signals when compared to constitutive or viral promoters (e.g., ubiquitin and CMV promoters). Other and our laboratories have been actively working on different strategies to amplify PSP signals (see Chapter 7). It is not clear which enhancer strategy will work better with which PSP, thus different enhancer strategies should be investigated with the PSP of interest. In our experience, the use of enhancer strategies commonly results in increased "signal leakage" or background, what should be separated from the true tissue-specific signal. The different amplification strategies should be tested to fully evaluate which one provides the best signal-to-background ratio.

In the previous section, some of the applications of reporter genes in the field of stem cell imaging have been discussed. In the next section, focus will be placed on the translation of these strategies to large-animal models and patients.

MOVING TOWARD CLINICAL APPLICATIONS

Over the last few years the biomedical research community has obtained a significant amount of information on the potential benefit of cell-based therapies. However, most of this knowledge was obtained indirectly (either by extrapolation of acquired knowledge from preclinical studies or by assessing the long-term effect of these therapies on organ function) or invasively (e.g., biopsy). These approaches, although important, fail to provide all the information needed to understand the biology of transplanted stem cells and assess their effects on patients. The ability to directly monitor and assess cell-based therapies in patients will be invaluable as it will allow us for the first time to investigate these therapies in the subject for whom they were intended. In many pathophysiological states, large-animal models have been shown to be similar to humans with respect to disease progression [126]. These similarities have led researchers to use large-animal models for diagnosis of disease as well as evaluation of different therapies.

Large-Animal Models

Large-animal models have been commonly used as a last preclinical step before the use of different therapies in humans. The use of large-animal models also depends on the area of research: for example, the field of cardiovascular diseases uses large-animal models more extensively than oncology. That is due to the similarities in disease progression between some large-animal models and the human disease. For example, it has been previously shown that the swine is a great model to study the cardiovascular system [126], in large part due to similarities in the development and progression of atherosclerosis in these animals. Whenever applicable, the use of large-animal models is critical for optimizing therapeutic and imaging strategies before they are applied in humans. In addition to pathophysiological similarities, many large animals have comparable weight, size, and anatomy to humans. These similarities also allow us to better optimize different imaging strategies for human applications. In diseases where the use of large-animal models is not pertinent, the translation from small-animal models to humans may be more difficult.

Our laboratory and others have used the swine model for imaging reporter genes in the myocardium [127, 128] and showed the feasibility of applying these imaging strategies for gene monitoring in large animals (Figure 12.8) and eventually humans. A similar strategy can be used for the imaging of stem cells (previously labeled to carry the reporter gene, as we have done in small animals). It is important to realize that when cells are exogenously delivered to a host, a myriad of mechanisms may play a role in survival of the transplanted stem cells: a relatively hypoxic scenario, lack of constant and efficient cell–cell contact with the microenvironment, activated immune response, and different degrees of rejection. In addition to the biological variables already mentioned, imaging of large animals (similar to what we will encounter in patients) has a few technical aspects that need to be kept in mind: on one side, the sensitivity of clinical systems is lower than that of small-animal

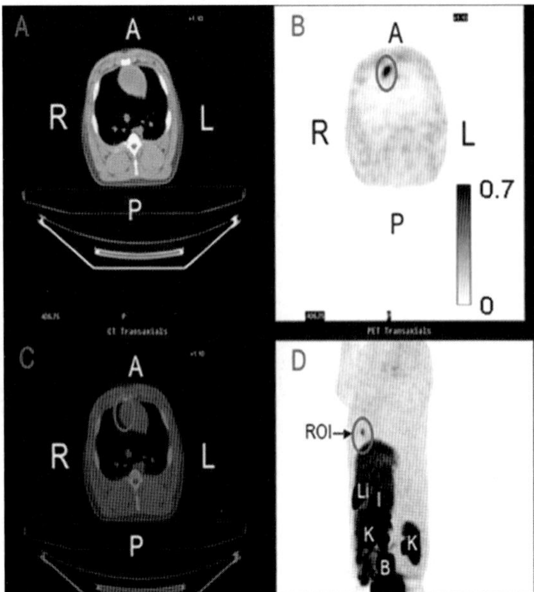

Figure 12.8. Noninvasive reporter gene myocardial delivery and imaging. A steerable percutaneous catheter with a helicoidal infusion catheter (left) is used to deliver reporter genes percutaneously under fluoroscopic guidance (middle), showing the delivery of a PET reporter gene to the antero-septal myocardium. Gene expression is imaged using PET-CT (right), clearly showing the increased uptake of ^{18}F-FHBG in the antero-septal myocardium (where the reporter gene was delivered). From Rodriguez-Porcel et al. *J Am Coll Cardiol* 2008 Feb 5; 51(5): 595–7 with permission.

imaging systems, what will make the imaging and further quantitation more challenging. In addition and depending on the imaging modality used, the amounts/strengths of the reporter gene/cell system to be delivered may need to be adjusted depending on the weight and other characteristics of the subject, what in some cases may represent a limitation. Studies in large animals are logistically more complicated and definitely require a team effort (in addition to what is usually required for small animals), and the quantities of reporter genes and imaging probe (PET or SPECT) are significantly greater, which increases cost significantly. However, it is necessary to reinforce that performance of these studies is critical to advance the field of cell-based therapies.

Clinical Imaging

Reporter gene imaging allows the possibility of studying the biology of genes or cells when they are delivered to patients. Reporter gene imaging in humans presents most of the unsolved issues described in the section of large animals. In addition, there are concerns regarding their safety and long-term effects, which are not present in animal studies as in the latter the survival is relatively short (animals are eventually used for complementary *ex vivo* studies). In addition, the variable of rejection takes

even further importance. In Chapter 13 of this book, Peñuelas et al. discuss the different strategies that can be used to image cells in patients.

IMAGING OF STEM CELLS: PUTTING IT ALL TOGETHER

In previous sections of this chapter, the major advantages and drawbacks of each imaging modality were discussed. In the next section this information will be integrated, providing a framework for the different investigators to choose the best imaging modality that will answer the proposed research question. In addition, a few potential scenarios will be presented, together with a discussion on the different imaging modalities that can be used.

The ideal imaging modality should have excellent spatial resolution and cell detection sensitivity. However, currently no such technology exists. Each imaging modality should be chosen depending on the question being addressed. For example, if the goal is to image the delivery and short-term homing of stem cells in different organs, one needs a modality that has sufficient spatial resolution and enough cell detection sensitivity to provide that information. Probably a direct labeling approach will suffice in this regard, always taking into consideration the potential toxicity it may have. Out of the molecular imaging modalities, magnetic resonance imaging provides the highest spatial resolution at all depths, albeit with significantly lower molecular sensitivity compared to other modalities like PET/SPECT or optical imaging. These variables should be taken into perspective of the limitations of MRI for molecular imaging. Similarly, ultrasound can be used in certain cases for the same purpose. In the case of ultrasound, cells could be "labeled" with microbubbles that can be identified and aid in the localization of the delivery area. MRI and ultrasound have the advantage that they can provide almost real-time image guidance for cell delivery. Nuclear medicine modalities (PET/SPECT) have been successfully and extensively used, although depending on the application they may not provide sufficient spatial resolution.

Another research question that can be posed is the long-term monitoring of stem cells. The drawbacks of a direct labeling strategy for long-term stem cell monitoring have been previously discussed. Conversely, reporter gene imaging appears more suited for the long-term monitoring of stem cells, and that can be achieved using optical imaging (bioluminescence, fluorescence) and nuclear techniques (PET/SPECT). Novel developments in MR reporter genes are discussed later in this chapter and in Chapter 4. Although optical imaging is more molecularly sensitive than PET or SPECT, it provides lesser anatomical localization and is limited to small animals. On the other hand, nuclear techniques

(PET/SPECT) provide good sensitivity with the advantage of reasonable anatomical localization and potential translation to human applications.

Similarly, when the goal is to study the biology of transplanted cells and whether these cells express a certain gene or perform a certain function, the imaging strategy to be used has to be driven by the action being studied, and as such the detected signal will correlate with the cell activity under study. At the present time, reporter gene imaging provides the best tool to address these questions. Again, the imaging modality (PET vs. SPECT vs. optical imaging) should be decided based on variables previously described.

With these concepts in mind, the following clinical scenario can be envisioned: A 60-year-old gentleman with history of hypertension, hypercholesterolemia, and previous myocardial infarction presents with symptoms of congestive heart failure (cardiac pump failure) that are refractory to maximal standard medical therapy. Echocardiogram shows left ventricular ejection fraction of 30% with akinesis of the left ventricular anterior wall. A decision is made to transplant mesenchymal stem cells (MSCs) for myocardial regeneration. Before transplantation, cells are labeled with reporter gene (e.g., HSV1-TK, hNIS) to monitor stem cell viability. A second reporter gene can be used to study cell biology (e.g., cell differentiation). To ensure that cells are delivered to myocardial areas of akinesis, MSCs are delivered using ultrasound guidance and cell survival and biology will then be monitored noninvasively using molecular imaging (likely using PET or SPECT). Concomitantly with cell monitoring, cardiac function can be monitored using echocardiography. In this way, cardiac function can be correlated with cell viability and with changes in cell biology (e.g., differentiation). Use of strategies like the ones presented here will determine, for the first time, whether the biology of stem cells are associated with changes (or lack thereof) in organ function.

In summary, the field of stem cell imaging arose due to a need to answer critical biological questions. As such, to decide which modality should be used, we need to focus on the research question posed, and based on it we will know what we need from the imaging modality (e.g., spatial resolution, molecular sensitivity, and clinical application) and use different molecular imaging modalities accordingly.

NOVEL DEVELOPMENTS IN REPORTER GENE IMAGING

Magnetic Resonance Reporter Genes

In previous sections of this chapter we have discussed the advantages of MRIs, in terms of high spatial resolution and the advantages of a reporter gene approach for the monitoring of transplanted stem cells in different diseases. Over the last few years, significant efforts have been devoted to combining these modalities to develop MR reporter genes (Figure 12.4) [129].

MRI-based reporter gene system development efforts are mostly based on production of intracellular metalloproteins (transferrin, ferritin, tyrosinase) [130]. As previously described, iron is a paramagnetic substance that induces changes in relaxivity (i.e., T2* effect) that can be detected using specific imaging sequences so the goal of these strategies is to accumulate large quantities of iron intracellularly for its detection. Physiologically, iron enters cells through the transferrin receptor (TfR) that binds the transferrin protein containing two iron atoms and internalizes iron molecules. Another approach for cell imaging uses ferritin. Ferritin is a metalloprotein that functions as the body's iron depot and can contain up to 4000 iron atoms. Native ferritin is in essence an antimagnetic particle and much weaker than SPIOs (several orders of magnitude weaker). Considerable efforts are being devoted to improving its relaxivity by removing its native core (oxyhydroxide) and reconstituting the protein shell with a superparamagnetic core. Figure 12.9 shows how a vector expressing the ferritin reporter gene (left panel) can be used to image cells (in a tumor model in this case) using MRI [129]. Briefly, cells transduced with the ferritin reporter gene were delivered to the hindlimb of mice. After several days, enough iron signal was "collected" inside the transplanted cells (due to the overexpression of ferritin) that made possible the noninvasive monitoring of transplanted cells using MRI.

Tyrosinase has also been used as an MR reporter gene [131]. Briefly, tyrosinase participates in the production of melanin, and melanin has high affinity for iron, which we know leads to increased relaxivity. Tyrosinase has been transfected into fibroblasts and embryonic kidney cells as well as breast cancer cells and has resulted in increased signal. However, there have been issues related to iron toxicity. Melanin production also produces reactive oxygen species (an important and deleterious component of the oxidative stress cascade) and thus exhibits significant toxic effects.

There are a few drawbacks in the use of metalloproteins as MR reporter genes that deserve attention [130]: First, the signal is dependent on the accumulation of iron inside the cells and for how long iron particles can be retained inside cells. Furthermore, when cells divide the signal gets diluted and the "clock" starts again, as cells need to start again to accumulate enough iron to make them detectable by MRI. From the imaging standpoint, the relaxation is dynamic and dependent on the iron loading conditions: R2 relaxivity is high at low iron doses and decreases at intermediate iron loading conditions. When iron conditions are high, T2 relaxation remains constant, what may preclude accurate quantification of the obtained signal. As previously described for direct cell labeling, MR signal

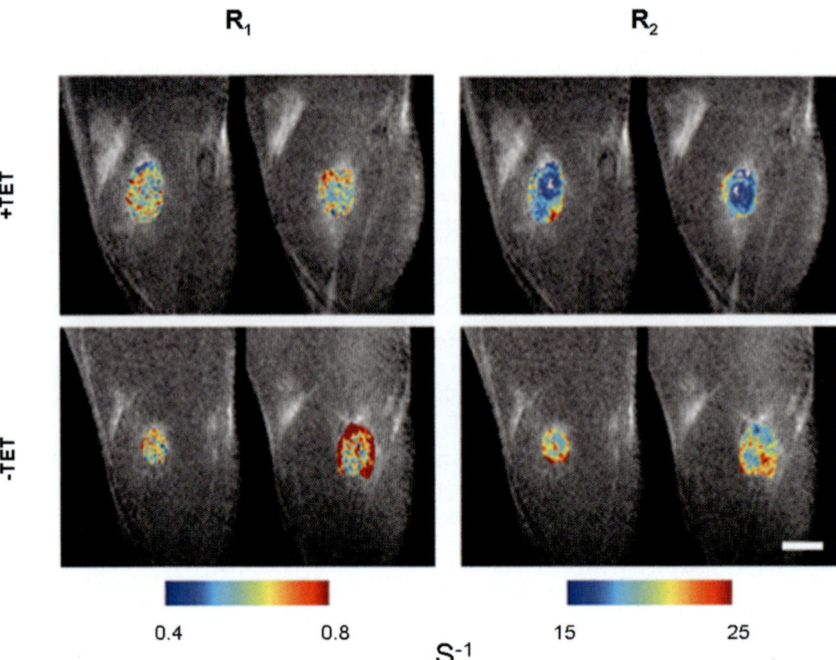

Figure 12.9. *In vivo* MRI detection of switchable ferritin expression in C6 tumors. MRI of ferritin-expressing tumors at different times after inoculation of C6-TET-eGFP-HA-ferritin tumor cells (clone 1) in the hind limb of nude mice. R1 and R2 maps of tumor regions overlaid on the MR images are shown for two representative mice from each group. Addition of tetracycline suppressed the overexpression of ferritin, leading to decreased iron uptake and detected signal by MR. Scalebar = 2.5 mm. From Cohen B et al. *Neoplasia* 2005 Feb 7(2): 109–117 with permission.

works on the basis of iron accumulation. Similar to what happens with SPIOs, when cells die the accumulated iron continues to be present inside the cells for some time (until cells are dissolved or eaten by macrophages), and thus it continues to induce a change in MR relaxivity (even if cells are not alive), and thus the signal is not representative nor linear of cell viability.

Improvements in Reporter Gene Technology

As previously described, most strategies for integration of reporter gene into the cell genome are based on random integration, signifying that there is the potential for deletions and mutations of genomic DNA. Significant efforts are being devoted to the development of site-specific integration strategies [132, 133]. For site-specific integration to be successful, it is critical to first identify the location of the critical genes within the DNA genome. Completion of the human genome constitutes a critical step toward site-specific integration into human stem cells. In this strategy, the researcher knows precisely where the reporter gene is being introduced and thus minimizes the chances of altering areas of the genome that can have significant implications for the transfected cells.

The development of reporter genes for different modalities has made a tremendous impact on the use of imaging as a research tool. Over the last few years,

significant advances have been made in the imaging of reporter genes. Furthermore, many different strategies of reporter gene imaging have been developed and have been described in other sections of this book. Over the next few years these strategies will permit a better understanding of the biology of stem cells and how they interact with the microenvironment in the living subject. In addition, the development of novel reporter genes as well improvement of existing ones will further broaden the spectrum of research questions that can be answered with these tools. For example, improvement in these novel imaging strategies will provide further insight on the timing, route, and amount of stem cells that will result in improved therapies, which will be critical to move the field of cell-based therapies into the clinics.

At the same time, novel and more sensitive tracers are being developed and will result in stronger signals. Lastly, improvements in imaging hardware for different imaging modalities will play an important role in the advancement of this field: from MR scanners with higher magnetic force to more sensitive and tomographic optical cameras, as well as improvements in PET and SPECT.

Another imaging area becoming more important is the use of nanotechnology and nanostructures (e.g., nanoparticles and nanotubes) for the study of stem cells. While sharing some characteristics with other direct cell labeling methodologies, nanoparticles/nanotubes can be used as carriers of different substances (e.g., drugs), or even as "smart nanotubes" (biologically prepared to interact with the stem cell in question), which may play a role in optimizing cell survival while providing information about the biology of these cells. In addition, nanoparticles and nanotubes have specific characteristics that make them suitable for different imaging applications and allow detection of different cell populations. So, one can envision a multimodality therapeutic and imaging strategy, where different drugs are delivered to the tissue of interest; at the same time one can monitor them while they are being delivered (using imaging).

SUMMARY

Over the last decade we have seen a revolution in gene expression imaging and more recently in cell imaging. In this chapter we have outlined some of the most important characteristics of direct- and indirect-cell labeling focusing on reporter gene technology, which will likely be the preferred methodology for long-term monitoring of stem cell biology. Currently no one technique answers all the questions posed, but use of a multimodality imaging approach presents a more appropriate approach to address the myriad of aspects posed in this exciting and rapidly evolving field.

REFERENCES

1   Gafni, Y., Turgeman, G., Liebergal, M., Pelled, G., Gazit, Z., Gazit, D. (2004). Stem cells as vehicles for orthopedic gene therapy. *Gene Ther* 11(4): 417–426.

2   Jorgensen, C., Noel, D., Apparailly, F., Sany, J. (2001). Stem cells for repair of cartilage and bone: the next challenge in osteoarthritis and rheumatoid arthritis. *Ann Rheum Dis* 60(4): 305–309.

3   Lumelsky, N., Blondel, O., Laeng, P., Velasco, I., Ravin, R., McKay, R. (2001). Differentiation of embryonic stem cells to insulin-secreting structures similar to pancreatic islets. *Science* 292(5520): 1389–1394.

4   Nawrot, I., Wozniewicz, B., Tolloczko, T., Sawicki, A., Gorski, A., Chudzinski, W., Wojtaszek, M., Grzesiuk, W., Sladowski, D., Karwacki, J., Zawitkowska, T., Szmidt, J. (2007). Allotransplantation of cultured parathyroid progenitor cells without immunosuppression: clinical results. *Transplantation* 83(6): 734–740.

5   Connor, B., van Roon-Mom, W. M., Curtis, M. A., Dragunow, M., Faull, R. L. (2001). Stem cells and neurodegenerative diseases. *N Z Med J* 114(1142): 477–479.

6   Orlic, D., Kajstura, J., Chimenti, S., Jakoniuk, I., Anderson, S. M., Li, B., Pickel, J., McKay, R., Nadal-Ginard, B., Bodine, D. M., Leri, A., Anversa, P. (2001). Bone marrow cells regenerate infarcted myocardium. *Nature* 410(6829): 701–705.

7   Forrester, J. S., Price, M. J., Makkar, R. R. (2003). Stem cell repair of infarcted myocardium: an overview for clinicians. *Circulation* 108(9): 1139–1145.

8   Kereiakes, D. J. Stem cells: the chameleon fountain of youth. *Circulation* 107(7): 939–940.

9   Hassink, R. J., Brutel de la Riviere, A., Mummery, C. L., Doevendans, P. A. (2003). Transplantation of cells for cardiac repair. *J Am Coll Cardiol* 41(5): 711–717.

10   Papadaki, H. A., Marsh, J. C., Eliopoulos, G. D. (2002). Bone marrow stem cells and stromal cells in autoimmune cytopenias. *Leuk Lymphoma* 43(4): 753–760.

11   Smith-Berdan S, Gille D, Weissman IL, Christensen JL (2007). Reversal of autoimmune disease in lupus-prone New Zealand black/New Zealand white mice by nonmyeloablative transplantation of purified allogeneic hematopoietic stem cells. *Blood* 110(4): 1370–8.

12   van Bekkum, D. W. (2000). Stem cell transplantation in experimental models of autoimmune disease. *J Clin Immunol* 20(1): 10–16.

13   Marmont, A. M. (1998). Stem cell transplantation for severe autoimmune diseases: progress and problems. *Haematologica* 83(8): 733–743.

14   Mutis, T., Goulmy, E. (2002). Targeting alloreactive T cells to hematopoietic system specific minor histocompatibility antigens for cellular immunotherapy of hematological malignancies after stem cell transplantation. *Ann Hematol* 81(Suppl 2): S38–39.

15   Nakanishi, M., Hamazaki, T. S., Komazaki, S., Okochi, H., Asashima, M. (2007). Pancreatic tissue formation from murine embryonic stem cells in vitro. *Differentiation* 75(1): 1–11.

16   Djouad, F., Mrugala, D., Noel, D., Jorgensen, C. (2006). Engineered mesenchymal stem cells for cartilage repair. *Regen Med* 1(4): 529–537.

17   Kuroda, R., Usas, A., Kubo, S., Corsi, K., Peng, H., Rose, T., Cummins, J., Fu, F. H., Huard, J. (2006). Cartilage repair using bone morphogenetic protein 4 and muscle-derived stem cells. *Arthritis Rheum* 54(2): 433–442.

18   Grande, D. A., Mason, J., Light, E., Dines, D. (2003). Stem cells as platforms for delivery of genes to enhance cartilage repair. *J Bone Joint Surg Am* 85-A(Suppl 2): 111–116.

19   Bruder, S. P., Fink, D. J., Caplan, A. I. (1994). Mesenchymal stem cells in bone development, bone repair, and skeletal regeneration therapy. *J Cell Biochem* 56(3): 283–294.

20   Orlic, D., Kajstura, J., Chimenti, S., Bodine, D. M., Leri, A., Anversa, P. (2003). Bone marrow stem cells regenerate infarcted myocardium. *Pediatr Transplant* 7(Suppl 3): 86–88.

21   Miyahara, Y., Nagaya, N., Kataoka, M., Yanagawa, B., Tanaka, K., Hao, H., Ishino, K., Ishida, H., Shimizu, T., Kangawa, K., Sano, S., Okano, T., Kitamura, S., Mori, H. (2006). Monolayered mesenchymal stem cells repair scarred myocardium after myocardial infarction. *Nat Med* 12(4): 459–465.

22   Yoshioka, T., Ageyama, N., Shibata, H., Yasu, T., Misawa, Y., Takeuchi, K., Matsui, K., Yamamoto, K., Terao, K., Shimada, K., Ikeda, U., Ozawa, K., Hanazono, Y. (2005). Repair of infarcted myocardium mediated by transplanted bone marrow-derived CD34+ stem cells in a nonhuman primate model. *Stem Cells* (Dayton, Ohio) 23(3): 355–364.

23   Genre, D., Viens, P., Bertucci, F., Chabannon, C., Gravis, G., Braud, A. C., Camerlo, J., Houvenaeghel, G., Moutardier, V., Goncalvez, A., Protiere, C., Bardou, V. J., Maraninchi, D. (2002). Modulations of dose intensity of doxorubicin and

cyclophosphamide in association with G-CSF and peripheral blood stem cells in adjuvant chemotherapy for breast cancer: comparative evaluation of completion and safety of three intensive regimens. *Bone Marrow Transplant* 29(11): 881–886.

24 Pittenger, M. F., Martin, B. J. (2004). Mesenchymal stem cells and their potential as cardiac therapeutics. *Circ Res* 95(1): 9–20.

25 Bhatia, R., Hare, J. M. (2005). Mesenchymal stem cells: future source for reparative medicine. *Congest Heart Fail* 11(2): 87–91.

26 Pittenger, M. F., Mackay, A. M., Beck, S. C., Jaiswal, R. K., Douglas, R., Mosca, J. D., Moorman, M. A., Simonetti, D. W., Craig, S., Marshak, D. R. (1999). Multilineage potential of adult human mesenchymal stem cells. *Science* 284(5411): 143–147.

27 Lee, J. P., Jeyakumar, M., Gonzalez, R., Takahashi, H., Lee, P. J., Baek, R. C., Clark, D., Rose, H., Fu, G., Clarke, J., McKercher, S., Meerloo, J., Muller, F. J., Park, K. I., Butters, T. D., Dwek, R. A., Schwartz, P., Tong, G., Wenger, D., Lipton, S. A., Seyfried, T. N., Platt, F. M., Snyder, E. Y. (2007). Stem cells act through multiple mechanisms to benefit mice with neurodegenerative metabolic disease. *Nat Med* 13(4): 439–447.

28 Rafii, S., Lyden, D. (2003). Therapeutic stem and progenitor cell transplantation for organ vascularization and regeneration. *Nat Med* 9(6): 702–712.

29 Roy, N. S., Cleren, C., Singh, S. K., Yang, L., Beal, M. F., Goldman, S. A. (2006). Functional engraftment of human ES cell-derived dopaminergic neurons enriched by coculture with telomerase-immortalized midbrain astrocytes. *Nat Med* 12(11): 1259–1268.

30 Balsam, L. B., Robbins, R. C. (2005). Haematopoietic stem cells and repair of the ischaemic heart. *Clin Sci (Lond)* 109(6): 483–492.

31 Stamm, C., Westphal, B., Kleine, H. D., Petzsch, M., Kittner, C., Klinge, H., Schumichen, C., Nienaber, C. A., Freund, M., Steinhoff, G. (2003). Autologous bone-marrow stem-cell transplantation for myocardial regeneration. *Lancet* 361(9351): 45–46.

32 Sorrentino, S. A., Bahlmann, F. H., Besler, C., Muller, M., Schulz, S., Kirchhoff, N., Doerries, C., Horvath, T., Limbourg, A., Limbourg, F., Fliser, D., Haller, H., Drexler, H., Landmesser, U. (2007). Oxidant stress impairs in vivo reendothelialization capacity of endothelial progenitor cells from patients with type 2 diabetes mellitus: restoration by the peroxisome proliferator-activated receptor-gamma agonist rosiglitazone. *Circulation* 116(2): 163–173.

33 Ii, M., Nishimura, H., Iwakura, A., Wecker, A., Eaton, E., Asahara, T., Losordo, D. W. (2005). Endothelial progenitor cells are rapidly recruited to myocardium and mediate protective effect of ischemic preconditioning via "imported" nitric oxide synthase activity. *Circulation* 111(9): 1114–1120.

34 Rookmaaker, M. B., Vergeer, M., van Zonneveld, A. J., Rabelink, T. J., Verhaar, M. C. (2003). Endothelial progenitor cells: mainly derived from the monocyte/macrophage-containing CD34- mononuclear cell population and only in part from the hematopoietic stem cell-containing CD34+ mononuclear cell population. *Circulation* 108(21): e150; author reply e150.

35 Imitola, J., Raddassi, K., Park, K. I., Mueller, F. J., Nieto, M., Teng, Y. D., Frenkel, D., Li, J., Sidman, R. L., Walsh, C. A., Snyder, E. Y., Khoury, S. J. (2004). Directed migration of neural stem cells to sites of CNS injury by the stromal cell-derived factor 1alpha/CXC chemokine receptor 4 pathway. *Proceedings of the National Academy of Sciences of the United States of America* 101(52): 18117–18122.

36 Song, H. J., Stevens, C. F., Gage, F. H. (2002). Neural stem cells from adult hippocampus develop essential properties of functional CNS neurons. *Nat Neurosci* 5(5): 438–445.

37 Massoud, T. F., Gambhir, S. S. (2003). Molecular imaging in living subjects: seeing fundamental biological processes in a new light. *Genes Dev* 17(5): 545–580.

38 Herschman, H. R., MacLaren, D. C., Iyer, M., Namavari, M., Bobinski, K., Green, L. A., Wu, L., Berk, A. J., Toyokuni, T., Barrio, J. R., Cherry, S. R., Phelps, M. E., Sandgren, E. P., Gambhir, S. S. (2000). Seeing is believing: non-invasive, quantitative and repetitive imaging of reporter gene expression in living animals, using positron emission tomography. *J Neurosci Res* 59(6): 699–705.

39 Thakur, M. L., Coleman, R. E., Mayhall, C. G., Welch, M. J., Jr. (1976). Preparation and evaluation of 111In-labeled leukocytes as an abscess imaging agent in dogs. *Radiology* 119(3): 731.

40 Thakur, M. L., Lavender, J. P., Arnot, R. N., Silvester, D. J., Segal, A. W. (1977). Indium-111-labeled autologous leukocytes in man. *J Nucl Med* 18(10): 1014–1021.

41 Ju, S., Teng, G. J., Lu, H., Zhang, Y., Zhang, A., Chen, F., Ni, Y. (2007). In vivo MR tracking of mesenchymal stem cells in rat liver after intrasplenic transplantation. *Radiology* 245(1): 206–215.

42 Guzman, R., Uchida, N., Bliss, T. M., He, D., Christopherson, K. K., Stellwagen, D., Capela, A., Greve, J., Malenka, R. C., Moseley, M. E., Palmer, T. D., Steinberg, G. K. (2007). Long-term monitoring of transplanted human neural stem cells in developmental and pathological contexts with MRI. *Proceedings of the National Academy of Sciences of the United States of America* 104(24): 10211–10216.

43 Rice, H. E., Hsu, E. W., Sheng, H., Evenson, D. A., Freemerman, A. J., Safford, K. M., Provenzale, J. M., Warner, D. S., Johnson, G. A. (2007). Superparamagnetic iron oxide labeling and transplantation of adipose-derived stem cells in middle cerebral artery occlusion-injured mice. *AJR Am J Roentgenol* 188(4): 1101–1108.

44 Chakraborty, S. K., Fitzpatrick, J. A., Phillippi, J. A., Andreko, S., Waggoner, A. S., Bruchez, M. P., Ballou, B. (2007). Cholera toxin B conjugated quantum dots for live cell labeling. *Nano Lett* 7(9): 2618–2626.

45 Rosen, A. B., Kelly, D. J., Schuldt, A. J., Lu, J., Potapova, I. A., Doronin, S. V., Robichaud, K. J., Robinson, R. B., Rosen, M. R., Brink, P. R., Gaudette, G. R., Cohen, I. S. (2007). Finding fluorescent needles in the cardiac haystack: tracking human mesenchymal stem cells labeled with quantum dots for quantitative in vivo three-dimensional fluorescence analysis. *Stem Cells* (Dayton, Ohio) 25(8): 2128–2138.

46 Keppler, A., Arrivoli, C., Sironi, L., Ellenberg, J. (2006). Fluorophores for live cell imaging of AGT fusion proteins across the visible spectrum. *Biotechniques* 41(2): 167–175.

47 Hill, J. M., Dick, A. J., Raman, V. K., Thompson, R. B., Yu, Z. X., Hinds, K. A., Pessanha, B. S., Guttman, M. A.,

Varney, T. R., Martin, B. J., Dunbar, C. E., McVeigh, E. R., Lederman, R. J. (2003). Serial cardiac magnetic resonance imaging of injected mesenchymal stem cells. *Circulation* **108**(8): 1009–1014.

48 Rogers, W. J., Meyer, C. H., Kramer, C. M. (2006). Technology insight: in vivo cell tracking by use of MRI. *Nat Clin Pract Cardiovasc Med* **3**(10): 554–562.

49 Kang, W. J., Kang, H. J., Kim, H. S., Chung, J. K., Lee, M. C., Lee, D. S. (2006). Tissue distribution of 18F-FDG-labeled peripheral hematopoietic stem cells after intracoronary administration in patients with myocardial infarction. *J Nucl Med* **47**(8): 1295–1301.

50 Rini, J. N., Bhargava, K. K., Tronco, G. G., Singer, C., Caprioli, R., Marwin, S. E., Richardson, H. L., Nichols, K. J., Pugliese, P. V., Palestro, C. J. (2006). PET with FDG-labeled leukocytes versus scintigraphy with 111In-oxine-labeled leukocytes for detection of infection. *Radiology* **238**(3): 978–987.

51 Adonai, N., Nguyen, K. N., Walsh, J., Iyer, M., Toyokuni, T., Phelps, M. E., McCarthy, T., McCarthy, D. W., Gambhir, S. S. (2002). Ex vivo cell labeling with 64Cu-pyruvaldehyde-bis(N4-methylthiosemicarbazone) for imaging cell trafficking in mice with positron-emission tomography. *Proceedings of the National Academy of Sciences of the United States of America* **99**(5): 3030–3035.

52 Bindslev, L., Haack-Sorensen, M., Bisgaard, K., Kragh, L., Mortensen, S., Hesse, B., Kjaer, A., Kastrup, J. (2006). Labelling of human mesenchymal stem cells with indium-111 for SPECT imaging: effect on cell proliferation and differentiation. *Eur J Nucl Med Mol Imaging* **33**(10): 1171–1177.

53 Michalet, X., Pinaud, F. F., Bentolila, L. A., Tsay, J. M., Doose, S., Li, J. J., Sundaresan, G., Wu, A. M., Gambhir, S. S., Weiss, S. Quantum dots for live cells, in vivo imaging, and diagnostics. *Science* **307**(5709): 538–544.

54 Kalchenko, V., Shivtiel, S., Malina, V., Lapid, K., Haramati, S., Lapidot, T., Brill, A., Harmelin, A. (2006). Use of lipophilic near-infrared dye in whole-body optical imaging of hematopoietic cell homing. *J Biomed Opt* **11**(5): 050507.

55 Muller-Borer, B. J., Collins, M. C., Gunst, P. R., Cascio, W. E., Kypson, A. P. (2007). Quantum dot labeling of mesenchymal stem cells. *J Nanobiotechnology* **5**:9.

56 Shah, B. S., Clark, P. A., Moioli, E. K., Stroscio, M. A., Mao, J. J. (2007). Labeling of mesenchymal stem cells by bioconjugated quantum dots. *Nano Lett* **7**(10): 3071–3079.

57 Seleverstov, O., Zabirnyk, O., Zscharnack, M., Bulavina, L., Nowicki, M., Heinrich, J. M., Yezhelyev, M., Emmrich, F., O'Regan, R., Bader, A. (2006). Quantum dots for human mesenchymal stem cells labeling. A size-dependent autophagy activation. *Nano Lett* **6**(12): 2826–2832.

58 Lin, S., Xie, X., Patel, M. R., Yang, Y. H., Li, Z., Cao, F., Gheysens, O., Zhang, Y., Gambhir, S. S., Rao, J. H., Wu, J. C. (2007). Quantum dot imaging for embryonic stem cells. *BMC Biotechnol* **7**(1):67.

59 Montet, X., Figueiredo, J. L., Alencar, H., Ntziachristos, V., Mahmood, U., Weissleder, R. (2007). Tomographic fluorescence imaging of tumor vascular volume in mice. *Radiology* **242**(3): 751–758.

60 Montet, X., Ntziachristos, V., Grimm, J., Weissleder, R. (2005). Tomographic fluorescence mapping of tumor targets. *Cancer Res* **65**(14): 6330–6336.

61 Roy, R., Thompson, A. B., Godavarty, A., Sevick-Muraca, E. M. (2005). Tomographic fluorescence imaging in tissue phantoms: a novel reconstruction algorithm and imaging geometry. *IEEE Transactions on Medical Imaging* **24**(2): 137–154.

62 Bos, C., Delmas, Y., Desmouliere, A., Solanilla, A., Hauger, O., Grosset, C., Dubus, I., Ivanovic, Z., Rosenbaum, J., Charbord, P., Combe, C., Bulte, J. W., Moonen, C. T., Ripoche, J., Grenier, N. (2004). In vivo MR imaging of intravascularly injected magnetically labeled mesenchymal stem cells in rat kidney and liver. *Radiology* **233**(3): 781–789.

63 Di Tucci, A. A., Matta, G., Deplano, S., Gabbas, A., Depau, C., Derudas, D., Caocci, G., Agus, A., Angelucci, E. (2008). Myocardial iron overload assessment by T2* magnetic resonance imaging in adult transfusion dependent patients with acquired anemias. *Haematologica* **93**(9): 1385–1388.

64 Ittrich, H., Lange, C., Togel, F., Zander, A. R., Dahnke, H., Westenfelder, C., Adam, G., Nolte-Ernsting, C. (2007). In vivo magnetic resonance imaging of iron oxide-labeled, arterially-injected mesenchymal stem cells in kidneys of rats with acute ischemic kidney injury: detection and monitoring at 3T. *J Magn Reson Imaging* **25**(6): 1179–1191.

65 Kraitchman, D. L., Heldman, A. W., Atalar, E., Amado, L. C., Martin, B. J., Pittenger, M. F., Hare, J. M., Bulte, J. W. (2003). In vivo magnetic resonance imaging of mesenchymal stem cells in myocardial infarction. *Circulation* **107**(18): 2290–2293.

66 Dick, A. J., Guttman, M. A., Raman, V. K., Peters, D. C., Pessanha, B. S., Hill, J. M., Smith, S., Scott, G., McVeigh, E. R., Lederman, R. J. (2003). Magnetic resonance fluoroscopy allows targeted delivery of mesenchymal stem cells to infarct borders in Swine. *Circulation* **108**(23): 2899–2904.

67 Watson, D. J., Walton, R. M., Magnitsky, S. G., Bulte, J. W., Poptani, H., Wolfe, J. H. (2006). Structure-specific patterns of neural stem cell engraftment after transplantation in the adult mouse brain. *Hum Gene Ther* **17**(7): 693–704.

68 Bengel, F. M., Schachinger, V., Dimmeler, S. (2005). Cell-based therapies and imaging in cardiology. *Eur J Nucl Med Mol Imaging* **32**(Suppl 2): S404–416.

69 Bertini, I., Luchinat, C., Parigi, G., Quacquarini, G., Marzola, P., Cavagna, F. M. (1998). Off-resonance experiments and contrast agents to improve magnetic resonance imaging. *Magn Reson Med* **39**(1): 124–131.

70 Corot, C., Petry, K. G., Trivedi, R., Saleh, A., Jonkmanns, C., Le Bas, J. F., Blezer, E., Rausch, M., Brochet, B., Foster-Gareau, P., Baleriaux, D., Gaillard, S., Dousset, V. (2004). Macrophage imaging in central nervous system and in carotid atherosclerotic plaque using ultrasmall superparamagnetic iron oxide in magnetic resonance imaging. *Invest Radiol* **39**(10): 619–625.

71 Zhang, Y., Dodd, S. J., Hendrich, K. S., Williams, M., Ho, C. (2000). Magnetic resonance imaging detection of rat renal transplant rejection by monitoring macrophage infiltration. *Kidney Int* **58**(3): 1300–1310.

72 Li, Z., Suzuki, Y., Huang, M., Cao, F., Xie, X., Connolly, A. J., Yang, P. C., Wu, J. C. (2008). Comparison of reporter

gene and iron particle labeling for tracking fate of human embryonic stem cells and differentiated endothelial cells in living subjects. *Stem Cells* (Dayton, Ohio) 26(4): 864–873.

73 Thakur, M. L., Segal, A. W., Louis, L., Welch, M. J., Hopkins, J., Peters, T. J. (1977). Indium-111-labeled cellular blood components: mechanism of labeling and intracellular location in human neutrophils. *J Nucl Med* 18(10): 1022–1026.

74 Kraitchman, D. L., Tatsumi, M., Gilson, W. D., Ishimori, T., Kedziorek, D., Walczak, P., Segars, W. P., Chen, H. H., Fritzges, D., Izbudak, I., Young, R. G., Marcelino, M., Pittenger, M. F., Solaiyappan, M., Boston, R. C., Tsui, B. M., Wahl, R. L., Bulte, J. W. (2005). Dynamic imaging of allogeneic mesenchymal stem cells trafficking to myocardial infarction. *Circulation* 112(10): 1451–1461.

75 Zhou, R., Thomas, D. H., Qiao, H., Bal, H. S., Choi, S. R., Alavi, A., Ferrari, V. A., Kung, H. F., Acton, P. D. (2005). In vivo detection of stem cells grafted in infarcted rat myocardium. *J Nucl Med* 46(5): 816–822.

76 Doyle, B., Kemp, B. J., Chareonthaitawee, P., Reed, C., Schmeckpeper, J., Sorajja, P., Russell, S., Araoz, P., Riederer, S. J., Caplice, N. M. (2007). Dynamic tracking during intracoronary injection of 18F-FDG-labeled progenitor cell therapy for acute myocardial infarction. *J Nucl Med* 48(10): 1708–1714.

77 Hofmann, M., Wollert, K. C., Meyer, G. P., Menke, A., Arseniev, L., Hertenstein, B., Ganser, A., Knapp, W. H., Drexler, H. (2005). Monitoring of bone marrow cell homing into the infarcted human myocardium. *Circulation* 111(17): 2198–2202.

78 Troy, T., Jekic-McMullen, D., Sambucetti, L., Rice, B. (2004). Quantitative comparison of the sensitivity of detection of fluorescent and bioluminescent reporters in animal models. *Mol Imaging* 3(1): 9–23.

79 Beeres, S. L., Bengel, F. M., Bartunek, J., Atsma, D. E., Hill, J. M., Vanderheyden, M., Penicka, M., Schalij, M. J., Wijns, W., Bax, J. J. Role of imaging in cardiac stem cell therapy. *J Am Coll Cardiol* 49(11): 1137–1148.

80 Bengel, F. M., Gambhir, S. S. (2005). Clinical molecular imaging and therapy-moving ahead together. *Eur J Nucl Med Mol Imaging* 32(Suppl 2): S323.

81 Chen, I. Y., Greve, J. M., Gheysens, O., Willmann, J. K., Rodriguez-Porcel, M., Chu, P., Sheikh, A., Faranesh, T., Paulmurugan, R., Yang, P. C., Wu, J. C., and Gambhir, S. S. (2008). Comparison of optical bioluminescence reporter gene and superparamagnetic iron oxide MR contrast agent as cell markers for non-invasive imaging of cardiac cell transplantation. *Mol Imag Biol* 2008 11(3): 178–87.

82 Wu, J. C., Tseng, J. R., Gambhir, S. S. (2004). Molecular imaging of cardiovascular gene products. *J Nucl Cardiol* 11(4): 491–505.

83 Slotkin, J. R., Chakrabarti, L., Dai, H. N., Carney, R. S., Hirata, T., Bregman, B. S., Gallicano, G. I., Corbin, J. G., Haydar, T. F. (2007). In vivo quantum dot labeling of mammalian stem and progenitor cells. *Dev Dyn* 236(12): 3393-3401.

84 Keren, S., Gheysens, O., Levin, C. S., Gambhir, S. S. (2008). A comparison between a time domain and continuous wave small animal optical imaging system. *IEEE Transactions on Medical Imaging* 27(1): 58–63.

85 Kumar, A. T., Raymond, S. B., Dunn, A. K., Bacskai, B. J., Boas, D. A. (2008). A time domain fluorescence tomography system for small animal imaging. *IEEE Transactions on Medical Imaging* 27(8): 1152–1163.

86 Model, R., Orlt, M., Walzel, M., Hunlich, R. (1998). Optical imaging: three-dimensional approximation and perturbation approaches for time-domain data. *Applied Optics* 37(34): 7968–7976.

87 Ray, P., Tsien, R., Gambhir, S. S. (2007). Construction and validation of improved triple fusion reporter gene vectors for molecular imaging of living subjects. *Cancer Res* 67(7): 3085–3093.

88 Kesarwala, A. H., Prior, J. L., Sun, J., Harpstrite, S. E., Sharma, V., Piwnica-Worms, D. (2006). Second-generation triple reporter for bioluminescence, micro-positron emission tomography, and fluorescence imaging. *Mol Imaging* 5(4): 465–474.

89 Li, Z., Wu, J. C., Sheikh, A. Y., Kraft, D., Cao, F., Xie, X., Patel, M., Gambhir, S. S., Robbins, R. C., Cooke, J. P., Wu, J. C. (2007). Differentiation, survival, and function of embryonic stem cell derived endothelial cells for ischemic heart disease. *Circulation* 116(11 Suppl): I46–54.

90 Chen, I. Y., Wu, J. C., Min, J. J., Sundaresan, G., Lewis, X., Liang, Q., Herschman, H. R., Gambhir, S. S. (2004). Micro-positron emission tomography imaging of cardiac gene expression in rats using bicistronic adenoviral vector-mediated gene delivery. *Circulation* 109(11): 1415–1420.

91 Wu, J. C., Chen, I. Y., Sundaresan, G., Min, J. J., De, A., Qiao, J. H., Fishbein, M. C., Gambhir, S. S. Molecular imaging of cardiac cell transplantation in living animals using optical bioluminescence and positron emission tomography. *Circulation* 108(11): 1302–1305.

92 Rodriguez-Porcel, M., Gheysens, O., Chen, I. Y., Wu, J. C., Gambhir, S. S. (2005). Image-guided cardiac cell delivery using high-resolution small-animal ultrasound. *Mol Ther* 12(6): 1142–1147.

93 Loening, A. M., Wu, A. M., Gambhir, S. S. (2007). Red-shifted Renilla reniformis luciferase variants for imaging in living subjects. *Nat Methods* 4(8): 641–643.

94 Sheikh, A. Y., Lin, S. A., Cao, F., Cao, Y., van der Bogt, K. E., Chu, P., Chang, C. P., Contag, C. H., Robbins, R. C., Wu, J. C. (2007). Molecular imaging of bone marrow mononuclear cell homing and engraftment in ischemic myocardium. *Stem Cells* (Dayton, Ohio) 25(10): 2677–2684.

95 Cao, F., Lin, S., Xie, X., Ray, P., Patel, M., Zhang, X., Drukker, M., Dylla, S. J., Connolly, A. J., Chen, X., Weissman, I. L., Gambhir, S. S., Wu, J. C. (2006). In vivo visualization of embryonic stem cell survival, proliferation, and migration after cardiac delivery. *Circulation* 113(7): 1005–1014.

96 Roth, D. J., Jansen, E. D., Powers, A. C., Wang, T. G. (2006). A novel method of monitoring response to islet transplantation: bioluminescent imaging of an NF-kB transgenic mouse model. *Transplantation* 81(8): 1185–1190.

97 Lu, Y., Dang, H., Middleton, B., Zhang, Z., Washburn, L., Campbell-Thompson, M., Atkinson, M. A., Gambhir, S. S., Tian, J., Kaufman, D. L. (2004). Bioluminescent monitoring of islet graft survival after transplantation. *Mol Ther* 9(3): 428–435.

98 Soloviev, V. Y. (2007). Tomographic bioluminescence imaging with varying boundary conditions. *Applied Optics* 46(14): 2778–2784.

99 Lv, Y., Tian, J., Cong, W., Wang, G., Yang, W., Qin, C., Xu, M. (2007). Spectrally resolved bioluminescence tomography with adaptive finite element analysis: methodology and simulation. *Phys Med Biol* 52(15): 4497–4512.

100 Lu, Y., Dang, H., Middleton, B., Campbell-Thompson, M., Atkinson, M. A., Gambhir, S. S., Tian, J., Kaufman, D. L. (2006). Long-term monitoring of transplanted islets using positron emission tomography. *Mol Ther* 14(6): 851–856.

101 Lu, Y., Dang, H., Middleton, B., Zhang, Z., Washburn, L., Stout, D. B., Campbell-Thompson, M., Atkinson, M. A., Phelps, M., Gambhir, S. S., Tian, J., Kaufman, D. L. (2006). Noninvasive imaging of islet grafts using positron-emission tomography. *Proceedings of the National Academy of Sciences of the United States of America* 103(30): 11294–11299.

102 Kang, J. H., Lee, D. S., Paeng, J. C., Lee, J. S., Kim, Y. H., Lee, Y. J., Hwang, D. W., Jeong, J. M., Lim, S. M., Chung, J. K., Lee, M. C. (2005). Development of a sodium/iodide symporter (NIS)-transgenic mouse for imaging of cardiomyocyte-specific reporter gene expression. *J Nucl Med* 46(3): 479–483.

103 Kim, Y. H., Lee, D. S., Kang, J. H., Lee, Y. J., Chung, J. K., Roh, J. K., Kim, S. U., Lee, M. C. (2005). Reversing the silencing of reporter sodium/iodide symporter transgene for stem cell tracking. *J Nucl Med* 46(2): 305–311.

104 Miyagawa, M., Anton, M., Wagner, B., Haubner, R., Souvatzoglou, M., Gansbacher, B., Schwaiger, M., Bengel, F. M. (2005). Non-invasive imaging of cardiac transgene expression with PET: comparison of the human sodium/iodide symporter gene and HSV1-tk as the reporter gene. *Eur J Nucl Med Mol Imaging* 32(9): 1108–1114.

105 Hou, D., Youssef, E. A., Brinton, T. J., Zhang, P., Rogers, P., Price, E. T., Yeung, A. C., Johnstone, B. H., Yock, P. G., March, K. L. (2005). Radiolabeled cell distribution after intramyocardial, intracoronary, and interstitial retrograde coronary venous delivery: implications for current clinical trials. *Circulation* 112(9 Suppl): I150–156.

106 Karmarkar, P. V., Kraitchman, D. L., Izbudak, I., Hofmann, L. V., Amado, L. C., Fritzges, D., Young, R., Pittenger, M., Bulte, J. W., Atalar, E. (2004). MR-trackable intramyocardial injection catheter. *Magn Reson Med* 51(6): 1163–1172.

107 Springer, M. L., Sievers, R. E., Viswanathan, M. N., Yee, M. S., Foster, E., Grossman, W., Yeghiazarians, Y. (2005). Closed-chest cell injections into mouse myocardium guided by high-resolution echocardiography. *Am J Physiol Heart Circ Physiol* 289(3): H1307–1314.

108 Wu, J. C., Sundaresan, G., Iyer, M., Gambhir, S. S. (2001). Noninvasive optical imaging of firefly luciferase reporter gene expression in skeletal muscles of living mice. *Mol Ther* 4(4): 297–306.

109 Kutschka, I., Kofidis, T., Chen, I. Y., von Degenfeld, G., Zwierzchoniewska, M., Hoyt, G., Arai, T., Lebl, D. R., Hendry, S. L., Sheikh, A. Y., Cooke, D. T., Connolly, A., Blau, H. M., Gambhir, S. S., Robbins, R. C. (2006). Adenoviral human BCL-2 transgene expression attenuates early donor cell death after cardiomyoblast transplantation into ischemic rat hearts. *Circulation* 114(1 Suppl): I174–180.

110 Kutschka, I., Chen, I. Y., Kofidis, T., Arai, T., von Degenfeld, G., Sheikh, A. Y., Hendry, S. L., Pearl, J., Hoyt, G., Sista, R., Yang, P. C., Blau, H. M., Gambhir, S. S., Robbins, R. C. (2006). Collagen matrices enhance survival of transplanted cardiomyoblasts and contribute to functional improvement of ischemic rat hearts. *Circulation* 114(1 Suppl): I167–173.

111 Krishnan, M., Park, J. M., Cao, F., Wang, D., Paulmurugan, R., Tseng, J. R., Gonzalgo, M. L., Gambhir, S. S., Wu, J. C. (2006). Effects of epigenetic modulation on reporter gene expression: implications for stem cell imaging. *Faseb J* 20(1): 106–108.

112 Hu, K., Denk, L., de Vries, U., Minuth, W. W. (2007). Chemically defined medium environment for the development of renal stem cells into tubules. *Biotechnol J* 2(8): 992–995.

113 Kasper, G., Glaeser, J. D., Geissler, S., Ode, A., Tuischer, J., Matziolis, G., Perka, C., Duda, G. N. (2007). Matrix metalloprotease activity is an essential link between mechanical stimulus and mesenchymal stem cell behavior. *Stem Cells* (Dayton, Ohio) 25(8): 1985–1994.

114 Vats, A., Bielby, R. C., Tolley, N., Dickinson, S. C., Boccaccini, A. R., Hollander, A. P., Bishop, A. E., Polak, J. M. Chondrogenic differentiation of human embryonic stem cells: the effect of the micro-environment. *Tissue Eng* 12(6): 1687–1697.

115 An, Y. H., Wang, H. Y., Gao, Z. X., Wang, Z. C. (2004). Differentiation of rat neural stem cells and its relationship with environment. *Biomed Environ Sci* 17(1): 1–7.

116 Napoli, C., Williams-Ignarro, S., de Nigris, F., de Rosa, G., Lerman, L. O., Farzati, B., Matarazzo, A., Sica, G., Botti, C., Fiore, A., Byrns, R. E., Sumi, D., Sica, V., Ignarro, L. J. (2005). Beneficial effects of concurrent autologous bone marrow cell therapy and metabolic intervention in ischemia-induced angiogenesis in the mouse hindlimb. *Proceedings of the National Academy of Sciences of the United States of America* 102(47): 17202–17206.

117 Davis, M. E., Hsieh, P. C., Takahashi, T., Song, Q., Zhang, S., Kamm, R. D., Grodzinsky, A. J., Anversa, P., Lee, R. T. (2006). Local myocardial insulin-like growth factor 1 (IGF-1) delivery with biotinylated peptide nanofibers improves cell therapy for myocardial infarction. *Proceedings of the National Academy of Sciences of the United States of America* 103(21): 8155–8160.

118 Choi, J. H., Hur, J., Yoon, C. H., Kim, J. H., Lee, C. S., Youn, S. W., Oh, I. Y., Skurk, C., Murohara, T., Park, Y. B., Walsh, K., Kim, H. S. (2004). Augmentation of therapeutic angiogenesis using genetically modified human endothelial progenitor cells with altered glycogen synthase kinase-3beta activity. *J Biol Chem* 279(47): 49430–49438.

119 Sun, W., Wang, Q., Zhang, L., Pan, J., Zhang, M., Lu, G., Yao, H., Wang, J., Cao, X. (2002). TGF-beta(1) gene modified immature dendritic cells exhibit enhanced tolerogenicity but induce allograft fibrosis in vivo. *J Mol Med* 80(8): 514–523.

120 Suzuki, K., Murtuza, B., Beauchamp, J. R., Smolenski, R. T., Varela-Carver, A., Fukushima, S., Coppen, S. R., Partridge, T. A., Yacoub, M. H. (2004). Dynamics and mediators of acute graft attrition after myoblast transplantation to the heart. *Faseb J* 18(10): 1153–1155.

121 Mangi, A. A., Noiseux, N., Kong, D., He, H., Rezvani, M., Ingwall, J. S., Dzau, V. J. (2003). Mesenchymal stem cells

modified with Akt prevent remodeling and restore performance of infarcted hearts. *Nat Med* **9**(9): 1195–1201.

122 Heng, B. C., Cao, T., Lee, E. H. (2004). Directing stem cell differentiation into the chondrogenic lineage in vitro. *Stem Cells* (Dayton, Ohio) **22**(7): 1152–1167.

123 Keller, G. (2005). Embryonic stem cell differentiation: emergence of a new era in biology and medicine. *Genes Dev* **19**(10): 1129–1155.

124 Krencik, R., Zhang, S. C. (2006). Stem cell neural differentiation: A model for chemical biology. *Curr Opin Chem Biol* **10**(6): 592–597.

125 Odorico, J. S., Kaufman, D. S., Thomson, J. A. (2001). Multilineage differentiation from human embryonic stem cell lines. *Stem Cells* (Dayton, Ohio) **19**(3): 193–204.

126 Bloor, C. M., White, F. C., Roth, D. M. (1992). The pig as a model of myocardial ischemia and gradual coronary artery occlusion. In: Swindle, M. M., Moody, D. C., Phillips, L. D., eds. *Swine as models in biomedical research*. Ames, IA: Iowa State University Press: 163–175.

127 Rodriguez-Porcel, M., Brinton, T. J., Chen, I. Y., Gheysens, O., Lyons, J., Ikeno, F., Willman, J. K., Wu, L., Wu, J. C., Yeung, A. C., Yock, P. Y., Gambhir, S. S. (2007). Reporter gene imaging following percutaneous delivery in swine: moving towards clinical application. *J Am Coll Cardiol* (in press).

128 Bengel, F. M., Anton, M., Richter, T., Simoes, M. V., Haubner, R., Henke, J., Erhardt, W., Reder, S., Lehner, T., Brandau, W., Boekstegers, P., Nekolla, S. G., Gansbacher, B., Schwaiger, M. (2003). Noninvasive imaging of transgene expression by use of positron emission tomography in a pig model of myocardial gene transfer. *Circulation* **108**(17): 2127–2133.

129 Cohen, B., Dafni, H., Meir, G., Harmelin, A., Neeman, M. (2005). Ferritin as an endogenous MRI reporter for noninvasive imaging of gene expression in C6 glioma tumors. *Neoplasia* **7**(2): 109–117.

130 Gilad, A. A., Winnard, P. T., Jr., van Zijl, P. C., Bulte, J. W. (2007). Developing MR reporter genes: promises and pitfalls. *NMR Biomed* **20**(3): 275–290.

131 Weissleder, R., Simonova, M., Bogdanova, A., Bredow, S., Enochs, W. S., Bogdanov, A., Jr. (1997). MR imaging and scintigraphy of gene expression through melanin induction. *Radiology* **204**(2): 425–429.

132 Linden, R. M., Ward, P., Giraud, C., Winocour, E., Berns, K. I. (1996). Site-specific integration by adeno-associated virus. *Proceedings of the National Academy of Sciences of the United States of America* **93**(21): 11288–11294.

133 Recchia, A., Mavilio, F. (2006). Site-specific integration into the human genome: ready for clinical application? *Rejuvenation Research* **9**(4): 446–449.

13

Clinical Applications of Reporter Gene Technology

Iván Peñuelas, Shahriar S. Yaghoubi, Felipe Prósper, and Sanjiv Sam Gambhir

INTRODUCTION

For more than a decade, molecular imaging (MI) has increasingly been used to successfully image gene expression in living animals, thus making significant contributions to the field of gene and cellular gene therapy. However, there has been slow progress in translating these technologies into clinical application, even though there is a real need to develop, test, and validate sensitive and reproducible noninvasive imaging methods that could be repeatedly and safely performed in patients undergoing gene therapy.

New molecular biology technologies now permit rapid determination of expression levels of hundreds of genes from minute tissue samples. These technologies combined with the complete sequencing of the human genome have allowed establishment of a molecular signature for many diseases. Nonetheless, gene expression patterns can change during the course of the pathology and in response to therapy. Such modification of gene expression patterns can alter drug sensitivity. Hence, molecular imaging can play a role in monitoring variation of gene expression during treatment.

Molecular imaging is emerging as a noninvasive technology for *in vivo* mapping of gene expression and provides promising tools for accelerated progress of molecular medicine (for a review see [1–3] and references therein). Recently the importance of multimodality imaging has been recognized, allowing imaging of gene expression in intact cultured cells up to noninvasive whole-body imaging, both in animal models and in humans. It is evident that no single technique alone will provide the full spectrum of specifications needed to image all aspects of gene therapy from basic research to clinical application. The use of multimodality reporter gene systems (discussed in Chapter 5) should facilitate translational research in gene and cell therapy.

Developing a molecular imaging paradigm for a particular disease requires finding a molecular target relevant to the disease and finding a high-affinity ligand for such a target. If this basic principle is possible, a molecular imaging agent has to be found that can be used to detect the target *in vivo*. Unfortunately, in many cases such molecules might not be amenable for use in humans. In the case of gene expression imaging, the paradigm is somehow more complex, as reporter gene expression imaging represents an indirect imaging method that involves several different components. One such component is a gene whose expression is in some way amenable to imaging *in vivo*.

The ideal way to monitor the expression of a gene is to use a specific imaging probe that can directly detect its expression (by targeting the protein product of the gene or its mRNA), but this is not available in most cases. The development of a new probe and its *in vivo* validation for each single gene of interest would be a time-consuming and costly effort, and for many genes the design of an imaging paradigm may not be possible. Consequently, the most reasonable alternative would be to use a handful of well-characterized imaging reporter genes for indirect imaging of therapeutic gene expression by different technologies. Although indirect imaging of the expression of a therapeutic transgene (TG) may be feasible simply by coadministering the TG and an imaging reporter gene in identical vectors [4, 5], often more complex systems may be required. Such approaches would include the use of fusion gene constructs, bidirectional promoter vectors, bi-cistronic vectors and dual promoter constructs among others [6]. This is discussed further in Chapter 9.

In the case of reporter gene imaging, multimodality imaging vectors would probably be the best way to validate a procedure and even get different molecular information from different imaging technologies. For example, a vector comprising a triple-fusion imaging reporter gene amenable for bioluminescence, fluorescence, and PET imaging would have many advantages [7–10]. Bioluminescence imaging would permit fast and affordable testing of the construct in laboratory animals with high throughput but would lack tomographic information. The use of fluorescence, although less sensitive than bioluminescence for most applications, might permit fast kinetic analysis not possible with bioluminescence and would enable not only *in vivo* imaging in animals but also detailed histological examination of explanted tissues if required. However, none of the above technologies could be used in humans *in vivo*, but the PET reporter might permit tomographic transgene expression imaging in small animals and definitely transgene expression monitoring in humans in the clinical setting.

IN VIVO IMAGING OF REPORTER GENES

Reporter genes (RGs) are DNA sequences encoding for easily assayed proteins. They are efficient tools for monitoring transgene delivery and gene expression under the control of all types of gene regulatory elements. In addition, cell trafficking can be monitored by labeling cells with RGs. By linking the signaling gene (reporter gene) to the selected promoter construct, it is possible to monitor the expression of the reporter gene using a suitable reporter probe by different methodologies (colorimetric, luminometric, radioisotopic). The most widely used reporter genes have been chloramphenicol acetyltransferase, firefly and Renilla luciferase, β-galactosidase, and the green fluorescent protein (GFP). However, all of these systems have important limitations, as their use is restricted to cell extracts, tissue samples, or small animals, thus making impossible the sequential analysis of the studied event in humans under different conditions and its application to gene therapy studies in patients.

REPORTER GENE IMAGING MODALITIES FOR *IN VIVO* CLINICAL APPLICATIONS

Optical Imaging

Different optical imaging approaches rely on bioluminescence, fluorescence, absorption, or reflectance of the source of contrast. We will focus on bioluminescence and fluorescence, but more exhaustive description of different optical techniques can be found in detail in Chapters 1 and 2 of this book.

Bioluminescence-based approaches have the intrinsic advantage that photons are only generated when the reporter probe interacts with the reporter protein. This kind of reporter probe yields extraordinary signal-to-background ratios. Bioluminescence-based approaches have an excellent sensitivity and do not involve the noise that usually hampers interpretation of radionuclide-based imaging modalities. Signals are detected only from cells expressing luciferase. However, bioluminescence depends on the excess availability of the light-producing substrate, is only semiquantitative, and lacks

tomographic information, although recent developments may permit volumetric bioluminescence images [11]. Furthermore, acquisition times to achieve a sufficiently high signal must be long, and persistence of the signal precludes fast repetition of imaging studies. Additionally, optical signals are strongly attenuated by tissues; thus, optical imaging is limited for applications in large animals or humans. However, endoscopic devices should allow optical imaging of the internal organs during surgical procedures, and the use of near-infrared probes may allow near surface imaging in humans.

Another possibility would be the use of handheld scanning devices for *in vivo* imaging of processes taking place on the surface of the body or just below it. Several devices for optical imaging mammography are under development, and some of them have already been tested in the clinical setting. These devices are based on diffuse optical tomography and diffuse optical spectroscopy techniques, which use near-infrared light to measure the optical properties of tissue [12]. These are not bioluminescence-based approaches, as fiber optics are used to illuminate the tissue and an array of detectors are used to pick up the transmitted and scattered photons. Reconstruction software can then use the photon distribution to produce images or even quantitative distribution of different physiological parameters such as hemoglobin concentration, oxygen saturation, and scattering particle size – all of which are considered potential predictors of disease [13].

Although in these cases no images of reporter genes are obtained, once the technology is developed imaging of either bioluminescence or fluorescence reporters would in theory be feasible.

In any case, to be useful as large-animal or human reporters, the light emission of the selected system should have a significant red wavelength component to overcome greater absorption of blue and green light in tissues owing to hemoglobin, lipids, and water. High quantum yield and the time kinetics of light production are also important parameters to allow real-time monitoring of luciferase expression. The most widely used luciferases are codon optimized for mammalian cell expression versions of the firefly and the coral *Renilla reniformis* enzymes.

Novel probes for optical imaging of cancer by fusing a cancer-targeting engineered antibody to Renilla luciferase have been developed [14]. The election of Renilla luciferase (Rluc) in these cases was not done only owing to the smaller size of its gene compared with Firefly luciferase (fluc) but also on the fact that the Rluc catalyzed chemiluminescent reaction is not dependent on ATP, which is critical considering targets on the cell surface (where very low levels of ATP are present). Imaging of intracellular targets may eventually be possible by targeting internalizing receptors. Potential clinical applications of this new concept can be imagined, although careful evaluation of any potential toxicity of the substrate, along with potential immunogenicity issues of the luciferase, should be addressed first.

Fluorescence does not require a reporter substrate, as the reporter fluorescent protein is first excited by external illumination and then emits light at a different wavelength that can be externally detected. Different spectral shifted variants of green fluorescent protein (GFP), including cyan fluorescent protein (CFP) and yellow fluorescent protein (YFP), have long been used in biotechnology research [15]. Several red fluorescent proteins including mutagenized variants with emission maximums shifted up to values of 616 nm have also been described [16]. Red-shifted emitting fluorescence seems to be more suitable for *in vivo* applications, although other issues such as the wavelength of the excitation light, the duration of the light emission, and the quantum yield also have to be considered [17].

Endogenous autofluorescence usually results in substantial background emissions that limit the sensitivity and specificity of this technology for *in vivo* imaging in living animals. Exponentially decreasing intensity of emitted light with increasing depth of the target also has to be taken into account. In contrast, fluorescence emission is much more intense than bioluminescence, and there are reporters available in the whole spectral range of light that can also be used for fluorescence microscopy imaging.

Imaging in the near-infrared (NIR) spectrum maximizes tissue penetrance in addition to minimizing the autofluorescence from nontarget tissues. Advances in NIR fluorescence imaging have permitted the development of fluorochromes coupled to quenching peptides activated by specific proteases at the target site [18].

Novel catheter-based systems [19] are potentially directly transferable to human studies. Such catheter-based optical fluorescent imaging systems might allow endoscopic intraluminal imaging of vessel walls directly through the bloodstream and imaging of internal organs, whereas many other possibilities for imaging from inside the body also exist. A system with both visible and near-infrared capabilities would allow precise anatomic orientation with acquisition of the standard real-time white-light video image coupled with the optical fluorescent or bioluminescence image of the target of interest.

MRI Imaging

Magnetic resonance imaging (MRI) techniques have been developed to obtain images with voxel resolutions down to about 50 μm^3 for *in vivo* studies in small animals under strong magnetic fields and of a few hundreds of cubic microns for human studies in clinical MR devices.

Key requirement for MRI is the availability of high-relaxivity contrast agents that have a large effect on the MRI signal. MRI contrast in soft tissue is excellent owing to the different properties of water in different tissues, but despite this the application of contrast agents is still important for increasing the sensitivity and specificity of the technique. Although MRI provides exquisite anatomical resolution, temporal resolution is limited and fast kinetic studies cannot be performed.

Molecular probe detection by MRI is several orders of magnitude less sensitive as compared with optical and nuclear techniques, but signal amplification strategies to generate higher contrast between target and background could eventually help overcome such limitations. Delivering the metal complexes used as contrast to the target is challenging as they generally have poor penetrance into tissues and cells. New reporters that sequester endogenous iron from the organism have been developed [20] [21] and thus the cell constructs the MRI paramagnetic contrast *in situ* under reporter gene-driven instructions permitting MRI imaging of gene expression in mice without the need for delivery of a contrast agent.

With the exception of reporter genes, which generate contrast themselves by accumulating MR-detectable iron, contrast materials must be administered. Recently, a new type of contrast was suggested in which radiofrequency irradiation is used to noninvasively label agent protons that exchange with water [22]. The contrast produced by these so-called chemical-exchange saturation transfer agents (CESTs), unlike that produced by paramagnetic metals, can be switched on and off by selectively irradiating at the exchangeable proton resonance frequency. A problem of this system's contrast agents is that very high magnetic fields are required, although future developments might permit its application in low field (<3 T) clinical MRI devices.

A very interesting difference of CEST-based reporters is that for all reporter genes to date, a naturally occurring protein is selected and given a new application to report cellular activity or localization. However, CEST reporters can be designed *de novo* based on desired properties. Future reporter targets with reduced charge and increased exchange rate should allow increased detection sensitivity, or even frequency-specific reporter genes may be designed. We expect that the concept of CEST reporter genes could have applications in the future in the study of gene delivery and gene expression even in humans.

In any case, it would be desirable to have a multi-reporter system similar to the trireporter genes used for bioluminescent, PET, and fluorescent imaging, in which one of the genes is replaced with an MR reporter [23]. In this way, the high resolution of MRI could be combined with the sensitivity and straightforward whole-body application of either PET or bioluminescent imaging, as well as potential histological validation with fluorescence. MRI reporter genes have been discussed in much greater detail in Chapter 4.

Radionuclide Imaging

Molecular imaging has its roots in radionuclide imaging techniques used in nuclear medicine procedures. PET, single photon emission computed tomography (SPECT), and planar scintigraphy have long been used to detect radionuclide-labeled probes. The probe is somehow concentrated at specific sites after interaction with its target, but in all cases a significant background signal exists as the probe emits radioactivity irrespective of probe/target interaction.

Gamma ray-emitting isotopes (such as ^{99m}Tc, ^{123}I, or ^{111}In) used in clinical nuclear medicine as part of radiopharmaceuticals can readily be used for SPECT, but the chemistry of these elements is somehow complex and the availability of different compounds is limited. By contrast, imaging in small animals with radiopharmaceuticals readily available for human use permits faster translation of imaging paradigms into clinical applications.

PET is an analytical imaging tool in which compounds labeled with positron-emitting radioisotopes are used to measure biological processes using trace amounts of the radiolabeled probe. Molecules labeled with positron-emitting radionuclides injected intravenously in trace quantities are retained in tissues as a result of binding to a receptor or cell entrapment owing to enzyme-catalyzed conversion after uptake by a cell membrane transporter. PET offers several advantages compared with SPECT and planar imaging. It is more sensitive as it can detect picomolar or even nanomolar concentrations, which are generally several orders of magnitude below pharmacological levels. It is more quantitative, which allows kinetic modeling for analysis of rate constants of the underlying biochemical process studied. Owing to the short half-lives of the radionuclides most widely used in PET (carbon-11 and fluorine-18 with 20 and 110 min, respectively), repetitive imaging at short time intervals is also possible. In addition, PET radiochemistry has access to many compounds that can be labeled with carbon-11, thus obtaining a chemically indistinguishable molecule from the unlabeled natural one. Finally, PET is fully tomographic.

CURRENT RESULTS OF REPORTER GENE IMAGING IN HUMANS

Many crucial questions regarding the mechanisms of gene therapy remain unanswered. Little is known about issues as important as those related to the efficiency of vector delivery to target and nontarget tissues, the levels of transgene expression, duration of transgene expression, or even the therapeutic efficacy of the overall gene therapy procedure. In this scenario, imaging of living subjects could play an essential role both in preclinical and clinical research in gene therapy. A noninvasive clinically applicable method that could help answer all these questions would be of enormous value [24, 25]. It would not only make possible monitoring and evaluation of gene therapy in human subjects by defining the location, magnitude, and time variation of gene expression over time, it would also lead to a better understanding of vector biology and pharmacology.

Noninvasive molecular imaging in gene therapy can be used in several different ways [26, 27]. These technologies can be used to monitor gene expression both locally and for the whole body, evaluate the treatment by imaging the functional effects of gene therapy at the biochemical level, and determine prognosis. Furthermore, they would help improve the design of more efficient, specific, and safer vectors, determine the appropriate dose to be used in each case, and consequently facilitate their approval for human use, thus speeding up the entire process.

The immunogenicity of non-human–derived reporter proteins is a major limitation for translation of reporter gene imaging into clinical practice. When repetitive administration of a reporter gene or long-term monitoring of transgene expression is required, this constraint is especially important due to the potential immune reaction against the foreign protein. A number of human-derived reporter genes have been proposed to circumvent this limitation. Dopamine receptor type 2 [28, 29], somatostatin receptor [30], sodium iodide symporter [31–33], norepinephrine transporter [34], and more recently the hTK2 [35] have been proposed as potential human-derived reporter genes for nuclear imaging. Endogenous expression of the corresponding genes might pose a problem in certain cases. Background human sodium iodide symporter (hNIS) expression is very high in thyrocytes and stomach but is lower in mammary, salivary, and lacrimal glands. The expression of the 2 receptors is more limited within the body; hD2R expression is largely limited to the striatal–nigral system of the brain, and hSSTR2 expression is largely limited to carcinoid tumors. Both transporters and receptors have clinically approved radiopharmaceuticals for imaging: 123I- or 131I-iodide, 111In-diethylenetriaminepentaaceticacid-octreotide, and 3-(2'-^{18}F-fluoroethyl)spiperone ([^{18}F]FESP), respectively.

Although transporter or receptor–ligand-based systems might be hindered by a relatively low sensitivity caused by the saturability of receptors in some cases, it has been demonstrated that taking into account the overall process (including the potential difficulty of accessing intracellular proteins due to the need for transmembrane transportation of the probe) of the sensitivity of HSV1-tk- and hD2R-based systems is similar [5]. An example of a transporter-based reporter gene limitation is the

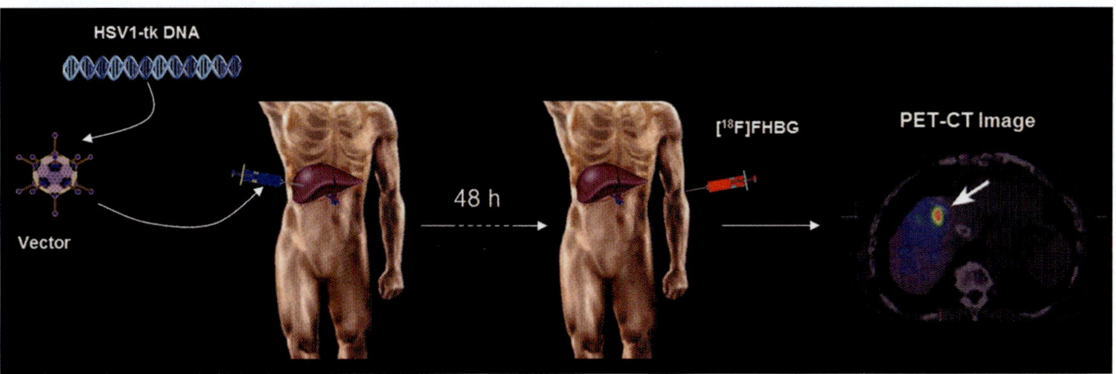

Figure 13.1. Gene therapy procedure and *in vivo* imaging of HSV1-tk gene expression in liver tumors as described by Peñuelas et al. [38]. The gene of interest (HSV1-tk in this case) is introduced into an appropriate vector. The vector is injected directly into the tumoral lesion in the liver of the patient. Two days after vector administration (the expected time for maximal transgene expression) the radioactive substrate of the HSV1-TK enzyme is intravenously administered to the patients and PET imaging carried out. Specific transgene expression (arrow) is in this case limited to the transduced tumoral nodule where the vector is injected.

well-known case of the hNIS. In this particular case, the radiolabeled substrate is not organified and trapped; hence specific signal can only be observed during a short time frame.

The novel hTK2 gene has some advantages that might hopefully circumvent limitations of most of the currently available reporter genes for use in humans. Such human-derived PET reporter genes have been developed to enable studies that require long-term monitoring of therapeutic gene coexpression in normal and neoplastic tissues, including adoptive immunotherapies, stem cell therapies, and other transplantation therapies. In addition, in contrast to HSV1-TK, hTK2 should in principle be nonimmunogenic and sensitize transduced cells to certain prodrugs for elimination of any undesired consequences of genetic and transplantation therapies [35]. However, hTK2 is approximately ten times less sensitive than HSV1-tk reporter genes.

NONINVASIVE MONITORING OF TRANSGENE EXPRESSION IN HUMANS

To the best of our knowledge, only three groups have been able to show reporter transgene expression in humans. In 2001, Jacobs et al. [36] showed in a single patient with glioblastoma after intratumoral infusion of a liposomal vector encoding HSV1-tk an increase in the accumulation rate of the radioactive uracil derivative 5-[^{124}I]iodo-2′-fluoro-2′-deoxy-1β-Darabinofuranosyl-5-iodouracil ([^{124}I]FIAU) while utilizing PET. Unfortunately, in this report the authors were not able to image the specific accumulation of [^{124}I]FIAU in the treated lesions in any of the four additional patients imaged. Furthermore, in this clinical trial, PET studies could not be serially repeated because of the long

half-life of ^{124}I (4 days). More recently, Dempsey et al. [37] were not able to obtain evidence of specific transgene expression in eight glioblastoma patients treated by direct intratumoral injection of a liposomal vector containing the selectively replication-competent mutant HSV1-tk 1716. In this case, imaging was done using ^{123}I labeled FIAU and SPECT imaging. The reasons for these negative results might be related to the lower sensitivity of SPECT imaging as compared with PET, the degree of disruption of the blood–brain barrier in individual gliomas, and transduction efficiency.

However, in 2005 Peñuelas et al. [38] demonstrated HSV-1-tk expression in tumoral hepatocarcinoma nodules treated by adenoviral-mediated suicide gene therapy in a group of patients (see Figures 13.1 and 13.2). Transgene expression was vector dose-dependent, although a direct relationship between the dose and the level of transgene expression could not be established. The authors also showed that PET reporter gene imaging can be used in humans for whole-body monitoring of transgene expression in nontarget sites. It was also observed that the signal-to-background ratio between the lesion and the nontreated liver increases with time after injection of the FHBG tracer. During the study, the progressive accumulation of the tracer in the transduced cells together with the metabolic elimination of the tracer to bile by the nontransduced liver parenchyma increases the specificity of the signal, making the transduced lesion more clearly distinguishable 5–6 h after tracer administration. In some cases, early images (1 h) are not able to show transgene expression, whereas it is clearly demonstrated on delayed ones (5–6 h). Specific [18F]FHBG accumulation (due to transgene expression) was not observed within the livers of any of the patients 9 days after adenoviral delivery of HSV1-tk to the tumors. This may have been due to the

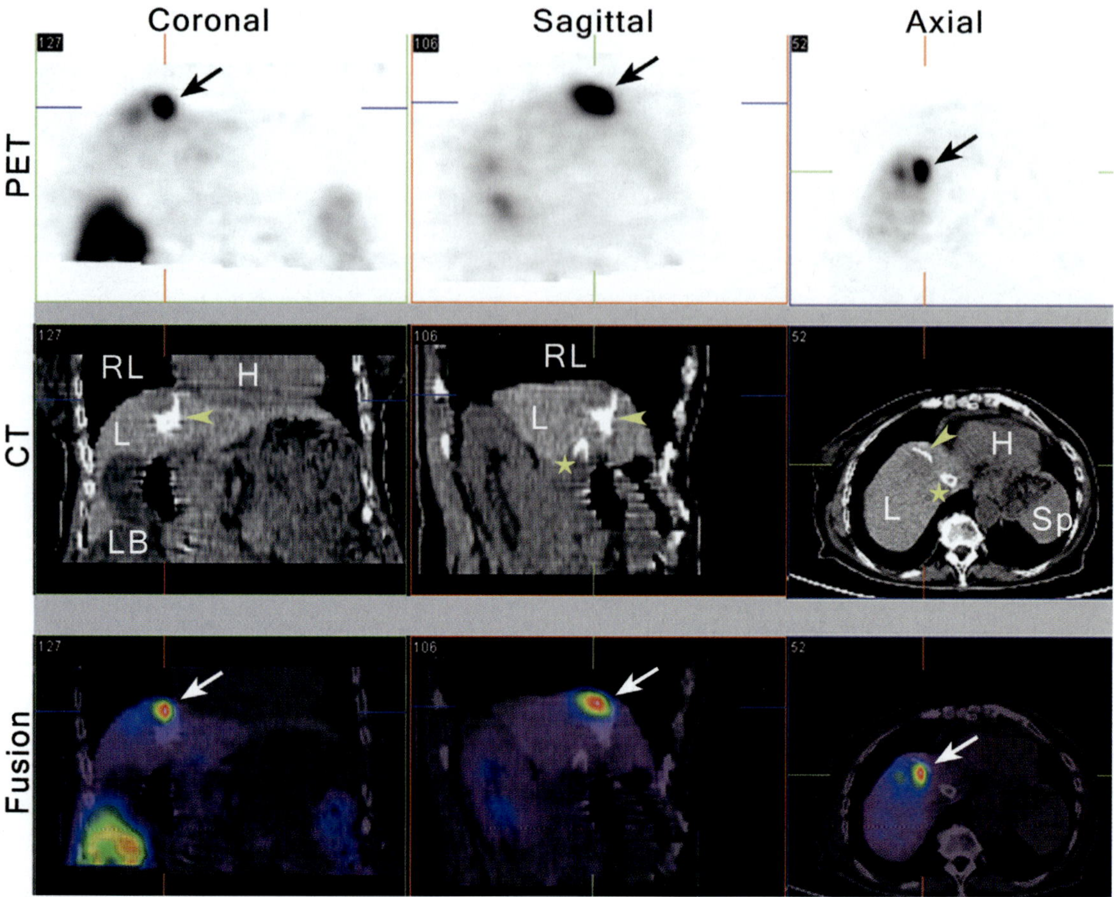

Figure 13.2. PET-CT imaging of HSV1-tk transgene expression in humans. PET images show [^{18}F]FHBG accumulation at the tumor site (arrows). Anatomo-metabolic correlation can be obtained by fused PET-CT imaging. Tracer signal can be seen in the treated lesion (arrows), whereas no specific accumulation of the tracer can be seen in necrotic, lipiodol-retaining regions around it. H: heart, L: liver, LB: large bowel, RL: right lung, Sp: spleen. Adapted from reference [38].

eradication of HSV1-tk-expressing cells following valganciclovir administration, transgene silencing, or immune reaction to HSV1-TK protein.

MOLECULAR IMAGING OF REPORTER GENE EXPRESSION IN HUMANS

The lack of therapeutic efficiency of many gene therapy protocols in the clinical setting is probably due to inefficient distribution of the vector being used and/or heterogeneous transduction of the desired tissue. Jacobs et al. [39] have recently demonstrated in animal models that to address these challenges it is important to develop assays that allow accurate noninvasive determination of viable target tissue and assessment of the transduced tissue dose of a therapeutic gene. Consequently, gene therapy clinical protocols should ideally be designed considering molecular imaging technology from the very beginning.

It should not only comprise a way to assess the location, magnitude, and duration of vector-mediated gene expression *in vivo* but also the precise identification of the tissue susceptible for treatment. Proper identification of target tissue permits directed vector application and enables a correlation of the induced therapeutic effect.

The only group that so far has demonstrated in a set of patients the feasibility of noninvasive gene expression monitoring in humans [26, 38] has shown that molecular imaging technologies for transgene expression monitoring can be used in humans in several different ways:

a) To avoid invasive procedures for gene transfer monitoring
b) To evaluate treatment by imaging the functional effects of gene therapy at the biochemical level, and to determine prognosis
c) To monitor gene expression both locally and for the whole body

d) To help determine the appropriate dose of the vector
e) To help improve vector design

IMAGING AS A NONINVASIVE PROCEDURE FOR GENE TRANSFER MONITORING

Until now, tissue sampling followed by molecular analysis of explanted tissues offered the only feasible way to know whether a vector had reached the desired target organ and/or whether the transgene is being expressed. However, the biopsy procedure has an inherent morbidity/mortality and can only provide limited results to determine gene transfer efficacy or its effects. In addition, several biopsy samples should be taken for each evaluation point to have a sufficiently representative overview, and the sampling procedure itself can alter the underlying tissue. As minute samples are usually taken, tissue heterogeneity may lead to a false idea of transgene expression in the target site. Finally, it is evident that it is not possible to take biopsy samples of every single organ of interest and that such invasive procedures are not likely to be used repeatedly owing to the clinical complications they can produce. On the other hand, radionuclide-based molecular imaging can obtain quantitative whole-body images of molecular events, noninvasively and repetitively from the patient, and can be the best method for monitoring transgene expression.

IDENTIFICATION OF TARGET TISSUE AND EVALUATION OF RESPONSE TO GENE THERAPY

Needless to say, noninvasive molecular imaging procedures are nowadays essential for proper identification of multiple diseases (see [2, 40, 41] and references therein). Detailed delineation of lesions by the use of fused images of both metabolic (PET and SPECT) and anatomic imaging modalities (CT and MRI) provide accurate identification of target tissue as a first step for targeted delivery of transgene expression vectors.

Radionuclide imaging was originally applied for tumor detection, but its application has been broadened to monitoring the efficacy of therapy. In fact, ^{18}F-fluorodeoxyglucose (FDG) PET has already been used in cancer patients to evaluate response to radio- or chemotherapy, and in many cases metabolic-based molecular imaging procedures such as PET and other radionuclide imaging modalities are increasingly gaining acceptance over anatomical-based imaging such as CT or MRI as the most appropriate procedures for evaluating response to therapy [42]. Regarding gene therapy, PET has been used in experimental models to evaluate response to HSV1-tk mediated suicide gene therapy [43].

Early gene therapy response evaluation or prediction is of great interest. We have recently shown that transgene expression monitoring by [^{18}F]FHBG can be used to predict the response to the gene therapy procedure in cancer patients [38]. In this case, only in those patients in whom we observed a positive accumulation of the PET reporter probe in the treated nodule when evaluated 30 d after the gene therapy procedure could clinical status be considered as stable disease. In contrast, those patients with [^{18}F]FHBG PET-negative studies had progression of the disease. Taking into account that the [^{18}F]FHBG PET study for evaluation of transgene expression was performed just 2 d after injection of the vector (and before starting ganciclovir (GCV) treatment), we could demonstrate that PET can in fact be used as a valuable tool for early prediction of response to the gene therapy procedure.

Our observation in humans is further supported by results in laboratory animals in which we have shown that only tumors accumulating [^{18}F]FHBG regress after GCV treatment [43]. In addition, these authors showed that because [^{18}F]FHBG (%ID) declines as GCV-exposed tumors regress, [^{18}F]FHBG PET can be used to monitor effectiveness of GCV treatment. Unnecessary prodrug treatments could hence be avoided and patients could be switched to other treatment regimens at an early stage.

An additional issue to be considered is how the gene therapy procedure can alter the uptake of widely used radiopharmaceuticals for early response monitoring (such as [^{18}F]FDG). Little information is available on this particular topic, and this is something that should be carefully considered. Could metabolic changes induced in the transduced cells by vector delivery modify the radiopharmaceutical uptake pattern in the lesion, thus giving either false positive or false negative results? Rigorous methodological PET protocols must be followed, different specific radiopharmaceuticals should be used whenever possible, and the results should be evaluated taking into account the aforementioned considerations. In any case, all noninvasive diagnostic technologies have their intrinsic limitations. Hence, gene therapy response and follow-up should ideally be done by combining several imaging modalities along with detailed biochemical and genetic analysis as the best way to gather as much information as possible.

NONINVASIVE MONITORING OF TRANSGENE EXPRESSION IN HUMANS

Transgene expression imaging in animals showing the feasibility of noninvasive quantitative evaluation of gene expression in combination with an appropriate PET reporter gene/PET reporter probe system has sufficiently been demonstrated and detailed in published protocols written by Yaghoubi et al. [44, 45]. In addition, a detailed tracer kinetic model that describes the pharmacokinetics of FHBG has been published [46], and we have shown that [18F]FHBG has the desirable *in vivo* characteristics of stability, rapid blood clearance, low background signal, biosafety, and acceptable radiation dosimetry in humans [47]. This study forms the foundation for using [18F]FHBG in applications to monitor HSV1-tk reporter gene expression. Furthermore, we have demonstrated the pharmacological safety of intravenously injecting 14 μg/kg of cold FHBG into male and female rats and rabbits [48] (14 μg/kg is 100 × the estimated injected dose of FHBG in humans, assuming a specific activity of 1.48×10^{10} Bq/μmole – lowest reported specific activity – and a tracer-injected dose of 555 MBq). Based on these preclinical and clinical data, the U.S. Food and Drug Administration has approved [18F]FHBG as an investigational new imaging agent ((IND #61,880)). A detailed protocol for PET imaging of HSV1-tk or mutant HSV1-sr39tk reporter gene expression in mice and humans using [18F]FHBG has also been published by our group [45].

In comparison with other fluorine-18 labeled acycloguanosine analogs, [18F]FHBG is a better substrate for both HSV1-TK and HSV1-sr39TK [49] (*Note: tk refers to the gene and TK refers to the enzyme*). Compared to [124I]FIAU, [18F]FHBG's sensitivity depends on the mode of HSV1-tk delivery into cells. Retrovirally transduced RG2 rat glioma cells (RG2-HSV1-tk), stable transfected C6 rat glioma cells, and RG2-HSV1-tk subcutaneous tumors accumulate greater amounts of FIAU [49, 50]. However, FHBG accumulation is significantly higher in HSV1-tk adenovirus-transduced ($>5 \times 10^6$ pfu) or HSV1-sr39tk-expressing (regardless of gene delivery technique) C6 cells [49]. Furthermore, [18F]FHBG entrapment is less than that of [124I]FIAU in control cells due to less efficient phosphorylation of [18F]FHBG by mammalian TK [49]. This characteristic improves [18F]FHBG's signal-to-noise ratio and may decrease the chance of toxicity in nontarget tissues. Furthermore, and albeit the hepatobiliary clearance of [18F]FHBG, it has been shown that it can even be used for *in vivo* imaging of HSV1-tk expression in the liver [38].

Using [18F]FHBG in clinical trials seems to have several advantages over [124I]FIAU. Despite the shorter half-life of Fluorine-18, we have shown that it is possible to obtain positive images of HSV1- tk expression using this tracer and that such images can be acquired even as soon as 50–60 min after injection when a high tumor-to-liver ratio is achieved in patients treated with sufficiently high doses of the vector. In any case, and albeit the dose of vector does not correlate with reporter gene expression, a certain threshold dose must be achieved.

Several investigators, including us, are in the process of investigating other PET reporter probes with improved sensitivity and specificity for HSV1-tk or HSV1-sr39tk. Our most recent findings indicate that the tritiated form

of 2′-fluoro-2′-deoxyarabinofuranosyl-5-ethyluracil ([3H]FEAU) has a significantly greater selective uptake in HSV1-tk-expressing cells than do [3H]penciclovir, [14C]FIAU or [14C] FMAU [1-(2′-fluoro-2′-deoxy-D-arabinofuranosyl)-5-methyluracil][51]. [18F]FEAU has recently been compared with several other HSV1-tk tracers in *in vitro* studies [52]. This investigation concluded that [18F]FEAU had the highest specific uptake in HSV1-tk-expressing C6 glioma cells relative to [18F]FHBG, [18F]FHPG, [18F]FIAU, [18F]FMAU, and [18F]FLT.

IMPORTANCE OF KINETICS STUDIES OF RADIOTRACER ACCUMULATION

Previous studies in cell culture and laboratory animals have shown that the *in vivo* [18F]FHBG-PET signal closely correlates with transgene expression as determined by messenger RNA abundance and protein levels. Regarding the kinetics of [18F]FHBG interaction with HSV1-tk enzyme, the published value for Km is 0.94 μmol/L (kcat = 0.04/s) [41]. The reported Km of thymidine for HSV1-TK is in the range 0.2–8.5 μmol/L, whereas the thymidine plasma concentration is 0.05–0.5 μmol/L. Taking into account that under physiological conditions the [18F]FHBG concentration is much lower than Km, the rate constant for the interaction of the labeled tracer with the enzyme (kcat/Km) can be used as a measure of catalytic efficiency and, hence, can compare an enzyme's preference for different substrates. Under these conditions, thymidine is a formidable competitor for HSV1-TK catalytic sites *in vivo*. Hence, for high signal-to-noise ratios in the PET images, an imaging probe with high affinity and catalytic efficiency is required. When compared with other similar compounds, FHBG has the highest kcat/Km known for acycloguanosines, and, hence, [18F]FHBG might be considered the most adequate and efficient probe for *in vivo* imaging of HSV1-tk gene expression. Nonetheless, the use of FHBG for imaging transgene expression in the liver or the abdomen might be considered impossible, as the tracer has hepatobiliary and intestinal clearance [47]. However, the only example of transgene expression imaging in humans with this tracer published thus far [38] was done in patients with hepatocellular carcinoma and transgene expression quantitatively determined in hepatocarcinoma nodules. PET images of gene expression were clear in this case, but quantitative data revealed additional results of great importance (Figure 13.3). This example shows that the use of appropriate imaging protocols and proper study design are critical for these kind of studies.

When a gene therapy procedure is considered, it is critical to demonstrate not only that it may be helpful for cure/remission of the considered disease but also that it is safe for the patient. Great concerns were raised over

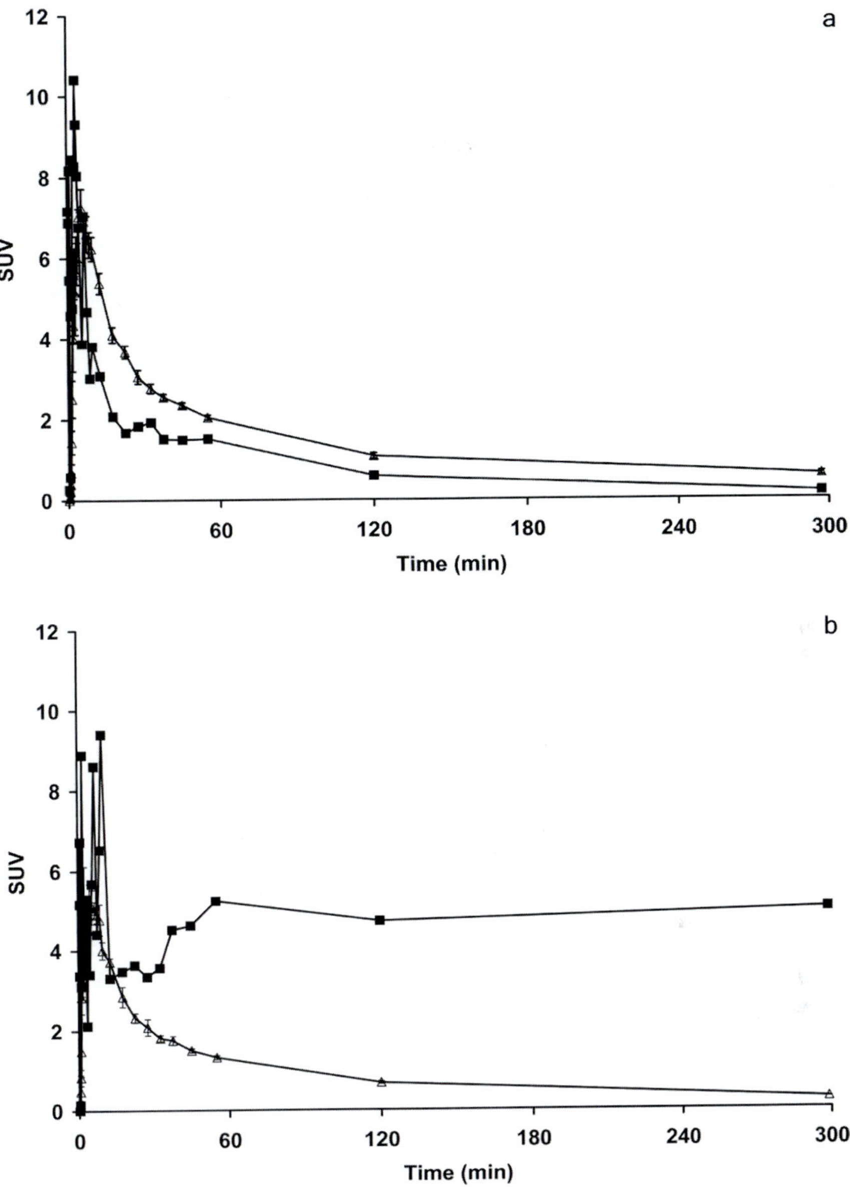

Figure 13.3. **Time course of [¹⁸F]FHBG accumulation in tumor nodules of hepatocarcinoma patients.** Panels *a* and *b* illustrate the time course of [¹⁸F]FHBG accumulation (expressed as SUV) in two different areas of the liver after injection of an adenoviral vector encoding the HSV1-tk in hepatocarcinoma patients: the treated tumor lesion (■) and the nontumoral parenchyma (△). For a patient treated with 10^{11} vp, (panel *a*) the two curves depict a similar pattern for both regions, whereas a clear difference between them can be observed for a patient treated with 10^{12} vp (panel *b*).

gene therapy at its beginning owing to the fact that genetically modified DNA sequences are being administered to humans. Undesired transduction of nontarget organs by viral vectors and nonspecific expression of the transgene in these organs could pose serious safety risks. The use of inducible or tissue-specific promoters to permit the expression of the transgene only in the desired locations is a good approach for prevention of side effects. The induction of mutations due to integration of the transduced

DNA into the host's genome in nontarget cells is another great concern [53, 54], and the best way to avoid adverse effects would be to obtain extremely specific vectors only able to transduce specific cells [55]. Local administration of the vector (e.g., intratumoral injection) would sometimes be preferable. We must always consider that there might be a certain level of gene expression that could not be detected by PET imaging, but with the available data it seems that the technique is so sensitive (PET can detect picomolar tracer concentrations in tissue) that such values would be so low they would not be harmful if they existed in unintended sites.

WHOLE-BODY MONITORING OF TRANSGENE EXPRESSION IN NONTARGET SITES

Whole-body PET imaging of therapeutic transgene expression in cancer patients not only reveals transgene expression at the target site but can also be used to confirm absence of transgene expression at nontarget sites. In our study, delivering HSV1-tk into the tumors of patients with hepatocellular cancer using an adenovirus, we observed accumulation of [^{18}F]FHBG within the Ad-CMV-tk-injected tumors, but only background [^{18}F]FHBG accumulation in other parts of the liver, including the surrounding cirrhotic liver, as well as distant tissues [38]. This finding is notable if we consider the marked hepatotropism of adenoviral vectors. It seems possible that the adenoviral particles released into the general circulation may not gain access to the hepatocytes in the context of liver cirrhosis owing to the fibrotic process and capillarization of sinusoids, which may form a physical barrier preventing infection of the cirrhotic liver. The observed sparing of nontumoral cirrhotic tissue is fortunate because there is no risk of damaging normal hepatocytes via the effect of the prodrug (e.g., ganciclovir). Thus, this finding argues in favor of the safety of gene therapy for this kind of tumor based on direct intratumoral injection of an adenoviral vector.

In addition, PET whole-body imaging can provide valuable information showing that the expression of the transgene is circumscribed to the treated lesion (Figure 13.4) while no other organs – and especially the liver, spleen, lungs, heart, and testes – show specific accumulation of the tracer. However, the use of [^{18}F]FHBG or other metabolically similar tracers with hepatobiliary and renal clearance might impede the visualization of positively transduced tissues in the gastrointestinal and urinary tracts, and also in the brain, unless the blood–brain barrier is compromised, because the tracer does not cross the blood–brain barrier.

It is also important to trace not only transgene expression but also vector biodistribution, as in many gene therapy protocols this is already a question of paramount importance, especially in relation to safety trials for new vectors. Scintigraphy has been used to detect enveloped viral particles labeled with ^{111}In, allowing the viruses to be traced *in vivo* [56]. The labeling procedure did not significantly reduce the infectivity of the herpes simplex virus, and the virus did not release any significant amounts of the radionuclide within 12 h after labeling. Sequential imaging of animals after intravenous administration of the ^{111}In-virus showed fast accumulation in

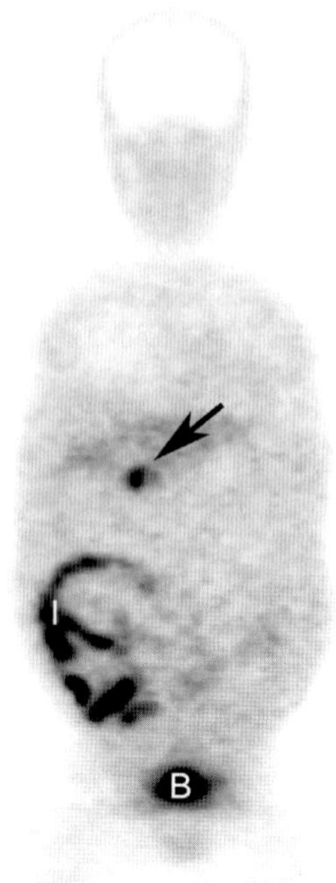

Figure 13.4. Whole-body *in vivo* monitoring of transgene expression by PET. Whole-body image, showing HSV1-tk gene expression only in the treated tumoral nodule (arrow), whereas no accumulation of [^{18}F]FHBG can be seen in any other site. The tracer accumulates in the bladder (**B**) and the intestines (**I**) due to physiological elimination. Adapted from reference [38].

the liver and redistribution from the blood pool to liver and spleen. These data demonstrated that *in vivo* imaging might be a sensitive tool for measuring changes to liver tropism. These approaches could be used for labeling the gene delivery vectors with a positron emitter and *in vivo* imaging with PET.

EVALUATION OF THE NECESSARY VECTOR DOSE

In patients administered different adenoviral doses, we have found that detectable tumor transduction was only observed when the dose of vector administered reached a certain threshold (Figures 13.3 and 13.5). However, an increase in the vector dose above this threshold was not associated with enhanced [^{18}F]FHBG accumulation in the treated tumor. In fact, patients who received 10^{12} viral particles (vp's) showed greater accumulation of the tracer in the tumor than those treated with 2×10^{12} vp (Figure 13.6). This variation in the accumulation of the radiolabel in different neoplasic lesions treated may reflect dissimilarities in tumor biology leading to different transduction by adenoviruses or different ability of the tumor cells to transport the tracer. Alternatively, this phenomenon may be due to impaired gene expression

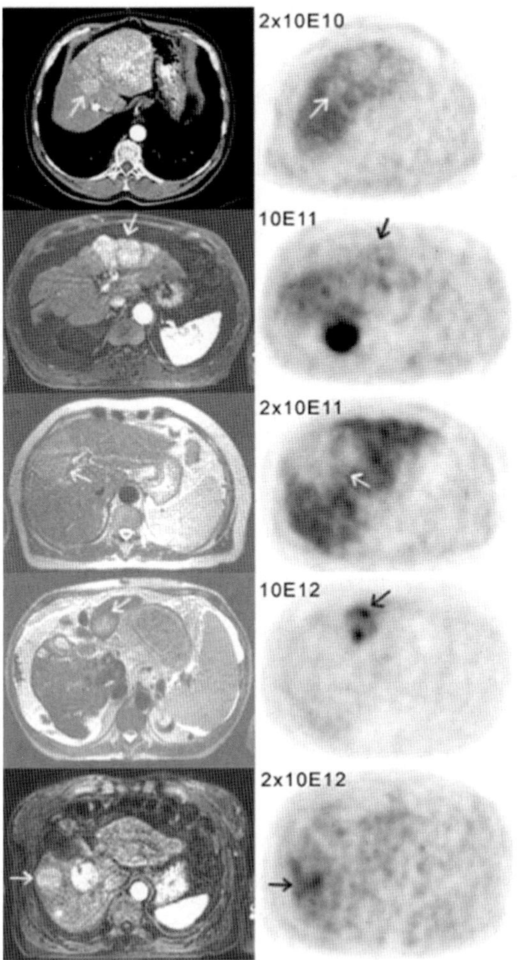

Figure 13.5. Viral dose-dependent imaging of HSV1-tk transgene expression with PET. Transaxial anatomical images (MRI or CT) showing the treated tumor nodules and the corresponding [^{18}F]FHBG-PET images obtained 60 min after injection of [^{18}F]FHBG for five patients enrolled in five consecutive adenoviral dose steps (2×10^{10}, 10^{11}, 2×10^{11}, 10^{12} and 2×10^{12} vp, respectively). Patients received the vector by direct intratumoral injection into the tumor nodule (as schematically shown in Figure 13.1) 2 days before. Detectable tumor transduction can only be observed when the dose of vector administered reaches a certain threshold (10^{12} in this particular case). Diffuse accumulation of the tracer in nontumoral hepatic tissue for the patients enrolled in the first three dose steps shows metabolic clearance of the tracer, whereas absent metabolism of the tracer in the tumor probably reflects cellular changes leading to dedifferentiation and loss of hepatocyte phenotype. Adapted from reference [38].

when the cells are infected with excess viral particles or to differences in perfusion and shunting volume. These results further support the benefits of imaging to measure levels of gene expression directly, as it seems that it is not easy to predict them based exclusively on the dose of vector administered to the patient. We found

that the magnitude of [^{18}F]FHBG accumulation in the treated lesion varies from patient to patient and cannot be directly correlated with the adenoviral dose used in each particular case. Thus in a patient treated with a lower vector dose, the transduced tumor was visible at 1.5 h after [^{18}F]FHBG injection due to very active accumulation of the tracer in the neoplastic nodule, while in another case, despite a higher vector dose, the treated tumor was not readily visible at 1.5 h. In this case the transduced tumor only became apparent later (6.5 h), when the tracer cleared from the nontumoral liver but remained trapped in the cells expressing HSV1-tk (Figure 13.4)

The ability of PET imaging to provide semiquantitative data permits the pharmacokinetic analysis of the tracer in the treated tumor and in nontreated tissues. We have found that while in all patients who received a vector dose below a certain threshold, the standard uptake value (SUV) decreased with time in parallel in both tumor and nontumor tissue; in patients who received 10^{12} or more vp, the SUV in the treated tumor remained high during the entire study period but decreased rapidly in the nontumoral liver (Figure 13.3). Accordingly, the ratio of radioactivity concentration in the tumor and nontumor tissue increased steadily during the study period in subjects treated with a sufficiently high adenovirus dose. In animal models, it has been demonstrated that for FHBG, better quantitation can be obtained when tracer kinetic modeling to quantitate rates of phosphorylation is applied [46]. Measurements such as SUV or %ID/g only convey accumulated activity in a region of interest, normalized to tissue and body weight. This measurement is not only affected by reporter gene expression but also tracer pharmacokinetics. Tracer kinetic modeling will allow more accurate measurement of reporter protein activity.

DURATION OF TRANSGENE EXPRESSION

In a gene therapy procedure the duration of the effect of genetic modification should be sufficiently long to achieve the desired therapeutic effect. Once more, there is little information regarding how long the expression of a transgene lasts in patient tissues. This information can also be obtained using PET. In hepatocarcinoma patients treated by suicide gene therapy with HSV1-tk, a [^{18}F]FHBG PET study performed at 9 d after the intratumoral injection of the vector showed no transgene expression in those patients who had shown positive tracer accumulation at just 2 d after tumor transduction (Figure 13.7) [38]. These results indicate that the level of HSV1-TK expression was likely reduced to undetectable levels after this time. This was not surprising because the tumor cells transduced with AdCMVtk are expected to be eliminated by the therapy with GCV and also by the

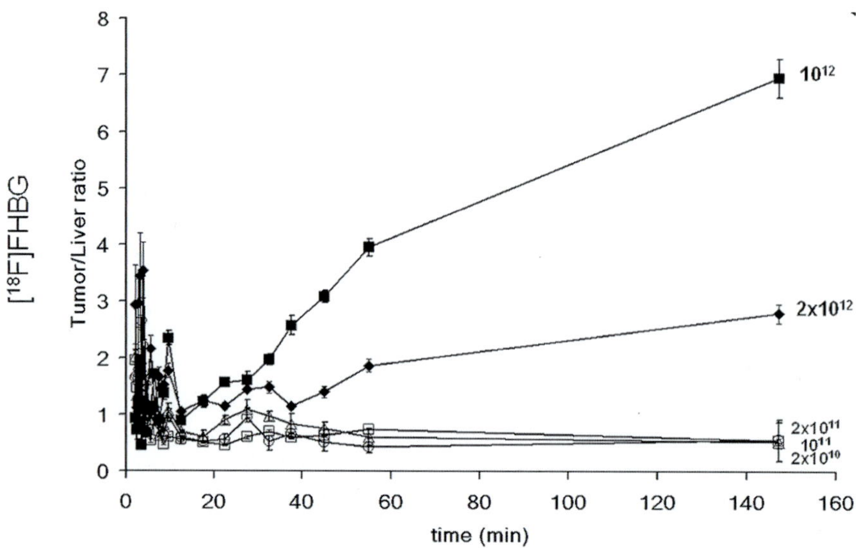

Figure 13.6. Relationship between transgene expression levels and the dose of gene therapy vector used for transduction of hepatocarcinoma nodules. Ratio between the radioactivity concentration in the treated tumor and in the nontumoral liver parenchyma as a function of time after injection of [^{18}F]FHBG was calculated from patients treated with five different doses of the vector. After Minute 40, ratios were equal to or lower than one for the three lower doses: 2×10^{10} (○), 10^{11} (□), and 2×10^{11} (△) vp. On the other hand, for vector doses of 10^{12} (■) and 2×10^{12} (♦) vp, radioactivity was concentrated in the treated lesion, likely due to HSV1-tk expression. Error bars show the error associated to the SUV ratios. Adapted from reference [38].

immune response against the viral antigens expressed by the infected cells, although other possible explanations can be considered, such as loss of the genetic information, which is episomal for adenoviral vectors. Yaghoubi et al. [47] state that the decrease in [^{18}F]FHBG accumulation after GCV treatment can be attributed to the following factors: (i) decrease in the level of HSV1-TK enzyme due to the elimination of transduced cells by GCV-induced cell death; (ii) decrease in the level of HSV1-TK enzyme due to attenuated expression of the gene; (iii) decrease in HSV1-TK enzymatic activity *in vivo*; and (iv) decreased uptake of the tracer. Based on these data, we conclude that the major reason is in fact likely the decrease in HSV1-TK enzyme due to the elimination of transduced cells.

These results might permit better planning of the prodrug treatment schedule and very early demonstration of the efficacy of treatment. In this particular case in which transduced cells are hopefully killed by GCV treatment, the results have to be carefully considered, but they undoubtedly serve as a proof of concept that the duration of transgene expression can be noninvasively monitored

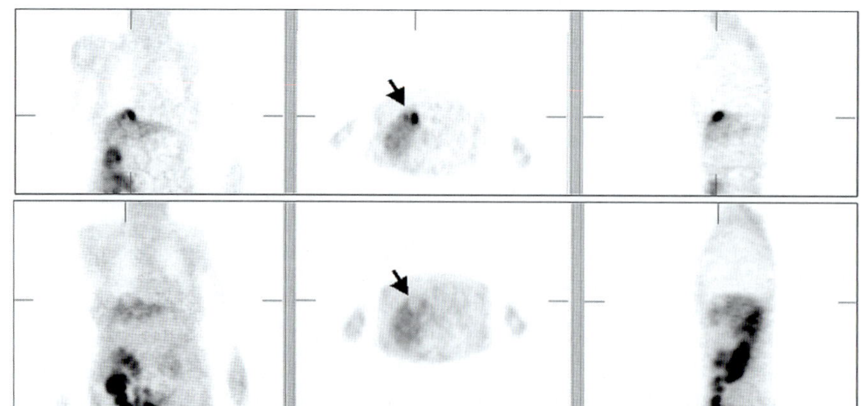

Figure 13.7. PET can be used to monitor duration of transgene in humans. [^{18}F]FHBG-PET images were obtained 2 and 9 d after intratumoral injection of a HSV1-tk encoding vector. Transgene expression can readily be observed at Day 2 in the treated tumoral nodule, but not at Day 9. Adapted from reference [26].

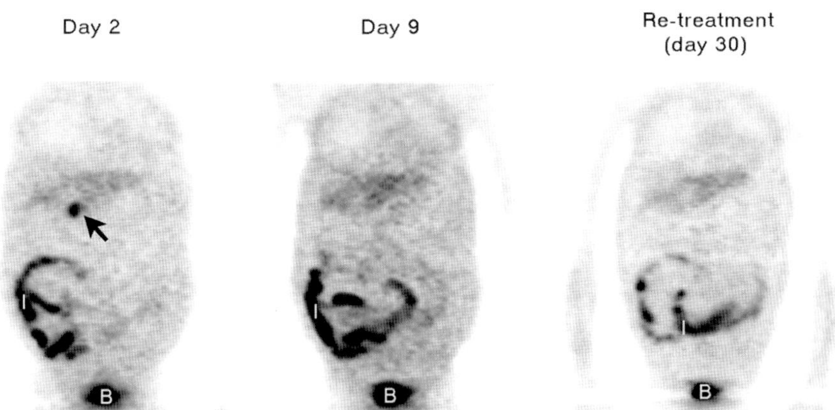

Figure 13.8. Monitoring of HSV1-tk transgene expression time variation by PET. [^{18}F]FHBG-PET images obtained 2 and 9 d after the first dose of AdCMVtk, and 2 d after readministration (Day 30) of the adenoviral vector on the same lesion. Specific [^{18}F]FHBG accumulation in the treated tumor can only be seen at Day 2 (arrow) whereas no specific accumulation can be seen either at Day 9 (as also shown in Figure 13.7) nor after retreatment. Radioactivity can be observed in all cases in the intestine (I) and the bladder (B) as a result of physiological clearance of the tracer. Adapted from reference [38].

in humans by PET. Serially acquired PET studies in the same subject, when using a tracer labeled with a very short-lived isotope (such as [^{18}F]FHBG), are hence feasible and might provide valuable information for better gene therapy planning.

Once the duration of transgene expression is ascertained, one should consider the possibility of retreatment of the same individual with a new dose of the vector to maintain high levels of transgene expression and to try to kill any cells not eliminated by the first treatment.

FEASIBILITY OF TUMOR TRANSDUCTION AFTER A SECOND DOSE OF THE VECTOR

PET imaging can also be used to determine feasibility of repeated therapeutic transgene transduction. In one patient who showed specific accumulation of [^{18}F]FHBG in the treated nodule after the first administration of AdCMVtk [38], additional [^{18}F]FHBG PET studies were performed after an additional dose of the vector 1 month later. This second dose of the vector was not followed by accumulation of [^{18}F]FHBG in the retreated nodule (Figure 13.8); a finding that may have been caused by several mechanisms. The first injection may have resulted in eradication of highly permissive hepatocellular carcinoma cells and selection of cells more resistant to adenoviral infection. Also, elicitation of a potent immune response after the first injection may have favored rapid clearance of the newly transduced cells after the second injection. But more likely, neutralization of the adenovirus by antibodies against the viral capsid strongly reduced the intensity of tumor transduction, as seen in animal models. In fact, in this patient the authors observed that the titer of neutralizing antiadenovirus

antibodies increased more than two logs after the first vector treatment.

IMPROVED VECTOR DESIGN/ PROTOCOL DESIGN

It is generally accepted that the Achilles' heels of gene therapy are the vectors utilized for gene delivery. Each new vector requires extensive safety testing before it is approved for human administration and further testing once it has been used in humans [57]. The overall process is not only very expensive but also time-consuming. Recent progress in vector production and multimodality imaging with optical/PET reporter genes will undoubtedly aid in faster development of new vectors. Molecular imaging can help trace not only transgene expression but also vector biodistribution, as discussed previously.

REPORTER GENE-BASED IMAGING OF THERAPEUTIC CELLS FOLLOWING ADOPTIVE TRANSFER INTO PATIENTS

Adoptive cellular gene therapy (ACGT) has the potential of curing, halting the progression of, or at least alleviating the symptoms of many diseases, such as cancer and autoimmune diseases, for a longer duration than currently available conventional treatments. However, success of ACGT is dependent on targeted homing of an adequate number of therapeutic cells that will remain functional at the target site for a minimum duration to be effective. Furthermore, often the trafficking of therapeutic cells to certain nontarget tissues should be avoided to prevent clinical complications. Therefore,

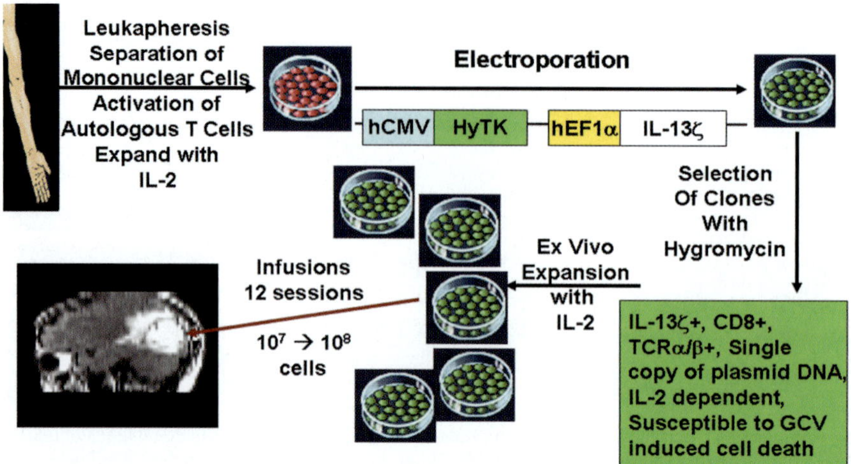

Figure 13.9. Diagram illustrating the procedure involved in cytolytic T-cell therapy of glioma tumors in a human patient. Cytolytic T cells (CTLs) are prepared by activation of isolated peripherals blood mononuclear cells and they are expanded by adding IL-2 to their growth media. CTLs are then electroporated to express hygromycin resistance gene/HSV1-tk and IL-13, then selected with hygromycin to make stable clones, which are then again expanded. These autologous cells are then infused during multiple sessions at increasing doses (depending on patient tolerance) into the tumor resection site.

noninvasive monitoring of cell trafficking is necessary because whole-body biopsy sampling is not practically feasible in patients.

Yaghoubi et al. have recently performed the first reported clinical trial of reporter gene-based imaging of therapeutic cells in a glioma cancer patient [58]. This patient had been treated on twelve sessions in a 4-week period by administration of genetically engineered autologous cytolytic T cells (CTLs), expressing the IL-13ζ and HSV1-tk genes, through a catheter into the tumor resection site (Figure 13.9). The expression of IL-13ζ was necessary to specifically target these cells to the residual glioma tumor cells that may have remained within the brain of the patient after surgical removal of his recurrent tumor. The week after completion of CTL infusions the patient was scanned for whole-body distribution of [18F]FHBG approximately 2 hours after its intravenous injection. The brain [18F]FHBG PET images showed greater than background accumulation of [18F]FHBG at the site of CTL infusions (which was also the tumor resection site) as well as a remote site near the corpus callosum of the patient's brain (Figure 13.10). Shortly thereafter, the patient's physician discovered that another tumor had recurred near the corpus callosum of the patient, which corresponded with the site of increased [18F]FHBG accumulation (Figure 13.10). Biopsy also confirmed the presence of infused CTLs after the [18F]FHBG PET scan at both sites of increased [18F]FHBG accumulation. When a region of interest was done to quantitatively compare [18F]FHBG activity at the tumor resection site and at the intact tumor of the CTL-infused patient relative to control glioma patients, the ratio of [18F]FHBG accumulation in the tumor resection

site over brain background for the CTL infused patient was 2.6 times higher than the ratio for control glioma patients who had not been administered CTLs. The ratio of [18F]FHBG accumulation in the intact tumor over brain background for the CTL-infused patient was 2.8 times higher than the ratio for control glioma patients who had not been administered CTLs. This is an ongoing study, and in the future patients receiving CTLs can be scanned for [18F]FHBG biodistribution before and after CTL infusions.

FINAL CONSIDERATIONS AND FUTURE PROSPECTS

Nearly all of the present knowledge regarding imaging of therapeutic transgenes comes from animal studies. However, the results obtained cannot be directly extrapolated to humans, and many important questions remain unanswered. For each particular patient, there are some crucial questions that should be answered: Has targeted gene transfer been successful? Have any other tissues been transduced? Is the transferred gene expressed in sufficient extent and for sufficient duration? The availability of these data is critical for the clinical success of gene therapy, which relies on our ability to develop methodologies able to address these questions. There is no doubt that the most exciting possibility for the use of PET in gene therapy is to monitor transgene expression. The availability of a quantitative, noninvasive tool to detect gene transfer and expression *in vivo* will probably change the way gene therapy protocols are designed and followed. Results with antitumoral HSV1-tk-based

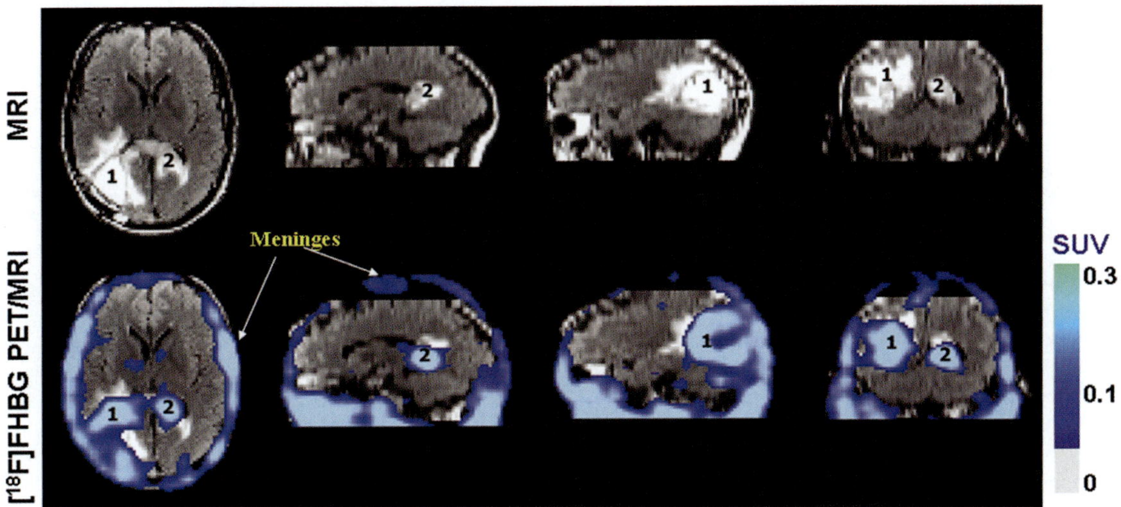

Figure 13.10. Brain [18F]FHBG PET/MRI coregistration in a patient infused with genetically engineered cytolytic T cells expressing the HSV1-tk PET reporter gene. The procedure is illustrated in Figure 13.9. Images were acquired about 2 h after intravenous [18F]FHBG injection (∼ 7 mCi). Areas of [18F]FHBG hyperintensity can be observed at the tumor resection site (Tumor 1), into where the CTLs had been infused and at a remote intact tumor site (Tumor 2) near the corpus callosum. Reprinted from reference [58].

therapy in humans have demonstrated the feasibility of such an approach. Much needs to be done in this field, as it is necessary to monitor not only the HSV1-tk gene but also any other therapeutic genes of interest amenable to use in a gene therapy protocol. Coupling of therapeutic and reporter genes may be one solution [7]. However, it remains to be demonstrated which of the therapeutic gene to imaging reporter gene linking approaches is best (fusion gene technology, bi-cistronic transcription units, bidirectional vectors, vector coadministration, etc.). The uncertainties in clinical gene therapy are related not only to problems of gene delivery and gene expression but also to the lack of objective clinical endpoints to evaluate therapy response. In many cases the clinical benefit of the therapeutic procedure is hardly measurable or requires complex invasive procedures that give only a partial picture of the situation as it is mostly based on molecular and histopathological analysis of biopsies. Hence there is a need for a technology that allows routine clinical use for a quantitative, whole-body spatiotemporal evaluation of gene delivery and expression as a way to determine and even predict the clinical outcome. The merger of both these ideas is important.

If molecular imaging by means of [18F]FHBG PET can be used to predict therapy response and serially performed [18F]FHBG PET studies can be used to monitor the effectiveness of the prodrug treatment in those cases in which response is foreseeable, we would have at our disposal one of the most desired tools pursued by researchers in the field of suicide gene therapy. PET might show very early on which patients must undergo the prodrug treatment, and in whom this will be ineffective. PET imaging, ideally combined with other molecular

imaging modalities, can be used as an unsurpassable tool to aid in the design of improved vectors that would more efficiently and precisely deliver the genes of interest to the target cells and enhance our understanding of the molecular biology, biochemistry, genetics, and pharmacokinetics of gene therapy vectors.

In summary, transgene expression can be monitored by PET in cancer patients. This noninvasive methodology is a valuable tool for the assessment of gene expression in clinical gene therapy and could be useful for revealing the transduction efficiency of a given vector in specific tissues or lesions. Therefore, PET imaging can allow us to study transgene expression distribution, determine its duration, and thereby help in evaluating new vectors and the design of novel therapeutic strategies.

Future studies will be influenced by the development of improved gene therapy vectors with higher transduction or transfection efficiencies and specificities. Development of new tracers with higher affinities for reporter proteins (enzymes, receptors, or transporters) will improve the sensitivity of imaging. Furthermore, the use of mutants of HSV1-tk and perhaps other reporter genes may yield better imaging reporter gene/probe systems, which can help in imaging aimed at improving gene therapy protocols. The transfer of nucleoside transporters can be used to enhance the tracer influx into the genetically modified cells and thereby increase suicide enzyme detection as well as therapeutic efficiency. The development of receptor-based reporter gene/probe systems with a lower background than the D_2R/[18F]fluoroethyl-spiperone and using a commercially available ligand will increase their application in gene therapy clinical trials. Finally, reporter genes encoding proteins secreted into the blood, such as humanized Gaussia luciferase, provide an alternative to imaging or can complement imaging for the detection of expression of the therapeutic gene.

We recommend that whenever possible molecular imaging be included in the protocols of gene therapy clinical trials. The use of different imaging surrogates would provide valuable information on many different topics. Indeed, it is important to consider the possibility of including imaging in the protocol from the very beginning, when the trial is being designed. This would allow inclusion of both the desired therapeutic gene and the appropriate imaging reporter gene in the gene therapy vector prior to obtaining approval for the clinical trial from regulatory agencies. Although this would require additional effort to determine the correlation between the expression of the two transgenes, imaging will yield essential information that would not otherwise have been gathered during the clinical trial. As discussed in this chapter, the inclusion of an imaging reporter gene would permit not only monitoring of the time variation of transgene expression but also noninvasive whole-body imaging of gene expression within and outside the target region, or even of the biodistribution of the vector on a whole-body basis and correlation with the sites of expression. The data obtained from the images would be important in a phase I trial and could speed up considerably the approval of further phase II/III protocols. In addition, molecular imaging should be considered as a way to monitor and evaluate response to the gene therapy procedure, or even as a prognostic indicator that would avoid unnecessary invasive sampling techniques and expensive and ineffective treatments.

The same principles developed for gene therapy imaging can also be applied to cell-based therapies. Cells have been labeled *in vitro* with [^{18}F]FDG for short-term follow-up of cell trafficking or with other longer-lived nuclides such as ^{64}Cu [59]. Nonetheless, such methods do not permit long-term follow-up of cell trafficking, a major issue that can be resolved by tracing cells *in vivo* with cell-specific labeled antibodies [60] or in a much better way by modifying the cells to carry an appropriate imaging reporter gene. Monitoring of cell trafficking with genetically modified cells [61] will permit grafting experiments and long-term follow-up of the cell fate. For imaging of cell trafficking using reporter genes one must keep in mind the minimum number of cells that need to accumulate at one site to be detectable by PET, SPECT, or MRI as well as the possibility that the cells may shut down the expression of the reporter gene. Finally, when the reporter and the therapeutic gene are different, one must consider the issue of reporter gene silencing, even though the therapeutic gene is still being expressed, and optimize the gene constructs to minimize the possibility of false negative results. False positive results by selective silencing of the therapeutic gene should also be taken into consideration.

REFERENCES

1 Hoffman, J. M., Gambhir, S. S. (2007). Molecular imaging: the vision and opportunity for radiology in the future. *Radiology* **244**(1): 39–47.

2 Massoud, T. F., Gambhir, S. S. (2003). Molecular imaging in living subjects: seeing fundamental biological processes in a new light. *Genes Dev* **17**(5): 545–580.

3 Massoud, T. F., Gambhir, S. S. (2007). Integrating non-invasive molecular imaging into molecular medicine: an evolving paradigm. *Trends Mol Med* **13**(5): 183–191.

4 Penuelas, I., Boan, J., Marti-Climent, J. M., Sangro, B., Mazzolini, G., Prieto, J. et al. (2004). Positron emission tomography and gene therapy: basic concepts and experimental approaches for in vivo gene expression imaging. *Mol Imaging Biol* **6**(4): 225–238.

5 Yaghoubi, S. S., Wu, L., Liang, Q., Toyokuni, T., Barrio, J. R., Namavari, M. et al. (2001). Direct correlation between positron emission tomographic images of two reporter genes delivered by two distinct adenoviral vectors. *Gene Ther* **8**(14): 1072–1080.

6 Ray, P., Bauer, E., Iyer, M., Barrio, J. R., Satyamurthy, N., Phelps, M. E. et al. (2001). Monitoring gene therapy with reporter gene imaging. *Semin Nucl Med* **31**(4): 312–320.

7 Ray, P., Tsien, R., Gambhir, S. S. (2007). Construction and validation of improved triple fusion reporter gene vectors for molecular imaging of living subjects. *Cancer Res* **67**(7): 3085–3093.

8 Deroose, C. M., De, A., Loening, A. M., Chow, P. L., Ray, P., Chatziioannou, A. F. et al. (2007). Multimodality imaging of tumor xenografts and metastases in mice with combined small-animal PET, small-animal CT, and bioluminescence imaging. *J Nucl Med* **48**(2): 295–303.

9 Ponomarev, V., Doubrovin, M., Serganova, I., Vider, J., Shavrin, A., Beresten, T. et al. (2004). A novel triple-modality reporter gene for whole-body fluorescent, bioluminescent, and nuclear noninvasive imaging. *Eur J Nucl Med Mol Imaging* **31**(5): 740–751.

10 Ray, P., De, A., Min, J. J., Tsien, R. Y., Gambhir, S. S. (2004). Imaging tri-fusion multimodality reporter gene expression in living subjects. *Cancer Res* **64**(4): 1323–1330.

11 Hielscher, A. H. (2005). Optical tomographic imaging of small animals. *Curr Opin Biotechnol* **16**(1): 79–88.

12 Gibson, A. P., Hebden, J. C., Arridge, S. R. (2005). Recent advances in diffuse optical imaging. *Phys Med Biol* **50**(4): R1–43.

13 Hawrysz, D. J., Sevick-Muraca, E. M. (2000). Developments toward diagnostic breast cancer imaging using near-infrared optical measurements and fluorescent contrast agents. *Neoplasia* **2**(5): 388–417.

14 Venisnik, K. M., Olafsen, T., Gambhir, S. S., Wu, A. M. (2007). Fusion of Gaussia luciferase to an engineered anti-carcinoembryonic antigen (CEA) antibody for in vivo optical imaging. *Mol Imaging Biol* **9**(5): 267–277.

15 Labas, Y. A., Gurskaya, N. G., Yanushevich, Y. G., Fradkov, A. F., Lukyanov, K. A., Lukyanov, S. A. et al. (2002).

Diversity and evolution of the green fluorescent protein family. *Proc Natl Acad Sci U S A* **99**(7): 4256–4261.

16 Matz, M. V., Fradkov, A. F., Labas, Y. A., Savitsky, A. P., Zaraisky, A. G., Markelov, M. L. et al. (1999). Fluorescent proteins from nonbioluminescent Anthozoa species. *Nat Biotechnol* **17**(10): 969–973.

17 Weissleder, R., Ntziachristos, V. (2003). Shedding light onto live molecular targets. *Nat Med* **9**(1): 123–128.

18 Ntziachristos, V., Tung, C. H., Bremer, C., Weissleder, R. (2002). Fluorescence molecular tomography resolves protease activity in vivo. *Nat Med* **8**(7): 757–760.

19 Funovics, M. A., Weissleder, R., Mahmood, U. (2004). Catheter-based in vivo imaging of enzyme activity and gene expression: feasibility study in mice. *Radiology* **231**(3): 659–666.

20 Genove, G., DeMarco, U., Xu, H., Goins, W. F., Ahrens, E. T. (2005). A new transgene reporter for in vivo magnetic resonance imaging. *Nat Med* **11**(4): 450–454.

21 Cohen, B., Dafni, H., Meir, G., Harmelin, A., Neeman, M. (2005). Ferritin as an endogenous MRI reporter for noninvasive imaging of gene expression in C6 glioma tumors. *Neoplasia* **7**(2): 109–117.

22 Gilad, A. A., McMahon, M. T., Walczak, P., Winnard, P. T., Jr., Raman, V., van Laarhoven, H. W. et al. (2007). Artificial reporter gene providing MRI contrast based on proton exchange. *Nat Biotechnol* **25**(2): 217–219.

23 Gilad, A. A., Winnard, P. T., Jr., van Zijl, P. C., Bulte, J. W. (2007). Developing MR reporter genes: promises and pitfalls. *NMR Biomed* **20**(3): 275–290.

24 Beanlands, R., Roberts, R. (2007). Positron molecular imaging, an in vivo glimpse of the genome. *J Mol Cell Cardiol* **43**(1): 11–14.

25 Rome, C., Couillaud, F., Moonen, C. T. (2007). Gene expression and gene therapy imaging. *Eur Radiol* **17**(2): 305–319.

26 Penuelas, I., Haberkorn, U., Yaghoubi, S., Gambhir, S. S. (2005). Gene therapy imaging in patients for oncological applications. *Eur J Nucl Med Mol Imaging* **32**(Suppl 2): S384–403.

27 Penuelas, I., Gambhir, S. S. (2005). Imaging studies for evaluating gene therapy in translational research. *Drug Discovery Today: Technologies* **2**(4): 335–343.

28 Liang, Q., Satyamurthy, N., Barrio, J. R., Toyokuni, T., Phelps, M. P., Gambhir, S. S. et al. (2001). Noninvasive, quantitative imaging in living animals of a mutant dopamine D2 receptor reporter gene in which ligand binding is uncoupled from signal transduction. *Gene Ther* **8**(19): 1490–1498.

29 MacLaren, D. C., Gambhir, S. S., Satyamurthy, N., Barrio, J. R., Sharfstein, S., Toyokuni, T. et al. (1999). Repetitive, non-invasive imaging of the dopamine D2 receptor as a reporter gene in living animals. *Gene Ther* **6**(5): 785–791.

30 Zinn, K. R., Chaudhuri, T. R. (2002). The type 2 human somatostatin receptor as a platform for reporter gene imaging. *Eur J Nucl Med Mol Imaging* **29**(3): 388–399.

31 Chung, J. K. (2002). Sodium iodide symporter: its role in nuclear medicine. *J Nucl Med* **43**(9): 1188–1200.

32 Groot-Wassink, T., Aboagye, E. O., Glaser, M., Lemoine, N. R., Vassaux, G. (2002). Adenovirus biodistribution and noninvasive imaging of gene expression in vivo by positron emission tomography using human sodium/iodide symporter as reporter gene. *Hum Gene Ther* **13**(14): 1723–1735.

33 Haberkorn, U., Henze, M., Altmann, A., Jiang, S., Morr, I., Mahmut, M. et al. (2001). Transfer of the human NaI symporter gene enhances iodide uptake in hepatoma cells. *J Nucl Med* **42**(2): 317–325.

34 Anton, M., Wagner, B., Haubner, R., Bodenstein, C., Essien, B. E., Bonisch, H. et al. (2004). Use of the norepinephrine transporter as a reporter gene for non-invasive imaging of genetically modified cells. *J Gene Med* **6**(1): 119–126.

35 Ponomarev, V., Doubrovin, M., Shavrin, A., Serganova, I., Beresten, T., Ageyeva, L. et al. (2007). A human-derived reporter gene for noninvasive imaging in humans: mitochondrial thymidine kinase type 2. *J Nucl Med* **48**(5): 819–826.

36 Jacobs, A., Voges, J., Reszka, R., Lercher, M., Gossmann, A., Kracht, L. et al. (2001). Positron-emission tomography of vector-mediated gene expression in gene therapy for gliomas. *Lancet* **358**(9283): 727–729.

37 Dempsey, M. F., Wyper, D., Owens, J., Pimlott, S., Papanastassiou, V., Patterson, J. et al. (2006). Assessment of 123I-FIAU imaging of herpes simplex viral gene expression in the treatment of glioma. *Nucl Med Commun* **27**(8): 611–617.

38 Penuelas, I., Mazzolini, G., Boan, J. F., Sangro, B., Marti-Climent, J., Ruiz, M. et al. (2005). Positron emission tomography imaging of adenoviral-mediated transgene expression in liver cancer patients. *Gastroenterology* **128**(7): 1787–1795.

39 Jacobs, A. H., Rueger, M. A., Winkeler, A., Li, H., Vollmar, S., Waerzeggers, Y. et al. (2007). Imaging-guided gene therapy of experimental gliomas. *Cancer Res* **67**(4): 1706–1715.

40 Weissleder, R. (2006). Molecular imaging in cancer. *Science* **312**(5777): 1168–1171.

41 Phelps, M. E. (2004). *Molecular Imaging and Its Biological Applications.* 1st ed. New York: Springer-Verlag.

42 Weber, W. A. (2006). Positron emission tomography as an imaging biomarker. *J Clin Oncol* **24**(20): 3282–3292.

43 Yaghoubi, S. S., Barrio, J. R., Namavari, M., Satyamurthy, N., Phelps, M. E., Herschman, H. R. et al. (2005). Imaging progress of herpes simplex virus type 1 thymidine kinase suicide gene therapy in living subjects with positron emission tomography. *Cancer Gene Ther* **12**(3): 329–339.

44 Yaghoubi, S. S., Berger, F., Gambhir, S. S. (2007). Studying the biodistribution of positron emission tomography reporter probes in mice. *Nat Protoc* **2**(7): 1752–1755.

45 Yaghoubi, S. S., Gambhir, S. S. (2006). PET imaging of herpes simplex virus type 1 thymidine kinase (HSV1-tk) or mutant HSV1-sr39tk reporter gene expression in mice and humans using [18F]FHBG. *Nat Protoc* **1**(6): 3069–3075.

46 Green, L. A., Nguyen, K., Berenji, B., Iyer, M., Bauer, E., Barrio, J. R. et al. (2004). A tracer kinetic model for 18F-FHBG for quantitating herpes simplex virus type 1 thymidine kinase reporter gene expression in living animals using PET. *J Nucl Med* **45**(9): 1560–1570.

47 Yaghoubi, S., Barrio, J. R., Dahlbom, M., Iyer, M., Namavari, M., Satyamurthy, N et al. (2001). Human pharmacokinetic and dosimetry studies of [(18)F]FHBG: a reporter probe for imaging herpes simplex virus type-1 thymidine kinase reporter gene expression. *J Nucl Med* **42**(8): 1225–1234.

48 Yaghoubi, S. S., Couto, M. A., Chen, C. C., Polavaram, L., Cui, G., Sen, L. et al. (2006). Preclinical safety evaluation of 18F-FHBG: a PET reporter probe for imaging herpes simplex virus type 1 thymidine kinase (HSV1-tk) or mutant HSV1-sr39tk's expression. *J Nucl Med* **47**(4): 706–715.

49 Min, J. J., Iyer, M., Gambhir, S. S. (2003). Comparison of [18F]FHBG and [14C]FIAU for imaging of HSV1-tk reporter gene expression: adenoviral infection vs stable transfection. *Eur J Nucl Med Mol Imaging* **30**(11): 1547–1560.

50 Tjuvajev, J. G., Doubrovin, M., Akhurst, T., Cai, S., Balatoni, J., Alauddin, M. M. et al. (2002). Comparison of radiolabeled nucleoside probes (FIAU, FHBG, and FHPG) for PET imaging of HSV1-tk gene expression. *J Nucl Med* **43**(8): 1072–1083.

51 Kang, K. W., Min, J. J., Chen, X., Gambhir, S. S. (2005). Comparison of [14C]FMAU, [3H]FEAU, [14C]FIAU, and [3H]PCV for monitoring reporter gene expression of wild type and mutant herpes simplex virus type 1 thymidine kinase in cell culture. *Mol Imaging Biol* **7**(4): 296–303.

52 Buursma, A. R., Rutgers, V., Hospers, G. A., Mulder, N. H., Vaalburg, W., de Vries, E. F. (2006). 18F-FEAU as a radiotracer for herpes simplex virus thymidine kinase gene expression: in-vitro comparison with other PET tracers. *Nucl Med Commun* **27**(1): 25–30.

53 Williams, D. A., Baum, C. (2003). Medicine. Gene therapy–new challenges ahead. *Science* **302**(5644): 400–401.

54 Hacein-Bey-Abina, S., von Kalle, C., Schmidt, M., Le Deist, F., Wulffraat, N., McIntyre, E. et al. (2003). A serious adverse event after successful gene therapy for X-linked severe combined immunodeficiency. *N Engl J Med* **348**(3): 255–256.

55 Thomas, C. E., Ehrhardt, A., Kay, M. A. (2003). Progress and problems with the use of viral vectors for gene therapy. *Nat Rev Genet* **4**(5): 346–358.

56 Schellingerhout, D., Bogdanov, A., Jr., Marecos, E., Spear, M., Breakefield, X., Weissleder, R. (1998). Mapping the in vivo distribution of herpes simplex virions. *Hum Gene Ther* **9**(11): 1543–1549.

57 Raty, J. K., Lesch, H. P., Wirth, T., Yla-Herttuala, S. (2008). Improving safety of gene therapy. *Curr Drug Saf* **3**(1): 46–53.

58 Yaghoubi, S., Jensen, M. C., Satyamurthy, N., Budhiraja, S., Paik, D., Czernin, J. et al. (2008). Non-invasive detection of therapeutic cytolytic T cells with [^{18}F]FHBG positron emission tomography in a glioma patient. *Nature Clinical Practice Oncology* **6**(1): 53–58.

59 Adonai, N., Nguyen, K. N., Walsh, J., Iyer, M., Toyokuni, T., Phelps, M. E. et al. (2002). Ex vivo cell labeling with 64Cu-pyruvaldehyde-bis(N4-methylthiosemicarbazone) for imaging cell trafficking in mice with positron-emission tomography. *Proc Natl Acad Sci U S A* **99**(5): 3030–3035.

60 Tamura, M., Unno, K., Yonezawa, S., Hattori, K., Nakashima, E., Tsukada, H. et al. (2004). In vivo trafficking of endothelial progenitor cells their possible involvement in the tumor neovascularization. *Life Sci* **75**(5): 575–584.

61 De, A., Lewis, X. Z., Gambhir, S. S. (2003). Noninvasive imaging of lentiviral-mediated reporter gene expression in living mice. *Mol Ther* **7**(5 Pt 1): 681–691.

Index